6 FET Biasing   Fixed-bias: $V_{GS} = -V_{GG}$, $V_{DS} = V_{DD} - I_D R_D$; self-bias: $V_{GS} = -I_D R_S$, $V_{DS} = V_{DD} - I_D(R_S + R_D)$, $V_S = I_S R_S$; voltage-divider: $V_G = R_2 V_{DD}/(R_1 + R_2)$, $V_{GS} = V_G - I_D R_S$, $V_{DS} = V_{DD} - I_D(R_D + R_S)$; enhancement-type MOSFET: $I_D = k(V_{GS} - V_{GS(Th)})^2$, $k = I_{D(on)}/(V_{GS(on)} - V_{GS(Th)})^2$; feedback bias: $V_{DS} = V_{GS}$, $V_{GS} = V_{DD} - I_D R_D$; voltage-divider: $V_G = R_2 V_{DD}/(R_1 + R_2)$, $V_{GS} = V_G - I_D R_S$; universal curve: $m = |V_P|/I_{DSS} R_S$, $M = m \times V_G/|V_P|$, $V_G = R_2 V_{DD}/(R_1 + R_2)$

7 BJT Transistor Modeling $Z_i = V_i/I_i$, $I_i = (V_s - V_i)/R_{sense}$, $I_o = (V_s - V_o)/R_{sense}$, $Z_o = V_o/I_o$, $A_v = V_o/V_i$, $A_{v_s} = Z_i A_{v_{NL}}/(Z_i + R_s)$, $A_i = -A_v Z_i/R_L$, $r_e = 26$ mV$/I_E$; common-base: $Z_i = r_e$, $Z_o \simeq \infty$ Ω, $A_v \simeq R_L/r_e$, $A_i \simeq -1$; common-emitter: $Z_i = \beta r_e$, $Z_o = r_o$, $A_v = -R_L/r_e$, $A_i \simeq \beta$, $h_{ie} = \beta r_e$, $h_{fe} = \beta_{ac}$, $h_{ib} = r_e$, $h_{fb} = -\alpha$.

8 BJT Small-Signal Analysis   Common-emitter: $A_v = -R_C/r_e$, $Z_i = R_B \| \beta r_e$, $Z_o = R_C$, $A_i = \beta$; voltage-divider: $R' = R_1 \| R_2$, $A_v = -R_C/r_e$, $Z_i = R' \| \beta r_e$, $Z_o = R_C$; emitter-bias: $Z_b = \beta(r_e + R_E) \simeq \beta R_E$, $A_v = -\beta R_C/Z_b = -R_C/(r_e + R_E) \simeq -R_C/R_E$; emitter-follower: $Z_b \simeq \beta(r_e + R_E)$, $A_v \simeq 1$, $Z_o \simeq r_e$; common-base: $A_v \simeq R_C/r_e$, $Z_i = R_E \| r_e$, $Z_o = R_C$; collector feedback: $A_v = -R_C/r_e$, $Z_i = \beta r_e \| R_F/|A_v|$, $Z_o \simeq R_C \| R_F$; collector dc feedback: $A_v = -(R_{F_2} \| R_C)/r_e$, $Z_i = R_{F_1} \| \beta r_e$, $Z_o = R_C \| R_{F_2}$; hybrid parameters: $A_i = h_f/(1 + h_o R_L)$, $A_v = -h_f R_L/[h_i + (h_i h_o - h_f h_r)R_L]$, $Z_i = h_i - h_f h_r R_L/(1 + h_o R_L)$, $Z_o = 1/[h_o - (h_f h_r/(h_i + R_s))]$

9 FET Small-Signal Analysis $g_m = g_{mo}(1 - V_{GS}/V_P)$, $g_{mo} = 2I_{DSS}/|V_P|$; basic configuration: $A_v = -g_m R_D$; unbypassed source resistance: $A_v = -g_m R_D/(1 + g_m R_S)$; source follower: $A_v = g_m R_S/(1 + g_m R_S)$; common gate: $A_v = g_m(R_D \| r_d)$

10 Systems Approach—Effect of R_s and R_L   BJT: $A_v = R_L A_{v_{NL}}/(R_L + R_o)$, $A_i = -A_v Z_i/R_L$, $V_i = R_i V_s/(R_i + R_s)$; fixed-bias: $A_v = -(R_C \| R_L)/r_e$, $A_{v_s} = Z_i A_v/(Z_i + R_s)$, $Z_i = \beta r_e$, $Z_o = R_C$; voltage-divider: $A_v = -(R_C \| R_L)/r_e$, $A_{v_s} = Z_i A_v/(Z_i + R_s)$, $Z_i \simeq R_1 \| R_2 \| \beta r_e$, $Z_o = R_C$; emitter-bias: $A_v = -(R_C \| R_L)/R_E$, $A_{v_s} = Z_i A_v/(Z_i + R_s)$, $Z_i \simeq R_B \| \beta R_E$, $Z_o = R_C$; collector-feedback: $A_v = -(R_C \| R_L)/r_e$, $A_{v_s} = Z_i A_v/(Z_i + R_s)$, $Z_i = \beta r_e \| R_F/|A_v|$, $Z_o \simeq R_C \| R_F$; emitter-follower: $R'_E = R_E \| R_L$, $A_v = R'_E/(R'_E + r_e)$, $A_{v_s} = R'_E/(R'_E + R_s/\beta + r_e)$, $Z_i = R_B \| \beta(r_e + R'_E)$, $Z_o = R_E \| (R_s/\beta + r_e)$; common-base: $A_v \simeq (R_C \| R_L)/r_e$, $A_i \simeq -1$, $Z_i \simeq r_e$, $Z_o = R_C$; FET: bypassed R_S: $A_v = -g_m(R_D \| R_L)$, $Z_i = R_G$, $Z_o = R_D$; unbypassed R_S: $A_v = -g_m(R_D \| R_L)/(1 + g_m R_S)$, $Z_i = R_G$, $Z_o = R_D$; source-follower: $A_v = g_m(R_S \| R_L)/[1 + g_m(R_S \| R_L)]$, $Z_i = R_G$, $Z_o = R_S \| r_d \| 1/g_m$; common gate: $A_v = g_m(R_D \| R_L)$, $Z_i = R_S \| 1/g_m$, $Z_o = R_D$; cascaded: $A_{v_T} = A_{v_1} \cdot A_{v_2} \cdot A_{v_3} \cdots A_{v_n}$, $A_{i_T} = \pm A_{v_T} Z_{i_1}/R_L$

ELECTRONIC DEVICES
AND CIRCUIT THEORY

FIFTH EDITION

ELECTRONIC DEVICES AND CIRCUIT THEORY

ROBERT BOYLESTAD
LOUIS NASHELSKY

PRENTICE HALL, Englewood Cliffs, NJ 07632

Library of Congress Cataloging-in-Publication Data

Boylestad, Robert L.
   Electronic devices and circuit theory/Robert Boylestad, Louis
Nashelsky.—5th ed.
    p. cm.
   ISBN 0-13-250994-6
   1. Electronic circuits.   2. Electronic apparatus and appliances.
I. Nashelsky, Louis. II. Title.
TK7867.B66 1992
621.3815—dc20   91-14454
                                CIP

Editorial/production supervision: Marcia Krefetz
Managing Editor: Mary Carnis
Design Directors: Janet Schmid and Chris Wolf
Interior design: Circa '86
Cover design: Rosemarie Paccione
Page Layout: Diane Koromhas
Manufacturing buyer: Ed O'Dougherty
Prepress Buyers: Mary McCartney and Ilene Levy
Editorial Assistant: Cathy Frank

Cover Photograph © Joseph Drivas/The Image Bank

 © 1992, 1987, 1982, 1978, 1972 by Prentice-Hall, Inc.
A Simon & Schuster Company
Englewood Cliffs, New Jersey 07632

PSpice is a registered trademark of MicroSim Corporation.

Printed in the United States of America

10 9 8 7 6 5 4 3 2 1

ISBN 0-13-250994-6

ISBN 0-13-251000-6 (Special Edition)

Prentice-Hall International (UK) Limited, *London*
Prentice-Hall of Australia Pty. Limited, *Sydney*
Prentice-Hall Canada Inc., *Toronto*
Prentice-Hall Hispanoamericana, S.A., *Mexico*
Prentice-Hall of India Private Limited, *New Delhi*
Prentice-Hall of Japan, Inc., *Tokyo*
Prentice-Hall of Southeast Asia Pte. Ltd., *Singapore*
Editora Prentice-Hall do Brasil, Ltda., *Rio de Janeiro*

Dedicated to:
ELSE MARIE, ERIC, ALISON, STACEY, and JOHANNA
and to
KATRIN, KIRA, LARREN, THOMAS, and JUSTIN

Contents

2 DIODE APPLICATIONS 51

3 BIPOLAR JUNCTION TRANSISTOR 108

4 DC BIASING BJTS 138

5

FIELD EFFECT TRANSISTORS

207

6

FET BIASING

248

7

BJT TRANSISTOR MODELING

301

8 BJT SMALL SIGNAL ANALYSIS

9 FET SMALL SIGNAL ANALYSIS

10 SYSTEMS APPROACH— EFFECTS OF R_S AND R_L

11 BJT AND JFET FREQUENCY RESPONSE   475

12 COMPOUND CONFIGURATIONS   526

13 DISCRETE AND IC MANUFACTURING TECHNIQUES   571

22 OSCILLOSCOPE AND OTHER MEASURING INSTRUMENTS 872

APPENDIX A HYBRID PARAMETERS— CONVERSION EQUATIONS (EXACT AND APPROXIMATE) 890

APPENDIX B RIPPLE FACTOR AND VOLTAGE CALCULATIONS 892

APPENDIX C CHARTS AND TABLES 899

APPENDIX D PSPICE 901

APPENDIX E SOLUTIONS TO SELECTED ODD-NUMBERED PROBLEMS 903

INDEX 909

Preface

As we approached the 20th anniversary of the text it became increasingly clear that this 5th edition should represent a major revision of the work. The growing use of computer software, packaged IC units, and the expanded range of coverage necessary in the basic courses all were contributing factors in defining the content of this edition. Our continued teaching experience with the subject matter, feedback from numerous educators and reviewers and comments from students have helped define an improved pedagogy for the text. The general appearance of the text needed to be enhanced for improved readability—making the text material appear ''friendlier'' to the broad range of students. Accuracy is obviously very important, and considerable effort was directed toward careful reviewing of all examples, artwork, problems, and technical development of concepts. Inconsistencies in the artwork developed over past editions have been removed by a totally new rendering of all the artwork using computer generated figures. A broad range of ancillary material will accompany the text to support the educational process.

PEDAGOGY

Without question one of the most important improvements in this text is the manner in which the content lends itself to the typical course syllabus. Not only has the order of chapters and internal sections been changed but sections of a chapter that were primarily reading material has been moved to later chapters to insure a continuous, logical, sequential presentation of important concepts and methods of analysis. Our teaching experience with the new presentation has reinforced our belief that the material now has an improved pedagogy to support the instructor's lecture and help the student build the foundation necessary for his/her future studies. The number of examples has been substantially increased and isolated bold-faced (''bullet'') statements have been introduced to identify important statements and conclusions. The format has been revamped to establish a friendlier appearance to the student and insure that the artwork is as close to the reference as possible. An additional color was employed in a manner that helps define important characteristics or isolate specific quantities in a network or on a characteristic. Icons have been developed for each chapter of the text to facilitate referencing a particular area of the text as quickly as possible. Problems have been developed for each section of the text that progress from the simple to the more complex. In addition, an asterisk has been added to identify the more difficult exercises. The title of each section is also reproduced in the problem section to clearly identify the exercises of interest for a particular topic of study.

SYSTEMS APPROACH

On numerous visits to other schools, technical institutes and meetings of various societies it was noted that a more "systems approach" should be developed to support a student's need to become adept in the application of packaged systems. Chapters 8, 9, and 10 were specifically reorganized to develop the foundation of systems analysis to the degree possible at this introductory level. Although it may be easier to consider the effects of R_s and R_L with each configuration when first introduced, the effects of R_s and R_L also provide an opportunity to apply some of the fundamental concepts of system analysis. The later chapters on Op-amps and IC units will further develop the concepts introduced in these early chapters.

ACCURACY

There is no question that a primary goal of any publication is that it be as free of errors as possible. Certainly, the intent is not to challenge the instructor or student with planned inconsistencies. In fact, there is nothing more distressing to an author than to hear of errors in a text. To insure the highest level of accuracy for this edition there were three technical reviewers in addition to the efforts of both authors. In addition, solutions to many problems and examples were checked using the computer. We now feel certain that this text will enjoy the highest level of accuracy obtainable for a publication of this kind.

TRANSISTOR MODELLING

BJT transistor modelling is area that is approached in various ways. Some institutions employ the r_e model exclusively while others lean toward the hybrid approach or a combination of these two. This edition will emphasize the r_e model with sufficient coverage of the hybrid model to permit comparison between models and the application of both. An entire chapter (Chapter 7) has been devoted to the introduction of the models to insure a clear, correct understanding of each and the relationships that exist between the two.

PSpice AND BASIC

The last few years have seen a continuing growth of the computer content in introductory courses. Not only is the use of word-processing appearing in the first semester, but spreadsheets and the use of a software analysis package such as PSpice is also being introduced in numerous educational institutions.

PSpice was chosen as the package to appear throughout this text because recent surveys suggest that it is most frequently employed. Other possible packages include Micro-Cap III and Breadboard. The coverage of PSpice provides sufficient content to permit writing the input file for the majority of networks analyzed in this text. No prior knowledge of computer software packages is presumed.

There are a number of BASIC programs still included in the text to demonstrate the advantages of knowing a computer language and the additional benefits that arise from its use.

TROUBLESHOOTING

Troubleshooting is undoubtedly one of the most difficult to introduce, develop and demonstrate in a text mode. It is an art that can be introduced using a variety of techniques but experience and exposure are obviously the key elements in developing the necessary skills. The content is essentially a review of situations that frequently occur in the laboratory environment. Some general hints as to how to isolate a problem area are introduced along with a list of typical causes. This is not to suggest that the student will become proficient in the debugging of networks introduced in this text, but at the very least the reader will have some understanding of what is involved with the troubleshooting process.

NEW MATERIAL

Specific areas of the text have been expanded and new material has been introduced to satisfy the changing requirements of the basic electronics courses. The application of various devices has been increased in number with an increased emphasis on the frequency response. Op-amps are a particularly important component in today's market and have received the full treatment with two expanded chapters of coverage.

The chapter on instrumentation has reappeared in this edition in response to a survey of current users and the obvious need to provide some reading material for those students with limited laboratory experience. The user guides provided with laboratory equipment is seldom at the reading level of new electronics students.

ANCILLARIES

The range of ancillary material has grown considerably. In addition to a completely revised laboratory manual with an associated instructor's manual (with typical data) there is a disk with all the input files of the PSpice programs and more than 250 transparency masters. The Instructor's Solutions Manual for the text has been carefully prepared and reviewed to insure the highest level of accuracy. In fact, a majority of the solutions were tested using PSpice.

USE OF TEST

In general the text is broken down into two main components, the dc analysis and the ac or frequency response. For some schools the dc section is sufficient for a one semester sequence while for others the entire text may be covered in one semester by choosing specific topics. In any event the text is one that "builds" from the early chapters. Superfluous material is relegated to the later chapters to avoid excessive content on a particular subject early in the development stage. For each device the text has covered a majority of the important configurations and applications. By choosing specific examples and applications the content of a course can be reduced without losing the progressive building characteristics of the text. Then again, if an instructor feels a specific area particularly important the detail is provided for a more extensive review.

ROBERT BOYLESTAD
WESTPORT, CONN.

LOUIS NASHELSKY
GREAT NECK, N.Y.

Acknowledgments

Our sincerest appreciation must be extended to the instructors who have used the text and sent in comments, corrections, and suggestions. Particular thanks is owed to Dr. Domingo Uy, Prof. Arthur Birch, and Prof. Scott Bisland for carrying the text material from manuscript stage, to galley proofs and finally into page form insuring a high level of accuracy throughout the text. We also want to thank Marcia Krefetz, Production Editor at Prentice Hall, for her heroic effort in keeping together the many detailed aspects of production. Our appreciation and thanks must also be extended to Ron Weickart of Network Graphics for his diligent attention to the art program through its many stages of development. Our sincerest thanks to Holly Hodder, Editor at Prentice Hall, for her commitment to producing a high quality, comprehensive, and error-free text. In addition, we want to thank Prof. Arthur Birch for his assistance with the instructor's manual.

Reviewer List

Saeed A. Shaikh	Miami-Dade Community College, Miami, FL
Thomas K. Grady	Western Washington University, Bellingham, WA
*Kenneth E. Kent	DeKalb Technical Institute, Clarkston, GA
Katherine L. Usik	Mohawk College of Applied Art & Technology, Hamilton, Ontario, CANADA
Phil Golden	DeVry Institute of Technology, Irving, TX
*Joe Baker	University of Southern California, Los Angeles, CA
Joseph Grabinski	Hartford State Technical College, Hartford, CT
*Richard J. Walters	DeVry Technical Institute, Woodbridge, NJ
*Jeffrey Bowe	Bunker Hill Community College, Charlestown, MA
Alfred D. Buerosse	Waukesha County Technical College, Pewaukee, WI
Gary C. Bocksch	Charles S. Mott Community College, Flint, MI
*A. Duane Bailey	Southern Alberta Institute of Technology, Calgary, Alberta, CANADA
*Jean Younes	ITT Technical Institute, Troy, MI
Ulrich E. Zeisler	Salt Lake Community College, Salt Lake City, UT
*John Darlington	Humber College, Ontario, CANADA
George T. Mason	Indiana Vocational Technical College, South Bend, IN
*Donald E. McMillan	Southwest State University, Marshall, MN
*Domingo Uy	Hampton University, Hampton, VA
Lucius B. Day	Metropolitan State College, Denver, CO
*Alan H. Czarapata	Montgomery College, Rockville, MD
Arthur Birch	Hartford State Technical College, Hartford, CT
E.F. Rockafellow	Southern Alberta Institute of Technology, Calgary, Alberta, CANADA
Scott Bisland	SEMATECH, Austin, TX

Questionnaire List

Richard J. Walters	DeVry Technical Institute, Woodbridge, NJ
Dr. Domingo L. Uy	Hampton University, Hampton, VA
Julian Wilson	Southern College of Technology, Marietta, GA
Jeffrey J. Bowe	Bunker Hill Community College, Charlestown, MA
Jean Younes	ITT Technical Institute, Troy, MI
Donald E. McMillan	Southwest State University, Marshall, MN
Alan H. Czarapata	Montgomery College, Rockville, MD
John Darlington	Humber College, Ontario, CANADA
Kenneth E. Kent	DeKalb Technical Institute, Clarkston, GA
John MacDougall	University of Western Ontario, London, Ontario, CANADA
F.D. Fuller	Humber College, Ontario, CANADA
Donald P. Szymanski	Owens Technical College, Toledo, OH
Ernest Lee Abbott	Napa Valley College, Napa, CA
Mike Durren	Indiana Vocational Technical College, South Bend, IN
Phillip D. Anderson	Muskegon Community College, Muskegon, MI
Charles E. Yunghans	Western Washington University, Bellingham, WA
Jeng-Nan Juang	Mercer University, Macon, GA
Joe Baker	University of Southern California, Los Angeles, CA
Donald E. King	ITT Technical Institute, Youngstown, OH
Albert L. Ickstadt	San Diego Mesa College, San Diego, CA
Parker M. Tabor	Greenville Technical College, Greenville, SC
A. Duane Bailey	Southern Alberta Institute of Technology, Calgary, Alberta, CANADA
Thomas E. Newman	L.H. Bates Vocational-Technical Institute, Tacoma, WA
Peter Tampas	Michigan Technological University, Houghton, MI

*THESE PEOPLE ARE ALSO LISTED ON THE QUESTIONNAIRE LIST.

List of Industrial Contributors

Mr. Al Anthony	EG&G VACTEC Inc.
Mr. Edward Bloch	The Perkin-Elmer Corporation
Ms. Sunny Carlson	MicroSim Corporation
Mr. Robert Casiano	International Rectifier Corporation
Mr. Jeff Gorin	Motorola Inc.
Dr. Jeff Bukhman	Motorola Inc.
Mrs. Karen Karger	Tektronix Inc.
Mr. Charles Lewis	APPLIED MATERIALS, Inc.
Mr. Elmer Smalling III	Texas Instruments Inc.
Mr. Eric Sung	Computronics Technology Inc.

Galley Reviewers

Scott Bisland	SEMATECH, Austin, TX
Arthur Birch	Hartford State Technical College, Hartford, CT
Domingo Uy	Hampton University, Hampton, VA

ELECTRONIC DEVICES AND CIRCUIT THEORY

Semiconductor Diodes

1.1 INTRODUCTION

The few decades following the introduction of the transistor in the late 1940s have seen a very dramatic change in the electronics industry. The miniaturization that has resulted leaves us to wonder about its limits. Complete systems now appear on a wafer thousands of times smaller than the single element of earlier networks. The advantages associated with current systems as compared to the tube networks of prior years are, for the most part, immediately obvious: smaller and lightweight, no heater requirement or heater loss (as required for tubes), more rugged construction, more efficient, and not requiring a warm-up period.

The miniaturization of recent years has resulted in systems so small that the primary purpose of the container is simply to provide some means of handling the device and ensuring that the leads remain properly fixed to the semiconductor wafer. The limits of miniaturization appear to be limited by three factors: the quality of the semiconductor material itself, the network design technique, and the limits of the manufacturing and processing equipment.

1.2 IDEAL DIODE

The first electronic device to be introduced is called the *diode*. It is the simplest of semiconductor devices but plays a very vital role in electronic systems, with its characteristics that closely match those of a simple switch. It will appear in a range of applications, extending from the simple to the very complex. In addition to the details of its construction and characteristics, the very important data and graphs to be found on specification sheets will also be covered to ensure an understanding of the terminology employed and to demonstrate the wealth of information typically available from manufacturers.

Before examining the construction and characteristics of an actual device, we first consider the ideal device, to provide a basis for comparison. The *ideal diode* is a *two-terminal* device having the symbol and characteristics shown in Fig. 1.1a and b, respectively.

Ideally, a diode will conduct current in the direction defined by the arrow in the symbol and act like an open circuit to any attempt to establish current in the opposite direction. In essence:

The characteristics of an ideal diode are those of a switch that can conduct current in only one direction.

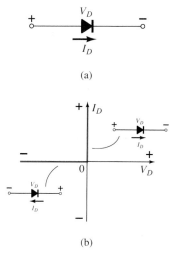

Figure 1.1 Ideal diode: (a) symbol; (b) characteristics.

1

In the description of the elements to follow, it is critical that the various *letter symbols, voltage polarities,* and *current directions* be defined. If the polarity of the applied voltage is consistent with that shown in Fig. 1.1a, the portion of the characteristics to be considered in Fig. 1.1b, is to the right of the vertical axis. If a reverse voltage is applied, the characteristics to the left are pertinent. If the current through the diode has the direction indicated in Fig. 1.1a, the portion of the characteristics to be considered is above the horizontal axis, while a reversal in direction would require the use of the characteristics below the axis. For the majority of the device characteristics to appear in this book the *ordinate* (or "y" axis) will be the *current* axis, while the *abscissa* (or "x"axis) will be the *voltage axis*.

One of the important parameters for the diode is the resistance at the point or region of operation. If we consider the conduction region defined by the direction of I_D and polarity of V_D in Fig. 1.1a (upper-right quadrant of Fig. 1.1b), we will find that the value of the forward resistance, R_F, as defined by Ohm's law is

$$R_F = \frac{V_F}{I_F} = \frac{0 \text{ V}}{2, 3, \text{ mA}, \ldots, \text{ or any positive value}} = 0 \text{ }\Omega \quad \text{(short circuit)}$$

where V_F is the forward voltage across the diode and I_F is the forward current through the diode.

The ideal diode, therefore, is a short circuit for the region of conduction.

If we now consider the region of negatively applied potential (third quandrant) of Fig. 1.1b,

$$R_R = \frac{V_R}{I_R} = \frac{-5, -20, \text{ or any reverse-bias potential}}{0 \text{ mA}} = \infty \text{ }\Omega \quad \text{(open-circuit)}$$

where V_R is the reverse voltage across the diode and I_R is the reverse current in the diode.

The ideal diode, therefore, is an open circuit in the region of nonconduction.

In review, the conditions depicted in Fig. 1.2 are applicable.

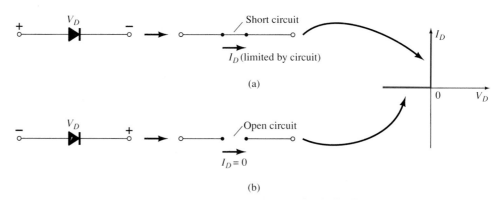

Figure 1.2 (a) Conduction and (b) nonconduction states of the ideal diode as determined by the applied bias.

In general, it is relatively simple to determine whether a diode is in the region of conduction or nonconduction simply by noting the direction of the current I_D established by an applied voltage. For conventional flow (opposite to that of electron flow), if the resultant diode current has the same direction as the arrowhead of the diode symbol, the diode is operating in the conducting region as depicted in Fig. 1.3a. If the resulting current has the opposite direction, as shown in Fig. 1.3b, the open-circuit equivalent is appropriate.

Chapter 1 Semiconductor Diodes

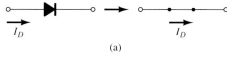

(a)

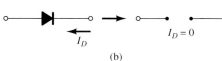

(b)

Figure 1.3 (a) Conduction and (b) nonconduction states of the ideal diode as determined by the direction of conventional current established by the network.

As indicated earlier, the primary purpose of this section is to introduce the characteristics of an ideal device for comparison with the characteristics of the commercial variety. As we progress through the next few sections, keep the following questions in mind:

How close will the forward or "on" resistance of a practical diode compare with the desired 0-Ω level?

Is the reverse-bias resistance sufficiently large to permit an open-circuit approximation?

1.3 SEMICONDUCTOR MATERIALS

The label *semiconductor* itself provides a hint as to its characteristics. The prefix *semi-* is normally applied to a range of levels midway between two limits.

The term conductor is applied to any material that will support a generous flow of charge when a voltage source of limited magnitude is applied across its terminals.

An insulator is a material that offers a very low level of conductivity under pressure from an applied voltage source.

A semiconductor, therefore, is a material that has a conductivity level somewhere between the extremes of an insulator and a conductor.

Inversely related to the conductivity of a material is its resistance to the flow of charge, or current. That is, the higher the conductivity level, the lower the resistance level. In tables, the term *resistivity* (ρ, Greek letter rho) is often used when comparing the resistance levels of materials. In metric units, the resistivity of a material is measured in Ω-cm or Ω-m. The units of Ω-cm is derived from the substitution of the units for each quantity of Fig. 1.4 into the following equation (derived from the basic resistance equation $R = \rho l/A$):

$$\rho = \frac{RA}{l} = \frac{(\Omega)(\text{cm}^2)}{\text{cm}} \Rightarrow \Omega\text{-cm} \qquad (1.1)$$

In fact, if the area of Fig. 1.4 is 1 cm² and the length 1 cm, the magnitude of the resistance of the cube of Fig. 1.4 is equal to the magnitude of the resistivity of the material as demonstrated below:

$$|R| = \rho \frac{l}{A} = \rho \frac{(1 \text{ cm})}{(1 \text{ cm}^2)} = |\rho|\text{ohms}$$

This fact will be helpful to remember as we compare resistivity levels in the discussions to follow.

In Table 1.1, typical resistivity values are provided for three broad categories of materials. Although you may be familiar with the electrical properties of copper and mica from your past studies, the characteristics of the semiconductor materials of

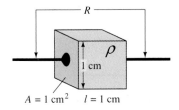

Figure 1.4 Defining the metric units of resistivity.

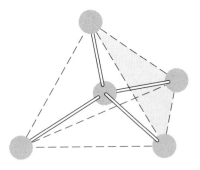

TABLE 1.1 Typical Resistivity Values		
Conductor	*Semiconductor*	*Insulator*
$\rho \cong 10^{-6}$ Ω-cm (copper)	$\rho \cong 50$ Ω-cm (germanium) $\rho \cong 50 \times 10^3$ Ω-cm (silicon)	$\rho \cong 10^{12}$ Ω-cm (mica)

germanium (Ge) and silicon (Si) may be relatively new. As you will find in the chapters to follow, they are certainly not the only two semiconductor materials. They are, however, the two materials that have received the broadest range of interest in the development of semiconductor devices. In recent years the shift has been steadily toward silicon and away from germanium, but germanium is still in modest production.

Note in Table 1.1 the extreme range between the conductor and insulating materials for the 1-cm length (1-cm^2 area) of the material. Eighteen places separate the placement of the decimal point for one number from the other. Ge and Si have received the attention they have for a number of reasons. One very important consideration is the fact that they can be manufactured to a very high purity level. In fact, recent advances have reduced impurity levels in the pure material to 1 part in 10 billion (1:10,000,000,000). One might ask if these low impurity levels are really necessary. They certainly are if you consider that the addition of one part impurity (of the proper type) per million in a wafer of silicon material can change that material from a relatively poor conductor to a good conductor of electricity. We are obviously dealing with a whole new spectrum of comparison levels when we deal with the semiconductor medium. The ability to change the characteristics of the material significantly through this process, known as "doping," is yet another reason why Ge and Si have received such wide attention. Further reasons include the fact that their characteristics can be altered significantly through the application of heat or light—an important consideration in the development of heat- and light-sensitive devices.

Some of the unique qualities of Ge and Si noted above are due to their atomic structure. The atoms of both materials form a very definite pattern that is periodic in nature (i.e., continually repeats itself). One complete pattern is called a *crystal* and the periodic arrangement of the atoms a *lattice*. For Ge and Si the crystal has the three-dimensional diamond structure of Fig. 1.5. Any material composed solely of repeating crystal structures of the same kind is called a *single-crystal* structure. For semiconductor materials of practical application in the electronics field, this single-crystal feature exists, and, in addition, the periodicity of the structure does not change significantly with the addition of impurities in the doping process.

Let us now examine the structure of the atom itself and note how it might affect the electrical characteristics of the material. As you are aware, the atom is composed of three basic particles; the *electron*, the *proton*, and the *neutron*. In the atomic lattice, the neutrons and protons form the *nucleus*, while the electrons revolve around the nucleus in a fixed *orbit*. The Bohr models of the two most commonly used semiconductors, *germanium* and *silicon*, are shown in Fig. 1.6.

As indicated by Fig. 1.6a, the germanium atom has 32 orbiting electrons, while silicon has 14 orbiting electrons. In each case, there are 4 electrons in the outermost (*valence*) shell. The potential (*ionization potential*) required to remove any one of these 4 valence electrons is lower than that required for any other electron in the structure. In a pure germanium or silicon crystal these 4 valence electrons are bonded to 4 adjoining atoms, as shown in Fig. 1.7 for silicon. Both Ge and Si are referred to as *tetravalent atoms* because they each have four valence electrons.

A bonding of atoms, strengthened by the sharing of electrons, is called covalent bonding.

Figure 1.5 Ge and Si single-crystal structure.

Chapter 1 Semiconductor Diodes

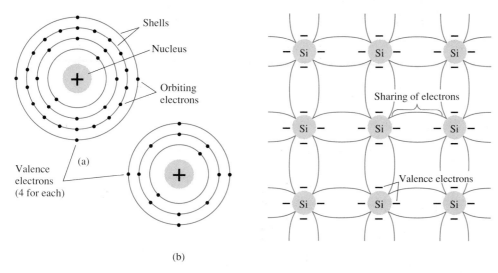

Figure 1.6 Atomic structure: (a) germanium; (b) silicon.

Figure 1.7 Covalent bonding of the silicon atom.

Although the covalent bond will result in a stronger bond between the valence electrons and their parent atom, it is still possible for the valence electrons to absorb sufficient kinetic energy from natural causes to break the covalent bond and assume the "free" state. The term "free" reveals that their motion is quite sensitive to applied electric fields such as established by voltage sources or any difference in potential. These natural causes include effects such as light energy in the form of photons and thermal energy from the surrounding medium. At room temperature there are approximately 1.5×10^{10} free carriers in a cubic centimeter of intrinsic silicon material.

Intrinsic materials are those semiconductors that have been carefully refined to reduce the impurities to a very low level—essentially as pure as can be made available through modern technology.

The free electrons in the material due only to natural causes are referred to as *intrinsic carriers*. At the same temperature, intrinsic germanium material will have approximately 2.5×10^{13} free carriers per cubic centimeter. The ratio of the number of carriers in germanium to that of silicon is greater than 10^3 and would indicate that germanium is a better conductor at room temperature. This may be true, but both are still considered poor conductors in the intrinsic state. Note in Table 1.1 that the resistivity also differs by a ratio of about 1000:1, with silicon having the larger value. This should be the case, of course, since resistivity and conductivity are inversely related.

An increase in temperature of a semiconductor can result in a substantial increase in the number of free electrons in the material.

As the temperature rises from absolute zero (0 K), an increasing number of valence electrons absorb sufficient thermal energy to break the covalent bond and contribute to the number of free carriers as described above. This increased number of carriers will increase the conductivity index and result in a lower resistance level.

Semiconductor materials such as Ge and Si that show a reduction in resistance with increase in temperature are said to have a negative temperature coefficient.

You will probably recall that the resistance of most conductors will increase with temperature. This is due to the fact that the numbers of carriers in a conductor will not increase significantly with temperature, but their vibration pattern about a relatively fixed location will make it increasingly difficult for electrons to pass through. An increase in temperature therefore results in an increased resistance level and a *positive temperature coefficient*.

1.4 ENERGY LEVELS

In the isolated atomic structure there are discrete (individual) energy levels associated with each orbiting electron, as shown in Fig. 1.8a. Each material will, in fact, have its own set of permissible energy levels for the electrons in its atomic structure.

The more distant the electron from the nucleus, the higher the energy state, and any electron that has left its parent atom has a higher energy state than any electron in the atomic structure.

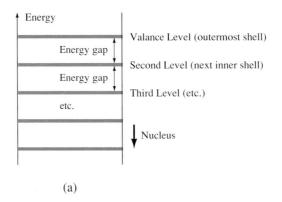

(a)

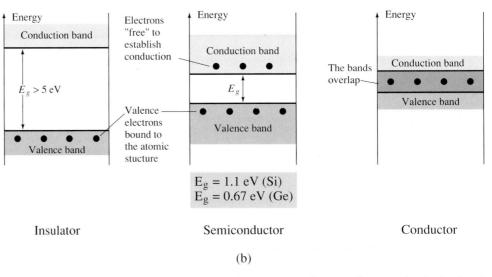

$$E_g = 1.1 \text{ eV (Si)}$$
$$E_g = 0.67 \text{ eV (Ge)}$$

Insulator   Semiconductor   Conductor

(b)

Figure 1.8 Energy levels: (a) discrete levels in isolated atomic structures; (b) conduction and valence bands of an insulator, semiconductor, and conductor.

Between the discrete energy levels are gaps in which no electrons in the isolated atomic structure can appear. As the atoms of a material are brought closer together to form the crystal lattice structure, there is an interaction between atoms that will result in the electrons in a particular orbit of one atom having slightly different energy levels from electrons in the same orbit of an adjoining atom. The net result is an expansion of the discrete levels of possible energy states for the valence electrons to that of bands as shown in Fig. 1.8b. Note that there are boundary levels and maximum energy states in which any electron in the atomic lattice can find itself, and there remains a *forbidden region* between the valence band and the ionization level. Recall

Chapter 1   Semiconductor Diodes

that ionization is the mechanism whereby an electron can absorb sufficient energy to break away from the atomic structure and enter the conduction band. You will note that the energy associated with each electron is measured in *electron volts* (eV). The unit of measure is appropriate, since

$$\boxed{W = QV} \qquad \text{eV} \qquad (1.2)$$

as derived from the defining equation for voltage $V = W/Q$. The charge Q is the charge associated with a single electron.

Substituting the charge of an electron and a potential difference of 1 volt into Eq. (1.2) will result in an energy level referred to as one *electron volt*. Since energy is also measured in joules and the charge of one electron $= 1.6 \times 10^{-19}$ coulomb,

$$W = QV = (1.6 \times 10^{-19} \text{ C})(1 \text{ V})$$

and

$$\boxed{1 \text{ eV} = 1.6 \times 10^{-19} \text{ J}} \qquad (1.3)$$

At 0 K or absolute zero ($-273.15°$C), all the valence electrons of semiconductor materials find themselves locked in their outermost shell of the atom with energy levels associated with the valence band of Fig. 1.8b. However, at room temperature (300 K, 25°C) a large number of valence electrons have acquired sufficient energy to leave the valence band, cross the energy gap defined by E_g in Fig. 1.8b and enter the conduction band. For silicon E_g is 1.1 eV, whereas for germanium it is 0.67 V. The obviously lower E_g for germanium accounts for the increased number of carriers in that material as compared to silicon at room temperature. Note for the insulator that the energy gap is typically 5 eV or more, which severely limits the number of electrons that can enter the conduction band at room temperature. The conductor has electrons in the conduction band even at 0 K. Quite obviously, therefore, at room temperature there are more than enough free carriers to sustain a heavy flow of charge, or current.

We will find in Section 1.5 that if certain impurities are added to the intrinsic semiconductor materials, energy states in the forbidden bands will occur which will cause a net reduction in E_g for both semiconductor materials—consequently, increased carrier density in the conduction band at room temperature!

1.5 EXTRINSIC MATERIALS—*n*- AND *p*-TYPE

The characteristics of semiconductor materials can be altered significantly by the addition of certain impurity atoms into the relatively pure semiconductor material. These impurities, although only added to perhaps 1 part in 10 million, can alter the band structure sufficiently to totally change the electrical properties of the material.

A semiconductor material that has been subjected to the doping process is called an extrinsic material.

There are two extrinsic materials of immeasurable importance to semiconductor device fabrication: *n*-type and *p*-type. Each will be described in some detail in the following paragraphs.

n-Type Material

Both the *n*- and *p*-type materials are formed by adding a predetermined number of impurity atoms into a germanium or silicon base. The *n*-type is created by introducing those impurity elements that have *five* valence electrons (*pentavalent*), such as *antimony, arsenic,* and *phosphorus*. The effect of such impurity elements is indicated in

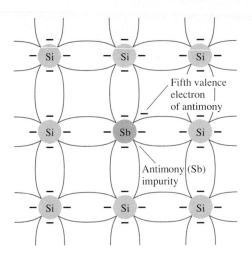

Figure 1.9 Antimony impurity in *n*-type material.

Fig. 1.9 (using antimony as the impurity in a silicon base). Note that the four covalent bonds are still present. There is, however, an additional fifth electron due to the impurity atom, which is *unassociated* with any particular covalent bond. This remaining electron, loosely bound to its parent (antimony) atom, is relatively free to move within the newly formed *n*-type material. Since the inserted impurity atom has donated a relatively "free" electron to the structure:

Diffused impurities with five valence electrons are called donor atoms.

It is important to realize that even though a large number of "free" carriers have been established in the *n*-type material, it is still electrically *neutral* since ideally the number of positively charged protons in the nuclei is still equal to the number of "free" and orbiting negatively charged electrons in the structure.

The effect of this doping process on the relative conductivity can best be described through the use of the energy-band diagram of Fig. 1.10. Note that a discrete energy level (called the *donor level*) appears in the forbidden band with an E_g significantly less than that of the intrinsic material. Those "free" electrons due to the added impurity sit at this energy level and have less difficulty absorbing a sufficient measure of thermal energy to move into the conduction band at room temperature. The result is that at room temperature, there are a large number of carriers (electrons) in the conduction level and the conductivity of the material increases significantly. At room temperature in an intrinsic Si material there is about one free electron for every 10^{12} atoms (1 to 10^9 for Ge). If our dosage level were 1 in 10 million (10^7), the ratio ($10^{12}/10^7 = 10^5$) would indicate that the carrier concentration has increased by a ratio of 100,000:1.

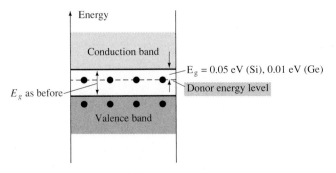

Figure 1.10 Effect of donor impurities on the energy band structure.

p-Type Material

The *p*-type material is formed by doping a pure germanium or silicon crystal with impurity atoms having *three* valence electrons. The elements most frequently used for this purpose are *boron, gallium,* and *indium.* The effect of one of these elements, boron, on a base of silicon is indicated in Fig. 1.11.

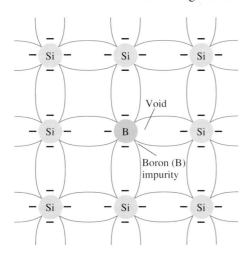

Figure 1.11 Boron impurity in *p*-type material.

Note that there is now an insufficient number of electrons to complete the covalent bonds of the newly formed lattice. The resulting vacancy is called a *hole* and is represented by a small circle or positive sign due to the absence of a negative charge. Since the resulting vacancy will readily *accept* a "free" electron:

The diffused impurities with three valence electrons are called acceptor atoms.

The resulting *p*-type material is electrically neutral, for the same reasons described for the *n*-type material.

Electron versus Hole Flow

The effect of the hole on conduction is shown in Fig. 1.12. If a valence electron acquires sufficient kinetic energy to break its covalent bond and fills the void created by a hole, then a vacancy, or hole, will be created in the covalent bond that released the electron. There is therefore a transfer of holes to the left and electrons to the right, as shown in Fig. 1.12. The direction to be used in this text is that of *conventional flow,* which is indicated by the direction of hole flow.

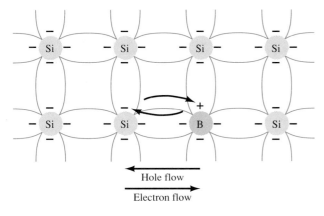

Figure 1.12 Electron versus hole flow.

Majority and Minority Carriers

In the intrinsic state, the number of free electrons in Ge or Si is due only to those few electrons in the valence band that have acquired sufficient energy from thermal or light sources to break the covalent bond or to the few impurities that could not be removed. The vacancies left behind in the covalent bonding structure represent our very limited supply of holes. In an *n*-type material, the number of holes has not changed significantly from this intrinsic level. The net result, therefore, is that the number of electrons far outweighs the number of holes. For this reason:

In an n-*type material (Fig. 1.13a) the electron is called the majority carrier and the hole the minority carrier.*

For the *p*-type material the number of holes far outweighs the number of electrons, as shown in Fig. 1.13b. Therefore:

In a p-*type material the hole is the majority carrier and the electron is the minority carrier.*

When the fifth electron of a donor atom leaves the parent atom, the atom remaining acquires a net positive charge: hence the positive sign in the donor-ion representation. For similar reasons, the negative sign appears in the acceptor ion.

The *n*- and *p*-type materials represent the basic building blocks of semiconductor devices. We will find in the next section that the "joining" of a single *n*-type material with a *p*-type material will result in a semiconductor element of considerable importance in electronic systems.

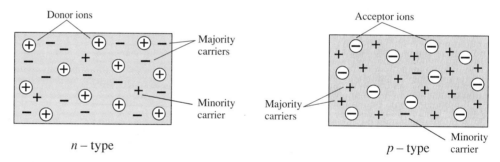

Figure 1.13 (a) *n*-type material; (b) *p*-type material.

1.6 SEMICONDUCTOR DIODE

In Section 1.5 both the *n*- and *p*-type materials were introduced. The semiconductor diode is formed by simply bringing these materials together (constructed from the same base—Ge or Si), as shown in Fig. 1.14, using techniques to be described in Chapter 20. At the instant the two materials are "joined" the electrons and holes in the region of the junction will combine resulting in a lack of carriers in the region near the junction.

This region of uncovered positive and negative ions is called the depletion region due to the depletion of carriers in this region.

Since the diode is a two-terminal device, the application of a voltage across its terminals leaves three possibilities: *no bias* ($V_D = 0$ V), *forward bias* ($V_D > 0$ V), and *reverse bias* ($V_D < 0$ V). Each is a condition that will result in a response that the user must clearly understand if the device is to be applied effectively.

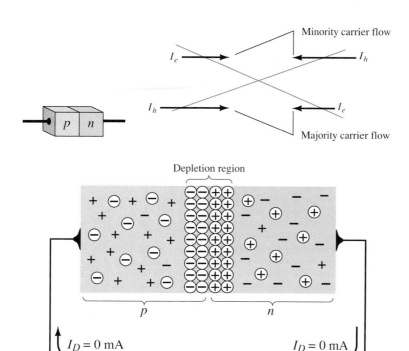

Figure 1.14 *p-n* junction with no external bias.

No Applied Bias ($V_D = 0$ V)

Under no-bias conditions, any minority carriers (holes) in the *n*-type material that find themselves within the depletion region will pass directly into the *p*-type material. The closer the minority carrier is to the junction, the greater the attraction for the layer of negative ions and the less the opposition of the positive ions in the depletion region of the *n*-type material. For the purposes of future discussions we shall assume that all the minority carriers of the *n*-type material that find themselves in the depletion region due to their random motion will pass directly into the *p*-type material. Similar discussion can be applied to the minority carriers (electrons) of the *p*-type material. This carrier flow has been indicated in Fig. 1.14 for the minority carriers of each material.

The majority carriers (electrons) of the *n*-type material must overcome the attractive forces of the layer of positive ions in the *n*-type material and the shield of negative ions in the *p*-type material in order to migrate into the area beyond the depletion region of the *p*-type material. However, the number of majority carriers is so large in the *n*-type material that there will invariably be a small number of majority carriers with sufficient kinetic energy to pass through the depletion region into the *p*-type material. Again, the same type of discussion can be applied to the majority carriers (holes) of the *p*-type material. The resulting flow due to the majority carriers is also shown in Fig. 1.14.

A close examination of Fig. 1.14 will reveal that the relative magnitudes of the flow vectors are such that the net flow in either direction is zero. This cancellation of vectors has been indicated by crossed lines. The length of the vector representing hole flow has been drawn longer than that for electron flow to demonstrate that the magnitude of each need not be the same for cancellation and that the doping levels for each material may result in an unequal carrier flow of holes and electrons. In summary, therefore:

In the absence of an applied bias voltage, the net flow of charge in any one direction for a semiconductor diode is zero.

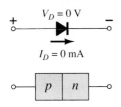

$V_D = 0 \text{ V}$

$I_D = 0 \text{ mA}$

Figure 1.15 No-bias conditions for a semiconductor diode.

The symbol for a diode is repeated in Fig. 1.15 with the associated *n*- and *p*-type regions. Note that the arrow is associated with the *p*-type component and the bar with the *n*-type region. As indicated, for $V_D = 0$ V, the current in any direction is 0 mA.

Reverse-Bias Condition ($V_D < 0$ V)

If an external potential of *V* volts is applied across the *p-n* junction such that the positive terminal is connected to the *n*-type material and the negative terminal is connected to the *p*-type material as shown in Fig. 1.16, the number of uncovered positive ions in the depletion region of the *n*-type material will increase due to the large number of "free" electrons drawn to the positive potential of the applied voltage. For similar reasons, the number of uncovered negative ions will increase in the *p*-type material. The net effect, therefore, is a widening of the depletion region. This widening of the depletion region will establish too great a barrier for the majority carriers to overcome, effectively reducing the majority carrier flow to zero as shown in Fig. 1.16.

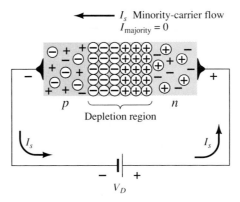

Figure 1.16 Reverse-biased *p-n* junction.

The number of minority carriers, however, that find themselves entering the depletion region will not change, resulting in minority-carrier flow vectors of the same magnitude indicated in Fig. 1.14 with no applied voltage.

The current that exists under reverse-bias conditions is called the reverse saturation current and is represented by I_s.

The reverse saturation current is seldom more than a few microamperes except for high-power devices. In fact, in recent years its level is typically in the nanoampere range for silicon devices and in the low-microampere range for germanium. The term *saturation* comes from the fact that it reaches its maximum level quickly and does not change significantly with increase in the reverse-bias potential, as shown on the diode characteristics of Fig. 1.19 for $V_D < 0$ V. The reverse-biased conditions are depicted in Fig. 1.17 for the diode symbol and *p-n* junction. Note, in particular, that the direction of I_s is against the arrow of the symbol. Note also that the negative potential is connected to the *p*-type material and the positive potential to the *n*-type material—the difference in underlined letters for each region revealing a reverse-bias condition.

Forward-Bias Condition ($V_D > 0$ V)

A *forward-bias* or "on" condition is established by applying the positive potential to the *p*-type material and the negative potential to the *n*-type material as shown in Fig. 1.18. For future reference, therefore:

A semiconductor diode is forward-biased when the association p-type and positive and n-type and negative has been established.

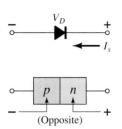

V_D

I_s

(Opposite)

Figure 1.17 Reverse-bias conditions for a semiconductor diode.

Chapter 1 Semiconductor Diodes

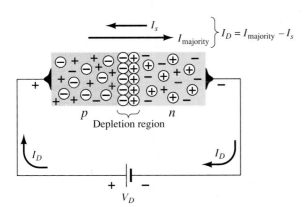

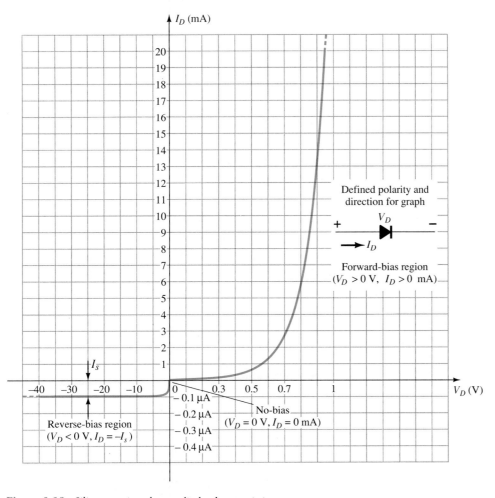

$$I_D = I_{majority} - I_s$$

Figure 1.18 Forward-biased *p-n* junction.

The application of a forward-bias potential V_D will ''pressure'' electrons in the *n*-type material and holes in the *p*-type material to recombine with the ions near the boundary and reduce the width of the depletion region as shown in Fig. 1.18. The resulting minority-carrier flow of electrons from the *p*-type material to the *n*-type material (and holes from the *n*-type material to the *p*-type material) has not changed in magnitude (since the conduction level is controlled primarily by the limited number of impurities in the material), but the reduction in the width of the depletion region has resulted in a heavy majority flow across the junction. An electron of the *n*-type

Figure 1.19 Silicon semiconductor diode characteristics.

material now "sees" a reduced barrier at the junction due to the reduced depletion region and a strong attraction for the positive potential applied to the *p*-type material. As the applied bias increases in magnitude the depletion region will continue to decrease in width until a flood of electrons can pass through the junction, resulting in an exponential rise in current as shown in the forward-bias region of the characteristics of Fig. 1.19. Note that the vertical scale of Fig. 1.19 is measured in milliamperes (although some semiconductor diodes will have a vertical scale measured in amperes) and the horizontal scale in the forward-bias region has a maximum of 1 V. Typically, therefore, the voltage across a forward-biased diode will be less than 1 V. Note also, how quickly the current rises beyond the knee of the curve.

It can be demonstrated through the use of solid-state physics that the general characteristics of a semiconductor diode can be defined by the following equation for the forward- and reverse-bias regions:

$$I_D = I_s(e^{kV_D/T_K} - 1) \tag{1.4}$$

where I_s = reverse saturation current

$k = 11,600/\eta$ with $\eta = 1$ for Ge and $\eta = 2$ for Si for relatively low levels of diode current (at or below the knee of the curve) and $\eta = 1$ for Ge and Si for higher levels of diode current (in the rapidly increasing section of the curve)

$T_K = T_C + 273°$

A plot of Eq. (1.4) is provided in Fig. 1.19. If we expand Eq. (1.4) into the following form, the contributing component for each region of Fig. 1.19 can easily be described:

$$I_D = I_s e^{kV_D/T_K} - I_s$$

For positive values of V_D the first term of the equation above will grow very quickly and overpower the effect of the second term. The result is that for positive values of V_D, I_D will be positive and grow as the function $y = e^x$ appearing in Fig. 1.20. At $V_D = 0$ V, Eq. (1.4) becomes $I_D = I_s(e^0 - 1) = I_s(1 - 1) = 0$ mA as appearing in Fig. 1.19. For negative values of V_D the first term will quickly drop off below I_s, resulting in $I_D = -I_s$, which is simply the horizontal line of Fig. 1.19. The break in the characteristics at $V_D = 0$ V is simply due to the dramatic change in scale from mA to μA.

It is important to note the change in scale for the vertical and horizontal axes. For positive values of I_D the scale is in milliamperes and the current scale below the axis is in microamperes (or possibly nanoamperes). For V_D the scale for positive values is in tenths of volts and for negative values the scale is in tens of volts.

Initially, Eq. (1.4) does appear somewhat complex and may develop an unwarranted fear that it will be applied for all the diode applications to follow. Fortunately, however, a number of approximations will be made in a later section that will negate the need to apply Eq. (1.4) and provide a solution with a minimum of mathematical difficulty.

Before leaving the subject of the forward-bias state the conditions for conduction (the "on" state) are repeated in Fig. 1.21 with the required biasing polarities and the resulting direction of majority-carrier flow. Note in particular how the direction of conduction matches the arrow in the symbol (as revealed for the ideal diode).

Zener Region

Even though the scale of Fig. 1.19 is in tens of volts in the negative region, there is a point where the application of too negative a voltage will result in a sharp change in the characteristics, as shown in Fig. 1.22. The current increases at a very rapid rate in

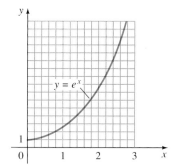

Figure 1.20 Plot of e^x.

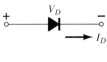
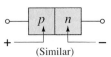

Figure 1.21 Forward-bias conditions for a semiconductor diode.

Chapter 1 Semiconductor Diodes

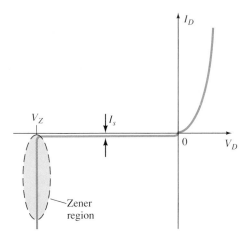

Figure 1.22 Zener region.

a direction opposite to that of the positive voltage region. The reverse-bias potential that results in this dramatic change in characteristics is called the *Zener potential* and is given the symbol V_Z.

As the voltage across the diode increases in the reverse-bias region, the velocity of the minority carriers responsible for the reverse saturation current I_s will also increase. Eventually, their velocity and associated kinetic energy ($W_K = \frac{1}{2}mv^2$) will be sufficient to release additional carriers through collisions with otherwise stable atomic structures. That is, an *ionization* process will result whereby valence electrons absorb sufficient energy to leave the parent atom. These additional carriers can then aid the ionization process to the point where a high *avalanche* current is established and the *avalanche breakdown* region determined.

The avalanche region (V_Z) can be brought closer to the vertical axis by increasing the doping levels in the *p*- and *n*-type materials. However, as V_Z decreases to very low levels, such as −5 V, another mechanism, called *Zener breakdown,* will contribute to the sharp change in the characteristic. It occurs because there is a strong electric field in the region of the junction that can disrupt the bonding forces within the atom and "generate" carriers. Although the Zener breakdown mechanism is only a significant contributor at lower levels of V_Z, this sharp change in the characteristic at any level is called the *Zener region* and diodes employing this unique portion of the characteristic of a *p-n* junction are called *Zener diodes*. They are described in detail in Section 1.14.

The Zener region of the semiconductor diode described must be avoided if the response of a system is not to be completely altered by the sharp change in character-istics in this reverse-voltage region.

The maximum reverse-bias potential that can be applied before entering the Zener region is called the peak inverse voltage (referred to simply as the PIV rating) or the peak reverse voltage (denoted by PRV rating).

If an application requires a PIV rating greater than that of a single unit, a number of diodes of the same characteristics can be connected in series. Diodes are also connected in parallel to increase the current-carrying capacity.

Silicon versus Germanium

Silicon diodes have, in general, higher PIV and current rating and wider temperature ranges than germanium diodes. PIV ratings for silicon can be in the neighborhood of 1000 V, whereas the maximum value for germanium is closer to 400 V. Silicon can be used for applications in which the temperature may rise to about 200°C (400°F), whereas germanium has a much lower maximum rating (100°C). The disadvantage of silicon, however, as compared to germanium, as indicated in Fig. 1.23, is the higher

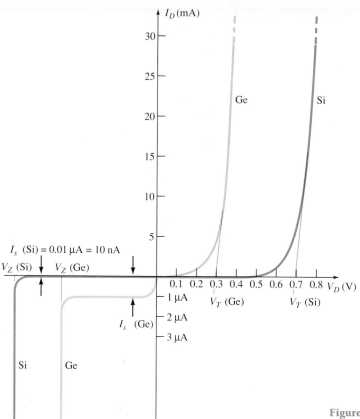

Figure 1.23 Comparison of Si and Ge semiconductor diodes.

forward-bias voltage required to reach the region of upward swing. It is typically of the order of magnitude of 0.7 V for *commercially* available silicon diodes and 0.3 V for germanium diodes when rounded off to the nearest tenths. The increased offset for silicon is due primarily to the factor η in Eq. (1.4). This factor plays a part in determining the shape of the curve only at very low current levels. Once the curve starts its vertical rise, the factor η drops to 1 (the continuous value for germanium). This is evidenced by the similarities in the curves once the offset potential is reached. The potential at which this rise occurs is commonly referred to as the *offset, threshold,* or *firing potential*. Frequently, the first letter of a term that describes a particular quantity is used in the notation for that quantity. However, to ensure a minimum of confusion with other terms, such as output voltage (V_o) and forward voltage (V_F), the notation V_T has been adopted for this book, from the word "threshold."

In review:

$$V_T = 0.7 \ (\text{Si})$$
$$V_T = 0.3 \ (\text{Ge})$$

Obviously, the closer the upward swing is to the vertical axis, the more "ideal" the device. However, the other characteristics of silicon as compared to germanium still make it the choice in the majority of commercially available units.

Temperature Effects

Temperature can have a marked effect on the characteristics of a silicon semiconductor diode as witnessed by a typical silicon diode in Fig. 1.24. It has been found experimentally that:

The reverse saturation current I_s will just about double in magnitude for every 10°C increase in temperature.

Chapter 1   Semiconductor Diodes

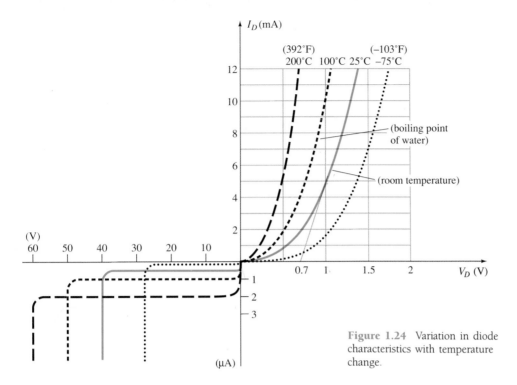

Figure 1.24 Variation in diode characteristics with temperature change.

It is not uncommon for a germanium diode with an I_s in the order of 1 or 2 μA at 25°C to have a leakage current of 100 μA = 0.1 mA at a temperature of 100°C. Current levels of this magnitude in the reverse-bias region would certainly question our desired open-circuit condition in the reverse-bias region. Typical values of I_s for silicon are much lower than that of germanium for similar power and current levels as shown in Fig. 1.23. The result is that even at high temperatures the levels of I_s for silicon diodes do not reach the same high levels obtained for germanium—a very important reason that silicon devices enjoy a significantly higher level of development and utilization in design. Fundamentally, the open-circuit equivalent in the reverse-bias region is better realized at any temperature with silicon than with germanium.

The increasing levels of I_s with temperature account for the lower levels of threshold voltage, as shown in Fig. 1.24. Simply increase the level of I_s in Eq. (1.4) and note the earlier rise in diode current. Of course, the level of T_K will also be increasing in the same equation, but the increasing level of I_s will overpower the smaller percent change in T_K. As the temperature increases the forward characteristics are actually becoming more "ideal," but we will find when we review the specifications sheets that temperatures beyond the normal operating range can have a very detrimental effect on the diode's maximum power and current levels. In the reverse-bias region the breakdown voltage is increasing with temperature, but note the undesirable increase in reverse saturation current.

1.7 RESISTANCE LEVELS

As the operating point of a diode moves from one region to another the resistance of the diode will also change due to the nonlinear shape of the characteristics curve. It will be demonstrated in the next few paragraphs that the type of applied voltage or signal will define the resistance level of interest. Three different levels will be introduced in this section that will appear again as we examine other devices. It is therefore paramount that their determination be clearly understood.

DC or Static Resistance

The application of a dc voltage to a circuit containing a semiconductor diode will result in an operating point on the characteristic curve that will not change with time. The resistance of the diode at the operating point can be found simply by finding the corresponding levels of V_D and I_D as shown in Fig. 1.25 and applying the following equation:

$$R_D = \frac{V_D}{I_D} \qquad\qquad (1.5)$$

The dc resistance levels at the knee and below will be greater than the resistance levels obtained for the vertical rise section of the characteristics. The resistance levels in the reverse-bias region will naturally be quite high. Since ohmmeters typically employ a relatively constant-current source, the resistance determined will be at a preset current level (typically, a few milliamperes).

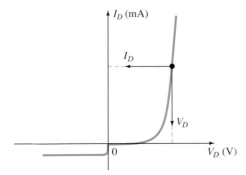

Figure 1.25 Determining the dc resistance of a diode at a particular operating point.

EXAMPLE 1.1

Determine the dc resistance levels for the diode of Fig. 1.26 at
(a) $I_D = 2$ mA
(b) $I_D = 20$ mA
(c) $V_D = -10$ V

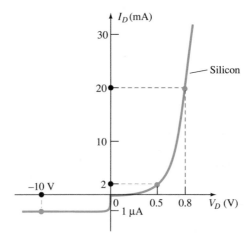

Figure 1.26 Example 1.1

Solution

(a) At $I_D = 2$ mA, $V_D = 0.5$ V (from the curve) and

$$R_D = \frac{V_D}{I_D} = \frac{0.5 \text{ V}}{2 \text{ mA}} = \mathbf{250 \ \Omega}$$

Chapter 1   Semiconductor Diodes

(b) At $I_D = 20$ mA, $V_D = 0.8$ V (from the curve) and

$$R_D = \frac{V_D}{I_D} = \frac{0.8\text{ V}}{20\text{ mA}} = \mathbf{40\ \Omega}$$

(c) At $V_D = -10$ V, $I_D = -I_s = -1$ μA (from the curve) and

$$R_D = \frac{V_D}{I_D} = \frac{10\text{ V}}{1\ \mu\text{A}} = \mathbf{10\ M\Omega}$$

clearly supporting some of the earlier comments regarding the dc resistance levels of a diode.

AC or Dynamic Resistance

It is obvious from Eq. 1.5 and Example 1.1 that the dc resistance of a diode is independent of the shape of the characteristic in the region surrounding the point of interest. If a sinusoidal rather than dc input is applied, the situation will change completely. The varying input will move the instantaneous operating point up and down a region of the characteristics and thus defines a specific change in current and voltage as shown in Fig. 1.27. With no applied varying signal, the point of operation would be the Q-point appearing on Fig. 1.27 determined by the applied dc levels. The designation Q-point is derived from the word *quiescent*, which means "still or un-varying level."

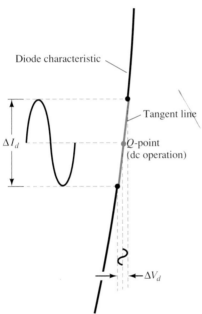

Diode characteristic

Tangent line

ΔI_d

Q-point
(dc operation)

ΔV_d

Figure 1.27 Defining the dynamic or ac resistance

A straight line drawn tangent to the curve through the Q-point as shown in Fig. 1.28 will define a particular change in voltage and current that can be used to determine the *ac* or *dynamic* resistance for this region of the diode characteristics. An effort should be made to keep the change in voltage and current as small as possible and equidistant to either side of the Q-point. In equation form,

$$\boxed{r_d = \frac{\Delta V_d}{\Delta I_d}}$$

where Δ signifies a finite change in the quantity. (1.6)

The steeper the slope, the less the value of ΔV_d for the same change in ΔI_d and the less the resistance. The ac resistance in the vertical-rise region of the characteristic is therefore quite small, while the ac resistance is much higher at low current levels.

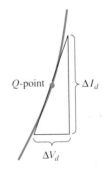

Q-point

ΔI_d

ΔV_d

Figure 1.28 Determining the ac resistance at a Q-point.

EXAMPLE 1.2

For the characteristics of Fig. 1.29:
(a) Determine the ac resistance at $I_D = 2$ mA.
(b) Determine the ac resistance at $I_D = 25$ mA.
(c) Compare the results of parts (a) and (b) to the dc resistances at each current level.

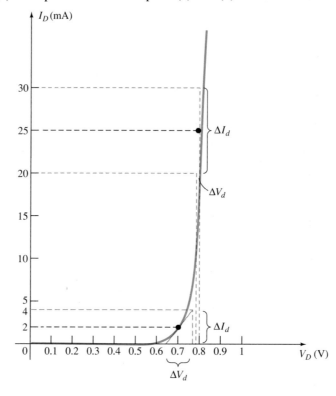

Figure 1.29 Example 1.2

Solution

(a) For $I_D = 2$ mA; the tangent line at $I_D = 2$ mA was drawn as shown in the figure and a swing of 2 mA above and below the specified diode current was chosen. At $I_D = 4$ mA, $V_D = 0.76$ V, and at $I_D = 0$ mA, $V_D = 0.65$ V. The resulting changes in current and voltage are

$$\Delta I_d = 4 \text{ mA} - 0 \text{ mA} = 4 \text{ mA}$$

and

$$\Delta V_d = 0.76 \text{ V} - 0.65 \text{ V} = 0.11 \text{ V}$$

and the ac resistance:

$$r_d = \frac{\Delta V_d}{\Delta I_d} = \frac{0.11 \text{ V}}{4 \text{ mA}} = \mathbf{27.5 \ \Omega}$$

(b) For $I_D = 25$ mA; the tangent line at $I_D = 25$ mA was drawn as shown on the figure and a swing of 5 mA above and below the specified diode current was chosen. At $I_D = 30$ mA, $V_D = 0.8$ V, and at $I_D = 20$ mA, $V_D = 0.78$ V. The resulting changes in current and voltage are

$$\Delta I_d = 30 \text{ mA} - 20 \text{ mA} = 10 \text{ mA}$$

and

$$\Delta V_d = 0.8 \text{ V} - 0.78 \text{ V} = 0.02 \text{ V}$$

and the ac resistance is

$$r_d = \frac{\Delta V_d}{\Delta I_d} = \frac{0.02 \text{ V}}{10 \text{ mA}} = \mathbf{2 \ \Omega}$$

(c) For $I_D = 2$ mA, $V_D = 0.7$ V and

$$R_D = \frac{V_D}{I_D} = \frac{0.7 \text{ V}}{2 \text{ mA}} = \mathbf{350 \ \Omega}$$

which far exceeds the r_d of 27.5 Ω.

For $I_D = 25$ mA, $V_D = 0.79$ V and

$$R_D = \frac{V_D}{I_D} = \frac{0.79 \text{ V}}{25 \text{ mA}} = \mathbf{31.62 \ \Omega}$$

which far exceeds the r_d of 2 Ω.

We have found the dynamic resistance graphically but there is a basic definition in differential calculus which states:

The derivative of a function at a point is equal to the slope of the tangent line drawn at that point.

Equation (1.6), as defined by Fig. 1.28, is, therefore, essentially finding the derivative of the function at the Q-point of operation. If we find the derivative of the general equation (1.4) for the semiconductor diode with respect to the applied forward bias and then invert the result, we will have an equation for the dynamic or ac resistance in that region. That is, taking the derivative of Eq. (1.4) with respect to the applied bias will result in

$$\frac{d}{dV_D}(I_D) = \frac{d}{dV}[I_s(e^{kV/T_K} - 1)]$$

and

$$\frac{dI_D}{dV_D} = \frac{k}{T_K}(I_D + I_s)$$

following a few basic maneuvers of differential calculus. In general, $I_D \gg I_s$ in the vertical slope section of the characteristics and

$$\frac{dI_D}{dV_D} \cong \frac{k}{T_K}I_D$$

Substituting $\eta = 1$ for Ge and Si in the vertical-rise section of the characteristics, we obtain

$$k = \frac{11,600}{\eta} = \frac{11,600}{1} = 11,600$$

and at room temperature,

$$T_K = T_C + 273° = 25° + 273° = 298°$$

so that

$$\frac{k}{T_K} = \frac{11,600}{298} \cong 38.93$$

and

$$\frac{dI_D}{dV_D} = 38.93 I_D$$

Flipping the result to define a resistance ratio ($R = V/I$) gives us

$$\frac{dV_D}{dI_D} \cong \frac{0.026}{I_D}$$

or

$$\boxed{r_d = \frac{26 \text{ mV}}{I_D}}_{\text{Ge,Si}}$$

(1.7)

The significance of Eq. (1.7) must be clearly understood. It implies that the dynamic resistance can be found simply by substituting the quiescent value of the diode current into the equation. There is no need to have the characteristics available or to worry about sketching tangent lines as defined by Eq. (1.6). It is important to keep in mind, however, that Eq. (1.7) is accurate only for values of I_D in the vertical-rise section of the curve. For lesser values of I_D, $\eta = 2$ (silicon) and the value of r_d obtained must be multiplied by a factor of 2. For small values of I_D below the knee of the curve, Eq. (1.7) becomes inappropriate.

All the resistance levels determined thus far have been defined by the p-n junction and do not include the resistance of the semiconductor material itself (called *body* resistance) and the resistance introduced by the connection between the semiconductor material and the external metallic conductor (called *contact* resistance). These additional resistance levels can be included in Eq. (1.7) by adding resistance denoted by r_B as appearing in Eq. (1.8). The resistance r'_d, therefore includes the dynamic resistance defined by Eq. 1.7 and the resistance r_B just introduced.

$$r'_d = \frac{26 \text{ mV}}{I_D} + r_B \qquad \text{ohms} \qquad (1.8)$$

The factor r_B can range from typically 0.1 Ω for high-power devices to 2 Ω for some low-power, general-purpose diodes. For Example 1.2 the ac resistance at 25 mA was calculated to be 2 Ω. Using Eq. (1.7), we have

$$r_d = \frac{26 \text{ mV}}{I_D} = \frac{26 \text{ mV}}{25 \text{ mA}} = \textbf{1.04 } \boldsymbol{\Omega}$$

The difference of about 1 Ω could be treated as the contribution of r_B.

For Example 1.2 the ac resistance at 2 mA was calculated to be 27.5 Ω. Using Eq. (1.7) but multiplying by a factor of 2 for this region (in the knee of the curve $\eta = 2$),

$$r_d = 2\left(\frac{26 \text{ mV}}{I_D}\right) = 2\left(\frac{26 \text{ mV}}{2 \text{ mA}}\right) = 2(13 \text{ } \Omega) = \textbf{26 } \boldsymbol{\Omega}$$

The difference of 1.5 Ω could be treated as the contribution due to r_B.

In reality, determining r_d to a high degree of accuracy from a characteristic curve using Eq. (1.6) is a difficult process at best and the results have to be treated with a grain of salt. At low levels of diode current the factor r_B is normally small enough compared to r_d to permit ignoring its impact on the ac diode resistance. At high levels of current the level of r_B may approach that of r_d, but since there will frequently be other resistive elements of a much larger magnitude in series with the diode we will assume in this book that the ac resistance is determined solely by r_d and the impact of r_B will be ignored unless otherwise noted. Technological improvements of recent years suggest that the level of r_B will continue to decrease in magnitude and eventually become a factor that can certainly be ignored in comparison to r_d.

The discussion above has centered solely on the forward-bias region. In the reverse-bias region we will assume that the change in current along the I_s line is nil from 0 V to the Zener region and the resulting ac resistance using Eq. (1.6) is sufficiently high to permit the open-circuit approximation.

Average AC Resistance

If the input signal is sufficiently large to produce a broad swing such as indicated in Fig. 1.30, the resistance associated with the device for this region is called the *average ac resistance*. The average ac resistance is, by definition, the resistance deter-

Chapter 1   Semiconductor Diodes

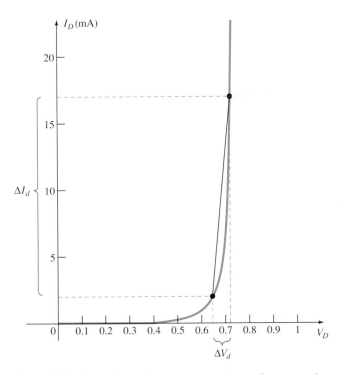

Figure 1.30 Determining the average ac resistance between indicated limits.

mined by a straight line drawn between the two intersections established by the maximum and minimum values of input voltage. In equation form (note Fig. 1.30)

$$r_{av} = \frac{\Delta V_d}{\Delta I_d}\bigg|_{pt. \ to \ pt.}$$

(1.9)

For the situation indicated by Fig. 1.30,

$$\Delta I_d = 17 \text{ mA} - 2 \text{ mA} = 15 \text{ mA}$$

and

$$\Delta V_d = 0.725 \text{ V} - 0.65 \text{ V} = 0.075 \text{ V}$$

with

$$r_{av} = \frac{\Delta V_d}{\Delta I_d} = \frac{0.075 \text{ V}}{15 \text{ mA}} = \mathbf{5 \ \Omega}$$

If the ac resistance (r_d) were determined at $I_D = 2$ mA its value would be more than 5 Ω, and if determined at 17 mA it would be less. In between the ac resistance would make the transition from the high value at 2 mA to the lower value at 17 mA. Equation (1.9) has defined a value that is considered the average of the ac values from 2 to 17 mA. The fact that one resistance level can be used for such a wide range of the characteristics will prove quite useful in the definition of equivalent circuits for a diode in a later section.

Summary Table

Table 1.2 was developed to reinforce the important conclusions of the last few pages and emphasize the difference between the various resistance levels. As indicated earlier, the content of this section is the foundation for a number of resistance calculations to be performed in later sections and chapters.

TABLE 1.2 Resistance Levels

Type	Equation	Special Characteristics	Graphical Determination
Dc or static	$R_D = \dfrac{V_D}{I_D}$	Defined as a *point* on the characteristics	
Ac or dynamic	$r_d = \dfrac{\Delta V_d}{\Delta I_d} = \dfrac{26 \text{ mV}}{I_D}$	Defined by a tangent line at the Q-point	
Average ac	$r_{av} = \dfrac{\Delta V_d}{\Delta I_d}\bigg|_{\text{pt. to pt.}}$	Defined by a straight line between limits of operation	

1.8 DIODE EQUIVALENT CIRCUITS

An equivalent circuit is a combination of elements properly chosen to best represent the actual terminal characteristics of a device, system, or such in a particular operating region.

In other words, once the equivalent circuit is defined, the device symbol can be removed from a schematic and the equivalent circuit inserted in its place without severely affecting the actual behavior of the system. The result is often a network that can be solved using traditional circuit analysis techniques.

Piecewise-Linear Equivalent Circuit

One technique for obtaining an equivalent circuit for a diode is to approximate the characteristics of the device by straight-line segments, as shown in Fig. 1.31. The resulting equivalent circuit is naturally called the *piecewise-linear equivalent circuit*. It should be obvious from Fig. 1.31 that the straight-line segments do not result in an exact duplication of the actual characteristics, especially in the knee region. However, the resulting segments are sufficiently close to the actual curve to establish an equivalent circuit that will provide an excellent first approximation to the actual behavior of the device. For the sloping section of the equivalence the average ac resistance as introduced in Section 1.7 is the resistance level appearing in the equivalent circuit of Fig. 1.32 next to the actual device. In essence, it defines the resistance level of the device when it is in the ''on'' state. The ideal diode is included to establish that there is only one direction of conduction through the device, and a

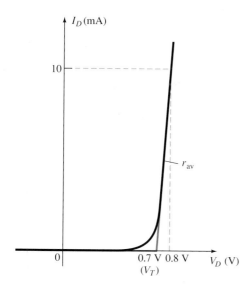

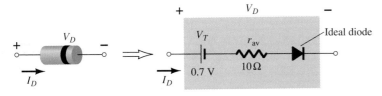

Figure 1.31 Defining the piece-wise-linear equivalent circuit using straight-line segments to approximate the characteristic curve.

Figure 1.32 Components of the piecewise-linear equivalent circuit.

reverse-bias condition will result in the open-circuit state for the device. Since a silicon semiconductor diode does not reach the conduction state until V_D reaches 0.7 V with a forward bias (as shown in Fig. 1.31), a battery V_T opposing the conduction direction must appear in the equivalent circuit as shown in Fig. 1.32. The battery simply specifies that the voltage across the device must be greater than the threshold battery voltage before conduction through the device in the direction dictated by the ideal diode can be established. When conduction is established the resistance of the diode will be the specified value of r_{av}.

Keep in mind, however, that V_T in the equivalent circuit is not an independent voltage source. If a voltmeter is placed across an isolated diode on the top of a lab bench, a reading of 0.7 V will not be obtained. The battery simply represents the horizontal offset of the characteristics that must be exceeded to establish conduction.

The approximate level of r_{av} can usually be determined from a specified operating point on the specification sheet (to be discussed in Section 1.9). For instance, for a silicon semiconductor diode, if $I_F = 10$ mA (a forward conduction current for the diode) at $V_D = 0.8$ V, we know for silicon that a shift of 0.7 V is required before the characteristics rise and

$$r_{av} = \frac{\Delta V_d}{\Delta I_d}\bigg|_{\text{pt. to pt.}} = \frac{0.8 \text{ V} - 0.7 \text{ V}}{10 \text{ mA} - 0 \text{ mA}} = \frac{0.1 \text{ V}}{10 \text{ mA}} = \mathbf{10\ \Omega}$$

as obtained for Fig. 1.30.

Simplified Equivalent Circuit

For most applications, the resistance r_{av} is sufficiently small to be ignored in comparison to the other elements of the network. The removal of r_{av} from the equivalent

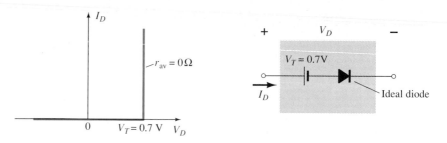

Figure 1.33 Simplified equivalent circuit for the silicon semiconductor diode.

circuit is the same as implying that the characteristics of the diode appear as shown in Fig. 1.33. Indeed, this approximation is frequently employed in semiconductor circuit analysis as demonstrated in Chapter 2. The reduced equivalent circuit appears in the same figure. It states that a forward-biased silicon diode in an electronic system under dc conditions has a drop of 0.7 V across it in the conduction state at any level of diode current (within rated values, of course).

Ideal Equivalent Circuit

Now that r_{av} has been removed from the equivalent circuit let us take it a step further and establish that a 0.7-V level can often be ignored in comparison to the applied voltage level. In this case the equivalent circuit will be reduced to that of an ideal diode as shown in Fig. 1.34 with its characteristics. In Chapter 2 we will see that this approximation is often made without a serious loss in accuracy.

In industry a popular substitution for the phrase "diode equivalent circuit" is diode *model*—a model by definition being a representation of an existing device, object, system, and so on. In fact, this substitute terminology will be used almost exclusively in the chapters to follow.

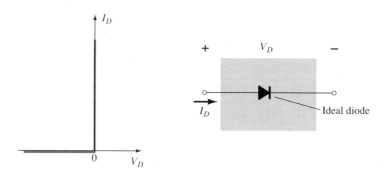

Figure 1.34 Ideal diode and its characteristics.

Summary Table

For clarity, the diode models employed for the range of circuit parameters and applications are provided in Table 1.3 with their piecewise-linear characteristics. Each will be investigated in greater detail in Chapter 2. There are always exceptions to the general rule but it is fairly safe to say that the simplified equivalent model will be employed most frequently in the analysis of electronic systems, while the ideal diode is frequently applied in the analysis of power supply systems where larger voltages are encountered.

TABLE 1.3 Diode Equivalent Circuits (Models)

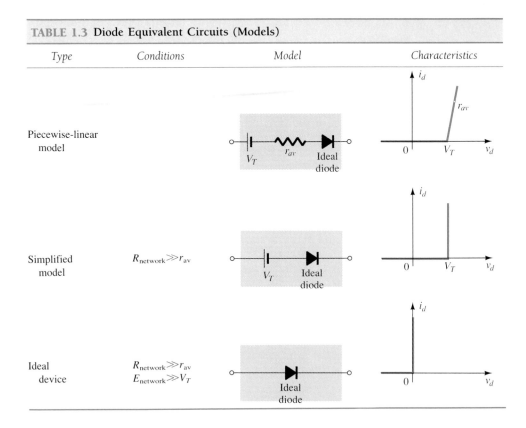

Type	Conditions	Model	Characteristics
Piecewise-linear model			
Simplified model	$R_{\text{network}} \gg r_{av}$		
Ideal device	$R_{\text{network}} \gg r_{av}$ $E_{\text{network}} \gg V_T$		

1.9 DIODE SPECIFICATION SHEETS

Data on specific semiconductor devices are normally provided by the manufacturer in one of two forms. Most frequently, it is a very brief description limited to perhaps one page. Otherwise, it is a thorough examination of the characteristics using graphs, artwork, tables, and so on. In either case, however, there are specific pieces of data that must be included for proper utilization of the device. They include:

1. The forward voltage V_F (at a specified current and temperature)
2. The maximum forward current I_F (at a specified temperature)
3. The reverse saturation current I_R (at a specified voltage and temperature)
4. The reverse-voltage rating [PIV or PRV or V(BR), where BR comes from the term "breakdown" (at a specified temperature)]
5. The maximum power dissipation level at a particular temperature
6. Capacitance levels (as defined in Section 1.10)
7. Reverse recovery time t_{rr} (as defined in Section 1.11)
8. Operating temperature range

Depending on the type of diode being considered, additional data may also be provided, such as frequency range, noise level, switching time, thermal resistance levels, and peak repetitive values. For the application in mind, the significance of the data will usually be self-apparent. If the maximum power or dissipation rating is also provided, it is understood to be equal to the following product:

$$P_{D_{\text{max}}} = V_D I_D \qquad (1.10)$$

where I_D and V_D are the diode current and voltage at a particular point of operation.

If we apply the simplified model for a particular application (a common occurrence), we can substitute $V_D = V_T = 0.7$ V for a silicon diode in Eq. (1.10) and determine the resulting power dissipation for comparison against the maximum power rating. That is,

$$P_{\text{dissipated}} \cong (0.7 \text{ V})I_D \qquad (1.11)$$

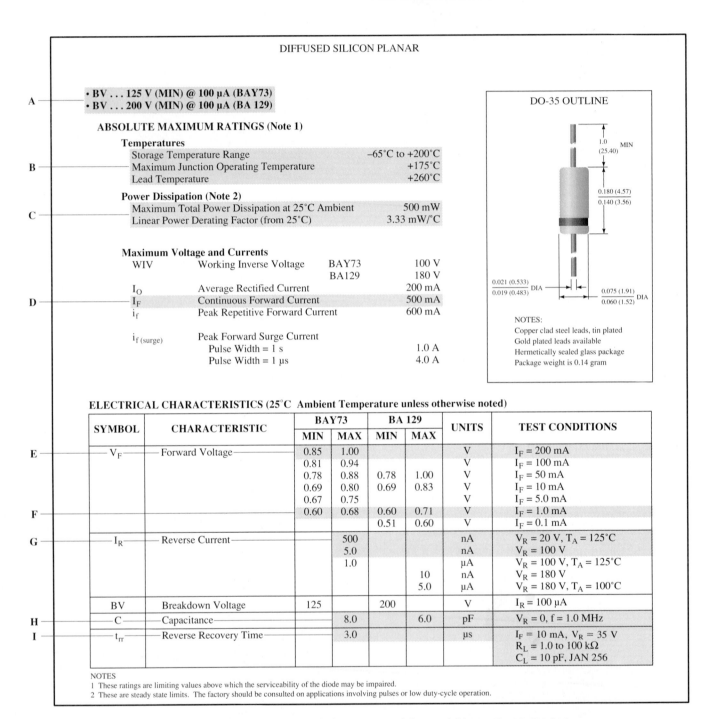

DIFFUSED SILICON PLANAR

A • **BV . . . 125 V (MIN) @ 100 μA (BAY73)**
• **BV . . . 200 V (MIN) @ 100 μA (BA 129)**

ABSOLUTE MAXIMUM RATINGS (Note 1)

Temperatures

B Storage Temperature Range	−65°C to +200°C
Maximum Junction Operating Temperature	+175°C
Lead Temperature	+260°C

Power Dissipation (Note 2)

C Maximum Total Power Dissipation at 25°C Ambient	500 mW
Linear Power Derating Factor (from 25°C)	3.33 mW/°C

Maximum Voltage and Currents

WIV	Working Inverse Voltage	BAY73	100 V
		BA129	180 V
I_O	Average Rectified Current		200 mA
D I_F	Continuous Forward Current		500 mA
i_f	Peak Repetitive Forward Current		600 mA
$i_{f\,(surge)}$	Peak Forward Surge Current		
	Pulse Width = 1 s		1.0 A
	Pulse Width = 1 μs		4.0 A

DO-35 OUTLINE

1.0 (25.40) MIN

0.180 (4.57)
0.140 (3.56)

$\frac{0.021\ (0.533)}{0.019\ (0.483)}$ DIA

$\frac{0.075\ (1.91)}{0.060\ (1.52)}$ DIA

NOTES:
Copper clad steel leads, tin plated
Gold plated leads available
Hermetically sealed glass package
Package weight is 0.14 gram

ELECTRICAL CHARACTERISTICS (25°C Ambient Temperature unless otherwise noted)

SYMBOL	CHARACTERISTIC	BAY73 MIN	BAY73 MAX	BA 129 MIN	BA 129 MAX	UNITS	TEST CONDITIONS
E V_F	Forward Voltage	0.85	1.00			V	$I_F = 200$ mA
		0.81	0.94			V	$I_F = 100$ mA
		0.78	0.88	0.78	1.00	V	$I_F = 50$ mA
		0.69	0.80	0.69	0.83	V	$I_F = 10$ mA
		0.67	0.75			V	$I_F = 5.0$ mA
F		0.60	0.68	0.60	0.71	V	$I_F = 1.0$ mA
				0.51	0.60	V	$I_F = 0.1$ mA
G I_R	Reverse Current		500			nA	$V_R = 20$ V, $T_A = 125$°C
			5.0			nA	$V_R = 100$ V
			1.0			μA	$V_R = 100$ V, $T_A = 125$°C
					10	nA	$V_R = 180$ V
					5.0	μA	$V_R = 180$ V, $T_A = 100$°C
BV	Breakdown Voltage	125		200		V	$I_R = 100$ μA
H C	Capacitance		8.0		6.0	pF	$V_R = 0$, f = 1.0 MHz
I t_{rr}	Reverse Recovery Time		3.0			μs	$I_F = 10$ mA, $V_R = 35$ V $R_L = 1.0$ to 100 kΩ $C_L = 10$ pF, JAN 256

NOTES
1 These ratings are limiting values above which the serviceability of the diode may be impaired.
2 These are steady state limits. The factory should be consulted on applications involving pulses or low duty-cycle operation.

Figure 1.35 Electrical characteristics of the Fairchild Bay 73 · BA 129 high-voltage, low-leakage diodes. (Courtesy Fairchild Camera and Instrument Corporation.)

An exact copy of the data provided by Fairchild Camera and Instrument Corporation for their BAY73 and BA129 high-voltage/low-leakage diodes appears in Figs. 1.35 and 1.36. This example would represent the expanded list of data and characteristics. The term *rectifier* is applied to a diode when it is frequently used in a *rectification* process to be described in Chapter 2.

TYPICAL ELECTRICAL CHARACTERISTIC CURVES
at 25°C ambient temperature unless otherwise noted

FORWARD VOLTAGE VERSUS FORWARD CURRENT

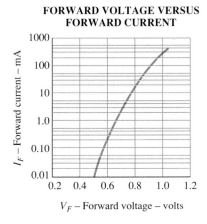

FORWARD CURRENT VERSUS TEMPERATURE COEFFICIENT

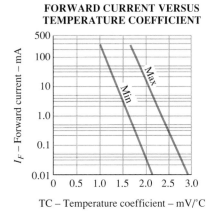

CAPACITANCE VERSUS REVERSE VOLTAGE

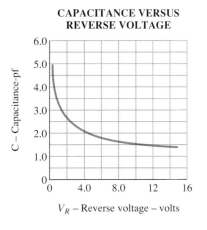

REVERSE VOLTAGE VERSUS REVERSE CURRENT

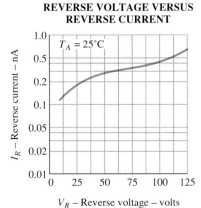

REVERSE CURRENT VERSUS TEMPERATURE COEFFICIENT

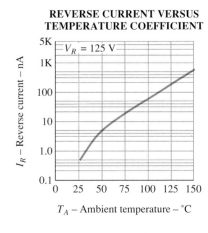

DYNAMIC IMPEDANCE VERSUS FORWARD CURRENT

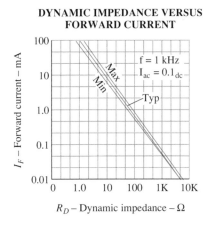

POWER DERATING CURVE

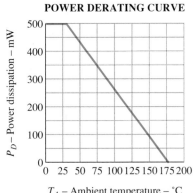

AVERAGE RECTIFIED CURRENT AND FORWARD CURRENT VERSUS AMBIENT TEMPERATURE

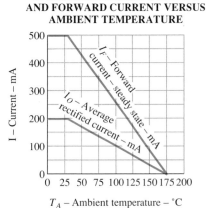

Figure 1.36 Terminal characteristics of the Fairchild Bay 73 · BA 129 high-voltage diodes. (Courtesy Fairchild Camera and Instrument Corporation.)

Specific areas of the specification sheet have been highlighted in blue with a letter identification corresponding with the following description:

A: The *minimum* reverse-bias voltages (PIVs) for each diode at a specified reverse saturation current.

B: Temperature characteristics as indicated. Note the use of the Celsius scale and the wide range of utilization [recall that $32°F = 0°C = $ freezing(H_2O) and $212°F = 100°C = $ boiling(H_2O)].

C: Maximum power dissipation level $P_D = V_D I_D = 500$ mW. The maximum power rating decreases at a rate of 3.33 mW per degree increase in temperature above room temperature (25°C), as clearly indicated by the *power derating curve* of Fig. 1.36.

D: Maximum continuous forward current $I_{F_{max}} = 500$ mA (note I_F versus temperature in Fig. 1.36).

E: Range of values of V_F at $I_F = 200$ mA. Note that it exceeds $V_T = 0.7$ V for both devices.

F: Range of values of V_F at $I_F = 1.0$ mA. Note in this case how the upper limits surround 0.7 V.

G: At $V_R = 20$ V and a typical operating temperature $I_R = 500$ nA $= 0.5$ μA, while at a higher reverse voltage I_R drops to 5 nA $= 0.005$ μA.

H: The capacitance level between terminals is about 8 pF for the BAY73 diode at $V_R = V_D = 0$ V (no-bias) and an applied frequency of 1 MHz.

I: The reverse recovery time is 3 μs for the list of operating conditions.

A number of the curves of Fig. 1.36 employ a log scale. A brief investigation of Section 11.2 should help with the reading of the graphs. Note in the top left figure how V_F increased from about 0.5 V to over 1 V as I_F increased from 10 μA to over 100 mA. In the figure below we find that the reverse saturation current does change slightly with increasing levels of V_R but remains at less than 1 nA at room temperature up to $V_R = 125$ V. As noted in the adjoining figure, however, note how quickly the reverse saturation current increases with increase in temperature (as forecasted earlier).

In the top right figure note how the capacitance decreases with increase in reverse-bias voltage, and in the figure below note that the ac resistance (r_d) is only about 1 Ω at 100 mA and increases to 100 Ω at currents less than 1 mA (as expected from the discussion of earlier sections).

The average rectified current, peak repetitive forward current, and peak forward surge current as they appear on the specification sheet are defined as follows:

1. *Average rectified current.* A half-wave-rectified signal (described in Section 2.8) has an average value defined by $I_{av} = 0.318 I_{peak}$. The average current rating is lower than the continuous or peak repetitive forward currents because a half-wave current waveform will have instantaneous values much higher than the average value.

2. *Peak repetitive forward current.* This is the maximum instantaneous value of repetitive forward current. Note that since it is at this level for a brief period of time, its level can be higher than the continuous level.

3. *Peak forward surge current.* On occasion during turn-on, malfunctions, and so on, there will be very high currents through the device for very brief intervals of time (that are not repetitive). This rating defines the maximum value and the time interval for such surges in current level.

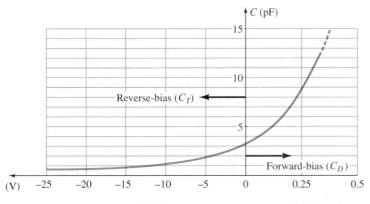

The more one is exposed to specification sheets, the "friendlier" they will become, especially when the impact of each parameter is clearly understood for the application under investigation.

1.10 TRANSITION AND DIFFUSION CAPACITANCE

Electronic devices are inherently sensitive to very high frequencies. Most shunt capacitive effects that can be ignored at lower frequencies because the reactance $X_C = 1/2\pi f C$ is very large (open-circuit equivalent). This, however, cannot be ignored at very high frequencies. X_C will become sufficiently small due to the high value of f to introduce a low-reactance "shorting" path. In the p-n semiconductor diode, there are two capacitive effects to be considered. Both types of capacitance are present in the forward- and reverse-bias regions, but one so outweighs the other in each region that we consider the effects of only one in each region.

In the reverse-bias region we have the transition- or depletion-region capacitance (C_T), while in the forward-bias region we have the diffusion (C_D) or storage capacitance.

Recall that the basic equation for the capacitance of a parallel-plate capacitor is defined by $C = \epsilon A/d$, where ϵ is the permittivity of the dielectric (insulator) between the plates of area A separated by a distance d. In the reverse-bias region there is a depletion region (free of carriers) that behaves essentially like an insulator between the layers of opposite charge. Since the depletion width (d) will increase with increased reverse-bias potential, the resulting transition capacitance will decrease, as shown in Fig. 1.37. The fact that the capacitance is dependent on the applied reverse-bias potential has application in a number of electronic systems. In fact, in Chapter 20 a diode will be introduced whose operation is wholly dependent on this phenomenon.

Although the effect described above will also be present in the forward-bias region, it is overshadowed by a capacitance effect directly dependent on the rate at which charge is injected into the regions just outside the depletion region. In other words, directly dependent on the resulting current of the diode. Increased levels of current will result in increased levels of diffusion capacitance. However, increased levels of current result in reduced levels of associated resistance (to be demonstrated shortly), and the resulting time constant ($\tau = RC$), which is very important in high-speed applications, does not become excessive.

Figure 1.37 Transition and diffusion capacitance versus applied bias for a silicon diode.

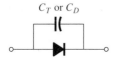

C_T or C_D

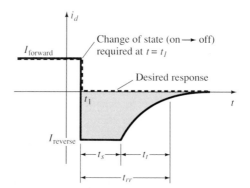

Figure 1.38 Including the effect of the transition or diffusion capacitance on the semiconductor diode.

The capacitive effects described above are represented by a capacitor in parallel with the ideal diode, as shown in Fig. 1.38. For low- or mid-frequency applications (except in the power area), however, the capacitor is normally not included in the diode symbol.

1.11   REVERSE RECOVERY TIME

There are certain pieces of data that are normally provided on diode specification sheets provided by manufacturers. One such quantity that has not been considered yet is the reverse recovery time, denoted by t_{rr}. In the forward-bias state it was shown earlier that there are a large number of electrons from the n-type material progressing through the p-type material and a large number of holes in the n-type—a requirement for conduction. The electrons in the p-type and holes progressing through the n-type material establish a large number of minority carriers in each material. If the applied voltage should be reversed to establish a reverse-bias situation, we would ideally like to see the diode change instantaneously from the conduction state to the nonconduction state. However, because of the large number of minority carriers in each material, the diode will simply reverse as shown in Fig. 1.39 and stay at this measurable level for the period of time t_s (storage time) required for the minority carriers to return to their majority-carrier state in the opposite material. In essence, the diode will remain in the short-circuit state with a current I_{reverse} determined by the network parameters. Eventually, when this storage phase has passed, the current will reduce in level to that associated with the nonconduction state. This second period of time is denoted by t_t (transition interval). The reverse recovery time is the sum of these two intervals: $t_{rr} = t_s + t_t$. Naturally, it is an important consideration in high-speed switching applications. Most commercially available switching diodes have a t_{rr} in the range of a few nanoseconds to 1 μs. Units are available, however, with a t_{rr} of only a few hundred picoseconds (10^{-12}).

Figure 1.39 Defining the reverse recovery time.

1.12   SEMICONDUCTOR DIODE NOTATION

The notation most frequently used for semiconductor diodes is provided in Fig. 1.40. For most diodes any marking such as a dot or band, as shown in Fig. 1.40, appears at the cathode end. The terminology anode and cathode is a carryover from vacuum-tube notation. The anode refers to the higher or positive potential and the cathode refers to the lower or negative terminal. This combination of bias levels will result in a forward-bias or "on" condition for the diode. A number of commercially available semiconductor diodes appear in Fig. 1.41. Some details of the actual construction of devices such as those appearing in Fig. 1.41 are provided in Chapters 12 and 20.

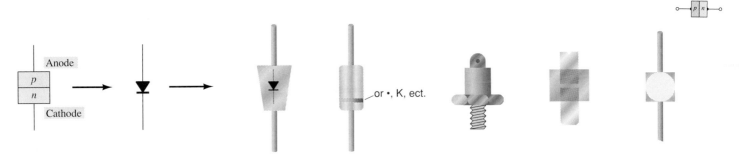

Figure 1.40 Semiconductor diode notation.

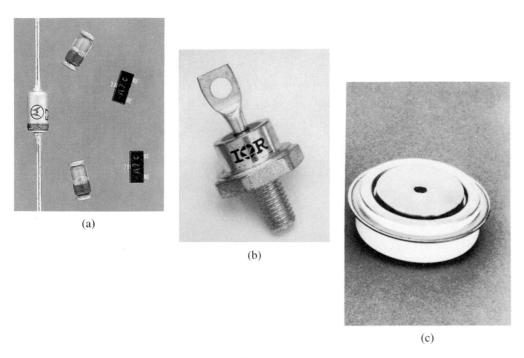

(a)

(b)

(c)

Figure 1.41 Various types of junction diodes. [(a) Courtesy of Motorola Inc.; and (b) and (c) Courtesy International Rectifier Corporation.]

1.13 DIODE TESTING

The condition of a semiconductor diode can be determined quickly using (1) a digital display meter (DDM) with a *diode checking function*, (2) the *ohmmeter section* of a multimeter, or (3) a *curve tracer*.

Diode Checking Function

A digital display meter with a diode checking capability appears in Fig. 1.42. Note the small diode symbol as the bottom option of the rotating dial. When set in this position and hooked up as shown in Fig. 1.43a the diode should be in the "on" state and the display will provide an indication of the forward-bias voltage such as 0.67 V. (for S*i*) The meter has an internal constant current source (about 2 mA) that will define the voltage level as indicated in Fig. 1.43b. An OL indication with the hookup of Fig. 1.43a reveals an open (defective) diode. If the leads are reversed, an OL indication should result due to the expected open-circuit equivalence for the diode. In general, therefore, an OL indication in both directions is an indication of an open or defective diode.

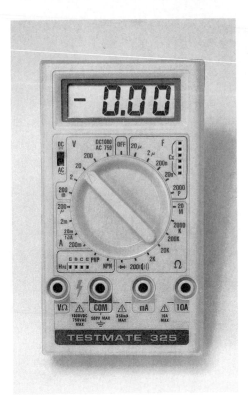

Figure 1.42 Digital display meter with diode checking capability. (Courtesy Computronics Technology, Inc.)

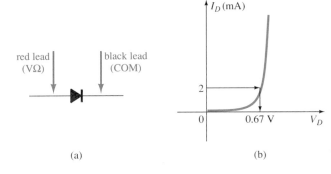

(a)

(b)

Figure 1.43 Checking a diode in the forward-bias state.

Ohmmeter Testing

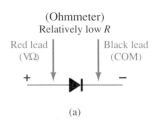

(a)

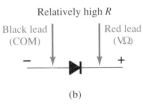

(b)

Figure 1.44 Checking a diode with an ohmmeter.

In Section 1.7 we found that the forward-bias resistance of a semiconductor diode is quite low compared to the reverse-bias level. Therefore, if we measure the resistance of a diode using the connections indicated in Fig. 1.44a, we can expect a relatively low level. The resulting ohmmeter indication will be a function of the current established through the diode by the internal battery (often 1.5 V) of the ohmmeter circuit. The higher the current, the less the resistance level. For the reverse-bias situation the reading should be quite high, requiring a high resistance scale on the meter, as indicated in Fig. 1.44b. A high resistance reading in both directions obviously indicates an open (defective device) condition, while a very low resistance reading in both directions will probably indicate a shorted device.

Curve Tracer

The curve tracer of Fig. 1.45 can display the characteristics of a host of devices, including the semiconductor diode. By properly connecting the diode to the test panel at the bottom center of the unit and adjusting the controls, the display of Fig. 1.46 can

Chapter 1   Semiconductor Diodes

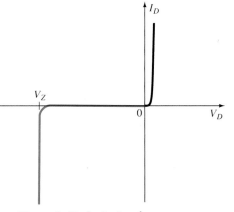

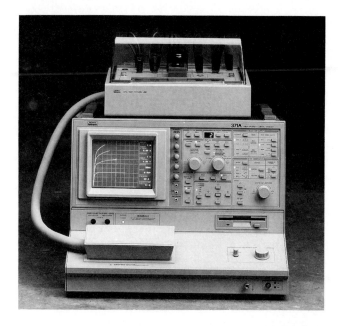

Figure 1.45 Curve tracer. (Courtesy of Tektronix, Inc.)

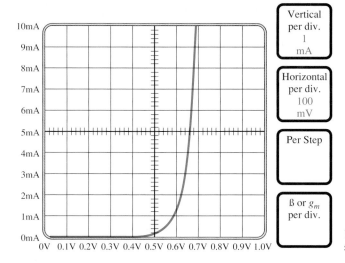

| Vertical per div. 1 mA |
| Horizontal per div. 100 mV |
| Per Step |
| ß or g_m per div. |

Figure 1.46 Curve tracer response to 1N4007 silicon diode.

be obtained. Note that the vertical scaling is 1 mA/div, resulting in the levels indicated. For the horizontal axis the scaling is 100 mV/div, resulting in the voltage levels indicated. For a 2-mA level as defined for a DDM, the resulting voltage would be about 625 mV = 0.625 V. Although the instrument initially appears quite complex, the instruction manual and a few moments of exposure will reveal that the desired results can usually be obtained without an excessive amount of effort and time. The same instrument will appear on more than one occasion in the chapters to follow as we investigate the characteristics of the variety of devices.

1.14 ZENER DIODES

The Zener region of Fig. 1.47 was discussed in some detail in Section 1.6. The characteristic drops in an almost vertical manner at a reverse-bias potential denoted V_Z. The fact that the curve drops down and away from the horizontal axis rather than up and away for the positive V_D region reveals that the current in the Zener region has a direction opposite to that of a forward-biased diode.

Figure 1.47 Reviewing the Zener region.

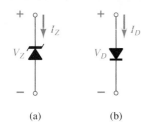

+ $\circ$ + $\circ$

V_Z I_Z V_D I_D

− $\circ$ − $\circ$

   (a)   (b)

Figure 1.48 Conduction direction: (a) Zener diode; (b) semiconductor diode.

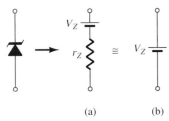

   (a)   (b)

Figure 1.49 Zener equivalent circuit: (a) complete; (b) approximate.

This region of unique characteristics is employed in the design of *Zener diodes* which have the graphic symbol appearing in Fig. 1.48a. Both the semiconductor diode and zener diode are presented side by side in Fig. 1.48 to ensure that the direction of conduction of each is clearly understood together with the required polarity of the applied voltage. For the semiconductor diode the "on" state will support a current in the direction of the arrow in the symbol. For the Zener diode the direction of conduction is opposite to that of the arrow in the symbol as pointed out in the introduction to this section. Note also that the polarity of V_D and V_Z are the same as would be obtained if each were a resistive element.

The location of the Zener region can be controlled by varying the doping levels. An increase in doping, producing an increase in the number of added impurities, will decrease the Zener potential. Zener diodes are available having Zener potentials of 1.8 to 200 V with power ratings from $\frac{1}{4}$ to 50 W. Because of its higher temperature and current capability, silicon is usually preferred in the manufacture of Zener diodes.

The complete equivalent circuit of the Zener diode in the Zener region includes a small dynamic resistance and dc battery equal to the Zener potential, as shown in Fig. 1.49. For all applications to follow, however, we shall assume as a first approximation that the external resistors are much larger in magnitude than the Zener-equivalent resistor and that the equivalent circuit is simply that indicated in Fig. 1.49b.

A larger drawing of the Zener region is provided in Fig. 1.50 to permit a description of the Zener nameplate data appearing in Table 1.4 for a 1N961, Fairchild, 500-mW, 20% diode. The term "nominal" associated with V_Z indicates that it is a typical average value. Since this is a 20% diode, the Zener potential can be expected

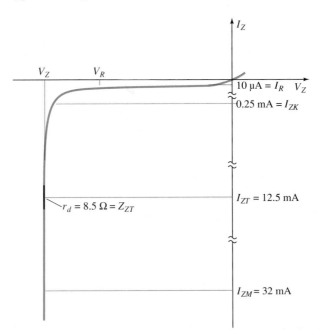

Figure 1.50 Zener test characteristics (Fairchild IN961).

TABLE 1.4 Electrical Characteristics (25°C Ambient Temperature Unless Otherwise Noted)

Jedec Type	Zener Voltage Nominal, V_Z (V)	Test Current, I_{ZT} (mA)	Max Dynamic Impedance, Z_{ZT} at I_{ZT} (Ω)	Maximum Knee Impedance, Z_{ZK} at I_{ZK} (Ω)	(mA)	Maximum Reverse Current, I_R at V_R (μA)	Test Voltage, V_R (V)	Maximum Regulator Current, I_{ZM} (mA)	Typical Temperature Coefficient (%/°C)
IN961	10	12.5	8.5	700	0.25	10	7.2	32	+0.072

to vary as 10 V ± 20% or from 8 to 12 V in its range of application. Also available are 10% and 5% diodes with the same specifications. The test current I_{ZT} is the current defined by the $\frac{1}{4}$ power level and Z_{ZT} is the dynamic impedance at this current level. The maximum knee impedance occurs at the knee current of I_{ZK}. The reverse saturation current is provided at a particular potential level and I_{ZM} is the maximum current for the 20% unit.

The temperature coefficient reflects the percent change in V_Z with temperature. It is defined by the equation

$$T_C = \frac{\Delta V_Z}{V_Z(T_1 - T_0)} \times 100\% \qquad \%/°C \qquad (1.12)$$

where ΔV_Z is the resulting change in Zener potential due to the temperature variation. Note in Fig. 1.51a that the temperature coefficient can be positive, negative, or even zero for different Zener levels. A positive value would reflect an increase in V_Z with an increase in temperature, while a negative value would result in a decrease in value with increase in temperature. The 24-V, 6.8-V, and 3.6-V levels refer to three Zener diodes having these nominal values within the same family of Zeners as the 1N961. The curve for the 10-V 1N961 Zener would naturally lie between the curves of the 6.8-V and 24-V devices. Returning to Eq. (1.12), T_0 is the temperature at which V_Z is provided (normally room temperature—25°C) and T_1 is the new level. Example 1.3 will demonstrate the use of Eq. (1.12).

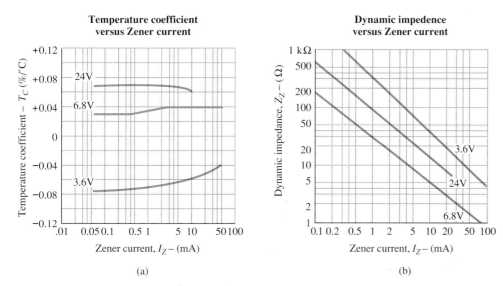

Figure 1.51 Electrical characteristics for a 500-mW Fairchild Zener diode. (Courtesy Fairchild Camera and Instrument Corporation.)

Determine the nominal voltage for a 1N961 Fairchild Zener diode of Table 1.4 at a temperature of 100°C.

EXAMPLE 1.3

Solution

From Eq. 1.12,

$$\Delta V_Z = \frac{T_C V_Z}{100}(T_1 - T_0)$$

1.14 Zener Diodes   37

Substitution values from Table 1.4 yield

$$\Delta V_Z = \frac{(0.072)(10 \text{ V})}{100}(100°C - 25°C)$$

$$= (0.0072)(75)$$

$$= 0.54 \text{ V}$$

and because of the positive temperature coefficient, the new Zener potential, defined by V_Z', is

$$V_Z' = V_Z + 0.54 \text{ V}$$

$$= \mathbf{10.54 \text{ V}}$$

The variation in dynamic impedance (fundamentally, its series resistance) with current appears in Fig. 1.51b. Again, the 10-V Zener appears between the 6.8-V and 24-V Zeners. Note that the heavier the current (or the farther up the vertical rise you are in Fig. 1.47, the less the resistance value. Also note that as you drop below the knee of the curve, the resistance increases to significant levels.

The terminal identification and the casing for a variety of Zener diodes appear in Fig. 1.52. Figure 1.53 is an actual photograph of a variety of Zener devices. Note that their appearance is very similar to the semiconductor diode. A few areas of application for the Zener diode will be examined in Chapter 2.

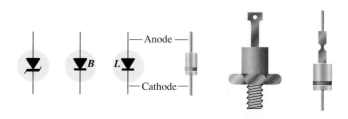

Figure 1.52 Zener terminal identification and symbols.

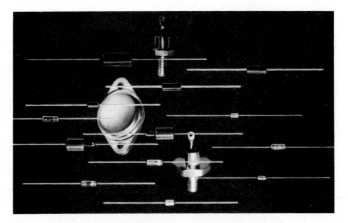

Figure 1.53 Zener diodes. (Courtesy Siemans Corporation.)

1.15 LIGHT-EMITTING DIODES

The increasing use of digital displays in calculators, watches, and all forms of instrumentation has contributed to the current extensive interest in structures that will emit light when properly biased. The two types in common use today to perform this function are the *light-emitting diode* (LED) and the *liquid-crystal display* (LCD). Since the LED falls within the family of *p-n* junction devices and will appear in some of the networks to appear in the next few chapters, it will be introduced in this chapter. The LCD display is described in Chapter 20.

As the name implies, the light-emitting diode (LED) is a diode that will give off visible light when it is energized. In any forward-biased *p-n* junction there is, within the structure and primarily close to the junction, a recombination of holes and electrons. This recombination requires that the energy possessed by the unbound free electron be transferred to another state. In all semiconductor *p-n* junctions some of this energy will be given off as heat and some in the form of photons. In silicon and germanium the greater percentage is given up in the form of heat and the emitted light is insignificant. In other materials, such as gallium arsenide phosphide (GaAsP) or gallium phosphide (GaP), the number of photons of light energy emitted is sufficient to create a very visible light source.

The process of giving off light by applying an electrical source of energy is called electroluminescence.

As shown in Fig. 1.54 with its graphic symbol, the conducting surface connected to the *p*-material is much smaller, to permit the emergence of the maximum number of photons of light energy. Note in the figure that the recombination of the injected carriers due to the forward-biased junction is resulting in emitted light at the site of recombination. There may, of course, be some absorption of the packages of photon energy in the structure itself, but a very large percentage are able to leave, as shown in the figure.

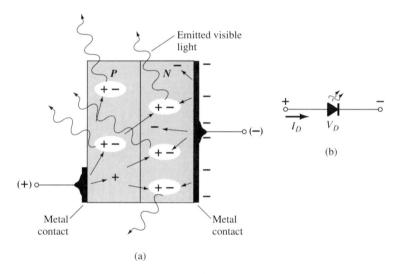

(a)

(b)

Figure 1.54 (a) Process of electroluminescence in the LED; (b) graphic symbol.

The appearance and characteristics of a subminiature high-efficiency solid-state lamp manufactured by Hewlett-Packard appears in Fig. 1.55. Note that the maximum forward current is 20 mA with 10 mA a typical operating level as indicated by the last column, labeled *Test Conditions*. The level of V_D under forward-bias conditions is listed as V_F and extends from 2.2 to 3 V. In other words, one can expect a typical operating current of about 10 mA at 2.5 V for good light emission.

Two quantities yet undefined appear under the heading electrical/optical characteristics at $T_A = 25°C$. They are the *axial luminous intensity* (I_V) and the *luminous efficacy* (η_v). Light intensity is measured in *candela*. One candela emits a light flux of 4π lumens and establishes an illumination of 1 footcandle on a 1-ft² area 1 ft from the light source. Even though this description may not provide a clear understanding of the candela as a unit of measure, its level can certainly be compared between similar devices. The term *efficacy* is, by definition, a measure of the ability of a device to produce a desired effect. For the LED this is the ratio of the number of lumens generated per applied watt of electrical energy. The relative efficiency is defined by the luminous intensity per unit current, as shown in Fig. 1.55g. The relative intensity of each color versus wavelength appears in Fig. 1.55d.

Since the LED is a *p-n* junction device, it will have a forward-biased characteristic (Fig. 1.55e) similar to the diode response curves. Note the almost linear increase in relative luminous intensity with forward current (Fig. 1.55f). Figure 1.55h reveals that the longer the pulse duration at a particular frequency, the lower the permitted peak current (after you pass the break value of t_p). Figure 1.55i simply reveals that the intensity is greater at 0° (or head on) and the least at 90° (when you view the device from the side).

(a)

Absolute Maximum Ratings at $T_A = 25°C$

Parameter	High Eff. Red 4160	Units
Power dissipation	120	mW
Average forward current	20[1]	mA
Peak forward current	60	mA
Operating and storage temperature range	−55°C to 100°C	
Lead soldering temperature [1.6 mm (0.063 in.) from body]	230°C for 3 seconds	

[1] Derate from 50°C at 0.2 mA/°C.

(b)

Electrical/Optical Characteristics at $T_A = 25°C$

Symbol	Description	High Eff. Red 4160 Min.	Typ.	Max.	Units	Test Conditions
						$I_F = 10$ mA
I_V	Axial luminous intensity	1.0	3.0		mcd	
$2\theta_{1/2}$	Included angle between half luminious intensity points		80		deg.	Note 1
λ_{peak}	Peak wavelength		635		nm	Measurement at peak
λ_d	Dominant wavelength		628		nm	Note 2
τ_s	Speed of response		90		ns	
C	Capacitance		11		pF	$V_F = 0$; $f = 1$ Mhz
θ_{JC}	Thermal resistance		120		°C/W	Junction to cathode lead at 0.79 mm (.031 in) from body
V_F	Forward voltage		2.2	3.0	V	$I_F = 10$ mA
BV_R	Reverse breakdown voltage	5.0			V	$I_R = 100$ µA
η_v	Luminous efficacy		147		lm/W	Note 3

NOTES:
1. $\theta_{1/2}$ is the off-axis angle at which the luminous intensity is half the axial luminous intensity.
2. The dominant wavelength, λ_d, is derived from the CIE chromaticity diagram and represents the single wavelength that defines the color of the device.
3. Radiant intensity, I_e, in watts/steradian, may be found from the equation $I_e = I_v/\eta_v$, where I_v is the luminous intensity in candelas and η_v is the luminous efficacy in lumens/watt.

(c)

Figure 1.55 Hewlett-Packard subminiature high-efficiency red solid-state lamp: (a) appearance; (b) absolute maximum ratings; (c) electrical/optical characteristics; (d) relative intensity versus wavelength; (e) forward current versus forward voltage; (f) relative luminous intensity versus forward current; (g) relative efficiency versus peak current; (h) maximum peak current versus pulse duration; (i) relative luminous intensity versus angular displacement. (Courtesy Hewlett-Packard Corporation.)

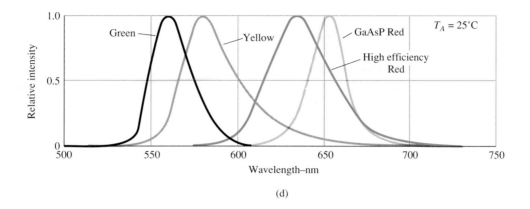

(d)

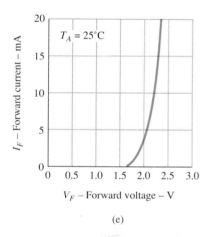

(e)

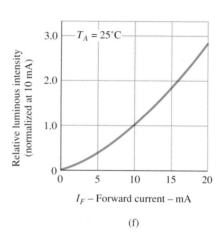

(f)

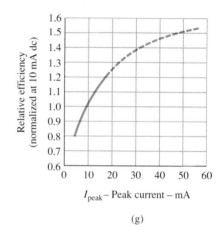

(g)

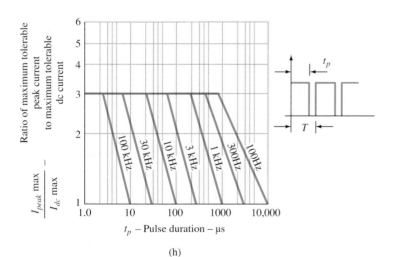

(h)

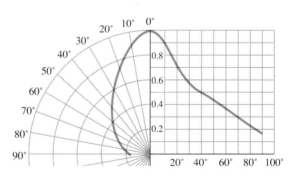

(i)

Figure 1.55 Continued.

1.15 **Light-Emitting Diodes**

41

LED displays are available today in many different sizes and shapes. The light-emitting region is available in lengths from 0.1 to 1 in. Numbers can be created by segments such as shown in Fig. 1.56. By applying a forward bias to the proper *p*-type material segment, any number from 0 to 9 can be displayed.

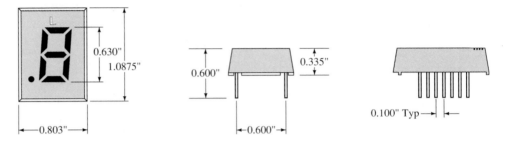

Figure 1.56 Litronix segment display.

The display of Fig. 1.57 is used in calculators and will provide eight digits. There are also two-lead LED lamps that contain two LEDs, so that a reversal in biasing will change the color from green to red, or vice versa. LEDs are presently available in red, green, yellow, orange, and white. It would appear that the introduction of the color blue is a possibility in the very near future. In general, LEDs operate at voltage levels from 1.7 to 3.3 V, which makes them completely compatible with solid-state circuits. They have a fast response time (nanoseconds) and offer good contrast ratios for visibility. The power requirement is typically from 10 to 150 mW with a lifetime of 100,000+ hours. Their semiconductor construction adds a significant ruggedness factor.

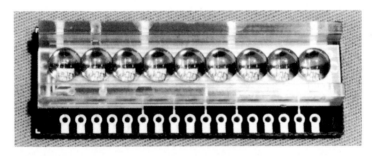

Figure 1.57 Eight-digit and a sign calculator display. (Courtesy Hewlett-Packard Corporation.)

1.16 DIODE ARRAYS—INTEGRATED CIRCUITS

The unique characteristics of integrated circuits will be introduced in Chapter 12. However, we have reached a plateau in our introduction to electronic circuits that permits at least a surface examination of diode arrays in the integrated-circuit package. You will find that the integrated circuit is not a unique device with characteristics totally different from those we examine in these introductory chapters. It is simply a packaging technique that permits a significant reduction in the size of electronic systems. In other words, internal to the integrated circuit are systems and discrete devices that were available long before the integrated circuit as we know it today became a reality.

One possible array appears in Fig. 1.58. Note that eight diodes are internal to the Fairchild FSA 1410M diode array. That is, in the container shown in Fig. 1.59 there are diodes set in a single silicon wafer that have all the anodes connected to pin 1 and the cathodes of each to pins 2 through 9. Note in the same figure that pin 1 can be determined as being to the left of the small projection in the case if we look from the bottom toward the case. The other numbers then follow in sequence. If only one diode is to be used, then only pins 1 and 2 (or any number from 3 to 9) would be used.

FSA1410M

PLANAR AIR–ISOLATED MONOLITHIC DIODE ARRAY

- C . . . 5.0 pF (MAX)
- ΔV_F . . . 15 mv (MAX) @ 10 mA

ABSOLUTE MAXIMUM RATINGS (Note 1)

Temperatures

Storage Temperature Range	$-55°C$ to $+200°C$
Maximum Junction Operating Temperature	$+150°C$
Lead Temperature	$+260°C$

Power Dissipation (Note 2)

Maximum Dissipation per Junction at 25°C Ambient	400 mW
per Package at 25°C Ambient	600 mW
Linear Derating Factor (from 25°C) Junction	3.2 mW/°C
Package	4.8 mW/°C

Maximum Voltage and Currents

WIV	Working Inverse Voltage	55 V
I_F	Continuous Forward Current	350 mA
$i_{f(surge)}$	Peak Forward Surge Current	
	Pulse Width = 1.0 s	1.0 A
	Pulse Width = 1.0 μs	2.0 A

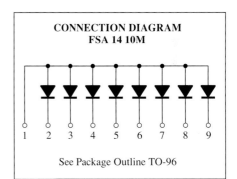

CONNECTION DIAGRAM
FSA 14 10M

See Package Outline TO-96

ELECTRICAL CHARACTERISTICS (25°C Ambient Temperature unless otherwise noted)

SYMBOL	CHARACTERISTIC	MIN	MAX	UNITS	TEST CONDITIONS
B_V	Breakdown Voltage	60		V	$I_R = 10$ μA
V_F	Forward Voltage (Note 3)		1.5	V	$I_F = 500$ mA
			1.1	V	$I_F = 200$ mA
			1.0	V	$I_F = 100$ mA
I_R	Reverse Current		100	nA	$V_R = 40$ V
	Reverse Current ($T_A = 150°C$)		100	μA	$V_R = 40$ V
C	Capacitance		5.0	pF	$V_R = 0$, f = 1 MHz
V_{FM}	Peak Forward Voltage		4.0	V	$I_f = 500$ mA, $t_r < 10$ ns
t_{fr}	Forward Recovery Time		40	ns	$I_f = 500$ mA, $t_r < 10$ ns
t_{rr}	Reverse Recovery Time		10	ns	$I_f = I_r = 10 - 200$ mA $R_L = 100\,\Omega$, Rec. to 0.1 I_r
			50	ns	$I_f = 500$ mA, $I_r = 50$ mA $R_L = 100\,\Omega$, Rec. to 5 mA
ΔV_F	Forward Voltage Match		15	mV	$I_F = 10$ mA

NOTES
1 These ratings are limiting values above which life or satisfactory performance may be impaired.
2 These are steady state limits. The factory should be consulted on applications involving pulsed or low duty cycle operation.
3 V_F is measured using an 8 ms pulse.

Figure 1.58 Monolithic diode array. (Courtesy Fairchild Camera and Instrument Corporation.)

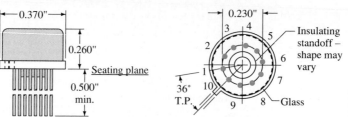

Figure 1.59 Package outline TO-96 for the FSA 1410M diode array. All dimensions are in inches. (Courtesy Fairchild Camera and Instrument Corporation.)

Notes:
Kovar leads, gold plated
Hermetically sealed package
Package weight is 1.32 grams

The remaining diodes would be left hanging and not affect the network to which pins 1 and 2 are connected.

Another diode array appears in Fig. 1.60. In this case the package is different but the numbering sequence appears in the outline. Pin 1 is the pin directly above the small indentation as you look down on the device.

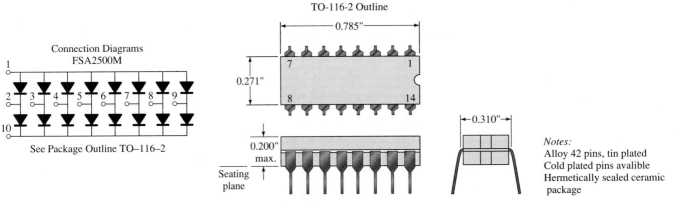

Connection Diagrams
FSA2500M

See Package Outline TO–116–2

Notes:
Alloy 42 pins, tin plated
Cold plated pins avalible
Hermetically sealed ceramic package

Figure 1.60 Monolithic diode array. All dimensions are in inches. (Courtesy Fairchild Camera and Instrument Corporation.)

1.17 COMPUTER ANALYSIS

The computer has now become such an integral part of the electronics industry that the capabilities of this working "tool" must be introduced at the earliest possible opportunity. For those students with no prior computer experience there is a common initial fear of this seemingly complicated powerful system. With this in mind the computer analysis of this book was designed to make the computer system more "friendly" by revealing the relative ease with which it can be applied to perform some very helpful and special tasks in a minimum amount of time with a high degree of accuracy. The content was written with the assumption that the reader has no prior computer experience or exposure to the terminology to be applied. There is also no suggestion that the content of this book is sufficient to permit a complete understanding of the "hows" and "whys" that will surface. The purpose here is solely to introduce some of the terminology, discuss a few of its capabilities, reveal the possibilities available, touch on some of its limitations, and demonstrate its versatility with a number of carefully chosen examples.

In general, the computer analysis of electronic systems can take one of two approaches: using a *language* such as BASIC, Fortran, Pascal, or C, or utilizing a *software package* such as PSpice, MicroCap II, Breadboard, or Circuit Master, to name a few. A language, through its symbolic notation, forms a bridge between the user and the computer that permits a dialogue between the two for establishing the operations to be performed.

The language employed in this book is BASIC, chosen because it uses a number of familiar words and phrases from the English language that in themselves reveal the operation to be performed. When a language is employed to analyze a system a *program* is developed that sequentially defines the operations to be performed—in much the same order in which we perform the same analysis in longhand. As with the longhand approach, one wrong step and the result obtained can be completely meaningless. Programs typically develop with time and application as more efficient paths toward a solution become obvious. Once established in its "best" form it can be cataloged for future use. The important advantage of the language approach is that a program can be tailored to meet all the special needs of the user. It permits innovative "moves" by the user that can result in printouts of data in an informative and interesting manner.

The alternative approach referred to above utilizes a software package to perform the desired investigation. A software package is a program written and tested over a period of time designed to perform a particular type of analysis or synthesis in an efficient manner with a high level of accuracy.

The package itself cannot be altered by the user and its application is limited to the operations built into the system. A user must adjust his or her desire for output information to the range of possibilities offered by the package. In addition, the user must input information exactly as requested by the package or the data may be misinterpreted. The software package chosen for this book is PSpice,* due to its growing popularity in the educational community. A photo of the disks and supporting material available for the PSpice software package appears in Fig. 1.61. A more sophisticated version referred to simply as SPICE is finding widespread application in industry.

In total, therefore, a software package is "packaged" to perform a specific series of calculations and operations and to provide the results in a defined format. A language permits an expanded level of flexibility but also fails to benefit from the extensive testing and research normally devoted to the development of a "trusted" package. The user must define which approach best fits the needs of the moment. Obviously, if a package exists for the desired analysis or synthesis, it should be considered before turning to the many hours required to develop a reliable, efficient program. In addition, one may acquire the data needed for a particular analysis from a software package and then turn to a language to define the format of the output. In many ways, the two approaches go hand in hand. If one is to depend on computer analysis on a continuing basis, knowledge of the use and limits of both languages and

Figure 1.61 PSpice software package. (Copyright 1990 MicroSim Corporation).

---

*PSpice is a registered trademark of MicroSim Corporation.

software packages is a necessity. The choice of which language or software package to become familiar with is primarily a function of the area of investigation. Fortunately, however, a fluent knowledge of one language or a particular software package will usually help the user become familiar with other languages and software packages. There is a similarity in purpose and procedures that ease the transition from one approach to another.

Some comment regarding computer analysis will be made in each chapter. In some cases a BASIC program and PSpice application will appear, while in other situations only one of the two will be applied. As the need for detail arises, sufficient background will be provided to permit at least a surface understanding of the analysis.

This chapter deals specifically with the characteristics of the semiconductor diode. In Chapter 2 the diode is investigated using the PSpice software package. As a first step toward that analysis the "model" for the semiconductor diode will now be introduced. The description in the PSpice manual includes a total of 14 parameters to define its terminal characteristics. They include the saturation current, series resistance, terminal capacitance, reverse breakdown voltage, reverse breakdown current, and a host of other factors that can be specified if necessary for the design or analysis to be performed.

The specification of a diode in a network has two components. The first specifies the location and model name and the other includes the parameters referred to above. The format for defining the location and model name of the diode is the following for the diode of Fig. 1.62:

Figure 1.62 PSpice labels for diode entry into a network description.

$$\underset{\text{name}}{\underline{D1}} \quad \underset{\substack{\text{node}\\+}}{\underline{2}} \quad \underset{\substack{\text{node}\\-}}{\underline{3}} \quad \underset{\substack{\text{model}\\\text{name}}}{\underline{DI}}$$

Note that a diode is specified by the uppercase letter D at the beginning of the line followed by the label applied to the diode on the schematic. The sequence of the nodes (connection points for the diode) defines the potential at each node and the direction of conduction for the diode of Fig. 1.62. In other words, conduction is specified from the positive to the negative node. The *model name* is the name assigned to the parameter description to follow. The same model name can be applied to any number of other diodes in the network, such as D2, D3, and so on.

The parameters are specified using a *MODEL statement* that has the following format for a diode:

$$.\text{MODEL} \quad \underset{\substack{\text{model}\\\text{name}}}{\underline{DI}} \quad \underset{\substack{\text{parameter}\\\text{specifications}}}{\underline{D(IS = 2E-15)}}$$

The specification begins with the entry **.MODEL** followed by the model name as specified in the location description and the uppercase letter D to specify a diode. The parameter specifications appear within parentheses and must use the notation specified in the PSpice manual. The reverse saturation current is listed as IS and is assigned the value 2×10^{-15} A. This value was chosen because it typically results in a diode voltage of about 0.7 V for levels of diode current typically encountered in the applications discussed in Chapter 2. In this way the computer and longhand analysis will have results that are relatively close in magnitude. Although only one parameter was specified in the listing above, the list can include all 10 appearing in the manual. For both statements introduced above it is particularly important to follow the format as defined. The absence of a period before MODEL or forgetting the letter D in the same line will invalidate the entry completely.

The description above simply introduces the format for the entry of a diode in a network description. In Chapter 2 we use the material above to analyze a diode network using PSpice. The role of the foregoing statements and how they fit into the PSpice analysis will then be a great deal more obvious.

§ 1.2 Ideal Diode

1. Describe in your own words the meaning of the word *ideal* as applied to a device or system.

2. Describe in your own words the characteristics of the *ideal* diode and how they determine the on and off states of the device. That is, describe why the short-circuit and open-circuit equivalents are appropriate.

3. What is the one important difference between the characteristics of a simple switch and those of an ideal diode?

§ 1.3 Semiconductor Materials

4. In your own words, define *semiconductor, resistivity, bulk resistance*, and *ohmic contact resistance*.

5. (a) Using Table 1.1, determine the resistance of a silicon sample having an area of 1 cm^2 and a length of 3 cm.
   (b) Repeat part (a) if the length is 1 cm and the area 4 cm^2.
   (c) Repeat part (a) if the length is 8 cm and the area 0.5 cm^2.
   (d) Repeat part (a) for copper and compare the results.

6. Sketch the atomic structure of copper and discuss why it is a good conductor and how its structure is different from germanium and silicon.

7. Define, in your own words, an intrinsic material, a negative temperature coefficient, and covalent bonding.

8. Consult your reference library and list three materials that have a negative temperature coefficient and three that have a positive temperature coefficient.

§ 1.4 Energy Levels

9. How much energy in joules is required to move a charge of 6 C through a difference in potential of 3 V?

10. If 48 eV of energy is required to move a charge through a potential difference of 12 V, determine the charge involved.

11. Consult your reference library and determine the level of E_g for GaP and ZnS, two semiconductor materials of practical value. In addition, determine the written name for each material.

§ 1.5 Extrinsic Materials—*n*- and *p*-Type

12. Describe the difference between *n*-type and *p*-type semiconductor materials.

13. Describe the difference between donor and acceptor impurities.

14. Describe the difference between majority and minority carriers.

15. Sketch the atomic structure of silicon and insert an impurity of arsenic as demonstrated for silicon in Fig. 1.9.

16. Repeat Problem 15 but insert an impurity of indium.

17. Consult your reference library and find another explanation of hole versus electron flow. Using both descriptions, describe in your own words the process of hole conduction.

§ 1.6 Semiconductor Diode

18. Describe in your own words the conditions established by forward- and reverse-bias conditions on a *p-n* junction diode and how it affects the resulting current.

19. Describe how you will remember the forward- and reverse-bias states of the *p-n* junction diode. That is, how you will remember which potential (positive or negative) is applied to which terminal?

20. Using Eq. (1.4), determine the diode current at 20°C for a silicon diode with $I_s = 50$ nA and an applied forward bias of 0.6 V.

21. Repeat Problem 20 for $T = 100$°C (boiling point of water). Assume that I_s has increased to 5.0 μA.

22. (a) Using Eq. (1.4), determine the diode current at 20°C for a silicon diode with $I_s = 0.1$ μA at a reverse-bias potential of -10 V.
(b) Is the result expected? Why?

23. (a) Plot the function $y = e^x$ for x from 0 to 5.
(b) What is the value of $y = e^x$ at $x = 0$?
(c) Based on the results of part (b), why is the factor -1 important in Eq. (1.4)?

24. In the reverse-bias region the saturation current of a silicon diode is about 0.1 μA ($T = 20$°C). Determine its approximate value if the temperature is increased 40°C.

25. Compare the characteristics of a silicon and a germanium diode and determine which you would prefer to use for most practical applications. Give some detail. Refer to a manufacturer's listing and compare the characteristic of a germanium and a silicon diode of similar maximum ratings.

26. Determine the forward voltage drop across the diode whose characteristics appear in Fig. 1.24 at temperatures of -75°C, 25°C, 100°C, and 200°C and a current of 10 mA. For each temperature, determine the level of saturation current. Compare the extremes of each and comment on the ratio of the two.

§ 1.7 Resistance Levels

27. Determine the static or dc resistance of the diode of Fig. 1.19 at a forward current of 2 mA.

28. Repeat Problem 26 at a forward current of 15 mA and compare results.

29. Determine the static or dc resistance of the diode of Fig. 1.19 at a reverse voltage of -10 V. How does it compare to the value determined at a reverse voltage of -30 V?

30. (a) Determine the dynamic (ac) resistance of the diode of Fig. 1.29 at a forward current of 10 mA using Eq. (1.6).
(b) Determine the dynamic (ac) resistance of the diode of Fig. 1.29 at a forward current of 10 mA using Eq. (1.7).
(c) Compare solutions of parts (a) and (b).

31. Calculate the dc and ac resistance for the diode of Fig. 1.29 at a forward current of 10 mA and compare their magnitudes.

32. Using Eq. (1.6), determine the ac resistance at a current of 1 mA and 15 mA for the diode of Fig. 1.29. Compare the solutions and develop a general conclusion regarding the ac resistance and increasing levels of diode current.

33. Using Eq. (1.7), determine the ac resistance at a current of 1 mA and 15 mA for the diode of Fig. 1.19. Modify the equation as necessary for low levels of diode current. Compare to the solutions obtained in Problem 32.

34. Determine the average ac resistance for the diode of Fig. 1.19 for the region between 0.6 and 0.9 V.

35. Determine the ac resistance for the diode of Fig. 1.19 at 0.75 V and compare to the average ac resistance obtained in Problem 34.

§ 1.8 Diode Equivalent Circuits

36. Find the piecewise-linear equivalent circuit for the diode of Fig. 1.19. Use a straight line segment that intersects the horizontal axis at 0.7 V and best approximates the curve for the region greater than 0.7 V.

37. Repeat Problem 36 for the diode of Fig. 1.29.

Chapter 1   Semiconductor Diodes

§ 1.9 Diode Specification Sheets

* **38.** Plot I_F versus V_F using linear scales for the Fairchild diode of Fig. 1.36. Note that the provided graph employs a log scale for the vertical axis (log scales are covered in sections 11.2 and 11.3).

39. Comment on the change in capacitance level with increase in reverse-bias potential for the BAY73 diode.

40. Does the reverse saturation current of the BAY73 diode change significantly in magnitude for reverse-bias potentials in the range -25 V to -100 V?

* **41.** For the diode of Fig. 1.36 determine the level of I_R at room temperature (25°C) and the boiling point of water (100°C). Is the change significant? Does the level just about double for every 10°C increase in temperature?

42. For the diode of Fig. 1.36 determine the maximum ac (dynamic) resistance at a forward current of 0.1 mA, 1.5 mA, and 20 mA. Compare levels and comment on whether the results support conclusions derived in earlier sections of this chapter.

43. Using the characteristics of Fig. 1.36, determine the maximum power dissipation levels for the diode at room temperature (25°C) and 100°C. Assuming that V_F remains fixed at 0.7 V, how has the maximum level of I_F changed between the two temperature levels?

44. Using the characteristics of Fig. 1.36, determine the temperature at which the diode current will be 50% of its value at room temperature (25°C).

§ 1.10 Transition and Diffusion Capacitance

* **45.** (a) Referring to Fig. 1.37, determine the transition capacitance at reverse-bias potentials of -25 V and -10 V. What is the ratio of the change in capacitance to the change in voltage?
  (b) Repeat part (a) for reverse-bias potentials of -10 V and -1 V. Determine the ratio of the change in capacitance to the change in voltage.
  (c) How do the ratios determined in parts (a) and (b) compare? What does it tell you about which range may have more areas of practical application?

46. Referring to Fig. 1.37, determine the diffusion capacitance at 0 V and 0.25 V.

47. Describe in your own words how diffusion and transition capacitances differ.

48. Determine the reactance offered by a diode described by the characteristics of Fig. 1.37 at a forward potential of 0.2 V and a reverse potential of -20 V if the applied frequency is 6 MHz.

§ 1.11 Reverse Recovery Time

49. Sketch the waveform for i of the network of Fig. 1.63 if $t_t = 2t_s$ and the total reverse recovery time is 9 ns.

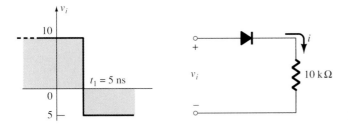

Figure 1.63 Problem 49

§ 1.14 Zener Diodes

50. The following characteristics are specified for a particular Zener diode: $V_Z = 29$ V, $V_R = 16.8$ V, $I_{ZT} = 10$ mA, $I_R = 20$ μA, and $I_{ZM} = 40$ mA. Sketch the characteristic curve in the manner displayed in Fig. 1.50.

* **51.** At what temperature will the 1N961 10-V Fairchild Zener diode have a nominal voltage of 10.75 V? (*Hint:* Note the data in Table 1.4.)

52. Determine the temperature coefficient of a 5-V Zener diode (rated 25°C value) if the nominal voltage drops to 4.8 V at a temperature of 100°C.

53. Using the curves of Fig. 1.51a, what level of temperature coefficient would you expect for a 20-V diode? Repeat for a 5-V diode. Assume a linear scale between nominal voltage levels and a current level of 0.1 mA.

54. Determine the dynamic impedance for the 24-V diode at $I_Z = 10$ mA for Fig. 1.51b. Note that it is a log scale.

* **55.** Compare the levels of dynamic impedance for the 24-V diode of Fig. 1.51b at current levels of 0.2 mA, 1 mA, and 10 mA. How do the results relate to the shape of the characteristics in this region?

§ 1.15 Light-Emitting Diodes

56. Referring to Fig. 1.55e, what would appear to be an appropriate value of V_T for this device? How does it compare to the value of V_T for silicon and germanium?

57. Using the information provided in Fig. 1.55, determine the forward voltage across the diode if the relative luminous intensity is 1.5.

* **58.** (a) What is the percent increase in relative efficiency of the device of Fig. 1.55 if the peak current is increased from 5 to 10 mA?
  (b) Repeat part (a) for 30 to 35 mA (the same increase in current).
  (c) Compare the percent increase from parts (a) and (b). At what point on the curve would you say there is little gained by further increasing the peak current?

* **59.** (a) Referring to Fig. 1.55h, determine the maximum tolerable peak current if the period of the pulse duration is 1 ms, the frequency is 300 Hz, and the maximum tolerable dc current is 20 mA.
  (b) Repeat part (a) for a frequency of 100 Hz.

60. (a) If the luminous intensity at 0° angular displacement is 3.0 mcd for the device of Fig. 1.55, at what angle will it be 0.75 mcd?
  (b) At what angle does the loss of luminous intensity drop below the 50% level?

* **61.** Sketch the current derating curve for the average forward current of the high-efficiency red LED of Fig. 1.55 as determined by temperature. (Note the absolute maximum ratings.)

---

*Please Note: Asterisks indicate more difficult problems.

Diode Applications

2.1 INTRODUCTION

The construction, characteristics, and models of semiconductor diodes were introduced in Chapter 1. The primary goal of this chapter is to develop a working knowledge of the diode in a variety of configurations using models appropriate for the area of application. By chapter's end, the fundamental behavior pattern of diodes in dc and ac networks should be clearly understood. The concepts learned in this chapter will have significant carryover in the chapters to follow. For instance, diodes are frequently employed in the description of the basic construction of transistors and in the analysis of transistor networks in the dc and ac domains.

The content of this chapter will reveal an interesting and very positive side of the study of a field such as electronic devices and systems—once the basic behavior of a device is understood, its function and response in an infinite variety of configurations can be determined. The range of applications is endless, yet the characteristics and models remain the same. The analysis will proceed from one that employs the actual diode characteristic to one that utilizes the approximate models almost exclusively. It is important that the role and response of various elements of an electronic system be understood without continually having to resort to lengthy mathematical procedures. This is usually accomplished through the approximation process, which can develop into an art itself. Although the results obtained using the actual characteristics may be slightly different from those obtained using a series of approximations, keep in mind that the characteristics obtained from a specification sheet may in themselves be slightly different from the device in actual use. In other words, the characteristics of a 1N4001 semiconductor diode may vary from one element to the next in the same lot. The variation may be slight, but it will often be sufficient to validate the approximations employed in the analysis. Also consider the other elements of the network: Is the resistor labeled 100 Ω exactly 100 Ω? Is the applied voltage exactly 10 V or perhaps 10.08 V? All these tolerances contribute to the general belief that a response determined through an appropriate set of approximations can often be "as accurate" as one that employs the full characteristics. In this book the emphasis is toward developing a working knowledge of a device through the use of appropriate approximations, thereby avoiding an unnecessary level of mathematical complexity. Sufficient detail will normally be provided, however, to permit a detailed mathematical analysis if desired.

2.2 LOAD-LINE ANALYSIS

The applied load will normally have an important impact on the point or region of operation of a device. If the analysis is performed in a graphical manner, a line can be drawn on the characteristics of the device that represents the applied load. The intersection of the load line with the characteristics will determine the point of operation of the system. Such an analysis is, for obvious reasons, called *load-line analysis*. Although the majority of the diode networks analyzed in this chapter do not employ the load-line approach, the technique is one used quite frequently in subsequent chapters, and this introduction offers the simplest application of the method. It also permits a validation of the approximate technique described throughout the remainder of this chapter.

Consider the network of Fig. 2.1a employing a diode having the characteristics of Fig. 2.1b. Note in Fig. 2.1a that the ''pressure'' established by the battery is to establish a current through the series circuit in the clockwise direction. The fact that this current and the defined direction of conduction of the diode are a ''match'' reveals that the diode is in the ''on'' state and conduction has been established. The resulting polarity across the diode will be as shown and the first quadrant (V_D and I_D positive) of Fig. 2.1b will be the region of interest—the forward-bias region.

Applying Kirchhoff's voltage law to the series circuit of Fig. 2.1a will result in

$$E - V_D - V_R = 0$$

or

$$\boxed{E = V_D + I_D R} \tag{2.1}$$

The two variables of Eq. (2.1) (V_D and I_D) are the same as the diode axis variables of Fig. 2.1b. This similarity permits a plotting of Eq. (2.1) on the same characteristics of Fig. 2.1b.

The intersections of the load line on the characteristics can easily be determined if one simply employs the fact that anywhere on the horizontal axis $I_D = 0$ A and anywhere on the vertical axis $V_D = 0$ V.

If we *set* $V_D = 0$ V in Eq. (2.1) and solve for I_D, we have the magnitude of I_D *on* the vertical axis. Therefore, with $V_D = 0$ V, Eq. (2.1) becomes

$$E = V_D + I_D R$$

$$= 0 \text{ V} + I_D R$$

and

$$\boxed{I_D = \left.\frac{E}{R}\right|_{V_D = 0 \text{ V}}} \tag{2.2}$$

as shown in Fig. 2.2. If we *set* $I_D = 0$ A in Eq. (2.1) and solve for V_D, we have the magnitude of V_D *on* the horizontal axis. Therefore, with $I_D = 0$ A, Eq. (2.1) becomes

$$E = V_D + I_D R$$

$$= V_D + (0 \text{ A})R$$

and

$$\boxed{V_D = E|_{I_D = 0 \text{ A}}} \tag{2.3}$$

as shown in Fig. 2.2. A straight line drawn between the two points will define the load line as depicted in Fig. 2.2. Change the level of R (the load) and the intersection on the vertical axis will change. The result will be a change in the slope of the load line and a different point of intersection between the load line and the device characteristics.

We now have a load line defined by the network and a characteristic curve defined by the device. The point of intersection between the two is the point of operation for

Figure 2.1 Series diode configuration: (a) circuit; (b) characteristics.

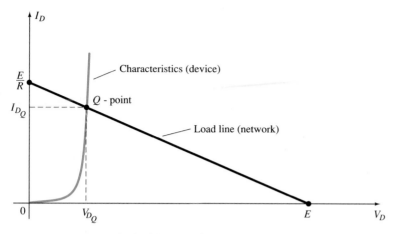

Figure 2.2 Drawing the load line and finding the point of operation.

this circuit. By simply drawing a line down to the horizontal axis the diode voltage V_{D_Q} can be determined, whereas a horizontal line from the point of intersection to the vertical axis will provide the level of I_{D_Q}. The current I_D is actually the current through the entire series configuration of Fig. 2.1a. The point of operation is usually called the *quiescent point* (abbreviated "Q pt.") to reflect its "still, unmoving" qualities as defined by a dc network.

The solution obtained at the intersection of the two curves is the same that would be obtained by a simultaneous mathematical solution of Eqs. (2.1) and (1.4) $[I_D = I_s(e^{kV_D/T_K} - 1)]$. Since the curve for a diode has nonlinear characteristics the mathematics envolved would require the use of nonlinear techniques that are beyond the needs and scope of this book. The load-line analysis described above provides a solution with a minimum of effort and provides a "pictorial" description of why the levels of solution for V_{D_Q} and I_{D_Q} were obtained. The next two examples will demonstrate the techniques introduced above and reveal the relative ease with which the load line can be drawn using Eqs. (2.2) and (2.3).

For the series diode configuration of Fig. 2.3a employing the diode characteristics of Fig. 2.3b determine:
(a) V_{D_Q} and I_{D_Q}.
(b) V_R.

EXAMPLE 2.1

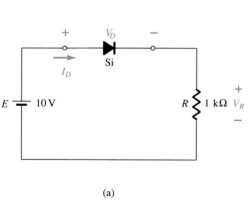

(a)

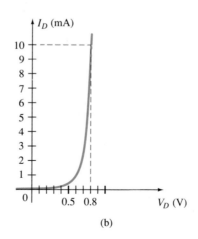

(b)

Figure 2.3 (a) Circuit; (b) characteristics.

2.2 **Load-line Analysis**   53

Solution

(a) Eq. (2.2): $I_D = \dfrac{E}{R}\bigg|_{V_D = 0\ \text{V}} = \dfrac{10\ \text{V}}{1\ \text{k}\Omega} = 10\ \text{mA}$

Eq. (2.3): $V_D = E\big|_{I_D = 0\ \text{A}} = 10\ \text{V}$

The resulting load line appears in Fig. 2.4. The intersection between the load line and the characteristic curve defines the Q point as

$$V_{D_Q} \cong \mathbf{0.78\ V}$$

$$I_{D_Q} \cong \mathbf{9.25\ mA}$$

The level of V_D is certainly an estimate and the accuracy of I_D is limited by the chosen scale. A higher degree of accuracy would require a plot that would be much larger and perhaps unwieldy.

(b) $V_R = I_R R = I_{D_Q} R = (9.25\ \text{mA})(1\ \text{k}\Omega) = \mathbf{9.25\ V}$

or $V_R = E - V_D = 10\ \text{V} - 0.78\ \text{V} = \mathbf{9.22\ V}$

The difference in results is due to the accuracy with which the graph can be read. Ideally, the results obtained either way should be the same.

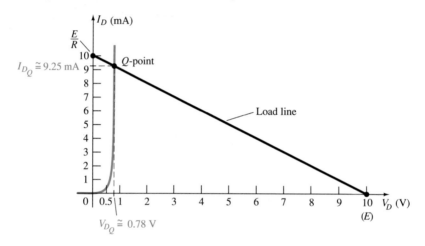

Figure 2.4 Solution to Example 2.1.

EXAMPLE 2.2

Repeat the analysis of Example 2.1 with $R = 2\ \text{k}\Omega$.

Solution

(a) Eq. (2.2): $I_D = \dfrac{E}{R}\bigg|_{V_D = 0\ \text{V}} = \dfrac{10\ \text{V}}{2\ \text{k}\Omega} = 5\ \text{mA}$

Eq. (2.3): $V_D = E\big|_{I_D = 0\ \text{A}} = 10\ \text{V}$

The resulting load line appears in Fig. 2.5. Note the reduced slope and levels of diode current for increasing loads. The resulting Q point is defined by

$$V_{D_Q} \cong \mathbf{0.7\ V}$$

$$I_{D_Q} \cong \mathbf{4.6\ mA}$$

(b) $V_R = I_R R = I_{D_Q} R = (4.6\ \text{mA})(2\ \text{k}\Omega) = \mathbf{9.2\ V}$

with $V_R = E - V_D = 10\ \text{V} - 0.7\ \text{V} = \mathbf{9.3\ V}$

The difference in levels is again due to the accuracy with which the graph can be read. Certainly, however, the results provide an expected magnitude for the voltage V_R.

Chapter 2 Diode Applications

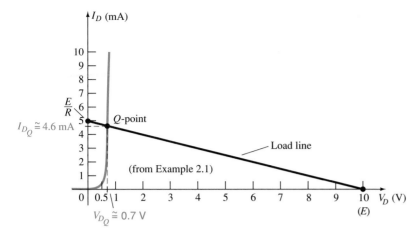

Figure 2.5 Solution to Example 2.2.

As noted in the examples above, the load line is determined solely by the applied network while the characteristics are defined by the chosen device. If we turn to our approximate model for the diode and do not disturb the network, the load line will be exactly the same as obtained in the examples above. In fact, the next two examples repeat the analysis of Examples 2.1 and 2.2 using the approximate model to permit a comparison of the results.

Repeat Example 2.1 using the approximate equivalent model for the silicon semiconductor diode.

EXAMPLE 2.3

Solution

The load line is redrawn as shown in Fig. 2.6 with the same intersections as defined in Example 2.1. The characteristics of the approximate equivalent circuit for the diode have also been sketched on the same graph. The resulting Q-point:

$$V_{D_Q} = \textbf{0.7 V}$$
$$I_{D_Q} = \textbf{9.25 mA}$$

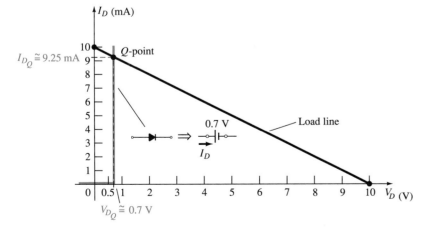

Figure 2.6 Solution to Example 2.1 using the diode approximate model.

The results obtained in Example 2.3 are quite interesting. The level of I_{D_Q} is exactly the same as obtained in Example 2.1 using a characteristic curve that is a great deal easier to draw than that appearing in Fig. 2.4. The level of $V_D = 0.7$ V versus 0.78 V from Example 2.1 is of a different magnitude to the hundredths place, but they are certainly in the same neighborhood if we compare their magnitudes to the magnitude of the other voltages of the network.

EXAMPLE 2.4

Repeat Example 2.2 using the approximate equivalent model for the silicon semiconductor diode.

Solution

The load line is redrawn as shown in Fig. 2.7 with the same intersections defined in Example 2.2. The characteristics of the approximate equivalent circuit for the diode have also been sketched on the same graph. The resulting Q-point:

$$V_{D_Q} = \textbf{0.7 V}$$

$$I_{D_Q} = \textbf{4.6 mA}$$

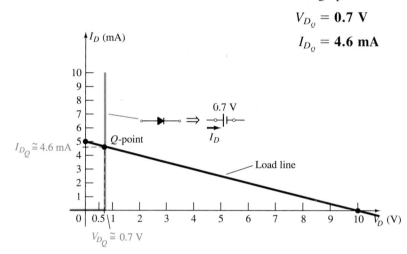

Figure 2.7 Solution to Example 2.2 using the diode approximate model.

In Example 2.4 the results obtained for both V_{D_Q} and I_{D_Q} are the same as those obtained using the full characteristics in Example 2.2. The examples above have demonstrated that the current and voltage levels obtained using the approximate model have been very close to those obtained using the full characteristics. It suggests, as will be applied in the sections to follow, that the use of appropriate approximations can result in solutions that are very close to the actual response with a reduced level of concern about properly reproducing the characteristics and choosing a large enough scale. In the next example we go a step further and substitute the ideal model. The results will reveal the conditions that must be satisfied to apply the ideal equivalent properly.

EXAMPLE 2.5

Repeat Example 2.1 using the ideal diode model.

Solution

As shown in Fig. 2.8 the load line continues to be the same, but the ideal characteristics now intersect the load line on the vertical axis. The Q-point is therefore defined by

$$V_{D_Q} = \textbf{0 V}$$

$$I_{D_Q} = \textbf{10 mA}$$

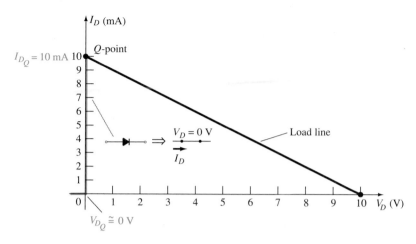

Figure 2.8 Solution to Example 2.1 using the ideal diode model.

The results are sufficiently different from the solutions of Example 2.1 to cause some concern about their accuracy. Certainly, they do provide some indication of the level of voltage and current to be expected relative to the other voltage levels of the network, but the additional effort of simply including the 0.7-V offset suggests that the approach of Example 2.3 is more appropriate.

Use of the ideal diode model therefore should be reserved for those occasions when the role of a diode is more important than voltage levels that differ by tenths of a volt and in those situations where the applied voltages are considerably larger than the threshold voltage V_T. In the next few sections the approximate model will be employed exclusively since the voltage levels obtained will be sensitive to variations that approach V_T. In later sections the ideal model will be employed more frequently since the applied voltages will frequently be quite a bit larger than V_T and the authors want to ensure that the role of the diode is correctly and clearly understood.

2.3 DIODE APPROXIMATIONS

In Section 2.2 we revealed that the results obtained using the approximate piecewise-linear equivalent model were quite close, if not equal, to the response obtained using the full characteristics. In fact, if one considers all the variations possible due to tolerances, temperature, and so on, one could certainly consider one solution to be ''as accurate'' as the other. Since the use of the approximate model normally results in a reduced expenditure of time and effort to obtain the desired results, it is the approach that will be employed in this book unless otherwise specified. Recall the following:

The primary purpose of this book is to develop a general knowledge of the behavior, capabilities, and possible areas of application of a device in a manner that will minimize the need for extensive mathematical developments.

The complete piecewise-linear equivalent model introduced in Chapter 1 was not employed in the load-line analysis because r_{av} is typically much less than the other series elements of the network. If r_{av} should be close in magnitude to the other series elements of the network, the complete equivalent model can be applied in much the same manner as described in Section 2.2.

In preparation for the analysis to follow, Table 2.1 was developed to review the important characteristics, models, and conditions of application for the approximate and ideal diode models. Although the silicon diode is used almost exclusively due to

TABLE 2.1 Approximate and Ideal Semiconductor Diode Models

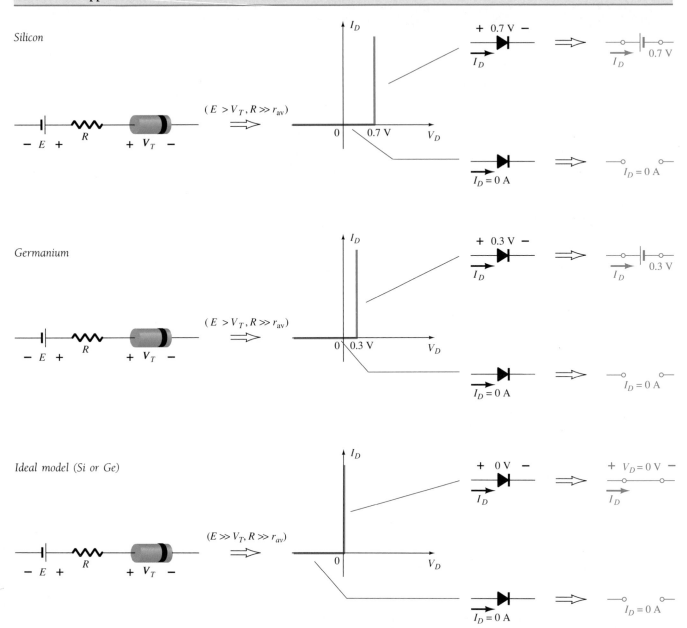

its temperature characteristics, the germanium diode is still employed and is therefore included in Table 2.1. As with the silicon diode a germanium diode is approximated by an open-circuit equivalent for voltages less than V_T. It will enter the "on" state when $V_D \geq V_T = 0.3$ V.

Keep in mind that the 0.7 V and 0.3 V in the equivalent circuits are not *independent* sources of energy but are there simply to remind us that there is a "price to pay" to turn on a diode. An isolated diode on a laboratory table will not indicate 0.7 V or 0.3 V if a voltmeter is placed across its terminals. The supplies specify the voltage drop across each when the device is "on" and specify that the diode voltage must be at least the indicated level before conduction can be established.

In the next few sections we demonstrate the impact of the models of Table 2.1 on the analysis of diode configurations. For those situations where the approximate equivalent circuit will be employed, the diode symbol will appear as shown in Fig. 2.9a for the silicon and germanium diodes. If conditions are such that the ideal diode model can be employed, the diode symbol will appear as shown in Fig. 2.9b.

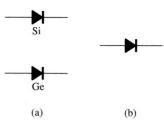

Figure 2.9 (a) Approximate model notation; (b) ideal diode notation.

2.4 SERIES DIODE CONFIGURATIONS WITH DC INPUTS

In this section the approximate model is utilized to investigate a number of series diode configurations with dc inputs. The content will establish a foundation in diode analysis that will carry over into the sections and chapters to follow. The procedure described can, in fact, be applied to networks with any number of diodes in a variety of configurations.

For each configuration the state of each diode must first be determined. Which diodes are "on" and which are "off"? Once determined, the appropriate equivalent as defined in Section 2.3 can be substituted and the remaining parameters of the network determined.

In general, a diode is in the "on" state if the current established by the applied sources is such that its direction matches that of the arrow in the diode symbol, and $V_D \geq 0.7$ V for silicon and $V_D \geq 0.3$ V for germanium.

For each configuration, *mentally* replace the diodes with resistive elements and note the resulting current direction as established by the applied voltages ("pressure"). If the resulting direction is a "match" with the arrow in the diode symbol, conduction through the diode will occur and the device is in the "on" state. The description above is, of course, contingent on the supply having a voltage greater than the "turn-on" voltage (V_T) of each diode.

If a diode is in the "on" state, one can either place a 0.7-V drop across the element, or the network can be redrawn with the V_T equivalent circuit as defined in Table 2.1. In time the preference will probably simply be to include the 0.7-V drop across each "on" diode and draw a line through each diode in the "off" or open state. Initially, however, the substitution method will be utilized to ensure that the proper voltage and current levels are determined.

The series circuit of Fig. 2.10 described in some detail in Section 2.2 will be used to demonstrate the approach described in the paragraphs above. The state of the diode is first determined by mentally replacing the diode by a resistive element as shown in Fig. 2.11. The resulting direction of I is a match with the arrow in the diode symbol, and since $E > V_T$ the diode is in the "on" state. The network is then redrawn as shown in Fig. 2.12 with the appropriate equivalent model for the forward-biased silicon diode. Note for future reference that the polarity of V_D is the same as would result if in fact the diode were a resistive element. The resulting voltage and current levels are the following:

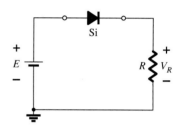

Figure 2.10 Series diode configuration.

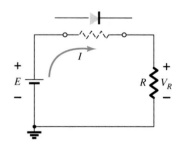

Figure 2.11 Determining the state of the diode of Fig. 2.10.

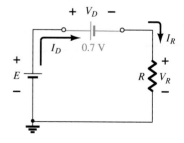

Figure 2.12 Substituting the equivalent model for the "on" diode of Fig. 2.10.

$$V_D = V_T \tag{2.4}$$

$$V_R = E - V_T \tag{2.5}$$

$$I_D = I_R = \frac{V_R}{R} \tag{2.6}$$

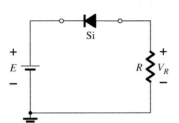

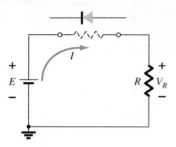

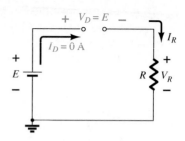

Figure 2.13 Reversing the diode of Fig. 2.10.

Figure 2.14 Determining the state of the diode of Fig. 2.13.

Figure 2.15 Substituting the equivalent model for the "off" diode of Figure 2.13.

In Fig. 2.13 the diode of Fig. 2.10 has been reversed. Mentally replacing the diode with a resistive element as shown in Fig. 2.14 will reveal that the resulting current direction does not match the arrow in the diode symbol. The diode is in the "off" state, resulting in the equivalent circuit of Fig. 2.15. Due to the open circuit, the diode current is 0 A and the voltage across the resistor R is the following:

$$V_R = I_R R = I_D R = (0 \text{ A})R = \mathbf{0 \text{ V}}$$

The fact that $V_R = 0$ V will establish E volts across the open circuit as defined by Kirchhoff's voltage law. Always keep in mind that under any circumstances—dc, ac instantaneous values, pulses, and so on—Kirchhoff's voltage law must be satisfied!

EXAMPLE 2.6

For the series diode configuration of Fig. 2.16, determine V_D, V_R, and I_D.

Solution

Since the applied voltage establishes a current in the clockwise direction to match the arrow of the symbol and the diode is in the "on" state,

$$V_D = \mathbf{0.7 \text{ V}}$$

$$V_R = E - V_T = 8 \text{ V} - 0.7 \text{ V} = \mathbf{7.3 \text{ V}}$$

$$I_D = I_R = \frac{V_R}{R} = \frac{7.3 \text{ V}}{2.2 \text{ k}\Omega} \cong \mathbf{3.32 \text{ mA}}$$

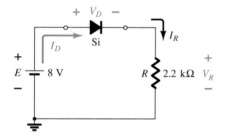

Figure 2.16 Circuit for Example 2.6.

EXAMPLE 2.7

Repeat Example 2.6 with the diode reversed.

Solution

Removing the diode, we find that the direction of I is opposite to the arrow in the diode symbol and the diode equivalent is the open circuit no matter which model is employed. The result is the network of Fig. 2.17, where $I_D = \mathbf{0 \text{ A}}$ due to the open circuit. Since $V_R = I_R R$, $V_R = (0)R = 0$ V. Applying Kirchhoff's voltage law around the closed loop yields

$$E - V_D - V_R = 0$$

and

$$V_D = E - V_R = E - 0 = E = \mathbf{8 \text{ V}}$$

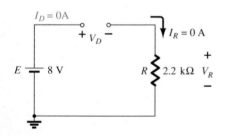

Figure 2.17 Determining the unknown quantities for Example 2.7.

In particular, note in Example 2.7 the high voltage across the diode even though it is an "off" state. The current is zero, but the voltage is significant. For review purposes, keep the following in mind for the analysis to follow:

1. An open circuit can have any voltage across its terminals, but the current is always 0 A.

2. A short circuit has a 0-V drop across its terminals, but the current is limited only by the surrounding network.

In the next example the notation of Fig. 2.18 will be employed for the applied voltage. It is a common industry notation and one with which the reader should become very familiar. Such notation and other defined voltage levels are treated further in Chapter 4.

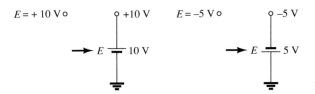

Figure 2.18 Source notation.

For the series diode configuration of Fig. 2.19, determine V_D, V_R, and I_D.

EXAMPLE 2.8

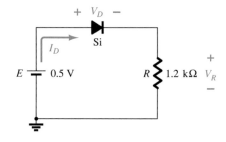

Figure 2.19 Series diode circuit for Example 2.8.

Solution

Although the "pressure" establishes a current with the same direction as the arrow symbol the level of applied voltage is insufficient to turn the silicon diode "on." The point of operation on the characteristics is shown in Fig. 2.20, establishing the open-circuit equivalent as the appropriate approximation. The resulting voltage and current levels are therefore the following:

$$I_D = \mathbf{0\ A}$$
$$V_R = I_R R = I_D R = (0\ A)1.2\ k\Omega = \mathbf{0\ V}$$

and
$$V_D = E = \mathbf{0.5\ V}$$

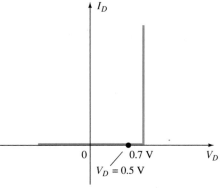

Figure 2.20 Operating point with $E = 0.5$ V.

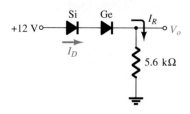

EXAMPLE 2.9   Determine V_o and I_D for the series circuit of Fig. 2.21.

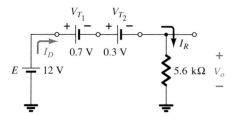

Figure 2.21 Circuit for Example 2.9.

Solution

An attack similar to that applied in Example 2.6 will reveal that the resulting current has the same direction as the arrowheads of the symbols of both diodes, and the network of Fig. 2.22 results because $E = 12\ V > (0.7\ V + 0.3\ V) = 1\ V$. Note the redrawn supply of 12 V and the polarity of V_o across the 5.6-kΩ resistor. The resulting voltage

$$V_o = E - V_{T_1} - V_{T_2} = 12\ V - 0.7\ V - 0.3\ V = \mathbf{11\ V}$$

and

$$I_D = I_R = \frac{V_R}{R} = \frac{V_o}{R} = \frac{11\ V}{5.6\ k\Omega} \cong \mathbf{1.96\ mA}$$

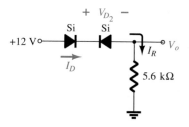

Figure 2.22 Determining the unknown quantities for Example 2.9.

EXAMPLE 2.10   Determine I_D, V_{D_2}, and V_o for the circuit of Fig. 2.23.

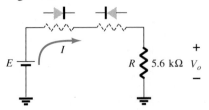

Figure 2.23 Circuit for Example 2.10.

Solution

Removing the diodes and determining the direction of the resulting current I will result in the circuit of Fig. 2.24. There is a match in current direction for the silicon diode but not for the germanium diode. The combination of a short circuit in series with an open circuit always results in an open circuit and $I_D = \mathbf{0\ A}$, as shown in Fig. 2.25.

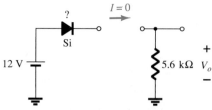

Figure 2.24 Determining the state of the diodes of Figure 2.23.

Figure 2.25 Substituting the equivalent state for the open diode.

Chapter 2 Diode Applications

The question remains as to what to substitute for the silicon diode. For the analysis to follow in this and succeeding chapters, simply recall for the actual practical diode that when $I_D = 0$ A, $V_D = 0$ V (and vice versa), as described for the no-bias situation in Chapter 1. The conditions described by $I_D = 0$ A and $V_{D_1} = 0$ V are indicated in Fig. 2.26.

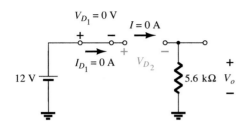

Figure 2.26 Determining the unknown quantities for the circuit of Example 2.10.

$$V_o = I_R R = I_D R = (0 \text{ A})R = 0 \text{ V}$$

and

$$V_{D_2} = V_{\text{open circuit}} = E = \textbf{12 V}$$

Applying Kirchhoff's voltage law in a clockwise direction gives us

$$E - V_{D_1} - V_{D_2} - V_o = 0$$

and

$$V_{D_2} = E - V_{D_1} - V_o = 12 \text{ V} - 0 - 0$$

$$= \textbf{12 V}$$

with

$$V_o = \textbf{0 V}$$

Determine I, V_1, V_2, and V_o for the series dc configuration of Fig. 2.27.

EXAMPLE 2.11

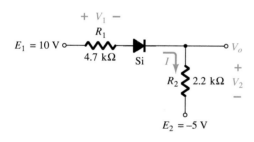

Figure 2.27 Circuit for Example 2.11.

Solution

The sources are drawn and the current direction indicated as shown in Fig. 2.28. The diode is in the "on" state and the notation appearing in Fig. 2.29 is included to indicate this state. Note that the "on" state is noted simply by the additional $V_D =$

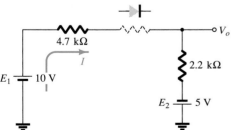
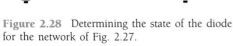

Figure 2.28 Determining the state of the diode for the network of Fig. 2.27.

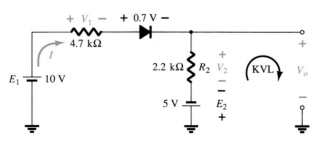

Figure 2.29 Determining the unknown quantities for the network of Fig. 2.27.

0.7 V on the figure. This eliminates the need to redraw the network and avoids any confusion that may result from the appearance of another source. As indicated in the introduction to this section, this is probably the path and notation that one will take when a level of confidence has been established in the analysis of diode configurations. In time the entire analysis will be performed simply by referring to the original network. Recall that a reverse-biased diode can simply be indicated by a line through the device.

The resulting current through the circuit,

$$I = \frac{E_1 + E_2 - V_D}{R_1 + R_2} = \frac{10 \text{ V} + 5 \text{ V} - 0.7 \text{ V}}{4.7 \text{ k}\Omega + 2.2 \text{ k}\Omega} = \frac{14.3 \text{ V}}{6.9 \text{ k}\Omega}$$

$$\cong \mathbf{2.07 \text{ mA}}$$

$$V_1 = IR_1 = (2.07 \text{ mA})(4.7 \text{ k}\Omega) = \mathbf{9.73 \text{ V}}$$

$$V_2 = IR_2 = (2.07 \text{ mA})(2.2 \text{ k}\Omega) = \mathbf{4.55 \text{ V}}$$

Applying Kirchhoff's voltage law to the output section in the clockwise direction will result in

$$-E_2 + V_2 - V_o = 0$$

and

$$V_o = V_2 - E_2 = 4.55 \text{ V} - 5 \text{ V} = \mathbf{-0.45 \text{ V}}$$

The minus sign indicates that V_o has a polarity opposite to that appearing in Fig. 2.27.

2.5 PARALLEL AND SERIES–PARALLEL CONFIGURATIONS

The methods applied in Section 2.4 can be extended to the analysis of parallel and series–parallel configurations. For each area of application, simply match the sequential series of steps applied to series diode configurations.

EXAMPLE 2.12

Determine V_o, I_1, I_{D_1}, and I_{D_2} for the parallel diode configuration of Fig. 2.30.

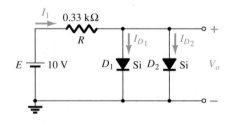

Figure 2.30 Network for Example 2.12.

Solution

For the applied voltage the "pressure" of the source is to establish a current through each diode in the same direction as shown in Fig. 2.31. Since the resulting current direction matches that of the arrow in each diode symbol and the applied voltage is greater than 0.7 V, both diodes are in the "on" state. The voltage across parallel elements is always the same and

$$V_o = \mathbf{0.7 \text{ V}}$$

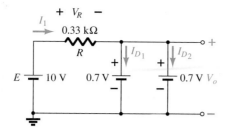

Figure 2.31 Determining the unknown quantities for the network of Example 2.12.

The current

$$I_1 = \frac{V_R}{R} = \frac{E - V_D}{R} = \frac{10\text{ V} - 0.7\text{ V}}{0.33\text{ k}\Omega} = \mathbf{28.18\text{ mA}}$$

Assuming diodes of similar characteristics, we have

$$I_{D_1} = I_{D_2} = \frac{I_1}{2} = \frac{28.18\text{ mA}}{2} = \mathbf{14.09\text{ mA}}$$

Example 2.12 demonstrated one reason for placing diodes in parallel. If the current rating of the diodes of Fig. 2.30 is only 20 mA, a current of 28.18 mA would damage the device if it appeared alone in Fig. 2.30. By placing two in parallel, the current is limited to a safe value of 14.09 mA with the same terminal voltage.

Determine the current I for the network of Fig. 2.32.

EXAMPLE 2.13

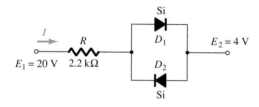

Figure 2.32 Network for Example 2.13.

Solution

Redrawing the network as shown in Fig. 2.33 reveals that the resulting current direction is such as to turn on diode D_1 and turn off diode D_2. The resulting current I is then

$$I = \frac{E_1 - E_2 - V_D}{R} = \frac{20\text{ V} - 4\text{ V} - 0.7\text{ V}}{2.2\text{ k}\Omega} \cong \mathbf{6.95\text{ mA}}$$

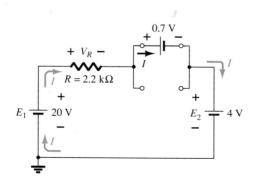

Figure 2.33 Determining the unknown quantities for the network of Example 2.13.

EXAMPLE 2.14

Determine the voltage V_o for the network of Fig. 2.34.

Solution

Initially, it would appear that the applied voltage will turn both diodes "on." However, if both were "on," the 0.7-V drop across the silicon diode would not match the 0.3 V across the germanium diode as required by the fact that the voltage across parallel elements must be the same. The resulting action can be explained simply by realizing that when the supply is turned on it will increase from 0 V to 12 V over a period of time—although probably measurable in milliseconds. At the instant during the rise that 0.3 V is established across the germanium diode it will turn "on" and maintain a level of 0.3 V. The silicon diode will never have the opportunity to capture its required 0.7 V and therefore remains in its open-circuit state as shown in Fig. 2.35. The result:

$$V_o = 12 \text{ V} - 0.3 \text{ V} = \textbf{11.7 V}$$

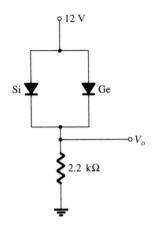

Figure 2.34 Network for Example 2.14.

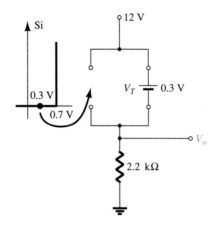

Figure 2.35 Determining V_o for the network of Fig. 2.34.

EXAMPLE 2.15

Determine the currents I_1, I_2, and I_{D_2} for the network of Fig. 2.36.

Solution

The applied voltage (pressure) is such as to turn both diodes on, as noted by the resulting current directions in the network of Fig. 2.37. Note the use of the abbreviated notation for "on" diodes and that the solution is obtained through an application of techniques applied to dc series-parallel networks.

$$I_1 = \frac{V_{T_2}}{R_1} = \frac{0.7 \text{ V}}{3.3 \text{ k}\Omega} = \textbf{0.212 mA}$$

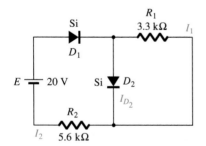

Figure 2.36 Network for Example 2.15.

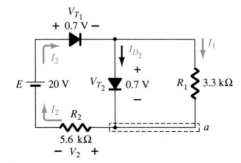

Figure 2.37 Determining the unknown quantities for Example 2.15.

Applying Kirchhoff's voltage law around the indicated loop in the clockwise direction yields

$$-V_2 + E - V_{T_1} - V_{T_2} = 0$$

and

$$V_2 = E - V_{T_1} - V_{T_2} = 20 \text{ V} - 0.7 \text{ V} - 0.7 \text{ V} = 18.6 \text{ V}$$

with

$$I_2 = \frac{V_2}{R_2} = \frac{18.6 \text{ V}}{5.6 \text{ k}\Omega} = \textbf{3.32 mA}$$

At the bottom node (a),

$$I_{D_2} + I_1 = I_2$$

and

$$I_{D_2} = I_2 - I_1 = 3.32 \text{ mA} - 0.212 \text{ mA} = \textbf{3.108 mA}$$

2.6 AND/OR GATES

The tools of analysis are now at our disposal and the opportunity to investigate a computer configuration is one that will demonstrate the range of applications of this relatively simple device. Our analysis will be limited to determining the voltage levels and will not include a detailed discussion of Boolean algebra or positive and negative logic.

The network to be analyzed in Example 2.16 is an OR gate for positive logic. That is, the 10-V level of Fig. 2.38 is assigned a "1" for Boolean algebra while the 0-V input is assigned a "0." An OR gate is such that the output voltage level will be a 1 if either *or* both inputs is a 1. The output is a 0 if both inputs are at the 0 level.

The analysis of AND/OR gates is made measurably easier by using the approximate equivalent for a diode rather than the ideal because we can stipulate that the voltage across the diode must be 0.7 V (0.3 V for Ge) positive for the silicon diode to switch to the "on" state.

In general, the best approach is simply to establish a "gut" feeling for the state of the diodes by noting the direction and the "pressure" established by the applied potentials. The analysis will then verify or negate your initial assumptions.

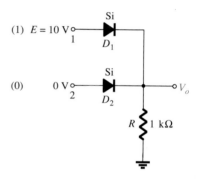

Figure 2.38 Positive logic OR gate.

Determine V_o for the network of Fig. 2.38.

EXAMPLE 2.16

Solution

First note that there is only one applied potential; 10 V at terminal 1. Terminal 2 with a 0 V input is essentially at ground potential, as shown in the redrawn network of Fig. 2.39. Figure 2.39 "suggests" that D_1, is probably in the "on" state due to the applied 10 V while D_2 with its "positive" side at 0 V is probably "off." Assuming these states will result in the configuration of Fig. 2.40.

The next step is simply to check that there is no contradiction to our assumptions. That is, note that the polarity across D_1 is such as to turn it on and the polarity across D_2 is such as to turn it off. For D_1 the "on" state establishes V_o at $V_o = E - V_D = 10 \text{ V} - 0.7 \text{ V} = \textbf{9.3 V}$. With 9.3 V at the cathode (−) side of D_2 and 0 V at the anode (+) side, D_2 is definitely in the "off" state. The current direction and the resulting continuous path for conduction further confirm our assumption that D_1 is conducting. Our assumptions seem confirmed by the resulting voltages and current, and our initial analysis can be assumed to be correct. The output voltage level is not 10 V as defined for an input of 1, but the 9.3 V is sufficiently large to be considered a 1 level. The output is therefore at a 1 level with only one input, which suggests that

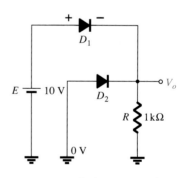

Figure 2.39 Redrawn network of Fig. 2.38.

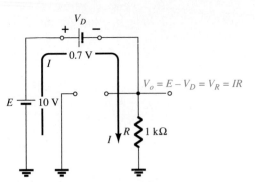

Figure 2.40 Assumed diode states for Fig. 2.38.

the gate is an OR gate. An analysis of the same network with two 10-V inputs will result in both diodes being in the "on" state and an output of 9.3 V. A 0-V input at both inputs will not provide the 0.7 V required to turn the diodes on and the output will be a 0 due to the 0-V output level. For the network of Fig. 2.40 the current level is determined by

$$I = \frac{E - V_D}{R} = \frac{10 \text{ V} - 0.7 \text{ V}}{1 \text{ k}\Omega} = \mathbf{9.3 \text{ mA}}$$

EXAMPLE 2.17

Determine the output level for the positive logic AND gate of Fig. 2.41.

Solution

Note in this case that an independent source appears in the grounded leg of the network. For reasons soon to become obvious it is chosen at the same level as the input logic level. The network is redrawn in Fig. 2.42 with our initial assumptions regarding the state of the diodes. With 10 V at the cathode side of D_1 it is assumed that D_1 is in the "off" state even though there is a 10-V source connected to the anode of D_1 through the resistor. However, recall that we mentioned in the introduction to this section that the use of the approximate model will be an aid to the analysis. For D_1, where will the 0.7 V come from if the input and source voltages are at the same level and creating opposing "pressures"? D_2 is assumed to be in the "on" state due to the low voltage at the cathode side and the availability of the 10-V source through the 1-kΩ resistor.

For the network of Fig. 2.42 the voltage at V_o is 0.7 V, due to the forward-biased diode D_2. With 0.7 V at the anode of D_1 and 10 V at the cathode, D_1 is definitely in the "off" state. The current I will have the direction indicated in Fig. 2.42 and a magnitude equal to

$$I = \frac{E - V_D}{R} = \frac{10 \text{ V} - 0.7 \text{ V}}{1 \text{ k}\Omega} = \mathbf{9.3 \text{ mA}}$$

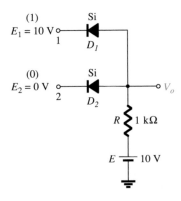

Figure 2.41 Positive logic AND gate.

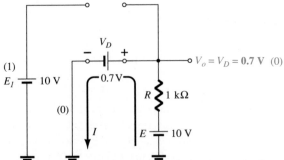

Figure 2.42 Substituting the assumed states for the diodes of Fig. 2.41.

Chapter 2 **Diode Applications**

The state of the diodes is therefore confirmed and our earlier analysis was correct. Although not 0 V as earlier defined for the 0 level, the output voltage is sufficiently small to be considered a 0 level. For the AND gate, therefore, a single input will result in a 0-level output. The remaining states of the diodes for the possibilities of two inputs and no inputs will be examined in the problems at the end of the chapter.

2.7 SINUSOIDAL INPUTS; HALF-WAVE RECTIFICATION

The diode analysis will now be expanded to include time-varying functions such as the sinusoidal waveform and the square wave. There is no question that the degree of difficulty will increase, but once a few fundamental maneuvers are understood, the analysis will be fairly direct and follow a common thread.

The simplest of networks to examine with a time-varying signal appears in Fig. 2.43. For the moment we will use the ideal model (note the absence of the Si or Ge label to denote ideal diode) to ensure that the approach is not clouded by additional mathematical complexity.

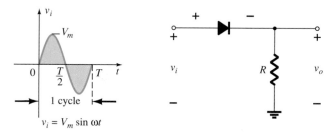

Figure 2.43 Half-wave rectifier.

Over one full cycle, defined by the period T of Fig. 2.43, the average value (the algebraic sum of the areas above and below the axis) is zero. The circuit of Fig. 2.43 called a *half-wave rectifier*, will generate a waveform v_o that will have an average value of particular use in the ac-to-dc conversion process. When employed in the rectification process a diode is typically referred to as a *rectifier*. Its power and current ratings are typically much higher than that of diodes employed in other applications, such as computers and communication systems.

During the interval $t = 0 \rightarrow T/2$ in Fig. 2.43 the polarity of the applied voltage v_i is such as to establish "pressure" in the direction indicated and turn on the diode with the polarity appearing above the diode. Substituting the short-circuit equivalence for the ideal diode will result in the equivalent circuit of Fig. 2.44, where it is fairly obvious that the output signal is an exact replica of the applied signal. The two terminals defining the output voltage are connected directly to the applied signal via the short-circuit equivalence of the diode.

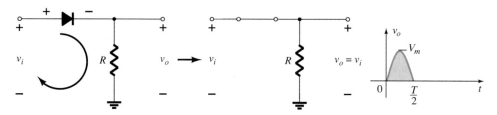

Figure 2.44 Conduction region $(0 \rightarrow T/2)$.

For the period $T/2 \rightarrow T$, the polarity of the input v_i is as shown in Fig. 2.45 and the resulting polarity across the ideal diode produces an "off" state with an open-circuit equivalent. The result is the absence of a path for charge to flow and $v_o = iR = (0)R = 0$ V for the period $T/2 \rightarrow T$. The input v_i and the output v_o were sketched together in Fig. 2.46 for comparison purposes. The output signal v_o now has a net positive area above the axis over a full period and an average value determined by

$$V_{dc} = 0.318V_m \quad \text{half-wave}$$ (2.7)

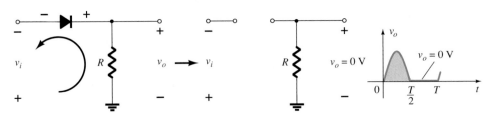

Figure 2.45 Nonconduction region $(T/2 \rightarrow T)$.

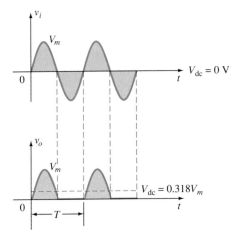

Figure 2.46 Half-wave rectified signal.

The process of removing one-half the input signal to establish a dc level is aptly called *half-wave rectification*.

The effect of using a silicon diode with $V_T = 0.7$ V is demonstrated in Fig. 2.47 for the forward-bias region. The applied signal must now be at least 0.7 V before the diode can turn "on." For levels of v_i less than 0.7 V the diode is still in an open-circuit state and $v_o = 0$ V as shown in the same figure. When conducting, the difference between v_o and v_i is a fixed level of $V_T = 0.7$ V and $v_o = v_i - V_T$, as shown in the figure. The net effect is a reduction in area above the axis, which naturally

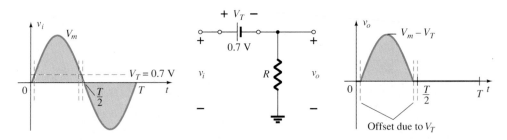

Figure 2.47 Effect of V_T on half-wave rectified signal.

Chapter 2 **Diode Applications**

reduces the resulting dc voltage level. For situations where $V_m \gg V_T$, equation 2.8 can be applied to determine the average value with a relatively high level of accuracy.

$$V_{dc} \cong 0.318(V_m - V_T) \tag{2.8}$$

In fact, if V_m is sufficiently greater than V_T, Eq. 2.7 is often applied as a first approximation for V_{dc}.

EXAMPLE 2.18

(a) Sketch the output v_o and determine the dc level of the output for the network of Fig. 2.48.
(b) Repeat part (a) if the ideal diode is replaced by a silicon diode.
(c) Repeat parts (a) and (b) if V_m is increased to 200 V and compare solutions using Eqs. (2.7) and (2.8).

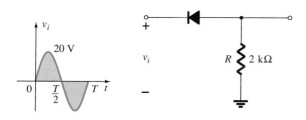

Figure 2.48 Network for Example 2.18.

Solution

(a) In this situation the diode will conduct during the negative part of the input as shown in Fig. 2.49, and v_o will appear as shown in the same figure. For the full period, the dc level is

$$V_{dc} = -0.318V_m = -0.318(20 \text{ V}) = \mathbf{-6.36 \text{ V}}$$

The negative sign indicates that the polarity of the output is opposite to the defined polarity of Fig. 2.48.

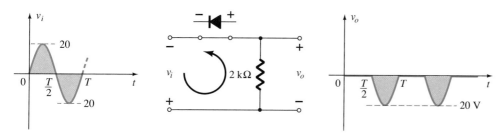

Figure 2.49 Resulting v_o for the circuit of Example 2.18.

(b) Using a silicon diode, the output has the appearance of Fig. 2.50 and

$$V_{dc} \cong -0.318(V_m - 0.7 \text{ V}) = -0.318(19.3 \text{ V}) \cong \mathbf{-6.14 \text{ V}}$$

The resulting drop in dc level is 0.22 V or about 3.5%.
(c) Eq. (2.7): $V_{dc} = -0.318V_m = -0.318(200 \text{ V}) = \mathbf{-63.6 \text{ V}}$

Eq. (2.8): $V_{dc} = -0.318(V_m - V_T) = -0.318(200 \text{ V} - 0.7 \text{ V})$
$$= -(0.318)(199.3 \text{ V}) = \mathbf{-63.38 \text{ V}}$$

which is a difference that can certainly be ignored for most applications. For part c the offset and drop in amplitude due to V_T would not be discernible on a typical oscilloscope if the full pattern is displayed.

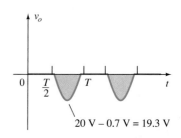

Figure 2.50 Effect of V_T on output of Fig. 2.49.

PIV (PRV)

The peak-inverse-voltage (PIV) [or PRV (peak reverse voltage)] rating of the diode is of primary importance in the design of rectification systems. Recall that it is voltage rating that must not be exceeded in the reverse-bias region or the diode will enter the Zener avalanche region. The required PIV rating for the half-wave rectifier can be determined from Fig. 2.51, which displays the reverse-biased diode of Fig. 2.43 with maximum applied voltage. Applying Kirchhoff's voltage law, it is fairly obvious that the PIV rating of the diode must equal or exceed the peak value of the applied voltage. Therefore,

$$\boxed{\text{PIV rating} \geqq V_m}$$
half-wave rectifier
(2.9)

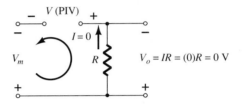

Figure 2.51 Determining the required PIV rating for the half-wave rectifier.

2.8 FULL-WAVE RECTIFICATION

Bridge Network

The dc level obtained from a sinusoidal input can be improved 100% using a process called *full-wave rectification*. The most familiar network for performing such a function appears in Fig. 2.52 with its four diodes in a *bridge* configuration. During the period $t = 0$ to $T/2$ the polarity of the input is as shown in Fig. 2.53. The resulting polarities across the ideal diodes are also shown in Fig. 2.53 to reveal that D_2 and D_3 are conducting while D_1 and D_4 are in the "off" state. The net result is the configuration of Fig. 2.54, with its indicated current and polarity across R. Since the diodes are ideal the load voltage $v_o = v_i$, as shown in the same figure.

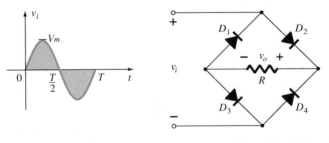

Figure 2.52 Full-wave bridge rectifier.

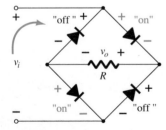

Figure 2.53 Network of Fig. 2.52 for the period $0 \rightarrow T/2$ of the input voltage v_i.

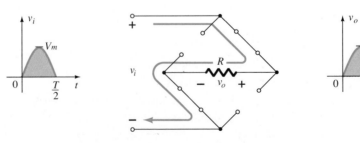

Figure 2.54 Conduction path for the positive region of v_i.

Chapter 2   Diode Applications

For the negative region of the input the conducting diodes are D_1 and D_4, resulting in the configuration of Fig. 2.55. The important result is that the polarity across the load resistor R is the same as in Fig. 2.53, establishing a second positive pulse, as shown in Fig. 2.55. Over one full cycle the input and output voltages will appear as shown in Fig. 2.56.

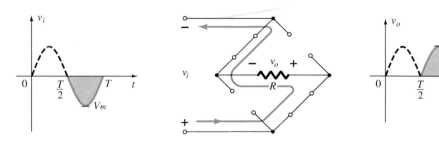

Figure 2.55 Conduction path for the negative region of v_i.

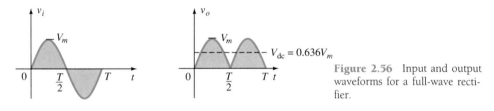

Figure 2.56 Input and output waveforms for a full-wave rectifier.

Since the area above the axis for one full cycle is now twice that obtained for a half-wave system, the dc level has also been doubled and

$$V_{dc} = 2(\text{Eq. } 2.7) = 2(0.318V_m)$$

or

$$\boxed{V_{dc} = 0.636V_m} \quad \text{full-wave} \tag{2.10}$$

If silicon rather than ideal diodes are employed as shown in Fig. 2.57, an application of Kirchhoff's voltage law around the conduction path would result in

$$v_i - V_T - v_o - V_T = 0$$

and

$$v_o = v_i - 2V_T$$

The peak value of the output voltage v_o is therefore

$$V_{o_{\max}} = V_m - 2V_T$$

For situations where $V_m \gg 2V_T$, Eq. 2.11 can be applied for the average value with a relatively high level of accuracy.

$$\boxed{V_{dc} \cong 0.636(V_m - 2V_T)} \tag{2.11}$$

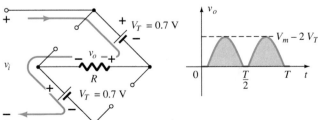

Figure 2.57 Determining $V_{o_{\max}}$ for silicon diodes in the bridge configuration.

Then again, if V_m is sufficiently greater than $2V_T$ then Eq. 2.10 is often applied as a first approximation for V_{dc}.

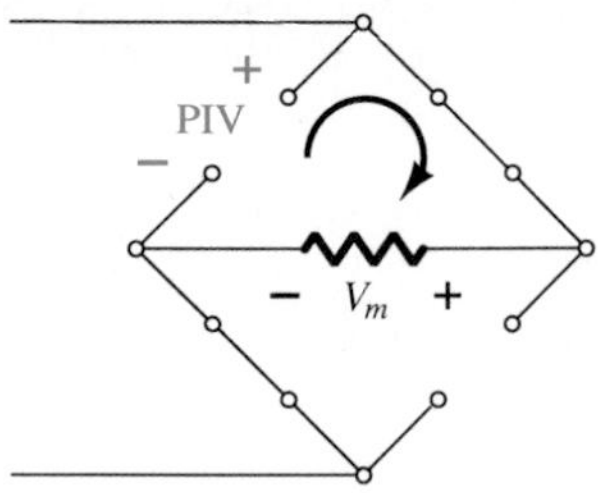

Figure 2.58 Determining the required PIV for the bridge configuration.

PIV

The required PIV of each diode (ideal) can be determined from Fig. 2.58 obtained at the peak of the positive region of the input signal. For the indicated loop the maximum voltage across R is V_m and the PIV rating is defined by

$$\boxed{\text{PIV} \geqq V_m} \quad \text{full-wave bridge rectifier} \tag{2.12}$$

Center-Tapped Transformer

A second popular full-wave rectifier appears in Fig. 2.59 with only two diodes but requiring a center-tapped (CT) transformer to establish the input signal across each section of the secondary of the transformer. During the positive portion of v_i applied to the primary of the transformer, the network will appear as shown in Fig. 2.60. D_1 assumes the short-circuit equivalent and D_2 the open-circuit equivalent, as determined by the secondary voltages and the resulting current directions. The output voltage appears as shown in Fig. 2.60.

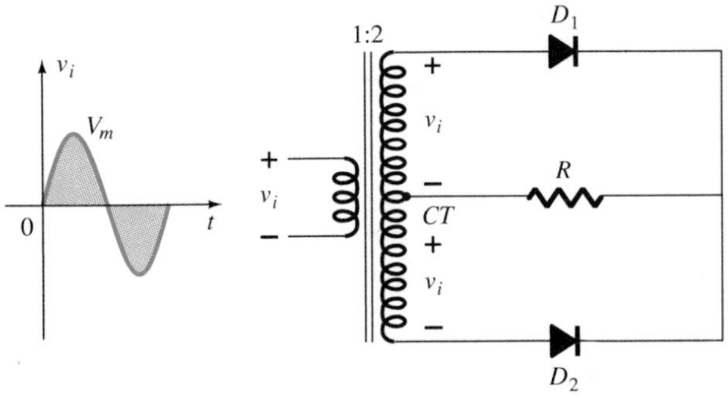

Figure 2.59 Center-tapped transformer full-wave rectifier.

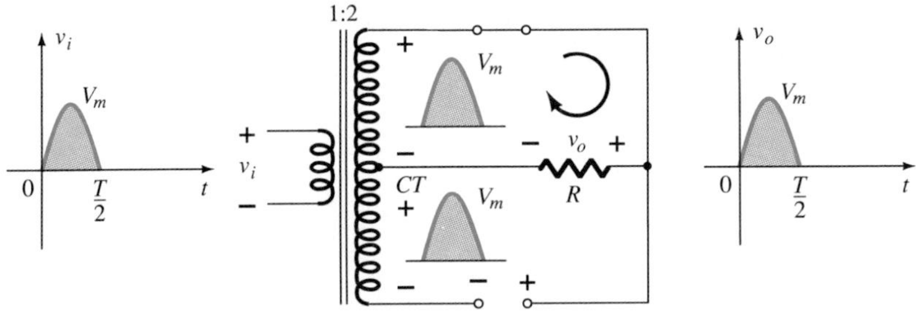

Figure 2.60 Network conditions for the positive region of v_i.

During the negative portion of the input the network appears as shown in Fig. 2.61, reversing the roles of the diodes but maintaining the same polarity for the

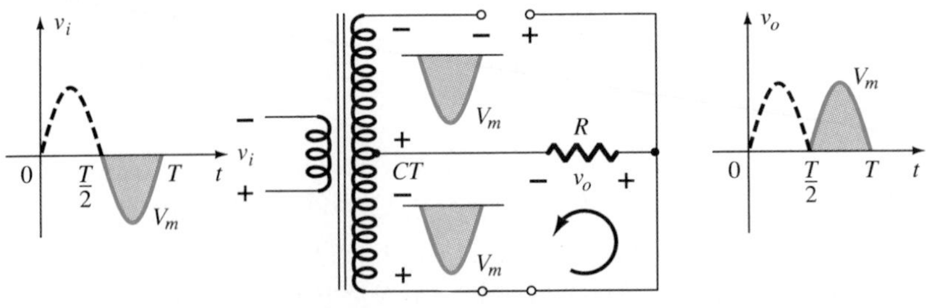

Figure 2.61 Network conditions for the negative region of v_i.

voltage across the load resistor R. The net effect is the same output as that appearing in Fig. 2.56 with the same dc levels.

PIV

The network of Fig. 2.62, will help us determine the net PIV for each diode for this full-wave rectifier. Inserting the maximum voltage for the secondary voltage and V_m as established by the adjoining loop will result in

$$\text{PIV} = V_{\text{secondary}} + V_R$$

$$= V_m + V_m$$

and

$$\boxed{\text{PIV} \cong 2V_m} \quad \text{CT transformer, full-wave rectifier} \tag{2.13}$$

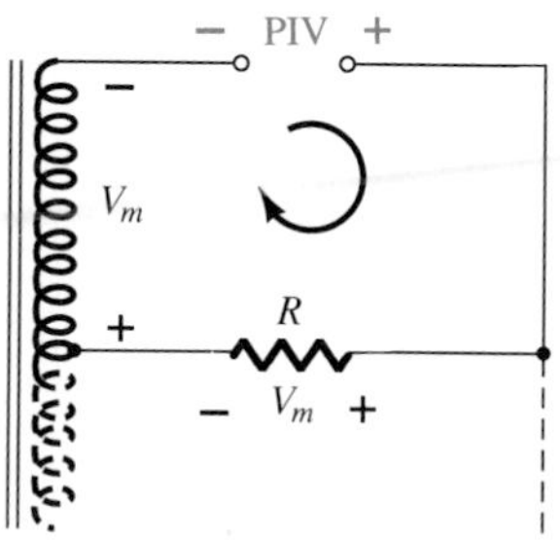

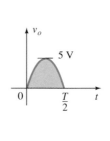

Figure 2.62 Determining the PIV level for the diodes of the CT transformer full-wave rectifier.

EXAMPLE 2.19

Determine the output waveform for the network of Fig. 2.63 and calculate the output dc level and the required PIV of each diode.

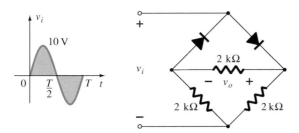

Figure 2.63 Bridge network for Example 2.19.

Solution

The network will appear as shown in Fig. 2.64 for the positive region of the input voltage. Redrawing the network will result in the configuration of Fig. 2.65, where $v_o = \frac{1}{2}v_i$ or $V_{o_{\text{max}}} = \frac{1}{2}V_{i_{\text{max}}} = \frac{1}{2}(10 \text{ V}) = 5 \text{ V}$, as shown in Fig. 2.65. For the negative part of the input the roles of the diodes will be interchanged and v_o will appear as shown in Fig. 2.66.

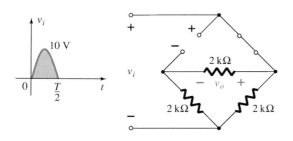

Figure 2.64 Network of Fig. 2.63 for the positive region of v_i.

Figure 2.65 Redrawn network of Fig. 2.64.

The effect of removing two diodes from the bridge configuration was therefore to reduce the available dc level to the following:

$$V_{\text{dc}} = 0.636(5 \text{ V}) = \textbf{3.18 V}$$

or that available from a half-wave rectifier with the same input. However, the PIV as determined from Fig. 2.58 is equal to the maximum voltage across R, which is 5 V or half of that required for a half-wave rectifier with the same input.

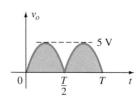

Figure 2.66 Resulting output for Example 2.19.

2.8 **Full-Wave Rectification**

75

2.9 CLIPPERS

There are a variety of diode networks called *clippers* that have the ability to "clip" off a portion of the input signal without distorting the remaining part of the alternating waveform. The half-wave rectifier of Section 2.7 is an example of the simplest form of diode clipper—one resistor and diode. Depending on the orientation of the diode, the positive or negative region of the input signal is "clipped" off.

There are two general categories of clippers: *series* and *parallel*. The series configuration is defined as one where the diode is in series with the load, while the parallel variety has the diode in a branch parallel to the load.

Series

The response of the series configuration of Fig. 2.67a to a variety of alternating waveforms is provided in Fig. 2.67b. Although first introduced as a half-wave rectifier (for sinusoidal waveforms), there are no boundaries on the type of signals that can be applied to a clipper. The addition of a dc supply such as shown in Fig. 2.68 can have a pronounced effect on the output of a clipper. Our initial discussion will be limited to ideal diodes, with the effect of V_T reserved for a concluding example.

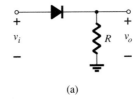

(a)

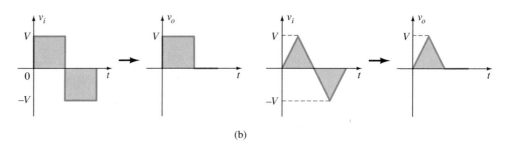

(b)

Figure 2.67 Series clipper.

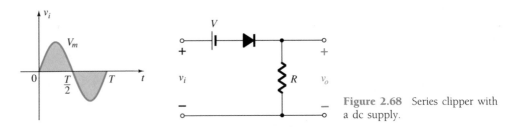

Figure 2.68 Series clipper with a dc supply.

There is no general procedure for analyzing networks such as the type in Fig. 2.68, but there are a few thoughts to keep in mind as you work toward a solution.

1. Make a mental sketch of the response of the network based on the direction of the diode and the applied voltage levels.

For the network of Fig. 2.68, the direction of the diode suggests that the signal v_i must be positive to turn it on. The dc supply further requires that the voltage v_i be greater than V volts to turn the diode on. The negative region of the input signal is "pressuring" the diode into the "off" state, supported further by the dc supply. In general, therefore, we can be quite sure that the diode is an open circuit ("off" state) for the negative region of the input signal.

> *2. Determine the applied voltage (transition voltage) that will cause a change in state for the diode.*

For the ideal diode the transition between states will occur at the point on the characteristics where $v_d = 0$ V and $i_d = 0$ A. Applying the condition $i_d = 0$ at $v_d = 0$ to the network of Fig. 2.68 will result in the configuration of Fig. 2.69, where it is recognized that the level of v_i that will cause a transition in state is

$$\boxed{v_i = V} \tag{2.14}$$

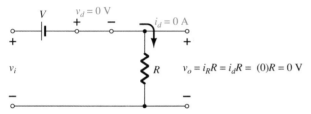

Figure 2.69 Determining the transition level for the circuit of Fig. 2.68.

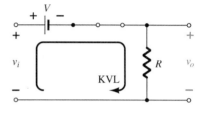

Figure 2.70 Determining v_o.

For an input voltage greater than V volts the diode is in the short-circuit state, while for input voltages less than V volts it is in the open-circuit or "off" state.

> *3. Be continually aware of the defined terminals and polarity of v_o.*

When the diode is in the short-circuit state, such as shown in Fig. 2.70, the output voltage v_o can be determined by applying Kirchhoff's voltage law in the clockwise direction:

$$v_i - V - v_o = 0 \text{ (CW direction)}$$

and

$$\boxed{v_o = v_i - V} \tag{2.15}$$

> *4. It can be helpful to sketch the input signal above the output and determine the output at instantaneous values of the input.*

It is then possible that the output voltage can be sketched from the resulting data points of v_o as demonstrated in Fig. 2.71. Keep in mind that at an instantaneous value of v_i the input can be treated as a dc supply of that value and the corresponding dc (the instantaneous value) value of the output determined. For instance, at $v_i = V_m$ for the network of Fig. 2.68, the network to be analyzed appears in Fig. 2.72. For $V_m > V$ the diode is in the short-circuit state and $v_o = V_m - V$, as shown in Fig. 2.71.

At $v_i = V$ the diodes change state and at $v_i = -V_m$, $v_o = $ OV, and the complete curve for v_o can be sketched as shown in Fig. 2.73.

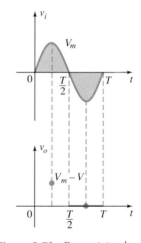

Figure 2.71 Determining levels of v_o.

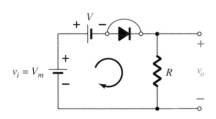

Figure 2.72 Determining v_o when $v_i = V_m$.

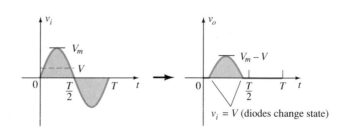

Figure 2.73 Sketching v_o.

EXAMPLE 2.20

Determine the output waveform for the network of Fig. 2.74.

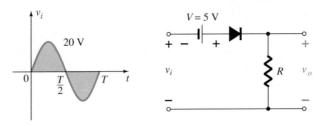

Figure 2.74 Series clipper for Example 2.20.

Solution

Past experience suggests that the diode will be in the "on" state for the positive region of v_i—especially when we note the aiding effect of $V = 5$ V. The network will then appear as shown in Fig. 2.75 and $v_o = v_i + 5$ V. Substituting $i_d = 0$ at $v_d = 0$ for the transition levels, we obtain the network of Fig. 2.76 and $v_i = -5$ V.

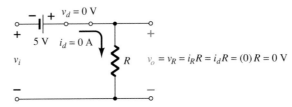

Figure 2.75 v_o with diode in the "on" state.

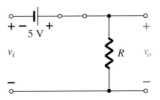

$v_o = v_R = i_R R = i_d R = (0)R = 0$ V

Figure 2.76 Determining the transition level for the clipper of Fig. 2.74.

For v_i more negative than -5 V the diode will enter its open-circuit state, while for voltages more positive than -5 V the diode is in the short-circuit state. The input and output voltages appear in Fig. 2.77.

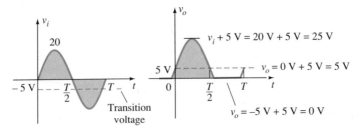

$v_i + 5$ V $= 20$ V $+ 5$ V $= 25$ V

$v_o = 0$ V $+ 5$ V $= 5$ V

$v_o = -5$ V $+ 5$ V $= 0$ V

Figure 2.77 Sketching v_o for Example 2.20.

The analysis of clipper networks with square-wave inputs is actually easier to analyze than with sinusoidal inputs because only two levels have to be considered. In other words, the network can be analyzed as if it had two dc level inputs with the resulting output v_o plotted in the proper time frame.

Repeat Example 2.20 for the square-wave input of Fig. 2.78.

EXAMPLE 2.21

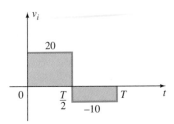

Figure 2.78 Applied signal for Example 2.21.

Solution

For $v_i = 20$ V $(0 \rightarrow T/2)$ the network of Fig. 2.79 will result. The diode is in the short-circuit state and $v_o = 20$ V $+ 5$ V $= 25$ V. For $v_i = -10$ V the network of Fig. 2.80 will result, placing the diode in the "off" state and $v_o = i_R R = (0)R = 0$ V. The resulting output voltage appears in Fig. 2.81.

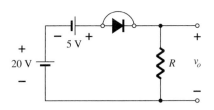

Figure 2.79 v_o at $v_i = +20$ V.

Figure 2.80 v_o at $v_i = -10$ V.

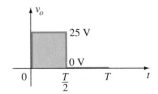

Figure 2.81 Sketching v_o for Example 2.21.

Note in Example 2.21 that the clipper not only clipped off 5 V from the total swing but raised the dc level of the signal by 5 V.

Parallel

The network of Fig. 2.82 is the simplest of parallel diode configurations with the output for the same inputs of Fig. 2.67. The analysis of parallel configurations is very similar to that applied to series configurations, as demonstrated in the next example.

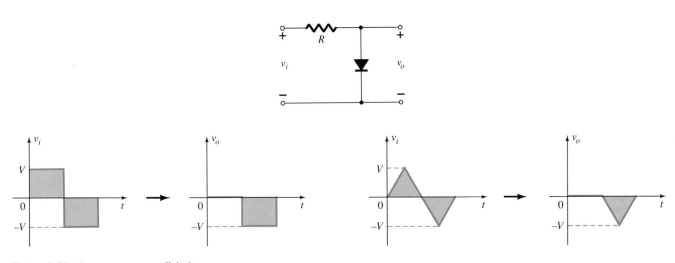

Figure 2.82 Response to a parallel clipper.

EXAMPLE 2.22   Determine v_o for the network of Fig. 2.83.

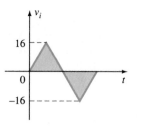

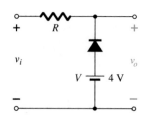

Figure 2.83 Example 2.22.

Solution

The polarity of the dc supply and the direction of the diode strongly suggest that the diode will be in the "on" state for the negative region of the input signal. For this region the network will appear as shown in Fig. 2.84, where the defined terminals for v_o require that $v_o = V = 4$ V.

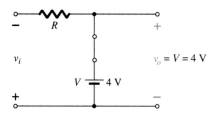

Figure 2.84 v_o for the negative region of v_i.

The transition state can be determined from Fig. 2.85, where the condition $i_d = 0$ A at $v_d = 0$ V has been imposed. The result is v_i (transition) $= V = 4$ V.

Since the dc supply is obviously "pressuring" the diode to stay in the short-circuit state, the input voltage must be greater than 4 V for the diode to be in the "off" state. Any input voltage less than 4 V will result in a short-circuited diode.

For the open-circuit state the network will appear as shown in Fig. 2.86, where $v_o = v_i$. Completing the sketch of v_o results in the waveform of Fig. 2.87.

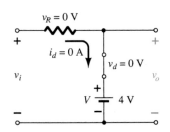

Figure 2.85 Determining the transition level for Example 2.22.

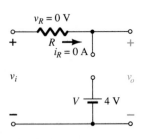

Figure 2.86 Determining v_o for the open state of the diode.

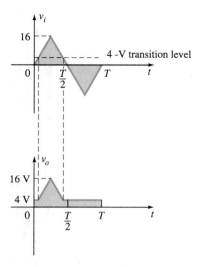

Figure 2.87 Sketching v_o for Example 2.22.

To examine the effects of V_T on the output voltage, the next example will specify a silicon diode rather than an ideal diode equivalent.

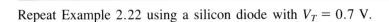

Repeat Example 2.22 using a silicon diode with $V_T = 0.7$ V.

EXAMPLE 2.23

Solution

The transition voltage can first be determined by applying the condition $i_d = 0$ A at $v_d = V_D = 0.7$ V and obtaining the network of Fig. 2.88. Applying Kirchhoff's voltage law around the output loop in the clockwise direction, we find that

$$v_i + V_T - V = 0$$

and
$$v_i = V - V_T = 4 \text{ V} - 0.7 \text{ V} = 3.3 \text{ V}$$

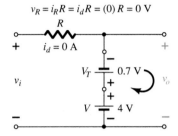

Figure 2.88 Determining the transition level for the network of Fig. 2.83.

For input voltages greater than 3.3 V, the diode will be an open circuit and $v_o = v_i$. For input voltages of less than 3.3 V, the diode will be in the "on" state and the network of Fig. 2.89 results, where

$$v_o = 4 \text{ V} - 0.7 \text{ V} = 3.3 \text{ V}$$

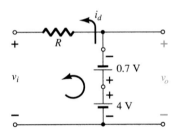

Figure 2.89 Determining v_o for the diode of Fig. 2.83 in the "on" state.

The resulting output waveform appears in Fig. 2.90. Note that the only effect of V_T was to drop the transition level to 3.3 V from 4 V.

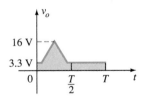

Figure 2.90 Sketching v_o for Example 2.23.

There is no question that including the effects of V_T will complicate the analysis somewhat, but once the analysis is understood with the ideal diode, the procedure, including the effects of V_T, will not be that difficult.

Summary

A variety of series and parallel clippers with the resulting output for the sinusoidal input are provided in Fig. 2.91. In particular, note the response of the last configuration, with its ability to clip off a positive and a negative section as determined by the magnitude of the dc supplies.

Simple Series Clippers (Ideal Diodes)

POSITIVE   **NEGATIVE**

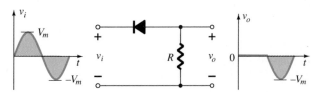

Biased Series Clippers (Ideal Diodes)

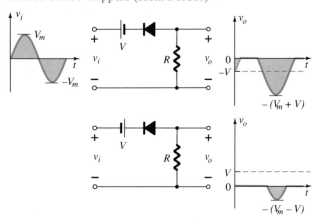

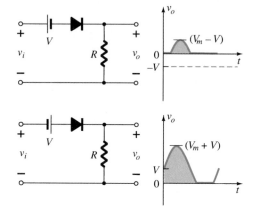

Simple Parallel Clippers (Ideal Diodes)

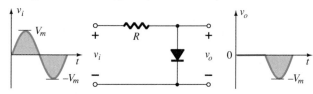

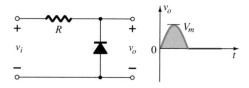

Biased Parallel Clippers (Ideal Diodes)

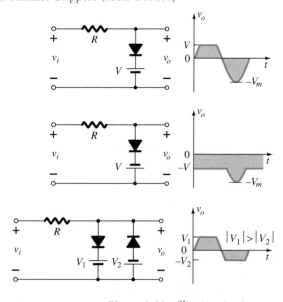

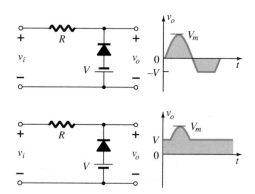

Figure 2.91 Clipping circuits.

2.10 CLAMPERS

The *clamping* network is one that will "clamp" a signal to a different dc level. The network must have a capacitor, a diode, and a resistive element, but it can also employ an independent dc supply to introduce an additional shift. The magnitude of R and C must be chosen such that the time constant $\tau = RC$ is large enough to ensure that the voltage across the capacitor does not discharge significantly during the interval the diode is nonconducting. Throughout the analysis we will assume that for all practical purposes the capacitor will fully charge or discharge in five time constants.

The network of Fig. 2.92 will clamp the input signal to the zero level (for ideal diodes). The resistor R can be the load resistor or a parallel combination of the load resistor and a resistor designed to provide the desired level of R.

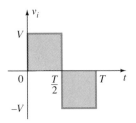

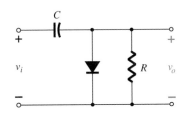

Figure 2.92 Clamper.

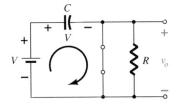

Figure 2.93 Diode "on" and the capacitor charging to V volts.

During the interval $0 \rightarrow T/2$ the network will appear as shown in Fig. 2.93, with the diode in the "on" effectively state "shorting out" the effect of the resistor R. The resulting RC time constant is so small (R determined by the inherent resistance of the network) that the capacitor will charge to V volts very quickly. During this interval the output voltage is directly across the short circuit and $v_o = 0$ V.

When the input switches to the $-V$ state, the network will appear as shown in Fig. 2.94, with the open-circuit equivalent for the diode determined by the applied signal and stored voltage across the capacitor—both "pressuring" current through the diode from cathode to anode. Now that R is back in the network the time constant determined by the RC product is sufficiently large to establish a discharge period 5τ much greater than the period $T/2 \rightarrow T$ and it can be assumed on an approximate basis that the capacitor holds onto all its charge and therefore voltage (since $V = Q/C$) during this period.

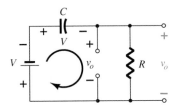

Figure 2.94 Determining v_o with the diode "off."

Since v_o is in parallel with the diode and resistor, it can also be drawn in the alternative position shown in Fig. 2.94. Applying Kirchhoff's voltage law around the input loop will result in

$$-V - V - v_o = 0$$

and

$$v_o = -2V$$

The negative sign resulting from the fact that the polarity of $2V$ is opposite to the polarity defined for v_o. The resulting output waveform appears in Fig. 2.95 with the input signal. The output signal is clamped to 0 V for the interval 0 to $T/2$ but maintains the same total swing ($2V$) as the input.

For a clamping network:

The total swing of the output is equal to the total swing of the input signal.

This fact is an excellent checking tool for the result obtained.

In general, the following steps may be helpful when analyzing clamping networks:

1. *Start the analysis of clamping networks by considering that part of the input signal that will forward bias the diode.*

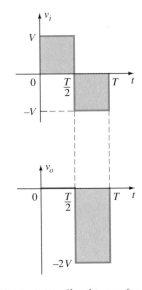

Figure 2.95 Sketching v_o for the network of Fig. 2.92.

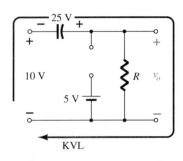

The statement above may require skipping an interval of the input signal (as demonstrated in an example to follow), but the analysis will not be extended by an unnecessary measure of investigation.

2. *During the period that the diode is in the "on" state, assume that the capacitor will charge up instantaneously to a voltage level determined by the network.*

3. *Assume that during the period when the diode is in the "off" state the capacitor will hold on to its established voltage level.*

4. *Throughout the analysis maintain a continual awareness of the location and reference polarity for v_o to ensure that the proper levels for v_o are obtained.*

5. *Keep in mind the general rule that the total swing of the total output must match the swing of the input signal.*

EXAMPLE 2.24

Determine v_o for the network of Fig. 2.96 for the input indicated.

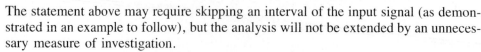

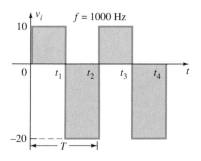

Figure 2.96 Applied signal and network for Example 2.24.

Solution

Note that the frequency is 1000 Hz, resulting in a period of 1 ms and an interval of 0.5 ms between levels. The analysis will begin with the period $t_1 \rightarrow t_2$ of the input signal since the diode is in its short-circuit state as recommended by comment 1. For this interval the network will appear as shown in Fig. 2.97. The output is across R, but it is also directly across the 5-V battery if you follow the direct connection between the defined terminals for v_o and the battery terminals. The result is $v_o = 5$ V for this interval. Applying Kirchhoff's voltage law around the input loop will result in

$$-20 \text{ V} + V_C - 5 \text{ V} = 0$$

and

$$V_C = 25 \text{ V}$$

The capacitor will therefore charge up to 25 V, as stated in comment 2. In this case the resistor R is not shorted out by the diode but a Thévenin equivalent circuit of that portion of the network which includes the battery and the resistor will result in $R_{Th} = 0 \ \Omega$ with $E_{Th} = V = 5$ V. For the period $t_2 \rightarrow t_3$ the network will appear as shown in Fig. 2.98.

The open-circuit equivalent for the diode will remove the 5-V battery from having any effect on v_o, and applying Kirchhoff's voltage law around the outside loop of the network will result in

$$+10 \text{ V} + 25 \text{ V} - v_o = 0$$

and

$$v_o = 35 \text{ V}$$

Figure 2.97 Determining v_o and v_C with the diode in the "on" state.

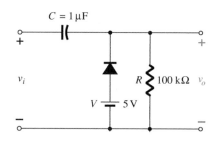

Figure 2.98 Determining v_o with the diode in the "off" state.

Chapter 2   Diode Applications

The time constant of the discharging network of Fig. 2.98 is determined by the product RC and has the magnitude

$$\tau = RC = (100 \text{ k}\Omega)(0.1 \ \mu\text{F}) = 0.01 \text{ s} = 10 \text{ ms}$$

The total discharge time is therefore $5\tau = 5(10 \text{ ms}) = 50 \text{ ms}$.

Since the interval $t_2 \rightarrow t_3$ will only last for 0.5 ms, it is certainly a good approximation that the capacitor will hold its voltage during the discharge period between pulses of the input signal. The resulting output appears in Fig. 2.99 with the input signal. Note that the output swing of 30 V matches the input swing as noted in step 5.

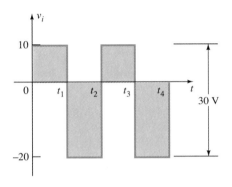

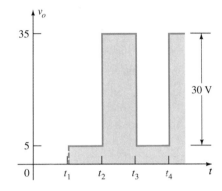

Figure 2.99 v_i and v_o for the clamper of Fig. 2.96.

Repeat Example 2.24 using a silicon diode with $V_T = 0.7$ V.

EXAMPLE 2.25

Solution

For the short-circuit state the network now takes on the appearance of Fig. 2.100 and v_o can be determined by Kirchhoff's voltage law in the output section.

$$+5 \text{ V} - 0.7 \text{ V} - v_o = 0$$

and

$$v_o = 5 \text{ V} - 0.7 \text{ V} = 4.3 \text{ V}$$

For the input section Kirchhoff's voltage law will result in

$$-20 \text{ V} + V_C + 0.7 \text{ V} - 5 \text{ V} = 0$$

and

$$V_C = 25 \text{ V} - 0.7 \text{ V} = 24.3 \text{ V}$$

For the period $t_2 \rightarrow t_3$ the network will now appear as in Fig. 2.101, with the only change being the voltage across the capacitor. Applying Kirchhoff's voltage law yields

$$+10 \text{ V} + 24.3 \text{ V} - v_o = 0$$

and

$$v_o = 34.3 \text{ V}$$

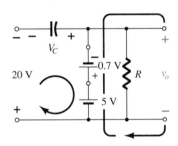

Figure 2.100 Determining v_o and v_C with the diode in the "on" state.

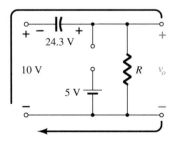

Figure 2.101 Determining v_o with the diode in the open state.

2.10 Clampers

85

The resulting output appears in Fig. 2.102 verifying the statement that the input and output swings are the same.

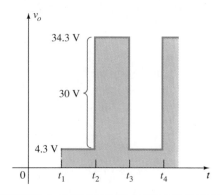

Figure 2.102 Sketching v_o for the clamper of Fig. 2.96 with a silicon diode.

A number of clamping circuits and their effect on the input signal are shown in Fig. 2.103. Although all the waveforms appearing in Fig. 2.103 are square waves, clamping networks work equally well for sinusoidal signals. In fact, one approach to the analysis of clamping networks with sinusoidal inputs is to replace the sinusoidal signal by a square wave of the same peak values. The resulting output will then form an envelope for the sinusoidal response as shown in Fig. 2.104 for a network appearing in the bottom right of Fig. 2.103.

Clamping Networks

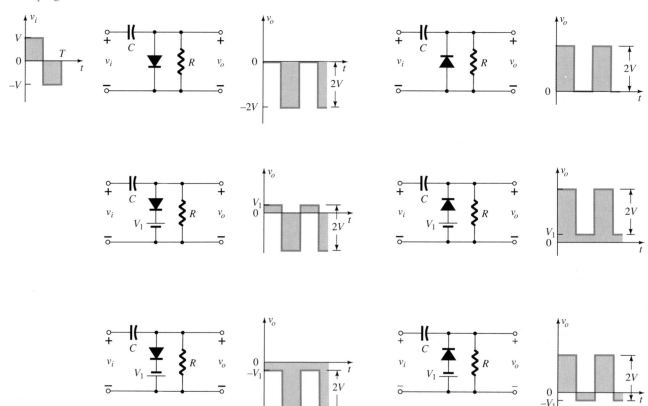

Figure 2.103 Clamping circuits with ideal diodes ($5\tau = 5RC \gg T/2$).

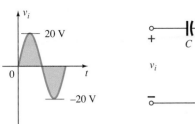

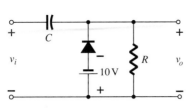

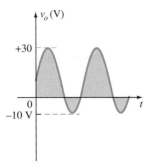

Figure 2.104 Clamping network with a sinusoidal input.

2.11 ZENER DIODES

The analysis of networks employing Zener diodes is quite similar to that applied to the analysis of semiconductor diodes in previous sections. First the state of the diode must be determined followed by a substitution of the appropriate model and a determination of the other unknown quantities of the network. Unless otherwise specified, the Zener model to be employed for the "on" state will be as shown in Fig. 2.105a. For the "off" state as defined by a voltage less than V_Z but greater than 0 V with the polarity indicated in Fig. 2.105b, the Zener equivalent is the open circuit that appears in the same figure.

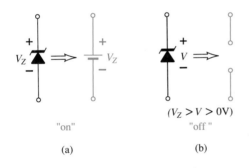

$(V_Z > V > 0V)$

"on"

"off"

(a)

(b)

Figure 2.105 Zener diode equivalents for the (a) "on" and (b) "off" states.

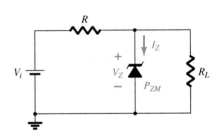

Figure 2.106 Basic Zener regulator.

V_i and R Fixed

The simplest of Zener diode networks appears in Fig. 2.106. The applied dc voltage is fixed, as is the load resistor. The analysis can fundamentally be broken down into two steps.

1. Determine the state of the Zener diode by removing it from the network and calculating the voltage across the resulting open circuit.

Applying step 1 to the network of Fig. 2.106 will result in the network of Fig. 2.107, where an application of the voltage divider rule will result in

$$V = V_L = \frac{R_L V_i}{R + R_L}$$ (2.16)

If $V \geq V_Z$, the Zener diode is "on" and the equivalent model of Fig. 2.105a can be substituted. If $V < V_Z$, the diode is "off" and the open-circuit equivalence of Fig. 2.105b is substituted.

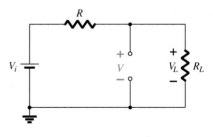

Figure 2.107 Determining the state of the Zener diode.

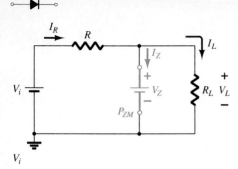

Figure 2.108 Substituting the Zener equivalent for the "on" situation.

2. *Substitute the appropriate equivalent circuit and solve for the desired unknowns.*

For the network of Fig. 2.106 the "on" state will result in the equivalent network of Fig. 2.108. Since voltages across parallel elements must be the same, we find that

$$V_L = V_Z \tag{2.17}$$

The Zener diode current must be determined by an application of Kirchhoff's current law. That is,

$$I_R = I_Z + I_L$$

and

$$I_Z = I_R - I_L \tag{2.18}$$

where

$$I_L = \frac{V_L}{R_L} \quad \text{and} \quad I_R = \frac{V_R}{R} = \frac{V_i - V_L}{R}$$

The power dissipated by the Zener diode is determined by

$$P_Z = V_Z I_Z \tag{2.19}$$

which must be less than the P_{ZM} specified for the device.

Before continuing, it is particularly important to realize that the first step was employed only to determine the *state of the Zener diode*. If the Zener diode is in the "on" state, the voltage across the diode is not V volts. When the system is turned on the Zener diode will turn "on" as soon as the voltage across the Zener diode is V_Z volts. It will then "lock in" at this level and never reach the higher level of V volts.

Zener diodes are most frequently used in *regulator* networks or as a *reference* voltage. Figure 2.106 is a simple regulator designed to maintain a fixed voltage across the load R_L. For values of applied voltage greater than required to turn the Zener diode "on," the voltage across the load will be maintained at V_Z volts. If the Zener diode is employed as a reference voltage, it will provide a level for comparison against other voltages.

EXAMPLE 2.26

(a) For the Zener diode network of Fig. 2.109, determine V_L, V_R, I_Z, and P_Z.
(b) Repeat part (a) with $R_L = 3\ k\Omega$.

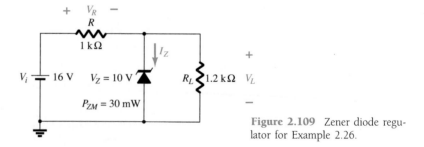

Figure 2.109 Zener diode regulator for Example 2.26.

Solution

(a) Following the suggested procedure the network is redrawn as shown in Fig. 2.110. Applying Eq. (2.16) gives

$$V = \frac{R_L V_i}{R + R_L} = \frac{1.2\ k\Omega(16\ V)}{1\ k\Omega + 1.2\ k\Omega} = 8.73\ V$$

Chapter 2 Diode Applications

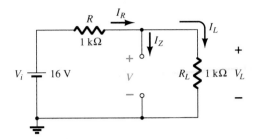

Figure 2.110 Determining V for the regulator of Fig. 2.109.

Since $V = 8.73$ V is less than $V_Z = 10$ V the diode is in the "off" state as shown on the characteristics of Fig. 2.111. Substituting the open-circuit equivalent will result in the same network as in Fig. 2.110, where we find that

$$V_L = V = \textbf{8.73 V}$$

$$V_R = V_i - V_L = 16 \text{ V} - 8.73 \text{ V} = \textbf{7.27 V}$$

$$I_Z = \textbf{0 A}$$

and

$$P_Z = V_Z I_Z = V_Z(0 \text{ A}) = \textbf{0 W}$$

(b) Applying Eq. (2.16) will now result in

$$V = \frac{R_L V_i}{R + R_L} = \frac{3 \text{ k}\Omega(16 \text{ V})}{1 \text{ k}\Omega + 3 \text{ k}\Omega} = 12 \text{ V}$$

Since $V = 12$ V is greater than $V_Z = 10$ V, the diode is in the "on" state and the network of Fig. 2.112 will result. Applying Eq. (2.17) yields

$$V_L = V_Z = \textbf{10 V}$$

and

$$V_R = V_i - V_L = 16 \text{ V} - 10 \text{ V} = \textbf{6 V}$$

with

$$I_L = \frac{V_L}{R_L} = \frac{10 \text{ V}}{3 \text{ k}\Omega} = 3.33 \text{ mA}$$

and

$$I_R = \frac{V_R}{R} = \frac{6 \text{ V}}{1 \text{ k}\Omega} = 6 \text{ mA}$$

so that

$$I_Z = I_R - I_L \text{ [Eq. (2.18)]}$$

$$= 6 \text{ mA} - 3.33 \text{ mA}$$

$$= \textbf{2.67 mA}$$

The power dissipated,

$$P_Z = V_Z I_Z = (10 \text{ V})(2.67 \text{ mA}) = \textbf{26.7 mW}$$

which is less than the specified $P_{ZM} = 30$ mW.

Figure 2.111 Resulting operating point for the network of Fig. 2.109.

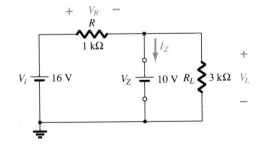

Figure 2.112 Network of Fig. 2.109 in the "on" state.

Fixed V_i, Variable R_L

Due to the offset voltage V_Z, there is a specific range of resistor values (and therefore load current) which will ensure that the Zener is in the "on" state. Too small a load resistance R_L will result in a voltage V_L across the load resistor less than V_Z and the Zener device will be in the "off" state.

To determine the minimum load resistance of Fig. 2.106 that will turn the Zener diode on, simply calculate the value of R_L that will result in a load voltage $V_L = V_Z$. That is,

$$V_L = V_Z = \frac{R_L V_i}{R_L + R}$$

Solving for R_L, we have

$$R_{L_{min}} = \frac{R V_Z}{V_i - V_Z} \qquad (2.20)$$

Any load resistance value greater than the R_L obtained from Eq. (2.20) will ensure that the Zener diode is in the "on" state and the diode can be replaced by its V_Z source equivalent.

The condition defined by Eq. (2.20) establishes the minimum R_L, but in turn specifies the maximum I_L as

$$I_{L_{max}} = \frac{V_L}{R_L} = \frac{V_Z}{R_{L_{min}}} \qquad (2.21)$$

Once the diode is in the "on" state, the voltage across R remains fixed at

$$V_R = V_i - V_Z \qquad (2.22)$$

and I_R remains fixed at

$$I_R = \frac{V_R}{R} \qquad (2.23)$$

The Zener current

$$I_Z = I_R - I_L \qquad (2.24)$$

resulting in a minimum I_Z when I_L is a maximum and a maximum I_Z when I_L is a minimum value since I_R is constant.

Since I_Z is limited to I_{ZM} as provided on the data sheet, it does affect the range of R_L and therefore I_L. Substituting I_{ZM} for I_Z establishes the minimum I_L as

$$I_{L_{min}} = I_R - I_{ZM} \qquad (2.25)$$

and the maximum load resistance as

$$R_{L_{max}} = \frac{V_Z}{I_{L_{min}}} \qquad (2.26)$$

(a) For the network of Fig. 2.113, determine the range of R_L and I_L that will result in V_{R_L} being maintained at 10 V.

(b) Determine the maximum wattage rating of the diode.

EXAMPLE 2.27

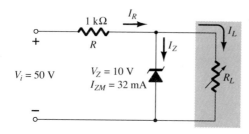

Figure 2.113 Voltage regulator for Example 2.27.

Solution

(a) To determine the value of R_L that will turn the Zener diode on, apply Eq. (2.20):

$$R_{L_{min}} = \frac{RV_Z}{V_i - V_Z} = \frac{(1\ k\Omega)(10\ V)}{50\ V - 10\ V} = \frac{10\ k\Omega}{40} = \mathbf{250\ \Omega}$$

The voltage across the resistor R is then determined by Eq. (2.22):

$$V_R = V_i - V_Z = 50\ V - 10\ V = \mathbf{40\ V}$$

and Eq. (2.23) provides the magnitude of I_R:

$$I_R = \frac{V_R}{R} = \frac{40\ V}{1\ k\Omega} = \mathbf{40\ mA}$$

The minimum level of I_L is then determined by Eq. (2.25):

$$I_{L_{min}} = I_R - I_{ZM} = 40\ mA - 32\ mA = \mathbf{8\ mA}$$

with Eq. (2.26) determining the maximum value of R_L:

$$R_{L_{max}} = \frac{V_Z}{I_{L_{min}}} = \frac{10\ V}{8\ mA} = \mathbf{1.25\ k\Omega}$$

A plot of V_L versus R_L appears in Fig. 2.114a and for V_L versus I_L in Fig. 2.114b.

(b) $P_{max} = V_Z I_{ZM}$

$\qquad = (10\ V)(32\ mA) = \mathbf{320\ mW}$

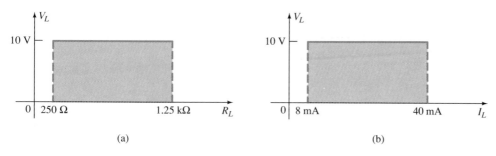

Figure 2.114. V_L versus R_L and I_L for the regulator of Fig. 2.113.

Fixed R_L, Variable V_i

For fixed values of R_L in Fig. 2.106, the voltage V_i must be sufficiently large to turn the Zener diode on. The minimum turn-on voltage $V_i = V_{i_{min}}$ is determined by

$$V_L = V_Z = \frac{R_L V_i}{R_L + R}$$

and

$$\boxed{V_{i_{min}} = \frac{(R_L + R)V_Z}{R_L}} \tag{2.27}$$

The maximum value of V_i is limited by the maximum Zener current I_{ZM}. Since $I_{ZM} = I_R - I_L$,

$$\boxed{I_{R_{max}} = I_{ZM} + I_L} \tag{2.28}$$

Since I_L is fixed at V_Z/R_L and I_{ZM} is the maximum value of I_Z, the maximum V_i is defined by

$$V_{i_{max}} = V_{R_{max}} + V_Z$$

$$\boxed{V_{i_{max}} = I_{R_{max}}R + V_Z} \tag{2.29}$$

EXAMPLE 2.28

Determine the range of values of V_i that will maintain the Zener diode of Fig. 2.115 in the "on" state.

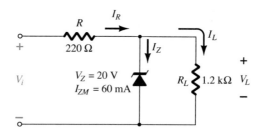

Figure 2.115 Regulator for Example 2.28.

Solution

Eq. (2.27): $V_{i_{min}} = \dfrac{(R_L + R)V_Z}{R_L} = \dfrac{(1200\ \Omega + 220\ \Omega)(20\ \text{V})}{1200\ \Omega} = \mathbf{23.67\ V}$

$I_L = \dfrac{V_L}{R_L} = \dfrac{V_Z}{R_L} = \dfrac{20\ \text{V}}{1.2\ \text{k}\Omega} = 16.67\ \text{mA}$

Eq. (2.28): $I_{R_{max}} = I_{ZM} + I_L = 60\ \text{mA} + 16.67\ \text{mA}$

$\qquad\qquad\quad = 76.67\ \text{mA}$

Eq. (2.29): $V_{i_{max}} = I_{R_{max}}R + V_Z$

$\qquad\qquad\quad = (76.67\ \text{mA})(0.22\ \text{k}\Omega) + 20\ \text{V}$

$\qquad\qquad\quad = 16.87\ \text{V} + 20\ \text{V}$

$\qquad\qquad\quad = \mathbf{36.87\ V}$

A plot of V_L versus V_i is provided in Fig. 2.116.

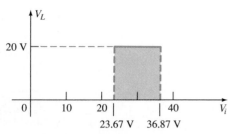

Figure 2.116 V_L versus V_i for the regulator of Fig. 2.115.

Chapter 2 Diode Applications

The results of Example 2.28 reveal that for the network of Fig. 2.115 with a fixed R_L, the output voltage will remain fixed at 20 V for a range of input voltage that extends from 23.67 to 36.87 V.

In fact, the input could appear as shown in Fig. 2.117 and the output would remain constant at 20 V, as shown in Fig. 2.116. The waveform appearing in Fig. 2.117 is obtained by *filtering* a half-wave- or full-wave-rectified output—a process described in detail in a later chapter. The net effect, however, is to establish a steady dc voltage (for a defined range of V_i) such as that shown in Fig. 2.116 from a sinusoidal source with 0 average value.

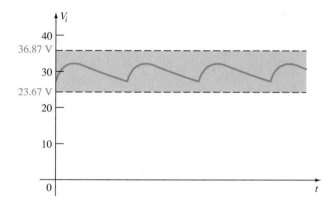

Figure 2.117 Waveform generated by a filtered rectified signal.

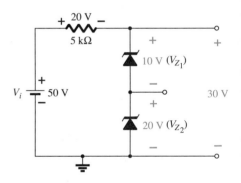

Figure 2.118 Establishing three reference voltage levels.

Two or more reference levels can be established by placing Zener diodes in series as shown in Fig. 2.118. As long as V_i is greater than the sum of V_{Z_1} and V_{Z_2}, both diodes will be in the "on" state and the three reference voltages will be available.

Two back-to-back Zeners can also be used as an ac regulator as shown in Fig. 2.119a. For the sinusoidal signal v_i the circuit will appear as shown in Fig. 2.119b at the instant $v_i = 10$ V. The region of operation for each diode is indicated in the adjoining figure. Note that Z_1 is in a low-impedance region, while the impedance of Z_2 is quite large, corresponding with the open-circuit representation. The result is that $v_o = v_i$ when $v_i = 10$ V. The input and output will continue to duplicate each other until v_i reaches 20 V. Z_2 will then "turn on" (as a Zener diode) while Z_1 will be in

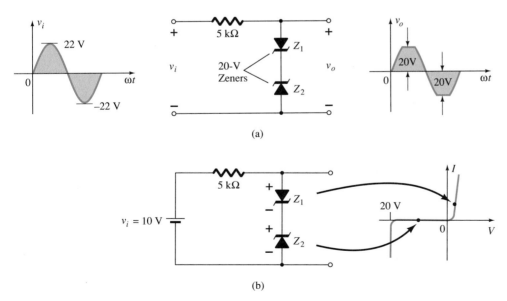

Figure 2.119 Sinusoidal ac regulation: (a) 40-V peak-to-peak sinusoidal ac regulator; (b) circuit operation at $v_i = 10$ V.

a region of conduction with a resistance level sufficiently small compared to the series 5-kΩ resistor to be considered a short circuit. The resulting output for the full range of v_i is provided in Fig. 2.119(a). Note that the waveform is not purely sinusoidal, but its rms value is lower than that associated with a full 22-V peak signal. The network is effectively limiting the rms value of the available voltage. The network of Fig. 2.119a can be extended to that of a simple square-wave generator (due to the clipping action) if the signal v_i is increased to perhaps a 50-V peak with 10-V Zeners as shown in Fig. 2.120 with the resulting output waveform.

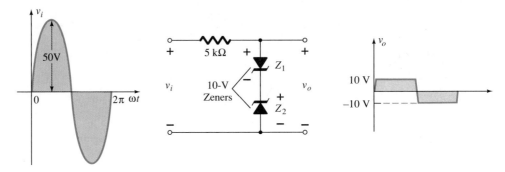

Figure 2.120 Simple square-wave generator.

2.12 VOLTAGE-MULTIPLIER CIRCUITS

Voltage-multiplier circuits are employed to maintain a relatively low transformer peak voltage while stepping up the peak output voltage to two, three, four, or more times the peak rectified voltage.

Voltage Doubler

The network of Figure 2.121 is a half-wave voltage doubler. During the positive-voltage half-cycle across the transformer, secondary diode D_1 conducts (and diode D_2 is cut off), charging capacitor C_1 up to the peak rectified voltage (V_m). Diode D_1 is ideally a short during this half-cycle and the input voltage charges capacitor C_1 to V_m with the polarity shown in Fig. 2.122a. During the negative half-cycle of the secondary voltage, diode D_1 is cut off and diode D_2 conducts charging capacitor C_2. Since diode D_2 acts as a short during the negative half-cycle (and diode D_1 is open), we can sum the voltages around the outside loop (see Fig. 2.122b):

$$-V_{C_2} + V_{C_1} + V_m = 0$$
$$-V_{C_2} + V_m + V_m = 0$$

from which

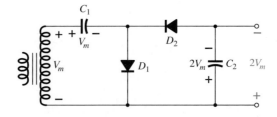

Figure 2.121 Half-wave voltage doubler.

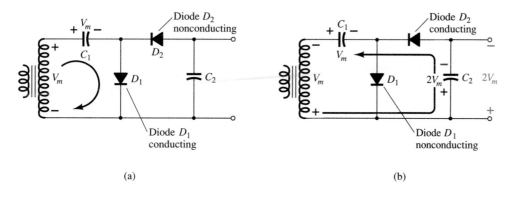

Figure 2.122 Double operation, showing each half-cycle of operation: (a) positive half-cycle; (b) negative half cycle.

(a)   (b)

On the next positive half-cycle, diode D_2 is nonconducting and capacitor C_2 will discharge through the load. If no load is connected across capacitor C_2, both capacitors stay charged—C_1 to V_m and C_2 to $2V_m$. If, as would be expected, there is a load connected to the output of the voltage doubler, the voltage across capacitor C_2 drops during the positive half-cycle (at the input) and the capacitor is recharged up to $2V_m$ during the negative half-cycle. The output waveform across capacitor C_2 is that of a half-wave signal filtered by a capacitor filter. The peak inverse voltage across each diode is $2V_m$.

Another doubler circuit is the full-wave doubler of Fig. 2.123. During the positive half-cycle of transformer secondary voltage (see Fig. 2.124a) diode D_1 conducts charging capacitor C_1 to a peak voltage V_m. Diode D_2 is nonconducting at this time.

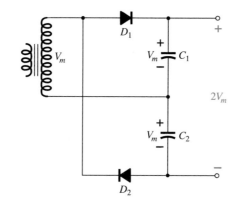

Figure 2.123 Full-wave voltage doubler.

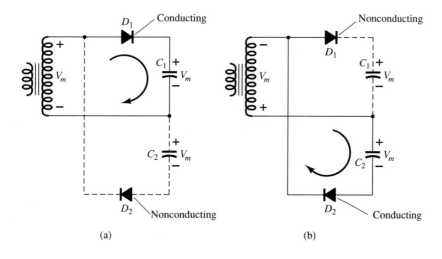

(a)   (b)

Figure 2.124 Alternate half-cycles of operation for full-wave voltage doubler.

During the negative half-cycle (see Fig. 2.124b) diode D_2 conducts charging capacitor C_2 while diode D_1 is nonconducting. If no load current is drawn from the circuit, the voltage across capacitors C_1 and C_2 is $2V_m$. If load current is drawn from the circuit, the voltage across capacitors C_1 and C_2 is the same as that across a capacitor fed by a full-wave rectifier circuit. One difference is that the effective capacitance is that of C_1 and C_2 in series, which is less than the capacitance of either C_1 or C_2 alone. The lower capacitor value will provide poorer filtering action than the single-capacitor filter circuit.

The peak inverse voltage across each diode is $2V_m$ as it is for the filter capacitor circuit. In summary, the half-wave or full-wave voltage-doubler circuits provide twice the peak voltage of the transformer secondary while requiring no center-tapped transformer and only $2V_m$ PIV rating for the diodes.

Voltage Tripler and Quadrupler

Figure 2.125 shows an extension of the half-wave voltage doubler, which develops three and four times the peak input voltage. It should be obvious from the pattern of the circuit connection how additional diodes and capacitors may be connected so that the output voltage may also be five, six, seven, and so on, times the basic peak voltage (V_m).

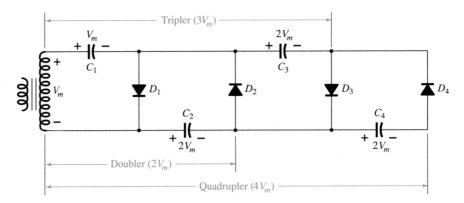

Figure 2.125 Voltage tripler and quadrupler.

In operation capacitor C_1 charges through diode D_1 to a peak voltage, V_m, during the positive half-cycle of the transformer secondary voltage. Capacitor C_2 charges to twice the peak voltage $2V_m$ developed by the sum of the voltages across capacitor C_1 and the transformer, during the negative half-cycle of the transformer secondary voltage.

During the positive half-cycle, diode D_3 conducts and the voltage across capacitor C_2 charges capacitor C_3 to the same $2V_m$ peak voltage. On the negative half-cycle, diodes D_2 and D_4 conduct with capacitor C_3, charging C_4 to $2V_m$.

The voltage across capacitor C_2 is $2V_m$, across C_1 and C_3 it is $3V_m$, and across C_2 and C_4 it is $4V_m$. If additional sections of diode and capacitor are used, each capacitor will be charged to $2V_m$. Measuring from the top of the transformer winding (Fig. 2.125) will provide odd multiples of V_m at the output, whereas measuring from the bottom of the transformer the output voltage will provide even multiples of the peak voltage, V_m.

The transformer rating is only V_m, maximum, and each diode in the circuit must be rated at $2V_m$ PIV. If the load is small and the capacitors have little leakage, extremely high dc voltages may be developed by this type of circuit, using many sections to step up the dc voltage.

Chapter 2 Diode Applications

2.13 COMPUTER ANALYSIS

The computer analysis of this chapter will employ PSpice to determine the unknown quantities for the network of Fig. 2.27 (Example 2.11). The first step would be to redraw the network as shown in Fig. 2.126, identify the nodes, and label them in a logical order. Ground is chosen as the reference level and assigned the label 0. The silicon diode is specified between nodes 2 and 3. The output voltage of Example 2.11 is from node 3 to ground. The voltage V_1 is between nodes 1 and 2 and V_2 between nodes 3 and 4.

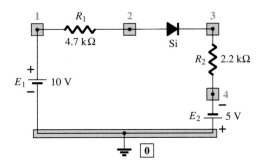

Figure 2.126 Figure 2.27 redrawn for PSpice analysis.

The network information is entered into the computer in an *input* file as shown in block form in Fig. 2.127. The first entry *must* be a title line to identify the analysis to be performed. The next set of entries is a description of the network using the chosen nodes and the specific format required by PSpice for each element. The Analysis Commands define the quantities to be determined. The last entry *must* be the **.END** statement in exactly the form indicated. Leaving out the period will invalidate the entire input file.

The input file for the network of Fig. 2.126 is provided in Fig. 2.128. The title line specifies the "Diode circuit of Fig. 2.126" as the circuit to be analyzed. The first line of the network description specifies the 10-V dc supply. For all dc supplies the first letter of the line must be the uppercase letter V followed by the *name* of the supply. The name is simply a choice of letters and/or numbers to identify the source in the network structure. Next the node with the positive side of the supply is entered followed by the negative polarity. The magnitude of the supply is then entered as indicated.

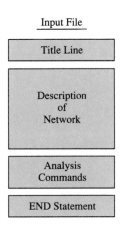

Figure 2.127 Components of an input file.

```
Diode circuit for network of Fig. 2.126
VE1 1 0 10V
R1  1 2 4.7K
D1  2 3 DI
R2  3 4 2.2K
VE2 0 4 5V
.MODEL DI D(IS=2E-15)
.DC VE1 10V 10V 1V
.PRINT DC V(3) I(D1) V(1,2) V(3,4) V(2,3)
.OPTIONS NOPAGE
.END
```

Figure 2.128 Input file for the network of Fig. 2.126.

The next entry of the input file is a resistive element requiring a capital letter R to start the line followed by its chosen name (in this case simply the number 1 to relate to its subscript in the network of Fig. 2.126). The "pressure" of the 10-V supply suggests that the resulting current will establish node 1 as positive with respect to node 2—hence the order of the nodes in the input file. The magnitude of the resistor is specified as 4.7 kΩ.

The format for the diode entry was introduced in Chapter 1. Note the entry on line 3 of the network description and the diode model description on line 6. Recall that IS was specified as 2E-15 to obtain a 0.7-V drop (or as close as possible to this level) across silicon diodes in the "on" state with current levels typical for electronic systems.

The next two entries are the second resistor and power supply. Note in each case an attempt to define the positive and negative nodes by the order of the node entries. An incorrect assumption, however, will simply result in a negative sign for the voltage across a particular element.

The .DC entry specifies a dc analysis with supply E_1 at 10 V. The .DC analysis can be specified for a range of values, hence the repeat of the 10-V level on the entry line. If the level is repeated, as in this case, the analysis will be performed only at the level indicated. If the second level were different, the analysis would be performed from the first to the second level at levels defined by the increment specified as the next entry of the line. Even though our analysis is at only one level, an entry for the increment is required as indicated by the 1 V typically used for this purpose. Hence, once the program is run and the computer system notes a repeat of the 10-V level, it will perform the analysis at only one level (10 V) and ignore the impact of the increment entry. It is unnecessary to include the second dc supply in this statement. The .DC entry specifies the type of analysis at a level of $E_1 = 10$ V with all the other elements as specified in the network description.

The .PRINT statement defines those quantities to be included in the output data. The quantity V(3) is the voltage from node 3 to ground—the output voltage of Fig. 2.126. Next is the current through the diode followed by the voltages between the indicated nodes.

The .OPTIONS NOPAGE entry is a command to "save paper" in that it limits the data provided in the *output file* unless specifically requested. The input file ends with the required .END statement.

Once the input file is entered *properly,* the PSpice program can be "run" and the desired information obtained in the format of the output file appearing in Fig. 2.129. Note in the figure the location of the title line and the repeat of the entire network description. Next the specified model parameters are listed followed by the desired results. VE1 is simply a repeat of the level of E_1 (1.000E + 1 = 10) is controlled by the computer to specify the conditions under which the calculations were made (recall the .DC statement), while $V(3) = V_o = -4.455\text{E}{-}01 = -0.4455$ V, which compares very favorably with the -0.45 V obtained in Example 2.11. The diode current $I(D1) = I_D = 2.07$ mA, which is an exact match of Example 2.11. The voltage $V(1,2) = V_1 = 9.73$ V to compare with 9.73 V for Example 2.11 and $V(3,4) =$

Figure 2.129 Output file for the network of Fig. 2.126.

```
**** 02/24/90 ******* Evaluation PSpice (January 1989) ******* 18:27:13 ****

Diode circuit for network of Fig. 2.126

****      CIRCUIT DESCRIPTION

*********************************************************************************

VE1 1 0 10V
R1   1 2 4.7K
D1   2 3 DI
R2   3 4 2.2K
VE2 0 4 5V
.MODEL DI D(IS=2E-15)
.DC VE1 10V 10V 1V
.PRINT DC V(3) I(D1) V(1,2) V(3,4) V(2,3)
.OPTIONS NOPAGE
.END

****      Diode MODEL PARAMETERS
             DI
      IS     2.000000E-15


****      DC TRANSFER CURVES                TEMPERATURE =    27.000 DEG C
  VE1           V(3)          I(D1)      V(1,2)       V(3,4)       V(2,3)
   1.000E+01   -4.455E-01    2.070E-03   9.730E+00    4.554E+00    7.155E-01
```

$V_2 = 4.554$ V to compare with 4.55 V in the same example. The last element of the output file is the voltage across the diode, which for the current level and IS chosen is 0.715 V, compared to the 0.7 V employed in Example 2.11. Recall from Chapter 1 that the diode voltage is a function of a variety of parameters, such as the reverse-saturation current, current level, temperature, and so on, and cannot simply be specified as 0.7 V unless we drop the use of the model altogether.

All in all, the results are an excellent match with those obtained in Example 2.11, as they should be if the *proper* care were applied to both approaches. The first exposure to any new technique, such as the PSpice analysis introduced in this section, will naturally leave questions and concerns about its application. However, be aware that our primary intent in this book is simply to expose the reader to various computer methods—not necessarily the detail required to perform the analysis on your own for a variety of configurations. This is not to say that the description above may not be sufficient to approach a number of diode configurations but only that questions may arise that require a class lecture on the subject or at the very least the availability of a PSpice manual. The above is a sample of the type of PSpice analysis that will be provided throughout the book. Be aware that PSpice is one of the most frequently applied packages in the educational community and any knowledge of its application will carry over into whatever computer analysis approach you may eventually choose.

§ 2.2 Load-Line Analysis

PROBLEMS

1. (a) Using the characteristics of Fig. 2.130b, determine I_D, V_D, and V_R for the circuit of Fig. 2.130a.
   (b) Repeat part (a) using the approximate model for the diode and compare results.
   (c) Repeat part (a) using the ideal model for the diode and compare results.

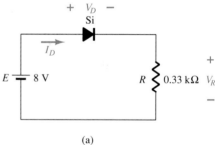

(a)

Figure 2.130 Problems 1, 2

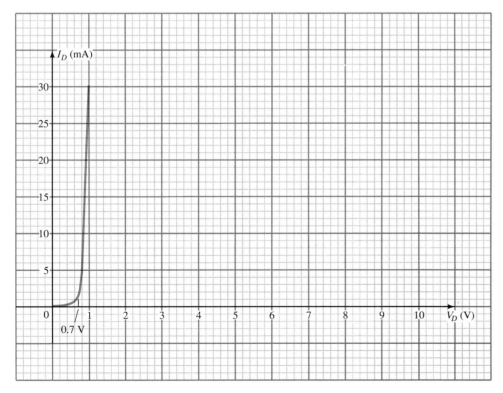

(b)

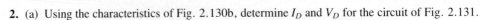

2. (a) Using the characteristics of Fig. 2.130b, determine I_D and V_D for the circuit of Fig. 2.131.
   (b) Repeat part (a) with $R = 0.47$ kΩ.
   (c) Repeat part (a) with $R = 0.18$ kΩ.
   (d) Is the level of V_D relatively close to 0.7 V in each case?

   How do the resulting levels of I_D compare? Comment accordingly.

3. Determine the value of R for the circuit of Fig. 2.131 that will result in a diode current of 10 mA if $E = 7$ V. Use the characteristics of Fig. 2.130b for the diode.

4. (a) Using the approximate characteristics for the Si diode, determine the level of V_D, I_D, and V_R for the circuit of Fig. 2.132.
   (b) Perform the same analysis as part (a) using the ideal model for the diode.
   (c) Do the results obtained in parts (a) and (b) suggest that the ideal model can provide a good approximation for the actual response under some conditions?

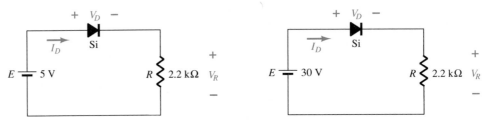

Figure 2.131 Problems 2, 3   Figure 2.132 Problem 4

§ 2.4 Series Diode Configurations with DC Inputs

5. Determine the current I for each of the configurations of Fig. 2.133 using the approximate equivalent model for the diode.

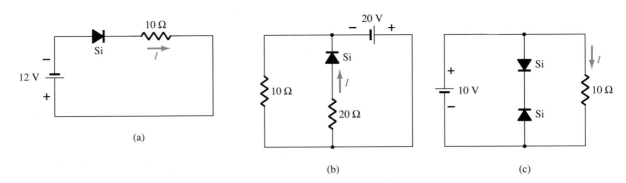

(a)   (b)   (c)

Figure 2.133 Problem 5

6. Determine V_o and I_D for the networks of Fig. 2.134.

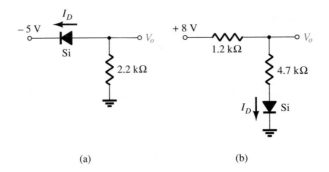

(a)   (b)   Figure 2.134 Problem 6

* **7.** Determine the level of V_o for each network of Fig. 2.135.

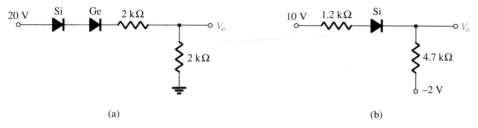

(a)   (b)

Figure 2.135   Problems 7, 51

* **8.** Determine V_o and I_D for the networks of Fig. 2.136.

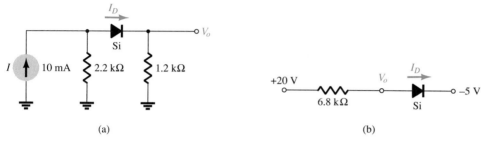

(a)   (b)

Figure 2.136   Problem 8

* **9.** Determine V_{o_1} and V_{o_2} for the networks of Fig. 2.137.

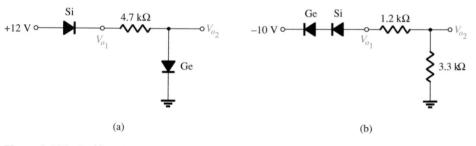

(a)   (b)

Figure 2.137   Problem 9

§ 2.5 Parallel and Series–Parallel Configurations

10. Determine V_o and I_D for the networks of Fig. 2.138.

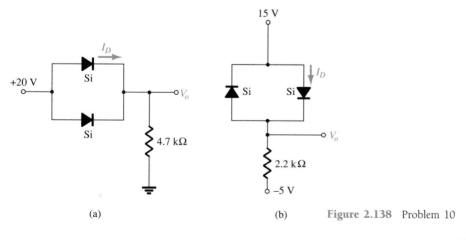

(a)   (b)   **Figure 2.138**   Problem 10

Problems   **101**

* **11.** Determine V_o and I for the networks of Fig. 2.139.

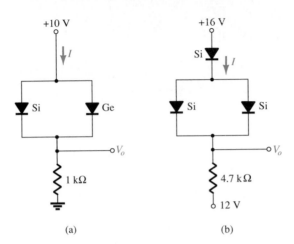

(a)   (b)   **Figure 2.139**   Problem 11

12. Determine V_{o_1}, V_{o_2}, and I for the network of Fig. 2.140.

* **13.** Determine V_o and I_D for the network of Fig. 2.141.

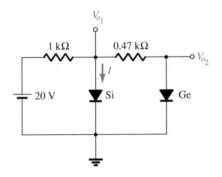

Figure 2.140   Problem 12   **Figure 2.141**   Problem 13

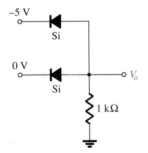

Figure 2.142   Problem 18

§ 2.6 AND/OR Gates

14. Determine V_o for the network of Fig. 2.38 with 0 V on both inputs.

15. Determine V_o for the network of Fig. 2.38 with 10 V on both inputs.

16. Determine V_o for the network of Fig. 2.41 with 0 V on both inputs.

17. Determine V_o for the network of Fig. 2.41 with 10 V on both inputs.

18. Determine V_o for the negative logic OR gate of Fig. 2.142.

19. Determine V_o for the negative logic AND gate of Fig. 2.143.

20. Determine the level of V_o for the gate of Fig. 2.144.

21. Determine V_o for the configuration of Fig. 2.145.

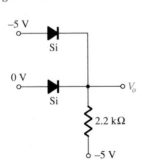

Figure 2.143   Problem 19

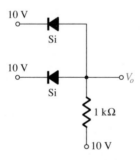

Figure 2.144   Problem 20

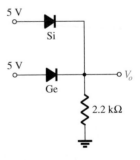

Figure 2.145   Problem 21

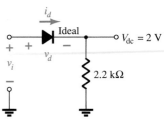

Figure 2.146 Problems 22, 23, 24

§ 2.7 Sinusoidal Inputs; Half-Wave Rectification

22. Assuming an ideal diode, sketch v_i, v_d, and i_d for the half-wave rectifier of Fig. 2.146. The input is a sinusoidal waveform with a frequency of 60 Hz.

* **23.** Repeat Problem 22 with a silicon diode ($V_T = 0.7$ V).

* **24.** Repeat Problem 22 with a 6.8-kΩ load applied as shown in Fig. 2.147. Sketch v_L and i_L.

25. For the network of Fig. 2.148, sketch v_o and determine V_{dc}.

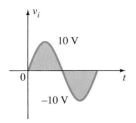

Figure 2.147 Problem 24

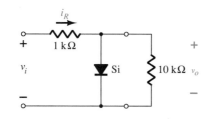

Figure 2.148 Problem 25

* **26.** For the network of Fig. 2.149, sketch v_o and i_R.

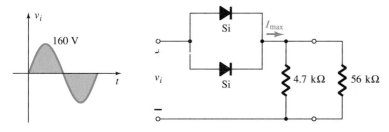

Figure 2.149 Problem 26

* **27.** (a) Given $P_{max} = 14$ mW for each diode of Fig. 2.150, determine the maximum current rating of each diode.
  (b) Determine I_{max} for $V_{i_{max}} = 160$ V.
  (c) Determine the current through each diode for $V_m = 160$ V.
  (d) Is the current determined in part (c) less than the maximum rating determined in part (a)?
  (e) If only one diode were present, determine the diode current and compare it to the maximum rating.

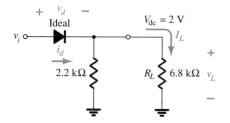

Figure 2.150 Problem 27

§ 2.8 Full-Wave Rectification

28. A full-wave bridge rectifier with a 120-V rms sinusoidal input has a load resistor of 1 kΩ.
  (a) If silicon diodes are employed, what is the dc voltage available at the load?
  (b) Determine the required PIV rating of each diode.
  (c) Find the maximum current through each diode during conduction.
  (d) What is the required power rating of each diode?

29. Determine v_o and the required PIV rating of each diode for the configuration of Fig. 2.151.

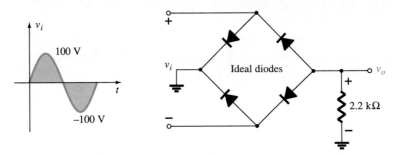

Figure 2.151 Problem 29

* **30.** Sketch v_o for the network of Fig. 2.152 and determine the dc voltage available.

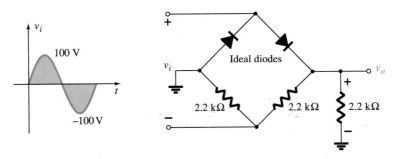

Figure 2.152 Problem 30

* **31.** Sketch v_o for the network of Fig. 2.153 and determine the dc voltage available.

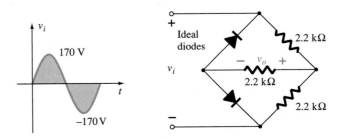

Figure 2.153 Problem 31

§ 2.9 Clippers

32. Determine v_o for each network of Fig. 2.154 for the input shown.

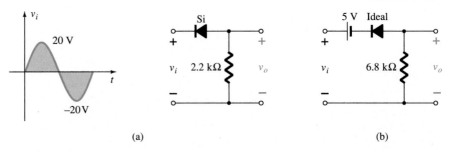

(a)   (b)

Figure 2.154 Problem 32

Chapter 2 Diode Applications

33. Determine v_o for each network of Fig. 2.155 for the input shown.

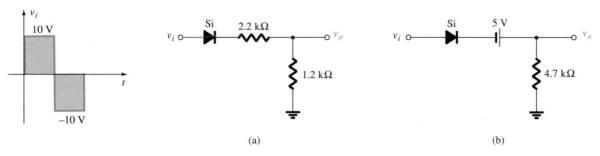

(a)   (b)

Figure 2.155   Problem 33

* **34.** Determine v_o for each network of Fig. 2.156 for the input shown.

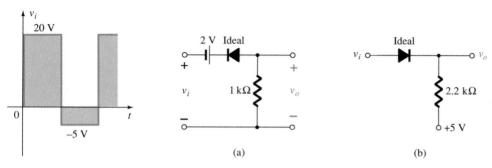

(a)   (b)

Figure 2.156   Problem 34

* **35.** Determine v_o for each network of Fig. 2.157 for the input shown.

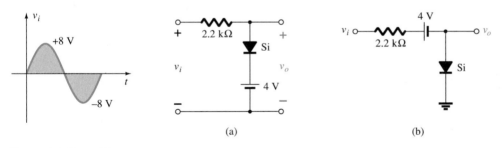

(a)   (b)

Figure 2.157   Problem 35

36. Sketch i_R and v_o for the network of Fig. 2.158 for the input shown.

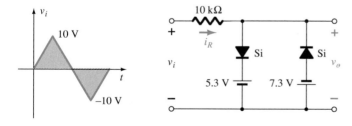

Figure 2.158   Problem 36

37. Sketch v_o for each network of Fig. 2.159 for the input shown.

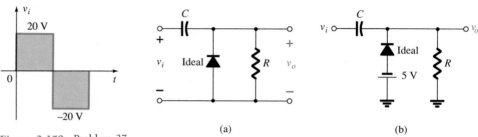

Figure 2.159 Problem 37

38. Sketch v_o for each network of Fig. 2.160 for the input shown. Would it be a good approximation to consider the diode to be ideal for both configurations? Why?

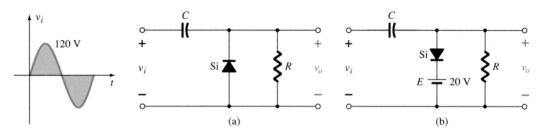

Figure 2.160 Problem 38

* **39.** For the network of Fig. 2.161:
   (a) Calculate 5τ.
   (b) Compare 5τ to half the period of the applied signal.
   (c) Sketch v_o.

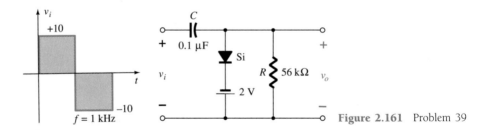

Figure 2.161 Problem 39

* **40.** Design a clamper to perform the function indicated in Fig. 2.162.

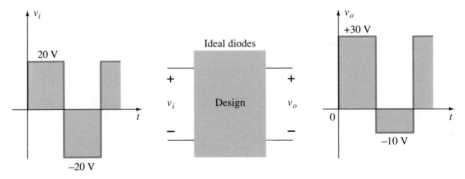

Figure 2.162 Problem 40

* **41.** Design a clamper to perform the function indicated in Fig. 2.163.

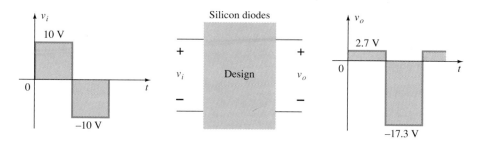

Figure 2.163 Problem 41

§ **2.11 Zener Diodes**

* **42.** (a) Determine V_L, I_L, I_Z, and I_R for the network Fig. 2.164 if $R_L = 180$ Ω
   (b) Repeat part (a) if $R_L = 470$ Ω.
   (c) Determine the value of R_L that will establish maximum power conditions for the Zener diode.
   (d) Determine the minimum value of R_L to ensure that the Zener diode is in the "on" state.

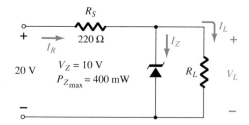

Figure 2.164 Problem 42

* **43.** (a) Design the network of Fig. 2.165 to maintain V_L at 12 V for a load variation (I_L) from 0 to 200 mA. That is, determine R_s and V_Z.
   (b) Determine $P_{Z_{max}}$ for the Zener diode of part (a).

* **44.** For the network of Fig. 2.166, determine the range of V_i that will maintain V_L at 8 V and not exceed the maximum power rating of the Zener diode.

45. Design a voltage regulator that will maintain an output voltage of 20 V across a 1-kΩ load with an input that will vary between 30 and 50 V. That is, determine the proper value of R_s and the maximum current I_{ZM}.

46. Sketch the output of the network of Fig. 2.120 if the input is a 50-V square wave. Repeat for a 5-V square wave.

§ **2.12 Voltage-Multiplier Circuits**

47. Determine the voltage available from the voltage doubler of Fig. 2.121 if the secondary voltage of the transformer is 120 V (rms).

48. Determine the required PIV ratings of the diodes of Fig. 2.121 in terms of the peak secondary voltage V_m.

§ **2.13 Computer Analysis**

49. Write the PSpice input file to determine the currents I_1, I_2, and I_{D_2} of Fig. 2.36 (Example 2.15).

50. Using PSpice, write the input file to determine V_o for the network of Fig. 2.38.

51. Write the PSpice input file to determine V_o for the network of Fig. 2.135b.

52. Repeat Problem 49 using BASIC.

53. Repeat Problem 50 using BASIC.

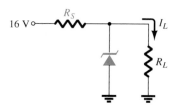

Figure 2.165 Problem 43

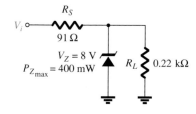

Figure 2.166 Problem 44

*Please Note: Asterisks indicate more difficult problems.

3

Bipolar Junction Transistors

β

3.1 INTRODUCTION

During the period 1904–1947, the vacuum tube was undoubtedly the electronic device of interest and development. In 1904, the vacuum-tube diode was introduced by J. A. Fleming. Shortly thereafter, in 1906, Lee De Forest added a third element, called the *control grid,* to the vacuum diode, resulting in the first amplifier, the *triode.* In the following years, radio and television provided great stimulation to the tube industry. Production rose from about 1 million tubes in 1922 to about 100 million in 1937. In the early 1930s the four-element tetrode and five-element pentode gained prominence in the electron-tube industry. In the years to follow, the industry became one of primary importance and rapid advances were made in design, manufacturing techniques, high-power and high-frequency applications, and miniaturization.

On December 23, 1947, however, the electronics industry was to experience the advent of a completely new direction of interest and development. It was on the afternoon of this day that Walter H. Brattain and John Bardeen demonstrated the amplifying action of the first transistor at the Bell Telephone Laboratories. The original transistor (a point-contact transistor) is shown in Fig. 3.1. The advantages of this three-terminal solid-state device over the tube were immediately obvious: It was

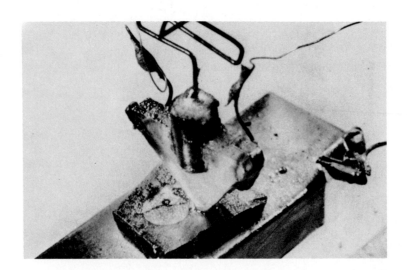

Figure 3.1 The first transistor. (Courtesy Bell Telephone Laboratories.)

108

smaller and lightweight; had no heater requirement or heater loss; had rugged construction; and was more efficient since less power was absorbed by the device itself; it was instantly available for use, requiring no warm-up period; and lower operating voltages were possible. Note in the discussion above that this chapter is our first discussion of devices with three or more terminals. You will find that all amplifiers (devices that increase the voltage, current, or power level) will have at least three terminals with one controlling the flow between two other terminals.

3.2 TRANSISTOR CONSTRUCTION

The transistor is a three-layer semiconductor device consisting of either two *n*- and one *p*-type layers of material or two *p*- and one *n*-type layers of material. The former is called an *npn transistor,* while the latter is called a *pnp transistor.* Both are shown in Fig. 3.2 with the proper dc biasing. We will find in Chapter 4 that the dc biasing is necessary to establish the proper region of operation for ac amplification. The outer layers of the transistor are heavily doped semiconductor materials having widths much greater than those of the sandwiched *p*- or *n*-type material. For the transistors shown in Fig. 3.2 the ratio of the total width to that of the center layer is $0.150/0.001 = 150:1$. The doping of the sandwiched layer is also considerably less than that of the outer layers (typically, $10:1$ or less). This lower doping level decreases the conductivity (increases the resistance) of this material by limiting the number of "free" carriers.

For the biasing shown in Fig. 3.2 the terminals have been indicated by the capital letters *E* for *emitter, C* for *collector,* and *B* for *base.* An appreciation for this choice of notation will develop when we discuss the basic operation of the transistor. The abbreviation BJT, from *bipolar junction transistor,* is often applied to this three-terminal device. The term *bipolar* reflects the fact that holes *and* electrons participate in the injection process into the oppositely polarized material. If only one carrier is employed (electron or hole), it is considered a *unipolar* device. The Schottky diode of Chapter 20 is such a device.

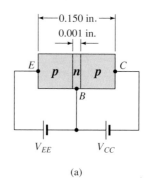

(a)

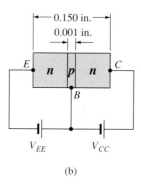

(b)

Figure 3.2 Types of transistors: (a) *pnp;* (b) *npn.*

3.3 TRANSISTOR OPERATION

The basic operation of the transistor will now be described using the *pnp* transistor of Fig. 3.2a. The operation of the *npn* transistor is exactly the same if the roles played by the electron and hole are interchanged. In Fig. 3.3 the *pnp* transistor has been redrawn without the base-to-collector bias. Note the similarities between this situation and that of the *forward-biased* diode in Chapter 1. The depletion region has been reduced in width due to the applied bias, resulting in a heavy flow of majority carriers from the *p*- to the *n*-type material.

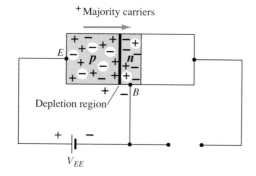

Figure 3.3 Forward-biased junction of a *pnp* transistor.

Let us now remove the base-to-emitter bias of the *pnp* transistor of Fig. 3.2a as shown in Fig. 3.4. Consider the similarities between this situation and that of the *reverse-biased* diode of Section 1.6. Recall that the flow of majority carriers is zero, resulting in only a minority-carrier flow, as indicated in Fig. 3.4. In summary, therefore:

One p-n junction of a transistor is reverse biased, while the other is forward biased.

In Fig. 3.5 both biasing potentials have been applied to a *pnp* transistor, with the resulting majority- and minority-carrier flow indicated. Note in Fig. 3.5 the widths of the depletion regions, indicating clearly which junction is forward-biased and which is reverse-biased. As indicated in Fig. 3.5, a large number of majority carriers will diffuse across the forward-biased p-n junction into the n-type material. The question then is whether these carriers will contribute directly to the base current I_B or pass directly into the p-type material. Since the sandwiched n-type material is very thin and has a low conductivity, a very small number of these carriers will take this path of high resistance to the base terminal. The magnitude of the base current is typically on the order of microamperes as compared to milliamperes for the emitter and collector currents. The larger number of these majority carriers will diffuse across the reverse-biased junction into the p-type material connected to the collector terminal as indicated in Fig. 3.5. The reason for the relative ease with which the majority carriers can cross the reverse-biased junction is easily understood if we consider that for the reverse-biased diode the injected majority carriers will appear as minority carriers in the n-type material. In other words, there has been an *injection* of minority carriers into the n-type base region material. Combining this with the fact that all the minority carriers in the depletion region will cross the reverse-biased junction of a diode accounts for the flow indicated in Fig. 3.5.

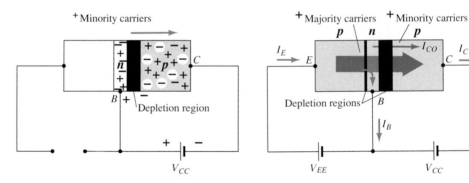

Figure 3.4 Reverse-biased junction of a *pnp* transistor.

Figure 3.5 Majority and minority carrier flow of a *pnp* transistor.

Applying Kirchhoff's current law to the transistor of Fig. 3.5 as if it were a single node, we obtain

$$I_E = I_C + I_B \tag{3.1}$$

and find that the emitter current is the sum of the collector and base currents. The collector current, however, is comprised of two components—the majority and minority carriers as indicated in Fig. 3.5. The minority-current component is called the *leakage current* and is given the symbol I_{CO} (I_C current with emitter terminal *O*pen). The collector current, therefore, is determined in total by Eq. 3.2).

$$I_C = I_{C_{\text{majority}}} + I_{CO_{\text{minority}}} \tag{3.2}$$

For general-purpose transistors, I_C is measured in milliamperes, while I_{CO} is measured in microamperes or nanoamperes. I_{CO}, like I_s for a reverse-biased diode, is temperature sensitive and must be examined carefully when applications of wide temperature ranges are considered. It can severely affect the stability of a system at high temperature if not considered properly. Improvements in construction techniques have resulted in significantly lower levels of I_{CO}, to the point where its effect can often be ignored.

3.4 COMMON-BASE CONFIGURATION

The notation and symbols used in conjunction with the transistor in the majority of texts and manuals published today are indicated in Fig. 3.6 for the common-base configuration with *pnp* and *npn* transistors. The common-base terminology is derived from the fact that the base is common to both the input and output sides of the configuration. In addition, the base is usually the terminal closest to, or at, ground potential. Throughout this book all current directions will refer to conventional (hole) flow rather than electron flow. This choice was based primarily on the fact that the vast amount of literature available at educational and industrial institutions employs conventional flow and the arrows in all electronic symbols have a direction defined by this convention. Recall that the arrow in the diode symbol defined the direction of conduction for conventional current. For the transistor:

The arrow in the graphic symbol defines the direction of emitter current (conventional flow) through the device.

All the current directions appearing in Fig. 3.6 are the actual directions as defined by the choice of conventional flow. Note in each case that $I_E = I_C + I_B$. Note also that the applied biasing (voltage sources) are such as to establish current in the direction indicated for each branch. That is, compare the direction of I_E to the polarity or V_{EE} for each configuration and the direction of I_C to the polarity of V_{CC}.

To fully describe the behavior of a three-terminal device such as the common-base amplifiers of Fig. 3.6 requires two sets of characteristics—one for the *driving point* or *input* parameters and the other for the *output* side. The input set for the common-base amplifier as shown in Fig. 3.7 will relate an input current (I_E) to an input voltage (V_{BE}) for various levels of output voltage (V_{CB}).

The output set will relate an output current (I_C) to an output voltage (V_{CB}) for various levels of input current (I_E) as shown in Fig. 3.8. The output or *collector* set of characteristics has three basic regions of interest, as indicated in Fig. 3.8: the *active*,

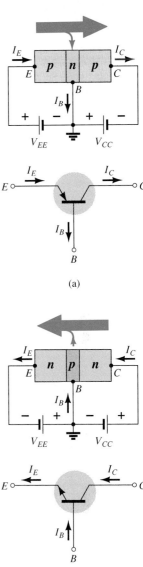

Figure 3.6 Notation and symbols used with the common-base configuration: (a) *pnp* transistor; (b) *npn* transistor.

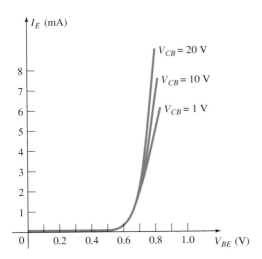

Figure 3.7 Input or driving point characteristics for a common-base silicon transistor amplifier.

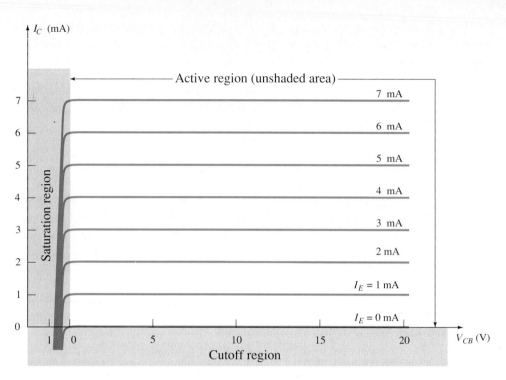

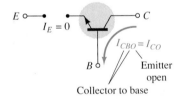

Figure 3.8 Output or collector characteristics for a common-base transistor amplifier.

Figure 3.9 Reverse saturation current.

cutoff, and *saturation* regions. The active region is the region normally employed for linear (undistorted) amplifiers. In particular:

In the active region the collector-base junction is reverse-biased, while the base-emitter junction is forward-biased.

The active region is defined by the biasing arrangements of Fig. 3.6. At the lower end of the active region the emitter current (I_E) is zero, the collector current is simply that due to the reverse saturation current I_{CO}, as indicated in Fig. 3.8. The current I_{CO} is so small (microamperes) in magnitude compared to the vertical scale of I_C (milliamperes) that it appears on virtually the same horizontal line as $I_C = 0$. The circuit conditions that exist when $I_E = 0$ for the common-base configuration are shown in Fig. 3.9. The notation most frequently used for I_{CO} on data and specification sheets is, as indicated in Fig. 3.9, I_{CBO}. Because of improved construction techniques, the level of I_{CBO} for general-purpose transistors (especially silicon) in the low- and mid-power ranges is usually so low that its effect can be ignored. However, for higher power units I_{CBO} will still appear in the microampere range. In addition, keep in mind that I_{CBO}, like I_s, for the diode (both reverse leakage currents) is temperature sensitive. At higher temperatures the effect of I_{CBO} may become an important factor since it increases so rapidly with temperature.

Note in Fig. 3.8 that as the emitter current increases above zero, the collector current increases to a magnitude essentially equal to that of the emitter current as determined by the basic transistor-current relations. Note also the almost negligible effect of V_{CB} on the collector current for the active region. The curves clearly indicate that *a first approximation to the relationship between I_E and I_C in the active region is given by*

$$I_C \cong I_E \qquad (3.3)$$

As inferred by its name, the cutoff region is defined as that region where the collector current is 0 A, as revealed on Fig. 3.8. In addition:

In the cutoff region the collector-base and base-emitter junctions of a transistor are both reverse-biased.

The saturation region is defined as that region of the characteristics to the left of $V_{CB} = 0$ V. The horizontal scale in this region was expanded to clearly show the dramatic change in characteristics in this region. Note the exponential increase in collector current as the voltage V_{CB} increases toward 0 V.

In the saturation region the collector-base and base-emitter junctions are forward-biased.

The input characteristics of Fig. 3.7 reveal that for fixed values of collector voltage (V_{CB}), as the base-to-emitter voltage increases, the emitter current increases in a manner that closely resembles the diode characteristics. In fact, increasing levels of V_{CB} have such a small effect on the characteristics that as a first approximation the change due to changes in V_{CB} can be ignored and the characteristics drawn as shown in Fig. 3.10a. If we then apply the piecewise-linear approach, the characteristics of Fig. 3.10b will result. Taking it a step further and ignoring the slope of the curve and therefore the resistance associated with the forward-biased junction will result in the characteristics of Fig. 3.10c. For the analysis to follow in this book the equivalent model of Fig. 3.10c will be employed for all dc analysis of transistor networks. That is, once a transistor is in the "on" state, the base-to-emitter voltage will be assumed to be the following:

$$V_{BE} = 0.7 \text{ V} \qquad (3.4)$$

In other words, the effect of variations due to V_{CB} and the slope of the input characteristics will be ignored as we strive to analyze transistor networks in a manner that will provide a good approximation to the actual response without getting too involved with parameter variations of less importance.

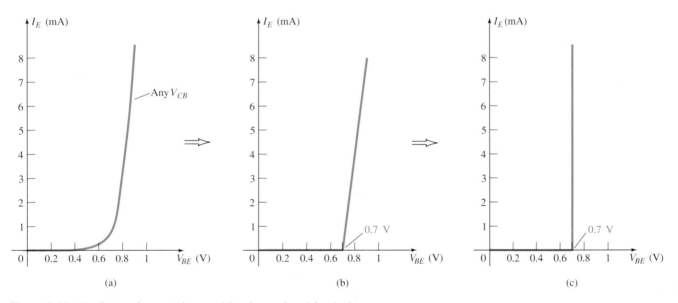

Figure 3.10 Developing the equivalent model to be employed for the base-to-emitter region of an amplifier in the dc mode.

It is important to fully appreciate the statement made by the characteristics of Fig. 3.10c. They specify that with the transistor in the "on" or active state the voltage from base to emitter will be 0.7 V at *any* level of emitter current as controlled by the external network. In fact, at the first encounter of any transistor configuration in the dc mode, one can now immediately specify that the voltage from base to emitter is 0.7 V if the device is in the active region—a very important conclusion for the dc analysis to follow.

EXAMPLE 3.1

(a) Using the characteristics of Fig. 3.8, determine the resulting collector current if $I_E = 3$ mA and $V_{CB} = 10$ V.
(b) Using the characteristics of Fig. 3.8, determine the resulting collector current if I_E remains at 3 mA but V_{CB} is reduced to 2 V.
(c) Using the characteristics of Figs. 3.7 and 3.8, determine V_{BE} if $I_C = 4$ mA and $V_{CB} = 20$ V.
(d) Repeat part (c) using the characteristics of Figs. 3.8 and 3.10c.

Solution

(a) The characteristics clearly indicate that $I_C \cong I_E = $ **3 mA**.
(b) The effect of changing V_{CB} is negligible and I_C continues to be **3 mA**.
(c) From Fig. 3.8, $I_E \cong I_C = 4$ mA. On Fig. 3.7 the resulting level of V_{BE} is about **0.74 V**.
(d) Again from Fig. 3.8, $I_E \cong I_C = 4$ mA. However, on Fig. 3.10c V_{BE} is **0.7 V** for any level of emitter current.

Alpha (α)

In the dc mode the levels of I_C and I_E due to the majority carriers are related by a quantity called *alpha* and defined by the following equation:

$$\alpha_{dc} = \frac{I_C}{I_E}$$

(3.5)

where I_C and I_E are the levels of current at the point of operation. Even though the characteristics of Fig. 3.8 would suggest that $\alpha = 1$, for practical devices the level of alpha typically extends from 0.90 to 0.998, with most approaching the high end of the range. Since alpha is defined solely for the majority carriers, Eq. (3.2) becomes

$$I_C = \alpha I_E + I_{CBO}$$

(3.6)

For the characteristics of Fig. 3.8 when $I_E = 0$ mA, I_C is therefore equal to I_{CBO}, but as mentioned earlier, the level of I_{CBO} is usually so small that it is virtually undetectable on the graph of Fig. 3.8. In other words, when $I_E = 0$ mA on Fig. 3.8, I_C also appears to be 0 mA for the range of V_{CB} values.

For ac situations where the point of operation moves on the characteristic curve, an ac alpha is defined by

$$\alpha_{ac} = \frac{\Delta I_C}{\Delta I_E}\bigg|_{V_{CB}=\text{constant}}$$

(3.7)

The ac alpha is formally called the *common-base, short-circuit, amplification factor,* for reasons that will be more obvious when we examine transistor equivalent circuits in Chapter 7. For the moment, recognize that Eq. (3.7) specifies that a relatively small change in collector current is divided by the corresponding change in I_E with the collector-to-base voltage held constant. For most situations the magnitudes of α_{ac} and α_{dc} are quite close, permitting the use of the magnitude of one for the other. The use of an equation such as (3.7) will be demonstrated in Section 3.6.

Biasing

The proper biasing of the common-base configuration in the active region can be determined quickly using the approximation $I_C \cong I_E$ and assuming for the moment

Chapter 3   Bipolar Junction Transistors

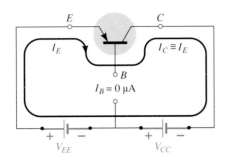

Figure 3.11 Establishing the proper biasing management for a common-base *pnp* transistor in the active region.

that $I_B \cong 0 \; \mu$A. The result is the configuration of Fig. 3.11 for the *pnp* transistor. The arrow of the symbol defines the direction of conventional flow for $I_E \cong I_C$. The dc supplies are then inserted with a polarity that will support the resulting current direction. For the *npn* transistor the polarities will be reversed.

Some students feel that they can remember whether the arrow of the device symbol in pointing in or out by matching the letters of the transistor type with the appropriate letters of the phrases "pointing in" or "not pointing in." For instance, there is a match between the letters *npn* and the italic letters of *not pointing in* and the letters *pnp* with *pointing in*.

3.5 TRANSISTOR AMPLIFYING ACTION

Now that the relationship between I_C and I_E has been established in Section 3.4, the basic amplifying action of the transistor can be introduced on a surface level using the network of Fig. 3.12. The dc biasing does not appear in the figure since our interest will be limited to the ac response. For the common-base configuration the ac input resistance determined by the characteristics of Fig. 3.7 is quite small and typically varies from 10 to 100 Ω. The output resistance as determined by the curves of Fig. 3.8 is quite high (the more horizontal the curves the higher the resistance) and typically varies from 50 kΩ to 1 MΩ. The difference in resistance is due to the forward-biased junction at the input (base to emitter) and the reverse-biased junction at the output (base to collector). Using a common value of 20 Ω for the input resistance, we find that

$$I_i = \frac{V_i}{R_i} = \frac{200 \text{ mV}}{20 \; \Omega} = 10 \text{ mA}$$

If we assume for the moment that $\alpha_{\text{ac}} = 1 \; (I_c = I_e)$,

$$I_L = I_i = 10 \text{ mA}$$

and

$$V_L = I_L R$$

$$= (10 \text{ mA})(5 \text{ k}\Omega)$$

$$= 50 \text{ V}$$

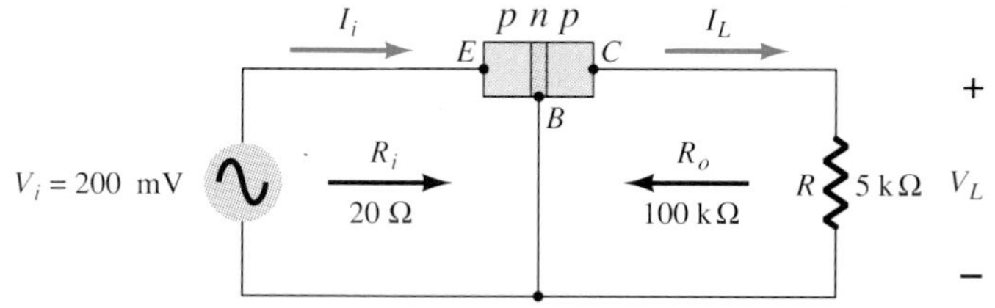

Figure 3.12 Basic voltage amplification action of the common-base configuration.

The voltage amplification is

$$A_v = \frac{V_L}{V_i} = \frac{50 \text{ V}}{200 \text{ mV}} = \mathbf{250}$$

Typical values of voltage amplification for the common-base configuration vary from 50 to 300. The current amplification (I_C/I_E) is always less than 1 for the common-base configuration. This latter characteristic should be obvious since $I_C = \alpha I_E$ and α is always less than 1.

The basic amplifying action was produced by *transferring* a current I from a low- to a high-*resistance* circuit. The combination of the two terms in italics results in the label *transistor;* that is,

*trans*fer + re*sistor* → *transistor*

3.6 COMMON-EMITTER CONFIGURATION

The most frequently encountered transistor configuration appears in Fig. 3.13 for the *pnp* and *npn* transistors. It is called the *common-emitter configuration* since the emitter is common or reference to both the input and output terminals (in this case common to both the base and collector terminals). Two sets of characteristics are again necessary to describe fully the behavior of the common-emitter configuration: one for the *input* or *base-emitter* circuit and one for the *output* or *collector-emitter* circuit. Both are shown in Fig. 3.14.

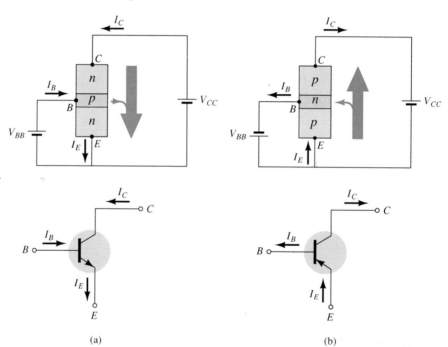

Figure 3.13 Notation and symbols used with the common-emitter configuration: (a) *npn* transistor; (b) *pnp* transistor.

The emitter, collector, and base currents are shown in their actual conventional current direction. Even though the transistor configuration has changed, the current relations developed earlier for the common-base configuration are still applicable. That is, $I_E = I_C + I_B$ and $I_C = \alpha I_E$.

For the common-emitter configuration the output characteristics are a plot of the output current (I_C) versus output voltage (V_{CE}) for a range of values of input current (I_B). The input characteristics are a plot of the input current (I_B) versus the input voltage (V_{BE}) for a range of values of output voltage (V_{CE}).

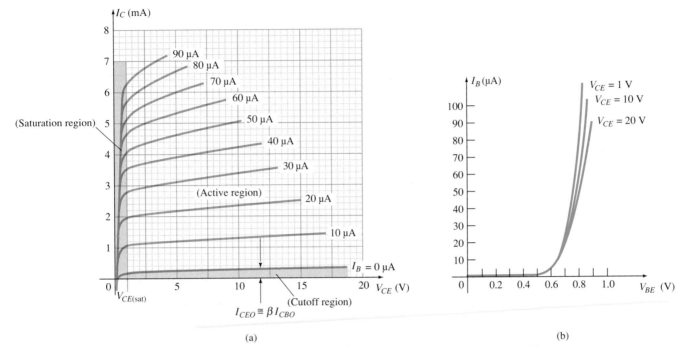

Figure 3.14 Characteristics of a silicon transistor in the common-emitter configuration: (a) collector characteristics; (b) base characteristics.

Note that on the characteristics of Fig. 3.14 the magnitude of I_B is in microamperes, compared to milliamperes of I_C. Consider also that the curves of I_B are not as horizontal as those obtained for I_E in the common-base configuration, indicating that the collector-to-emitter voltage will influence the magnitude of the collector current.

The active region for the common-emitter configuration is that portion of the upper-right quadrant that has the greatest linearity, that is, that region in which the curves for I_B are nearly straight and equally spaced. In Fig. 3.14a this region exists to the right of the vertical dashed line at $V_{CE_{sat}}$ and above the curve for I_B equal to zero. The region to the left of $V_{CE_{sat}}$ is called the saturation region.

In the active region of a common-emitter amplifier the collector-base junction is reverse-biased, while the base-emitter junction is forward-biased.

You will recall that these were the same conditions that existed in the active region of the common-base configuration. The active region of the common-emitter configuration can be employed for voltage, current, or power amplification.

The cutoff region for the common-emitter configuration is not as well defined as for the common-base configuration. Note on the collector characteristics of Fig. 3.14 that I_C is not equal to zero when I_B is zero. For the common-base configuration, when the input current I_E was equal to zero, the collector current was equal only to the reverse saturation current I_{CO}, so that the curve $I_E = 0$ and the voltage axis were, for all practical purposes, one.

The reason for this difference in collector characteristics can be derived through the proper manipulation of Eqs. (3.3) and (3.6). That is,

$$\text{Eq. (3.6):} \quad I_C = \alpha I_E + I_{CBO}$$

Substitution gives $\quad$ Eq. (3.3): $\quad I_C = \alpha(I_C + I_B) + I_{CBO}$

Rearranging yields
$$I_C = \frac{\alpha I_B}{1-\alpha} + \frac{I_{CBO}}{1-\alpha} \qquad (3.8)$$

3.6 Common-Emitter Configuration $\qquad$ **117**

If we consider the case discussed above, where $I_B = 0$ A, and substitute a typical value of α such as 0.996, the resulting collector current is the following:

$$I_C = \frac{\alpha(0 \text{ A})}{1 - \alpha} + \frac{I_{CBO}}{1 - 0.996}$$

$$= \frac{I_{CBO}}{0.004} = 250 \, I_{CBO}$$

If I_{CBO} were 1 μA, the resulting collector current with $I_B = 0$ A would be $250(1 \ \mu A) = 0.25$ mA, as reflected in the characteristics of Fig. 3.14.

For future reference, the collector current defined by the condition $I_B = 0 \ \mu$A will be assigned the notation indicated by Eq. (3.9).

$$I_{CEO} = \left.\frac{I_{CBO}}{1 - \alpha}\right|_{I_B = 0 \ \mu A} \tag{3.9}$$

In Fig. 3.15 the conditions surrounding this newly defined current are demonstrated with its assigned reference direction.

For linear (least distortion) amplification purposes, cutoff for the common-emitter configuration will be defined by $I_C = I_{CEO}$.

In other words, the region below $I_B = 0 \ \mu$A is to be avoided if an undistorted output signal is required.

When employed as a switch in the logic circuitry of a computer, a transistor will have two points of operation of interest: one in the cutoff and one in the saturation region. The cutoff condition should ideally be $I_C = 0$ mA for the chosen V_{CE} voltage. Since I_{CEO} is typically low in magnitude for silicon materials, *cutoff will exist for switching purposes when $I_B = 0 \ \mu A$ or $I_C = I_{CEO}$ for silicon transistors only. For germanium transistors, however, cutoff for switching purposes will be defined as those conditions that exist when $I_C = I_{CBO}$.* This condition can normally be obtained for germanium transistors by reverse-biasing the base-to-emitter junction a few tenths of a volt.

Recall for the common-base configuration that the input set of characteristics was approximated by a straight-line equivalent that resulted in $V_{BE} = 0.7$ V for any level of I_E greater than 0 mA. For the common-emitter configuration the same approach can be taken, resulting in the approximate equivalent of Fig. 3.16. The result supports our earlier conclusion that for a transistor in the "on" or active region the base-to-emitter voltage is 0.7 V. In this case the voltage is fixed for any level of base current.

Figure 3.15 Circuit conditions related to I_{CEO}.

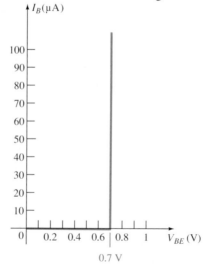

Figure 3.16 Piecewise-linear equivalent for the diode characteristics of Fig. 3.14b.

EXAMPLE 3.2

(a) Using the characteristics of Fig. 3.14, determine I_C at $I_B = 30\ \mu\text{A}$ and $V_{CE} = 10$ V.
(b) Using the characteristics of Fig. 3.14, determine I_C at $V_{BE} = 0.7$ V and $V_{CE} = 15$ V.

Solution

(a) At the intersection of $I_B = 30\ \mu\text{A}$ and $V_{CE} = 10$ V, $I_C =$ **3.4 mA**.
(b) Using Fig. 3.14b, $I_B = 20\ \mu\text{A}$ at $V_{BE} = 0.7$ V. From Fig. 3.14a we find that $I_C =$ **2.5 mA** at the intersection of $I_B = 20\ \mu\text{A}$ and $V_{CE} = 15$ V.

Beta (β)

In the dc mode the levels of I_C and I_B are related by a quantity called *beta* and defined by the following equation:

$$\beta_{\text{dc}} = \frac{I_C}{I_B} \qquad (3.10)$$

where I_C and I_B are determined at a particular operating point on the characteristics. For practical devices the level of β typically ranges from about 50 to over 400, with most in the midrange. As for α, β certainly reveals the relative magnitude of one current to the other. For a device with a β of 200, the collector current is 200 times the magnitude of the base current.

On specification sheets β_{dc} is usually included as h_{FE} with the h derived from an ac *h*ybrid equivalent circuit to be introduced in Chapter 7. The subscripts FE are derived from *f*orward-current amplification and common-*e*mitter configuration, respectively.

For ac situations an ac beta has been defined as follows:

$$\beta_{\text{ac}} = \left. \frac{\Delta I_C}{\Delta I_B} \right|_{V_{CE} = \text{constant}} \qquad (3.11)$$

The formal name for β_{ac} is *common-emitter forward-current amplification factor*. Since the collector current is usually the output current for a common-emitter configuration and the base current the input current, the term *amplification* is included in the nomenclature above.

Equation (3.11) is similar in format to the equation for α_{ac} in Section 3.4. The procedure for obtaining α_{ac} from the characteristic curves was not described because of the difficulty of actually measuring changes of I_C and I_E on the characteristics. Equation (3.11), however, is one that can be described with some clarity, and in fact, the result can be used to find α_{ac} using an equation to be derived shortly.

On specification sheets β_{ac} is normally referred to as h_{fe}. Note that the only difference between the notation used for the dc beta, specifically, $\beta_{\text{dc}} = h_{FE}$, is the type of lettering for each subscript quantity. The lowercase letter h continues to refer to the hybrid equivalent circuit to be described in Chapter 7 and the *fe* to the *f*orward current gain in the common-*e*mitter configuration.

The use of Eq. (3.11) is best described by a numerical example using an actual set of characteristics such as appearing in Fig. 3.14a and repeated in Fig. 3.17. Let us determine β_{ac} for a region of the characteristics defined by an operating point of $I_B = 25\ \mu\text{A}$ and $V_{CE} = 7.5$ V as indicated on Fig. 3.17. The restriction of $V_{CE} =$ constant requires that a vertical line be drawn through the operating point at $V_{CE} = 7.5$ V. At any location on this vertical line the voltage V_{CE} is 7.5 V, a constant. The

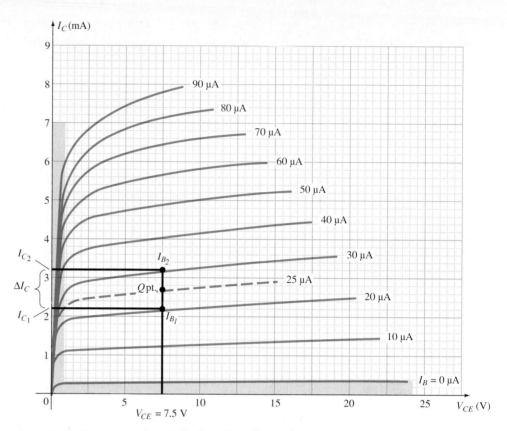

Figure 3.17 Determining β_{ac} and β_{dc} from the collector characteristics.

change in I_B (ΔI_B) as appearing in Eq. (3.11) is then defined by choosing two points on either side of the Q-point along the vertical axis of about equal distances to either side of the Q-point. For this situation the $I_B = 20$ μA and 30 μA curves meet the requirement without extending too far from the Q-point. They also define levels of I_B that are easily defined rather than have to interpolate the level of I_B between the curves. It should be mentioned that the best determination is usually made by keeping the chosen ΔI_B as small as possible. At the two intersections of I_B and the vertical axis, the two levels of I_C can be determined by drawing a horizontal line over to the vertical axis and reading the resulting values of I_C. The resulting β_{ac} for the region can then be determined by

$$\beta_{ac} = \frac{\Delta I_C}{\Delta I_B}\bigg|_{V_{CE}=\text{constant}} = \frac{I_{C_2} - I_{C_1}}{I_{B_2} - I_{B_1}}$$

$$= \frac{3.2 \text{ mA} - 2.2 \text{ mA}}{30 \ \mu\text{A} - 20 \ \mu\text{A}} = \frac{1 \text{ mA}}{10 \ \mu\text{A}}$$

$$= \mathbf{100}$$

The solution above reveals that for an ac input at the base, the collector current will be about 100 times the magnitude of the base current.

If we determine the dc beta at the Q-point:

$$\beta_{dc} = \frac{I_C}{I_B} = \frac{2.7 \text{ mA}}{25 \ \mu\text{A}} = \mathbf{108}$$

Although not exactly equal, the levels of β_{ac} and β_{dc} are usually reasonably close and are often used interchangeably. That is, if β_{ac} is known, it is assumed to be about the same magnitude as β_{dc}, and vice versa. Keep in mind that in the same lot, the value of β_{ac} will vary somewhat from one transistor to the next even though each transistor has the same number code. The variation may not be significant but for the majority of applications, it is certainly sufficient to validate the approximate approach above. Generally, the smaller the level of I_{CEO}, the closer the magnitude of the two betas. Since the trend is toward lower and lower levels of I_{CEO}, the validity of the foregoing approximation is further substantiated.

If the characteristics had the appearance of those appearing in Fig. 3.18, the level of β_{ac} would be the same in every region of the characteristics. Note that the step in I_B is fixed at 10 μA and the vertical spacing between curves is the same at every point in the characteristics—namely, 2 mA. Calculating the β_{ac} at the Q-point indicated will result in

$$\beta_{ac} = \frac{\Delta I_C}{\Delta I_B}\bigg|_{V_{CE}=\text{constant}} = \frac{9 \text{ mA} - 7 \text{ mA}}{45 \text{ } \mu\text{A} - 35 \text{ } \mu\text{A}} = \frac{2 \text{ mA}}{10 \text{ } \mu\text{A}} = \mathbf{200}$$

Determining the dc beta at the same Q-point will result in

$$\beta_{dc} = \frac{I_C}{I_B} = \frac{8 \text{ mA}}{40 \text{ } \mu\text{A}} = \mathbf{200}$$

revealing that if the characteristics have the appearance of Fig. 3.18, the magnitude of β_{ac} and β_{dc} *will be the same* at every point on the characteristics. In particular, note that $I_{CEO} = 0$ μA.

Although a true set of transistor characteristics will never have the exact appearance of Fig. 3.18, it does provide a set of characteristics for comparison with those obtained from a curve tracer (to be described shortly).

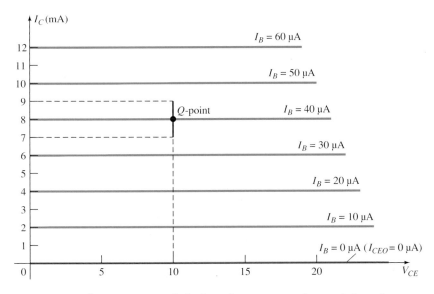

Figure 3.18 Characteristics in which β_{ac} is the same everywhere and $\beta_{ac} = \beta_{dc}$.

For the analysis to follow the subscript dc or ac will not be included with β to avoid cluttering the expressions with unnecessary labels. For dc situations it will simply be recognized as β_{dc} and for any ac analysis as β_{ac}. If a value of β is specified for a particular transistor configuration, it will normally be used for both the dc and ac calculations.

A relationship can be developed between β and α using the basic relationships introduced thus far. Using $\beta = I_C/I_B$ we have $I_B = I_C/\beta$, and from $\alpha = I_C/I_E$ we have $I_E = I_C/\alpha$. Substituting into

$$I_E = I_C + I_B$$

we have

$$\frac{I_C}{\alpha} = I_C + \frac{I_C}{\beta}$$

and dividing both sides of the equation by I_C will result in

$$\frac{1}{\alpha} = 1 + \frac{1}{\beta}$$

or

$$\beta = \alpha\beta + \alpha = (\beta + 1)\alpha$$

so that

$$\boxed{\alpha = \frac{\beta}{\beta + 1}} \qquad (3.12a)$$

or

$$\boxed{\beta = \frac{\alpha}{1-\alpha}} \qquad (3.12b)$$

In addition, recall that

$$I_{CEO} = \frac{I_{CBO}}{1 - \alpha}$$

but using an equivalence of

$$\frac{1}{1 - \alpha} = \beta + 1$$

derived from the above, we find that

$$I_{CEO} = (\beta + 1)I_{CBO}$$

or

$$\boxed{I_{CEO} \cong \beta I_{CBO}} \qquad (3.13)$$

as indicated on Fig. 3.14a. Beta is a particularly important parameter because it provides a direct link between current levels of the input and output circuits for a common-emitter configuration. That is,

$$\boxed{I_C = \beta I_B} \qquad (3.14)$$

and since

$$I_E = I_C + I_B$$
$$= \beta I_B + I_B$$

we have

$$\boxed{I_E = (\beta + 1)I_B} \qquad (3.15)$$

Both of the equations above play a major role in the analysis in Chapter 4.

Biasing

The proper biasing of a common-emitter amplifier can be determined in a manner similar to that introduced for the common-base configuration. Let us assume that we are presented with an *npn* transistor such as shown in Fig. 3.19a and asked to apply the proper biasing to place the device in the active region.

The first step is to indicate the direction of I_E as established by the arrow in the transistor symbol as shown in Fig. 3.19b. Next, the other currents are introduced as

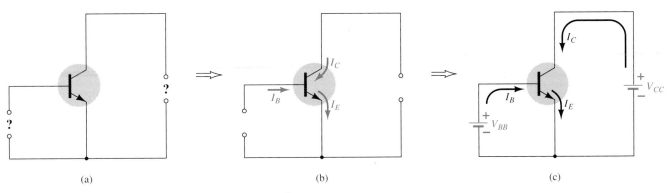

(a)   (b)   (c)

Figure 3.19 Determining the proper biasing arrangement for a common-emitter *npn* transistor configuration.

shown, keeping in mind the Kirchhoff's current law relationship: $I_C + I_B = I_E$. Finally, the supplies are introduced with polarities that will support the resulting directions of I_B and I_C as shown in Fig. 3.19c to complete the picture. The same approach can be applied to *pnp* transistors. If the transistor of Fig. 3.19 was a *pnp* transistor, all the currents and polarities of Fig. 3.19c would be reversed.

3.7 COMMON-COLLECTOR CONFIGURATION

The third and final transistor configuration is the *common-collector configuration*, shown in Fig. 3.20 with the proper current directions and voltage notation. The common-collector configuration is used primarily for impedance-matching purposes since it has a high input impedance and low output impedance, opposite to that of the common-base and common-emitter configurations.

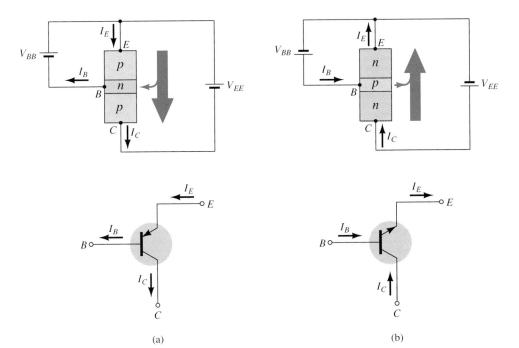

(a)   (b)

Figure 3.20 Notation and symbols used with the common-collector configuration: (a) *pnp* transistor; (b) *npn* transistor.

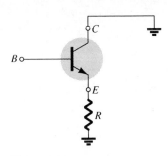

Figure 3.21 Common-collector configuration used for impedance-matching purposes.

A common-collector circuit configuration is provided in Fig. 3.21 with the load resistor connected from emitter to ground. Note that the collector is tied to ground even though the transistor is connected in a manner similar to the common-emitter configuration. From a design viewpoint, there is no need for a set of common-collector characteristics to choose the parameters of the circuit of Fig. 3.21. It can be designed using the common-emitter characteristics of Section 3.6. For all practical purposes, the output characteristics of the common-collector configuration are the same as for the common-emitter configuration. For the common-collector configuration the output characteristics are a plot of I_E versus V_{EC} for a range of values of I_B. The input current, therefore, is the same for both the common-emitter and common-collector characteristics. The horizontal voltage axis for the common-collector configuration is obtained by simply changing the sign of the collector-to-emitter voltage of the common-emitter characteristics. Finally, there is an almost unnoticeable change in the vertical scale of I_C of the common-emitter characteristics if I_C is replaced by I_E for the common-collector characteristics (since $\alpha \cong 1$). For the input circuit of the common-collector configuration the common-emitter base characteristics are sufficient for obtaining the required information.

3.8 LIMITS OF OPERATION

For each transistor there is a region of operation on the characteristics which will ensure that the maximum ratings are not being exceeded and the output signal exhibits minimum distortion. Such a region has been defined for the transistor characteristics of Fig. 3.22. All of the limits of operation are defined on a typical transistor specification sheet described in Section 3.9.

Some of the limits of operation are self-explanatory, such as maximum collector current (normally referred to on the specification sheet as *continuous* collector current) and maximum collector-to-emitter voltage (often abbreviated as V_{CEO} or $V_{(BR)CEO}$ on the specification sheet). For the transistor of Fig. 3.22, $I_{C_{max}}$ was specified as 50 mA and V_{CEO} as 20 V. The vertical line on the characteristics defined as $V_{CE_{sat}}$

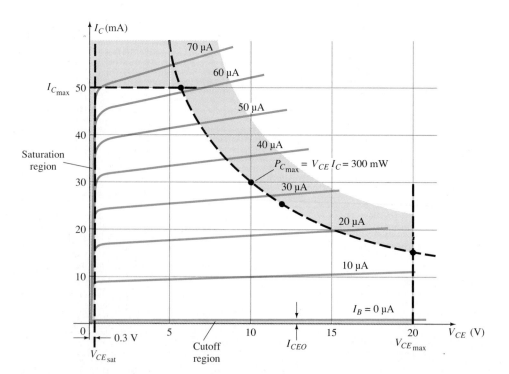

Figure 3.22 Defining the linear (undistorted) region of operation for a transistor.

Chapter 3 Bipolar Junction Transistors

specifies the minimum V_{CE} that can be applied without falling into the nonlinear region labeled the *saturation* region. The level of $V_{CE_{sat}}$ is typically in the neighborhood of the 0.3 V specified for this transistor.

The maximum dissipation level is defined by the following equation:

$$\boxed{P_{C_{max}} = V_{CE}I_C}\qquad(3.16)$$

For the device of Fig. 3.22, the collector power dissipation was specified as 300 mW. The question then arises of how to plot the collector power dissipation curve specified by the fact that

$$P_{C_{max}} = V_{CE}I_C = 300 \text{ mW}$$

or
$$V_{CE}I_C = 300 \text{ mW}$$

At any point on the characteristics the product of V_{CE} and I_C must be equal to 300 mW. If we choose I_C to be the maximum value of 50 mA and substitute into the relationship above, we obtain

$$V_{CE}I_C = 300 \text{ mW}$$

$$V_{CE}(50 \text{ mA}) = 300 \text{ mW}$$

$$V_{CE} = \frac{300 \text{ mW}}{50 \text{ mA}} = \mathbf{6\ V}$$

As a result we find that if $I_C = 50$ mA, then $V_{CE} = 6$ V on the power dissipation curve as indicated in Fig. 3.22. If we now choose V_{CE} to be its maximum value of 20 V, the level of I_C is the following:

$$(20 \text{ V})I_C = 300 \text{ mW}$$

$$I_C = \frac{300 \text{ mW}}{20 \text{ V}} = \mathbf{15\ mA}$$

defining a second point on the power curve.

If we now choose a level of I_C in the midrange such as 25 mA, and solve for the resulting level of V_{CE}, we obtain

$$V_{CE}(25 \text{ mA}) = 300 \text{ mW}$$

and
$$V_{CE} = \frac{300 \text{ mW}}{25 \text{ mA}} = \mathbf{12\ V}$$

as also indicated on Fig. 3.22.

A rough estimate of the actual curve can usually be drawn using the three points defined above. Of course, the more points you have, the more accurate the curve, but a rough estimate is normally all that is required.

The *cutoff* region is defined as that region below $I_C = I_{CEO}$. This region must also be avoided if the output signal is to have minimum distortion. On some specification sheets only I_{CBO} is provided. One must then use the equation $I_{CEO} = \beta I_{CBO}$ to establish some idea of the cutoff level if the characteristic curves are unavailable. Operation in the resulting region of Fig. 3.22 will ensure minimum distortion of the output signal and current and voltage levels that will not damage the device.

If the characteristic curves are unavailable or do not appear on the specification sheet (as is often the case), one must simply be sure that I_C, V_{CE}, and their product $V_{CE}I_C$ fall into the range appearing in Eq. (3.17).

$$\boxed{\begin{aligned} I_{CEO} &\leqq I_C \leqq I_{C_{max}} \\ V_{CE_{sat}} &\leqq V_{CE} \leqq V_{CE_{max}} \\ V_{CE} I_C &\leqq P_{C_{max}} \end{aligned}}$$
(3.17)

For the common-base characteristics the maximum power curve is defined by the following product of output quantities:

$$\boxed{P_{C_{max}} = V_{CB} I_C}$$
(3.18)

3.9 TRANSISTOR SPECIFICATION SHEET

Since the specification sheet is the communication link between the manufacturer and user, it is particularly important that the information provided be recognized and correctly understood. Although all the parameters have not been introduced, a broad number will now be familiar. The remaining parameters will be introduced in the chapters that follow. Reference will then be made to this specification sheet to review the manner in which the parameter is presented.

The information provided as Fig. 3.23 is taken directly from the *Small-Signal Transistors, FETs, and Diodes* publication prepared by Motorola Inc. The 2N4123 is a general-purpose *npn* transistor with the casing and terminal identification appearing in the top-right corner of Fig. 3.23a. Most specification sheets are broken down into *Maximum ratings, thermal characteristics, and electrical characteristics*. The electrical characteristics are further broken down into ''on,'' ''off,'' and small-signal characteristics. The ''on'' and ''off'' characteristics refer to dc limits, while the small-signal characteristics include the parameters of importance to ac operation.

Note in the maximum rating list that $V_{CE_{max}} = V_{CEO} = 30$ V with $I_{C_{max}} = 200$ mA. The maximum collector dissipation $P_{C_{max}} = 625$ mW. The derating factor under the maximum rating specifies that the maximum rating must be decreased 5 mW for every 1° rise in temperature above 25°C. In the ''off'' characteristics I_{CBO} is specified as 50 nA and in the ''on'' characteristics $V_{CE_{sat}} = 0.3$ V. The level of h_{FE} has a range of 50 to 150 at $I_C = 2$ mA and $V_{CE} = 1$ V and a minimum value of 25 at a higher current of 50 mA at the same voltage.

The limits of operation have now been defined for the device and are repeated below in the format of Eq. (3.17) using $h_{FE} = 150$ (the upper limit) and $I_{CEO} \cong \beta I_{CBO} = (150)\,(50\text{ nA}) = 7.5\ \mu\text{A}$. Certainly, for many applications the 7.5 μA $= 0.0075$ mA can be considered to be 0 mA on an approximate basis.

Limits of Operation
$$7.5\ \mu\text{A} \leqq I_C \leqq 200\text{ mA}$$

$$0.3\text{ V} \leqq V_{CE} \leqq 30\text{ V}$$

$$V_{CE} I_C \leqq 650\text{ mW}$$

In the small-signal characteristics the level of h_{fe} (β_{ac}) is provided along with a plot of how it varies with collector current in Fig. 3.23f. In Fig. 3.23j the effect of temperature and collector current on the level of h_{FE} (β_{ac}) is demonstrated. At room temperature (25°C), note that h_{FE} (β_{dc}) is a maximum value of 1 in the neighborhood of about 8 mA. As I_C increased beyond this level, h_{FE} drops off to one-half the value with I_C equal to 50 mA. It also drops to this level if I_C decreases to the low level of 0.15 mA. Since this is a *normalized* curve, if we have a transistor with $\beta_{dc} = h_{FE} = 50$ at room temperature, the maximum value at 8 mA is 50. At $I_C = 50$ mA it has dropped to $50/2 = 25$. In other words, normalizing reveals that the actual level of h_{FE}

at any level of I_C has been divided by the maximum value of h_{FE} at that temperature and $I_C = 8$ mA. Note also that the horizontal scale of Fig. 3.23j is a log scale. Log scales are examined in depth in Chapter 11. You may want to look back at the plots of this section when you find time to review the first few sections of Chapter 11.

MAXIMUM RATINGS

Rating	Symbol	2N4123	Unit
Collector-Emitter Voltage	V_{CEO}	30	Vdc
Collector-Base Voltage	V_{CBO}	40	Vdc
Emitter-Base Voltage	V_{EBO}	5.0	Vdc
Collector Current – Continuous	I_C	200	mAdc
Total Device Dissipation @ $T_A = 25°C$ Derate above 25°C	P_D	625 5.0	mW mW°C
Operating and Storage Junction Temperature Range	T_J, T_{stg}	−55 to +150	°C

THERMAL CHARACTERISTICS

Characteristic	Symbol	Max	Unit
Thermal Resistance, Junction to Case	$R_{\theta JC}$	83.3	°C W
Thermal Resistance, Junction to Ambient	$R_{\theta JA}$	200	°C W

2N4123

CASE 29-04, STYLE 1
TO-92 (TO-226AA)

3 Collector

2
Base

1 Emitter

1
2 3

**GENERAL PURPOSE
TRANSISTOR**

NPN SILICON

ELECTRICAL CHARACTERISTICS ($T_A = 25°C$ unless otherwise noted)

Characteristic	Symbol	Min	Max	Unit
OFF CHARACTERISTICS				
Collector-Emitter Breakdown Voltage (1) ($I_C = 1.0$ mAdc, $I_E = 0$)	$V_{(BR)CEO}$	30		Vdc
Collector-Base Breakdown Voltage ($I_C = 10$ µAdc, $I_E = 0$)	$V_{(BR)CBO}$	40		Vdc
Emitter-Base Breakdown Voltage ($I_E = 10$ µAdc, $I_C = 0$)	$V_{(BR)EBO}$	5.0	–	Vdc
Collector Cutoff Current ($V_{CB} = 20$ Vdc, $I_E = 0$)	I_{CBO}	–	50	nAdc
Emitter Cutoff Current ($V_{BE} = 3.0$ Vdc, $I_C = 0$)	I_{EBO}	–	50	nAdc
ON CHARACTERISTICS				
DC Current Gain(1) ($I_C = 2.0$ mAdc, $V_{CE} = 1.0$ Vdc) ($I_C = 50$ mAdc, $V_{CE} = 1.0$ Vdc)	h_{FE}	50 25	150 –	–
Collector-Emitter Saturation Voltage(1) ($I_C = 50$ mAdc, $I_B = 5.0$ mAdc)	$V_{CE(sat)}$	–	0.3	Vdc
Base-Emitter Saturation Voltage(1) ($I_C = 50$ mAdc, $I_B = 5.0$ mAdc)	$V_{BE(sat)}$	–	0.95	Vdc
SMALL-SIGNAL CHARACTERISTICS				
Current-Gain – Bandwidth Product ($I_C = 10$ mAdc, $V_{CE} = 20$ Vdc, f = 100 MHz)	f_T	250	–	MHz
Output Capacitance ($V_{CB} = 5.0$ Vdc, $I_E = 0$, f = 100 MHz)	C_{obo}	–	4.0	pF
Input Capacitance ($V_{BE} = 0.5$ Vdc, $I_C = 0$, f = 100 kHz)	C_{ibo}	–	8.0	pF
Collector-Base Capacitance ($I_E = 0$, $V_{CB} = 5.0$ V, f = 100 kHz)	C_{cb}	–	4.0	pF
Small-Signal Current Gain ($I_C = 2.0$ mAdc, $V_{CE} = 10$ Vdc, f = 1.0 kHz)	h_{fe}	50	200	–
Current Gain – High Frequency ($I_C = 10$ mAdc, $V_{CE} = 20$ Vdc, f = 100 MHz) ($I_C = 2.0$ mAdc, $V_{CE} = 10$ V, f = 1.0 kHz)	h_{fe}	2.5 50	– 200	–
Noise Figure ($I_C = 100$ µAdc, $V_{CE} = 5.0$ Vdc, $R_S = 1.0$ k ohm, f = 1.0 kHz)	NF	–	6.0	dB

(1) Pulse Test: Pulse Width = 300 µs, Duty Cycle = 2.0%

(a)

Figure 3.23 Transistor specification sheet.

Before leaving this description of the characteristics, take note of the fact that the actual collector characteristics are not provided. In fact, most specification sheets as provided by the range of manufacturers fail to provide the full characteristics. It is expected that the data provided are sufficient to use the device effectively in the design process.

As noted in the introduction to this section, all the parameters of the specification sheet have not been defined in the preceding sections or chapters. However, the specification sheet provided in Fig. 3.23 will be referenced continually in the chapters to follow as parameters are introduced. The specification sheet can be a very valuable tool in the design or analysis mode, and every effort should be made to be aware of the importance of each parameter and how it may vary with changing levels of current, temperature, and so on.

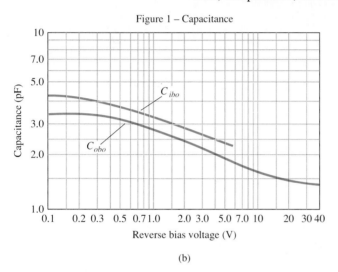

(b)

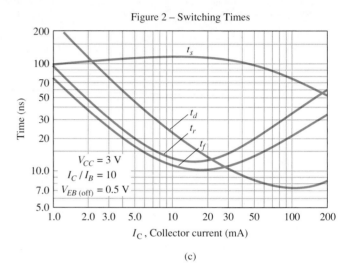

(c)

AUDIO SMALL SIGNAL CHARACTERISTICS

NOISE FIGURE

(V_{CE} = 5 Vdc, T_A = 25°C)
Bandwidth = 1.0 Hz

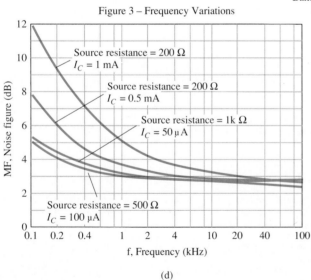

(d)

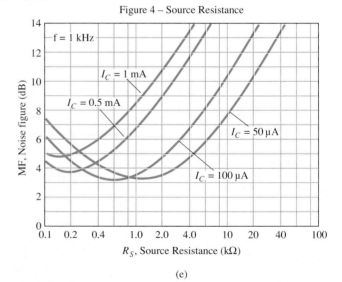

(e)

Figure 3.23 Continued.

h PARAMETERS

V_{CE} = 10 V, f = 1 kHz, T_A = 25°C

Figure 5 – Current Gain

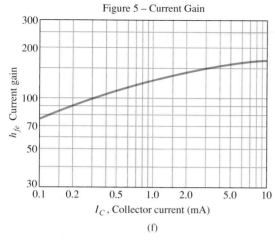

(f)

Figure 6 – Output Admittance

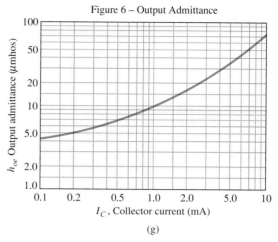

(g)

Figure 7 – Input Impedance

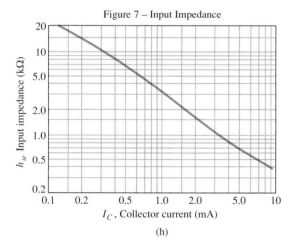

(h)

Figure 8 – Voltage Feedback Ratio

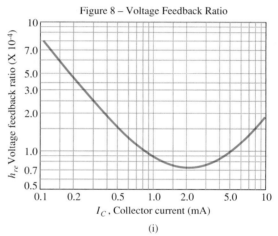

(i)

STATIC CHARACTERISTICS

Figure 9 – DC Current Gain

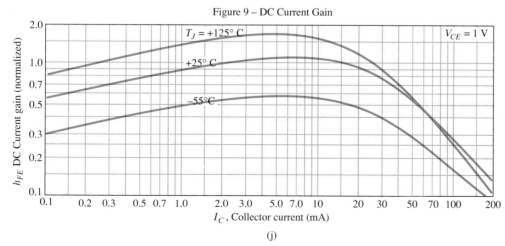

(j)

Figure 3.23 Continued.

3.10 TRANSISTOR TESTING

As with diodes, there are three routes one can take to check a transistor: *curve tracer, digital meter,* and *ohmmeter.*

Curve Tracer

The curve tracer of Fig. 1.45 will provide the display of Fig. 3.24 once all the controls have been properly set. The smaller displays to the right reveal the scaling to be applied to the characteristics. The vertical sensitivity is 2 mA/div, resulting in the scale shown to the left of the monitor's display. The horizontal sensitivity is 1 V/div, resulting in the scale shown below the characteristics. The step function reveals that the curves are separated by a difference of 10 μA, starting at 0 μA for the bottom curve. The last scale factor provided can be used to quickly determine the β_{ac} for any region of the characteristics. Simply multiply the displayed factor by the number of divisions between I_B curves in the region of interest. For instance, let us determine β_{ac} at a Q-point of $I_C = 7$ mA and $V_{CE} = 5$ V. In this region of the display, the distance between I_B curves is $\frac{9}{10}$ of a division, as indicated on Fig. 3.25. Using the factor specified, we find that

$$\beta_{ac} = \frac{9}{10} \text{ div} \left(\frac{200}{\text{div}} \right) = \mathbf{180}$$

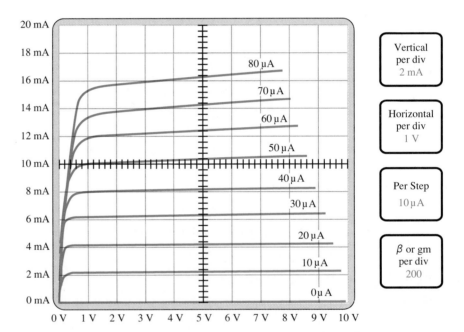

Figure 3.24 Curve tracer response to 2N3904 npn transistor.

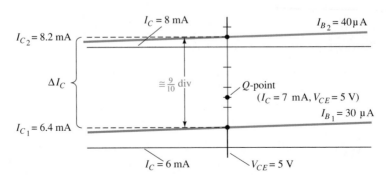

Figure 3.25 Determining β_{ac} for the transistor characteristics of Fig. 3.24 at $I_C = 7$ mA and $V_{CE} = 5$ V.

Chapter 3 Bipolar Junction Transistors

Using Eq. (3.11) gives us

$$\beta_{ac} = \frac{\Delta I_C}{\Delta I_B}\bigg|_{V_{CE}=\text{constant}} = \frac{I_{C_2} - I_{C_1}}{I_{B_2} - I_{B_1}} = \frac{8.2 \text{ mA} - 6.4 \text{ mA}}{40 \text{ } \mu\text{A} - 30 \text{ } \mu\text{A}}$$

$$= \frac{1.8 \text{ mA}}{10 \text{ } \mu\text{A}} = \mathbf{180}$$

verifying the determination above.

Advanced Digital Meters

Advanced digital meters such as that shown in Fig. 3.26 are now available that can provide the level of h_{FE} using the lead sockets appearing at the bottom left of the dial. Note the choice of *pnp* or *npn* and the availability of two emitter connections which used to handle the sequence of leads as connected to the casing. The level of h_{FE} is determined at a collector current of 2 mA for the Testmate 175A which is also provided on the digital display. Note that this versatile instrument can also check a diode. It can measure capacitance and frequency in addition to the normal functions of voltage, current, and resistance measurements.

In fact, in the diode testing mode it can be used to check the *p-n* junctions of a transistor. With the collector open the base-to-emitter junction should result in a low voltage of about 0.7 V with the red (positive) lead connected to the base and the black (negative) lead connected to the emitter. A reversal of the leads should result in an O.L. indication to represent the reverse-biased junction. Similarly, with the emitter open, the forward- and reverse-bias states of the base-to-collector junction can be checked.

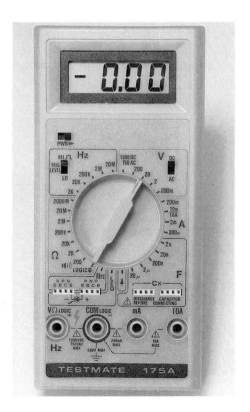

Figure 3.26 Transistor tester. (Courtesy Computronics Technology, Inc.)

β

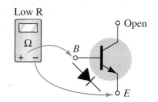

Figure 3.27 Checking the forward-biased base-to-emitter junction of an *npn* transistor.

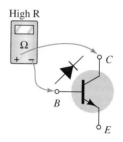

Figure 3.28 Checking the reverse-biased base-to-collector junction of an *npn* transistor.

Ohmmeter

An ohmmeter or the resistance scales of a DMM can be used to check the state of a transistor. Recall that for a transistor in the active region the base-to-emitter junction is forward-biased and the base-to-collector junction is reverse-biased. Essentially, therefore, the forward-biased junction should register a relatively low resistance while the reverse-biased junction a much higher resistance. For an *npn* transistor, the forward-biased junction (biased by the internal supply in the resistance mode) from base to emitter should be checked as shown in Fig. 3.27 and result in a reading that will typically fall in the range of 100Ω to a few kilohms. The reverse-biased base-to-collector junction (again reverse-biased by the internal supply) should be checked as shown in Fig. 3.28 with a reading typically exceeding 100 kΩ. For a *pnp* transistor the leads are reversed for each junction. Obviously, a large or small resistance in both directions (reversing the leads) for either junction of an *npn* or *pnp* transistor indicates a faulty device.

If both junctions of a transistor result in the expected readings the type of transistor can also be determined by simply noting the polarity of the leads as applied to the base-emitter junction. If the positive (+) lead is connected to the base and the negative lead (−) to the emitter a low resistance reading would indicate an *npn* transistor. A high resistance reading would indicate a *pnp* transistor. Although an ohmmeter can also be used to determine the leads (base, collector and emitter) of a transistor it is assumed that this determination can be made by simply looking at the orientation of the leads on the casing.

3.11 TRANSISTOR CASING AND TERMINAL IDENTIFICATION

After the transistor has been manufactured using one of the techniques described in Chapter 12, leads of, typically, gold, aluminum, or nickel are then attached and the entire structure is encapsulated in a container such as that shown in Fig. 3.29. Those with the heavy duty construction are high-power devices, while those with the small can (top hat) or plastic body are low- to medium-power devices.

Whenever possible, the transistor casing will have some marking to indicate which leads are connected to the emitter, collector, or base of a transistor. A few of the methods commonly used are indicated in Fig. 3.30.

(a)   (b)   (c)   (d)

Figure 3.29 Various types of transistors. [(a) Courtesy General Electric Company; (b) and (c) Courtesy of Motorola Inc.; (d) Courtesy International Rectifier Corporation.

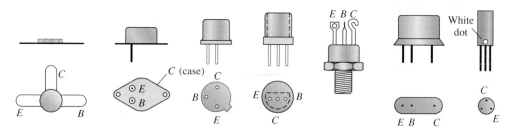

Figure 3.30 Transistor terminal identification.

The internal construction of a TO-92 package in the Fairchild line appears in Fig. 3.31. Note the very small size of the actual semiconductor device. There are gold bond wires, a copper frame, and an epoxy encapsulation.

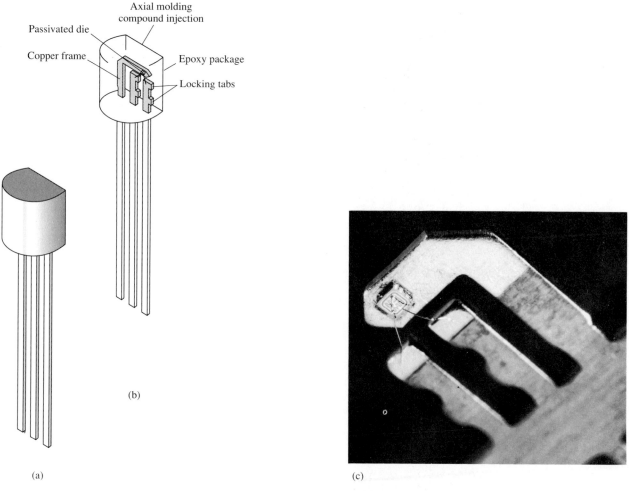

Figure 3.31 Internal construction of a Fairchild transistor in a TO-92 package. (Courtesy Fairchild Camera and Instrument Corporation.)

Four (quad) individual *pnp* silicon transistors can be housed in the 14-pin plastic dual-in-line package appearing in Fig. 3.32a. The internal pin connections appear in Fig. 3.32b. As with the diode IC package, the indentation in the top surface reveals the number 1 and 14 pins.

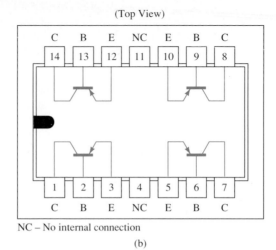

(Top View)

NC – No internal connection

Figure 3.32 Type Q2T2905 Texas Instruments quad *pnp* silicon transistors: (a) appearance; (b) pin connections. (Courtesy Texas Instruments Incorporated.)

(a)   (b)

3.12 COMPUTER ANALYSIS

In Chapter 4 a transistor network will be investigated using BASIC and PSpice. The approach using BASIC will parallel an analysis performed in the longhand fashion, while the PSpice analysis will employ a transistor model to be introduced in the next few paragraphs.

The PSpice statement for the introduction of transistor elements has the following format:

$$Q1 \quad 3 \quad 1 \quad 4 \quad QN$$

name   C   B   E   model name

The Q is required to identify the device as a transistor. The number 1 is the chosen name for the transistor, although it can include up to seven characters (numbers and letters). Next, the terminals are entered *in the order* appearing above. The last entry is the *model name,* to direct the software package (program) to the location of the parameters that define the transistor.

The model statement has the following format:

$$.\text{MODEL} \quad QN \quad NPN \; (BF = 140 \; IS = 2E-15)$$

model   type   parameters to be specified
name

As indicated, the statement must begin with .MODEL and be followed by the transistor's model name as specified in the earlier statement. Next, the type of transistor is indicated and the parameter values to be specified are included within the parentheses. The parameter list as appearing in the PSpice manual is quite extensive and in fact includes 40 terms. For our present needs there are only two parameters that need to be specified. They include the value of Beta, referred to as BF, and the reverse saturation current IS at a level to result in a base-to-emitter voltage of about 0.7 V when the device is "on."

Both statements listed above will appear in the analysis provided in the computer analysis section of Chapter 4. In fact, they will be the only statements different from those appearing in the diode analysis of Chapter 2. In other words, new elements can be introduced into the PSpice library without modifying the procedures already described. In this sense of use of the PSpice package is a true "building experience" with the possibility of analyzing some very complicated networks only a few statements away.

§ 3.2 Transistor Construction

1. What names are applied to the two types of BJT transistors? Sketch the basic construction of each and label the various minority and majority carriers in each. Draw the graphic symbol next to each. Is any of this information altered by changing from a silicon to a germanium base?

2. What is the major difference between a bipolar and a unipolar device?

§ 3.3 Transistor Operation

3. How must the two transistor junctions be biased for proper transistor amplifier operation?

4. What is the source of the leakage current in a transistor?

5. Sketch a figure similar to Fig. 3.3 for the forward-biased junction of an *npn* transistor. Describe the resulting carrier motion.

6. Sketch a figure similar to Fig. 3.4 for the reverse-biased junction of an *npn* transistor. Describe the resulting carrier motion.

7. Sketch a figure similar to Fig. 3.5 for the majority- and minority-carrier flow of an *npn* transistor. Describe the resulting carrier motion.

8. Which of the transistor currents is always the largest? Which is always the smallest? Which two currents are relatively close in magnitude?

9. If the emitter current of a transistor is 8 mA and I_B is 1/100 of I_C, determine the levels of I_C and I_B.

§ 3.4 Common-Base Configuration

10. From memory, sketch the transistor symbol for a *pnp* and an *npn* transistor, and then insert the conventional flow direction for each current.

11. Using the characteristics of Fig. 3.7, determine V_{BE} at $I_E = 5$ mA for $V_{CB} = 1$ V, 10 V, and 20 V. Is it reasonable to assume on an approximate basis that V_{CB} has only a slight effect on the relationship between V_{BE} and I_E?

12. (a) Determine the average ac resistance for the characteristics of Fig. 3.10b.
    (b) For networks in which the magnitude of the resistive elements is typically in kilohms, is the approximation of Fig. 3.10c a valid one [based on the results of part (a)]?

13. (a) Using the characteristics of Fig. 3.8, determine the resulting collector current if $I_E = 4.5$ mA and $V_{CB} = 4$ V.
    (b) Repeat part (a) for $I_E = 4.5$ mA and $V_{CB} = 16$ V.
    (c) How have the changes in V_{CB} affected the resulting level of I_C?
    (d) On an approximate basis, how are I_E and I_C related based on the results above?

14. (a) Using the characteristics of Figs. 3.7 and 3.8, determine I_C if $V_{CB} = 10$ V and $V_{BE} = 800$ mV.
    (b) Determine V_{BE} if $I_C = 5$ mA and $V_{CB} = 10$ V.
    (c) Repeat part (b) using the characteristics of Fig. 3.10b.
    (d) Repeat part (b) using the characteristics of Fig. 3.10c.
    (e) Compare the solutions for V_{BE} for parts (b), (c), and (d). Can the difference be ignored if voltage levels greater than a few volts are typically encountered?

15. (a) Given an α_{dc} of 0.998, determine I_C if $I_E = 4$ mA.
    (b) Determine α_{dc} if $I_E = 2.8$ mA and $I_B = 20$ μA.
    (c) Find I_E if $I_B = 40$ μA and α_{dc} is 0.98.

16. From memory, and memory only, sketch the common-base BJT transistor configuration (for *npn* and *pnp*) and indicate the polarity of the applied bias and resulting current directions.

§ 3.5 Transistor Amplifying Action

17. Calculate the voltage gain ($A_v = V_L/V_i$) for the network of Fig. 3.12 if $V_i = 500$ mV and $R = 1$ kΩ. (The other circuit values remain the same.)

18. Calculate the voltage gain ($A_v = V_L/V_i$) for the network of Fig. 3.12 if the source has an internal resistance of 100 Ω in series with V_i.

§ 3.6 Common-Emitter Configuration

19. Define I_{CBO} and I_{CEO}. How are they different? How are they related? Are they typically close in magnitude?

20. Using the characteristics of Fig. 3.14:
(a) Find the value of I_C corresponding to $V_{BE} = +750$ mV and $V_{CE} = +5$ V.
(b) Find the value of V_{CE} and V_{BE} corresponding to $I_C = 3$ mA and $I_B = 30$ μA.

* 21. (a) For the common-emitter characteristics of Fig. 3.14, find the dc beta at an operating point of $V_{CE} = +8$ V and $I_C = 2$ mA.
(b) Find the value of α corresponding to this operating point.
(c) At $V_{CE} = +8$ V, find the corresponding value of I_{CEO}.
(d) Calculate the approximate value of I_{CBO} using the dc beta value obtained in part (a).

* 22. (a) Using the characteristics of Fig. 3.14a, determine I_{CEO} at $V_{CE} = 10$ V.
(b) Determine β_{dc} at $I_B = 10$ μA and $V_{CE} = 10$ V.
(c) Using the β_{dc} determined in part (b), calculate I_{CBO}.

23. (a) Using the characteristics of Fig. 3.14a, determine β_{dc} at $I_B = 80$ μA and $V_{CE} = 5$ V.
(b) Repeat part (a) at $I_B = 5$ μA and $V_{CE} = 15$ V.
(c) Repeat part (a) at $I_B = 30$ μA and $V_{CE} = 10$ V.
(d) Reviewing the results of parts (a) through (c), does the value of β_{dc} change from point to point on the characteristics? Where were the higher values found? Can you develop any general conclusions about the value of β_{dc} on a set of characteristics such as those provided in Fig. 3.14a?

* 24. (a) Using the characteristics of Fig. 3.14a, determine β_{ac} at $I_B = 80$ μA and $V_{CE} = 5$ V.
(b) Repeat part (a) at $I_B = 5$ μA and $V_{CE} = 15$ V.
(c) Repeat part (a) at $I_B = 30$ μA and $V_{CE} = 10$ V.
(d) Reviewing the results of parts (a) through (c), does the value of β_{ac} change from point to point on the characteristics? Where are the high values located? Can you develop any general conclusions about the value of β_{ac} on a set of collector characteristics?
(e) The chosen points in this exercise are the same as those employed in Problem 23. If Problem 23 was performed, compare the levels of β_{dc} and β_{ac} for each point and comment on the trend in magnitude for each quantity.

25. Using the characteristics of Fig. 3.14a, determine β_{dc} at $I_B = 25$ μA and $V_{CE} = 10$ V. Then calculate α_{dc} and the resulting level of I_E. (Use the level of I_C determined by $I_C = \beta_{dc}I_B$.)

26. (a) Given that $\alpha_{dc} = 0.987$, determine the corresponding value of β_{dc}.
(b) Given $\beta_{dc} = 120$, determine the corresponding value of α.
(c) Given that $\beta_{dc} = 180$ and $I_C = 2.0$ mA, find I_E and I_B.

27. From memory, and memory only, sketch the common-emitter configuration (for *npn* and *pnp*) and insert the proper biasing arrangement with the resulting current directions for I_B, I_C, and I_E.

§ 3.7 Common-Collector Configuration

28. An input voltage of 2 V rms (measured from base to ground) is applied to the circuit of Fig. 3.21. Assuming that the emitter voltage follows the base voltage exactly and that V_{be} (rms) = 0.1 V, calculate the circuit voltage amplification ($A_v = V_o/V_i$) and emitter current for $R_E = 1$ kΩ.

29. For a transistor having the characteristics of Fig. 3.14, sketch the input and output characteristics of the common-collector configuration.

§ 3.8 Limits of Operation

30. Determine the region of operation for a transistor having the characteristics of Fig. 3.14 if $I_{C_{max}} = 7$ mA, $V_{CE_{max}} = 17$ V, and $P_{C_{max}} = 40$ mW.

31. Determine the region of operation for a transistor having the characteristics of Fig. 3.8 if $I_{C_{max}} = 6$ mA, $V_{CB_{max}} = 15$ V, and $P_{C_{max}} = 30$ mW.

§ 3.9 Transistor Specification Sheet

32. Referring to Fig. 3.23, determine the temperature range for the device in degrees Fahrenheit.

33. Using the information provided in Fig. 3.23 regarding $P_{D_{max}}$, $V_{CE_{max}}$, $I_{C_{max}}$ and $V_{CE_{sat}}$, sketch the boundaries of operation for the device.

34. Based on the data of Fig. 3.23, what is the expected value of I_{CEO} using the average value of β_{dc}?

35. How does the range of h_{FE} (Fig. 3.23(j), normalized from $h_{FE} = 100$) compare with the range of h_{fe} (Fig. 3.23(f)) for the range of I_C from 0.1 mA to 10 mA?

36. Using the characteristics of Fig. 3.23b, determine whether the input capacitance in the common-base configuration increases or decreases with increasing levels of reverse bias potential. Can you explain why?

* **37.** Using the characteristics of Fig. 3.23f, determine how much the level of h_{fe} has changed from its value at 1mA to its value at 10 mA. Note that the vertical scale is a log scale that may require reference to Section 11.2. Is the change one that should be considered in a design situation?

* **38.** Using the characteristics of Fig. 3.23j, determine the level of β_{dc} at $I_C = 10$ mA at the three levels of temperature appearing in the figure. Is the change significant for the specified temperature range? Is it an element to be concerned about in the design process?

§ 3.10 Transistor Testing

39. (a) Using the characteristics of Fig. 3.24, determine β_{ac} at $I_C = 14$ mA and $V_{CE} = 3$ V.
(b) Determine β_{dc} at $I_C = 1$ mA and $V_{CE} = 8$ V.
(c) Determine β_{ac} at $I_C = 14$ mA and $V_{CE} = 3$ V.
(d) Determine β_{dc} at $I_C = 1$ mA and $V_{CE} = 8$ V.
(e) How does the level of β_{ac} and β_{dc} compare in each region?
(f) Is the approximation $\beta_{dc} \cong \beta_{ac}$ a valid one for this set of characteristics?

*Please note: Asterisks indicate more difficult problems.

4 DC Biasing—BJTs

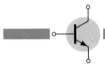

4.1 INTRODUCTION

The analysis or design of a transistor amplifier requires a knowledge of both the dc and ac response of the system. Too often it is assumed that the transistor is a magical device that can raise the level of the applied ac input without the assistance of an external energy source. In actuality, the improved output ac power level is the result of a transfer of energy from the applied dc supplies. The analysis or design of any electronic amplifier, therefore, has two components: the dc portion and the ac portion. Fortunately, the superposition theorem is applicable and the investigation of the dc conditions can be totally separated from the ac response. However, one must keep in mind that during the design or synthesis stage the choice of parameters for the required dc levels will affect the ac response, and vice versa.

The dc level of operation of a transistor is controlled by a number of factors, including the range of possible operating points on the device characteristics. In Section 4.2 we specify the range for the BJT amplifier. Once the desired dc current and voltage levels have been defined, a network must be constructed that will establish the desired operating point—a number of these networks are analyzed in this chapter. Each design will also determine the stability of the system. That is, how sensitive the system is to temperature variations—another topic to be investigated in a later section of this chapter.

Although a number of networks are analyzed in this chapter, there is an underlying similarity between the analysis of each configuration due to the recurring use of the following important basic relationships for a transistor:

$$V_{BE} = 0.7 \text{ V} \tag{4.1}$$

$$I_E = (\beta + 1)I_B \cong I_C \tag{4.2}$$

$$I_C = \beta I_B \tag{4.3}$$

In fact, once the analysis of the first few networks is clearly understood, the path toward the solution of the networks to follow will begin to become quite apparent. In most instances the base current I_B is the first quantity to be determined. Once I_B is known the relationships of Eqs. (4.1) through (4.3) can be applied to find the remaining quantities of interest. The similarities in analysis will be immediately obvious as we progress through the chapter. The equations for I_B are so similar for a number of configurations that one equation can be derived from another simply by dropping or adding a term or two. The primary function of this chapter is to develop a level of familiarity with the BJT transistor that would permit a dc analysis of any system that might employ the BJT amplifier.

4.2   OPERATING POINT

The term *biasing* appearing in the title of this chapter is an all-inclusive term for the application of dc voltages to establish a fixed level of current and voltage. For transistor amplifiers the resulting dc current and voltage establish an *operating point* on the characteristics that define the region that will be employed for amplification of the applied signal. Since the operating point is a fixed point on the characteristics it is also called the *quiescent point* (abbreviated Q-point). By definition, *quiescent* means quiet, still, inactive. Figure 4.1 shows a general output device characteristic with four operating points indicated. The biasing circuit can be designed to set the device operation at any of these points or others within the *active region*. The maximum ratings are indicated on the characteristics of Fig. 4.1 by a horizontal line for the maximum collector current $I_{C_{max}}$ and a vertical line at the maximum collector-to-emitter voltage $V_{CE_{max}}$. The maximum power constraint is defined by the curve $P_{C_{max}}$ in the same figure. At the lower end of the scales are the *cutoff region*, defined by $I_B \leq 0\ \mu A$, and the *saturation region*, defined by $V_{CE} \leq V_{CE_{sat}}$.

The BJT device could be biased to operate outside these maximum limits, but the result of such operation would be either a considerable shortening of the lifetime of the device or destruction of the device. Confining ourselves to the *active* region one can select many different operating areas or points. The chosen Q pt often depends on

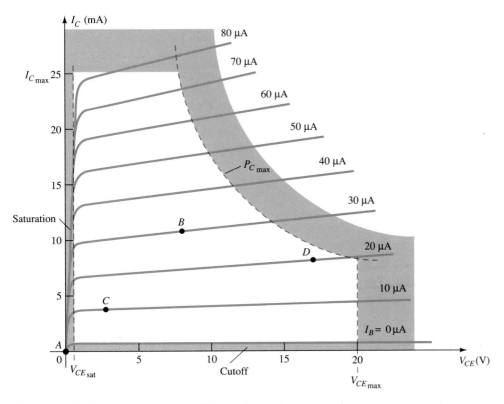

Figure 4.1   Various operating points within the limits of operation of a transistor.

the intended use of the circuit. Still, we can consider some differences between the various points shown in Fig. 4.1 to present some basic ideas about the operating point and, thereby, the bias circuit.

If no bias were used, the device would initially be completely off, resulting in a Q pt at A—namely, zero current through the device (and zero voltage across it). Since it is necessary to bias a device so that it can respond to the entire range of an input signal point A would not be suitable. For point B if a signal is applied to the circuit, the device will vary in current and voltage from operating point, allowing the device to react to (and possibly amplify) both the positive and negative excursions of the input signal. If, the input signal is properly chosen, the voltage and current of the device will vary but not enough to drive the device into *cutoff* or *saturation*. Point C would allow some positive and negative variation of the output signal but the peak-to-peak value would be limited by the proximity of $V_{CE} = 0V/I_C = 0$ mA. Operating at point C also raises some concern about the nonlinearities introduced by the fact that the spacing between I_B curves is rapidly changing in this region. In general, it is preferable to operate where the gain of the device is fairly constant (or linear) to insure that the amplification over the entire swing of input signal is the same. Point *B* is a region of more linear spacing and, therefore, more linear operation, as shown in Fig. 4.1. Point *D* sets the device operating point near the maximum voltage and power level. The output voltage swing in the positive direction is thus limited if the maximum voltage is not to be exceeded. Point *B*, therefore, seems the best operating point in terms of linear gain and largest possible voltage and current swing. This is usually the desired condition for small-signal amplifiers (Chapter 8) but not the case necessarily for power amplifiers which will be considered in Chapter 16. In this discussion, we will be concentrating primarily on biasing the transistor for *small-signal* amplification operation.

One other very important biasing factor must be considered. Having selected and biased the BJT at a desired operating point, the effect of temperature must also be taken into account. Temperature causes the device parameters such as the transistor current gain (β_{ac}) and the transistor leakage current (I_{CEO}) to change. Higher temperatures result in increased leakage currents in the device, thereby changing the operating condition set by the biasing network. The result is that the network design must also provide a degree of *temperature stability* so that temperature changes result in minimum changes in the operating point. This maintenance of operating point can be specified by a *stability factor, S,* which indicates the degree of change in operating-point due to a temperature variation. A highly stable circuit is desirable and the stability of a few basic bias circuits will be compared.

For the BJT to be biased in its linear or active operating region the following must be true:

1. The base–emitter junction *must* be forward-biased (*p*-region voltage more *positive*) with a resulting forward-bias voltage of about 0.6 to 0.7 V.

2. The base–collector junction *must* be reverse-biased (*n*-region more *positive*), with the reverse-bias voltage being any value within the maximum limits of the device.

[Note that for forward bias the voltage across the *p-n* junction is *p-positive*, while for reverse bias it is opposite (reverse) with *n-positive*. This emphasis on the initial letter should provide a means of helping memorize the necessary voltage polarity.]

Operation in the cutoff, saturation, and linear regions of the BJT characteristic are provided as follows:

1. *Linear-region operation:*
   Base–emitter junction forward biased
   Base–collector junction reverse biased

2. *Cutoff-region operation:*
   Base–emitter junction reverse biased

3. *Saturation-region operation:*
   Base–emitter junction forward biased
   Base–collector junction forward biased

4.3 FIXED-BIAS CIRCUIT

The fixed-bias circuit of Fig. 4.2 provides a relatively straightforward and simple introduction to transistor dc bias analysis. Even though the network employs an *npn* transistor, the equations and calculations apply equally well to a *pnp* transistor configuration merely by changing all current directions and voltage polarities. The current directions of Fig. 4.2 are the actual current directions and the voltages are defined by the standard double-subscript notation. For the dc analysis the network can be isolated from the indicated ac levels by replacing the capacitors by an open-circuit equivalent. In addition, the dc supply V_{CC} can be separated into two supplies (for analysis purposes only) as shown in Fig. 4.3 to permit a separation of input and output circuits. It also reduces the linkage between the two to the base current I_B. The separation is certainly valid, as we note in Fig. 4.3 that V_{CC} is connected directly to R_B and R_C just as in Fig. 4.2.

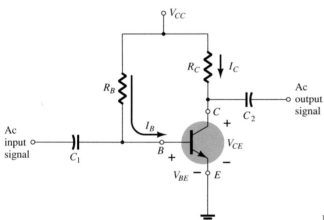

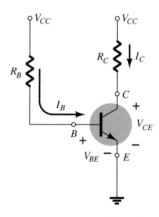

Figure 4.2   Fixed-bias circuit.

Figure 4.3   Dc equivalent of Fig. 4.2.

Forward Bias of Base–Emitter

Consider first the base–emitter circuit loop of Fig. 4.4. Writing Kirchhoff's voltage equation in the clockwise direction for the loop, we obtain

$$+V_{CC} - I_B R_B - V_{BE} = 0$$

Note the polarity of the voltage drop across R_B as established by the indicated direction of I_B. Solving the equation for the current I_B will result in the following:

$$\boxed{I_B = \frac{V_{CC} - V_{BE}}{R_B}} \qquad (4.4)$$

Equation (4.4) is certainly not a difficult one to remember if one simply keeps in mind that the base current is the current through R_B and by Ohm's law that current is the voltage across R_B divided by the resistance R_B. The voltage across R_B is the applied voltage V_{CC} at one end less the drop across the base-to-emitter junction (V_{BE}).

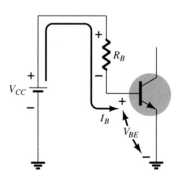

Figure 4.4   Base–emitter loop.

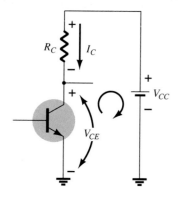

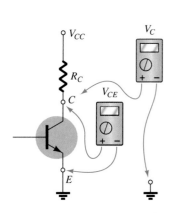

R_C I_C

$+$ V_{CC} $-$

V_{CE}

Figure 4.5 Collector–emitter loop.

V_{CC} V_C

R_C

V_{CE}

C

E

Figure 4.6 Measuring V_{CE} and V_C.

In addition, since the supply voltage V_{CC} and the base–emitter voltage V_{BE} are constants, the selection of a base resistor, R_B, sets the level of base current for the operating point.

Collector–Emitter Loop

The collector–emitter section of the network appears in Fig. 4.5 with the indicated direction of current I_C and the resulting polarity across R_C. The magnitude of the collector current is related directly to I_B through

$$I_C = \beta I_B \qquad (4.5)$$

It is interesting to note that since the base current is controlled by the level of R_B and I_C is related to I_B by a constant β the magnitude of I_C is not a function of the resistance R_C. Change R_C to any level and it will not affect the level of I_B or I_C as long as we remain in the active region of the device. However, as we shall see, the level of R_C will determine the magnitude of V_{CE}, which is an important parameter.

Applying Kirchhoff's voltage law in the clockwise direction around the indicated closed loop of Fig. 4.5 will result in the following:

$$V_{CE} + I_C R_C - V_{CC} = 0$$

and

$$V_{CE} = V_{CC} - I_C R_C \qquad (4.6)$$

which states in words that the voltage across the collector-emitter region of a transistor in the fixed-bias configuration is the supply voltage less the drop across R_C.

As a brief review of single- and double-subscript notation recall that

$$V_{CE} = V_C - V_E \qquad (4.7)$$

where V_{CE} is the voltage from collector to emitter and V_C and V_E are the voltages from collector and emitter to ground respectively. But *in this case* since $V_E = 0$ V, we have

$$V_{CE} = V_C \qquad (4.8)$$

In addition, since

$$V_{BE} = V_B - V_E \qquad (4.9)$$

and $V_E = 0$ V, then

$$V_{BE} = V_B \qquad (4.10)$$

Keep in mind that voltage levels such as V_{CE} are determined by placing the red (positive) lead of the voltmeter at the collector terminal with the black (negative) lead at the emitter terminal as shown in Fig. 4.6. V_C is the voltage from collector to ground and is measured as shown in the same figure. In this case the two readings are identical, but in the networks to follow the two can be quite different. Clearly understanding the difference between the two measurements can prove to be quite important in the troubleshooting of transistor networks.

EXAMPLE 4.1

Determine the following for the fixed-bias configuration of Fig. 4.7.
(a) I_{B_Q} and I_{C_Q}.
(b) V_{CE_Q}.
(c) V_B and V_C.
(d) V_{BC}.

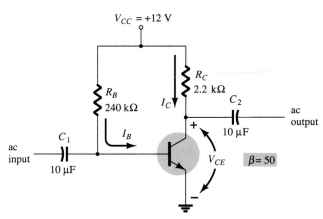

Figure 4.7 Dc fixed-bias circuit for Example 4.1.

Solution

(a) Eq. (4.4): $I_{B_Q} = \dfrac{V_{CC} - V_{BE}}{R_B} = \dfrac{12 \text{ V} - 0.7 \text{ V}}{240 \text{ k}\Omega} = \mathbf{47.08 \mu A}$

Eq. (4.5): $I_{C_Q} = \beta I_{B_Q} = (50)(47.08 \; \mu A) = \mathbf{2.35 \text{ mA}}$

(b) Eq. (4.6): $V_{CE_Q} = V_{CC} - I_C R_C$

$$= 12 \text{ V} - (2.35 \text{ mA})(2.2 \text{ k}\Omega)$$

$$= \mathbf{6.83 \text{ V}}$$

(c) $V_B = V_{BE} = \mathbf{0.7 \text{ V}}$

$V_C = V_{CE} = \mathbf{6.83 \text{ V}}$

(d) Using double-subscript notation yields

$$V_{BC} = V_B - V_C = 0.7 \text{ V} - 6.83 \text{ V}$$

$$= \mathbf{-6.13 \text{ V}}$$

with the negative sign revealing that the junction is reversed-biased, as it should be for linear amplification.

Transistor Saturation

The term *saturation* is applied to any system where levels have reached their maximum values. A saturated sponge is one that cannot hold another drop of liquid. For a transistor operating in the saturation region the current is a maximum value *for the particular design*. Change the design and the corresponding saturation level may rise or drop. Of course, the highest saturation level is defined by the maximum collector current as provided by the specification sheet.

Saturation conditions are normally avoided because the base-collector junction is no longer reverse-biased and the output amplified signal will be distorted. An operating point in the saturation region is depicted in Fig. 4.8a. Note that it is in a region where the characteristic curves join and the collector-to-emitter voltage is at or below $V_{CE_{sat}}$. In addition, the collector current is relatively high on the characteristics.

If we approximate the curves of Fig. 4.8a by those appearing in Fig. 4.8b, a quick direct method for determining the saturation level becomes apparent. In Fig. 4.8b the current is relatively high and the voltage V_{CE} is assumed to be zero volts. Applying Ohm's law the resistance between collector and emitter terminals can be determined as follows:

$$R_{CE} = \frac{V_{CE}}{I_C} = \frac{0 \text{ V}}{I_{C_{sat}}} = 0 \; \Omega$$

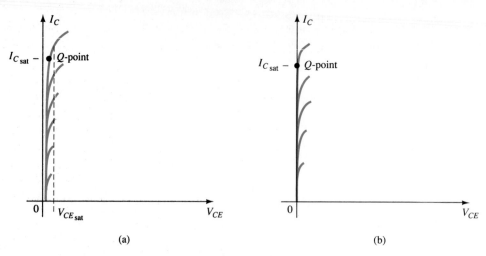

Figure 4.8 Saturation region (a) actual (b) approximate.

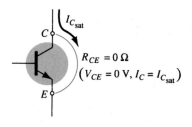

Figure 4.9 Determining $I_{C_{sat}}$.

Applying the results to the network schematic would result in the configuration of Fig. 4.9.

For the future, therefore, if there were an immediate need to know the approximate maximum collector current (saturation level) for a particular design, simply insert a short-circuit equivalent between collector and emitter of the transistor and calculate the resulting collector current. In short, set $V_{CE} = 0$V. For the fixed-bias configuration of Fig. 4.10 the short circuit has been applied, causing the voltage across R_C to be the applied voltage V_{CC}. The resulting saturation current for the fixed-bias configuration:

$$I_{C_{sat}} = \frac{V_{CC}}{R_C} \qquad (4.11)$$

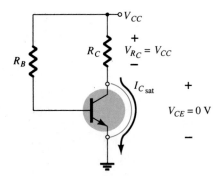

Figure 4.10 Determining $I_{C_{sat}}$ for the fixed-bias configuration.

Once $I_{C_{sat}}$ is known we have some idea of the maximum possible collector current for the chosen design and the level to stay below if we expect linear amplification.

EXAMPLE 4.2

Determine the saturation level for the network of Fig. 4.7.

Solution

$$I_{C_{sat}} = \frac{V_{CC}}{R_C} = \frac{12 \text{ V}}{2.2 \text{ k}\Omega} = \textbf{5.45 mA}$$

Chapter 4   DC Biasing—BJTs

The design of Example 4.1 resulted in $I_{C_Q} = 2.35$ mA, which is far from the saturation level and about one-half the maximum value for the design.

Load-Line Analysis

The analysis thus far has been performed using a level of β corresponding with the resulting Q-point. We will now investigate how the network parameters define the possible range of Q-points and how the actual Q-point is determined. The network of Fig. 4.11a establishes an output equation that relates the variables I_C and V_{CE} in the following manner:

$$V_{CE} = V_{CC} - I_C R_C \qquad (4.12)$$

The output characteristics of the transistor also relate the same two variables I_C and V_{CE} as shown in Fig. 4.11b.

In essence, therefore, we have a network equation and a set of characteristics that employ the same variables. The common solution of the two occurs where the constraints established by each are satisfied simultaneously. In other words, this is similar to finding the solution of two simultaneous equations: one established by the network and the other by the device characteristics.

The device characteristics of I_C versus V_{CE} are provided in Fig. 4.11b. We must now superimpose the straight line defined by Eq. 4.12 on the characteristics. The most direct method of plotting Eq. (4.12) on the output characteristics is to use the fact that a straight line is defined by two points. If we *choose* I_C to be 0 mA, we are specifying the horizontal axis as the line on which one point is located. By substituting $I_C = 0$ mA into Eq. (4.12), we find that

$$V_{CE} = V_{CC} - (0)R_C$$

and

$$V_{CE} = V_{CC}|_{I_C=0 \text{ mA}} \qquad (4.13)$$

defining one point for the straight line as shown in Fig. 4.12.

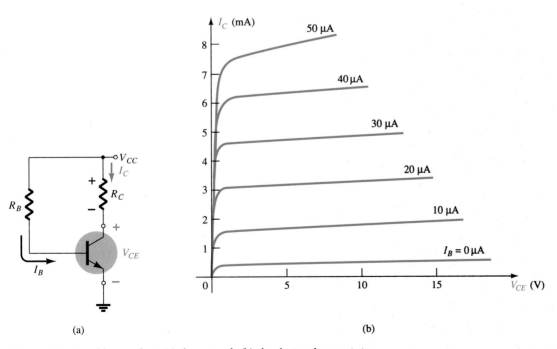

(a)

(b)

Figure 4.11 Load-line analysis (a) the network (b) the device characteristics.

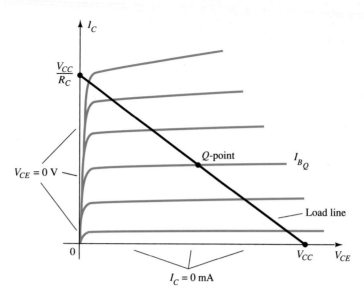

Figure 4.12 Fixed-bias load line.

If we now *choose* V_{CE} to be 0 V, which establishes the vertical axis as the line on which the second point will be defined, we find that I_C is determined by the following equation:

$$0 = V_{CC} - I_C R_C$$

and

$$I_C = \frac{V_{CC}}{R_C}\bigg|_{V_{CE}=0\text{ V}} \tag{4.14}$$

as appearing on Fig. 4.12.

By joining the two points defined by Eqs. (4.13) and (4.14) the straight line established by Eq. (4.12) can be drawn. The resulting line on the graph of Fig. 4.12 is called the *load line* since it is defined by the load resistor R_C. By solving for the resulting level of I_B the actual Q-point can be established as shown in Fig. 4.12.

If the level of I_B is changed by varying the value of R_B the Q-point moves up or down the load line as shown in Fig. 4.13. If V_{CC} is held fixed and R_C changed, the load line will shift as shown in Fig. 4.14. If I_B is held fixed, the Q-point will move as shown in the same figure. If R_C is fixed and V_{CC} varied, the load line shifts as shown in Fig. 4.15.

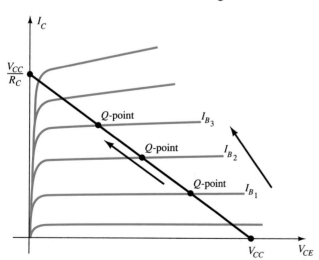

Figure 4.13 Movement of Q-point with increasing levels of I_B.

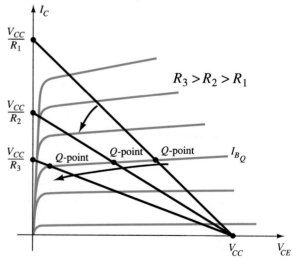

Figure 4.14 Effect of increasing levels of R_C on the load line and Q-point.

Chapter 4 DC Biasing—BJTs

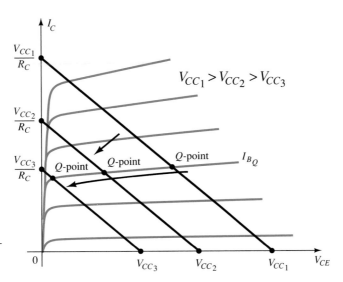

Figure 4.15 Effect of lower values of V_{CC} on the load line and Q-point.

Given the load line of Fig. 4.16 and the defined Q-point, determine the required values of V_{CC}, R_C, and R_B for a fixed-bias configuration.

EXAMPLE 4.3

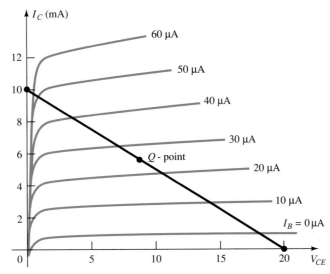

Figure 4.16 Example 4.3

Solution

From Fig. 4.16,

$$V_{CE} = V_{CC} = \mathbf{20\ V} \text{ at } I_C = 0 \text{ mA}$$

$$I_C = \frac{V_{CC}}{R_C} \text{ at } V_{CE} = 0 \text{ V}$$

and

$$R_C = \frac{V_{CC}}{I_C} = \frac{20\ V}{10\ mA} = \mathbf{2\ k\Omega}$$

$$I_B = \frac{V_{CC} - V_{BE}}{R_B}$$

and

$$R_B = \frac{V_{CC} - V_{BE}}{I_B} = \frac{20\ V - 0.7\ V}{25\ \mu A} = \mathbf{772\ k\Omega}$$

4.4 EMITTER-STABILIZED BIAS CIRCUIT

The dc bias network of Fig. 4.17 contains an emitter resistor to improve the stability level over that of the fixed-bias configuration. The improved stability will be demonstrated through a numerical example later in the section. The analysis will be performed by first examining the base–emitter loop and then using the results to investigate the collector–emitter loop.

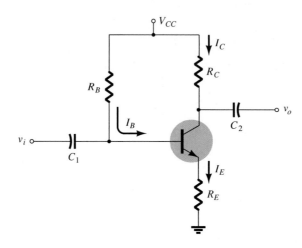

Figure 4.17 BJT bias circuit with emitter resistor.

Base–Emitter Loop

The base–emitter loop of the network of Fig. 4.17 can be redrawn as shown in Fig. 4.18. Writing Kirchhoff's voltage law around the indicated loop in the clockwise direction will result in the following equation:

$$+V_{CC} - I_B R_B - V_{BE} - I_E R_E = 0 \qquad (4.15)$$

Recall from Chapter 3 that

$$I_E = (\beta + 1)I_B \qquad (4.16)$$

Substituting for I_E in Eq. (4.15) will result in

$$V_{CC} - I_B R_B - V_{BE} - (\beta + 1)I_B R_E = 0$$

Grouping terms will then provide the following:

$$-I_B(R_B + (\beta + 1)R_E) + V_{CC} - V_{BE} = 0$$

Multiplying through by (-1) we have

$$I_B(R_B + (\beta + 1)R_E) - V_{CC} + V_{BE} = 0$$

with

$$I_B(R_B + (\beta + 1)R_E) = V_{CC} - V_{BE}$$

and solving for I_B gives

$$\boxed{I_B = \frac{V_{CC} - V_{BE}}{R_B + (\beta + 1)R_E}} \qquad (4.17)$$

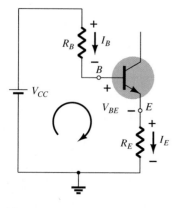

Figure 4.18 Base–emitter loop.

Note that the only difference between this equation for I_B and that obtained for the fixed-bias configuration is the term $(\beta + 1)R_E$.

There is an interesting result that can be derived from Eq. (4.17) if the equation is used to sketch a series network that would result in the same equation. Such is the

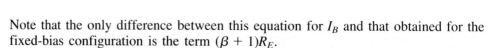

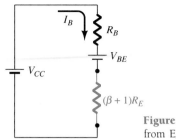

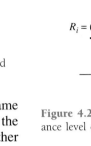

Figure 4.19 Network derived from E$_q$. (4.17).

Figure 4.20 Reflected impedance level of R_E.

case for the network of Fig. 4.19. Solving for the current I_B will result in the same equation obtained above. Note that aside from the base-to-emitter voltage V_{BE} the resistor R_E is *reflected* back to the input base circuit by a factor $(\beta + 1)$. In other words, the emitter resistor, which is part of the collector–emitter loop, "appears as" $(\beta + 1)R_E$ in the base–emitter loop. Since β is typically 50 or more, the emitter resistor appears to be a great deal larger in the base circuit. In general, therefore, for the configuration of Fig. 4.20,

$$\boxed{R_i = (\beta + 1)R_E} \tag{4.18}$$

Equation (4.18) is one that will prove useful in the analysis to follow. In fact, it provides a fairly easy way to remember Eq. (4.17). Using Ohm's law, we know that the current through a system is the voltage divided by the resistance of the circuit. For the base–emitter circuit the net voltage is $V_{CC} - V_{BE}$. The resistance levels are R_B plus R_E reflected by $(\beta + 1)$. The result is Eq. (4.17).

Collector–Emitter Loop

The collector–emitter loop is redrawn in Fig. 4.21. Writing Kirchhoff's voltage law for the indicated loop in the clockwise direction will result in

$$+I_E R_E + V_{CE} + I_C R_C - V_{CC} = 0$$

Substituting $I_E \cong I_C$ and grouping terms gives

$$V_{CE} - V_{CC} + I_C(R_C + R_E) = 0$$

and

$$\boxed{V_{CE} = V_{CC} - I_C(R_C + R_E) = 0} \tag{4.19}$$

The single-subscript voltage V_E is the voltage from emitter to ground and is determined by

$$\boxed{V_E = I_E R_E} \tag{4.20}$$

while the voltage from collector to ground can be determined from

$$V_{CE} = V_C - V_E$$

and

$$\boxed{V_C = V_{CE} + V_E} \tag{4.21}$$

or

$$\boxed{V_C = V_{CC} - I_C R_C} \tag{4.22}$$

The voltage at the base with respect to ground can be determined from

$$\boxed{V_B = V_{CC} - I_B R_B} \tag{4.23}$$

or

$$\boxed{V_B = V_{BE} + V_E} \tag{4.24}$$

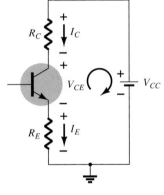

Figure 4.21 Collector–emitter loop.

EXAMPLE 4.4

For the emitter bias network of Fig. 4.22, determine:
(a) I_B.
(b) I_C.
(c) V_{CE}.
(d) V_C.
(e) V_E.
(f) V_B.
(g) V_{BC}.

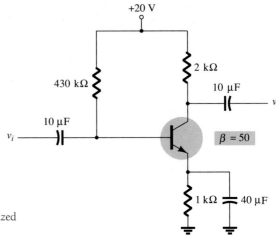

Figure 4.22 Emitter-stabilized bias circuit for Example 4.4.

Solution

(a) Eq. (4.17): $I_B = \dfrac{V_{CC} - V_{BE}}{R_B + (\beta + 1)R_E} = \dfrac{20\ \text{V} - 0.7\ \text{V}}{430\ \text{k}\Omega + (51)(1\ \text{k}\Omega)}$

$= \dfrac{19.3\ \text{V}}{481\ \text{k}\Omega} = \mathbf{40.1\ \mu A}$

(b) $I_C = \beta I_B$

$= (50)(40.1\ \mu A)$

$\cong \mathbf{2.01\ mA}$

(c) Eq. (4.19): $V_{CE} = V_{CC} - I_C(R_C + R_E)$

$= 20\ \text{V} - (2.01\ \text{mA})(2\ \text{k}\Omega + 1\ \text{k}\Omega) = 20\ \text{V} - 6.03\ \text{V}$

$= \mathbf{13.97\ V}$

(d) $V_C = V_{CC} - I_C R_C$

$= 20\ \text{V} - (2.01\ \text{mA})(2\ \text{k}\Omega) = 20\ \text{V} - 4.02\ \text{V}$

$= \mathbf{15.98\ V}$

(e) $V_E = V_C - V_{CE}$

$= 15.98\ \text{V} - 13.97\ \text{V}$

$= \mathbf{2.01\ V}$

or $V_E = I_E R_E \cong I_C R_E$

$= (2.01\ \text{mA})(1\ \text{k}\Omega)$

$= \mathbf{2.01\ V}$

(f) $V_B = V_{BE} + V_E$

$= 0.7\ \text{V} + 2.01\ \text{V}$

$= \mathbf{2.71\ V}$

(g) $V_{BC} = V_B - V_C$

$= 2.71\ \text{V} - 15.98\ \text{V}$

$= \mathbf{-13.27\ V}$ (reverse-biased as required)

Improved Bias Stability

The addition of the emitter resistor to the dc bias of the BJT provides improved stability; that is, the dc bias currents and voltages remain closer to where they were set by the circuit when outside conditions, such as temperature, and transistor beta, change. While a mathematical analysis is provided in Section 4.12, some comparison of the improvement can be obtained as demonstrated by Example 4.5.

Prepare a table and compare the bias voltage and currents of the circuits of Figs. 4.7 and Fig. 4.22 for the given value of $\beta = 50$ and for a new value of $\beta = 100$. Compare the changes in I_C and V_{CE} for same increase in β.

EXAMPLE 4.5

Solution

Using the results calculated in Example 4.1 and then repeating for a value of $\beta = 100$ yields the following:

β	I_B (μA)	I_C (mA)	V_{CE} (V)
50	47.08	2.35	6.83
100	47.08	4.71	1.64

The BJT collector current is seen to change by 100% due to the 100% change in the value of β. I_B is the same and V_{CE} decreased by 76%.

Using the results calculated in Example 4.4 and then repeating for a value of $\beta = 100$, we have the following:

β	I_B (μA)	I_C (mA)	V_{CE} (V)
50	40.1	2.01	13.97
100	36.3	3.63	9.11

Now the BJT collector current increases by about 81% due to the 100% increase in β. Notice that I_B decreased, helping maintain the value of I_C—or at least reducing the overall change in I_C due to the change in β. The change in V_{CE} has dropped to about 35%. The network of Fig. 4.22 is therefore more stable than that of Fig. 4.7 for the same change in β.

Saturation Level

The collector saturation level or maximum collector current for an emitter-bias design can be determined using the same approach applied to the fixed-bias configuration: Apply a short circuit between the collector–emitter terminals as shown in Fig. 4.23 and calculate the resulting collector current. For Fig. 4.23:

$$I_{C_{sat}} = \frac{V_{CC}}{R_C + R_E} \qquad (4.25)$$

The addition of the emitter resistor reduces the collector saturation level below that obtained with a fixed-bias configuration using the same collector resistor.

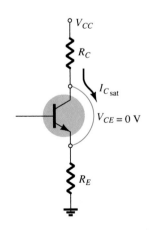

Figure 4.23 Determining $I_{C_{sat}}$ for the emitter–stabilized bias circuit.

EXAMPLE 4.6

Determine the saturation current for the network of Example 4.4.

Solution

$$I_{C_{sat}} = \frac{V_{CC}}{R_C + R_E}$$

$$= \frac{20 \text{ V}}{2 \text{ k}\Omega + 1 \text{ k}\Omega} = \frac{20 \text{ V}}{3 \text{ k}\Omega}$$

$$= \mathbf{6.67 \text{ mA}}$$

which is about twice the level of I_{C_Q} for Example 4.4.

Load-Line Analysis

The load-line analysis of the emitter-bias network is only slightly different from that encountered for the fixed-bias configuration. The level of I_B as determined by Eq. (4.17) defines the level of I_B on the characteristics of Fig. 4.24 (denoted I_{B_Q}).

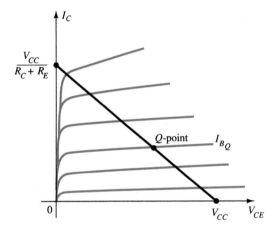

Figure 4.24 Load line for the emitter bias configuration.

The collector–emitter loop equation that defines the load line is the following:

$$V_{CE} = V_{CC} - I_C(R_C + R_E)$$

Choosing $I_C = 0$ mA gives

$$\boxed{V_{CE} = V_{CC}|_{I_C=0 \text{ mA}}} \tag{4.26}$$

as obtained for the fixed-bias configuration. Choosing $V_{CE} = 0$ V gives

$$\boxed{I_C = \frac{V_{CC}}{R_C + R_E}\bigg|_{V_{CE}=0 \text{ V}}} \tag{4.27}$$

as shown in Fig. 4.24. Different levels of I_{B_Q} will, of course, move the Q-point up or down the load line.

4.5 VOLTAGE-DIVIDER BIAS

In the previous bias configurations the bias current I_{C_Q} and voltage V_{CE_Q} were a function of the current gain (β) of the transistor. However, since β is temperature sensitive, especially for silicon transistors, and the actual value of beta is usually not well defined, it would be desirable to develop a bias circuit that is less dependent, or

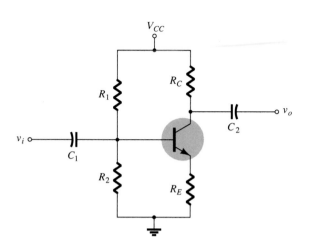

Figure 4.25 Voltage-divider bias configuration.

Figure 4.26 Defining the Q point for the voltage-divider bias configuration.

in fact, independent of the transistor beta. The voltage-divider bias configuration of Fig. 4.25 is such a network. If analyzed on an exact basis the sensitivity to changes in beta is quite small. If the circuit parameters are properly chosen, the resulting levels of I_{C_Q} and V_{CE_Q} can be almost totally independent of beta. Recall from previous discussions that a Q-point is defined by a fixed level of I_{C_Q} and V_{CE_Q} as shown in Fig. 4.26. The level of I_{B_Q} will change with the change in beta, but the operating point on the characteristics defined by I_{C_Q} and V_{CE_Q} can remain fixed if the proper circuit parameters are employed.

As noted above, there are two methods that can be applied to analyze the voltage-divider configuration. The reason for the choice of names for this configuration will become obvious in the analysis to follow. The first to be demonstrated is the *exact method* that can be applied to *any* voltage-divider configuration. The second is referred to as the *approximate method* and can be applied only if specific conditions are satisfied. The approximate approach permits a more direct analysis with a savings in time and energy. It is also particularly helpful in the design mode to be described in a later section. All in all, the approximate approach can be applied to the majority of situations and therefore should be examined with the same interest as the exact method.

Exact Analysis

The input side of the network of Fig. 4.25 can be redrawn as shown in Fig. 4.27 for the dc analysis. The Thévenin equivalent network for the network to the left of the base terminal can then be found in the following manner:

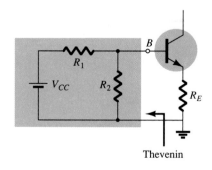

Thevenin

Figure 4.27 Redrawing the input side of the network of Fig. 4.25.

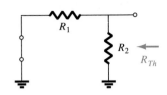

Figure 4.28 Determining R_{Th}.

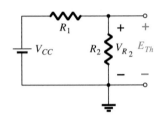

Figure 4.29 Determining E_{Th}.

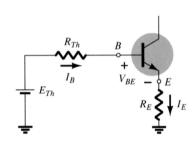

Figure 4.30 Inserting the Thévenin equivalent circuit.

R_{Th}: The voltage source is replaced by a short-circuit equivalent as shown in Fig. 4.28.

$$R_{Th} = R_1 \| R_2 \tag{4.28}$$

E_{Th}: The voltage source V_{CC} is returned to the network and the open-circuit Thévenin voltage of Fig. 4.29 determined as follows:

Applying the voltage-divider rule:

$$E_{Th} = V_{R_2} = \frac{R_2 V_{CC}}{R_1 + R_2} \tag{4.29}$$

The Thévenin network is then redrawn as shown in Fig. 4.30 and I_{B_Q} can be determined by first applying Kirchhoff's voltage law in the clockwise direction for the loop indicated:

$$E_{Th} - I_B R_{Th} - V_{BE} - I_E R_E = 0$$

Substituting $I_E = (\beta + 1)I_B$ and solving for I_B yields

$$I_B = \frac{E_{Th} - V_{BE}}{R_{Th} + (\beta + 1)R_E} \tag{4.30}$$

Although Eq. (4.30) initially appears different from those developed earlier, note that the numerator is again a difference of two voltage levels and the denominator is the base resistance plus the emitter resistor reflected by $(\beta + 1)$—certainly very similar to Eq. (4.17).

Once I_B is known the remaining quantities of the network can be found in the same manner as developed for the emitter-bias configuration. That is,

$$V_{CE} = V_{CC} - I_C(R_C + R_E) \tag{4.31}$$

which is exactly the same as Eq. (4.19). The remaining equations for V_E, V_C, and V_B are also the same as obtained for the emitter-bias configuration.

EXAMPLE 4.7

Determine the dc bias voltage V_{CE} and the current I_C for the voltage-divider configuration of Fig. 4.31.

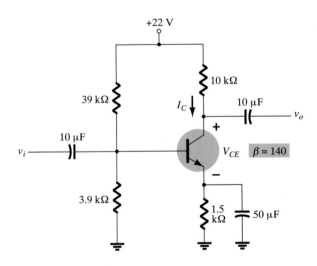

Figure 4.31 Beta-stabilized circuit for Example 4.7.

Solution

Eq. (4.28): $R_{Th} = R_1 \| R_2$

$$= \frac{(39\ k\Omega)(3.9\ k\Omega)}{39\ k\Omega + 3.9\ k\Omega} = 3.55\ k\Omega$$

Eq. (4.29): $E_{Th} = \dfrac{R_2 V_{CC}}{R_1 + R_2}$

$$= \frac{(3.9\ k\Omega)(22\ V)}{39\ k\Omega + 3.9\ k\Omega} = 2\ V$$

Eq. (4.30): $I_B = \dfrac{E_{Th} - V_{BE}}{R_{Th} + (\beta + 1)R_E}$

$$= \frac{2\ V - 0.7\ V}{3.55\ k\Omega + (141)(1.5\ k\Omega)} = \frac{1.3\ V}{3.55\ k\Omega + 211.5\ k\Omega}$$

$$= 6.05\ \mu A$$

$$I_C = \beta I_B$$

$$= (140)(6.05\ \mu A)$$

$$= \mathbf{0.85\ mA}$$

Eq. (4.31): $V_{CE} = V_{CC} - I_C(R_C + R_E)$

$$= 22\ V - (0.85\ mA)(10\ k\Omega + 1.5\ k\Omega)$$

$$= 22\ V - 9.78\ V$$

$$= \mathbf{12.22\ V}$$

Approximate Analysis

The input section of the voltage-divider configuration can be represented by the network of Fig. 4.32. The resistance R_i is the equivalent resistance between base and ground for the transistor with an emitter resistor R_E. Recall from Section 4.4 [Eq. (4.18)] that the reflected resistance between base and emitter is defined by $R_i = (\beta + 1)R_E$. If R_i is much larger than the resistance R_2, the current I_B will be much smaller than I_2 (current always seeks the path of least resistance) and I_2 will be approximately equal to I_1. If we accept the approximation that I_B is essentially zero amperes compared to I_1 or I_2, then $I_1 = I_2$ and R_1 and R_2 can be considered series

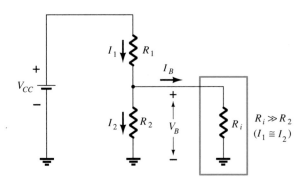

Figure 4.32 Partial bias circuit for calculating the approximate base voltage V_B.

elements. The voltage across R_2, which is actually the base voltage, can be determined using the voltage-divider rule (hence the name for the configuration). That is,

$$V_B = \frac{R_2 V_{CC}}{R_1 + R_2} \qquad (4.32)$$

Since $R_i = (\beta + 1)R_E \cong \beta R_E$ the condition that will define whether the approximate approach can be applied will be the following:

$$\beta R_E \geq 10 \, R_2 \qquad (4.33)$$

In other words, if beta times the value of R_E is at least 10 times the value of R_2, the approximate approach can be applied with a high degree of accuracy.

Once V_B is determined, the level of V_E can be calculated from

$$V_E = V_B - V_{BE} \qquad (4.34)$$

and the emitter current can be determined from

$$I_E = \frac{V_E}{R_E} \qquad (4.35)$$

and

$$I_{C_Q} \cong I_E \qquad (4.36)$$

The collector-to-emitter voltage is determined by

$$V_{CE} = V_{CC} - I_C R_C - I_E R_E$$

but since $I_E \cong I_C$,

$$V_{CE_Q} = V_{CC} - I_C(R_C + R_E) \qquad (4.37)$$

Note in the sequence of calculations from Eq. (4.33) through Eq. (4.37) that beta does not appear and I_B was not calculated. The Q-point (as determined by I_{C_Q} and V_{CE_Q}) is therefore independent of the value of beta.

EXAMPLE 4.8   Repeat the analysis of Fig. 4.31 using the approximate technique and compare solutions for I_{C_Q} and V_{CE_Q}.

Solution

Testing:

$$\beta R_E \geq 10 \, R_2$$

$$(140)(1.5 \text{ k}\Omega) \geq 10(3.9 \text{ k}\Omega)$$

$$210 \text{ k}\Omega \geq 39 \text{ k}\Omega \; (\textit{satisfied})$$

Eq. (4.32): $V_B = \dfrac{R_2 V_{CC}}{R_1 + R_2}$

$$= \frac{(3.9 \text{ k}\Omega)(22 \text{ V})}{39 \text{ k}\Omega + 3.9 \text{ k}\Omega}$$

$$= 2 \text{ V}$$

Note that the level of V_B is the same as E_{Th} determined in Example 4.7. Essentially, therefore, the primary difference between the exact and approximate techniques is the effect of R_{Th} in the exact analysis that separates E_{Th} and V_B.

$$\text{Eq. (4.33):} \quad V_E = V_B - V_{BE}$$

$$= 2\text{ V} - 0.7\text{ V}$$

$$= 1.3\text{ V}$$

$$I_{C_Q} \cong I_E = \frac{V_E}{R_E} = \frac{1.3\text{ V}}{1.5\text{ k}\Omega} = \mathbf{0.867\ mA}$$

compared to 0.85mA with the exact analysis. Finally,

$$V_{CE_Q} = V_{CC} - I_C(R_C + R_E)$$

$$= 22\text{ V} - (0.867\text{ mA})(10\text{ k}\Omega + 1.5\text{ k}\Omega)$$

$$= 22\text{ V} - 9.97\text{ V}$$

$$= \mathbf{12.03\ V}$$

versus 12.22 V obtained in Example 4.7.

The results for I_{C_Q} and V_{CE_Q} are certainly close and considering the actual variation in parameter values one can certainly be considered as accurate as the other. The larger the level of R_i compared to R_2, the closer the approximate to the exact solution. Example 4.10 will compare solutions at a level well below the condition established by Eq. (4.33).

Repeat the exact analysis of Example 4.7 if β is reduced to 70 and compare solutions for I_{C_Q} and V_{CE_Q}.

EXAMPLE 4.9

Solution

This example is not a comparison of exact versus approximate methods but a testing of how much the Q-point will move if the level of β is cut in half. R_{Th} and E_{Th} are the same:

$$R_{Th} = 3.55\text{ k}\Omega, \qquad E_{Th} = 2\text{ V}$$

$$I_B = \frac{E_{Th} - V_{BE}}{R_{Th} + (\beta + 1)R_E}$$

$$= \frac{2\text{ V} - 0.7\text{ V}}{3.55\text{ k}\Omega + (71)(1.5\text{ k}\Omega)} = \frac{1.3\text{ V}}{3.55\text{ k}\Omega + 106.5\text{ k}\Omega}$$

$$= 11.81\ \mu A$$

$$I_{C_Q} = \beta I_B$$

$$= (70)(11.81\ \mu A)$$

$$= 0.83\text{ mA}$$

$$V_{CE_Q} = V_{CC} - I_C(R_C + R_E)$$

$$= 22\text{ V} - (0.83\text{ mA})(10\text{ k}\Omega + 1.5\text{ k}\Omega)$$

$$= 12.46\text{ V}$$

Tabulating the results, we have:

β	I_{C_Q}	V_{CE_Q}
140	0.85 mA	12.22 V
70	0.83 mA	12.46 V

The results clearly show the relative insensitivity of the circuit to the change in β. Even though β is drastically cut in half, from 140 to 70, the levels of I_{C_Q} and V_{CE_Q} are essentially the same.

EXAMPLE 4.10

Determine the levels of I_{C_Q} and V_{CE_Q} for the voltage-divider configuration of Fig. 4.33 using the exact and approximate techniques and compare solutions. In this case the conditions of Eq. (4.33) will not be satisfied, but the results will reveal the difference in solution if the criterion of Eq. (4.33) is ignored.

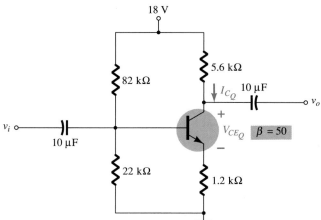

Figure 4.33 Voltage-divider configuration for Example 4.10.

Solution

Exact Analysis:

$$\text{Eq. (4.33):} \quad \beta R_E \geq 10\, R_2$$

$$(50)(1.2 \text{ k}\Omega) \geq 10(22 \text{ k}\Omega)$$

$$60 \text{ k}\Omega \neq 220 \text{ k}\Omega \ (\textit{not satisfied})$$

$$R_{Th} = R_1 \| R_2 = 82 \text{ k}\Omega \| 22 \text{ k}\Omega = 17.35 \text{ k}\Omega$$

$$E_{Th} = \frac{R_2 V_{CC}}{R_1 + R_2} = \frac{22 \text{ k}\Omega (18 \text{ V})}{82 \text{ k}\Omega + 22 \text{ k}\Omega} = 3.81 \text{ V}$$

$$I_B = \frac{E_{Th} - V_{BE}}{R_{Th} + (\beta + 1)R_E} = \frac{3.81 \text{ V} - 0.7 \text{ V}}{17.35 \text{ k}\Omega + (51)(1.2 \text{ k}\Omega)} = \frac{3.11 \text{ V}}{78.55 \text{ k}\Omega}$$

$$= 39.6 \text{ } \mu\text{A}$$

$$I_{C_Q} = \beta I_B = (50)(39.6 \text{ } \mu\text{A}) = \mathbf{1.98 \text{ mA}}$$

$$V_{CE_Q} = V_{CC} - I_C(R_C + R_E)$$

$$= 18 \text{ V} - (1.98 \text{ mA})(5.6 \text{ k}\Omega + 1.2 \text{ k}\Omega)$$

$$= \mathbf{4.54 \text{ V}}$$

Approximate Analysis:

$$V_B = E_{Th} = 3.81 \text{ V}$$

$$V_E = V_B - V_{BE} = 3.81 \text{ V} - 0.7 \text{ V} = 3.11 \text{ V}$$

$$I_{C_Q} \cong I_E = \frac{V_E}{R_E} = \frac{3.11 \text{ V}}{1.2 \text{ k}\Omega} = \mathbf{2.59 \text{ mA}}$$

$$V_{CE_Q} = V_{CC} - I_C(R_C + R_E)$$
$$= 18 \text{ V} - (2.59 \text{ mA})(5.6 \text{ k}\Omega + 1.2 \text{ k}\Omega)$$
$$= \mathbf{3.88 \text{ V}}$$

Tabulating the results, we have:

	I_{C_Q}	V_{CE_Q}
Exact	1.98 mA	4.54 V
Approximate	2.59 mA	3.88 V

The results reveal the difference between exact and approximate solutions. I_{C_Q} is about 30% greater with the approximate solution, while V_{CE_Q} is about 10% less. The results are notably different in magnitude, but even though βR_E is only about three times larger than R_2, the results are still relatively close to each other. For the future, however, our analysis will be dictated by Eq. (4.33) to ensure a close similarity between exact and approximate solutions.

Transistor Saturation

The output collector–emitter circuit for the voltage-divider configuration has the same appearance as the emitter-biased circuit analyzed in Section 4.4. The resulting equation for the saturation current (when V_{CE} is set to zero volts on the schematic) is therefore the same as obtained for the emitter-biased configuration. That is,

$$I_{C_{sat}} = I_{C_{max}} = \frac{V_{CC}}{R_C + R_E} \qquad (4.38)$$

Load-Line Analysis

The similarities with the output circuit of the emitter-biased configuration result in the same intersections for the load line of the voltage-divider configuration. The load line will therefore have the same appearance as that of Fig. 4.24, with

$$I_C = \frac{V_{CC}}{R_C + R_E}\bigg|_{V_{CE}=0 \text{ V}} \qquad (4.39)$$

and

$$V_{CE} = V_{CC}\big|_{I_C=0 \text{ mA}} \qquad (4.40)$$

The level of I_B is of course determined by a different equation for the voltage-divider bias and the emitter-bias configurations.

4.6 DC BIAS WITH VOLTAGE FEEDBACK

An improved level of stability can also be obtained by introducing a feedback path from collector to base as shown in Fig. 4.34. Although the Q-point is not totally independent of beta (even under approximate conditions), the sensitivity to changes in beta or temperature variations is normally less than encountered for the fixed-bias or emitter-biased configurations. The analysis will again be performed by first analyzing the base–emitter loop with the results applied to the collector–emitter loop.

Base–Emitter Loop

Figure 4.35 shows the base–emitter loop for the voltage feedback configuration. Writing Kirchhoff's voltage law around the indicated loop in the clockwise direction will result in

$$V_{CC} - I_C'R_C - I_BR_B - V_{BE} - I_ER_E = 0$$

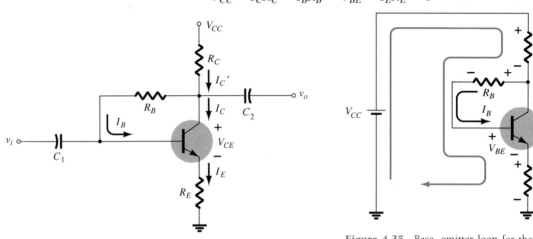

Figure 4.34 Dc bias circuit with voltage feedback.

Figure 4.35 Base–emitter loop for the network of Fig. 4.34.

It is important to note that the current through R_C is not I_C but I_C' (where $I_C' = I_C + I_B$). However, the level of I_C and I_C' far exceeds the usual level of I_B and the approximation $I_C' \cong I_C$ is normally employed. Substituting $I_C' \cong I_C = \beta I_B$ and $I_E \cong I_C$ will result in

$$V_{CC} - \beta I_B R_C - I_B R_B - V_{BE} - \beta I_B R_E = 0$$

Gathering terms, we have

$$V_{CC} - V_{BE} - \beta I_B(R_C + R_E) - I_B R_B = 0$$

and solving for I_B yields

$$I_B = \frac{V_{CC} - V_{BE}}{R_B + \beta(R_C + R_E)} \tag{4.41}$$

The result is quite interesting in that the format is very similar to equations for I_B obtained for earlier configurations. The numerator is again the difference of available voltage levels, while the denominator is the base resistance plus the collector and emitter resistors reflected by beta. In general, therefore, the feedback path results in a reflection of the resistance R_C back to the input circuit, much like the reflection of R_E.

In general, the equation for I_B has had the following format:

$$I_B = \frac{V'}{R_B + \beta R'}$$

with the absence of R' for the fixed-bias configuration, $R' = R_E$ for the emitter-bias setup (with $(\beta + 1) \cong \beta$), and $R' = R_C + R_E$ for the collector-feedback arrangement. The voltage V' is the difference between two voltage levels.

Since $I_C = \beta I_B$,

$$I_{C_Q} = \frac{\beta V'}{R_B + \beta R'}$$

In general, the larger $\beta R'$ is compared to R_B, the less the sensitivity of I_{C_Q} to variations in beta. Obviously, if $\beta R' \gg R_B$ and $R_B + \beta R' \cong \beta R'$, then

$$I_{C_Q} = \frac{\beta V'}{R_B + \beta R'} \cong \frac{\beta V'}{\beta R'} = \frac{V'}{R'}$$

and I_{C_Q} is independent of the value of beta. Since R' is typically larger for the voltage-feedback configuration than for the emitter-bias configuration, the sensitivity to variations in beta is less. Of course, R' is zero ohms for the fixed-bias configuration and is therefore quite sensitive to variations in beta.

Collector–Emitter Loop

The collector-emitter loop for the network of Fig. 4.34 is provided in Fig. 4.36. Applying Kirchhoff's voltage law around the indicated loop in the clockwise direction will result in

$$I_E R_E + V_{CE} + I'_C R_C - V_{CC} = 0$$

Since $I'_C \cong I_C$ and $I_E \cong I_C$, we have

$$I_C(R_C + R_E) + V_{CE} - V_{CC} = 0$$

and

$$\boxed{V_{CE} = V_{CC} - I_C(R_C + R_E)} \qquad (4.42)$$

which is exactly as obtained for the emitter-bias and voltage-divider bias configurations.

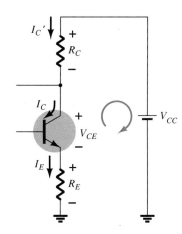

Figure 4.36 Collector–emitter loop for the network of Fig. 4.34.

Determine the quiescent levels of I_{C_Q} and V_{CE_Q} for the network of Fig. 4.37.

EXAMPLE 4.11

Solution

Eq. (4.41): $I_B = \dfrac{V_{CC} - V_{BE}}{R_B + \beta(R_C + R_E)}$

$$= \frac{10\text{ V} - 0.7\text{ V}}{250\text{ k}\Omega + (90)(4.7\text{ k}\Omega + 1.2\text{ k}\Omega)}$$

$$= \frac{9.3\text{ V}}{250\text{ k}\Omega + 531\text{ k}\Omega} = \frac{9.3\text{ V}}{781\text{ k}\Omega}$$

$$= 11.91\ \mu\text{A}$$

$$I_{C_Q} = \beta I_B = (90)(11.91\ \mu\text{A})$$

$$= \mathbf{1.07\ mA}$$

$$V_{CE_Q} = V_{CC} - I_C(R_C + R_E)$$

$$= 10\text{ V} - (1.07\text{ mA})(4.7\text{ k}\Omega + 1.2\text{ k}\Omega)$$

$$= 10\text{ V} - 6.31\text{ V}$$

$$= \mathbf{3.69\ V}$$

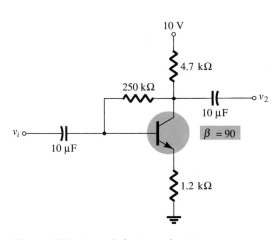

Figure 4.37 Network for Example 4.11.

EXAMPLE 4.12

Repeat Example 4.11 using a beta of 135 (50% more than Example 4.11).

Solution

It is important to note in the solution for I_B in Example 4.11 that the second term in the denominator of the equation is larger than the first. Recall in a recent discussion that the larger this second term is compared to the first, the less the sensitivity to changes in beta. In this example the level of beta is increased by 50%, which will increase the magnitude of this second term even more compared to the first. It is more important to note in these examples, however, that once the second term is relatively large compared to the first, the sensitivity to changes in beta is significantly less.

Solving for I_B gives

$$I_B = \frac{V_{CC} - V_{BE}}{R_B + \beta(R_C + R_E)} = \frac{10 \text{ V} - 0.7 \text{ V}}{250 \text{ k}\Omega + (135)(4.7 \text{ k}\Omega + 1.2 \text{ k}\Omega)}$$

$$= \frac{9.3 \text{ V}}{250 \text{ k}\Omega + 796.5 \text{ k}\Omega} = \frac{9.3 \text{ V}}{1046.5 \text{ k}\Omega}$$

$$= 8.89 \ \mu\text{A}$$

and

$$I_{C_Q} = \beta I_B$$

$$= (135)(8.89\mu\text{A})$$

$$= \mathbf{1.2mA}$$

with

$$V_{CE_Q} = V_{CC} - I_C(R_C + R_E)$$

$$= 10 \text{ V} - (1.2\text{mA})(4.7 \text{ k}\Omega + 1.2 \text{ k}\Omega)$$

$$= 10 \text{ V} - 7.08 \text{ V}$$

$$= \mathbf{2.92 \ V}$$

Even though the level of β increased 50%, the level of I_{C_Q} only increased 12.1% while the level of V_{CE_Q} decreased about 20.9%. If the network were a fixed-bias design, a 50% increase in β would have resulted in a 50% increase in I_{C_Q} and a dramatic change in the location of the Q-point.

EXAMPLE 4.13

Determine the dc level of I_B and V_C for the network of Fig. 4.38.

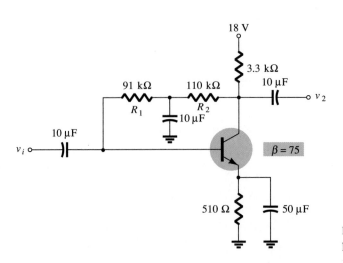

Figure 4.38 Network for Example 4.13.

Solution

In this case the base resistance for the dc analysis is composed of two resistors with a capacitor connected from their junction to ground. For the dc mode, the capacitor assumes the open-circuit equivalence and $R_B = R_1 + R_2$.

Solving for I_B gives

$$I_B = \frac{V_{CC} - V_{BE}}{R_B + \beta(R_C + R_E)}$$

$$= \frac{18 \text{ V} - 0.7 \text{ V}}{(91 \text{ k}\Omega + 110 \text{ k}\Omega) + (75)(3.3 \text{ k}\Omega + 0.51 \text{ k}\Omega)}$$

$$= \frac{17.3 \text{ V}}{201 \text{ k}\Omega + 285.75 \text{ k}\Omega} = \frac{17.3 \text{ V}}{486.75 \text{ k}\Omega}$$

$$= \mathbf{35.5 \ \mu A}$$

$$I_C = \beta I_B$$

$$= (75)(35.5 \ \mu A)$$

$$= 2.66 \text{ mA}$$

$$V_C = V_{CC} - I_C' R_C \cong V_{CC} - I_C R_C$$

$$= 18 \text{ V} - (2.66 \text{ mA})(3.3 \text{ k}\Omega)$$

$$= 18 \text{ V} - 8.78 \text{ V}$$

$$= \mathbf{9.22 \text{ V}}$$

Saturation Conditions

Using the approximation $I_C' = I_C$ the equation for the saturation current is the same as obtained for the voltage-divider and emitter-bias configurations. That is,

$$\boxed{I_{C_{\text{sat}}} = I_{C_{\text{max}}} = \frac{V_{CC}}{R_C + R_E}} \qquad (4.43)$$

Load-Line Analysis

Continuing with the approximation $I_C' = I_C$ will result in the same load line defined for the voltage-divider and emitter-biased configurations. The level of I_{B_Q} will be defined by the chosen bias configuration.

4.7 MISCELLANEOUS BIAS CONFIGURATIONS

There are a number of BJT bias configurations that do not match the basic mold of those analyzed in the previous sections. In fact, there are variations in design that would require many more pages than is possible in a book of this type. However, the primary purpose here is to emphasize those characteristics of the device that permit a dc analysis of the configuration and to establish a general procedure toward the desired solution. For each configuration discussed thus far the first step has been the derivation of an expression for the base current. Once the base current is known, the collector current and voltage levels of the output circuit can be determined quite directly. This is not to imply that all solutions will take this path, but it does suggest a possible route to follow if a new configuration is encountered.

The first example is simply one where the emitter resistor has been dropped from the voltage-feedback configuration of Fig. 4.34. The analysis is quite similar but does require dropping R_E from the applied equation.

EXAMPLE 4.14

For the network of Fig. 4.39:
(a) Determine I_{C_Q} and V_{CE_Q}.
(b) Find V_B, V_C, V_E, and V_{BC}.

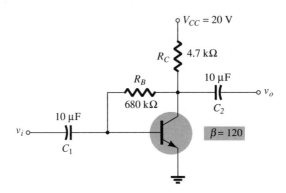

Figure 4.39 Collector feedback with $R_E = 0\Omega$.

Solution

(a) The absence of R_E reduces the reflection of resistive levels to simply that of R_C and the equation for I_B reduces to

$$I_B = \frac{V_{CC} - V_{BE}}{R_B + \beta R_C}$$

$$= \frac{20 \text{ V} - 0.7 \text{ V}}{680 \text{ k}\Omega + (120)(4.7 \text{ k}\Omega)} = \frac{19.3 \text{ V}}{1.244 \text{ M}\Omega}$$

$$= 15.51 \ \mu\text{A}$$

$$I_{C_Q} = \beta I_B = (120)(15.51 \ \mu\text{A})$$

$$= \textbf{1.86 mA}$$

$$V_{CE_Q} = V_{CC} - I_C R_C$$

$$= 20 \text{ V} - (1.86 \text{ mA})(4.7 \text{ k}\Omega)$$

$$= \textbf{11.26 V}$$

(b)
$$V_B = V_{BE} = \textbf{0.7 V}$$
$$V_C = V_{CE} = \textbf{11.26 V}$$
$$V_E = \textbf{0 V}$$
$$V_{BC} = V_B - V_C = 0.7 \text{ V} - 11.26 \text{ V}$$
$$= \textbf{-10.56 V}$$

In the next example the applied voltage is connected to the emitter leg and R_C is connected directly to ground. Initially, it appears somewhat unorthodox and quite different from those encountered thus far, but one application of Kirchhoff's voltage law to the base circuit will result in the desired base current.

Determine V_C and V_B for the network of Fig. 4.40.

EXAMPLE 4.15

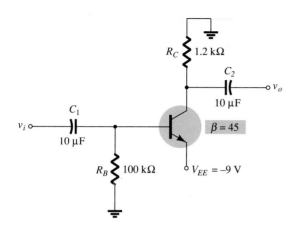

Figure 4.40 Example 4.15

Solution

Applying Kirchhoff's voltage law in the clockwise direction for the base–emitter loop will result in

$$-I_B R_B - V_{BE} + V_{EE} = 0$$

and

$$I_B = \frac{V_{EE} - V_{BE}}{R_B}$$

Substitution yields

$$I_B = \frac{9 \text{ V} - 0.7 \text{ V}}{100 \text{ k}\Omega}$$

$$= \frac{8.3 \text{ V}}{100 \text{ k}\Omega}$$

$$= 83 \text{ } \mu\text{A}$$

$$I_C = \beta I_B$$

$$= (45)(83 \text{ } \mu\text{A})$$

$$= 3.735 \text{ mA}$$

$$V_C = -I_C R_C$$

$$= -(3.735 \text{ mA})(1.2 \text{ k}\Omega)$$

$$= \mathbf{-4.48 \text{ V}}$$

$$V_B = -I_B R_B$$

$$= -(83 \text{ } \mu\text{A})(100 \text{ k}\Omega)$$

$$= \mathbf{-8.3 \text{ V}}$$

The next example employs a network referred to as an *emitter-follower* configuration. When the same network is analyzed on an ac basis we will find that the output and input signals are in phase (one following the other) and the output voltage is slightly less than the applied signal. For the dc analysis the collector is grounded and the applied voltage is in the emitter leg.

EXAMPLE 4.16   Determine V_{CE_Q} and I_E for the network of Fig. 4.41.

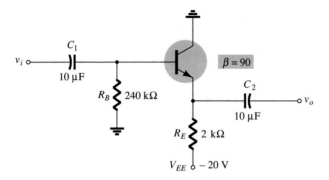

Figure 4.41   Common-collector (emitter-follower) configuration.

Solution

Applying Kirchhoff's voltage law to the input circuit will result in

$$-I_B R_B - V_{BE} - I_E R_E + V_{EE} = 0$$

but

$$I_E = (\beta + 1)I_B$$

and

$$V_{EE} - V_{BE} - (\beta + 1)I_B R_E - I_B R_B = 0$$

with

$$I_B = \frac{V_{EE} - V_{BE}}{R_B + (\beta + 1)R_E}$$

Substituting values yields

$$I_B = \frac{20\ V - 0.7\ V}{240\ k\Omega + (91)(2\ k\Omega)}$$

$$= \frac{19.3\ V}{240\ k\Omega + 182\ k\Omega} = \frac{19.3\ V}{422\ k\Omega}$$

$$= 45.73\ \mu A$$

$$I_C = \beta I_B$$

$$= (90)(45.73\ \mu A)$$

$$= 4.12\ mA$$

Applying Kirchhoff's voltage law to the output circuit, we have

$$-V_{EE} + I_E R_E + V_{CE} = 0$$

but

$$I_E = (\beta + 1)I_B$$

and

$$V_{CE_Q} = V_{EE} - (\beta + 1)I_B R_E$$

$$= 20\ V - (91)(45.73\ \mu A)(2\ k\Omega)$$

$$= \mathbf{11.68\ V}$$

$$I_E = \mathbf{4.16\ mA}$$

All of the examples thus far have employed a common-emitter or common-collector configuration. In the next example we investigate the common-base configuration. In this situation the input circuit will be employed to determine I_E rather than I_B. The collector current is then available to perform an analysis of the output circuit.

Determine the voltage V_{CB} and the current I_B for the common-base configuration of Fig. 4.42.

EXAMPLE 4.17

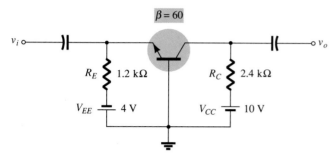

Figure 4.42 Common-base configuration.

Solution

Applying Kirchhoff's voltage law to the input circuit yields

$$-V_{EE} + I_E R_E + V_{BE} = 0$$

and

$$I_E = \frac{V_{EE} - V_{BE}}{R_E}$$

Substituting values, we obtain

$$I_E = \frac{4\text{ V} - 0.7\text{ V}}{1.2\text{ k}\Omega} = 2.75\text{ mA}$$

Applying Kirchhoff's voltage law to the output circuit gives

$$-V_{CB} + I_C R_C - V_{CC} = 0$$

and

$$V_{CB} = V_{CC} - I_C R_C \text{ with } I_C \cong I_E$$

$$= 10\text{V} - (2.75\text{mA})(2.4\text{ k}\Omega)$$

$$= \mathbf{3.4\ V}$$

$$I_B = \frac{I_C}{\beta}$$

$$= \frac{2.75\text{ mA}}{60}$$

$$= \mathbf{45.8\ \mu A}$$

Example 4.18 employs a split supply and will require the application of Thévenin's theorem to determine the desired unknowns.

EXAMPLE 4.18

Determine V_C and V_B for the network of Fig. 4.43.

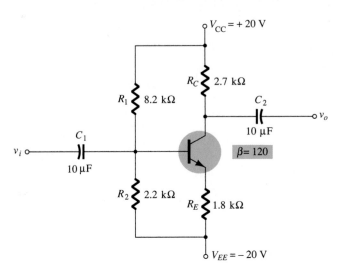

Figure 4.43 Example 4.18

Solution

The Thévenin resistance and voltage are determined for the network to the left of the base terminal as shown in Figs. 4.44 and 4.45.

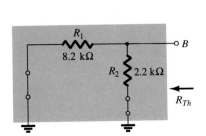

Figure 4.44 Determining R_{Th}.

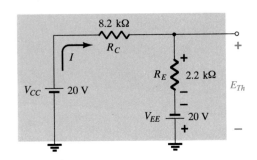

Figure 4.45 Determining E_{Th}.

R_{Th}:

$$R_{Th} = 8.2 \text{ k}\Omega \| 2.2 \text{ k}\Omega = 1.73 \text{ k}\Omega$$

E_{Th}:

$$I = \frac{V_{CC} + V_{EE}}{R_1 + R_2} = \frac{20 \text{ V} + 20 \text{ V}}{8.2 \text{ k}\Omega + 2.2 \text{ k}\Omega} = \frac{40 \text{ V}}{10.4 \text{ k}\Omega}$$

$$= 3.85 \text{ mA}$$

$$E_{Th} = IR_2 - V_{EE}$$

$$= (3.85 \text{ mA})(2.2 \text{ k}\Omega) - 20 \text{ V}$$

$$= -11.53 \text{ V}$$

The network can then be redrawn as shown in Fig. 4.46, where the application of Kirchhoff's voltage law will result in

$$-E_{Th} - I_B R_{Th} - V_{BE} - I_E R_E + V_{EE} = 0$$

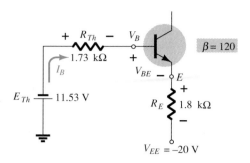

Figure 4.46 Substituting the Thévenin equivalent circuit.

Substituting $I_E = (\beta + 1)I_B$ gives

$$V_{EE} - E_{Th} - V_{BE} - (\beta + 1)I_B R_E - I_B R_{Th} = 0$$

and

$$I_B = \frac{V_{EE} - E_{Th} - V_{BE}}{R_{Th} + (\beta + 1)R_E}$$

$$= \frac{20 \text{ V} - 11.53 \text{ V} - 0.7 \text{ V}}{1.73 \text{ k}\Omega + (121)(1.8 \text{ k}\Omega)}$$

$$= \frac{7.77 \text{ V}}{219.53 \text{ k}\Omega}$$

$$= 35.39 \text{ } \mu\text{A}$$

$$I_C = \beta I_B$$

$$= (120)(35.39 \text{ } \mu\text{A})$$

$$= 4.25 \text{ mA}$$

$$V_C = V_{CC} - I_C R_C$$

$$= 20 \text{ V} - (4.25 \text{ mA})(2.7 \text{ k}\Omega)$$

$$= \mathbf{8.53 \text{ V}}$$

$$V_B = -E_{Th} - I_B R_{Th}$$

$$= -(11.53 \text{ V}) - (35.39 \text{ } \mu\text{A})(1.73 \text{ k}\Omega)$$

$$= \mathbf{-11.59 \text{ V}}$$

•

4.8 DESIGN OPERATIONS

Discussions thus far have focused on the analysis of existing networks. All the elements are in place and it is simply a matter of solving for the current and voltage levels of the configuration. The design process is one where a current and/or voltage may be specified and the elements required to establish the designated levels must be determined. This synthesis process requires a clear understanding of the characteristics of the device, the basic equations for the network, and a firm understanding of the basic laws of circuit analysis, such as Ohm's law, Kirchhoff's voltage law, and so on. In most situations the thinking process is challenged to a higher degree in the design process than in the analysis sequence. The path toward a solution is less defined and in fact may require a number of basic assumptions that do not have to be made when simply analyzing a network.

The design sequence is obviously sensitive to the components that are already specified and the elements to be determined. If the transistor and supplies are specified, the design process will simply determine the required resistors for a particular design. Once the theoretical value of the resistors are determined, the nearest standard commercial value is normally chosen and any variation due to not using the exact resistance value accepted as part of the design. This is certainly a valid approximation considering the tolerances normally associated with resistive elements and the transistor parameters.

If resistive values are to be determined, one of the most powerful equations is simply Ohm's law in the following form:

$$R_{unk} = \frac{V_R}{I_R} \tag{4.44}$$

In a particular design the voltage across a resistor can often be determined from specified levels. If additional specifications define the current level equation (4.44) can then be used to calculate the required resistance level. The first few examples will demonstrate how particular elements can be determined from specified levels. A complete design procedure will then be introduced for two popular configurations.

EXAMPLE 4.19

Given the device characteristics of Fig. 4.47a, determine V_{CC}, R_B, and R_C for the fixed-bias configuration of Fig. 4.47b.

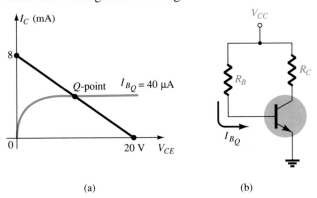

(a) (b) **Figure 4.47** Example 4.19

Solution

From the load line $V_{CC} = \mathbf{20\ V}$

$$I_C = \frac{V_{CC}}{R_C}\bigg|_{V_{CE} = 0\ V}$$

and

$$R_C = \frac{V_{CC}}{I_C} = \frac{20\ V}{8\ mA} = \mathbf{2.5\ k\Omega}$$

$$I_B = \frac{V_{CC} - V_{BE}}{R_B}$$

with

$$R_B = \frac{V_{CC} - V_{BE}}{I_B}$$

$$= \frac{20\ V - 0.7\ V}{40\ \mu A} = \frac{19.3\ V}{40\ \mu A}$$

$$= \mathbf{482.5\ k\Omega}$$

Standard resistor values:

$$R_C = 2.4 \text{ k}\Omega$$

$$R_B = 470 \text{ k}\Omega$$

Using standard resistor values gives

$$I_B = 41.1 \ \mu A$$

which is well within 5% of the value specified.

EXAMPLE 4.20

Given that $I_{C_Q} = 2$ mA and $V_{CE_Q} = 10$ V, determine R_1 and R_C for the network of Fig. 4.48.

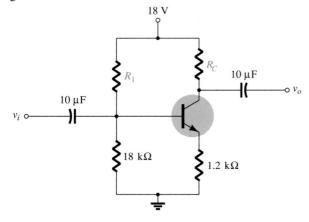

18 V

Figure 4.48 Example 4.20

Solution

$$V_E = I_E R_E \cong I_C R_E$$

$$= (2 \text{ mA})(1.2 \text{ k}\Omega) = 2.4 \text{ V}$$

$$V_B = V_{BE} + V_E = 0.7 \text{ V} + 2.4 \text{ V} = 3.1 \text{ V}$$

$$V_B = \frac{R_2 V_{CC}}{R_1 + R_2} = 3.1 \text{ V}$$

and

$$\frac{(18 \text{ k}\Omega)(18 \text{ V})}{R_1 + 18 \text{ k}\Omega} = 3.1 \text{ V}$$

$$324 \text{ k}\Omega = 3.1 \, R_1 + 55.8 \text{ k}\Omega$$

$$3.1 \, R_1 = 268.2 \text{ k}\Omega$$

$$R_1 = \frac{268.2 \text{ k}\Omega}{3.1} = \mathbf{86.52 \text{ k}\Omega}$$

Eq. (4.44):
$$R_C = \frac{V_{R_C}}{I_C} = \frac{V_{CC} - V_C}{I_C}$$

with
$$V_C = V_{CE} + V_E = 10 \text{ V} + 2.4 \text{ V} = 12.4 \text{ V}$$

and
$$R_C = \frac{18 \text{ V} - 12.4 \text{ V}}{2 \text{ mA}}$$

$$= \mathbf{2.8 \text{ k}\Omega}$$

The nearest standard commercial values to R_1 are 82 kΩ and 91 kΩ. However, using the series combination of standard values of 82 kΩ and 4.7 kΩ = 86.7 kΩ would result in a value very close to the design level.

EXAMPLE 4.21

The emitter-bias configuration of Fig. 4.49 has the following specifications: $I_{C_Q} = \frac{1}{2}I_{C_{\text{sat}}}$, $I_{C_{\text{sat}}} = 8$ mA, $V_C = 18$ V, and $\beta = 110$. Determine R_C, R_E, and R_B.

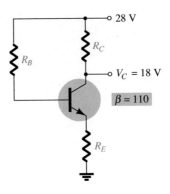

Figure 4.49 Example 4.21

Solution

$$I_{C_Q} = \tfrac{1}{2}I_{C_{\text{sat}}} = 4 \text{ mA}$$

$$R_C = \frac{V_{R_C}}{I_{C_Q}} = \frac{V_{CC} - V_C}{I_{C_Q}}$$

$$= \frac{28 \text{ V} - 18 \text{ V}}{4 \text{ mA}} = \mathbf{2.5 \text{ k}\Omega}$$

$$I_{C_{\text{sat}}} = \frac{V_{CC}}{R_C + R_E}$$

and

$$R_C + R_E = \frac{V_{CC}}{I_{C_{\text{sat}}}} = \frac{28 \text{ V}}{8 \text{ mA}} = 3.5 \text{ k}\Omega$$

$$R_E = 3.5 \text{ k}\Omega - R_C$$

$$= 3.5 \text{ k}\Omega - 2.5 \text{ k}\Omega$$

$$= \mathbf{1 \text{ k}\Omega}$$

$$I_{B_Q} = \frac{I_{C_Q}}{\beta} = \frac{4 \text{ mA}}{110} = 36.36 \text{ }\mu\text{A}$$

$$I_{B_Q} = \frac{V_{CC} - V_{BE}}{R_B + (\beta + 1)R_E}$$

and

$$R_B + (\beta + 1)R_E = \frac{V_{CC} - V_{BE}}{I_{B_Q}}$$

with

$$R_B = \frac{V_{CC} - V_{BE}}{I_{B_Q}} - (\beta + 1)R_E$$

$$= \frac{28 \text{ V} - 0.7 \text{ V}}{36.36 \text{ }\mu\text{A}} - (111)(1 \text{ k}\Omega)$$

$$= \frac{27.3 \text{ V}}{36.36 \text{ }\mu\text{A}} - 111 \text{ k}\Omega$$

$$= \mathbf{639.8 \text{ k}\Omega}$$

For standard values:

$$R_C = 2.4 \text{ k}\Omega$$

$$R_E = 1 \text{ k}\Omega$$

$$R_B = 620 \text{ k}\Omega$$

The discussion to follow will introduce one technique for designing an entire circuit to operate at a specified bias point. Often the manufacturer's specification (spec) sheets provide information on a suggested operating point (or operating region) for a particular transistor. In addition, other system components connected to the given amplifier stage may also define the current swing, voltage swing, value of common supply voltage, and so on, for the design.

In actual practice, many other factors may have to be considered and which may affect the selection of the desired operating point. For the moment we shall concentrate, however, on determining the component values to obtain a specified operating point. The discussion will be limited to the emitter-bias and voltage-divider bias configurations, although the same procedure can be applied to a variety of other transistor circuits.

Design of a Bias Circuit with an Emitter Feedback Resistor

Consider first the design of the dc bias components of an amplifier circuit having emitter-resistor bias stabilization as shown in Fig. 4.50. The supply voltage and operating point were selected from the manufacturer's information on the transistor used in the amplifier.

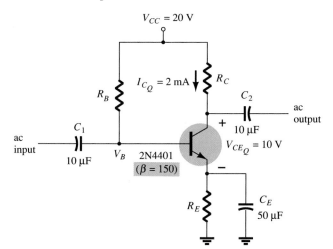

Figure 4.50 Emitter-stabilized bias circuit for design consideration.

The selection of collector and emitter resistors cannot proceed directly from the information just specified. The equation that relates the voltages around the collector–emitter loop has two unknown quantities present—the resistors R_C and R_E. At this point some engineering judgment must be made, such as the level of the emitter voltage compared to the applied supply voltage. Recall that the need for including a resistor from emitter to ground was to provide a means of dc bias stabilization so that the change of collector current due to leakage currents in the transistor and the transistor beta would not cause a large shift in the operating point. The emitter resistor cannot be unreasonably large because the voltage across it limits the range of voltage swing of the voltage from collector to emitter (to be noted when the ac response is discussed). The examples examined in this chapter reveal that the voltage from emitter to ground is typically around one-fourth to one-tenth of the supply voltage. Selecting the conservative case of one-tenth will permit calculating the emitter resistor R_E and the resistor R_C in a manner similar to the examples just completed. In the next example we perform a complete design of the network of Fig. 4.49 using the criteria just introduced for the emitter voltage.

EXAMPLE 4.22

Determine the resistor values for the network of Fig. 4.50 for the indicated operating point and supply voltage.

Solution

$$V_E = \tfrac{1}{10} V_{CC} = \tfrac{1}{10} (20 \text{ V}) = 2 \text{ V}$$

$$R_E = \frac{V_E}{I_E} \cong \frac{V_E}{I_C} = \frac{2 \text{ V}}{2 \text{ mA}} = \mathbf{1 \text{ k}\Omega}$$

$$R_C = \frac{V_{R_C}}{I_C} = \frac{V_{CC} - V_{CE} - V_E}{I_C} = \frac{20 \text{ V} - 10 \text{ V} - 2 \text{ V}}{2 \text{ mA}} = \frac{8 \text{ V}}{2 \text{ mA}}$$

$$= \mathbf{4 \text{ k}\Omega}$$

$$I_B = \frac{I_C}{\beta} = \frac{2 \text{ mA}}{150} = 13.33 \text{ } \mu A$$

$$R_B = \frac{V_{R_B}}{I_B} = \frac{V_{CC} - V_{BE} - V_E}{I_B} = \frac{20 \text{ V} - 0.7 \text{ V} - 2 \text{ V}}{13.33 \text{ } \mu A}$$

$$\cong \mathbf{1.3 \text{ M}\Omega}$$

Design of a Current-Gain-Stabilized (Beta-Independent) Circuit

The circuit of Fig. 4.51 provides stabilization both for leakage and current gain (beta) changes. The four resistor values shown must be obtained for the specified operating point. Engineering judgment in selecting a value of emitter voltage, V_E, as in the previous design consideration, leads to a direct straightforward solution for all the resistor values. The design steps are all demonstrated in the next example.

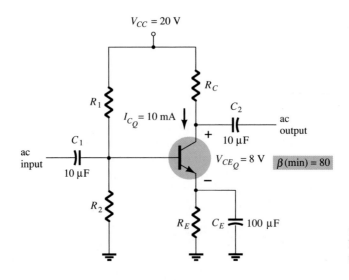

Figure 4.51 Current-gain-stabilized circuit for design considerations.

Determine the levels of R_C, R_E, R_1, and R_2 for the network of Fig. 4.51 for the operating point indicated.

EXAMPLE 4.23

Solution

$$V_E = \tfrac{1}{10} V_{CC} = \tfrac{1}{10} (20 \text{ V}) = 2 \text{ V}$$

$$R_E = \frac{V_E}{I_E} \cong \frac{V_E}{I_C} = \frac{2 \text{ V}}{10 \text{ mA}} = \textbf{200 } \mathbf{\Omega}$$

$$R_C = \frac{V_{R_C}}{I_C} = \frac{V_{CC} - V_{CE} - V_E}{I_C} = \frac{20 \text{ V} - 8 \text{ V} - 2 \text{ V}}{10 \text{ mA}} = \frac{10 \text{ V}}{10 \text{ mA}}$$

$$= \textbf{1 k}\mathbf{\Omega}$$

$$V_B = V_{BE} + V_E = 0.7 + 2 \text{ V} = 2.7 \text{ V}$$

The equations for the calculation of the base resistors R_1 and R_2 will require a little thought. Using the value of base voltage calculated above and the value of the supply voltage will provide one equation—but there are two unknowns, R_1 and R_2. An additional equation can be obtained from an understanding of the operation of these two resistors in providing the necessary base voltage. For the circuit to operate efficiently it is assumed that the current through R_1 and R_2 should be approximately equal and much larger than the base current (at least 10:1). This fact and the voltage-divider equation for the base voltage provide the two relationships necessary to determine the base resistors. That is,

$$R_2 \leq \tfrac{1}{10}\beta R_E$$

and

$$V_B = \frac{R_2}{R_1 + R_2} V_{CC}$$

Substitution yields

$$R_2 \leq \tfrac{1}{10} (80)(0.2 \text{ k}\Omega)$$

$$= \textbf{1.6 k}\mathbf{\Omega}$$

$$V_B = 2.7 \text{ V} = \frac{(1.6 \text{ k}\Omega)(20 \text{ V})}{R_1 + 1.6 \text{ k}\Omega}$$

and

$$2.7 R_1 + 4.32 \text{ k}\Omega = 32 \text{ k}\Omega$$

$$2.7 R_1 = 27.68 \text{ k}\Omega$$

$$R_1 = \textbf{10.25 k}\mathbf{\Omega} \quad \text{(use 10 k}\Omega\text{)}$$

4.9 TRANSISTOR SWITCHING NETWORKS

The application of transistors is not limited solely to the amplification of signals. Through proper design it can be used as a switch for computer and control applications. The network of Fig. 4.52a can be employed as an *inverter* in computer logic circuitry. Note that the output voltage V_C is opposite to that applied to the base or input terminal. In addition, note the absence of a dc supply connected to the base circuit. The only dc source is connected to the collector or output side and for computer applications is typically equal to the magnitude of the "high" side of the applied signal—in this case 5 V.

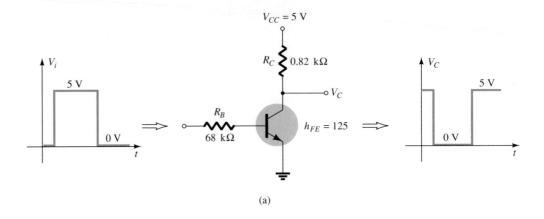

(a)

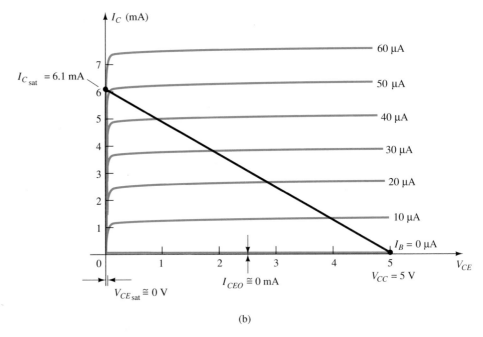

(b)

Figure 4.52 Transistor inverter.

Proper design for the inversion process requires that the operating point switch from cutoff to saturation along the load line depicted in Fig. 4.52b. For our purposes we will assume that $I_C = I_{CEO} = 0$ mA when $I_B = 0$ μA (an excellent approximation in light of improving construction techniques), as shown in Fig. 4.52b. In addition, we will assume that $V_{CE} = V_{CE_{sat}} = 0$ V rather than the typical 0.1- to 0.3-V level.

When $V_i = 5$ V the transistor will be "on" and the design must ensure that the network is heavily saturated by a level of I_B greater than that associated with the I_B curve appearing near the saturation level. In Fig. 4.52b this requires that $I_B > 50$ μA. The saturation level for the collector current for the circuit of Fig. 4.52a is defined by

$$I_{C_{sat}} = \frac{V_{CC}}{R_C} \qquad (4.45)$$

The level of I_B in the active region just before saturation results can be approximated by the following equation:

$$I_{B_{max}} \cong \frac{I_{C_{sat}}}{\beta_{dc}}$$

For the saturation level we must therefore ensure that the following condition is satisfied:

$$\boxed{I_B > \frac{I_{C_{sat}}}{\beta_{dc}}} \tag{4.46}$$

For the network of Fig. 4.52b when $V_i = 5$ V, the resulting level of I_B is the following:

$$I_B = \frac{V_i - 0.7 \text{ V}}{R_B} = \frac{5 \text{ V} - 0.7 \text{ V}}{68 \text{ k}\Omega} = 63 \ \mu A$$

and

$$I_{C_{sat}} = \frac{V_{CC}}{R_C} = \frac{5 \text{ V}}{0.82 \text{ k}\Omega} \cong 6.1 \text{ mA}$$

Testing Eq. (4.46) gives

$$I_B = 63 \ \mu A > \frac{I_{C_{sat}}}{\beta_{dc}} = \frac{6.1 \text{ mA}}{125} = 48.8 \ \mu A$$

which is satisfied. Certainly, any level of I_B greater than 60 μA will pass through a Q-point on the load line that is very close to the vertical axis.

For $V_i = 0$ V, $I_B = 0 \ \mu A$, and since we are assuming that $I_C = I_{CEO} = 0$ mA, the voltage drop across R_C as determined by $V_{R_C} = I_C R_C = 0$ V, resulting in $V_C = +5$ V for the response indicated in Fig. 4.52a.

In addition to its contribution to computer logic the transistor can also be employed as a switch using the same extremities of the load line. At saturation the current I_C is quite high and the voltage V_{CE} very low. The result is a resistance level between the two terminals determined by

$$R_{sat} = \frac{V_{CE_{sat}}}{I_{C_{sat}}}$$

and depicted in Fig. 4.53.

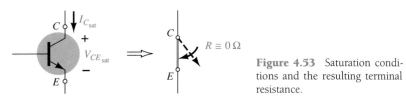

Figure 4.53 Saturation conditions and the resulting terminal resistance.

Using a typical average value of $V_{CE_{sat}}$ such as 0.15 V gives

$$R_{sat} = \frac{V_{CE_{sat}}}{I_{C_{sat}}} = \frac{0.15 \text{ V}}{6.1 \text{ mA}} = 24.6 \ \Omega$$

which is a relatively low value and $\cong 0 \ \Omega$ when placed in series with resistors in the kilohm range.

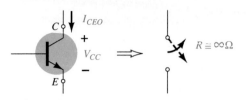

Figure 4.54 Cutoff conditions and the resulting terminal resistance.

For $V_i = 0$ V as shown in Fig. 4.54, the cutoff condition will result in a resistance level of the following magnitude:

$$R_{\text{cutoff}} = \frac{V_{CC}}{I_{CEO}} = \frac{5 \text{ V}}{0 \text{ mA}} = \infty \; \Omega$$

resulting in the open-circuit equivalence. For a typical value of $I_{CEO} = 10 \; \mu\text{A}$, the magnitude of the cutoff resistance is

$$R_{\text{cutoff}} = \frac{V_{CC}}{I_{CEO}} = \frac{5 \text{ V}}{10 \; \mu\text{A}} = \textbf{500 k}\boldsymbol{\Omega}$$

which certainly approaches an open-circuit equivalence for many situations.

EXAMPLE 4.24   Determine R_B and R_C for the transistor inverter of Fig. 4.55 if $I_{C_{\text{sat}}} = 10$ mA.

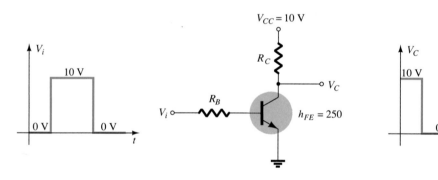

Figure 4.55 Inverter for Example 4.24.

Solution

At saturation:

$$I_{C_{\text{sat}}} = \frac{V_{CC}}{R_C}$$

and

$$10 \text{ mA} = \frac{10 \text{ V}}{R_C}$$

so that

$$R_C = \frac{10 \text{ V}}{10 \text{ mA}} = 1 \text{ k}\Omega$$

At saturation:

$$I_B \cong \frac{I_{C_{\text{sat}}}}{\beta_{\text{dc}}} = \frac{10 \text{ mA}}{250} = 40 \; \mu\text{A}$$

Choosing $I_B = 60 \; \mu$A to ensure saturation, and using

$$I_B = \frac{V_i - 0.7 \text{ V}}{R_B}$$

we obtain
$$R_B = \frac{V_i - 0.7 \text{ V}}{I_B} = \frac{10 \text{ V} - 0.7 \text{ V}}{60 \ \mu\text{A}} = 155 \text{ k}\Omega$$

Choose $R_B = 150 \text{ k}\Omega$ which is a standard value. Then
$$I_B = \frac{V_i - 0.7 \text{ V}}{R_B} = \frac{10 \text{ V} - 0.7 \text{ V}}{150 \text{ k}\Omega} = 62 \ \mu\text{A}$$

and
$$I_B = 62 \ \mu\text{A} > \frac{I_{C_{\text{sat}}}}{\beta_{\text{dc}}} = 40 \ \mu\text{A}$$

Therefore, use $R_B = \mathbf{150 \text{ k}\Omega}$ and $R_C = \mathbf{1 \text{ k}\Omega}$.

There are transistors that are referred to as *switching transistors* due to the speed with which they can switch from one voltage level to the other. In Fig. 3.23c the periods of time defined as t_s, t_d, t_r, and t_f are provided versus collector current. Their impact on the speed of response of the collector output is defined by the collector current response of Fig. 4.56. The total time required for the transistor to switch from the "off" to the "on" state is designated as t_{on} and defined by

$$\boxed{t_{\text{on}} = t_r + t_d} \tag{4.47}$$

with t_d the delay time between the changing state of the input and the beginning of a response at the output. The time element t_r is the rise time from 10% to 90% of the final value.

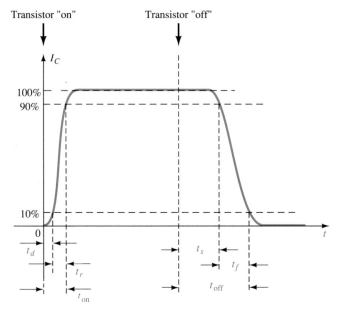

Figure 4.56 Defining the time intervals of a pulse waveform.

The total time required for a transistor to switch from the "on" to the "off" state is referred to as t_{off} and is defined by

$$\boxed{t_{\text{off}} = t_s + t_f} \tag{4.48}$$

where t_s is the storage time and t_f the fall time from 90% to 10% of the initial value.

For the general-purpose transistor of Fig. 3.23c at $I_C = 10$ mA, we find that

$$t_s = 120 \text{ ns}$$

$$t_d = 25 \text{ ns}$$

$$t_r = 13 \text{ ns}$$

and

$$t_f = 12 \text{ ns}$$

so that

$$t_{\text{on}} = t_r + t_d = 13 \text{ ns} + 25 \text{ ns} = \textbf{38 ns}$$

and

$$t_{\text{off}} = t_s + t_f = 120 \text{ ns} + 12 \text{ ns} = \textbf{132 ns}$$

Comparing the values above with the following parameters of a BSV52L switching transistor reveals one of the reasons for choosing a switching transistor when the need arises.

$$t_{\text{on}} = \textbf{12 ns} \qquad \text{and} \qquad t_{\text{off}} = \textbf{18 ns}$$

4.10 TROUBLESHOOTING TECHNIQUES

The art of troubleshooting is such a broad topic that a full range of possibilities and techniques cannot be covered in a few sections of a book. However, the practitioner should be aware of a few basic maneuvers and measurements that can isolate the problem area and possibly identify a solution.

Quite obviously, the first step in being able to troubleshoot a network is to fully understand the behavior of the network and to have some idea of the expected voltage and current levels. For the transistor in the active region the most important measurable dc level is the base-to-emitter voltage.

For an "on" transistor the voltage V_{BE} should be in the neighborhood of 0.7 V.

The proper connections for measuring V_{BE} appear in Fig. 4.57. Note that the red (positive) lead is connected to the base terminal for an *npn* transistor and the black (negative) lead to the emitter terminal. Any reading totally different from the expected level of about 0.7 V, such as 0 V, 4 V, or 12 V, or negative in value would be suspect and the device or network connections should be checked. For a *pnp* transistor the same connections can be used but a negative reading should be expected.

A voltage level of equal importance is the collector-to-emitter voltage. Recall from the general characteristics of a BJT that levels of V_{CE} in the neighborhood of 0.3 V suggest a saturated device—a condition that should not exist unless being employed in a switching mode. However:

For the typical transistor amplifier in the active region, V_{CE} is usually about 25% to 75% of V_{CC}.

For $V_{CC} = 20$ V a reading of V_{CE} of 1 to 2 V or 18 to 20 V as measured in Fig. 4.58 is certainly an uncommon result, and unless knowingly designed for this response the design and operation should be investigated. If $V_{CE} = 20$ V (with $V_{CC} = 20$ V) at least two possibilities exist—either the device (BJT) is damaged and has the

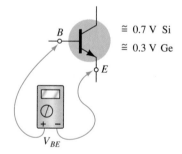

$\cong 0.7$ V Si

$\cong 0.3$ V Ge

Figure 4.57 Checking the dc level of V_{BE}.

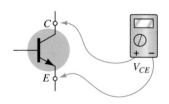

0.3 V = saturation

0 V = short-circuit state
    or poor connection

Normally a few volts
    or more

Figure 4.58 Checking the dc level of V_{CE}.

Chapter 4   DC Biasing—BJTs

characteristics of an open-circuit between collector and emitter terminals or a connection in the collector-emitter or base-emitter circuit loop is open as shown in Fig. 4.59, establishing I_C at 0 mA and $V_{R_C} = 0$ V. In Fig. 4.59 the black lead of the voltmeter is connected to the common ground of the supply and the red lead to the bottom terminal of the resistor. The absence of a collector current and a resulting drop across R_C will result in a reading of 20 V. If the meter is connected to the collector terminal of the BJT, the reading will be 0 V since V_{CC} is blocked from the active device by the open circuit. One of the most common errors in the laboratory experience is the use of the wrong resistance value for a given design. Imagine the impact of using a 680-Ω resistor for R_B rather than the design value of 680 kΩ. For $V_{CC} = 20$ V and a fixed-bias configuration, the resulting base current would be

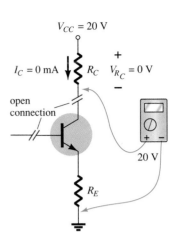

$$I_B = \frac{20 \text{ V} - 0.7 \text{ V}}{680 \ \Omega} = 28.4 \text{ mA}$$

rather than the desired 28.4 μA—a significant difference!

Figure 4.59 Effect of a poor connection or damaged device.

A base current of 28.4 mA would certainly place the design in a saturation region and possibly damage the device. Since actual resistor values are often different from the nominal color-code value (recall the common tolerance levels for resistive elements), it is time well spent to measure a resistor before inserting it in the network. The result is actual values closer to theoretical levels and some insurance that the correct resistance value is being employed.

There are times when frustration will develop. You have checked the device on a curve tracer or other BJT testing instrumentation and it looks good. All resistor levels seem correct, the connections appear solid, and the proper supply voltage has been applied—what next? Now the troubleshooter must strive to attain a higher level of sophistication. Could it be that the internal connection between the wire and the end connection of a lead is faulty? How often has simply touching a lead at the proper point created a "make or break" situation between connections? Perhaps the supply was turned on and set at the proper voltage but the current-limiting knob was left in the zero position, preventing the proper level of current as demanded by the network design. Obviously, the more sophisticated the system, the broader the range of possibilities. In any case, one of the most effective methods of checking the operation of network is to check various voltage levels with respect to ground by hooking up the black (negative) lead of a voltmeter to ground and "touching" the important terminals with the red (positive) lead. In Fig. 4.60 if the red lead is connected directly to V_{CC}, it should read V_{CC} volts since the network has one common ground for the supply and network parameters. At V_C the reading should be less as determined by the drop across R_C and V_E should be less than V_C by the collector–emitter voltage V_{CE}. The failure of any of these points to register what would appear to be a reasonable level may be sufficient in itself to define the faulty connection or element. If V_{R_C} and V_{R_E} are reasonable values but V_{CE} is 0 V, the possibility exists that the BJT is damaged and displays a short-circuit equivalence between collector and emitter terminals. As noted earlier, if V_{CE} registers a level of about 0.3 V as defined by $V_{CE} = V_C - V_E$ (the difference of the two levels as measured above), the network may be in saturation with a device that may or may not be defective.

It should be somewhat obvious from the discussion above that the voltmeter section of the VOM or DMM is quite important in the troubleshooting process. Current levels are usually calculated from the voltage levels across resistors rather than "breaking" the network to insert the milliammeter section of a multimeter. On large schematics specific voltage levels are provided with respect to ground for easy checking and identification of possible problem areas. Of course, for the networks covered in this chapter one must simply be aware of typical levels within the system as defined by the applied potential and general operation of the network.

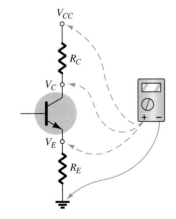

Figure 4.60 Checking voltage levels with respect to ground.

All in all, the troubleshooting process is a true test of your clear understanding of the proper behavior of a network and the ability to isolate problem areas using a few basic measurements with the appropriate instruments. Experience is the key, and that will come only with continued exposure to practical circuits.

EXAMPLE 4.25

Based on the readings provided in Fig. 4.61, determine whether the network is operating properly and if not, the probable cause.

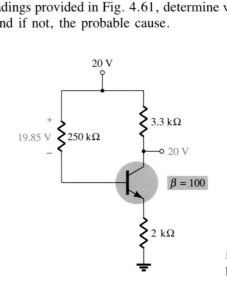

Figure 4.61 Network for Example 4.25.

Solution

The 20 V at the collector immediately reveals that $I_C = 0$ mA, due to an open circuit or a nonoperating transistor. The level of $V_{R_B} = 19.85$ V also reveals that the transistor is "off" since the difference of $V_{CC} - V_{R_B} = 0.15$ V is less than that required to turn "on" the transistor and provide some voltage for V_E. In fact, if we assume a short circuit condition from base to emitter, we obtain the following current through R_B.

$$I_{R_B} = \frac{V_{CC}}{R_B + R_E} = \frac{20 \text{ V}}{252 \text{ k}\Omega} = 79.4 \ \mu\text{A}$$

which matches that obtained from

$$I_{R_B} = \frac{V_{R_B}}{R_B} = \frac{19.85 \text{ V}}{250 \text{ k}\Omega} = 79.4 \ \mu\text{A}$$

If the network were operating properly, the base current should be

$$I_B = \frac{V_{CC} - V_{BE}}{R_B + (\beta + 1)R_E} = \frac{20 \text{ V} - 0.7 \text{ V}}{250 \text{ k}\Omega + (101)(2 \text{ k}\Omega)} = \frac{19.3 \text{ V}}{452 \text{ k}\Omega} = 42.7 \ \mu\text{A}$$

The result, therefore, is that the transistor is in a damaged state with a short-circuit condition between base and emitter.

EXAMPLE 4.26

Based on the readings appearing in Fig. 4.62, determine whether the transistor is "on" and the network is operating properly.

Solution

Based on the resistor values of R_1 and R_2 and the magnitude of V_{CC}, the voltage $V_B = 4$ V seems appropriate (and in fact it is). The 3.3 V at the emitter results in a 0.7-V drop across the base-to-emitter junction of the transistor suggesting an "on" transistor. However, the 20 V at the collector reveals that $I_C = 0$ mA, although the connection to the supply must be "solid" or the 20 V would not appear at the collector of the device. Two possibilities exist—there can be a poor connection between R_C and the collector terminal of the transistor or the transistor has an open base-to-collector junction. First check the continuity at the collector junction using an ohmmeter, and if okay, the transistor should be checked using one of the methods described in Chapter 3.

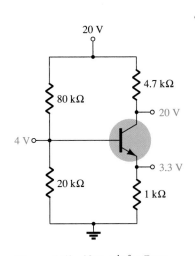

Figure 4.62 Network for Example 4.26.

4.11 PNP TRANSISTORS

The analysis thus far has been limited totally to *npn* transistors to ensure that the initial analysis of the basic configurations was as clear as possible and uncomplicated by switching between types of transistors. Fortunately, the analysis of *pnp* transistors follows the same pattern established for *npn* transistors. The level of I_B is first determined, followed by the application of the appropriate transistor relationships to determine the list of unknown quantities. In fact, the only difference between the resulting equations for a network in which an *npn* transistor has been replaced by a *pnp* transistor is the sign associated with particular quantities.

As noted in Fig. 4.63, the double-subscript notation continues as normally defined. The current directions, however, have been reversed to reflect the actual conduction directions. Using the defined polarities of Fig. 4.63, both V_{BE} and V_{CE} will be negative quantities.

Applying Kirchhoff's voltage law to the base–emitter loop will result in the following equation for the network of Fig. 4.63:

$$-I_E R_E + V_{BE} - I_B R_B + V_{CC} = 0$$

Substituting $I_E = (\beta + 1)I_B$ and solving for I_B yields

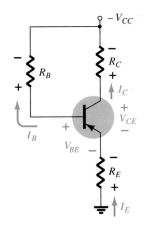

Figure 4.63 PNP transistor in an emitter-stabilized configuration.

$$I_B = \frac{V_{CC} + V_{BE}}{R_B + (\beta + 1)R_E} \tag{4.49}$$

The resulting equation is the same as Eq. (4.17) except for the sign for V_{BE}. However, in this case $V_{BE} = -0.7$ V and the substitution of values will result in the same sign for each term of Eq. (4.49) as Eq. (4.17). Keep in mind that the direction of I_B is now defined opposite of that for a *pnp* transistor as shown in Fig. 4.63.

For V_{CE} Kirchhoff's voltage law is applied to the collector–emitter loop, resulting in the following equation:

$$-I_E R_E + V_{CE} - I_C R_C + V_{CC} = 0$$

Substituting $I_E \cong I_C$ gives

$$V_{CE} = -V_{CC} + I_C(R_C + R_E) \tag{4.50}$$

The resulting equation has the same format as Eq. (4.19), but the sign in front of each term on the right of the equal sign has changed. Since V_{CC} will be larger than the magnitude of the succeeding term, the voltage V_{CE} will have a negative sign, as noted in an earlier paragraph.

EXAMPLE 4.27

Determine V_{CE} for the voltage-divider bias configuration of Fig. 4.64.

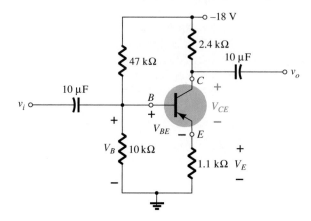

Figure 4.64 PNP transistor in a voltage-divider bias configuration.

Solution

Testing the condition

$$\beta R_E \geq 10\,R_2$$

results in $(120)(1.1\ \text{k}\Omega) \geq 10\ (10\ \text{k}\Omega)$

$$132\ \text{k}\Omega \geq 100\ \text{k}\Omega\ (\textit{satisfied})$$

Solving for V_B, we have

$$V_B = \frac{R_2 V_{CC}}{R_1 + R_2} = \frac{(10\ \text{k}\Omega)(-18\ \text{V})}{47\ \text{k}\Omega + 10\ \text{k}\Omega} = -3.16\ \text{V}$$

Note the similarity in format of the equation with the resulting negative voltage for V_B.

Applying Kirchhoff's voltage law around the base–emitter loop yields

$$+V_B - V_{BE} - V_E = 0$$

and
$$V_E = V_B - V_{BE}$$

Substituting values, we obtain

$$V_E = -3.16\ \text{V} - (-0.7\ \text{V})$$

$$= -3.16\ \text{V} + 0.7\ \text{V}$$

$$= -2.46\ \text{V}$$

Note in the equation above that the standard single- and double-subscript notation is employed. For an *npn* transistor the equation $V_E = V_B - V_{BE}$ would be exactly the same. The only difference surfaces when the values are substituted.

The current

$$I_E = \frac{V_E}{R_E} = \frac{2.46\ \text{V}}{1.1\ \text{k}\Omega} = 2.24\ \text{mA}$$

For the collector–emitter loop:

$$-I_E R_E + V_{CE} - I_C R_C + V_{CC} = 0$$

Substituting $I_E \cong I_C$ and gathering terms, we have

$$V_{CE} = -V_{CC} + I_C(R_C + R_E)$$

Substituting values, gives

$$V_{CE} = -18 \text{ V} + (2.24 \text{ mA})(2.4 \text{ k}\Omega + 1.1 \text{ k}\Omega)$$

$$= -18 \text{ V} + 7.84 \text{ V}$$

$$= \mathbf{-10.16 \text{ V}}$$

4.12 BIAS STABILIZATION

The stability of a system is a measure of the sensitivity of a network to variations in its parameters. In any amplifier employing a transistor the collector current I_C is sensitive to each of the following parameters:

β: *increases with increase in temperature*

$|V_{BE}|$ *decreases about 7.5 mV per degree Centigrade (°C) increase in temperature*

I_{CO} *(reverse saturation current): doubles in value for every 10°C increase in temperature*

Any or all of these factors can cause the bias point to drift from the designed point of operation. Table 4.1 reveals how the level of I_{CO} and V_{BE} changed with increase in temperature for a particular transistor. At room temperature (about 25°C) $I_{CO} = 0.1$ nA, while at 100°C (boiling point of water) I_{CO} is about 200 times larger at 20 nA. For the same temperature variation β increased from 50 to 80 and V_{BE} dropped from 0.65 V to 0.48 V. Recall that I_B is quite sensitive to the level of V_{BE}, especially for levels beyond the threshold value.

TABLE 4.1 Variation of Silicon Transistor Parameters with Temperature

T (°C)	I_{CO} (nA)	β	V_{BE} (V)
−65	0.2×10^{-3}	20	0.85
25	0.1	50	0.65
100	20	80	0.48
175	3.3×10^3	120	0.3

The effect of changes in leakage current (I_{CO}) and current gain (β) on the dc bias point is demonstrated by the common-emitter collector characteristics of Fig. 4.65a and b. Figure 4.65 shows how the transistor collector characteristics change from a temperature of 25°C to a temperature of 100°C. Note that the significant increase in leakage current not only causes the curves to rise but there is also an increase in beta as revealed by the larger spacing between curves.

An operating point may be specified by drawing the circuit dc load line on the graph of the collector characteristic and noting the intersection of the load line and the dc base current set by the input circuit. An arbitrary point is marked in Fig. 4.65a at $I_B = 30 \text{ } \mu\text{A}$. Since the fixed-bias circuit provides a base current whose value depends approximately on the supply voltage and base resistor, neither of which is affected by temperature or the change in leakage current or beta, the same base current magnitude will exist at high temperatures as indicated on the graph of Fig. 4.65b. As the figure shows, this will result in the dc bias point's shifting to a higher collector current and a lower collector–emitter voltage operating point. In the extreme, the transistor could be driven into saturation. In any case the new operating point may not be at all

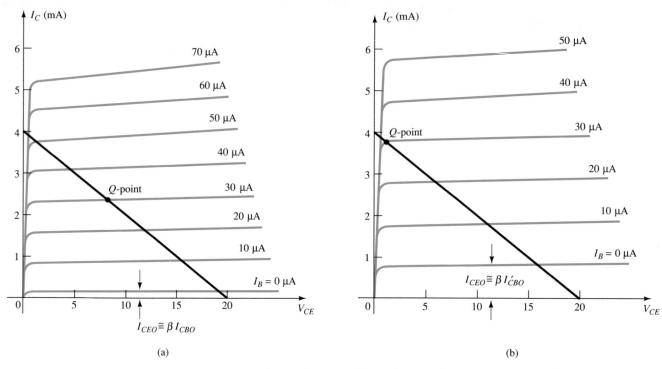

(a)   (b)

Figure 4.65 Shift in dc bias point (Q-point) due to change in temperature: (a) 25°C; (b) 100°C.

satisfactory and considerable distortion may result because of the bias-point shift. A better bias circuit is one that will stabilize or maintain the dc bias initially set, so that the amplifier can be used in a changing-temperature environment.

Stability Factors, $S(I_{CO})$, $S(V_{BE})$ and $S(\beta)$

A stability factor, S, is defined for each of the parameters affecting bias stability as listed below:

$$S(I_{CO}) = \frac{\Delta I_C}{\Delta I_{CO}} \qquad (4.51)$$

$$S(V_{BE}) = \frac{\Delta I_C}{\Delta V_{BE}} \qquad (4.52)$$

$$S(\beta) = \frac{\Delta I_C}{\Delta \beta} \qquad (4.53)$$

In each case the delta symbol (Δ) signifies change in that quantity. The numerator of each equation is the change in collector current as established by the change in the quantity in the denominator. For a particular configuration if a change in I_{CO} fails to produce a significant change in I_C, the stability factor defined by $S(I_{CO}) = \Delta I_C/\Delta I_{CO}$ will be quite small. In other words:

Networks that are quite stable and relatively insensitive to temperature variations have low stability factors.

In some ways it would seem more appropriate to consider the quantities defined by Eqs. 4.51–4.53 to be sensitivity factors because:

Chapter 4   DC Biasing—BJTs

The higher the stability factor, the more sensitive the network to variations in that parameter.

The study of stability factors requires the knowledge of differential caculus. Our purpose, here, however, is to review the results of the mathematical analysis and to form an overall assessment of the stability factors for a few of the most popular bias configurations. A great deal of literature is available on this subject and if time permits you are encouraged to read more on the subject.

$S(I_{CO})$:

EMITTER-BIAS CONFIGURATION

For the emitter-bias configuration an analysis of the network will result in

$$S(I_{CO}) = (\beta + 1)\frac{1 + R_B/R_E}{(\beta + 1) + R_B/R_E} \qquad (4.54)$$

For $R_B/R_E \gg (\beta + 1)$, Eq. (4.54) will reduce to the following:

$$S(I_{CO}) = \beta + 1 \qquad (4.55)$$

as shown on the graph of $S(I_{CO})$ versus R_B/R_E in Fig. 4.66.

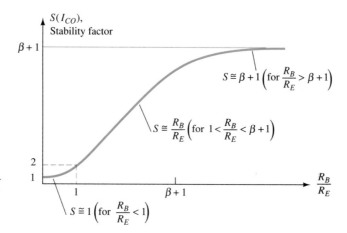

Figure 4.66 Variation of stability factor $S(I_{CO})$ with the resistor ratio R_B/R_E for the emitter-bias configuration.

For $R_B/R_E \ll 1$, Eq. (4.54) will approach the following level (as shown in Fig. 4.66):

$$S(I_{CO}) = (\beta + 1)\frac{1}{(\beta + 1)} = \rightarrow 1 \qquad (4.56)$$

revealing that the stability factor will approach its lowest level as R_E becomes sufficiently large. Keep in mind, however, that good bias control normally requires that R_B be greater than R_E. The result, therefore, is a situation where the best stability levels are associated with poor design criteria. Obviously, a trade-off must occur that will satisfy both the stability and bias specifications. It is interesting to note in Fig. 4.66 that the lowest value of $S(I_{CO})$ is 1, revealing that I_C will always increase at a rate equal to or greater than I_{CO}.

For the range where R_B/R_E ranges between 1 and $(\beta + 1)$, the stability factor will be determined by

$$S(I_{CO}) \cong \frac{R_B}{R_E} \qquad (4.57)$$

as shown in Fig. 4.66. The results reveal that the emitter-bias configuration is quite stable when the ratio R_B/R_E is as small as possible and the least stable when the same ratio approaches $(\beta + 1)$.

EXAMPLE 4.28

Calculate the stability factor and the change in I_C from 25°C to 100°C for the transistor defined by Table 4.1 for the following emitter-bias arrangements.
(a) $R_B/R_E = 250$ ($R_B = 250R_E$)
(b) $R_B/R_E = 10$ ($R_B = 10\ R_E$).
(c) $R_B/R_E = 0.01$ ($R_E = 100\ R_B$).

Solution

(a) $S(I_{CO}) = (\beta + 1)\dfrac{1 + R_B/R_E}{1 + \beta + R_B/R_E}$

$= 51\left(\dfrac{1 + 250}{51 + 250}\right) = 51\left(\dfrac{251}{301}\right)$

$\cong \mathbf{42.53}$

which begins to approach the level defined by $\beta + 1 = 51$

$\Delta I_C = [S(I_{CO})](\Delta I_{CO}) = (42.53)(19.9\text{ nA})$

$\cong \mathbf{0.85\ \mu A}$

(b) $S(I_{CO}) = (\beta + 1)\dfrac{1 + R_B/R_E}{1 + \beta + R_B/R_E}$

$= 51\left(\dfrac{1 + 10}{51 + 10}\right) = 51\left(\dfrac{11}{61}\right)$

$\cong \mathbf{9.2}$

$\Delta I_C = [S(I_{CO})](\Delta I_{CO}) = (9.2)(19.9\text{ nA})$

$\cong \mathbf{0.18\ \mu A}$

(c) $S(I_{CO}) = (\beta + 1)\dfrac{1 + R_B/R_E}{1 + \beta + R_B/R_E}$

$= 51\left(\dfrac{1 + 0.01}{51 + 0.01}\right) = 51\left(\dfrac{1.01}{51.01}\right)$

$\cong \mathbf{1.01}$

which is certainly very close to the level of 1 forecast if $R_B/R_E \ll 1$.

$\Delta I_C = [S(I_{CO})](\Delta I_{CO}) = 1.01(19.9\text{ nA})$

$= \mathbf{20.1\ nA}$

Example 4.28 reveals how lower and lower levels of I_{CO} for the modern-day BJT transistor have improved the stability level of the basic bias configurations. Even though the change in I_C is considerably different in a circuit having ideal stability ($S = 1$) from one having a stability factor of 34.1, the change in I_C is not that significant. For example, the amount of change in I_C from a dc bias current set at, say, 2 mA, would be from 2 mA to 2.085 mA in the worst case, which is obviously small enough to be ignored for most applications. Some power transistors exhibit larger leakage currents, but for most amplifier circuits the lower levels of I_{CO} have had a very positive impact on the stability question.

FIXED-BIAS CONFIGURATION

For the fixed-bias configuration if we multiply the top and bottom of Eq. 4.54 by R_E and then plug in $R_E = 0\Omega$ the following equation will result:

$$\boxed{S(I_{CO}) = \beta + 1} \tag{4.58}$$

Chapter 4 DC Biasing—BJTs

Note that the resulting equation matches the maximum value for the emitter-bias configuration. The result is a configuration with a poor stability factor and a high sensitivity to variations in I_{CO}.

Voltage-Divider Bias Configuration

Recall from Section 4.5 the development of the Thévenin equivalent network appearing in Fig. 4.67 for the voltage-divider bias configuration. For the network of Fig. 4.67 the equation for $S(I_{CO})$ the following:

$$S(I_{CO}) = (\beta + 1)\frac{1 + R_{Th}/R_E}{(\beta + 1) + R_{Th}/R_E} \qquad (4.59)$$

Note the similarities with Eq. (4.54), where it was determined that $S(I_{CO})$ had its lowest level and the network had its greatest stability when $R_E > R_B$. For Eq. (4.59) the corresponding condition is $R_E > R_{Th}$ or R_{Th}/R_E should be as small as possible. For the voltage-divider bias configuration R_{Th} can be much less than the corresponding R_B of the emitter-bias configuration and still have an effective design.

Feedback-Bias Configuration ($R_E = 0\ \Omega$)

In this case,

$$S(I_{CO}) = (\beta + 1)\frac{1 + R_B/R_C}{(\beta + 1) + R_B/R_C} \qquad (4.60)$$

Since the equation is similar in format to that obtained for the emitter-bias and voltage-divider bias configurations, the same conclusions regarding the ratio R_B/R_C can be applied here also.

Physical Impact

Equations of the type developed above often fail to provide a physical sense for why the networks perform as they do. We are now aware of the relative levels of stability and how the choice of parameters can affect the sensitivity of the network, but without the equations it may be difficult for us to explain in words why one network is more stable than another. The next few paragraphs attempt to fill this void through the use of some of the very basic relationships associated with each configuration.

For the fixed-bias configuration of Fig. 4.68a, the equation for the base current is the following:

$$I_B = \frac{V_{CC} - V_{BE}}{R_B}$$

with the collector current determined by

$$I_C = \beta I_B + (\beta + 1)I_{CO} \qquad (4.61)$$

If I_C as defined by Eq. (4.61) should increase due to an increase in I_{CO}, there is nothing in the equation for I_B that would attempt to offset this undesirable increase in current level (assuming V_{BE} remains constant). In other words, the level of I_C would continue to rise with temperature with I_B maintaining a fairly constant value—a very unstable situation.

For the emitter-bias configuration of Fig. 4.68b, however, an increase in I_C due to an increase in I_{CO} will cause the voltage $V_E = I_E R_E \cong I_C R_E$ to increase. The result is a drop in the level of I_B as determined by the following equation:

Figure 4.67 Equivalent circuit for the voltage-divider configuration.

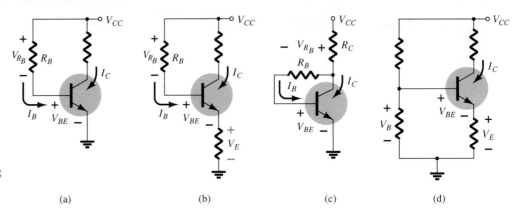

Figure 4.68 Review of biasing managements and the stability factors $S(I_{CO})$.

(a) (b) (c) (d)

$$I_B \downarrow \; = \; \frac{V_{CC} - V_{BE} - V_E \uparrow}{R_B} \qquad (4.62)$$

A drop in I_B will have the effect of reducing the level of I_C through transistor action and thereby offset the tendency of I_C to increase due to an increase in temperature. In total, therefore, the configuration is such that there is a reaction to an increase in I_C that will tend to oppose the change in bias conditions.

The feedback configuration of Fig. 4.68c operates in much the same way as the emitter-bias configuration when it comes to levels of stability. If I_C should increase due to an increase in temperature, the level of V_{R_C} will increase in the following equation:

$$I_B \downarrow \; = \; \frac{V_{CC} - V_{BE} - V_{R_C} \uparrow}{R_B} \qquad (4.63)$$

and the level of I_B will decrease. The result is a stabilizing effect as described for the emitter-bias configuration. One must be aware that the action described above does not happen in a step-by-step sequence. Rather, it is a simultaneous action to maintain the established bias conditions. In other words, the very instant I_C begins to rise the network will sense the change and the balancing effect described above will take place.

The most stable of the configurations is the voltage-divider bias network of Fig. 4.68d. If the condition $\beta R_E \gg 10 R_2$ is satisfied, the voltage V_B will remain fairly constant for changing levels of I_C. The base-to-emitter voltage of the configuration is determined by $V_{BE} = V_B - V_E$. If I_C should increase, V_E will increase as described above, and for a constant V_B the voltage V_{BE} will drop. A drop in V_{BE} will establish a lower level of I_B which will try to offset the increased level of I_C.

$S(V_{BE})$:

The stability factor defined by

$$S(V_{BE}) = \frac{\Delta I_C}{\Delta V_{BE}}$$

will result in the following equation for the emitter-bias configuration:

$$S(V_{BE}) = \frac{-\beta}{R_B + (\beta + 1)R_E} \qquad (4.64)$$

Substituting $R_E = 0 \; \Omega$ as occurs for the fixed-bias configuration will result in

$$S(V_{BE}) = -\frac{\beta}{R_B} \qquad (4.65)$$

Equation (4.64) can be written in the following form:

$$S(V_{BE}) = \frac{-\beta/R_E}{R_B/R_E + (\beta + 1)} \qquad (4.66)$$

Substituting the condition $(\beta + 1) \gg R_B/R_E$ will result in the following equation for $S(V_{BE})$:

$$S(V_{BE}) \cong \frac{-\beta/R_E}{\beta + 1} \cong \frac{-\beta/R_E}{\beta} = -\frac{1}{R_E} \qquad (4.67)$$

revealing that the larger the resistance R_E, the lower the stability factor and the more stable the system.

Determine the stability factor $S(V_{BE})$ and the change in I_C from 25°C to 100°C for the transistor defined by Table 4.1 for the following bias arrangements.
(a) Fixed-bias with $R_B = 240$ kΩ and $\beta = 100$.
(b) Emitter-bias with $R_B = 240$ kΩ, $R_E = 1$ kΩ, and $\beta = 100$.
(c) Emitter-bias with $R_B = 47$ kΩ, $R_E = 4.7$ kΩ, and $\beta = 100$.

EXAMPLE 4.29

Solution

(a) Eq. (4.69): $\quad S(V_{BE}) = -\dfrac{\beta}{R_B}$

$$= -\frac{100}{240 \text{ k}\Omega}$$

$$= -0.417 \times 10^{-3}$$

and $\qquad \Delta I_C = [S(V_{BE})](\Delta V_{BE})$

$$= (-0.417 \times 10^{-3})(0.48 \text{ V} - 0.65 \text{ V})$$

$$= (-0.417 \times 10^{-3})(-0.17 \text{ V})$$

$$= 70.9 \ \mu\text{A}$$

(b) In this case, $(\beta + 1) = 101$ and $R_B/R_E = 240$. The condition $(\beta + 1) \gg R_B/R_E$ is not satisfied, negating the use of Eq. (4.67) and requiring the use of Eq. (4.64).

Eq. (4.64): $\quad S(V_{BE}) = \dfrac{-\beta}{R_B + (\beta + 1)R_E}$

$$= \frac{-100}{240 \text{ k}\Omega + (101)1 \text{ k}\Omega} = -\frac{100}{341 \text{ k}\Omega}$$

$$= -0.293 \times 10^{-3}$$

which is about 30% less than the fixed-bias value due to the additional $(\beta + 1)R_E$ term in the denominator of the $S(V_{BE})$ equation.

$$\Delta I_C = [S(V_{BE})](\Delta V_{BE})$$

$$= (-0.293 \times 10^{-3})(-0.17 \text{ V})$$

$$\cong 50 \ \mu\text{A}$$

(c) In this case,

$$(\beta + 1) = 101 \gg \frac{R_B}{R_E} = \frac{47 \text{ k}\Omega}{4.7 \text{ k}\Omega} = 10 \text{ (satisfied)}$$

Eq. (4.67): $S(V_{BE}) = -\dfrac{1}{R_E}$

$$= -\dfrac{1}{4.7\ \text{k}\Omega}$$

$$= -0.212 \times 10^{-3}$$

and $\Delta I_C = [S(V_{BE})](\Delta V_{BE})$

$$= (-0.212 \times 10^{-3})(-0.17\ \text{V})$$

$$= \mathbf{36.04\ \mu A}$$

In Example 4.29 the increase of 70.9 μA will have some impact on the level of I_{C_Q}. For a situation where $I_{C_Q} = 2$ mA the resulting collector current will increase to

$$I_{C_Q} = 2\ \text{mA} + 70.9\ \mu\text{A}$$

$$= 2.0709\ \text{mA}$$

a 3.5% increase

For the voltage-divider configuration the level of R_B will be changed to R_{Th} in Eq. (4.64) (as defined by Fig. 4.67). In Example 4.29 the use of $R_B = 47$ kΩ is a questionable design. However, R_{Th} for the voltage-divider configuration can be this level or lower and still maintain good design characteristics. The resulting equation for $S(V_{BE})$ for the feedback network will be similar to that of Eq. (4.64) with R_E replaced by R_C.

$S(\beta)$:

The last stability factor to be investigated is that of $S(\beta)$. The mathematical development is more complex than that encountered for $S(I_{CO})$ and $S(V_{BE})$, as suggested by the following equation for the emitter-bias configuration:

$$S(\beta) = \frac{\Delta I_C}{\Delta \beta} = \frac{I_{C_1}(1 + R_B/R_E)}{\beta_1(1 + \beta_2 + R_B/R_E)} \tag{4.68}$$

The notation I_{C_1} and β_1 is used to define their values under one set of network conditions while the notation β_2 is used to define the new value of beta as established by such causes as temperature change, variation in β for the same transistor or a change in transistors.

EXAMPLE 4.30

Determine I_{C_Q} at a temperature of 100°C if $I_{C_Q} = 2$ mA at 25°C. Use the transistor described by Table 4.1, where $\beta_1 = 50$ and $\beta_2 = 80$ and a resistance ratio R_B/R_E of 20.

Solution

Eq. 4.68: $S(\beta) = \dfrac{I_{C_1}(1 + R_B/R_E)}{\beta_1(1 + \beta_2 + R_B/R_E)}$

$$= \frac{(2 \times 10^{-3})(1 + 20)}{(50)(1 + 80 + 20)} = \frac{42 \times 10^{-3}}{5050}$$

$$= \mathbf{8.32 \times 10^{-6}}$$

and $\Delta I_C = [S(\beta)][\Delta \beta]$

$$= (8.32 \times 10^{-6})(30)$$

$$\cong \mathbf{0.25\ mA}$$

In conclusion, therefore, the collector current changed from 2 mA at room temperature to 2.25 mA at 100°C, representing a change of 12.5%

The fixed-bias configuration is defined by $S(\beta) = I_{C_1}/\beta_1$ and R_B of Eq. 4.68 can be replaced by R_{Th} for the voltage-divider configuration.

For the collector feedback configuration with $R_E = 0\Omega$

$$S(\beta) = \frac{I_{C_1}(R_B + R_C)}{\beta_1(R_B + R_C(1 + \beta_2))} \qquad (4.69)$$

Summary

Now that the three stability factors of importance have been introduced, the total effect on the collector current can be determined using the following equation:

$$\Delta I_C = S(I_{CO})\Delta I_{CO} + S(V_{BE})\Delta V_{BE} + S(\beta)\Delta\beta \qquad (4.70)$$

The equation may initially appear quite complex, but take note that each component is simply a stability factor for the configuration multiplied by the resulting change in a parameter between the temperature limits of interest. In addition, the ΔI_C to be determined is simply the change in I_C from the level at room temperature.

For instance, if we examine the fixed-bias configuration, Eq. (4.70) becomes the following:

$$\Delta I_C = (\beta + 1)\Delta I_{CO} - \frac{\beta}{R_B}\Delta V_{BE} + \frac{I_{C_1}}{\beta_1}\Delta\beta \qquad (4.71)$$

after substituting the stability factors as derived in this section. Let us now use Table 4.1 to find the change in collector current for a temperature change from 25°C (room temperature) to 100°C (the boiling point of water). For this range the table reveals that

$$\Delta I_{CO} = 20 \text{ nA} - 0.1 \text{ nA} = 19.9 \text{ nA}$$

$$\Delta V_{BE} = 0.48 \text{ V} - 0.65 \text{ V} = -0.17 \text{ V} \quad \text{(note the sign)}$$

and $\qquad \Delta\beta = 80 - 50 = 30$

Starting with a collector current of 2 mA with an R_B of 240 kΩ, the resulting change in I_C due to an increase in temperature of 75°C is the following:

$$\Delta I_C = (50 + 1)(19.9 \text{ nA}) - \frac{50}{240 \text{ k}\Omega}(-0.17 \text{ V}) + \frac{2 \text{ mA}}{50}(30)$$

$$= 1.01 \text{ } \mu A + 35.42 \text{ } \mu A + 1200 \text{ } \mu A$$

$$= 1.236 \text{ mA}$$

which is a significant change due primarily to the change in β. The collector current has increased from 2 mA to 3.236 mA—but this was expected in the sense that we recognize from the content of this section that the fixed-bias configuration is the least stable.

If the more stable voltage-divider configuration were employed with a ratio $R_{Th}/R_E = 2$ and $R_E = 4.7$ kΩ, then

$$S(I_{CO}) = 2.89, \qquad S(V_{BE}) = -0.2 \times 10^{-3}, \qquad S(\beta) = 1.445 \times 10^{-6}$$

and $\qquad \Delta I_C = (2.89)(19.9 \text{ nA}) - 0.2 \times 10^{-3}(-0.17 \text{ V}) + 1.445 \times 10^{-6}(30)$

$$= 57.51 \text{ nA} + 34 \text{ } \mu A + 43.4 \text{ } \mu A$$

$$= 0.077 \text{ mA}$$

The resulting collector current is 2.077 mA or essentially 2.1 mA, compared to the 2.0 mA at 25°C. The network is obviously a great deal more stable then the fixed-bias configuration, as mentioned in earlier discussions. In this case, $S(\beta)$ did not override the other two factors and the effects of $S(V_{BE})$ and $S(I_{CO})$ were equally important. In fact, at higher temperatures the effects of $S(I_{CO})$ and $S(V_{BE})$ will be greater than $S(\beta)$ for the device of Table 4.1. For temperatures below 25°C I_C will decrease with increasingly negative temperature levels.

The effect of $S(I_{CO})$ in the design process is becoming a lesser concern because of improved manufacturing techniques that continue to lower the level of $I_{CO} = I_{CBO}$. It should also be mentioned that for a particular transistor the variation in levels of I_{CBO} and V_{BE} from one transistor to another in a lot is almost negligible compared to the variation in beta. In addition, the results of the analysis above support the fact that for a good stabilized design:

The ratio R_B/R_E or R_{Th}/R_E should be as small as possible with due considera- tion to all aspects of the design, including the ac response.

Although the analysis above may have been clouded by some of the complex equations for some of the sensitivities, the purpose here was to develop a higher level of awareness of the factors that go into a good design and to be more intimate with the transistor parameters and their impact on the network's performance. The analysis of the earlier sections was for idealized situations with nonvarying parameter values. We are now more aware of how the dc response of the design can vary with the parameter variations of a transistor.

4.13   COMPUTER ANALYSIS

The content of this section will include an analysis of the voltage-divider network of Example 4.7 using both BASIC and PSpice. It will provide an excellent opportunity to compare the relative advantages of each.

PSpice

The network of Example 4.7 is redrawn in Fig. 4.69 with the chosen nodes for the PSpice analysis. The input file for the network appears in Fig. 4.70. Note that all the parameters are defined between the indicated nodes, with the first node assumed to be at the higher potential. The format of the transistor statement and its .MODEL entry are as defined in Chapter 3. If specific quantities such as $I(RC) = I_{R_C} = I_C$ and $V(3,4) = V_{CE}$ are required rather than simply a listing of all the nodal voltages, a .DC control statement must be added as indicated. Following the .DC indication, the dc supply is specified at the level to be employed. If the 22 V is repeated as in this case, the analysis will only be performed at that level. If the second level is different, the software package will perform the analysis at every level at and between the two

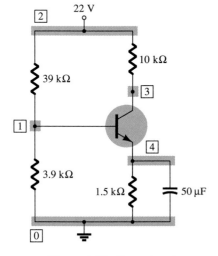

Figure 4.69 Network to be analyzed using PSpice.

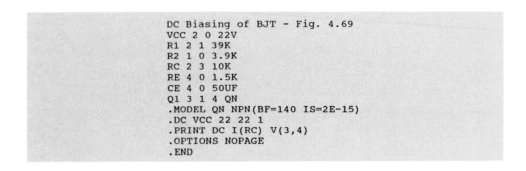

Figure 4.70 Input file for the PSpice analysis of the network of Fig. 4.69.

```
DC Biasing of BJT - Fig. 4.69
VCC 2 0 22V
R1 2 1 39K
R2 1 0 3.9K
RC 2 3 10K
RE 4 0 1.5K
CE 4 0 50UF
Q1 3 1 4 QN
.MODEL QN NPN(BF=140 IS=2E-15)
.DC VCC 22 22 1
.PRINT DC I(RC) V(3,4)
.OPTIONS NOPAGE
.END
```

```
**** 02/26/90 ******* Evaluation PSpice (January 1989) ******* 07:29:57 ****

  DC Biasing of BJT - Fig. 4.69

****         CIRCUIT DESCRIPTION

*************************************************************************

VCC 2 0 22V
R1 2 1 39K
R2 1 0 3.9K
RC 2 3 10K
RE 4 0 1.5K
CE 4 0 50UF
Q1 3 1 4 QN
.MODEL QN NPN(BF=140 IS=2E-15)
.DC VCC 22 22 1
.PRINT DC I(RC) V(3,4)
.OPTIONS NOPAGE
.END

****         BJT MODEL PARAMETERS
                 QN
                 NPN
       IS     2.000000E-15
       BF    140


****         DC TRANSFER CURVES            TEMPERATURE =    27.000 DEG C
    VCC          I(RC)         V(3,4)
    2.200E+01    8.512E-04    1.220E+01
```

Figure 4.71 Output file for the PSpice analysis of the network of Fig. 4.69.

levels using an increment defined as the next entry—in this case 1 V. However, since the 22 V is repeated in this .DC control statement, the 1 V is required to complete the format of the statement but is ignored in the operational sequence. The .PRINT statement can then be written to specify the quantities desired in the listing of the output file.

The output file appears in Fig. 4.71 with the list of specified model parameters and the desired output levels. For both I_{C_Q} and V_{CE_Q} the results obtained using PSpice are an exact match with the solutions of Example 4.7. That is, $I_{C_Q} = 8.512\text{E-}04 = 0.8512$ mA and $V_{CE_Q} = 1.220\text{E+}01 = 12.2$ V.

BASIC

The program to be developed using BASIC will perform the same analysis as listed above but will go a step further and permit changing the configuration by specifying an open circuit or a short circuit for particular parameters. For instance, if R_2 is set equal to 1E30 ohms, it is essentially an open circuit and the emitter-bias configuration will result. If R_E is set to zero ohms with R_2 at 1E30 ohms, the fixed-bias configuration will result. In this way the range of application is greatly expanded and limits the library of programs required to perform the analysis in a particular subject area.

A summary of the equations employed appears in Table 4.2 with a summary of variables in Table 4.3. A program module starting at line 10000 is written in BASIC to perform the calculations required for the dc analysis of the network of Fig. 4.69. Line 10010 calculates the Thévenin equivalent base resistance of R_1 in parallel with R_2. Line 10020 calculates the Thévenin equivalent voltage at the base. I_B is then determined on line 10030 using a base–emitter voltage of 0.7 V. Line 10040 then tests for a cutoff bias condition which occurs if the value of V_T is less than $V_{BE} = 0.7$ V, in which case I_B is set to zero; otherwise, I_B remains as calculated on line 10030. Lines 10060 and 10070 then calculate I_C and I_E, respectively.

TABLE 4.2 Equations and Computer Statements for DC Bias Calculations Module

Equation	Computer Statement
$R_{Th} = \dfrac{R_1 R_2}{R_1 + R_2}$	RT = (R1 * R2)/(R1 + R2)
$E_{Th} = \dfrac{R_2}{R_1 + R_2} V_{CC}$	VT = (R2 * CC)/(R1 + R2)
$I_B = \dfrac{E_{Th} - 0.7}{R_{Th} + (\beta + 1)R_E}$	IB = (VT − 0.7)/(RT + (BETA + 1) * RE)
$I_C = \beta I_B$	IC = BETA * IB
$I_E = (\beta + 1)I_B$	IE = (BETA + 1) * IB
$V_E = I_E R_E$	VE = IE * RE
$V_B = V_E + 0.7$	VB = VE + 0.7
$V_C = V_{CC} - I_C R_C$	VC = CC − IC * RC
$V_{CE} = V_C - V_E$	CE = VC − VE

TABLE 4.3 Equation and Computer Variables for DC Bias Calculations Module

Equation Variable	Computer Variable
R_1	R1
R_2	R2
R_{Th}	RT
V_{CC}	CC
V_{Th}	VT
I_B	IB
V_{BE}	BE
β	BETA
R_E	RE
I_C	IC
I_E	IE
V_E	VE
V_B	VB
V_C	VC
V_{CE}	CE

Line 10090 tests for the condition of circuit saturation, setting I_C (and I_E) to the saturation value in that case; otherwise, the values of I_C and I_E remain those calculated previously. Lines 10100 through 10120 then calculate V_E, V_B, and V_E, respectively. Line 10130 calculates V_{CE} and the program module then returns to the main program.

The main program, requesting input of all appropriate circuit data, module 10000 to do the dc bias calculations, and main program steps to print out the results are provided in Fig. 4.72. Again note the very close correspondence with the results obtained earlier for I_{C_Q} and V_{CE_Q}.

```
10 REM ********************************************************
20 REM
30 REM          DC BIAS CALCULATIONS OF STANDARD CIRCUIT
40 REM
50 REM ********************************************************
60 REM
100 PRINT "This program calculates the dc bias"
110 PRINT "for a standard circuit as shown in Figure 4.69."
120 PRINT
130 PRINT "First, enter the following circuit data:"
140 INPUT "RB1=";R1
150 INPUT "RB2(use 1E30 if 'open')=";R2
160 INPUT "RE=";RE
170 INPUT "RC=";RC
180 PRINT
190 INPUT "VCC=";CC
200 PRINT
210 INPUT "Transistor beta=";BETA
220 PRINT
230 REM Now do circuit calculations
240 GOSUB 10000
250 PRINT "The results of dc bias calculations are:"
260 PRINT
270 PRINT "Circuit currents:"
280 PRINT "IB=";IB*1000000!;"uA"
290 PRINT "IC=";IC*1000;"mA"
300 PRINT "IE=";IE*1000;"mA"
310 PRINT
320 PRINT "Circuit voltages:"
330 PRINT "VB=";VB;"volts"
340 PRINT "VE=";VE;"volts"
350 PRINT "VC=";VC;"volts"
360 PRINT "VCE=";CE;"volts"
370 PRINT :PRINT
380 END
10000 REM Module to calculate dc bias of BJT circuit
10010 RT=R1*(R2/(R1+R2))
10020 VT=CC*(R2/(R1+R2))
10030 IB=(VT-.7)/(RT+(BETA+1)*RE)
10040 REM Test for cutoff condition
10050 IF VT<=.7 THEN IB=0
10060 IC=BETA*IB
10070 IE=(BETA+1)*IB
10080 REM Test for saturation condition
10090 IF IC*(RC+RE)=CC THEN IC=CC/(RE+RC) :IE=IC
10100 VE=IE*RE
10110 VB=VE+.7
10120 VC=CC-IC*RC
10130 CE=VC-VE
10140 RETURN

RUN
This program calculates the dc bias
for a standard circuit as shown in Figure 4.69.

First, enter the following circuit data:
RB1=? 39E3
RB2(use 1E30 if 'open')=? 3.9E3
RE=? 1.5E3
RC=? 10E3

VCC=? 22

Transistor beta=? 140

The results of dc bias calculations are:

Circuit currents:
IB= 6.045233 uA
IC= .8463327 mA
IE= .8523779 mA

Circuit voltages:
VB= 1.978567 volts
VE= 1.278567 volts
VC= 13.53667 volts
VCE= 12.25811 volts
```

Figure 4.72 BASIC program for the analysis of the network of Fig. 4.69.

PROBLEMS

1. For the fixed-bias configuration of Fig. 4.73, determine:
(a) I_{B_Q}.
(b) I_{C_Q}.
(c) V_{CE_Q}.
(d) V_C.
(e) V_B.
(f) V_E.

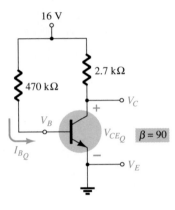

Figure 4.73 Problems 1, 4, 11, 47, 51, 52, 53, 57

2. Given the information appearing in Fig. 4.74, determine:
(a) I_C.
(b) R_C.
(c) R_B.
(d) V_{CE}.

3. Given the information appearing in Fig. 4.75, determine:
(a) I_C.
(b) V_{CC}.
(c) β.
(d) R_B.

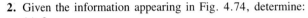

Figure 4.74 Problem 2

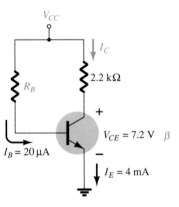

Figure 4.75 Problem 3

4. Find the saturation current ($I_{C_{sat}}$) for the fixed-bias configuration of Fig. 4.73.

* **5.** Given the BJT transistor characteristics of Fig. 4.76:
(a) Draw a load line on the characteristics determined by $E = 21$ V and $R_C = 3$ kΩ for a fixed-bias configuration.
(b) Choose an operating point midway between cutoff and saturation. Determine the value of R_B to establish the resulting operating point.
(c) What are the resulting values of I_{C_Q} and V_{CE_Q}?
(d) What is the value of β at the operating point?
(e) What is the value of α defined by the operating point?
(f) What is the saturation ($I_{C_{sat}}$) current for the design?
(g) Sketch the resulting fixed-bias configuration.
(h) What is the dc power dissipated by the device at the operating point?
(i) What is the power supplied by V_{CC}?
(j) Determine the power dissipated by the resistive elements by taking the difference between the results of parts (h) and (i).

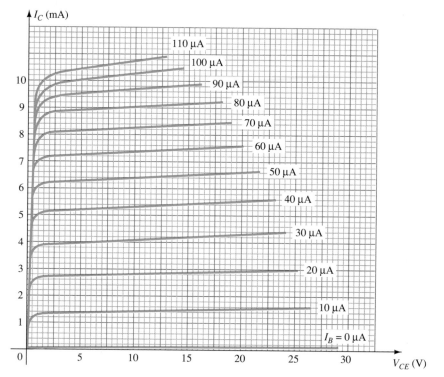

Figure 4.76 Problems 5, 10, 19, 35, 36

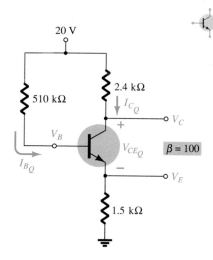

Figure 4.77 Problems 6, 9, 11, 20, 24, 48, 51, 54, 58

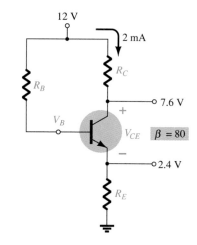

Figure 4.78 Problem 7

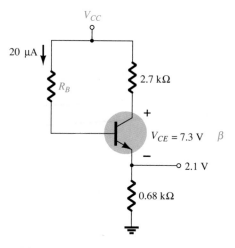

Figure 4.79 Problem 8

§ 4.4 Emitter-Stabilized Bias Circuit

6. For the emitter-stabilized bias circuit of Fig. 4.77, determine:
   (a) I_{B_Q}.
   (b) I_{C_Q}.
   (c) V_{CE_Q}.
   (d) V_C.
   (e) V_B.
   (f) V_E.

7. Given the information provided in Fig. 4.78, determine:
   (a) R_C.
   (b) R_E.
   (c) R_B.
   (d) V_{CE}.
   (e) V_B.

8. Given the information provided in Fig. 4.79, determine:
   (a) β.
   (b) V_{CC}.
   (c) R_B.

9. Determine the saturation current ($I_{C_{sat}}$) for the network of Fig. 4.77.

* 10. Using the characteristics of Fig. 4.76, determine the following for an emitter-bias configuration if a Q-point is defined at $I_{C_Q} = 4$ mA and $V_{CE_Q} = 10$ V.
   (a) R_C if $V_{CC} = 24$ V and $R_E = 1.2$ kΩ.
   (b) β at the operating point.
   (c) R_B.
   (d) Power dissipated by the transistor.
   (e) Power dissipated by the resistor R_C.

* **11.** (a) Determine I_C and V_{CE} for the network of Fig. 4.73.
   (b) Change β to 135 and determine the new value of I_C and V_{CE} for the network of Fig. 4.73.
   (c) Determine the magnitude of the percent change in I_C and V_{CE} using the following equations:

$$\%\Delta I_C = \left| \frac{I_{C(\text{part b})} - I_{C(\text{part a})}}{I_{C(\text{part a})}} \right| \times 100\%, \qquad \%\Delta V_{CE} = \left| \frac{V_{CE(\text{part b})} - V_{CE(\text{part a})}}{V_{CE(\text{part a})}} \right| \times 100\%$$

   (d) Determine I_C and V_{CE} for the network of Fig. 4.77.
   (e) Change β to 150 and determine the new value of I_C and V_{CE} for the network of Fig. 4.77.
   (f) Determine the magnitude of the percent change in I_C and V_{CE} using the following equations:

$$\%\Delta I_C = \left| \frac{I_{C(\text{part e})} - I_{C(\text{part d})}}{I_{C(\text{part d})}} \right| \times 100\%, \qquad \%\Delta V_{CE} = \left| \frac{V_{CE(\text{part e})} - V_{CE(\text{part d})}}{V_{CE(\text{part d})}} \right| \times 100\%$$

   (g) In each of the above the magnitude of β was increased 50%. Compare the percent change in I_C and V_{CE} for each configuration and comment on which seems to be less sensitive to changes in β.

§ **4.5 Voltage-Divider Bias**

12. For the voltage-divider bias configuration of Fig. 4.80, determine:
   (a) I_{B_Q}.
   (b) I_{C_Q}.
   (c) V_{CE_Q}.
   (d) V_C.
   (e) V_E.
   (f) V_B.

13. Given the information provided in Fig. 4.81, determine:
   (a) I_C.
   (b) V_E.
   (c) V_B.
   (d) R_1.

14. Given the information appearing in Fig. 4.82, determine:
   (a) I_C.
   (b) V_E.
   (c) V_{CC}.
   (d) V_{CE}.
   (e) V_B.
   (f) R_1.

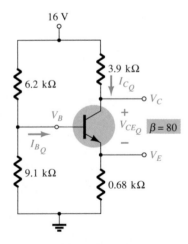

Figure 4.80 Problems 12, 15, 18, 20, 24, 49, 51, 52, 55, 59

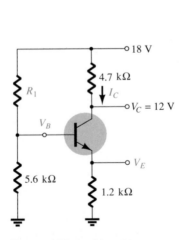

Figure 4.81 Problem 13

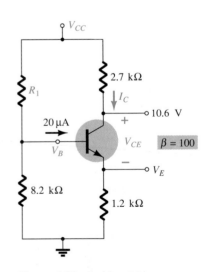

Figure 4.82 Problem 14

15. Determine the saturation current ($I_{C_{\text{sat}}}$) for the network of Fig. 4.80.

* **16.** Determine the following for the voltage-divider configuration of Fig. 4.83 using the approximate approach if the condition established by Eq. 4.33 is satisfied.

    (a) I_C.
    (b) V_{CE}.
    (c) I_B.
    (d) V_E.
    (e) V_B.

* **17.** Repeat Problem 16 using the exact (Thévenin) approach and compare solutions. Based on the results, is the approximate approach a valid analysis technique if Eq. (4.33) is satisfied?

18. (a) Determine I_{C_Q}, V_{CE_Q}, and I_{B_Q} for the network of Problem 12 (Fig. 4.80) using the approximate approach even though the condition established by Eq. (4.33) is not satisfied.
    (b) Determine I_{C_Q}, V_{CE_Q}, and I_{B_Q} using the exact approach.
    (c) Compare solutions and comment on whether the difference is sufficiently large to require standing by Eq. (4.33) when determining which approach to employ.

* **19.** (a) Using the characteristics of Fig. 4.76, determine R_C and R_E for a voltage-divider network having a Q-point of $I_{C_Q} = 5$ mA and $V_{CE_Q} = 8$ V. Use $V_{CC} = 24$ V and $R_C = 3R_E$.
    (b) Find V_E.
    (c) Determine V_B.
    (d) Find R_2 if $R_1 = 24$ kΩ assuming that $\beta R_E > 10R_2$.
    (e) Calculate β at the Q-point.
    (f) Test Eq. (4.33) and note whether the assumption of part (d) is correct.

* **20.** (a) Determine I_C and V_{CE} for the network of Fig. 4.80.
    (b) Change β to 120 (50% increase) and determine the new values of I_C and V_{CE} for the network of Fig. 4.80.
    (c) Determine the magnitude of the percent change in I_C and V_{CE} using the following equations:

$$\%\Delta I_C = \left| \frac{I_{C(\text{part b})} - I_{C(\text{part a})}}{I_{C(\text{part a})}} \right| \times 100\%, \qquad \%\Delta V_{CE} = \left| \frac{V_{CE(\text{part b})} - V_{CE(\text{part a})}}{V_{CE(\text{part a})}} \right| \times 100\%$$

    (d) Compare the solution to part (c) with the solutions obtained for parts (c) and (f) of Problem 11. If not performed, note the solutions provided in Appendix E.
    (e) Based on the results of part (d), which configuration is least sensitive to variations in β?

* **21.** I Repeat parts (a) through (e) of Problem 20 for the network of Fig. 4.83. Change β to 180 in part (b).
    II What general conclusions can be made about networks in which the condition $\beta R_E > 10R_2$ is satisfied and the quantities I_C and V_{CE} are to be determined in response to a change in β?

§ 4.6 DC Bias with Voltage Feedback

22. For the collector feedback configuration of Fig. 4.84, determine:
    (a) I_B.
    (b) I_C.
    (c) V_C.

23. For the voltage feedback network of Fig. 4.85, determine:
    (a) I_C.
    (b) V_C.
    (c) V_E.
    (d) V_{CE}.

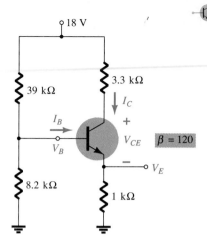

Figure 4.83 Problems 16, 17, 21

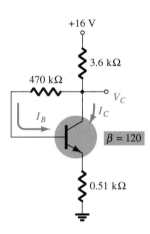

Figure 4.84 Problems 22, 50, 56, 60

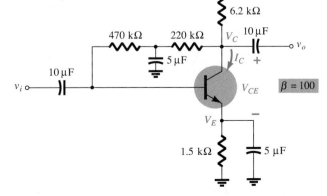

Figure 4.85 Problem 23

* **24.** (a) Determine the level of I_C and V_{CE} for the network of Fig. 4.86.
   (b) Change β to 135 (50% increase) and calculate the new levels of I_C and V_{CE}.
   (c) Determine the magnitude of the percent change in I_C and V_{CE} using the following equations:

$$\%\Delta I_C = \left| \frac{I_{C_{(part\ b)}} - I_{C_{(part\ a)}}}{I_{C_{(part\ a)}}} \right| \times 100\%, \qquad \%\Delta V_{CE} = \left| \frac{V_{CE_{(part\ b)}} - V_{CE_{(part\ a)}}}{V_{CE_{(part\ a)}}} \right| \times 100\%$$

   (d) Compare the results of part (c) with those of Problems 11 (c), 11 (f), and 20 (c). How does the collector-feedback network stack up against the other configurations in sensitivity to changes in β?

25. Determine the range of possible values for V_C for the network of Fig. 4.87 using the 1-MΩ potentiometer.

* **26.** Given $V_B = 4$ V for the network of Fig. 4.88, determine:
   (a) V_E.
   (b) I_C.
   (c) V_C.
   (d) V_{CE}.
   (e) I_B.
   (f) β.

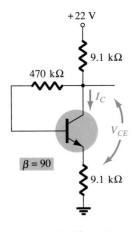

Figure 4.86 Problem 24

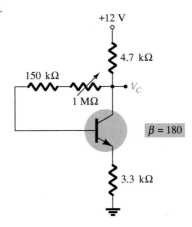

Figure 4.87 Problem 25

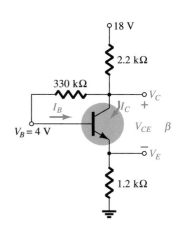

Figure 4.88 Problem 26

§ **4.7 Miscellaneous Bias Configurations**

27. Given $V_C = 8$ V for the network of Fig. 4.89, determine:
   (a) I_B.
   (b) I_C.
   (c) β.
   (d) V_{CE}.

* **28.** For the network of Fig. 4.90, determine:
   (a) I_B.
   (b) I_C.
   (c) V_{CE}.
   (d) V_C.

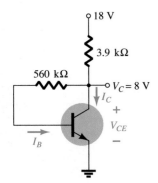

Figure 4.89 Problem 27

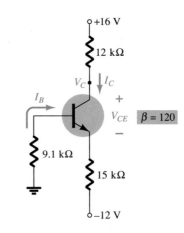

Figure 4.90 Problem 28

* **29.** For the network of Fig. 4.91, determine:
  (a) I_B.
  (b) I_C.
  (c) V_E.
  (d) V_{CE}.

* **30.** Determine the level of V_E and I_E for the network of Fig. 4.92.

* **31.** For the network of Fig. 4.93, determine:
  (a) I_E.
  (b) V_C.
  (c) V_{CE}.

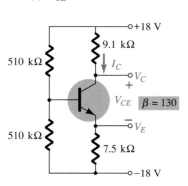

Figure 4.91 Problem 29

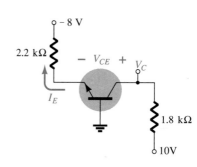

Figure 4.92 Problem 30

Figure 4.93 Problem 31

§ **4.8 Design Operations**

32. Determine R_C and R_B for a fixed-bias configuration if $V_{CC} = 12$ V, $\beta = 80$, and $I_{C_Q} = 2.5$ mA with $V_{CE_Q} = 6$ V. Use standard values.

33. Design an emitter-stabilized network at $I_{C_Q} = \frac{1}{2}I_{C_{sat}}$ and $V_{CE_Q} = \frac{1}{2}V_{CC}$. Use $V_{CC} = 20$ V, $I_{C_{sat}} = 10$ mA, $\beta = 120$, and $R_C = 4R_E$. Use standard values.

34. Design a voltage-divider bias network using a supply of 24 V, a transistor with a beta of 110, and an operating point of $I_{C_Q} = 4$ mA and $V_{CE_Q} = 8$ V. Choose $V_E = \frac{1}{8}V_{CC}$. Use standard values.

* **35.** Using the characteristics of Fig. 4.76, design a voltage-divider configuration to have a saturation level of 10 mA, and a Q-point one-half the distance between cutoff and saturation. The available supply is 28 V and V_E is to be one-fifth of V_{CC}. The condition established by Eq. (4.33) should also be met to provide a high stability factor. Use standard values.

§ **4.9 Transistor Switching Networks**

* **36.** Using the characteristics of Fig. 4.76, determine the appearance of the output waveform for the network of Fig. 4.94. Include the effects of $V_{CE_{sat}}$ and determine I_B, $I_{B_{max}}$, and $I_{C_{sat}}$ when $V_i = 10$ V. Determine the collector-to-emitter resistance at saturation and cutoff.

* **37.** Design the transistor inverter of Fig. 4.95 to operate with a saturation current of 8 mA using a transistor with a beta of 100. Use a level of I_B equal to 120% of $I_{B_{max}}$ and standard resistor values.

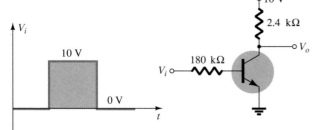

Figure 4.94 Problem 36

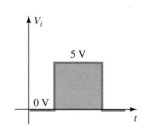

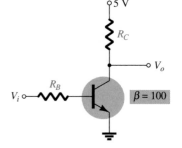

Figure 4.95 Problem 37

38. (a) Using the characteristics of Fig. 3.23c, determine t_{on} and t_{off} at a current of 2 mA. Note the use of log scales and the possible need to refer to Section 11.2.

(b) Repeat part (a) at a current of 10 mA. How have t_{on} and t_{off} changed with increase in collector current?

(c) For parts (a) and (b) sketch the pulse waveform of Fig. 4.56 and compare results.

§ **4.10 Troubleshooting Techniques**

* **39.** The measurements of Fig. 4.96 all reveal that the network is not functioning correctly. List as many reasons as you can for the measurements obtained.

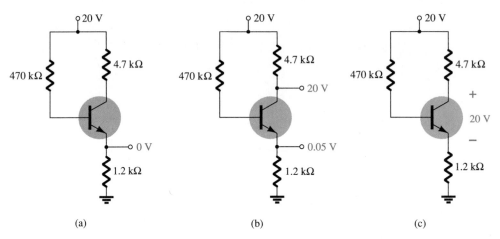

(a)　　　　　　　(b)　　　　　　　(c)

Figure 4.96 Problem 39

* **40.** The measurements appearing in Fig. 4.97 reveal that the networks are not operating properly. Be specific in describing why the levels obtained reflect a problem with the expected network behavior. In other words, the levels obtained reflect a very specific problem in each case.

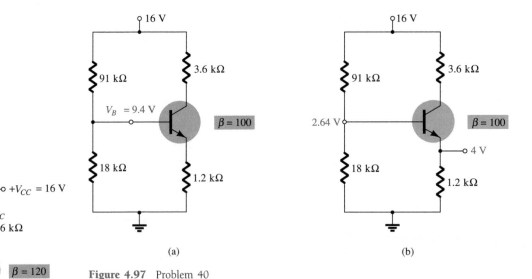

(a)　　　　　　　(b)

Figure 4.97 Problem 40

Figure 4.98 Problem 41

41. For the circuit of Fig. 4.98:

(a) Does V_C increase or decrease if R_B is increased?

(b) Does I_C increase or decrease if β is reduced?

(c) What happens to the saturation current if β is increased?

(d) Does the collector current increase or decrease if V_{CC} is reduced?

(e) What happens to V_{CE} if the transistor is replaced by one with smaller β?

42. Answer the following questions about the circuit of Fig. 4.99.
  (a) What happens to the voltage V_C if the transistor is replaced by one having a larger value of β?
  (b) What happens to the voltage V_{CE} if the ground leg of resistor R_{B_2} opens (does not connect to ground)?
  (c) What happens to I_C if the supply voltage is low?
  (d) What voltage V_{CE} would occur if the transistor base-emitter junction fails by becoming open?
  (e) What voltage V_{CE} would result if the transistor base-emitter junction fails by becoming a short?

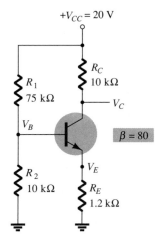

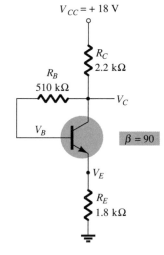

Figure 4.99 Problem 42

* **43.** Answer the following questions about the circuit of Fig. 4.100.
  (a) What happens to the voltage V_C if the resistor R_B is open?
  (b) What should happen to V_{CE} if β increases due to temperature?
  (c) How will V_E be affected when replacing the collector resistor with one whose resistance is at the lower end of the tolerance range?
  (d) If the transistor collector connection becomes open, what will happen to V_E?
  (e) What might cause V_{CE} to become nearly 18 V?

Figure 4.100 Problem 43

§ **4.11 PNP Transistors**

44. Determine V_C, V_{CE}, and I_C for the network of Fig. 4.101.

45. Determine V_C and I_B for the network of Fig. 4.102.

46. Determine I_E and V_C for the network of Fig. 4.103.

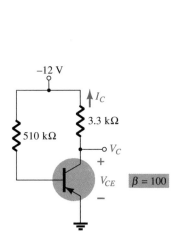

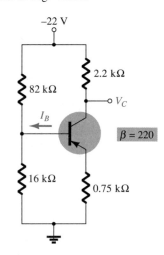

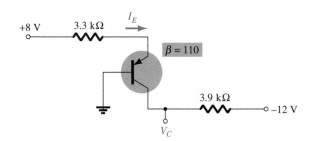

Figure 4.101 Problem 44   **Figure 4.102** Problem 45   **Figure 4.103** Problem 46

47. Determine the following for the network of Fig. 4.73.
  - (a) $S(I_{CO})$.
  - (b) $S(V_{BE})$.
  - (c) $S(\beta)$ using T_1 as the temperature at which the parameter values are specified and $\beta(T_2)$ as 25% more than $\beta(T_1)$.
  - (d) Determine the net change in I_C if a change in operating conditions results in I_{CO} increasing from 0.2 μA to 10 μA, V_{BE} drops from 0.7 V to 0.5 V, and β increases 25%.

* **48.** For the network of Fig. 4.77, determine:
  - (a) $S(I_{CO})$.
  - (b) $S(V_{BE})$.
  - (c) $S(\beta)$ using T_1 as the temperature at which the parameter values are specified and $\beta(T_2)$ as 25% more than $\beta(T_1)$.
  - (d) Determine the net change in I_C if a change in operating conditions results in I_{CO} increasing from 0.2 μA to 10 μA, V_{BE} drops from 0.7 V to 0.5 V, and β increases 25%.

* **49.** For the network of Fig. 4.80, determine:
  - (a) $S(I_{CO})$.
  - (b) $S(V_{BE})$.
  - (c) $S(\beta)$ using T_1 as the temperature at which the parameter values are specified and $\beta(T_2)$ as 25% more than $\beta(T_1)$.
  - (d) Determine the net change in I_C if a change in operating conditions results in I_{CO} increasing from 0.2 μA to 10 μA, V_{BE} drops from 0.7 V to 0.5 V, and β increases 25%.

* **50.** For the network of Fig. 4.89, determine:
  - (a) $S(I_{CO})$.
  - (b) $S(V_{BE})$.
  - (c) $S(\beta)$ using T_1 as the temperature at which the parameter values are specified and $\beta(T_2)$ as 25% more than $\beta(T_1)$.
  - (d) Determine the net change in I_C if a change in operating conditions results in I_{CO} increasing from 0.2 μA to 10 μA, V_{BE} drops from 0.7 V to 0.5 V, and β increases 25%.

* **51.** Compare the relative values of stability for Problems 47 through 50. The results for Exercises 47 and 49 can be found in Appendix E. Can any general conclusions be derived from the results?

* **52.** (a) Compare the levels of stability for the fixed-bias configuration of Problem 47.
  - (b) Compare the levels of stability for the voltage-divider configuration of Problem 49.
  - (c) Which factors of parts (a) and (b) seem to have the most influence on the stability of the system, or is there no general pattern to the results?

§ **4.13 Computer Analysis**

53. Perform a SPice analysis of the network of Fig. 4.73. That is, determine I_C, V_{CE}, and I_B.

54. Repeat Problem 53 for the network of Fig. 4.77.

55. Repeat Problem 53 for the network of Fig. 4.80.

56. Repeat Problem 53 for the network of Fig. 4.84.

57. Perform an analysis of the network of Fig. 4.73 using BASIC. That is, determine I_C, V_{CE}, and I_B.

58. Repeat Problem 57 for the network of Fig. 4.77.

59. Repeat Problem 57 for the network of Fig. 4.80.

60. Repeat Problem 57 for the network of Fig. 4.84.

---

*Please Note: Asterisks indicate more difficult problems.

Field-Effect Transistors

5.1 INTRODUCTION

The field-effect transistor (FET) is a three-terminal device used for a variety of applications that match, to a large extent, those of the BJT transistor described in Chapters 3 and 4. Although there are important differences between the two types of devices, there are also many similarities that will be pointed out in the sections to follow.

The primary difference between the two types of transistors is the fact that the BJT transistor is a *current-controlled* device as depicted in Fig. 5.1a, while the JFET transistor is a *voltage-controlled* device as shown in Fig. 5.1b. In other words, the current I_C in Fig. 5.1a is a direct function of the level of I_B. For the FET the current I will be a function of the voltage V_{GS} applied to the input circuit as shown in Fig. 5.1b. In each case the current of the output circuit is being controlled by a parameter of the input circuit—in one case a current level and in the other an applied voltage.

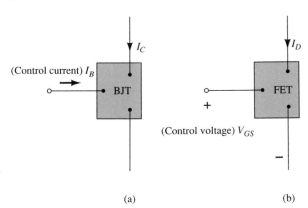

Figure 5.1 (a) Current-controlled and (b) voltage-controlled amplifiers.

Just as there are *npn* and *pnp* bipolar transistors, there are *n-channel* and *p-channel* field-effect transistors. However, it is important to keep in mind that the BJT transistor is a *bipolar* device—the prefix *bi-* revealing that the conduction level is a function of two charge carriers, electrons and holes. The FET is a *unipolar* device depending solely on either electron (*n*-channel) or hole (*p*-channel) conduction.

The term "field-effect" in the chosen name deserves some explanation. We are all familiar with the ability of a permanent magnet to draw metal filings to the magnet without the need for actual contact. The magnetic field of the permanent magnet has enveloped the filings and attracted them to the magnet through an effort on the part of the magnetic flux lines to be as short as possible. For the FET an *electric field* is established by the charges present that will control the conduction path of the output circuit without the need for direct contact between the controlling and controlled quantities.

There is a natural tendency when introducing a second device with a range of applications similar to one already introduced to compare some of the general characteristics of one versus the other. One of the most important characteristics of the FET is its *high input impedance*. At a level of 1 to several hundred megohms it far exceeds the typical input resistance levels of the BJT transistor configurations—a very important characteristic in the design of linear ac amplifier systems. On the other hand, the BJT transistor has a much higher sensitivity to changes in the applied signal. In other words, the variation in output current is typically a great deal more for BJTs then FETs for the same change in applied voltage. For this reason, typical ac voltage gains for BJT amplifiers are a great deal more than for FETs. In general, FETs are more temperature stable than BJTs, and FETs are usually smaller in construction than BJTs, making them particularly useful in *integrated-circuit (IC)* chips. The construction characteristics of some FETs, however, can make them more sensitive to handling than BJTs.

Two types of FETs will be introduced in this chapter: the *junction field-effect transistor* (JFET) and the *metal-oxide-semiconductor field-effect transistor (MOSFET)*. The MOSFET category is further broken down into depletion and enhancement types, which are both described. The MOSFET transistor has become one of the most important devices used in the design and construction of integrated circuits for digital computers. It's thermal stability, and other general characteristics make it extremely popular in computer circuit design. However, as a discrete element in a typical top-hat container, it must be handled with care (to be discussed in a later section).

Once the FET construction and characteristics have been introduced the biasing arrangements will be covered in Chapter 6. The analysis performed in Chapter 4 using BJT transistors will prove helpful in the derivation of the important equations and understanding the results obtained for FET circuits.

5.2 CONSTRUCTION AND CHARACTERISTICS OF JFETs

As indicated earlier, the JFET is a three-terminal device with one terminal capable of controlling the current between the other two. In our discussion of the BJT transistor the *npn* transistor was employed through the major part of the analysis and design sections with a section devoted to the impact of using a *pnp* transistor. For the JFET transistor the *n*-channel device will appear as the prominent device with paragraphs and sections devoted to the impact of using a *p*-channel JFET.

The basic construction of the *n*-channel JFET is shown in Fig. 5.2. Note that the major part of the structure is the *n*-type material that forms the channel between the embedded layers of *p*-type material. The top of the *n*-type channel is connected through an ohmic contact to the a terminal referred to as the *drain (D),* while the lower end of the same material is connected through an ohmic contact to a terminal referred to as the *source (S).* The two *p*-type materials are connected together and to the *gate (G)* terminal. In essence, therefore, the drain and source are connected to the ends of the *n*-type channel and the gate to the two layers of *p*-type material. In the absence of any applied potentials the JFET has two *p-n* junctions under no-bias conditions. The result is a depletion region at each junction as shown in Fig. 5.2 that resembles the same region of a diode under no-bias conditions. Recall also that a depletion region is that region void of free carriers and therefore unable to support conduction through the region.

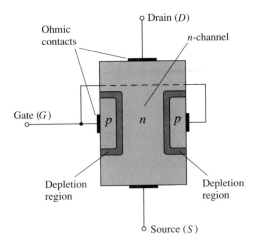

Figure 5.2 Junction field-effect transistor (JFET).

Analogies are seldom perfect and at times can be misleading, but the water analogy of Fig. 5.3 does provide a sense for the JFET control at the gate terminal and the appropriateness of the terminology applied to the terminals of the device. The source of water pressure can be likened to the applied voltage from drain to source that will establish a flow of water (electrons) from the spigot (source). The "gate," through an applied signal (potential), controls the flow of water (charge) to the "drain." The drain and source terminals are at opposite ends of the n-channel as introduced in Fig. 5.2 because the terminology is defined for electron flow.

Figure 5.3 Water analogy for the JFET control mechanism.

$V_{GS} = 0$ V, V_{DS} Some Positive Value

In Fig. 5.4 a positive voltage V_{DS} has been applied across the channel and the gate has been connected directly to the source to establish the condition $V_{GS} = 0$ V. The result is a gate and source terminal at the same potential and a depletion region in the low end of each p-material similar to the distribution of the no-bias conditions of Fig. 5.2. The instant the voltage V_{DD} ($= V_{DS}$) is applied, the electrons will be drawn to the drain terminal, establishing the conventional current I_D with the defined direction of Fig. 5.4. The path of charge flow clearly reveals that the drain and source currents are equivalent ($I_D = I_S$). Under the conditions appearing in Fig. 5.4, the flow of charge is relatively uninhibited and limited solely by the resistance of the n-channel between drain and source.

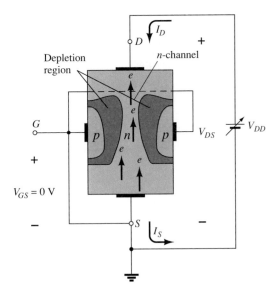

Figure 5.4 JFET in the $V_{GS} = 0$ V and $V_{DS} > 0$ V.

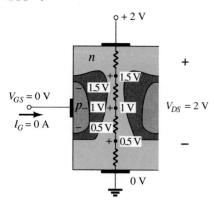

Figure 5.5 Varying reverse-bias potentials across the *p-n* junction of an *n*-channel JFET.

It is important to note that the depletion region is wider near the top of both *p*-type materials. The reason for the change in width of the region is best described through the help of Fig. 5.5. Assuming a uniform resistance in the *n*-channel, the resistance of the channel can be broken down to the divisions appearing in Fig. 5.5. The current I_D will establish the voltage levels through the channel as indicated on the same figure. The result is that the upper region of the *p*-type material will be reverse-biased by about 1.5 V, with the lower region only reverse-biased by 0.5 V. Recall from the discussion of the diode operation that the greater the applied reverse bias, the wider the depletion region—hence the distribution of the depletion region as shown in Fig. 5.5. The fact that the *p-n* junction is reverse-biased for the length of the channel results in a gate current of zero amperes as shown in the same figure. The fact that $I_G = 0$ A is an important characteristic of the JFET.

As the voltage V_{DS} is increased from 0 to a few volts, the current will increase as determined by Ohm's law and the plot of I_D versus V_{DS} will appear as shown in Fig. 5.6. The relative straightness of the plot reveals that for the region of low values of V_{DS}, the resistance is essentially constant. As V_{DS} increases and approaches a level referred to as V_P in Fig. 5.6, the depletion regions of Fig. 5.4 will widen, causing a noticeable reduction in the channel width. The reduced path of conduction causes the resistance to increase and the curve in the graph of Fig. 5.6 to occur. The more horizontal the curve, the higher the resistance, suggesting that the resistance is approaching "infinite" ohms in the horizontal region. If V_{DS} is increased to a level where it appears that the two depletion regions would "touch" as shown in Fig. 5.7, a condition referred to as *pinch-off* will result. The level of V_{DS} that establishes this condition is referred to as the *pinch-off voltage* and is denoted by V_P, as shown in Fig. 5.6. In actuality, the term "pinch-off" is a misnomer in that it suggests that the current I_D is pinched off and drops to 0 A. As shown in Fig. 5.6, however, this is hardly the case—I_D maintains a saturation level defined as I_{DSS} in Fig. 5.6. In reality a very small channel still exists, with a current of very high density. The fact that I_D does not drop off at pinch-off and maintains the saturation level indicated in Fig. 5.6 is verified by the following fact: The absence of a drain current would remove the possibility of different potential levels through the *n*-channel material to establish the varying levels of reverse bias along the *p-n* junction. The result would be a loss of the depletion region distribution that caused pinch-off in the first place.

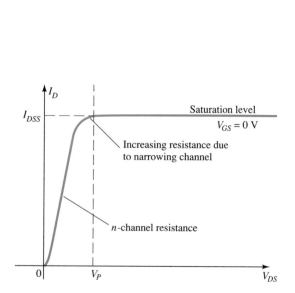

Figure 5.6 I_D versus V_{DS} for $V_{GS} = 0$ V.

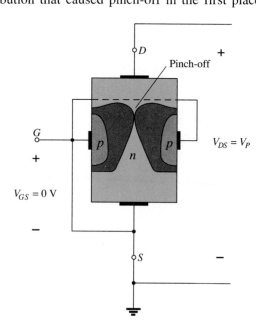

Figure 5.7 Pinch-off ($V_{GS} = 0$ V, $V_{DS} = V_P$).

As V_{DS} is increased beyond V_P, the region of close encounter between the two depletion regions will increase in length along the channel, but the level of I_D remains essentially the same. In essence, therefore, once $V_{DS} > V_P$ the JFET has the characteristics of a current source. As shown in Fig. 5.8, the current is fixed at $I_D = I_{DSS}$, but the voltage V_{DS} (for levels $> V_P$) is determined by the applied load.

The choice of notation I_{DSS} is derived from the fact that it is the *Drain-to-Source* current with a *Short*-circuit connection from gate to source. As we continue to investigate the characteristics of the device we will find that:

> *I_{DSS} is the maximum drain current for a JFET and is defined by the conditions $V_{GS} = 0$ V and $V_{DS} > |V_P|$.*

Note in Fig. 5.6 that $V_{GS} = 0$ V for the entire length of the curve. The next few paragraphs will describe how the characteristics of Fig. 5.6 are affected by changes in the level of V_{GS}.

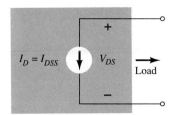

Figure 5.8 Current source equivalent for $V_{GS} = 0$ V, $V_{DS} > V_P$.

$V_{GS} < 0$ V

The voltage from gate to source, denoted V_{GS}, is the controlling voltage of the JFET. Just as various curves for I_C versus V_{CE} were established for different levels of I_B for the BJT transistor, curves of I_D versus V_{DS} for various levels of V_{GS} can be developed for the JFET. For the *n*-channel device the controlling voltage V_{GS} is made more and more negative from its $V_{GS} = 0$ V level. In other words, the gate terminal will be set at lower and lower potential levels as compared to the source.

In Fig. 5.9 a negative voltage of -1 V has been applied between the gate and source terminals for a low level of V_{DS}. The effect of the applied negative bias V_{GS} is to establish depletion regions similar to those obtained with $V_{GS} = 0$ V but at lower levels of V_{DS}. Therefore, the result of applying a negative bias to the gate is to reach the saturation level at a lower level of V_{DS} as shown in Fig. 5.10 for $V_{GS} = -1$ V. The resulting saturation level for I_D has been reduced and in fact will continue to decrease as V_{GS} is made more and more negative. Note also on Fig. 5.10 how the pinch-off voltage continues to drop in a parabolic manner as V_{GS} becomes more and more negative. Eventually, V_{GS} when $V_{GS} = -V_P$ will be sufficiently negative to establish a saturation level that is essentially 0 mA, and for all practical purposes the device has been "turned off." In summary:

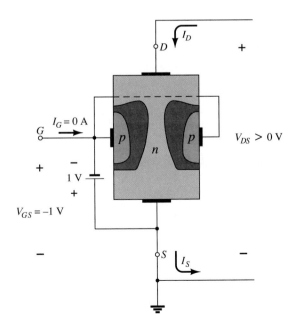

Figure 5.9 Application of a negative voltage to the gate of a JFET.

5.2 Construction and Characteristics of JFETs

211

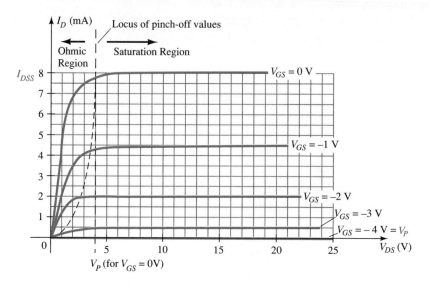

Figure 5.10 *n*-Channel JFET characteristics with $I_{DSS} = 8$ mA and $V_P = -4$ V.

The level of V_{GS} that results in $I_D = 0$ mA is defined by $V_{GS} = V_P$, with V_P being a negative voltage for n-channel devices and a positive voltage for p-channel JFETs.

On most specification sheets, the pinch-off voltage is specified as $V_{GS(off)}$ rather than V_P. A specification sheet will be reviewed later in the chapter when the primary elements of concern have been introduced. The region to the right of the pinch-off locus of Fig. 5.10 is the region typically employed in linear amplifiers (amplifiers with minimum distortion of the applied signal) and is commonly referred to as the *constant-current, saturation,* or *linear amplification region.*

Voltage-Controlled Resistor

The region to the left of the pinch-off locus of Fig. 5.10 is referred to as the *ohmic* or *voltage-controlled resistance region.* In this region the JFET can actually be employed as a variable resistor (possibly for an automatic gain control system) whose resistance is controlled by the applied gate-to-source voltage. Note in Fig. 5.10 that the slope of each curve and therefore the resistance of the device between drain and source for $V_{DS} < V_P$ is a function of the applied voltage V_{GS}. As V_{GS} becomes more and more negative, the slope of each curve becomes more and more horizontal, corresponding with an increasing resistance level. The following equation will provide a good first approximation to the resistance level in terms of the applied voltage V_{GS}.

$$r_d = \frac{r_o}{(1 - V_{GS}/V_P)^2} \tag{5.1}$$

where r_o is the resistance with $V_{GS} = 0$ V and r_d the resistance at a particular level of V_{GS}.

For an n-channel JFET with r_o equal to 10 kΩ ($V_{GS} = 0$ V, $V_P = -6$ V), Eq. (5.1) will result in 40 kΩ at $V_{GS} = -3$ V.

p-Channel Devices

The *p*-channel JFET is constructed in exactly the same manner as the *n*-channel device of Fig. 5.2 with a reversal of the *p*- and *n*-type materials as shown in Fig. 5.11.

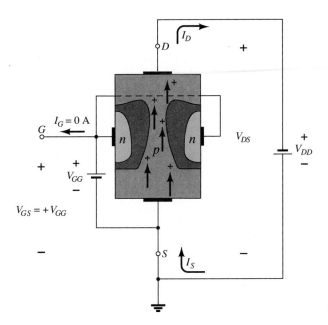

Figure 5.11 *p*-Channel JFET.

The defined current directions are reversed, as are the actual polarities for the voltages V_{GS} and V_{DS}. For the *p*-channel device the channel will be constricted by increasing positive voltages from gate to source and the double-subscript notation for V_{DS} will result in negative voltages for V_{DS} on the characteristics of Fig. 5.12, which has an I_{DSS} of 6 mA and a pinch-off voltage of $V_{GS} = +6$ V. Do not let the minus signs for V_{DS} throw you. They simply indicate that the source is at a higher potential than the drain.

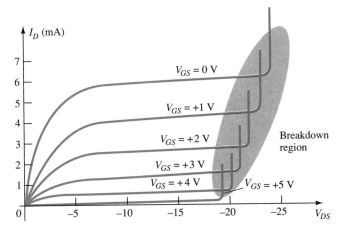

Figure 5.12 *p*-Channel JFET characteristics with $I_{DSS} = 6$ mA and $V_P = +6$ V.

Note at high levels of V_{DS} that the curves suddenly rise to levels that seem unbounded. The vertical rise is an indication that breakdown has occurred and the current through the channel (in the same direction as normally encountered) is now limited solely by the external circuit. Although not appearing in Fig. 5.10 for the *n*-channel device, they do occur for the *n*-channel device if sufficient voltage is applied. This region can be avoided if the level of $V_{DS_{max}}$ is noted on the specification sheet and the design is such that the actual level of V_{DS} is less than this value for *all* values of V_{GS}.

Symbols

The graphic symbols for the *n*-channel and *p*-channel JFETs are provided in Fig. 5.13. Note that the arrow is pointing in for the *n*-channel device of Fig. 5.13a to represent the direction I_G would flow if the *p-n* junction were forward-biased. For the *p*-channel device (Fig. 5.13b) the only difference in the symbol is the direction of the arrow.

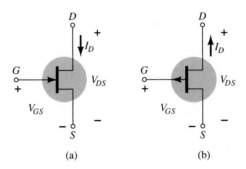

(a)   (b)

Figure 5.13 JFET symbols: (a) *n*-channel; (b) *p*-channel.

Summary

A number of important parameters and relationships were introduced in this section. A few that will surface frequently in the analysis to follow in this chapter and the next for *n*-channel JFETs include the following:

The maximum current is defined as I_{DSS} and occurs when $V_{GS} = 0$ V and $V_{DS} \geq |V_P|$ as shown in Fig. 5.14a.

For gate-to-source voltages V_{GS} less than (more negative than) the pinch-off level, the drain current is 0 A ($I_D = 0$ A), as appearing in Fig. 5.14b.

For all levels of V_{GS} between 0 V and the pinch-off level the current I_D will range between I_{DSS} and 0 A, respectively, as reviewed by Fig. 5.14c.

For p-channel JFETs a similar list can be developed.

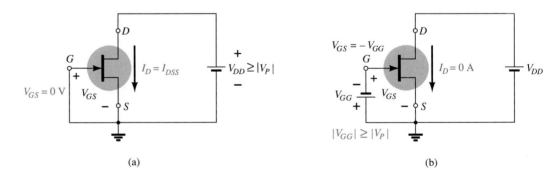

(a)   (b)

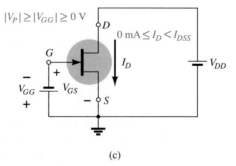

(c)

Figure 5.14 (a) $V_{GS} = 0$ V, $I_D = I_{DSS}$; (b) cutoff ($I_D = 0$ A) V_{GS} less than the pinch-off level; (c) I_D exists between 0 A and I_{DSS} for V_{GS} less than or equal to 0 V and greater than the pinch-off level.

5.3 TRANSFER CHARACTERISTICS

Derivation

For the BJT transistor the output current I_C and input controlling current I_B were related by beta, which was considered constant for the analysis to be performed. In equation form,

$$I_C = f(I_B) = \beta I_B \qquad \text{(5.2)}$$

control variable

constant

In Eq. (5.2) a linear relationship exists between I_C and I_B. Double the level of I_B and I_C will increase by a factor of 2 also.

Unfortunately, this linear relationship does not exist between the output and input quantities of a JFET. The relationship between I_D and V_{GS} is defined by *Shockley's equation:*

$$I_D = I_{DSS}\left(1 - \frac{V_{GS}}{V_P}\right)^2 \qquad \text{(5.3)}$$

control variable

constants

The squared term of the equation will result in a nonlinear relationship between I_D and V_{GS}, producing a curve that grows exponentially with increasing values of V_{GS}.

For the dc analysis to be performed in Chapter 6 a graphical rather than mathematical approach will in general be more direct and easier to apply. The graphical approach, however, will require a plot of Eq. (5.3) to represent the device and a plot of the network equation relating the same variables. The solution is defined by the point of intersection of the two curves. It is important to keep in mind when applying the graphical approach that the device characteristics will be *unaffected* by the network in which the device is employed. The network equation may change along with the intersection between the two curves, but the transfer curve defined by Eq. (5.3) is unaffected. In general, therefore:

> *The transfer characteristics defined by Shockley's equation are unaffected by the network in which the device is employed.*

The transfer curve can be obtained using Shockley's equation or from the output characteristics of Fig. 5.10. In Fig. 5.15 two graphs are provided with the vertical

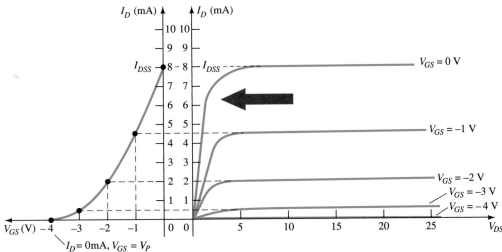

Figure 5.15 Obtaining the transfer curve from the drain characteristics.

scaling in milliamperes for each graph. One is a plot of I_D versus V_{DS} while the other is I_D versus V_{GS}. Using the drain characteristics on the right of the "y" axis a horizontal line can be drawn from the saturation region of the curve denoted $V_{GS} = 0$ V to the I_D axis. The resulting current level for both graphs is I_{DSS}. The point of intersection on the I_D versus V_{GS} curve will be as shown since the vertical axis is defined as $V_{GS} = 0$ V.

In review:

When $V_{GS} = 0$ V, $I_D = I_{DSS}$.

When $V_{GS} = V_P = -4$ V, the drain current is zero milliamperes, defining another point on the transfer curve. That is:

When $V_{GS} = V_P$, $I_D = 0$ mA.

Before continuing, it is important to realize that the drain characteristics relate one output (or drain) quantity to another output (or drain) quantity—both axes are defined by variables in the same region of the device characteristics. The transfer characteristics are a plot of an output (or drain) current versus an input-controlling quantity. There is therefore a direct "transfer" from input to output variables when employing the curve to the left of Fig. 5.15. If the relationship were a linear, the plot of I_D versus V_{GS} would result in straight line between I_{DSS} and V_P. However, a parabolic curve will result because the vertical spacing between steps of V_{GS} on the drain characteristics of Fig. 5.15 decreases noticeably as V_{GS} becomes more and more negative. Compare the spacing between $V_{GS} = 0$ V and $V_{GS} = -1$ V to that between $V_{GS} = -3$ V and pinch-off. The change in V_{GS} is the same, but the resulting change in I_D is quite different.

If a horizontal line is drawn from the $V_{GS} = -1$ V curve to the I_D axis and then extended to the other axis, another point on the transfer curve can be located. Note that $V_{GS} = -1$ V on the bottom axis of the transfer curve with $I_D = 4.5$ mA. Note in the definition of I_D at $V_{GS} = 0$ V and -1 V that the saturation levels of I_D are employed and the ohmic region ignored. Continuing with $V_{GS} = -2$ V and -3 V the transfer curve can be completed. It is the transfer curve of I_D versus V_{GS} that will receive extended use in the analysis of Chapter 6 and not the drain characteristics of Fig. 5.15. The next few paragraphs will introduce a quick, efficient method of plotting I_D versus V_{GS} given only the levels of I_{DSS} and V_P and Shockley's equation.

Applying Shockley's Equation

The transfer curve of Fig. 5.15 can also be obtained directly from Shockley's equation (5.3) given simply the values of I_{DSS} and V_P. The levels of I_{DSS} and V_P define the limits of the curve on both axes and leave only the necessity of finding a few intermediate plot points. The validity of Eq. (5.3) as a source of the transfer curve of Fig. 5.15 is best demonstrated by examining a few specific levels of one variable and finding the resulting level of the other as follows:

Substituting $V_{GS} = 0$ V gives

$$\text{Eq. (5.3):} \quad I_D = I_{DSS}\left(1 - \frac{V_{GS}}{V_P}\right)^2$$

$$= I_{DSS}\left(1 - \frac{0}{V_P}\right)^2 = I_{DSS}(1 - 0)^2$$

and
$$\boxed{I_D = I_{DSS}|_{V_{GS}=0V}} \tag{5.4}$$

Substituting $V_{GS} = V_P$ yields

$$I_D = I_{DSS}\left(1 - \frac{V_P}{V_P}\right)^2$$

$$= I_{DSS}(1 - 1)^2 = I_{DSS}(0)$$

$$\boxed{I_D = 0A\big|_{V_{GS} = V_P}} \tag{5.5}$$

For the drain characteristics of Fig. 5.15, if we substitute $V_{GS} = -1$ V,

$$I_D = I_{DSS}\left(1 - \frac{V_{GS}}{V_P}\right)^2$$

$$= 8 \text{ mA}\left(1 - \frac{-1 \text{ V}}{-4 \text{ V}}\right)^2 = 8 \text{ mA}\left(1 - \frac{1}{4}\right)^2 = 8 \text{ mA}(0.75)^2$$

$$= 8 \text{ mA}(0.5625)$$

$$= \mathbf{4.5 \text{ mA}}$$

as shown in Fig. 5.15. Note the care taken with the negative signs for V_{GS} and V_P in the calculations above. The loss of one sign would result in a totally erroneous result.

It should be obvious from the above that given I_{DSS} and V_P (as is normally provided on specification sheets) the level of I_D can be found for any level of V_{GS}. Conversely, by using basic algebra we can obtain [from Eq. (5.3)] an equation for the resulting level of V_{GS} for a given level of I_D. The derivation is quite straight forward and will result in

$$\boxed{V_{GS} = V_P\left(1 - \sqrt{\frac{I_D}{I_{DSS}}}\right)} \tag{5.6}$$

Let us test Eq. (5.6) by finding the level of V_{GS} that will result in a drain current of 4.5 mA for the device with the characteristics of Fig. 5.15.

$$V_{GS} = -4 \text{ V}\left(1 - \sqrt{\frac{4.5 \text{ mA}}{8 \text{ mA}}}\right)$$

$$= -4 \text{ V}(1 - \sqrt{0.5625}) = -4 \text{ V}(1 - 0.75)$$

$$= -4 \text{ V}(0.25)$$

$$= \mathbf{-1 \text{ V}}$$

as substituted in the above calculation and verified by Fig. 5.15.

Shorthand Method

Since the transfer curve must be plotted so frequently, it would be quite advantageous to have a shorthand method for plotting the curve in the quickest, most efficient manner while maintaining an acceptable degree of accuracy. The format of Eq. (5.3) is such that specific levels of V_{GS} will result in levels of I_D that can be memorized to provide the plot points needed to sketch the transfer curve. If we specify V_{GS} to be one-half the pinch-off value V_P, the resulting level of I_D will be the following, as determined by Shockley's equation:

$$I_D = I_{DSS}\left(1 - \frac{V_{GS}}{V_P}\right)^2$$

$$= I_{DSS}\left(1 - \frac{V_P/2}{V_P}\right)^2 = I_{DSS}\left(1 - \frac{1}{2}\right)^2 = I_{DSS}(0.5)^2$$

$$= I_{DSS}(0.25)$$

and

$$I_D = \frac{I_{DSS}}{4}\bigg|_{V_{GS} = V_P/2}$$

(5.7)

Now it is important to realize that Eq. (5.7) is not for a particular level of V_P. It is a general equation for any level of V_P as long as $V_{GS} = V_P/2$. The result specifies that the drain current will always be one-fourth of the saturation level I_{DSS}, as long as the gate-to-source voltage is one-half the pinch-off value. Note the level of I_D for $V_{GS} = V_P/2 = -4$ V/2 $= -2$ V in Fig. 5.15.

If we choose $I_D = I_{DSS}/2$ and plug into Eq. (5.6), we find that

$$V_{GS} = V_P\left(1 - \sqrt{\frac{I_D}{I_{DSS}}}\right)$$

$$= V_P\left(1 - \sqrt{\frac{I_{DSS}/2}{I_{DSS}}}\right) = V_P(1 - \sqrt{0.5}) = V_P(0.293)$$

and

$$V_{GS} \cong 0.3\, V_P\big|_{I_D = I_{DSS}/2}$$

(5.8)

Additional points can be determined, but the transfer curve can be sketched to a satisfactory level of accuracy simply using the four plot points defined above and reviewed in Table 5.1. In fact, in the analysis of Chapter 6, a maximum of four plot points are used to sketch the transfer curves. On most occasions using just the plot point defined by $V_{GS} = V_P/2$ and the axis intersections at I_{DSS} and V_P will provide a curve accurate enough for most calculations.

TABLE 5.1 V_{GS} versus I_D Using Shockley's Equation	
V_{GS}	I_D
0	I_{DSS}
0.3 V_P	$I_{DSS}/2$
0.5 V_P	$I_{DSS}/4$
V_P	0 mA

EXAMPLE 5.1

Sketch the transfer curve defined by $I_{DSS} = 12$ mA and $V_P = -6$ V.

Solution

Two plot points are defined by

$$I_{DSS} = \mathbf{12\ mA} \qquad \text{and} \qquad V_{GS} = \mathbf{0\ V}$$

and

$$I_D = \mathbf{0\ mA} \qquad \text{and} \qquad V_{GS} = V_P$$

At $V_{GS} = V_P/2 = -6$ V/2 $= \mathbf{-3\ V}$ the drain current will be determined by $I_D = I_{DSS}/4 = 12$ mA/4 $= \mathbf{3\ mA.}$ At $I_D = I_{DSS}/2 = 12$ mA/2 $= \mathbf{6\ mA}$ the gate-to-source voltage is determined by $V_{GS} \cong 0.3\, V_P = 0.3(-6$ V$) = \mathbf{-1.8\ V.}$ All four plot points are well defined on Fig. 5.16 with the complete transfer curve.

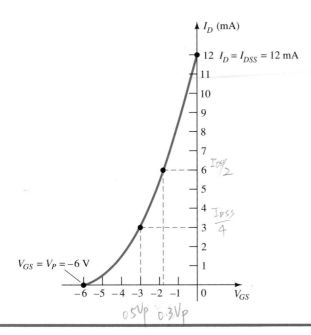

Figure 5.16 Transfer curve for Example 5.1.

For *p*-channel devices, Shockley's equation (5.3) can still be applied exactly as it appears. In this case, both V_P and V_{GS} will be positive and the curve will be the mirror image of the transfer curve obtained with an *n*-channel and the same limiting values.

Sketch the transfer curve for a *p*-channel device with $I_{DSS} = 4$ mA and $V_P = 3$ V.

EXAMPLE 5.2

Solution

At $V_{GS} = V_P/2 = 3$ V/2 = 1.5 V, $I_D = I_{DSS}/4 = 4$ mA/4 = 1 mA. At $I_D = I_{DSS}/2 = 4$ mA/2 = 2 mA, $V_{GS} = 0.3 V_P = 0.3(3$ V) = 0.9 V. Both plot points appear in Fig. 5.17 along with the points defined by I_{DSS} and V_P.

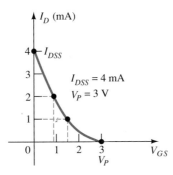

Figure 5.17 Transfer curve for the *p*-channel device of Example 5.2.

5.4 SPECIFICATION SHEETS (JFETs)

Although the general content of specification sheets may vary from the absolute minimum to an extensive display of graphs and charts, there are a few fundamental parameters that will be provided by all manufacturers. A few of the most important are discussed in the following paragraphs. The specification sheet for the 2N5457 *n*-channel JFET as provided by Motorola is provided as Fig. 5.18.

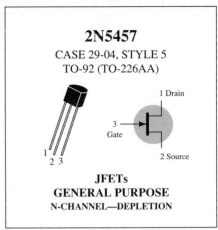

2N5457

CASE 29-04, STYLE 5
TO-92 (TO-226AA)

1 Drain

3
Gate

2 Source

JFETs
GENERAL PURPOSE
N-CHANNEL—DEPLETION

Refer to 2N4220 for graphs.

MAXIMUM RATINGS

Rating	Symbol	Value	Unit
Drain-Source Voltage	V_{DS}	25	Vdc
Drain-Gate Voltage	V_{DG}	25	Vdc
Reverse Gate-Source Voltage	V_{GSR}	−25	Vdc
Gate Current	I_G	10	mAdc
Total Device Dissipation @ $T_A = 25°C$ Derate above 25°C	P_D	310 2.82	mW mW/°C
Junction Temperature Range	T_J	125	°C
Storage Channel Temperature Range	T_{stg}	−65 to +150	°C

ELECTRICAL CHARACTERISTICS ($T_A = 25°C$ unless otherwise noted)

Characteristic		Symbol	Min	Typ	Max	Unit
OFF CHARACTERISTICS						
Gate-Source Breakdown Voltage ($I_G = -10\ \mu Adc, V_{DS} = 0$)		$V_{(BR)GSS}$	−25	–	–	Vdc
Gate Reverse Current ($V_{GS} = -15\ Vdc, V_{DS} = 0$) ($V_{GS} = -15\ Vdc, V_{DS} = 0, T_A = 100°C$)		I_{GSS}	– –	– –	−1.0 −200	nAdc
Gate Source Cutoff Voltage ($V_{DS} = 15\ Vdc, I_D = 10\ nAdc$)	2N5457	$V_{GS(off)}$	−0.5	–	−6.0	Vdc
Gate Source Voltage ($V_{DS} = 15\ Vdc, I_D = 100\ \mu Adc$)	2N5457	V_{GS}	–	−2.5	–	Vdc
ON CHARACTERISTICS						
Zero-Gate-Voltage Drain Current* ($V_{DS} = 15\ Vdc, V_{GS} = 0$)	2N5457	I_{DSS}	1.0	3.0	5.0	mAdc
SMALL-SIGNAL CHARACTERISTICS						
Forward Transfer Admittance Common Source* ($V_{DS} = 15\ Vdc, V_{GS} = 0, f = 1.0\ kHz$)	2N5457	$|y_{fs}|$	1000	–	5000	μmhos
Output Admittance Common Source* ($V_{DS} = 15\ Vdc, V_{GS} = 0, f = 1.0\ kHz$)		$|y_{os}|$	–	10	50	μmhos
Input Capacitance ($V_{DS} = 15\ Vdc, V_{GS} = 0, f = 1.0\ MHz$)		C_{iss}	–	4.5	7.0	pF
Reverse Transfer Capacitance ($V_{DS} = 15\ Vdc, V_{GS} = 0, f = 1.0\ MHz$)		C_{rss}	–	1.5	3.0	pF

*Pulse Test: Pulse Width ≤ 630 ms; Duty Cycle ≤ 10%

Figure 5.18 2N5457 Motorola *n*-channel JFET.

Maximum Ratings

The maximum rating list usually appears at the beginning of the specification sheet, with the maximum voltages between specific terminals, maximum current levels, and the maximum power dissipation level of the device. The specified maximum levels for V_{DS} and V_{DG} must not be exceeded at any point in the design operation of the device. The applied source V_{DD} can exceed these levels, but the actual level of voltage between these terminals must never exceed the level specified. Any good design will try to avoid these levels by a good margin of safety. The term *reverse* in V_{GSR} defines the maximum voltage with the source positive with respect to the gate (as normally biased for an *n*-channel device) before breakdown will occur. On some specification sheets it is referred to as BV_{DSS}—the *B*reakdown *V*oltage with the *D*rain-*S*ource *S*horted ($V_{DS} = 0$ V). Although normally designed to operate with $I_G = 0$ mA, if *forced* to accept a gate current it could withstand 10 mA before damage would occur. The total device dissipation at 25°C (room temperature) is the maximum power the device can dissipate under normal operating conditions and is defined by

$$P_D = V_{DS}I_D \qquad (5.9)$$

Note the similarity in format with the maximum power dissipation equation for the BJT transistor.

The derating factor is discussed in detail in Chapter 3, but for the moment recognize that the 2.82 mW/°C rating reveals that the dissipation rating *decreases* by 2.82 mW for each *increase* in temperature of 1°C above 25°C.

Electrical Characteristics

The electrical characteristics include the level of V_P in the OFF CHARACTERISTICS and I_{DSS} in the ON CHARACTERISTICS. In this case $V_P = V_{GS(off)}$ has a range from -0.5 to -6.0 V and I_{DSS} from 1 to 5 mA. The fact that both will vary from device to device with the same nameplate identification must be considered in the design process. The other quantities are defined under conditions appearing in parentheses. The small-signal characteristics are discussed in Chapter 9.

Case Construction and Terminal Identification

This particular JFET has the appearance provided on the specification sheet of Fig. 5.18. The terminal identification is also provided directly under the figure. JFETs are also available in top-hat containers, as shown in Fig. 5.19 with its terminal identification.

Operating Region

The specification sheet and the curve defined by the pinch-off levels at each level of V_{GS} define the region of operation for linear amplification on the drain characteristics as shown in Fig. 5.20. The ohmic region defines the minimum permissible values of V_{DS} at each level of V_{GS}, and $V_{DS_{max}}$ specifies the maximum value for this parameter.

2N2844

CASE 22-03, STYLE 12
TO-18 (TO-206AA)

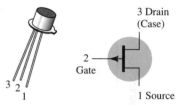

JFETs
GENERAL PURPOSE
P-CHANNEL

Figure 5.19 Top-hat container and terminal identification for a *p*-channel JFET.

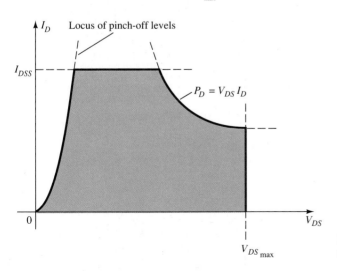

Figure 5.20 Normal operating region for linear amplifier design.

The saturation current I_{DSS} is the maximum drain current and the maximum power dissipation level defines the curve drawn in the same manner as described for BJT transistors. The resulting shaded region is the normal operating region for amplifier design.

5.5 INSTRUMENTATION

Recall from Chapter 3 that hand-held instruments are available to measure the level of β_{dc} for the BJT transistor. Similar instrumentation is not available to measure the levels of I_{DSS} and V_P. However, the curve tracer introduced for the BJT transistor can also display the drain characteristics of the JFET transistor through a proper setting of the various controls. The vertical scale (in milliamperes) and the horizontal scale (in volts) have been set to provide a full display of the characteristics, as shown in Fig. 5.21. For the JFET of Fig. 5.21 each vertical division (in centimeters) reflects a 1-mA change in I_C while each horizontal division has a value of 1 V. The step voltage is 500 mV/step (0.5 V/step), revealing that the top curve is defined by $V_{GS} = 0$ V and the next curve down −0.5 V for the n-channel device. Using the same step voltage the next curve is −1 V, then −1.5 V, and finally −2 V. By drawing a line from the top curve over to the I_D axis the level of I_{DSS} can be estimated to be about 9 mA. The level of V_P can be estimated by noting the V_{GS} value of the bottom curve and taking into account the shrinking distance between curves as V_{GS} becomes more and more negative. In this case, V_P is certainly more negative than −2 V and perhaps V_P is close to −2.5 V. However, keep in mind that the V_{GS} curves contract very quickly as they approach the cutoff condition, and perhaps $V_P = -3$ V is a better

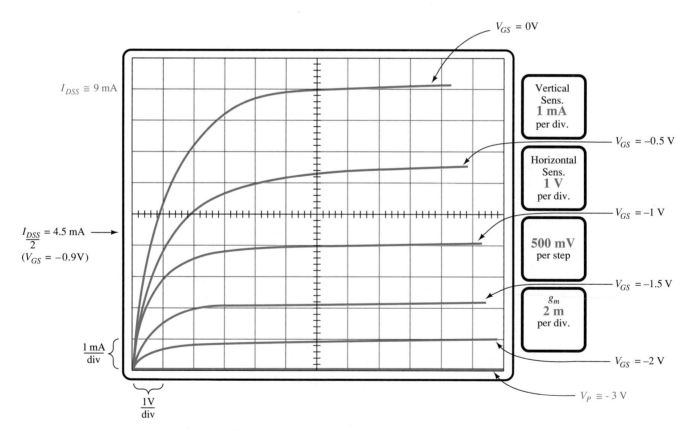

Figure 5.21 Drain characteristics for a 2N4416 JFET transistor as displayed on a curve tracer.

choice. It should also be noted that the step control is set for a five-step display limiting the displayed curves to $V_{GS} = 0$ V, -0.5 V, -1 V, -1.5 V, and -2 V. If the step control had been increased to 10, the voltage per step could be reduced to 250 mV = 0.25 V and the curve for $V_{GS} = -2.25$ V would have been included in addition to an additional curve between each step of Fig. 5.21. The $V_{GS} = -2.25$ V curve would reveal how quickly the curves are closing in on each other for the same step voltage. Fortunately, the level of V_P can be estimated to a reasonable degree of accuracy simply by applying a condition appearing in Table 5.1. That is, when $I_D = I_{DSS}/2$, then $V_{GS} = 0.3 V_P$. For the characteristics of Fig. 5.21, $I_D = I_{DSS}/2 = 9$ mA/2 = 4.5 mA, and as visible from Fig. 5.21 the corresponding level of V_{GS} is about -0.9 V. Using this information we find that $V_P = V_{GS}/0.3 = -0.9$ V/0.3 = -3 V, which will be our choice for this device. Using this value we find that at $V_{GS} = -2$ V,

$$
\begin{aligned}
I_D &= I_{DSS}\left(1 - \frac{V_{GS}}{V_P}\right)^2 \\
&= 9 \text{ mA}\left(1 - \frac{-2 \text{ V}}{-3 \text{ V}}\right)^2 \\
&\cong 1 \text{ mA}
\end{aligned}
$$

as supported by Fig. 5.21.

At $V_{GS} = -2.5$ V Schockley's equation will result in $I_D = 0.25$ mA, with $V_P = -3$ V clearly revealing how quickly the curves contract near V_P. The importance of the parameter g_m and how it is determined from the characteristics of Fig. 5.21 are described in Chapter 8 when small-signal ac conditions are examined.

5.6 IMPORTANT RELATIONSHIPS

A number of important equations and operating characteristics have been introduced in the last few sections that are of particular importance for the analysis to follow for the dc and ac configurations. In an effort to isolate and emphasize their importance they are repeated below next to a corresponding equation for the BJT transistor. The JFET equations are defined for the configuration of Fig. 5.22a while the BJT equations relate to Fig. 5.22b.

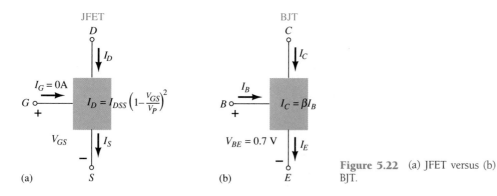

Figure 5.22 (a) JFET versus (b) BJT.

$$
\boxed{
\begin{array}{lcl}
\textbf{JFET} & & \textbf{BJT} \\
I_D = I_{DSS}\left(1 - \dfrac{V_{GS}}{V_P}\right)^2 & \Longleftrightarrow & I_C = \beta I_B \\
I_D = I_S & \Longleftrightarrow & I_C \cong I_E \\
I_G = 0 \text{ A} & \Longleftrightarrow & V_{BE} = 0.7 \text{ V}
\end{array}
}
\tag{5.10}
$$

A clear understanding of the impact of each of the equations above is sufficient background to approach the most complex of dc configurations. Recall that $V_{BE} = 0.7$ V was often the key to initiating an analysis of a BJT configuration. Similarly, the condition $I_G = 0$ A is often the starting point for the analysis of a JFET configuration. For the BJT configuration, I_B is normally the first parameter to be determined. For the JFET it is normally V_{GS}. The number of similarities between the analysis of BJT and JFET dc configurations will become quite apparent in Chapter 6.

5.7 DEPLETION-TYPE MOSFET

As noted in the chapter introduction, there are two types of FETs: JFETs and MOS-FETs. MOSFETs are further broken down into *depletion type* and *enhancement type*. The terms *depletion* and *enhancement* define their basic mode of operation, while the label MOSFET stands for *metal-oxide-semiconductor-field-effect transistor*. Since there are differences in the characteristics and operation of each type of MOSFET they are covered in separate sections. In this section we examine the depletion-type MOSFET, which happens to have characteristics similar to those of a JFET between cutoff and saturation at I_{DSS}, but then has the added feature of characteristics that extend into the region of opposite polarity for V_{GS}.

Basic Construction

The basic construction of the *n*-channel depletion-type MOSFET is provided in Fig. 5.23. A slab of *p*-type material is formed from a silicon base and is referred to as the *substrate*. It is the foundation upon which the device will be constructed. In some cases the substrate is internally connected to the source terminal. However, many discrete devices provide an additional terminal labeled SS, resulting in a four-terminal device, such as that appearing in Fig. 5.23. The source and drain terminals are connected through metallic contacts to *n*-doped regions linked by an *n*-channel as shown in the figure. The gate is also connected to a metal contact surface but remains insulated from the *n*-channel by a very thin silicon dioxide (SiO_2) layer. SiO_2 is a particular type of insulator referred to as a *dielectric* that sets up opposing (as revealed

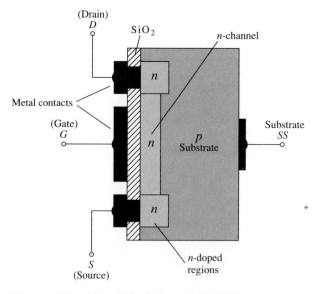

Figure 5.23 *n*-Channel depletion-type MOSFET.

by the prefix *di-*) electric fields within the dielectric when exposed to an externally applied field. The fact that the SiO$_2$ layer is an insulating layer reveals the following fact:

> *There is no direct electrical connection between the gate terminal and the channel of a MOSFET.*

In addition:

> *It is the insulating layer of SiO$_2$ in the MOSFET construction that accounts for the very desirable high input impedance of the device.*

In fact, the input resistance of a MOSFET is often that of the typical JFET, even though the input impedance of most JFETs is sufficiently high for most applications. The very high input impedance continues to fully support the fact that the gate current (I_G) is essentially zero amperes for dc-biased configurations.

The reason for the label metal-oxide-semiconductor FET is now fairly obvious. The *metal* for the drain, source, and gate connections to the proper surface—in particular, the gate terminal and the control to be offered by the surface area of the contact, the *oxide* for the silicon dioxide insulating layer, and the *semiconductor* for the basic structure on which the *n*- and *p*-type regions are diffused. The insulating layer between the gate and channel has resulted in another name for the device: *insulated-gate FET* or *IGFET*, although this label is used less and less in current literature.

Basic Operation and Characteristics

In Fig. 5.24 the gate-to-source voltage is set to zero volts by the direct connection from one terminal to the other, and a voltage V_{DS} is applied across the drain-to-source terminals. The result is an attraction for the positive potential at the drain by the *free* electrons of the *n*-channel and a current similar to that established through the channel of the JFET. In fact, the resulting current with $V_{GS} = 0$ V continues to be labeled I_{DSS}, as shown in Fig. 5.25.

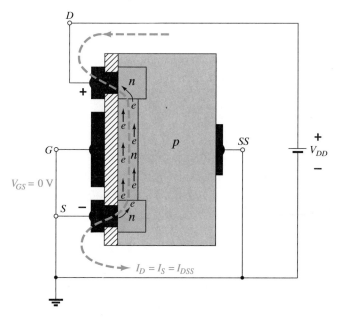

Figure 5.24 *n*-Channel depletion-type MOSFET with $V_{GS} = 0$ V and an applied voltage V_{DD}.

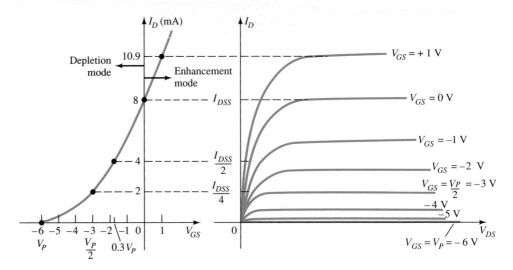

Figure 5.25 Drain and transfer characteristics for an *n*-channel depletion-type MOSFET.

In Fig. 5.26, V_{GS} has been set at a negative voltage such as -1 V. The negative potential at the gate will tend to pressure electrons toward the *p*-type substrate (like charges repel) and attract holes from the *p*-type substrate (opposite charges attract) as shown in Fig. 5.26. Depending on the magnitude of the negative bias established by V_{GS}, a level of recombination between electrons and holes will occur that will reduce the number of free electrons in the *n*-channel available for conduction. The more negative the bias, the higher the rate of recombination. The resulting level of drain current is therefore reduced with increasing negative bias for V_{GS} as shown in Fig. 5.25 for $V_{GS} = -1$ V, -2 V, and so on, to the pinch-off level of -6 V. The resulting levels of drain current and the plotting of the transfer curve proceeds exactly as described for the JFET.

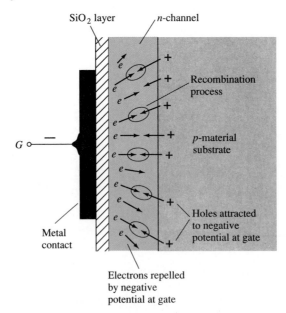

Figure 5.26 Reduction in free carriers in channel due to a negative potential at the gate terminal.

For positive values of V_{GS} the positive gate will draw additional electrons (free carriers) from the *p*-type substrate due to the reverse leakage current and establish new carriers through the collisions resulting between accelerating particles. As the gate-to-source voltage continues to increase in the positive direction, Fig. 5.25 reveals that the drain current will increase at a rapid rate for the reasons listed above.

The vertical spacing between the $V_{GS} = 0$ V and $V_{GS} = +1$ V curves of Fig. 5.25 is a clear indication of how much the current has increased for the 1 volt change in V_{GS}. Due to the rapid rise, the user must be aware of the maximum drain current rating since it could be exceeded with a positive gate voltage. That is, for the device of Fig. 5.25, the application of a voltage $V_{GS} = +4$ V would result in a drain current of 22.2 mA, which could possibly exceed the maximum rating (current or power) for the device. As revealed above, the application of a positive gate-to-source voltage has ''enhanced'' the level of free carriers in the channel compared to that encountered with $V_{GS} = 0$ V. For this reason the region of positive gate voltages on the drain or transfer characteristics is often referred to as the *enhancement region,* with the region between cutoff and the saturation level of I_{DSS} referred to as the *depletion region.*

It is particularly interesting and helpful that Shockley's equation will continue to be applicable for the depletion-type MOSFET characteristics in both the depletion and enhancement regions. For both regions it is simply necessary that the proper sign be included with V_{GS} in the equation and the sign be carefully monitored in the mathematical operations.

Sketch the transfer characteristics for an *n*-channel depletion-type MOSFET with $I_{DSS} = 10$ mA and $V_P = -4$ V.

EXAMPLE 5.3

Solution

At $V_{GS} = 0$ V, $\quad I_D = I_{DSS} = 10$ mA

$\quad V_{GS} = V_P = -4$ V, $\quad I_D = 0$ mA

$\quad V_{GS} = \dfrac{V_P}{2} = \dfrac{-4 \text{ V}}{2} = -2$ V, $\quad I_D = \dfrac{I_{DSS}}{4} = \dfrac{10 \text{ mA}}{4} = 2.5$ mA

and at $I_D = \dfrac{I_{DSS}}{2}$, $\quad V_{GS} = 0.3 \, V_P = 0.3(-4 \text{ V}) = -1.2$ V

all of which appear in Fig. 5.27.

Before plotting the positive region of V_{GS}, keep in mind that I_D increases very rapidly with increasing positive values of V_{GS}. In other words, be conservative with the choice of values to be substituted into Shockley's equation. In this case, we will try $+1$ V as follows:

$$I_D = I_{DSS}\left(1 - \frac{V_{GS}}{V_P}\right)^2$$

$$= 10 \text{ mA}\left(1 - \frac{+1 \text{ V}}{-4 \text{ V}}\right)^2 = 10 \text{ mA}(1 + 0.25)^2 = 10 \text{ mA}(1.5625)$$

$$\cong 15.63 \text{ mA}$$

which is sufficiently high to finish the plot.

---

p-Channel Depletion-Type MOSFET

The construction of a *p*-channel depletion-type MOSFET is exactly the reverse of that appearing in Fig. 5.23. That is, there is now an *n*-type substrate and a *p*-type channel, as shown in Fig. 5.28a. The terminals remain as identified, but all the voltage polarities and the current directions are reversed, as shown in the same figure. The drain characteristics would appear exactly as in Fig. 5.25 but with V_{DS} negative values, I_D

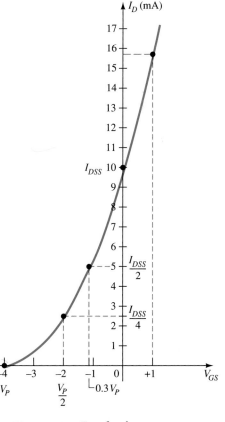

Figure 5.27 Transfer characteristics for an *n*-channel depletion-type MOSFET with $I_{DSS} = 10$ mA and $V_P = -4$ V.

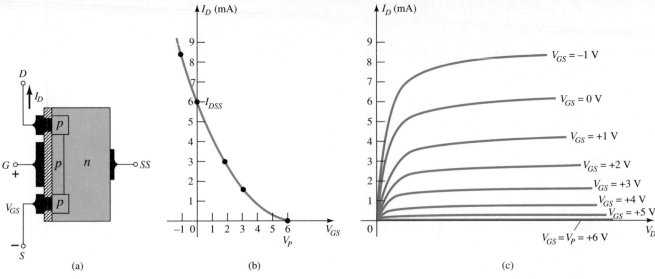

Figure 5.28 *p*-Channel depletion-type MOSFET with $I_{DSS} = 6$ mA and $V_P = +6$ V.

positive as indicated (since the defined direction is now reversed), and V_{GS} having the opposite polarities as shown in Fig. 5.28c. The reversal in V_{GS} will result in a mirror image (about the I_D axis) for the transfer characteristics as shown in Fig. 5.28b. In other words, the drain current will increase from cutoff at $V_{GS} = V_P$ in the positive V_{GS} region to I_{DSS} and then continue to increase for increasingly negative values of V_{GS}. Shockley's equation is still applicable and requires simply placing the correct sign for both V_{GS} and V_P in the equation.

Symbols, Specification Sheets, and Case Construction

The graphic symbols for an *n*- and *p*-channel depletion-type MOSFET are provided in Fig. 5.29. Note how the symbols chosen try to reflect the actual construction of the device. The lack of a direct connection (due to the gate insulation) between the gate and channel is represented by a space between the gate and the other terminals of the symbol. The vertical line representing the channel is connected between the drain and source and is "supported" by the substrate. Two symbols are provided for each type of channel to reflect the fact that in some cases the substrate is externally available, while in others it is not. For most of the analysis to follow in Chapter 6, the substrate and source will be connected and the lower symbols will be employed.

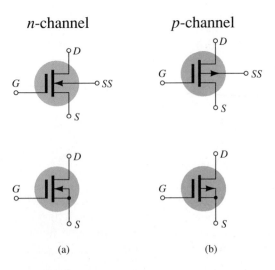

Figure 5.29 Graphic symbols for (a) *n*-channel depletion-type MOSFETs and (b) *p*-channel depletion-type MOSFETs.

The device appearing in Fig. 5.30 has three terminals with the terminal identification appearing in the same figure. The specification sheet for a depletion-type MOSFET is similar to that of a JFET. The levels of V_P and I_{DSS} are provided along with a list of maximum values and typical "on" and "off" characteristics. In addition,

2N3797

CASE 22-03, STYLE 2
TO-18 (TO-206AA)

MOSFETs
LOW POWER AUDIO

N-CHANNEL – DEPLETION

MAXIMUM RATINGS

Rating	Symbol	Value	Unit
Drain–Source Voltage 2N3797	V_{DS}	20	Vdc
Gate–Source Voltage	V_{GS}	±10	Vdc
Drain Current	I_D	20	mAdc
Total Device Dissipation @ $T_A = 25°C$ Derate above 25°C	P_D	200 1.14	mW mW/°C
Junction Temperature Range	T_J	+175	°C
Storage Channel Temperature Range	T_{stg}	−65 to +200	°C

ELECTRICAL CHARACTERISTICS ($T_A = 25°C$ unless otherwise noted)

Characteristic	Symbol	Min	Typ	Max	Unit
OFF CHARACTERISTICS					
Drain Source Breakdown Voltage ($V_{GS} = -7.0$ V, $I_D = 5.0 \mu A$) 2N3797	$V_{(BR)DSX}$	20	25	–	Vdc
Gate Reverse Current (1) ($V_{GS} = -10$ V, $V_{DS} = 0$) ($V_{GS} = -10$ V, $V_{DS} = 0$, $T_A = 150°C$)	I_{GSS}	– –	– –	1.0 200	pAdc
Gate Source Cutoff Voltage ($I_D = 2.0 \mu A$, $V_{DS} = 10$ V) 2N3797	$V_{GS(off)}$	–	−5.0	−7.0	Vdc
Drain-Gate Reverse Current (1) ($V_{DG} = 10$ V, $I_S = 0$)	I_{DGO}	–	–	1.0	pAdc
ON CHARACTERISTICS					
Zero-Gate-Voltage Drain Current ($V_{DS} = 10$ V, $V_{GS} = 0$) 2N3797	I_{DSS}	2.0	2.9	6.0	mAdc
On-State Drain Current ($V_{DS} = 10$ V, $V_{GS} = +3.5$ V) 2N3797	$I_{D(on)}$	9.0	14	18	mAdc
SMALL-SIGNAL CHARACTERISTICS					
Forward Transfer Admittance ($V_{DS} = 10$ V, $V_{GS} = 0$, f = 1.0 kHz) 2N3797	$\|y_{fs}\|$	1500	2300	3000	μmhos
($V_{DS} = 10$ V, $V_{GS} = 0$, f = 1.0 MHz) 2N3797		1500	–	–	
Output Admittance ($I_{DS} = 10$ V, $V_{GS} = 0$, f = 1.0 kHz) 2N3797	$\|y_{os}\|$	–	27	60	μmhos
Input Capacitance ($V_{DS} = 10$ V, $V_{GS} = 0$, f = 1.0 MHz) 2N3797	C_{iss}	–	6.0	8.0	pF
Reverse Transfer Capacitance ($V_{DS} = 10$ V, $V_{GS} = 0$, f = 1.0 MHz)	C_{rss}	–	0.5	0.8	pF
FUNCTIONAL CHARACTERISTICS					
Noise Figure ($V_{DS} = 10$ V, $V_{GS} = 0$, f = 1.0 kHz, $R_S = 3$ megohms)	NF	–	3.8	–	dB

(1) This value of current includes both the FET leakage current as well as the leakage current associated with the test socket and fixture when measured under best attainable conditions.

Figure 5.30 2N3797 Motorola n-Channel depletion-type MOSFET.

however, since I_D can extend beyond the I_{DSS} level, another point is normally provided that reflects a typical value of I_D for some positive voltage (for an n-channel device). For the unit of Fig. 5.30, I_D is specified as $I_{D\,(on)} = 9$ mA dc with $V_{DS} = 10$ V and $V_{GS} = 3.5$ V.

5.8 ENHANCEMENT-TYPE MOSFET

Although there are some similarities in construction and mode of operation between depletion-type and enhancement-type MOSFETs, the characteristics of the enhancement-type MOSFET are quite different from anything obtained thus far. The transfer curve is not defined by Shockley's equation and the drain current is now cut off until the gate-to-source voltage reaches a specific magnitude. In particular, current control in an n-channel device is now effected by a positive gate-to-source voltage rather than the range of negative voltages encountered for n-channel JFETs and n-channel depletion-type MOSFETs.

Basic Construction

The basic construction on the n-channel enhancement-type MOSFET is provided in Fig. 5.31. A slab of p-type material is formed from a silicon base and is again referred to as the substrate. As with the depletion-type MOSFET, the substrate is sometimes internally connected to the source terminal, while in other cases a fourth lead is made available for external control of its potential level. The source and drain terminals are again connected through metallic contacts to n-doped regions, but note in Fig. 5.31 the absence of a channel between the two n-doped regions. This is the primary difference between the construction of a depletion-type and enhancement-type MOSFETs— the absence of a channel as a constructed component of the device. The SiO_2 layer is still present to isolate the gate metallic platform from the region between the drain and source, but now it is simply separated from a section of the p-type material. In summary, therefore, the construction of an enhancement-type MOSFET is quite similar to that of the depletion-type MOSFET except for the absence of a channel between the drain and source terminals.

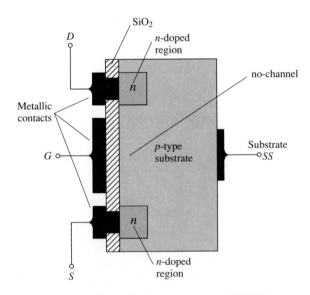

Figure 5.31 n-Channel enhancement-type MOSFET.

Basic Operation and Characteristics

If V_{GS} is set at 0 V and a voltage applied between the drain and source of the device of Fig. 5.31, the absence of an *n*-channel (with its generous number of free carriers) will result in a current of effectively zero amperes—quite different from the depletion-type MOSFET and JFET where $I_D = I_{DSS}$. It is not sufficient to have a large accumulation of carriers (electrons) at the drain and source (due to the *n*-doped regions) if a path fails to exist between the two. With V_{DS} some positive voltage, V_{GS} at 0 V, and terminal SS directly connected to the source, there are, in fact, two reverse-biased *p-n* junctions between the *n*-doped regions and the *p*-substrate to oppose any significant flow between drain and source.

In Fig. 5.32 both V_{DS} and V_{GS} have been set at some positive voltage greater than zero volts, establishing the drain and gate at a positive potential with respect to the source. The positive potential at the gate will pressure the holes (since like charges repel) in the *p*-substrate along the edge of the SiO_2 layer to leave the area and enter deeper regions of the *p*-substrate, as shown in the figure. The result is a depletion region near the SiO_2 insulating layer void of holes. However, the electrons in the *p*-substrate (the minority carriers of the material) will be attracted to the positive gate and accumulate in the region near the surface of the SiO_2 layer. The SiO_2 layer and its insulating qualities will prevent the negative carriers from being absorbed at the gate terminal. As V_{GS} increases in magnitude the concentration of electrons near the SiO_2 surface increases until eventually the induced *n*-type region can support a measurable flow between drain and source. The level of V_{GS} that results in the significant increase in drain current is called the *threshold voltage* and is given the symbol V_T. On specification sheets it is referred to as $V_{GS(Th)}$, although V_T is less unwieldy and will be used in the analysis to follow. Since the channel is nonexistent with $V_{GS} = 0$ V and "enhanced" by the application of a positive gate-to-source voltage, this type of MOSFET is called an *enhancement-type MOSFET*. Both depletion and enhancement-type MOSFETs have enhancement-type regions, but the label was applied to the latter since it is its only mode of operation.

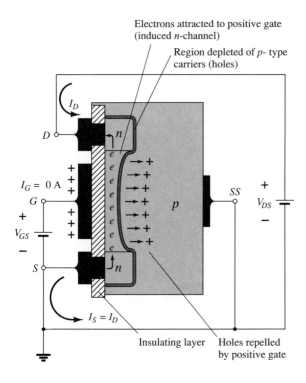

Figure 5.32 Channel formation in the *n*-channel enhancement-type MOSFET.

As V_{GS} is increased beyond the threshold level the density of free carriers in the induced channel will increase, resulting in an increased level of drain current. However, if we hold V_{GS} constant and increase the level of V_{DS}, the drain current will eventually reach a saturation level as occurred for the JFET and depletion-type MOSFET. The leveling off of I_D is due to a pinching-off process depicted by the narrower channel at the drain end of the induced channel as shown in Fig. 5.33. Applying Kirchhoff's voltage law to the terminal voltages of the MOSFET of Fig. 5.33, we find that

$$V_{DG} = V_{DS} - V_{GS} \tag{5.11}$$

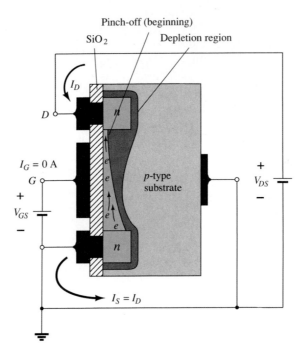

Figure 5.33 Change in channel and depletion region with increasing level of V_{DS} for a fixed value of V_{GS}.

If V_{GS} is held fixed at some value such as 8 V and V_{DS} is increased from 2 V to 5 V, the voltage V_{DG} [by Eq. (5.11)] will drop from −6 V to −3 V and the gate will become less and less positive with respect to the drain. This reduction in gate-to-drain voltage will in turn reduce the attractive forces for free carriers (electrons) in this region of the induced channel, causing a reduction in the effective channel width. Eventually, the channel will be reduced to the point of pinch-off and a saturation condition will be established as described earlier for the JFET and depletion-type MOSFET. In other words, any further increase in V_{DS} at the fixed value of V_{GS} will not affect the saturation level of I_D until breakdown conditions are encountered.

The drain characteristics of Fig. 5.34 reveal that for the device of Fig. 5.33 with $V_{GS} = 8$ V, saturation occurred at a level of $V_{DS} = 6$ V. In fact, the saturation level for V_{DS} is related to the level of applied V_{GS} by

$$V_{DS_{sat}} = V_{GS} - V_T \tag{5.12}$$

Obviously, therefore, for a fixed value of V_T the higher the level of V_{GS}, the more the saturation level for V_{DS}, as shown in Fig. 5.33 by the locus of saturation levels.

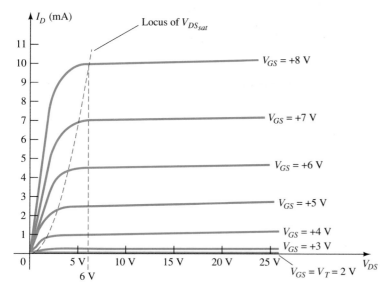

Figure 5.34 Drain characteristics of an *n*-channel enhancement-type MOSFET with $V_T = 2$ V and $k = 0.278 \times 10^{-3}$ A/V².

For the characteristics of Fig. 5.33 the level of V_T is 2 V, as revealed by the fact that the drain current has dropped to 0 mA. In general, therefore:

For values of V_{GS} less than the threshold level the drain current of an en-hancement-type MOSFET is 0 mA.

Figure 5.34 clearly reveals that as the level of V_{GS} increased from V_T to 8 V, the resulting saturation level for I_D also increased from a level of 0 mA to 10 mA. In addition, it is quite noticeable that the spacing between the levels of V_{GS} increased as the magnitude of V_{GS} increased, resulting in ever-increasing increments in drain current.

For levels of $V_{GS} > V_T$ the drain current is related to the applied gate-to-source voltage by the following nonlinear relationship:

$$I_D = k(V_{GS} - V_T)^2 \qquad (5.13)$$

Again, it is the squared term that results in the nonlinear (curved) relationship between I_D and V_{GS}. The k term is a constant that is a function of the construction of the device. The value of k can be determined from the following equation [derived from Eq. (5.13)] where $I_{D(\text{on})}$ and $V_{GS(\text{on})}$ are the values for each at a particular point on the characteristics of the device.

$$k = \frac{I_{D(\text{on})}}{(V_{GS(\text{on})} - V_T)^2} \qquad (5.14)$$

Substituting $I_{D(\text{on})} = 10$ mA when $V_{GS(\text{on})} = 8$ V from the characteristics of Fig. 5.34 yields

$$k = \frac{10 \text{ mA}}{(8 \text{ V} - 2 \text{ V})^2} = \frac{10 \text{ mA}}{(6 \text{ V})^2} = \frac{10 \text{ mA}}{36 \text{ V}^2}$$

$$= \mathbf{0.278 \times 10^{-3} \text{ A/V}^2}$$

and a general equation for I_D for the characteristics of Fig. 5.34 results:

$$I_D = 0.278 \times 10^{-3}(V_{GS} - 2 \text{ V})^2$$

5.8 Enhancement-Type MOSFET

Substituting $V_{GS} = 4$ V, we find that

$$I_D = 0.278 \times 10^{-3}(4\text{ V} - 2\text{ V})^2 = 0.278 \times 10^{-3}(2)^2$$

$$= 0.278 \times 10^{-3}(-4) = \mathbf{1.11\text{ mA}}$$

as verified by Fig. 5.34. At $V_{GS} = V_T$ the squared term is 0 and $I_D = 0$ mA.

For the dc analysis of enhancement-type MOSFETs to appear in Chapter 6, the transfer characteristics will again be the characteristics to be employed in the graphical solution. In Fig. 5.35 the drain and transfer characteristics have been set side by side to describe the transfer process from one to the other. Essentially, it proceeds as introduced earlier for the JFET and depletion-type MOSFETs. In this case, however, it must be remembered that the drain current is 0 mA for $V_{GS} \leq V_T$. At this point a measurable current will result for I_D and will increase as defined by Eq. (5.13). Note that in defining the points on the transfer characteristics from the drain characteristics, only the saturation levels are employed, thereby limiting the region of operation to levels of V_{DS} greater than the saturation levels as defined by Eq. (5.12).

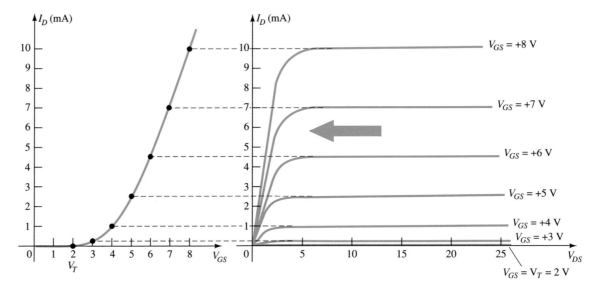

Figure 5.35 Sketching the transfer characteristics for an n-channel enhancement-type MOSFET from the drain characteristics.

The transfer curve of Fig. 5.35 is certainly quite different from those obtained earlier. For an n-channel (induced) device it is now totally in the positive V_{GS} region and does not rise until $V_{GS} = V_T$. The question now surfaces as to how to plot the transfer characteristics given the levels of k and V_T as included below for a particular MOSFET:

$$I_D = 0.5 \times 10^{-3}(V_{GS} - 4\text{ V})^2$$

First a horizontal line is drawn at $I_D = 0$ mA from $V_{GS} = 0$ V to $V_{GS} = 4$ V as shown in Fig. 5.36a. Next, a level of V_{GS} greater than V_T such as 5 V is chosen and substituted into Eq. (5.13) to determine the resulting level of I_D as follows:

$$I_D = 0.5 \times 10^{-3}(V_{GS} - 4\text{ V})^2$$

$$= 0.5 \times 10^{-3}(5\text{ V} - 4\text{ V})^2 = 0.5 \times 10^{-3}(1)^2$$

$$= \mathbf{0.5\text{ mA}}$$

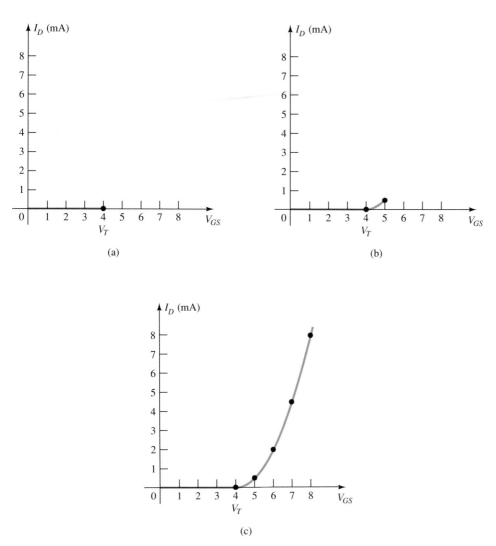

Figure 5.36 Plotting the transfer characteristics of an *n*-channel enhancement-type MOSFET with $k = 0.5 \times 10^{-3}$ A/V² and $V_T = 4$ V.

and a point on the plot is obtained as shown in Fig. 5.36b. Finally, additional levels of V_{GS} are chosen and the resulting levels of I_D obtained. In particular, at $V_{GS} = 6$ V, 7 V, and 8 V the level of I_D is 2 mA, 4.5 mA, and 8 mA, respectively, as shown on the resulting plot of Fig. 5.36c.

p-Channel Enhancement-Type MOSFETs

The construction of a *p*-channel enhancement-type MOSFET is exactly the reverse of that appearing in Fig. 5.31 as shown in Fig. 5.37a. That is, there is now an *n*-type substrate and *p*-doped regions under the drain and source connections. The terminals remain as identified, but all the voltage polarities and the current directions are reversed. The drain characteristics will appear as shown in Fig. 5.37c, with increasing levels of current resulting from increasingly negative values of V_{GS}. The transfer characteristics will be the mirror image (about the I_D axis) of the transfer curve of Fig. 5.35 with I_D increasing with increasingly negative values of V_{GS} beyond V_T, as shown in Fig. 5.37b. Equations (5.11) through (5.14) are equally applicable to *p*-channel devices.

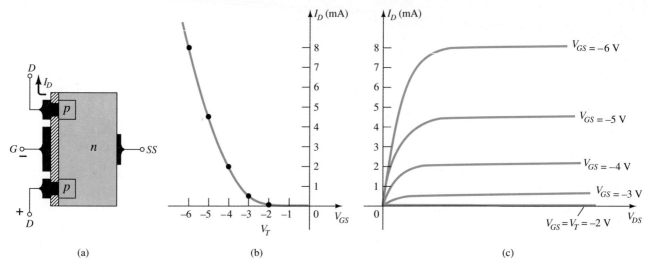

Figure 5.37 *p*-Channel enhancement-type MOSFET with $V_T = 2$ V and $k = 0.5 \times 10^{-3}$ A/V².

Symbols, Specification Sheets, and Case Construction

The graphic symbols for the *n*- and *p*-channel enhancement-type MOSFETs are provided as Fig. 5.38. Again note how the symbols try to reflect the actual construction of the device. The dashed line between drain and source was chosen to reflect the fact that a channel does not exist between the two under no-bias conditions. It is, in fact, the only difference between the symbols for the depletion-type and enhancement-type MOSFETs.

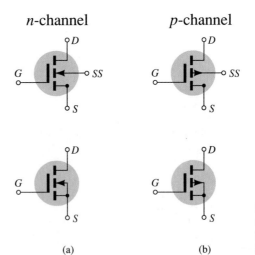

Figure 5.38 Symbols for (a) *n*-channel enhancement-type MOSFETs and (b) *p*-channel enhancement-type MOSFETs.

The specification sheet for a Motorola *n*-channel enhancement-type MOSFET is provided as Fig. 5.39. The case construction and terminal identification are provided next to the maximum ratings, which now include a maximum drain current of 30 mA dc. The specification sheet provides the level of I_{DSS} under "off" conditions, which is now simply 10 nA dc (at $V_{DS} = 10$ V and $V_{GS} = 0$ V) compared to the milliampere range for the JFET and depletion-type MOSFET. The threshold voltage

MAXIMUM RATINGS

Rating	Symbol	Value	Unit
Drain–Source Voltage	V_{DS}	25	Vdc
Drain–Gate Voltage	V_{DG}	30	Vdc
Gate–Source Voltage*	V_{GS}	30	Vdc
Drain Current	I_D	30	mAdc
Total Device Dissipation @ $T_A = 25°C$ Derate above 25°C	P_D	300 1.7	mW mW/°C
Junction Temperature Range	T_J	175	°C
Storage Temperature Range	T_{stg}	–65 to +175	°C

* Transient potentials of ± 75 Volt will not cause gate-oxide failure.

2N4351
CASE 20-03, STYLE 2
TO-72 (TO-206AF)

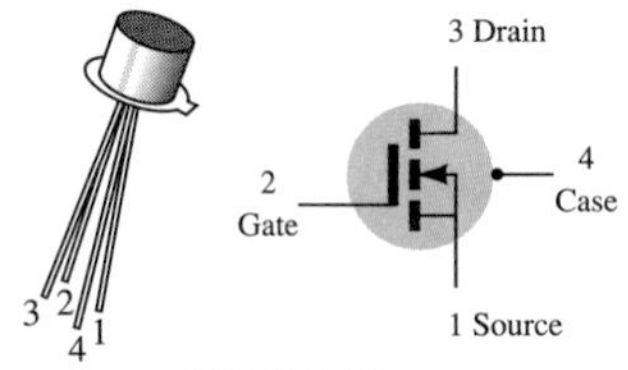

3 Drain
2 Gate
4 Case
1 Source

**MOSFET
SWITCHING
N-CHANNEL – ENHANCEMENT**

ELECTRICAL CHARACTERISTICS ($T_A = 25°C$ unless otherwise noted.)

Characteristic	Symbol	Min	Max	Unit
OFF CHARACTERISTICS				
Drain-Source Breakdown Voltage ($I_D = 10\ \mu A$, $V_{GS} = 0$)	$V_{(BR)DSX}$	25	–	Vdc
Zero-Gate-Voltage Drain Current ($V_{DS} = 10$ V, $V_{GS} = 0$) $T_A = 25°C$ $T_A = 150°C$	I_{DSS}	– –	10 10	nAdc μAdc
Gate Reverse Current ($V_{GS} = \pm 15$ Vdc, $V_{DS} = 0$)	I_{GSS}	–	± 10	pAdc
ON CHARACTERISTICS				
Gate Threshold Voltage ($V_{DS} = 10$ V, $I_D = 10\,\mu A$)	$V_{GS(Th)}$	1.0	5	Vdc
Drain-Source On-Voltage ($I_D = 2.0$ mA, $V_{GS} = 10$V)	$V_{DS(on)}$	–	1.0	V
On-State Drain Current ($V_{GS} = 10$ V, $V_{DS} = 10$ V)	$I_{D(on)}$	3.0	–	mAdc
SMALL-SIGNAL CHARACTERISTICS				
Forward Transfer Admittance ($V_{DS} = 10$ V, $I_D = 2.0$ mA, f = 1.0 kHz)	$\|y_{fs}\|$	1000	–	μmho
Input Capacitance ($V_{DS} = 10$ V, $V_{GS} = 0$, f = 140 kHz)	C_{iss}	–	5.0	pF
Reverse Transfer Capacitance ($V_{DS} = 0$, $V_{GS} = 0$, f = 140 kHz)	C_{rss}	–	1.3	pF
Drain-Substrate Capacitance ($V_{D(SUB)} = 10$ V, f = 140 kHz)	$C_{d(sub)}$	–	5.0	pF
Drain-Source Resistance ($V_{GS} = 10$ V, $I_D = 0$, f = 1.0 kHz)	$r_{ds(on)}$	–	300	ohms
SWITCHING CHARACTERISTICS				
Turn-On Delay (Fig. 5)	t_{d1}	–	45	ns
Rise Time (Fig. 6)	t_r	–	65	ns
Turn-Off Delay (Fig. 7)	t_{d2}	–	60	ns
Fall Time (Fig. 8)	t_f	–	100	ns

Switching: $I_D = 2.0$ mAdc, $V_{DS} = 10$ Vdc, ($V_{GS} = 10$ Vdc) (See Figure 9; Times Circuit Determined)

Figure 5.39 2N4351 Motorola n-Channel enhancement-type MOSFET.

is specified as $V_{GS(Th)}$ and has a range of 1 to 5 V dc, depending on the unit employed. Rather than provide a range of k in Eq. (5.13), a typical level of $I_{D(on)}$ (3 mA in this case) is specified at a particular level of $V_{GS(on)}$ (10 V for the specified I_D level). In other words, when $V_{GS} = 10$ V, $I_D = 3$ mA. The given levels of $V_{GS(Th)}$, $I_{D(on)}$, and $V_{GS(on)}$ permit a determination of k from Eq. (5.14) and a writing of the general equation for the transfer characteristics. The handling requirements of MOSFETs are reviewed in Section 5.9.

5.8 Enhancement-Type MOSFET

EXAMPLE 5.4

Using the data provided on the specification sheet of Fig. 5.39 and an average threshold voltage of $V_{GS(Th)} = 3$ V, determine:
(a) The resulting value of k for the MOSFET.
(b) The transfer characteristics.

Solution

(a) Eq. (5.14): $k = \dfrac{I_{D(on)}}{(V_{GS(on)} - V_{GS(Th)})^2}$

$$= \dfrac{3 \text{ mA}}{(10 \text{ V} - 3 \text{ V})^2} = \dfrac{3 \text{ mA}}{(7 \text{ V})^2} = \dfrac{3 \times 10^{-3}}{49} \text{ A/V}^2$$

$$= \mathbf{0.061 \times 10^{-3} \text{ A/V}^2}$$

(b) Eq. (5.13): $I_D = k(V_{GS} - V_T)^2$
$$= 0.061 \times 10^{-3}(V_{GS} - 3 \text{ V})^2$$

For $V_{GS} = 5$ V,

$$I_D = 0.061 \times 10^{-3}(5 \text{ V} - 3 \text{ V})^2 = 0.061 \times 10^{-3}(2)^2$$

$$= 0.061 \times 10^{-3}(4) = 0.244 \text{ mA}$$

For $V_{GS} = 8$ V, 10 V, and 12 V, I_D will be 1.525 mA, 3 mA (as defined), and 4.94 mA, respectively. The transfer characteristics are sketched in Fig. 5.40.

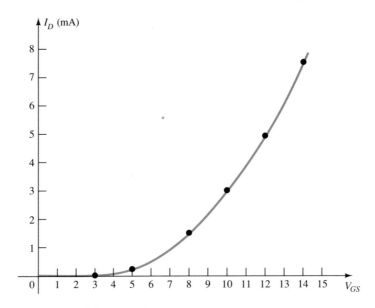

Figure 5.40 Solution to Example 5.4.

5.9 MOSFET HANDLING

The thin SiO_2 layer between the gate and channel of MOSFETs has the positive effect of providing a high-input-impedance characteristic for the device but because of its extremely thin layer introduces a concern for its handling that was not present for the BJT or JFET transistors. There is often sufficient accumulation of static charge (that we pick up from our surroundings) to establish a potential difference across the thin layer that can break down the layer and establish conduction through the layer. It is therefore imperative that we leave the shorting (or conduction) shipping foil (or ring) connecting the leads of the device together until the device is to be inserted in the system. The shorting ring prevents the possibility of applying a potential across any two terminals of the device. With the ring the potential difference between any two terminals is maintained at 0 V. At the very least always touch ground to permit discharge of the accumulated static charge before handling the device, and always pick up the transistor by the casing.

There are often transients (sharp changes in voltage or current) in a network when elements are removed or inserted if the power is on. The transient levels can often be more than the device can handle, and therefore the power should always be off when network changes are made.

The maximum gate-to-source voltage is normally provided in the list of maximum ratings of the device. One method of ensuring that this voltage is not exceeded (perhaps by transient effects) for either polarity is to introduce two Zener diodes, as shown in Fig. 5.41. The Zeners are back to back to ensure protection for either polarity. If both are 30-V Zeners and a positive transient of 40 V appears, the lower Zener will "fire" at 30 V and the upper will turn on with a zero-volt drop (ideally— for the positive "on" region of a semiconductor diode) across the other diode. The result is a maximum of 30 V for the gate-to-source voltage. One disadvantage introduced by the Zener protection is that the off resistance of a Zener diode is less than the input impedance established by the SiO$_2$ layer. The result is a reduction in input resistance, but even so it is still high enough for most applications. So many of the discrete devices now have the Zener protection that some of the concerns listed above are not as troublesome. However, it is still best to be somewhat cautious when handling discrete MOSFET devices.

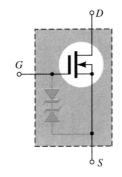

Figure 5.41 Zener-protected MOSFET.

5.10 VMOS

One of the disadvantages of the typical MOSFET is the reduced power-handling levels (typically, less than 1 W) compared to BJT transistors. This short fall for a device with so many positive characteristics can be softened by changing the construction mode from one of a planar nature such as shown in Fig. 5.23 to one with a vertical structure as shown in Fig. 5.42. All the elements of the planar MOSFET are present in the vertical metal-oxide-silicon FET (VMOS)—the metallic surface connection to the terminals of the device—the SiO$_2$ layer between the gate and the p-type region between the drain and source for the growth of the induced n-channel (en-

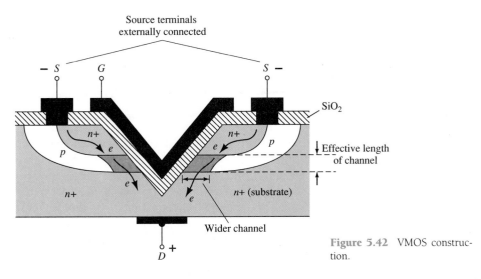

Figure 5.42 VMOS construction.

hancement-mode operation). The term *vertical* is due primarily to the fact that the channel is now formed in the vertical direction rather than the horizontal direction for the planar device. However, the channel of Fig. 5.42 also has the appearance of a V cut in the semiconductor base, which often stands out as a characteristic for mental memorization of the name of the device. The construction of Fig. 5.42 is somewhat simplistic in nature, leaving out some of the transition levels of doping, but it does permit a description of the most important facets of its operation.

The application of a positive voltage to the drain and a negative voltage to the source with the gate at 0 V or some typical positive "on" level as shown in Fig. 5.42 will result in the induced *n*-channel in the narrow *p*-type region of the device. The length of the channel is now defined by the vertical height of the *p*-region, which can be made significantly less than that of a channel using planar construction. On a horizontal plane the length of the channel is limited to 1 to 2 micrometers (μm) (1 μm = 10^{-6} m). Diffusion layers (such as the *p*-region of Fig. 5.42) can be controlled to small fractions of a micrometer. Since decreasing channel lengths result in reduced resistance levels, the power dissipation level of the device (power lost in the form of heat) at operating current levels will be reduced. In addition, the contact area between the channel and the n^+ region is greatly increased by the vertical mode construction, contributing to a further decrease in the resistance level and an increased area for current between the doping layers. There is also the existence of two conduction paths between drain and source, as shown in Fig. 5.42, to further contribute to a higher current rating. The net result is a device with drain currents that can reach the ampere levels with power levels exceeding 10 W.

In general:

Compared with commercially available planar MOSFETs, VMOS FETs have reduced channel resistance levels and higher current and power ratings.

An additional important characteristic of the vertical construction is:

VMOS FETs have a positive temperature coefficient that will combat the possibility of thermal runaway.

If the temperature of a device should increase due to the surrounding medium or currents of the device, the resistance levels will increase, causing a reduction in drain current rather than an increase as encountered for a conventional device. Negative temperature coefficients result in decreased levels or resistance with increase in temperature that fuel the growing current levels and result in further temperature instability and thermal runaway.

A further positive characteristic of the VMOS configuration is:

The reduced charge storage levels result in faster switching times for VMOS construction compared to those for conventional planar construction.

In fact, VMOS devices typically have switching times less than one-half that encountered for the typical BJT transistor.

5.11 CMOS

A very effective logic circuit can be established by constructing a *p*-channel and an *n*-channel MOSFET on the same substrate as shown in Fig. 5.43. Note the induced *p*-channel on the left and the induced *n*-channel on the right for the *p*- and *n*-channel devices, respectively. The configuration is referred to as a *complementary MOSFET* arrangement, abbreviated CMOS, that has extensive applications in computer logic design. The relatively high input impedance, fast switching speeds, and lower operating power levels of the CMOS configuration have resulted in a whole new discipline referred to as *CMOS logic design*.

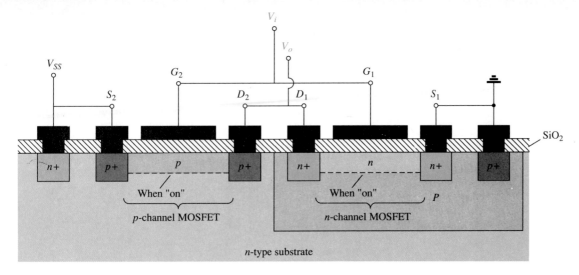

Figure 5.43 CMOS with the connections indicated in Fig. 5.44.

One very effective use of the complementary arrangement is as an inverter, as appearing in Fig. 5.44. As introduced for switching transistors, an inverter is a logic element that "inverts" the applied signal. That is, if the logic levels of operation are 0 V (0 state) and 5 V (1 state), an input level of 0 V will result in an output level of 5 V, and vice versa. Note in Fig. 5.44 that both gates are connected to the applied signal and both drains to the output V_o. The source of the p-channel MOSFET (Q_2) is connected directly to the applied voltage V_{SS} while the source of the n-channel MOS-FET (Q_1) is connected to ground. For the logic levels defined above, the application of 5 V at the input should result in approximately 0 V at the output. With 5 V at V_i (with respect to ground), $V_{GS_1} = V_i$ and Q_1 is "on," resulting in a relatively low resistance between drain and source as shown in Fig. 5.45. Since V_i and V_{SS} are at 5 V, $V_{GS_2} = 0$ V, which is less than the required V_T for the device resulting in an "off" state. The resulting resistance level between drain and source is quite high for Q_2, as shown in Fig. 5.45. A simple application of the voltage-divider rule will reveal that V_o is very close to 0 V or the 0 state, establishing the desired inversion process. For an applied voltage V_i of 0 V (0 state) $V_{GS_1} = 0$ V and Q_1 will be off with $V_{SS_1} = -5$ V, turning on the p-channel MOSFET. The result is that Q_2 will present a small resistance level and Q_1 a high resistance and $V_o = V_{SS} = 5$ V (the 1 state). Since the drain current that flows for either case is limited by the "off" transistor to the leakage value, the power dissipated by the device in either state is very low. Additional comment on the application of CMOS logic is presented in Chapter 17.

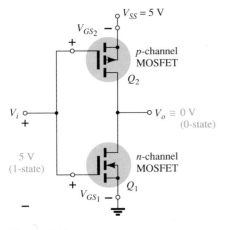

Figure 5.44 CMOS inverter.

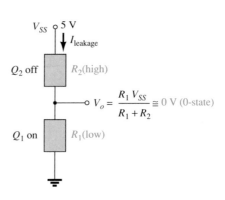

Figure 5.45 Relative resistance levels for $V_i = 5$ V (1-state).

5.11 CMOS   241

5.12 SUMMARY TABLE

Since the transfer curves and some important characteristics vary from one type of FET to another, Table 5.2 was developed to clearly display the differences from one device to the next. A clear understanding of all the curves and parameters of the table will provide a sufficient background for the dc and ac analysis to follow in Chapters 6 and 8. Take a moment to ensure that each curve is recognizable and their derivation understood and then establish a basis for comparison of the levels of the important parameters of R_i and C_i for each device.

TABLE 5.2 Field Effect Transistors

Type	-Symbol- Basic Relationships	Transfer Curve	Input Resistance and Capacitance
JFET (*n*-channel)	$I_G = 0$ A, $I_D = I_S$ $I_D = I_{DSS}(1 - \dfrac{V_{GS}}{V_P})^2$		$R_i > 100$ MΩ C_i: $(1 - 10)$ pF
MOSFET depletion-type (*n*-channel)	$I_G = 0$ A, $I_D = I_S$ $I_D = I_{DSS}(1 - \dfrac{V_{GS}}{V_P})^2$		$R_i > 10^{10}$ Ω C_i: $(1 - 10)$ pF
MOSFET enhancement-type (*n*-channel)	$I_G = 0$ A, $I_D = I_S$ $I_D = k(V_{GS} - V_T)^2$ $k = \dfrac{I_{D(on)}}{(V_{GS(on)} - V_T)^2}$		$R_i > 10^{10}$ Ω C_i: $(1 - 10)$ pF

Chapter 5   Field-Effect Transistors

5.13 COMPUTER ANALYSIS

The computer analysis of a FET amplifier in the dc mode using BASIC requires that the characteristic equation for the device be employed in conjunction with the network equations to obtain a mathematical solution. As noted for the BJT configuration, the analysis will proceed in much the same manner as a longhand approach. In Chapter 6, BASIC is employed to investigate one of the most common JFET amplifier configurations.

For PSpice there is a specific format that must be employed to properly enter the JFET parameters. For an *n*- or *p*-channel device the format is the following:

$$\underbrace{J1}_{name} \quad \underbrace{3}_{D} \quad \underbrace{1}_{G} \quad \underbrace{4}_{S} \quad \underbrace{JN}_{model\ name}$$

The format is very similar to that employed for the BJT transistor. The J is a JFET designator with the number 1 the chosen name. The nodes to which the terminals are connected are then listed *in the order* appearing in the example above. Finally, the model name must be entered to provide a location that will define the parameters of the JFET.

The format for the model description is the following:

$$.MODEL \quad \underbrace{JN}_{model\ name} \quad NJF\underbrace{(VTO = -4V,\ BETA = .5E-3)}_{parameter\ specifications}$$

The required .MODEL is followed by the model name as listed in the earlier statement. The NJF specifies an *n*-channel JFET, while PJF will specify a *p*-channel JFET. The choice of parameters to be specified can number as many as 14, but for our purposes it will usually be sufficient simply to specify VTO and BETA. VTO is the threshold voltage normally specified as V_P. BETA is not the β defined for BJT transistors but determined by the following equation:

$$BETA = \frac{I_{DSS}}{|V_P|^2} \tag{5.15}$$

For example if $V_P = -4$ V and $I_{DSS} = 8$ mA, the values appearing in the model statement above will result. That is, VTO $= -4$ V and BETA $= I_{DSS}/|V_P|^2 = 8$ mA/$(4$ V$)^2 = 8$ mA/16 V$^2 = 0.5 \times 10^{-3}$ A/V^2.

Both of the statements above will appear in a PSpice analysis performed in Chapter 6 on a voltage-divider configuration. Start to recognize the similarity of statements used for the entry of the parameters of a network. This similarity continues for a wide variety of devices, permitting a relatively easy adjustment to the analysis of networks with a variety of elements.

§ 5.2 Construction and Characteristics of JFETs

PROBLEMS

1. (a) Draw the basic construction of a *p*-channel JFET.
   (b) Apply the proper biasing between drain and source and sketch the depletion region for $V_{GS} = 0$ V.

2. Using the characteristics of Fig. 5.10, determine I_D for the following levels of V_{GS} (with $V_{DS} > V_P$).
   (a) $V_{GS} = 0$ V.
   (b) $V_{GS} = -1$ V.
   (c) $V_{GS} = -1.5$ V.
   (d) $V_{GS} = -1.8$ V.
   (e) $V_{GS} = -4$ V.
   (f) $V_{GS} = -6$ V.

3. (a) Determine V_{DS} for $V_{GS} = 0$ V and $I_D = 6$ mA using the characteristics of Fig. 5.10.
   (b) Using the results of part (a), calculate the resistance of the JFET for the region $I_D = 0$ mA to 6 mA for $V_{GS} = 0$ V.
   (c) Determine V_{DS} for $V_{GS} = -1$ V and $I_D = 3$ mA.
   (d) Using the results of part (c), calculate the resistance of the JFET for the region $I_D = 0$ mA to 3 mA for $V_{GS} = -1$ V.
   (e) Determine V_{DS} for $V_{GS} = -2$ V and $I_D = 1.5$ mA.
   (f) Using the results of part (e), calculate the resistance of the JFET for the region $I_D = 0$ mA to 1.5 mA for $V_{GS} = -2$ V.
   (g) Defining the result of part (b) as r_o, determine the resistance for $V_{GS} = -1$ V using Eq. (5.1) and compare with the results of part (d).
   (h) Repeat part (g) for $V_{GS} = -2$ V using the same equation, and compare the results with part (f).
   (i) Based on the results of parts (g) and (h), does Eq. (5.1) appear to be a valid approximation?

4. Using the characteristics of Fig. 5.10:
   (a) Determine the difference in drain current (for $V_{DS} > V_P$) between $V_{GS} = 0$ V and $V_{GS} = -1$ V.
   (b) Repeat part (a) between $V_{GS} = -1$ and -2 V.
   (c) Repeat part (a) between $V_{GS} = -2$ and -3 V.
   (d) Repeat part (a) between $V_{GS} = -3$ and -4 V.
   (e) Is there a marked change in the difference in current levels as V_{GS} becomes increasingly negative?
   (f) Is the relationship between the change in V_{GS} and the resulting change in I_D linear or nonlinear? Explain.

5. What are the major differences between the collector characteristics of a BJT transistor and the drain characteristics of a JFET transistor? Compare the units of each axis and the controlling variable. How does I_C react to increasing levels of I_B versus changes in I_D to increasingly negative values of V_{GS}? How does the spacing between steps of I_B compare to the spacing between steps of V_{GS}? Compare $V_{C_{sat}}$ to V_P in defining the nonlinear region at low levels of output voltage.

6. (a) Describe in your own words why I_G is effectively zero amperes for a JFET transistor.
   (b) Why is the input impedance to a JFET so high?
   (c) Why is the terminology *field effect* appropriate for this important three-terminal device?

7. Given $I_{DSS} = 12$ mA and $|V_P| = 6$ V, sketch a probable distribution of characteristic curves for the JFET (similar to Fig. 5.10).

8. In general, comment on the polarity of the various voltages and direction of the currents for an *n*-channel JFET versus a *p*-channel JFET.

§ 5.3 Transfer Characteristics

9. Given the characteristics of Fig. 5.46:
   (a) Sketch the transfer characteristics directly from the drain characteristics.
   (b) Using Fig. 5.46 to establish the values of I_{DSS} and V_P, sketch the transfer characteristics using Shockley's equation.
   (c) Compare the characteristics of parts (a) and (b). Are there any major differences?

10. (a) Given $I_{DSS} = 12$ mA and $V_P = -4$ V, sketch the transfer characteristics for the JFET transistor.
    (b) Sketch the drain characteristics for the device of part (a).

11. Given $I_{DSS} = 9$ mA and $V_P = -3.5$ V, determine I_D when:
    (a) $V_{GS} = 0$ V.
    (b) $V_{GS} = -2$ V.
    (c) $V_{GS} = -3.5$ V.
    (d) $V_{GS} = -5$ V.

12. Given $I_{DSS} = 16$ mA and $V_P = -5$ V, sketch the transfer characteristics using the data points of Table 5.1. Determine the value of I_D at $V_{GS} = -3$ V from the curve and compare to the value determined using Shockley's equation. Repeat the above for $V_{GS} = -1$ V.

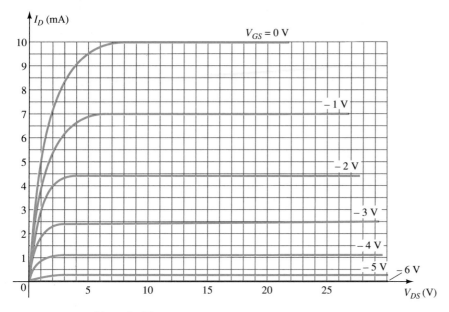

Figure 5.46 Problems 9, 17

13. A *p*-channel JFET has device parameters of $I_{DSS} = 7.5$ mA and $V_P = 4$ V. Sketch the transfer characteristics.

14. Given $I_{DSS} = 6$ mA and $V_P = -4.5$ V:
    (a) Determine I_D at $V_{GS} = -2$ V and -3.6 V.
    (b) Determine V_{GS} at $I_D = 3$ mA and 5.5 mA.

15. Given a Q point of $I_{D_Q} = 3$ mA and $V_{GS} = -3$ V, determine I_{DSS} if $V_P = -6$ V.

§ 5.4 Specification Sheets (JFETs)

16. Define the region of operation for the 2N5457 JFET of Fig. 5.18 using the range of I_{DSS} and V_P provided. That is, sketch the transfer curve defined by the maximum I_{DSS} and V_P and the transfer curve for the minimum I_{DSS} and V_P. Then shade in the resulting area between the two curves.

17. Define the region of operation for the JFET of Fig. 5.46 if $V_{DS_{max}} = 25$ V and $P_{D_{max}} = 120$ mW.

§ 5.5 Instrumentation

18. Using the characteristics of Fig. 5.21, determine I_D at $V_{GS} = -0.7$ V and $V_{DS} = 10$ V.

19. Referring to Fig. 5.21, is the locus of pinch-off values defined by the region of $V_{DS} < |V_P| = 3$ V?

20. Determine V_P for the characteristics of Fig. 5.21 using I_{DSS} and I_D at some value of V_{GS}. That is, simply substitute into Shockley's equation and solve for V_P. Compare the result to the assumed value of -3 V from the characteristics.

21. Using $I_{DSS} = 9$ mA and $V_P = -3$ V for the characteristics of Fig. 5.21, calculate I_D at $V_{GS} = -1$ V using Shockley's equation and compare to the level appearing in Fig. 5.21.

22. (a) Calculate the resistance associated with the JFET of Fig. 5.21 for $V_{GS} = 0$ V from $I_D = 0$ mA to 4 mA.
    (b) Repeat part (a) for $V_{GS} = -0.5$ V from $I_D = 0$ to 3 mA.
    (c) Assigning the label r_o to the result of part (a) and r_d to that of part (b), use Eq. (5.1) to determine r_d and compare to the result of part (b).

§ 5.7 Depletion-Type MOSFET

23. (a) Sketch the basic construction of a p-channel depletion-type MOSFET.
(b) Apply the proper drain-to-source voltage and sketch the flow of electrons for $V_{GS} = 0$ V.

24. In what ways is the construction of a depletion-type MOSFET similar to that of a JFET? In what ways is it different?

25. Explain in your own words why the application of a positive voltage to the gate of an n-channel depletion-type MOSFET will result in a drain current exceeding I_{DSS}.

26. Given a depletion-type MOSFET with $I_{DSS} = 6$ mA and $V_P = -3$ V, determine the drain current at $V_{GS} = -1$ V, 0 V, 1 V, and 2 V. Compare the difference in current levels between -1 and 0 V with the difference between 1 and 2 V. In the positive V_{GS} region, does the drain current increase at a significantly higher rate than for negative values? Does the I_D curve become more and more vertical with increasing positive values of V_{GS}? Is there a linear or a nonlinear relationship between I_D and V_{GS}? Explain.

27. Sketch the transfer and drain characteristics of an n-channel depletion-type MOSFET with $I_{DSS} = 12$ mA and $V_P = -8$ V for a range of $V_{GS} = -V_P$ to $V_{GS} = 1$ V.

28. Given $I_D = 14$ mA and $V_{GS} = 1$ V, determine V_P if $I_{DSS} = 9.5$ mA for a depletion-type MOSFET.

29. Given $I_D = 4$ mA at $V_{GS} = -2$ V, determine I_{DSS} if $V_P = -5$ V.

30. Using an average value of 2.9 mA for the I_{DSS} of the 2N3797 MOSFET of Fig. 5.30, determine the level of V_{GS} that will result in a maximum drain current of 20 mA if $V_P = -5$ V.

31. If the drain current for the 2N3797 MOSFET of Fig. 5.30 is 8 mA, what is the maximum permissible value of V_{DS} utilizing the maximum power rating?

§ 5.8 Enhancement-Type MOSFET

32. (a) What is the significant difference between the construction of an enhancement-type MOSFET and a depletion-type MOSFET?
(b) Sketch a p-channel enhancement-type MOSFET with the proper biasing applied ($V_{DS} > 0$ V, $V_{GS} > V_T$) and indicate the channel, the direction of electron flow, and the resulting depletion region.
(c) In your own words, briefly describe the basic operation of an enhancement-type MOSFET.

33. (a) Sketch the transfer and drain characteristics of an n-channel enhancement-type MOSFET if $V_T = 3.5$ V and $k = 0.4 \times 10^{-3}$ A/V^2.
(b) Repeat part (a) for the transfer characteristics if V_T is maintained at 3.5 V but k is increased by 100% to 0.8×10^{-3} A/V^2.

34. (a) Given $V_{GS(Th)} = 4$ V and $I_{D(on)} = 4$ mA at $V_{GS(on)} = 6$ V, determine k and write the general expression for I_D in the format of Eq. (5.13).
(b) Sketch the transfer characteristics for the device of part (a).
(c) Determine I_D for the device of part (a) at $V_{GS} = 2$ V, 5 V, and 10 V.

35. Given the transfer characteristics of Fig. 5.47, determine V_T and k and write the general equation for I_D.

36. Given $k = 0.4 \times 10^{-3}$ A/V^2 and $I_{D(on)} = 3$ mA with $V_{GS(on)} = 4$ V, determine V_T.

37. The maximum drain current for the 2N4351 n-channel enhancement-type MOSFET is 30 mA. Determine V_{GS} at this current level if $k = 0.06 \times 10^{-3}$ A/V^2 and V_T is the maximum value.

38. Does the current of an enhancement-type MOSFET increase at about the same rate as a depletion-type MOSFET for the conduction region? Carefully review the general format of the equations, and if your mathematics background includes differential calculus, calculate dI_D/dV_{GS} and compare its magnitude.

39. Sketch the transfer characteristics of a p-channel enhancement-type MOSFET if $V_T = -5$ V and $k = 0.45 \times 10^{-3}$ A/V^2.

40. Sketch the curve of $I_D = 0.5 \times 10^{-3}(V_{GS}^2)$ and $I_D = 0.5 \times 10^{-3}(V_{GS} - 4)^2$ for V_{GS} from 0 to 10 V. Does $V_T = 4$ V have a significant impact on the level of I_D for this region?

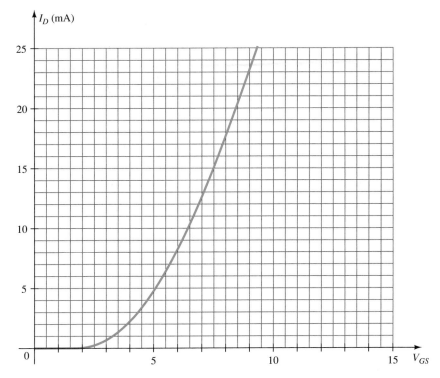

Figure 5.47 Problem 35

§ 5.10 VMOS

41. (a) Describe in your own words why the VMOS FET can withstand a higher current and power rating then the standard construction technique.
  (b) Why do VMOS FETs have reduced channel resistance levels?
  (c) Why is a positive temperature coefficient desirable?

§ 5.11 CMOS

* **42.** (a) Describe in your own words the operation of the network of Fig. 5.44 with $V_i = 0$ V.
  (b) If the "on" MOSFET of Fig. 5.44 (with $V_i = 0$ V) has a drain current of 4 mA with $V_{DS} = 0.1$ V, what is the approximate resistance level of the device? If $I_D = 0.5$ μA for the "off" transistor, what is the approximate resistance of the device? Do the resulting resistance levels suggest that the desired output voltage level will result?

43. Research CMOS logic at your local or college library and describe the range of applications and basic advantages of the approach.

---

*Please Note: Asterisks indicate more difficult problems.

6 FET Biasing

6.1 INTRODUCTION

In Chapter 5 we found that the biasing levels for a silicon transistor configuration can be obtained using the characteristic equations $V_{BE} = 0.7$ V, $I_C = \beta I_B$, and $I_C \cong I_E$. The linkage between input and output variables is provided by β, which is assumed to be fixed in magnitude for the analysis to be performed. The fact that beta is a constant establishes a *linear* relationship between I_C and I_B. Doubling the value of I_B will double the level of I_C, and so on.

For the field-effect transistor the relationship between input and output quantities is *nonlinear* due to the squared term in Shockley's equation. Linear relationships result in straight lines when plotted on a graph of one variable versus the other, while nonlinear functions result in curves as obtained for the transfer characteristics of a JFET. The nonlinear relationship between I_D and V_{GS} can complicate the mathematical approach to the dc analysis of FET configurations. A graphical approach may limit solutions to tenths place accuracy but it is a quicker method for most FET amplifiers. Since the graphical approach is in general the most popular the analysis of this chapter will have a graphical orientation rather than direct mathematical techniques.

Another distinct difference between the analysis of BJT and FET transistors is that the input controlling variable for a BJT transistor is a current level, while for the FET a voltage is the controlling variable. In both cases, however, the controlled variable on the output side is a current level that also defines the important voltage levels of the output circuit.

The general relationships that can be applied to the dc analysis of all FET amplifiers are

$$\boxed{I_G = 0 \text{ A}} \tag{6.1}$$

and

$$\boxed{I_D = I_S} \tag{6.2}$$

For JFETS and depletion-type MOSFETs Shockley's equation is applied to relate the input and output quantities:

$$\boxed{I_D = I_{DSS}\left(1 - \frac{V_{GS}}{V_P}\right)^2} \tag{6.3}$$

For enhancement-type MOSFETs the following equation is applicable:

$$\boxed{I_D = k(V_{GS} - V_T)^2} \qquad (6.4)$$

It is particularly important to realize that all of the equations above are for the *device only!* They do not change with each network configuration so long as the device is in the active region. The network simply defines the level of current and voltage associated with the operating point through its own set of equations. In reality the dc solution of BJT and FET networks is the solution of simultaneous equations established by the device and network. The solution can be determined using a mathematical or graphical approach—a fact to be demonstrated by the first few networks to be analyzed. However, as noted earlier, the graphical approach is the most popular for FET networks and is employed in this book.

The first few sections of this chapter are limited to JFETs and the graphical approach to analysis. The depletion-type MOSFET will then be examined with its increased range of operating points followed by the enhancement-type MOSFET. Finally, problems of a design nature are investigated to fully test the concepts and procedures introduced in the chapter.

6.2 FIXED-BIAS CONFIGURATION

The simplest of biasing arrangements for the *n*-channel JFET appears in Fig. 6.1. Referred to as the fixed-bias configuration, it is one of the few FET configurations that can be solved just as directly using a mathematical or graphical approach. Both methods are included in this section to demonstrate the difference between the two philosophies but also to establish the fact that the same solution can be obtained using either method.

The configuration of Fig. 6.1 includes the ac levels V_i and V_o and the coupling capacitors (C_1 and C_2). Recall that the coupling capacitors are "open circuits" for the dc analysis and low impedances (essentially short circuits) for the ac analysis. The resistor R_G is present to ensure that V_i appears at the input to the FET amplifier for the ac analysis (Chapter 9). For the dc analysis,

$$I_G = 0 \text{ A}$$

and
$$V_{R_G} = I_G R_G = (0 \text{ A})R_G = 0 \text{ V}$$

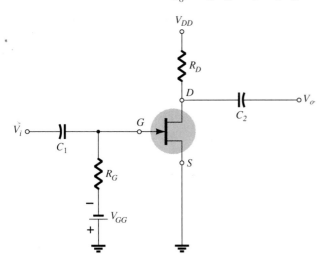

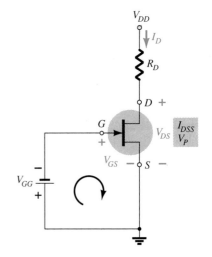

Figure 6.1 Fixed-bias configuration.

The zero-volt drop across R_G permits replacing R_G by a short-circuit equivalent, as appearing in the network of Fig. 6.2 specifically redrawn for the dc analysis.

Figure 6.2 Network for dc analysis.

The fact that the negative terminal of the battery is connected directly to the defined positive potential of V_{GS} clearly reveals that the polarity of V_{GS} is directly opposite to V_{GG}. Applying Kirchhoff's voltage law in the clockwise direction of the indicated loop of Fig. 6.2 will result in

$$-V_{GG} - V_{GS} = 0$$

and

$$\boxed{V_{GS} = -V_{GG}} \qquad (6.5)$$

Since V_{GG} is a fixed dc supply the voltage V_{GS} is fixed in magnitude, resulting in the notation "fixed-bias configuration."

The resulting level of drain current I_D is now controlled by Shockley's equation:

$$I_D = I_{DSS}\left(1 - \frac{V_{GS}}{V_P}\right)^2$$

Since V_{GS} is a fixed quantity for this configuration its magnitude and sign can simply be substituted into Shockley's equation and the resulting level of I_D calculated. This is one of the few instances in which a mathematical solution to a FET configuration is quite direct.

A graphical analysis would require a plot of Shockley's equation as shown in Fig. 6.3. Recall that choosing $V_{GS} = V_P/2$ will result in a drain current of $I_{DSS}/4$ when plotting the equation. For the analysis of this chapter the three points defined by I_{DSS}, V_P, and the intersection just described will be sufficient for plotting the curve.

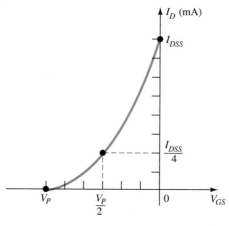

Figure 6.3 Plotting Shockley's equation.

In Fig. 6.4 the fixed level of V_{GS} has been superimposed as a vertical line at $V_{GS} = -V_{GG}$. At any point on the vertical line the level of V_{GS} is $-V_{GG}$—the level of I_D must simply be determined on this vertical line. The point where the

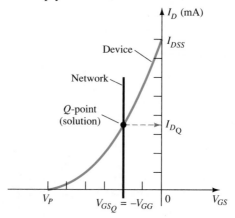

Figure 6.4 Finding the solution for the fixed-bias configuration.

two curves intersect is the common solution to the configuration—commonly referred to as the *quiescent* or *operating point*. The subscript Q will be applied to drain current and gate-to-source voltage to identify their levels at the Q-point. Note in Fig. 6.4 that the quiescent level of I_D is determined by drawing a horizontal line from the Q-point to the vertical I_D axis as shown in Fig. 6.4. It is important to realize that once the network of Fig. 6.1 is constructed and operating, the dc levels of I_D and V_{GS} that will be measured by the meters of Fig. 6.5 are the quiescent values defined by Fig. 6.4.

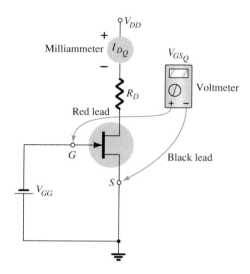

Figure 6.5 Measuring the quiescent values of I_D and V_{GS}.

The drain-to-source voltage of the output section can be determined by applying Kirchhoff's voltage law as follows:

$$+V_{DS} + I_D R_D - V_{DD} = 0$$

and
$$\boxed{V_{DS} = V_{DD} - I_D R_D} \qquad (6.6)$$

Recall that single-subscript voltages refer to the voltage at a point with respect to ground. For the configuration of Fig. 6.2,

$$\boxed{V_S = 0 \text{ V}} \qquad (6.7)$$

Using double-subscript notation:

$$V_{DS} = V_D - V_S$$

or
$$V_D = V_{DS} + V_S = V_{DS} + 0 \text{ V}$$

and
$$\boxed{V_D = V_{DS}} \qquad (6.8)$$

In addition,
$$V_{GS} = V_G - V_S$$

or
$$V_G = V_{GS} + V_S = V_{GS} + 0 \text{ V}$$

and
$$\boxed{V_G = V_{GS}} \qquad (6.9)$$

The fact that $V_D = V_{DS}$ and $V_G = V_{GS}$ is fairly obvious from the fact that $V_S = 0$ V, but the derivations above were included to emphasize the relationship that exists between double-subscript and single-subscript notation. Since the configuration requires two dc supplies, its use is limited and will not be included in the forthcoming list of the most common FET configurations.

EXAMPLE 6.1

Determine the following for the network of Fig. 6.6.
(a) V_{GS_Q}.
(b) I_{D_Q}.
(c) V_{DS}.
(d) V_D.
(e) V_G.
(f) V_S.

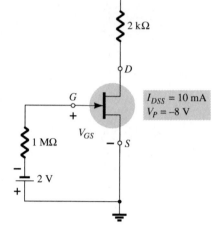

Figure 6.6 Example 6.1

Solution

Mathematical Approach:

(a) $V_{GS_Q} = -V_{GG} = \mathbf{-2\ V}$

(b) $I_{D_Q} = I_{DSS}\left(1 - \dfrac{V_{GS}}{V_P}\right)^2 = 10\ \text{mA}\left(1 - \dfrac{-2\ \text{V}}{-8\ \text{V}}\right)^2$

$\qquad = 10\ \text{mA}(1 - 0.25)^2 = 10\ \text{mA}(0.75)^2 = 10\ \text{mA}(0.5625)$

$\qquad = \mathbf{5.625\ mA}$

(c) $V_{DS} = V_{DD} - I_D R_D = 16\ \text{V} - (5.625\ \text{mA})(2\ \text{k}\Omega)$

$\qquad = 16\ \text{V} - 11.25\ \text{V} = \mathbf{4.75\ V}$

(d) $V_D = V_{DS} = \mathbf{4.75\ V}$

(e) $V_G = V_{GS} = \mathbf{-2\ V}$

(f) $V_S = \mathbf{0\ V}$

Graphical Approach: The resulting Shockley curve and the vertical line at $V_{GS} = -2$ V are provided in Fig. 6.7. It is certainly difficult to read beyond the second place

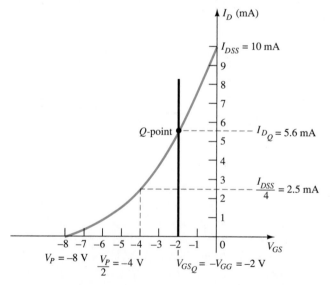

Figure 6.7 Graphical solution for the network of Fig. 6.6.

without significantly increasing the size of the figure but a solution of 5.6 mA from the graph of Fig. 6.7 is quite acceptable. Therefore, for part (a),

$$V_{GS_Q} = -V_{GG} = -2 \text{ V}$$

(b) $I_{D_Q} = \textbf{5.6 mA}$
(c) $V_{DS} = V_{DD} - I_D R_D = 16 \text{ V} - (5.6 \text{ mA})(2 \text{ k}\Omega)$

$\qquad = 16 \text{ V} - 11.2 \text{ V} = \textbf{4.8 V}$

(d) $V_D = V_{DS} = \textbf{4.8 V}$
(e) $V_G = V_{GS} = \textbf{-2 V}$
(f) $V_S = \textbf{0 V}$

The results clearly confirm the fact that the mathematical and graphical approach generate solutions that are quite close.

6.3 SELF-BIAS CONFIGURATION

The self-bias configuration eliminates the need for two dc supplies. The controlling gate-to-source voltage is now determined by the voltage across a resistor R_S introduced in the source leg of the configuration as shown in Fig. 6.8.

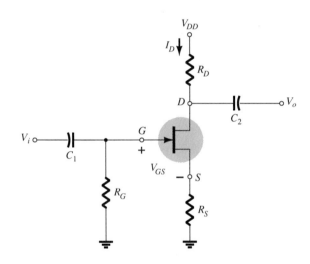

Figure 6.8 JFET self-bias configuration.

For the dc analysis the capacitors can again be replaced by "open circuits" and the resistor R_G replaced by a short-circuit equivalent since $I_G = 0$ A. The result is the network of Fig. 6.9 for the important dc analysis.

The current through R_S is the source current I_S, but $I_S = I_D$ and

$$V_{R_S} = I_D R_S$$

For the indicated closed loop of Fig. 6.9, we find that

$$-V_{GS} - V_{R_S} = 0$$

and

$$V_{GS} = -V_{R_S}$$

or

$$\boxed{V_{GS} = -I_D R_S} \qquad (6.10)$$

Note in this case that V_{GS} is a function of the output current I_D and not fixed in magnitude as occurred for the fixed-bias configuration.

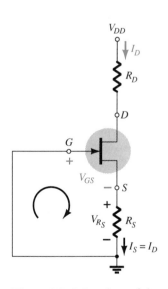

Figure 6.9 DC analysis of the self-bias configuration.

Equation (6.10) is defined by the network configuration, and Shockley's equation relates the input and output quantities of the device. Both equations relate the same two variables, permitting a mathematical or graphical solution.

A mathematical solution could be obtained simply by substituting Eq. (6.10) into Shockley's equation as shown below:

$$I_D = I_{DSS}\left(1 - \frac{V_{GS}}{V_P}\right)^2$$

$$= I_{DSS}\left(1 - \frac{-I_D R_S}{V_P}\right)^2$$

or

$$I_D = I_{DSS}\left(1 + \frac{I_D R_S}{V_P}\right)^2$$

By performing the squaring process indicated and rearranging terms, an equation of the following form can be obtained:

$$I_D^2 + K_1 I_D + K_2 = 0$$

The quadratic equation can then be solved for the appropriate solution for I_D.

The sequence above defines the mathematical approach. The graphical approach requires that we first establish the device transfer characteristics as shown in Fig. 6.10. Since Eq. (6.10) defines a straight line on the same graph, let us now identify two points on the graph that are on the line and simply draw a straight line between the two points. The most obvious condition to apply is $I_D = 0$ A since it results in $V_{GS} = -I_D R_S = (0 \text{ A})R_S = 0$ V. For Eq. (6.10), therefore, one point on the straight line is defined by $I_D = 0$ A and $V_{GS} = 0$ V, as appearing on Fig. 6.10.

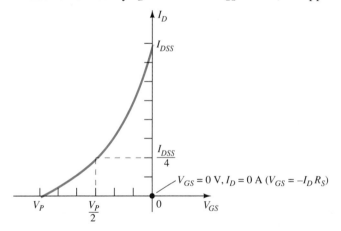

Figure 6.10 Defining a point on the self-bias line.

The second point for Eq. (6.10) requires that a level of V_{GS} or I_D be chosen and the corresponding level of the other quantity be determined using Eq. (6.10). The resulting levels of I_D and V_{GS} will then define another point on the straight line and permit an actual drawing of the straight line. Suppose, for example, that we choose a level of I_D equal to one-half the saturation level. That is,

$$I_D = \frac{I_{DSS}}{2}$$

then

$$V_{GS} = -I_D R_S = -\frac{I_{DSS} R_S}{2}$$

The result is a second point for the straight-line plot as shown in Fig. 6.11. The straight line as defined by Eq. (6.10) is then drawn and the quiescent point obtained at

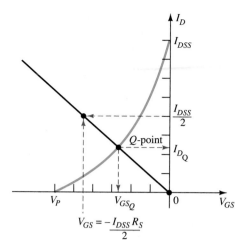

$$V_{GS} = -\frac{I_{DSS}R_S}{2}$$

Figure 6.11 Sketching the self-bias line.

the intersection of the straight line plot and the device characteristic curve. The quiescent values of I_D and V_{GS} can then be determined and used to find the other quantities of interest.

The level of V_{DS} can be determined by applying Kirchhoff's voltage law to the output circuit, with the result that

$$V_{R_S} + V_{DS} + V_{R_D} - V_{DD} = 0$$

and $\qquad V_{DS} = V_{DD} - V_{R_S} - V_{R_D} = V_{DD} - I_S R_S - I_D R_D$

but $\qquad I_D = I_S$

and $\qquad \boxed{V_{DS} = V_{DD} - I_D(R_S + R_D)} \qquad$ (6.11)

In addition:

$$\boxed{V_S = I_D R_S} \qquad (6.12)$$

$$\boxed{V_G = 0 \text{ V}} \qquad (6.13)$$

and $\qquad \boxed{V_D = V_{DS} + V_S = V_{DD} - V_{R_D}} \qquad$ (6.14)

Determine the following for the network of Fig. 6.12.
(a) V_{GS_Q}.
(b) I_{D_Q}.
(c) V_{DS}.
(d) V_S.
(e) V_G.
(f) V_D.

EXAMPLE 6.2

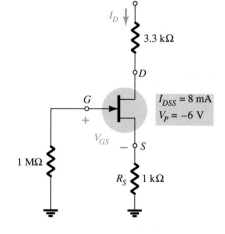

Figure 6.12 Example 6.2

6.3 Self-Bias Configuration

Solution

(a) The gate-to-source voltage is determined by

$$V_{GS} = -I_D R_S$$

Choosing $I_D = 4$ mA, we obtain

$$V_{GS} = -(4 \text{ mA})(1 \text{ k}\Omega) = -4 \text{ V}$$

The result is the plot of Fig. 6.13 as defined by the network.

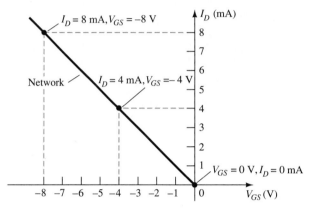

Figure 6.13   Sketching the self-bias line for the network of Fig. 6.12.

If we happen to choose $I_D = 8$ mA, the resulting value of V_{GS} would be -8 V, as shown on the same graph. In either case the same straight line will result, clearly demonstrating that any appropriate value of I_D can be chosen as long as the corresponding value of V_{GS} as determined by Eq. (6.10) is employed. In addition, keep in mind that the value of V_{GS} could be chosen and the value of I_D calculated with the same resulting plot.

For Shockley's equation if we choose $V_{GS} = V_P/2 = -3$ V, we find that $I_D = I_{DSS}/4 = 8$ mA/4 = 2 mA, and the plot of Fig. 6.14 will result, representing the characteristics of the device. The solution is obtained by superimposing the network characteristics defined by Fig. 6.13 on the device characteristics of Fig. 6.14 and finding the point of intersection of the two as indicated on Fig. 6.15. The resulting operating point results in a quiescent value of gate-to-source voltage of

$$V_{GS_Q} = -2.6 \text{ V}$$

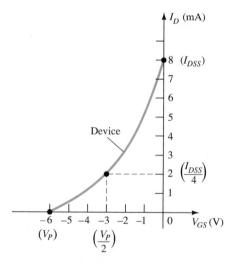

Figure 6.14   Sketching the device characteristics for the JFET of Fig. 6.12.

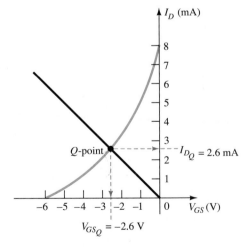

Figure 6.15   Determining the Q point for the network of Fig. 6.12.

(b) At the quiescent point:

$$I_{D_Q} = \textbf{2.6 mA}$$

(c) Eq. (6.11): $V_{DS} = V_{DD} - I_D(R_S + R_D)$

$$= 20 \text{ V} - (2.6 \text{ mA})(1 \text{ k}\Omega + 3.3 \text{ k}\Omega)$$

$$= 20 \text{ V} - 11.18 \text{ V}$$

$$= \textbf{8.82 V}$$

(d) Eq. (6.12): $V_S = I_D R_S$

$$= (2.6 \text{ mA})(1 \text{ k}\Omega)$$

$$= \textbf{2.6 V}$$

(e) Eq. (6.13): $V_G = \textbf{0 V}$

(f) Eq. (6.14): $V_D = V_{DS} + V_S = 8.82 \text{ V} + 2.6 \text{ V} = \textbf{11.42 V}$

or $\qquad\qquad V_D = V_{DD} - I_D R_D = 20 \text{ V} - (2.6 \text{ mA})(3.3 \text{ k}\Omega) = \textbf{11.42 V}$

Find the quiescent point for the network of Fig. 6.12 if:
(a) $R_S = 100 \ \Omega$.
(b) $R_S = 10 \text{ k}\Omega$.

EXAMPLE 6.3

Solution

Note Fig. 6.16.

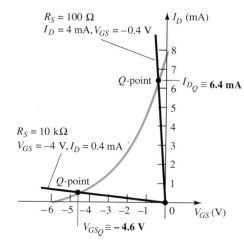

$R_S = 100 \ \Omega$
$I_D = 4 \text{ mA}, V_{GS} = -0.4 \text{ V}$

Q-point

$I_{D_Q} \cong \textbf{6.4 mA}$

$R_S = 10 \text{ k}\Omega$
$V_{GS} = -4 \text{ V}, I_D = 0.4 \text{ mA}$

Q-point

$V_{GS_Q} \cong \textbf{-4.6 V}$

Figure 6.16 Example 6.3

(a) With the I_D scale,

$$I_{D_Q} \cong \textbf{6.4 mA}$$

From Eq. (6.10),

$$V_{GS_Q} \cong \textbf{0.64 V}$$

(b) With the V_{GS} scale,

$$V_{GS_Q} \cong \textbf{-4.6 V}$$

From Eq. (6.10),

$$I_{D_Q} \cong \textbf{0.46 mA}$$

In particular, note how lower levels of R_S bring the load line of the network closer to the I_D axis, while increasing levels of R_S bring the load line closer to the V_{GS} axis.

EXAMPLE 6.4

Determine the following for the network of Fig. 6.17.
(a) V_{GS_Q}.
(b) I_{D_Q}.
(c) V_D.
(d) V_G.
(e) V_S.
(f) V_{DS}.

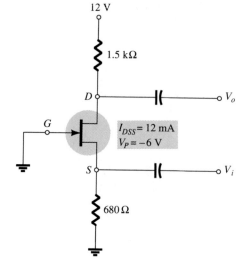

Figure 6.17 Example 6.4

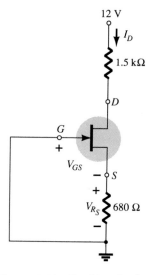

Figure 6.18 Sketching the dc equivalent of the network of Fig. 6.17.

Solution

The grounded gate terminal and the location of the input establish strong similarities with the common-base BJT amplifier. Although different in appearance from the basic structure of Fig. 6.8, the resulting dc network of Fig. 6.18 has the same basic structure as Fig. 6.9. The dc analysis can therefore proceed in the same manner as recent examples.

(a) The transfer characteristics and load line appear in Fig. 6.19. In this case the second point for the sketch of the load line was determined by choosing (arbitrarily) $I_D = 6$ mA and solving for V_{GS}. That is,

$$V_{GS} = I_D R_S = -(6 \text{ mA})(680 \text{ } \Omega) = -4.08 \text{ V}$$

as shown in Fig. 6.19. The device transfer curve was sketched using

$$I_D = \frac{I_{DSS}}{4} = \frac{12 \text{ mA}}{4} = 3 \text{ mA}$$

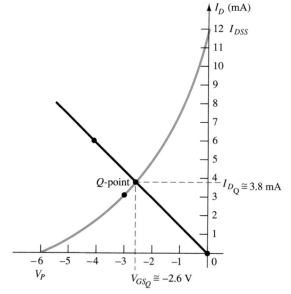

Figure 6.19 Determining the Q point for the network of Fig. 6.17.

and the associated value of V_{GS}:

$$V_{GS} = \frac{V_P}{2} = -\frac{6 \text{ V}}{2} = -3 \text{ V}$$

as shown on Fig. 6.19. Using the resulting quiescent point of Fig. 6.19 results in

$$V_{GS_Q} \cong -2.6 \text{ V}$$

(b) From Fig. 6.19,

$$I_{D_Q} \cong 3.8 \text{ mA}$$

(c) $V_D = V_{DD} - I_D R_D$

$\qquad = 12 \text{ V} - (3.8 \text{ mA})(1.5 \text{ k}\Omega) = 12 \text{ V} - 5.7 \text{ V}$

$\qquad = \mathbf{6.3 \text{ V}}$

(d) $V_G = \mathbf{0 \text{ V}}$

(e) $V_S = I_D R_S = (3.8 \text{ mA})(680 \text{ }\Omega)$

$\qquad = \mathbf{2.58 \text{ V}}$

(f) $V_{DS} = V_D - V_S$

$\qquad = 6.3 \text{ V} - 2.58 \text{ V}$

$\qquad = \mathbf{3.72 \text{ V}}$

6.4 VOLTAGE-DIVIDER BIASING

The voltage-divider bias arrangement applied to BJT transistor amplifiers is also applied to FET amplifiers as demonstrated by Fig. 6.20. The basic construction is exactly the same, but the dc analysis of each is quite different. $I_G = 0$ A for FET amplifiers, but the magnitude of I_B for common-emitter BJT amplifiers can affect the dc levels of current and voltage in both the input and output circuits. Recall that I_B provided the link between input and output circuits for the BJT voltage-divider configuration while V_{GS} will do the same for the FET configuration.

The network of Fig. 6.20 is redrawn as shown in Fig. 6.21 for the dc analysis. Note that all the capacitors, including the bypass capacitor C_S, have been replaced by an "open-circuit" equivalent. In addition, the source V_{DD} was separated into two

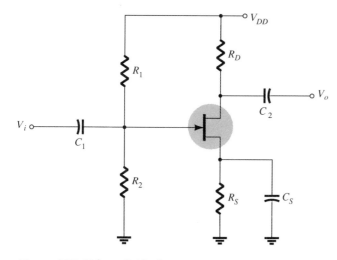

Figure 6.20 Voltage-divider bias arrangement.

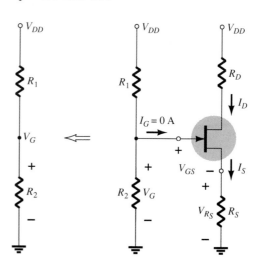

Figure 6.21 Redrawn network of Fig. 6.20 for dc analysis.

equivalent sources to permit a further separation of the input and output regions of the network. Since $I_G = 0$ A, Kirchhoff's current law requires that $I_{R_1} = I_{R_2}$ and the series equivalent circuit appearing to the left of the figure can be used to find the level of V_G. The voltage V_G, equal to the voltage across R_2, can be found using the voltage-divider rule as follows:

$$V_G = \frac{R_2 V_{DD}}{R_1 + R_2}$$ (6.15)

Applying Kirchhoff's voltage law in the clockwise direction to the indicated loop of Fig. 6.21 will result in

$$V_G - V_{GS} - V_{R_S} = 0$$

and

$$V_{GS} = V_G - V_{R_S}$$

Substituting $V_{R_S} = I_S R_S = I_D R_S$, we have

$$V_{GS} = V_G - I_D R_S$$ (6.16)

The result is an equation that continues to include the same two variables appearing in Shockley's equation: V_{GS} and I_D. The quantities V_G and R_S are fixed by the network construction. Equation (6.16) is still the equation for a straight line, but the origin is no longer a point in the plotting of the line. The procedure for plotting Eq. (6.16) is not a difficult one and will proceed as follows. Since any straight line requires two points to be defined, let us first use the fact that *anywhere on the horizontal axis* of Fig. 6.22 the current $I_D = 0$ mA. If we therefore *select* I_D to be 0 mA, we are in essence stating that we are somewhere on the horizontal axis. The exact location can be determined simply by substituting $I_D = 0$ mA into Eq. (6.16) and finding the resulting value of V_{GS} as follows:

$$V_{GS} = V_G - I_D R_S$$

$$= V_G - (0 \text{ mA}) R_S$$

and

$$V_{GS} = V_G|_{I_D = 0 \text{ mA}}$$ (6.17)

The result specifies that whenever we plot Eq. (6.16), if we choose $I_D = 0$ mA, the value of V_{GS} for the plot will be V_G volts. The point just determined appears in Fig. 6.22.

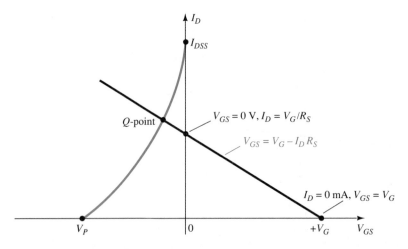

Figure 6.22 Sketching the network equation for the voltage-divider configuration.

Chapter 6 FET Biasing

For the other point let us now employ the fact that at any point on the vertical axis $V_{GS} = 0$ V and solve for the resulting value of I_D:

$$V_{GS} = V_G - I_D R_S$$

$$0 \text{ V} = V_G - I_D R_S$$

and
$$\boxed{I_D = \left.\frac{V_G}{R_S}\right|_{V_{GS}=0 \text{ V}}} \tag{6.18}$$

The result specifies that whenever we plot Eq. (6.16) if $V_{GS} = 0$ V, the level of I_D is determined by Eq. (6.18). This intersection also appears on Fig. 6.22.

The two points defined above permit the drawing of a straight line to represent Eq. (6.16). The intersection of the straight line with the transfer curve in the region to the left of the vertical axis will define the operating point and the corresponding levels of I_D and V_{GS}.

Since the intersection on the vertical axis is determined by $I_D = V_G/R_S$ and V_G is fixed by the input network, increasing values of R_S will reduce the level of the I_D intersection as shown in Fig. 6.23. It is fairly obvious from Fig. 6.23 that:

Increasing values of R_S result in lower quiescent values of I_D and more negative values of V_{GS}.

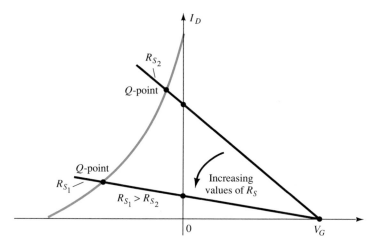

Figure 6.23 Effect of R_S on the resulting Q point.

Once the quiescent values of I_{D_Q} and V_{GS_Q} are determined, the remaining network analysis can be performed in the usual manner. That is,

$$\boxed{V_{DS} = V_{DD} - I_D(R_D + R_S)} \tag{6.19}$$

$$\boxed{V_D = V_{DD} - I_D R_D} \tag{6.20}$$

$$\boxed{V_S = I_D R_S} \tag{6.21}$$

$$\boxed{I_{R_1} = I_{R_2} = \frac{V_{DD}}{R_1 + R_2}} \tag{6.22}$$

EXAMPLE 6.5

Determine the following for the network of Fig. 6.24.
(a) I_{D_Q} and V_{GS_Q}.
(b) V_D.
(c) V_S.
(d) V_{DS}.
(e) V_{DG}.

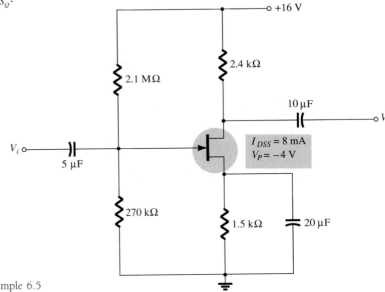

Figure 6.24 Example 6.5

Solution

(a) For the transfer characteristics, if $I_D = I_{DSS}/4 = 8$ mA/4 $= 2$ mA, then $V_{GS} = V_P/2 = -4$ V/2 $= -2$ V. The resulting curve representing Shockley's equation appears in Fig. 6.25. The network equation is defined by

$$V_G = \frac{R_2 V_{DD}}{R_1 + R_2}$$

$$= \frac{(270 \text{ k}\Omega)(16 \text{ V})}{2.1 \text{ M}\Omega + 0.27 \text{ M}\Omega}$$

$$= 1.82 \text{ V}$$

and
$$V_{GS} = V_G - I_D R_S$$

$$= 1.82 \text{ V} - I_D (1.5 \text{ k}\Omega)$$

$I_D = 0$ mA:

$$V_{GS} = +1.82 \text{ V}$$

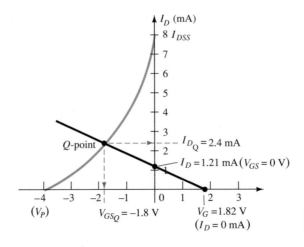

Figure 6.25 Determining the Q point for the network of Fig. 6.24.

Chapter 6   FET Biasing

$V_{GS} = 0$ V:

$$I_D = \frac{1.82 \text{ V}}{1.5 \text{ k}\Omega} = 1.21 \text{ mA}$$

The resulting bias line appears on Fig. 6.25 with quiescent values of

$$I_{D_Q} = \textbf{2.4 mA}$$

and

$$V_{GS_Q} = \textbf{-1.8 V}$$

(b) $V_D = V_{DD} - I_D R_D$

$\qquad = 16 \text{ V} - (2.4 \text{ mA})(2.4 \text{ k}\Omega)$

$\qquad = \textbf{10.24 V}$

(c) $V_S = I_D R_S = (2.4 \text{ mA})(1.5 \text{ k}\Omega)$

$\qquad = \textbf{3.6 V}$

(d) $V_{DS} = V_{DD} - I_D(R_D + R_S)$

$\qquad = 16 \text{ V} - (2.4 \text{ mA})(2.4 \text{ k}\Omega + 1.5 \text{ k}\Omega)$

$\qquad = \textbf{6.64 V}$

or $V_{DS} = V_D - V_S = 10.24 \text{ V} - 3.6 \text{ V}$

$\qquad = \textbf{6.64 V}$

(e) Although seldom requested, the voltage V_{DG} can easily be determined using

$$V_{DG} = V_D - V_G$$

$$= 10.24 \text{ V} - 1.82 \text{ V}$$

$$= \textbf{8.42 V}$$

Although the basic construction of the network in the next example is quite different from the voltage-divider bias arrangement, the resulting equations require a solution very similar to that just described. Note that the network employs a supply at the drain and source.

Determine the following for the network of Fig. 6.26.

(a) I_{D_Q} and V_{GS_Q}.

(b) V_{DS}.

(c) V_D.

(d) V_S.

EXAMPLE 6.6

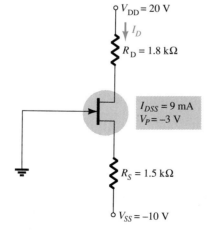

Figure 6.26   Example 6.6

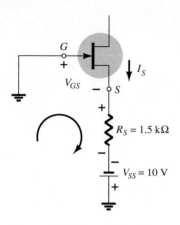

Figure 6.27 Determining the network equation for the configuration of Fig. 6.26.

Solution

(a) An equation for V_{GS} in terms of I_D is obtained by applying Kirchhoff's voltage law to the input section of the network as redrawn in Fig. 6.27.

$$-V_{GS} - I_S R_S + V_{SS} = 0$$

or

$$V_{GS} = V_{SS} - I_S R_S$$

but

$$I_S = I_D$$

and

$$\boxed{V_{GS} = V_{SS} - I_D R_S} \tag{6.23}$$

The result is an equation very similar in format to Eq. (6.16) that can be superimposed on the transfer characteristics using the procedure described for Eq. (6.16). That is, for this example,

$$V_{GS} = 10\ \text{V} - I_D(1.5\ \text{k}\Omega)$$

For $I_D = 0$ mA,

$$V_{GS} = V_{SS} = 10\ \text{V}$$

For $V_{GS} = 0$ V,

$$0 = 10\ \text{V} - I_D(1.5\ \text{k}\Omega)$$

and

$$I_D = \frac{10\ \text{V}}{1.5\ \text{k}\Omega} = 6.67\ \text{mA}$$

The resulting plot points are identified on Fig. 6.28.

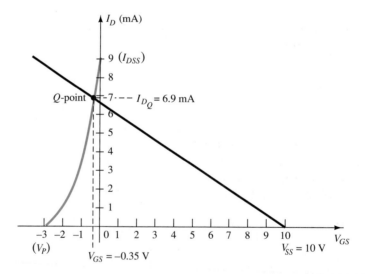

Figure 6.28 Determining the Q point for the network of Fig. 6.26.

The transfer characteristics are sketched using the plot point established by $V_{GS} = V_P/2 = -3\ \text{V}/2 = -1.5\ \text{V}$ and $I_D = I_{DSS}/4 = 9\ \text{mA}/4 = 2.25\ \text{mA}$, as also appearing on Fig. 6.28. The resulting operating point establishes the following quiescent levels:

$$I_{D_Q} = \textbf{6.9 mA}$$

$$V_{GS_Q} = \textbf{−0.35 V}$$

(b) Applying Kirchhoff's voltage law to the output side of Fig. 6.26 will result in

$$-V_{SS} + I_S R_S + V_{DS} + I_D R_D - V_{DD} = 0$$

Substituting $I_S = I_D$ and rearranging gives

$$\boxed{V_{DS} = V_{DD} + V_{SS} - I_D(R_D + R_S)} \qquad (6.24)$$

which for this example results in

$$V_{DS} = 20\ V + 10\ V - (6.9\ mA)(1.8\ k\Omega + 1.5\ k\Omega)$$

$$= 30\ V - 22.77\ V$$

$$= \mathbf{7.23\ V}$$

(c) $\quad V_D = V_{DD} - I_D R_D$

$$= 20\ V - (6.9\ mA)(1.8\ k\Omega) = 20\ V - 12.42\ V$$

$$= \mathbf{7.58\ V}$$

(d) $\quad V_{DS} = V_D - V_S$

$\quad$ or $V_S = V_D - V_{DS}$

$$= 7.58\ V - 7.23\ V$$

$$= \mathbf{0.35\ V}$$

6.5 DEPLETION-TYPE MOSFETs

The similarities in appearance between the transfer curves of JFETs and depletion-type MOSFETs permit a similar analysis of each in the dc domain. The primary difference between the two is the fact that depletion-type MOSFETs permit operating points with positive values of V_{GS} and levels of I_D that exceed I_{DSS}. In fact, for all the configurations discussed thus far, the analysis is the same if the JFET is replaced by a depletion-type MOSFET.

The only undefined part of the analysis is how to plot Shockley's equation for positive values of V_{GS}. How far into the region of positive values of V_{GS} and values of I_D greater than I_{DSS} does the transfer curve have to extend? For most situations this required range will be fairly well defined by the MOSFET parameters and the resulting bias line of the network. A few examples will reveal the impact of the change in device on the resulting analysis.

For the n-channel depletion-type MOSFET of Fig. 6.29, determine:
(a) I_{D_Q} and V_{GS_Q}.
(b) V_{DS}.

EXAMPLE 6.7

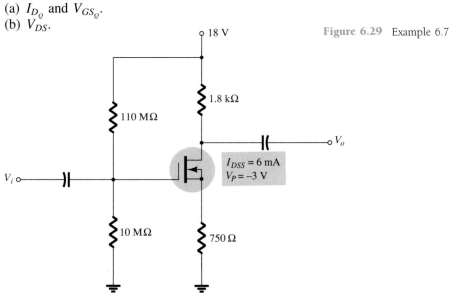

Figure 6.29 Example 6.7

Solution

(a) For the transfer characteristics a plot point is defined by $I_D = I_{DSS}/4 = 6$ mA$/4 =$ 1.5 mA and $V_{GS} = V_P/2 = -3$V$/2 = -1.5$ V. Considering the level of V_P and the fact that Shockley's equation defines a curve that rises more rapidly as V_{GS} becomes more positive, a plot point will be defined at $V_{GS} = +1$ V. Substituting into Shockley's equation yields

$$I_D = I_{DSS}\left(1 - \frac{V_{GS}}{V_P}\right)^2$$

$$= 6 \text{ mA}\left(1 - \frac{+1 \text{ V}}{-3 \text{ V}}\right)^2 = 6 \text{ mA}\left(1 + \frac{1}{3}\right)^2 = 6 \text{ mA}(1.778)$$

$$= 10.67 \text{ mA}$$

The resulting transfer curve appears in Fig. 6.30. Proceeding as described for JFETs, we have:

Eq. (6.15): $V_G = \dfrac{10 \text{ M}\Omega(18 \text{ V})}{10 \text{ M}\Omega + 110 \text{ M}\Omega} = 1.5$ V

Eq. (6.16): $V_{GS} = V_G - I_D R_S = 1.5 \text{ V} - I_D(750 \text{ }\Omega)$

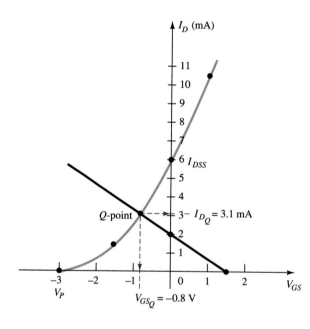

Figure 6.30 Determining the Q point for the network of Fig. 6.29.

Setting $I_D = 0$ mA, results in

$$V_{GS} = V_G = 1.5 \text{ V}$$

Setting $V_{GS} = 0$ V, yields

$$I_D = \frac{V_G}{R_S} = \frac{1.5 \text{ V}}{750 \text{ }\Omega} = 2 \text{ mA}$$

The plot points and resulting bias line appear in Fig. 6.30. The resulting operating point

$$I_{D_Q} = \textbf{3.1 mA}$$

$$V_{GS_Q} = \textbf{-0.8 V}$$

Chapter 6   FET Biasing

(b) Eq. (6.19): $V_{DS} = V_{DD} - I_D(R_D + R_S)$

$$= 18 \text{ V} - (3.1 \text{ mA})(1.8 \text{ k}\Omega + 750 \text{ }\Omega)$$

$$\cong \textbf{10.1 V}$$

Repeat Example 6.7 with $R_S = 150$ Ω.

Solution

(a) The plot points are the same for the transfer curve as shown in Fig. 6.31. For the bias line,

$$V_{GS} = V_G - I_D R_S = 1.5 \text{ V} - I_D \text{ (150 }\Omega)$$

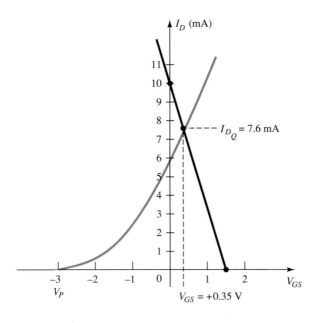

Figure 6.31 Example 6.8

Setting $I_D = 0$ mA results in

$$V_{GS} = 1.5 \text{ V}$$

Setting $V_{GS} = 0$ V yields

$$I_D = \frac{V_G}{R_S} = \frac{1.5 \text{ V}}{150 \text{ }\Omega} = 10 \text{ mA}$$

The bias line is included on Fig. 6.31. Note in this case that the quiescent point results in a drain current that exceeds I_{DSS} with a positive value for V_{GS}. The result:

$$I_{D_Q} = \textbf{7.6 mA}$$

$$V_{GS_Q} = \textbf{+0.35 V}$$

(b) Eq. (6.19): $V_{DS} = V_{DD} - I_D(R_D + R_S)$

$$= 18 \text{ V} - (7.6 \text{ mA})(1.8 \text{ k}\Omega + 150 \text{ }\Omega)$$

$$= \textbf{3.18 V}$$

6.5 Depletion-Type MOSFETs

EXAMPLE 6.9

Determine the following for the network of Fig. 6.32.
(a) I_{D_Q} and V_{GS_Q}.
(b) V_D.

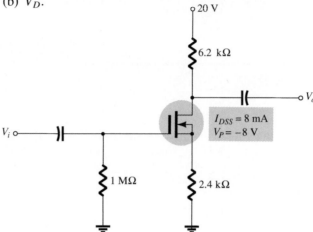

Figure 6.32 Example 6.9

Solution

(a) The self-bias configuration results in

$$V_{GS} = -I_D R_S$$

as obtained for the JFET configuration, establishing the fact that V_{GS} must be less than zero volts. There is therefore no requirement to plot the transfer curve for positive values of V_{GS}, although it was done on this occasion to complete the transfer characteristics. A plot point for the transfer characteristics for $V_{GS} < 0$ V is

$$I_D = \frac{I_{DSS}}{4} = \frac{8 \text{ mA}}{4} = 2 \text{ mA}$$

and

$$V_{GS} = \frac{V_P}{2} = \frac{-8 \text{ V}}{2} = -4 \text{ V}$$

and for $V_{GS} > 0$ V since $V_P = -8$ V we will choose

$$V_{GS} = +2 \text{ V}$$

and

$$I_D = I_{DSS}\left(1 - \frac{V_{GS}}{V_P}\right)^2 = 8 \text{ mA} \left(1 - \frac{+2 \text{ V}}{-8 \text{ V}}\right)^2$$

$$= 12.5 \text{ mA}$$

The resulting transfer curve appears in Fig. 6.33. For the network bias line, at $V_{GS} = 0$ V, $I_D = 0$ mA. Choosing $V_{GS} = -6$ V gives

$$I_D = -\frac{V_{GS}}{R_S} = -\frac{-6 \text{ V}}{2.4 \text{ k}\Omega} = 2.5 \text{ mA}$$

The resulting Q-point:

$$I_{D_Q} = \textbf{1.7 mA}$$

$$V_{GS_Q} = \textbf{-4.3 V}$$

(b) $V_D = V_{DD} - I_D R_D$
$\qquad = 20 \text{ V} - (1.7 \text{ mA})(6.2 \text{ k}\Omega)$
$\qquad = \textbf{9.46 V}$

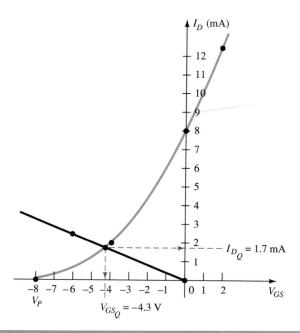

Figure 6.33 Determining the Q point for the network of Fig. 6.32.

The example to follow employs a design that can also be applied to JFET transistors. At first impression it appears rather simplistic, but in fact it often causes some confusion when first analyzed due to the special point of operation.

Determine V_{DS} for the network of Fig. 6.34.

EXAMPLE 6.10

Solution

The direct connection between the gate and source terminals requires that

$$V_{GS} = 0 \text{ V}$$

Since V_{GS} is fixed at 0 V the drain current must be I_{DSS} (by definition). In other words,

$$V_{GS_Q} = \mathbf{0 \text{ V}}$$

and

$$I_{D_Q} = \mathbf{10 \text{ mA}}$$

There is therefore no need to draw the transfer curve and

$$V_D = V_{DD} - I_D R_D = 20 \text{ V} - (10 \text{ mA})(1.5 \text{ k}\Omega)$$

$$= 20 \text{ V} - 15 \text{ V}$$

$$= \mathbf{5 \text{ V}}$$

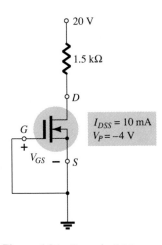

Figure 6.34 Example 6.10

6.6 ENHANCEMENT-TYPE MOSFETs

The transfer characteristics of the enhancement-type MOSFET are quite different from those encountered for the JFET and depletion-type MOSFETs resulting in a graphical solution quite different from the preceding sections. First and foremost, recall that for the n-channel enhancement-type MOSFET, the drain current is zero for levels of gate-to-source voltage less than the threshold level $V_{GS(Th)}$, as shown in Fig. 6.35. For levels of V_{GS} greater than $V_{GS(Th)}$, the drain current is defined by

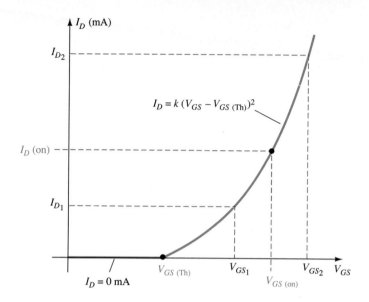

Figure 6.35 Transfer characteristics of an *n*-channel enhancement-type MOSFET.

$$I_D = k(V_{GS} - V_{GS(\text{Th})})^2 \qquad (6.25)$$

Since specification sheets typically provide the threshold voltage and a level of drain current ($I_{D(\text{on})}$) and its corresponding level of $V_{GS(\text{on})}$, two points are defined immediately as shown in Fig. 6.35. To complete the curve the constant k of Eq. (6.25) must be determined from the specification sheet data by substituting into Eq. (6.25) and solving for k as follows:

$$I_D = k(V_{GS} - V_{GS(\text{Th})})^2$$

$$I_{D(\text{on})} = k(V_{GS(\text{on})} - V_{GS(\text{Th})})^2$$

and
$$k = \frac{I_{D(\text{on})}}{(V_{GS(\text{on})} - V_{GS(\text{Th})})^2} \qquad (6.26)$$

Once k is defined, other levels of I_D can be determined for chosen values of V_{GS}. Typically, a point between $V_{GS(\text{Th})}$ and $V_{GS(\text{on})}$ and one just greater than $V_{GS(\text{on})}$ will provide a sufficient number of points to plot Eq. (6.25) (note I_{D_1} and I_{D_2} on Fig. 6.35).

Feedback Biasing Arrangement

A popular biasing arrangement for enhancement-type MOSFETs is provided in Fig. 6.36. The resistor R_G brings a suitably large voltage to the gate to drive the MOSFET "on." Since $I_G = 0$ mA and $V_{R_G} = 0$ V, the dc equivalent network appears as shown in Fig. 6.37.

A direct connection now exists between drain and gate, resulting in

$$V_D = V_G$$

and
$$V_{DS} = V_{GS} \qquad (6.27)$$

Chapter 6   FET Biasing

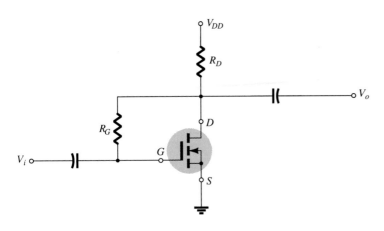

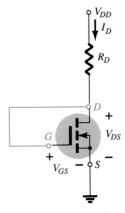

Figure 6.36 Feedback biasing arrangement.

Figure 6.37 DC equivalent of the network of Fig. 6.36.

For the output circuit,

$$V_{DS} = V_{DD} - I_D R_D$$

which becomes the following after substituting Eq. (6.27):

$$V_{GS} = V_{DD} - I_D R_D \qquad (6.28)$$

The result is an equation that relates the same two variables as Eq. (6.25), permitting the plot of each on the same set of axes.

Since Eq. (6.28) is that of a straight line, the same procedure described earlier can be employed to determine the two points that will define the plot on the graph. Substituting $I_D = 0$ mA into Eq. (6.28) gives

$$V_{GS} = V_{DD}|_{I_D=0\text{mA}} \qquad (6.29)$$

Substituting $V_{GS} = 0$ V into Eq. (6.28), we have

$$I_D = \left.\frac{V_{DD}}{R_D}\right|_{V_{GS}=0\text{V}} \qquad (6.30)$$

The plots defined by Eqs. (6.25) and (6.28) appear in Fig. 6.38 with the resulting operating point.

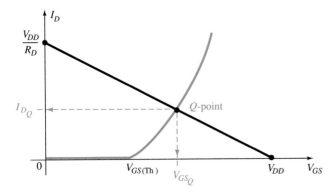

Figure 6.38 Determining the Q point for the network of Fig. 6.36.

EXAMPLE 6.11   Determine I_{D_Q} and V_{DS_Q} for the enhancement-type MOSFET of Fig. 6.39.

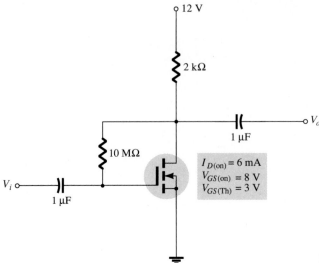

Figure 6.39   Example 6.11

Solution

Plotting the Transfer Curve: Two points are defined immediately as shown in Fig. 6.40. Solving for k:

$$\text{Eq. (6.26):} \quad k = \frac{I_{D(\text{on})}}{(V_{GS(\text{on})} - V_{GS(\text{Th})})^2}$$

$$= \frac{6 \text{ mA}}{(8 \text{ V} - 3 \text{ V})^2} = \frac{6 \times 10^{-3}}{25} \text{A/v}^2$$

$$= \mathbf{0.24 \times 10^{-3} \text{ A/V}^2}$$

For $V_{GS} = 6$ V (between 3 and 8 V):

$$I_D = 0.24 \times 10^{-3}(6 \text{ V} - 3 \text{ V})^2 = 0.24 \times 10^{-3}(9)$$

$$= 2.16 \text{ mA}$$

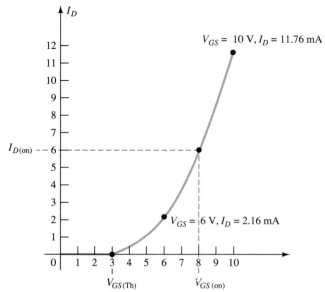

Figure 6.40   Plotting the transfer curve for the MOSFET of Fig. 6.39.

as shown on Fig. 6.40. For $V_{GS} = 10$ V (slightly greater than $V_{GS(Th)}$):

$$I_D = 0.24 \times 10^{-3}(10 \text{ V} - 3 \text{ V})^2 = 0.24 \times 10^{-3}(49)$$

$$= 11.76 \text{ mA}$$

as also appearing on Fig. 6.40. The four points are sufficient to plot the full curve for the range of interest as shown in Fig. 6.40.

For the Network Bias Line:

$$V_{GS} = V_{DD} - I_D R_D$$

$$= 12 \text{ V} - I_D(2 \text{ k}\Omega)$$

Eq. (6.29): $V_{GS} = V_{DD} = 12 \text{ V}|_{I_D = 0\text{mA}}$

Eq. (6.30): $I_D = \dfrac{V_{DD}}{R_D} = \dfrac{12 \text{ V}}{2 \text{ k}\Omega} = 6 \text{ mA}|_{V_{GS} = 0\text{V}}$

The resulting bias line appears in Fig. 6.41. The operating point:

$$I_{D_Q} = \mathbf{2.75 \text{ mA}}$$

and $V_{GS_Q} = 6.4 \text{ V}$

with $V_{DS_Q} = V_{GS_Q} = \mathbf{6.4 \text{ V}}$

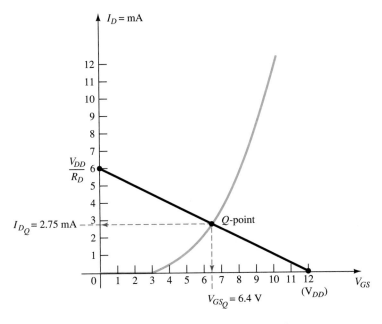

Figure 6.41   Determining the Q point for the network of Fig. 6.39.

Voltage-Divider Biasing Arrangement

A second popular biasing arrangement for the enhancement-type MOSFET appears in Fig. 6.42. The fact that $I_G = 0$ mA results in the following equation for V_{GG} as derived from an application of the voltage-divider rule:

$$\boxed{V_G = \frac{R_2 V_{DD}}{R_1 + R_2}}$$

(6.31)

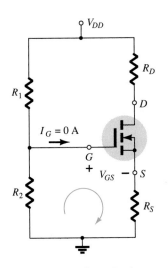

Figure 6.42   Voltage-divider biasing arrangement for an n-channel enhancement MOSFET.

Applying Kirchhoff's voltage law around the indicated loop of Fig. 6.42 will result in

$$+V_G - V_{GS} - V_{R_S} = 0$$

and

$$V_{GS} = V_G - V_{R_S}$$

or

$$\boxed{V_{GS} = V_G - I_D R_S} \qquad (6.32)$$

For the output section:

$$V_{R_S} + V_{DS} + V_{R_D} - V_{DD} = 0$$

and

$$V_{DS} = V_{DD} - V_{R_S} - V_{R_D}$$

or

$$\boxed{V_{DS} = V_{DD} - I_D(R_S + R_D)} \qquad (6.33)$$

Since the characteristics are a plot of I_D versus V_{GS} and Eq. (6.32) relates the same two variables, the two curves can be plotted on the same graph and a solution determined at their intersection. Once I_{D_Q} and V_{GS_Q} are known, all the remaining quantities of the network such as V_{DS}, V_D, and V_S can be determined.

EXAMPLE 6.12   Determine I_{D_Q}, V_{GS_Q}, and V_{DS} for the network of Fig. 6.43.

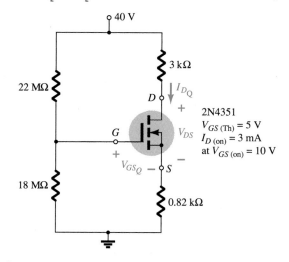

Figure 6.43 Example 6.12.

Solution

Network:

$$\text{Eq. (6.31):} \quad V_G = \frac{R_2 V_{DD}}{R_1 + R_2} = \frac{(18\ \text{M}\Omega)(40\ \text{V})}{22\ \text{M}\Omega + 18\ \text{M}\Omega} = 18\ \text{V}$$

$$\text{Eq. (6.32):} \quad V_{GS} = V_G - I_D R_S = 18\ \text{V} - I_D(0.82\ \text{k}\Omega)$$

When $I_D = 0$ mA,

$$V_{GS} = 18\ \text{V} - (0\ \text{mA})(0.82\ \text{k}\Omega) = 18\ \text{V}$$

as appearing on Fig. 6.44. When $V_{GS} = 0$ V,

$$V_{GS} = 18\ \text{V} - I_D(0.82\ \text{k}\Omega)$$

$$0 = 18\ \text{V} - I_D(0.82\ \text{k}\Omega)$$

$$I_D = \frac{18\ \text{V}}{0.82\ \text{k}\Omega} = 21.95\ \text{mA}$$

as appearing on Fig. 6.44.

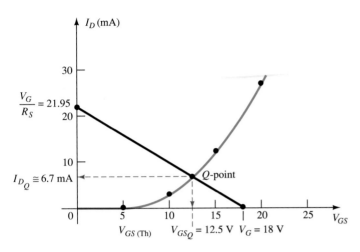

Figure 6.44 Determining the Q-point for the network of Example 6.12.

Device:

$$V_{GS(Th)} = 5 \text{ V}, \qquad I_{D(on)} = 3 \text{ mA with } V_{GS(on)} = 10 \text{ V}$$

Eq. (6.26): $\quad k = \dfrac{I_{D(on)}}{(V_{GS(on)} - V_{GS(Th)})^2}$

$$= \dfrac{3 \text{ mA}}{(10 \text{ V} - 5 \text{ V})^2} = 0.12 \times 10^{-3} \text{ A/V}^2$$

and $\qquad\qquad I_D = k(V_{GS} - V_{GS(Th)})^2$

$$= 0.12 \times 10^{-3}(V_{GS} - 5)^2$$

which is plotted on the same graph (Fig. 6.44). From Fig. 6.44,

$$I_{D_Q} \cong \textbf{6.7 mA}$$

$$V_{GS_Q} = \textbf{12.5 V}$$

Eq. (6.33): $\quad V_{DS} = V_{DD} - I_D(R_S + R_D)$

$$= 40 \text{ V} - (6.7 \text{ mA})(0.82 \text{ k}\Omega + 3.0 \text{ k}\Omega)$$

$$= 40 \text{ V} - 25.6 \text{ V}$$

$$= \textbf{14.4 V}$$

6.7 SUMMARY TABLE

Now that the most popular biasing arrangements for the various FETs have been introduced, Table 6.1 was developed to review the basic results and to demonstrate the similarity in approach for a number of configurations. It also reveals that the general analysis of dc configurations for FETs is not overly complex. Once the transfer characteristics are established, the network self-bias line can be drawn and the Q-point determined at the intersection of the device transfer characteristic and the network bias curve. The remaining analysis is simply an application of the basic laws of circuit analysis.

TABLE 6.1 FET Bias Configurations

Type	Configuration	Pertinent Equations	Graphical Solution
JFET Fixed-bias		$V_{GS_Q} = -V_{GG}$ $V_{DS} = V_{DD} - I_D R_S$	
JFET Self-bias		$V_{GS} = -I_D R_S$ $V_{DS} = V_{DD} - I_D(R_D + R_S)$	
JFET Voltage-divider bias		$V_G = \dfrac{R_2 V_{DD}}{R_1 + R_2}$ $V_{GS} = V_G - I_D R_S$ $V_{DS} = V_{DD} - I_D(R_D + R_S)$	
JFET Dual-supply		$V_{GS} = V_{SS} - I_D R_S$ $V_{DS} = V_{DD} + V_{SS} - I_D(R_D + R_S)$	
JFET $(V_{GS_Q} = 0\ \text{V})$		$V_{GS_Q} = 0\ \text{V}$ $I_{D_Q} = I_{DSS}$	
JFET $(R_D = 0\ \Omega)$		$V_{GS} = -I_D R_S$ $V_D = V_{DD}$ $V_S = I_D R_S$ $V_{DS} = V_{DD} - I_S R_S$	
Depletion-type MOSFET *(All configurations above plus cases where V_{GS_Q} = +voltage) Fixed-bias		$V_{GS_Q} = +V_{GG}$ $V_{DS} = V_{DD} - I_D R_S$	
Depletion-type MOSFET Voltage-divider bias		$V_G = \dfrac{R_2 V_{DD}}{R_1 + R_2}$ $V_{GS} = V_G - I_S R_S$ $V_{DS} = V_{DD} - I_D(R_D + R_S)$	
Enhancement-type MOSFET Feedback configuration		$V_{GS} = V_{DS}$ $V_{GS} = V_{DD} - I_D R_D$	
Enhancement-type MOSFET Voltage-divider bias		$V_G = \dfrac{R_2 V_{DD}}{R_1 + R_2}$ $V_{GS} = V_G - I_D R_S$	

6.8 COMBINATION NETWORKS

Now that the dc analysis of a variety of BJT and FET configurations is established, the opportunity to analyze networks with both types of devices presents itself. Fundamentally, the analysis simply requires that we *first* approach that device that will provide a terminal voltage or current level. The door is then usually open to calculate other quantities and concentrate on the remaining unknowns. These are usually particularly interesting problems, due to the challenge of finding the opening and then using the results of the past few sections and Chapter 5 to find the important quantities for each device. The equations and relationships used are simply those we have now employed on more than one occasion—no need to develop any new methods of analysis.

Determine the levels of V_D and V_C for the network of Fig. 6.45.

EXAMPLE 6.13

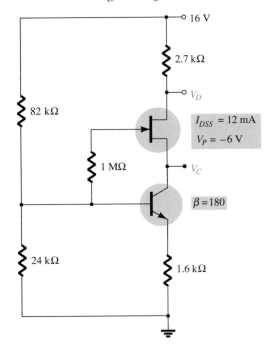

Figure 6.45 Example 6.13

Solution

From past experience we now realize that V_{GS} is typically an important quantity to determine or write an equation for when analyzing JFET networks. Since V_{GS} is a level for which an immediate solution is not obvious, let us turn our attention to the transistor configuration. The voltage-divider configuration is one where the approximate technique can be applied ($\beta R_E = (180 \times 1.6 \text{ k}\Omega) = 288 \text{ k}\Omega > 10R_2 = 240 \text{ k}\Omega$), permitting a determination of V_B using the voltage-divider rule on the input circuit.

For V_B:

$$V_B = \frac{24 \text{ k}\Omega \ (16 \text{ V})}{82 \text{ k}\Omega + 24 \text{ k}\Omega} = 3.62 \text{ V}$$

Using the fact that $V_{BE} = 0.7$ V results in

$$V_E = V_B - V_{BE} = 3.62 \text{ V} - 0.7 \text{ V}$$

$$= 2.92 \text{ V}$$

and
$$I_E = \frac{V_{R_E}}{R_E} = \frac{V_E}{R_E} = \frac{2.92 \text{ V}}{1.6 \text{ k}\Omega} = 1.825 \text{ mA}$$

with
$$I_C \cong I_E = 1.825 \text{ mA}$$

Continuing, we find for this configuration that

$$I_D = I_S = I_C$$

and
$$V_D = 16 \text{ V} - I_D(2.7 \text{ k}\Omega)$$

$$= 16 \text{ V} - (1.825 \text{ mA})(2.7 \text{ k}\Omega) = 16 \text{ V} - 4.93 \text{ V}$$

$$= \mathbf{11.07 \text{ V}}$$

The question of how to determine V_C is not as obvious. Both V_{CE} and V_{DS} are unknown quantities preventing us from establishing a link between V_D and V_C or from V_E to V_D. A more careful examination of Fig. 6.45 reveals that V_C is linked to V_B by V_{GS} (assuming that $V_{R_G} = 0$ V). Since we know V_B if we can find V_{GS}, V_C can be determined from

$$V_C = V_B - V_{GS}$$

The question then arises as to how to find the level of V_{GS_Q} from the quiescent value of I_D. The two are related by Shockley's equation:

$$I_{D_Q} = I_{DSS}\left(1 - \frac{V_{GS_Q}}{V_P}\right)^2$$

and V_{GS_Q} could be found mathematically by solving for V_{GS_Q} and substituting numerical values. However, let us turn to the graphical approach and simply work in the reverse order employed in the preceding sections. The JFET transfer characteristics are first sketched as shown in Fig. 6.46. The level of I_{D_Q} is then established by a horizontal line as shown in the same figure. V_{GS_Q} is then determined by dropping a line down from the operating point to the horizontal axis, resulting in

$$V_{GS_Q} = \mathbf{-3.7 \text{ V}}$$

The level of V_C:

$$V_C = V_B - V_{GS_Q} = 3.62 \text{ V} - (-3.7 \text{ V})$$

$$= \mathbf{7.32 \text{ V}}$$

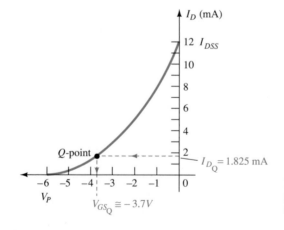

Figure 6.46 Determining the Q point for the network of Fig. 6.45.

Determine V_D for the network of Fig. 6.47.

EXAMPLE 6.14

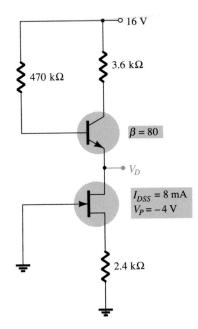

Figure 6.47 Example 6.14

Solution

In this case there is no obvious path to determine a voltage or current level for the transistor configuration. However, turning to the self-biased JFET, an equation for V_{GS} can be derived and the resulting quiescent point determined using graphical techniques. That is,

$$V_{GS} = -I_D R_S = -I_D(2.4 \text{ k}\Omega)$$

resulting in the self-bias line appearing in Fig. 6.48 that establishes a quiescent point at

$$V_{GS_Q} = -2.6 \text{ V}$$

$$I_{D_Q} = 1 \text{ mA}$$

For the transistor,

$$I_E \cong I_C = I_D = 1 \text{ mA}$$

and

$$I_B = \frac{I_C}{\beta} = \frac{1 \text{ mA}}{80} = 12.5 \text{ }\mu\text{A}$$

$$V_B = 16 \text{ V} - I_B(470 \text{ k}\Omega)$$

$$= 16 \text{ V} - (12.5 \text{ }\mu\text{A})(470 \text{ k}\Omega) = 16 \text{ V} - 5.875 \text{ V}$$

$$= 10.125 \text{ V}$$

and

$$V_E = V_D = V_B - V_{BE}$$

$$= 10.125 \text{ V} - 0.7 \text{ V}$$

$$= \mathbf{9.425 \text{ V}}$$

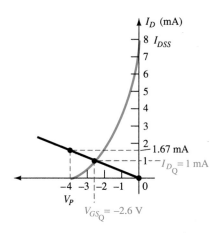

Figure 6.48 Determining the Q point for the network of Fig. 6.47.

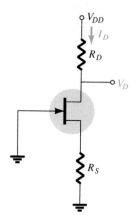

Figure 6.49 Self-bias configuration to be designed.

6.9 DESIGN

The design process is one that is not limited solely to dc conditions. The area of application, level of amplification desired, signal strength, and operating conditions are just a few conditions that enter into the total design process. However, we will first concentrate on establishing the chosen dc conditions.

For example, if the level of V_D and I_D is specified for the network of Fig. 6.49, the level of V_{GS_Q} can be determined from a plot of the transfer curve and R_S can then be determined from $V_{GS} = -I_D R_S$. If V_{DD} is specified, the level of R_D can then be calculated from $R_D = (V_{DD} - V_D)/I_D$. Of course, the value of R_S and R_D may not be standard commercial values requiring that the nearest commercial value be employed. However, with the tolerance (range of values) normally specified for the parameters of a network, the slight variation due to the choice of standard values will seldom cause a real concern in the design process.

The above is only one possibility for the design phase involving the network of Fig. 6.49. It is possible that only V_{DD} and R_D are specified together with the level of V_{DS}. The device to be employed may have to be specified along with the level of R_S. It appears logical that the device chosen should have a maximum V_{DS} greater than the specified value by a safe margin.

In general, it is good design practice for linear amplifiers to choose operating points that do not crowd the saturation level (I_{DSS}) or cutoff (V_P) regions. Levels of V_{GS_Q} close to $V_P/2$ or I_{D_Q} near $I_{DSS}/2$ are certainly reasonable starting points in the design. Of course, in every design procedure the maximum levels of I_D and V_{DS} as appearing on the specification sheet must not be considered exceeded.

The examples to follow have a design or synthesis orientation in that specific levels are provided and network parameters such as R_D, R_S, V_{DD}, and so on, must be determined. In any case the approach is in many ways the opposite of that described in previous sections. In some cases it is just a matter of applying Ohm's law in its appropriate form. In particular, if resistive levels are requested, the result is often obtained simply by applying Ohm's law in the following form:

$$R_{unknown} = \frac{V_R}{I_R} \qquad (6.34)$$

where V_R and I_R are often parameters that can be found directly from the specified voltage and current levels.

EXAMPLE 6.15

For the network of Fig. 6.50 the levels of V_{D_Q} and I_{D_Q} are specified. Determine the required values of R_D and R_S. What are the closest standard commercial values?

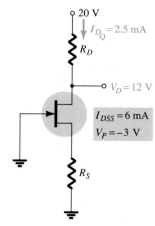

Figure 6.50 Example 6.15

Chapter 6   FET Biasing

Solution

As defined by Eq. (6.34),

$$R_D = \frac{V_{R_D}}{I_{D_Q}} = \frac{V_{DD} - V_{D_Q}}{I_{D_Q}}$$

and

$$= \frac{20\ V - 12\ V}{2.5\ mA} = \frac{8\ V}{2.5\ mA} = \mathbf{3.2\ k\Omega}$$

Plotting the transfer curve in Fig. 6.51 and drawing a horizontal line at $I_{D_Q} = 2.5$ mA will result in $V_{GS_Q} = -1$ V, and applying $V_{GS} = -I_D R_S$ will establish the level of R_S:

$$R_S = \frac{-(V_{GS_Q})}{I_{D_Q}} = \frac{-(-1\ V)}{2.5\ mA} = \mathbf{0.4\ k\Omega}$$

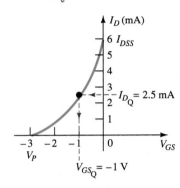

Figure 6.51 Determining V_{GS_Q} for the network of Fig. 6.50.

The nearest standard commercial values are

$$R_D = 3.2\ k\Omega \Rightarrow \mathbf{3.3\ k\Omega}$$

$$R_S = 0.4\ k\Omega \Rightarrow \mathbf{0.39\ k\Omega}$$

For the voltage-divider bias configuration of Fig. 6.52, if $V_D = 12$ V and $V_{GS_Q} = -2$ V, determine the value of R_S.

EXAMPLE 6.16

Solution

The level of V_G is determined as follows:

$$V_G = \frac{47\ k\Omega(16\ V)}{47\ k\Omega + 91\ k\Omega} = 5.44\ V$$

with

$$I_D = \frac{V_{DD} - V_D}{R_D}$$

$$= \frac{16\ V - 12\ V}{1.8\ k\Omega} = 2.22\ mA$$

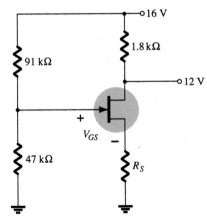

Figure 6.52 Example 6.16

The equation for V_{GS} is then written and the known values substituted:

$$V_{GS} = V_G - I_D R_S$$

$$-2\ V = 5.44\ V - (2.22\ mA)R_S$$

$$-7.44\ V = -(2.22\ mA)R_S$$

and

$$R_S = \frac{7.44\ V}{2.22\ mA} = \mathbf{3.35\ k\Omega}$$

The nearest standard commercial value is 3.3 kΩ.

EXAMPLE 6.17

The levels of V_{DS} and I_D are specified as $V_{DS} = \frac{1}{2}V_{DD}$ and $I_D = I_{D(on)}$ for the network of Fig. 6.53. Determine the level of V_{DD} and R_D.

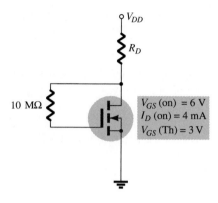

Figure 6.53 Example 6.17

Solution

Given $I_D = I_{D(on)} = 4$ mA and $V_{GS} = V_{GS(on)} = 6$ V, for this configuration,

$$V_{DS} = V_{GS} = \tfrac{1}{2}V_{DD}$$

and

$$6\text{ V} = \tfrac{1}{2}V_{DD}$$

so that

$$V_{DD} = \textbf{12 V}$$

Applying Eq. (6.34) yields

$$R_D = \frac{V_{R_D}}{I_D} = \frac{V_{DD} - V_{DS}}{I_{D(on)}} = \frac{V_{DD} - \frac{1}{2}V_{DD}}{I_{D(on)}} = \frac{\frac{1}{2}V_{DD}}{I_{D(on)}}$$

and

$$R_D = \frac{6\text{ V}}{4\text{ mA}} = \textbf{1.5 k}\Omega$$

which is a standard commercial value.

6.10 TROUBLESHOOTING

How often has a network been carefully constructed only to find that when the power is applied, the response is totally unexpected and fails to match the theoretical calculations. What is the next step? Is it a bad connection? A misreading of the color code for a resistive element or simply an error in the construction process? The range of possibilities seems vast and often frustrating. The troubleshooting process first described in the analysis of BJT transistor configurations should narrow down the list of possibilities and isolate the problem area following a definite plan of attack. In general, the process begins with a rechecking of the network construction and the terminal connections. This is usually followed by the checking of voltage levels between specific terminals and ground or between terminals of the network. Seldom are current levels measured since such maneuvers require disturbing the network structure to insert the meter. Of course, once the voltage levels are obtained, current levels can be calculated using Ohm's law. In any case, some idea of the expected voltage or current level must be known for the measurement to have any importance. In total, therefore, the troubleshooting process can only begin with some hope of success if the basic operation of the network is understood along with some expected levels of voltage or current. For the *n*-channel JFET amplifier it is clearly understood that the quiescent value of V_{GS_Q} is limited to 0 V or a negative voltage. For the network of Fig. 6.54, V_{GS_Q} is limited to negative values in the range 0 V to V_P. If a meter is hooked up as shown in Fig. 6.54, with the positive lead (normally red) to the gate and the negative lead (usually black) to the source, the resulting reading should have a negative sign and a magnitude of a few volts. Any other response should be considered suspicious and needs to be investigated.

The level of V_{DS} is typically between 25 and 75% of V_{DD}. A reading of 0 V for V_{DS} clearly indicates that either the output circuit has an "open" or the JFET is internally short-circuited between drain and source. If V_D is V_{DD} volts, there is obviously no drop across R_D due to the lack of current through R_D and the connections should be checked for continuity.

If the level of V_{DS} seems inappropriate, the continuity of the output circuit can easily be checked by grounding the negative lead of the voltmeter and measuring the voltage levels from V_{DD} to ground using the positive lead. If $V_D = V_{DD}$, the current through R_D may be zero, but there is continuity between V_D and V_{DD}. If $V_S = V_{DD}$, the device is not open between drain and source, but it is also not "on." The continuity through to V_S is confirmed, however. In this case it is possible that there is a poor ground connection between R_S and ground that may not be obvious. The internal connection between the wire of your lead and the terminal connector may have separated. Other possibilities also exist such as a shorted device from drain to source but the troubleshooter will simply have to narrow down the possible causes for the malfunction.

The continuity of a network can also be checked simply by measuring the voltage across any resistor of the network (except for R_G in the JFET configuration). An indication of 0 V immediately reveals the lack of current through the element due to an open circuit in the network.

The most sensitive element in the BJT and JFET configurations is the amplifier itself. The application of excessive voltage during the construction or testing phase or the use of incorrect resistor values resulting in high current levels can destroy the device. If you question the condition of the amplifier, the best test for the FET is the curve tracer since it not only reveals whether the device is operable but also its range of current and voltage levels. Some testers may reveal that the device is still fundamentally sound but do not reveal whether its range of operation has been severely reduced.

The development of good troubleshooting techniques comes primarily from experience and a level of confidence in what to expect and why. There are, of course, times when the reasons for a strange response seem to disappear mysteriously when you check a network. In such cases it is best not to breathe a sigh of relief and continue with the construction. The cause for such a sensitive "make or break" situation should be found and corrected or it may reoccur at the most inopportune moment.

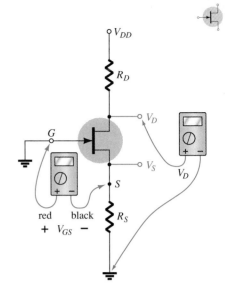

Figure 6.54 Checking the dc operation of the JFET self-bias configuration.

6.11 P-CHANNEL FETs

The analysis thus far has been limited solely to *n*-channel FETs. For *p*-channel FETs a mirror image of the transfer curves is employed and the defined current directions are reversed as shown in Fig. 6.55 for the various types of FETs.

Note for each configuration of Fig. 6.55 that each supply voltage is now a negative voltage drawing current in the indicated direction. In particular, note that the double-subscript notation for voltages continues as defined for the *n*-channel device: V_{GS}, V_{DS}, and so on. In this case, however, V_{GS} is positive (positive or negative for the depletion-type MOSFET) and V_{DS} negative.

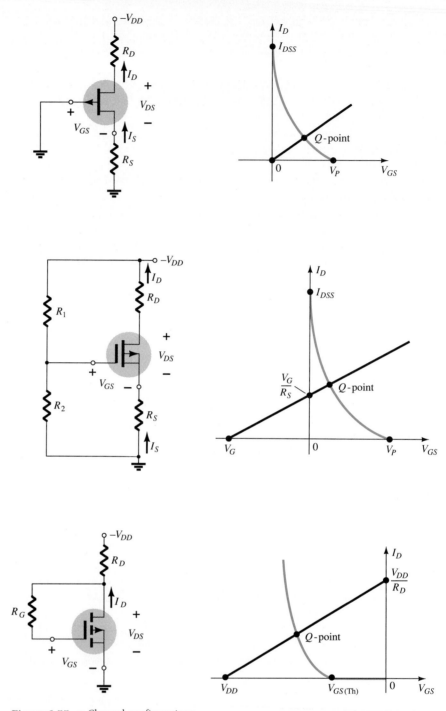

Figure 6.55 *p*-Channel configurations.

Due to the similarities between the analysis of *n*-channel and *p*-channel devices, one can actually assume an *n*-channel device and reverse the supply voltage and perform the entire analysis. When the results are obtained the magnitude of each quantity will be correct, although the current direction and voltage polarities will have to be reversed. However, the next example will demonstrate that with the experience gained through the analysis of *n*-channel devices the analysis of *p*-channel devices is quite straightforward.

Determine I_{D_Q}, V_{GS_Q}, and V_{DS} for the *p*-channel JFET of Fig. 6.56.

EXAMPLE 6.18

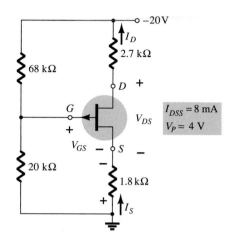

Figure 6.56 Example 6.18

Solution

$$V_G = \frac{20 \text{ k}\Omega(-20 \text{ V})}{20 \text{ k}\Omega + 68 \text{ k}\Omega} = -4.55 \text{ V}$$

Applying Kirchhoff's voltage law gives

$$V_G - V_{GS} + I_D R_S = 0$$

and

$$V_{GS} = V_G + I_D R_S$$

Choosing $I_D = 0$ mA yields

$$V_{GS} = V_G = -4.55 \text{ V}$$

as appearing in Fig. 6.57.

Choosing $V_{GS} = 0$ V, we obtain

$$I_D = -\frac{V_G}{R_S} = -\frac{-4.55 \text{ V}}{1.8 \text{ k}\Omega} = 2.53 \text{ mA}$$

as also appearing in Fig. 6.57. The resulting quiescent point from Fig. 6.57:

$$I_{D_Q} = \mathbf{3.4 \text{ mA}}$$

$$V_{GS_Q} = \mathbf{1.4 \text{ V}}$$

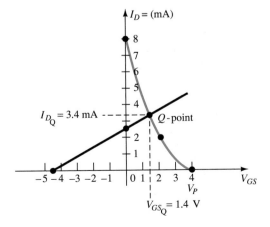

Figure 6.57 Determining the Q point for the JFET configuration of Fig. 6.56.

For V_{DS} Kirchhoff's voltage law will result in

$$-I_D R_S + V_{DS} - I_D R_D + V_{DD} = 0$$

and
$$V_{DS} = -V_{DD} + I_D(R_D + R_S)$$

$$= -20 \text{ V} + (3.4 \text{ mA})(2.7 \text{ k}\Omega + 1.8 \text{ k}\Omega)$$

$$= -20 \text{ V} + 15.3 \text{ V}$$

$$= \mathbf{-4.7 \text{ V}}$$

6.12 UNIVERSAL JFET BIAS CURVE

Since the dc solution of a FET configuration requires drawing the transfer curve for each analysis, a universal curve was developed that can be used for any level of I_{DSS} and V_P. The universal curve for an *n*-channel JFET or depletion-type MOSFET (for negative values of V_{GS_Q}) is provided in Fig. 6.58. Note that the horizontal axis is not that of V_{GS} but of a normalized level defined by $V_{GS}/|V_P|$, the $|V_P|$ indicating that only the magnitude of V_P is to be employed, not its sign. For the vertical axis the scale is also a normalized level of I_D/I_{DSS}. The result is that when $I_D = I_{DSS}$ the ratio is 1, and when $V_{GS} = V_P$ the ratio $V_{GS}/|V_P|$ is -1. Note also that the scale for I_D/I_{DSS} is on the left rather than on the right as encountered for I_D in past exercises. The additional two scales on the right need an introduction. The vertical scale labeled *m* can in itself be used to find the solution to fixed-bias configurations. The other scale, labeled *M*, is employed along with the *m* scale to find the solution to voltage-divider configura-

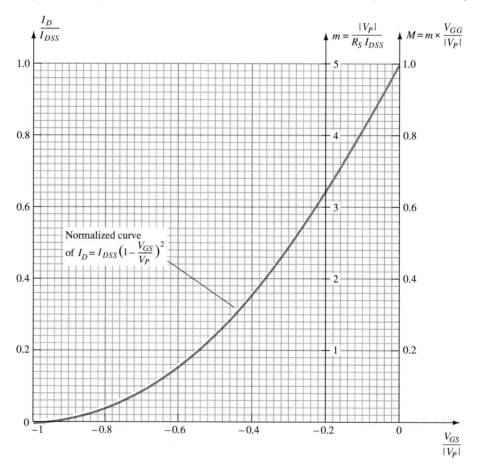

Figure 6.58 Universal JFET bias curve.

tions. The scaling for m and M come from a mathematical development involving the network equations and normalized scaling just introduced. The description to follow will not concentrate on why the m scale extends from 0 to 5 at $V_{GS}/|V_P| = -0.2$ and the M scale from 0 to 1 at $V_{GS}/|V_P| = 0$ but rather on how to use the resulting scales to obtain a solution for the configurations. The equations for m and M are the following, with V_G as defined by Eq. (6.18).

$$m = \frac{|V_P|}{I_{DSS}R_S} \qquad (6.35)$$

$$M = m \times \frac{V_G}{|V_P|} \qquad (6.36)$$

with $\qquad V_G = \dfrac{R_2 V_{DD}}{R_1 + R_2}$

Keep in mind that the beauty of this approach is the elimination of the need to sketch the transfer curve for each analysis, that the superposition of the bias line is a great deal easier, and that the calculations are fewer. The use of the m and M axes is best described by examples employing the scales. Once the procedure is clearly understood, the analysis can be quite rapid with a good measure of accuracy.

Determine the quiescent values of I_D and V_{GS} for the network of Fig. 6.59. **EXAMPLE 6.19**

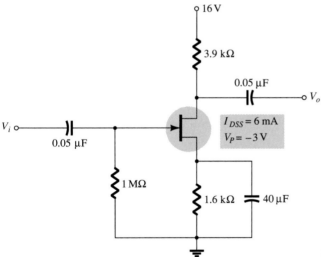

Figure 6.59 Example 6.19

Solution

Calculating the value of m, we obtain

$$m = \frac{|V_P|}{I_{DSS}R_S} = \frac{|-3V|}{(6 \text{ mA})(1.6 \text{ k}\Omega)} = 0.31$$

The self-bias line defined by R_S is plotted by drawing a straight line from the origin through a point defined by $m = 0.31$, as shown in Fig. 6.60.
The resulting Q-point:

$$\frac{I_D}{I_{DSS}} = 0.18 \qquad \text{and} \qquad \frac{V_{GS}}{|V_P|} = -0.575$$

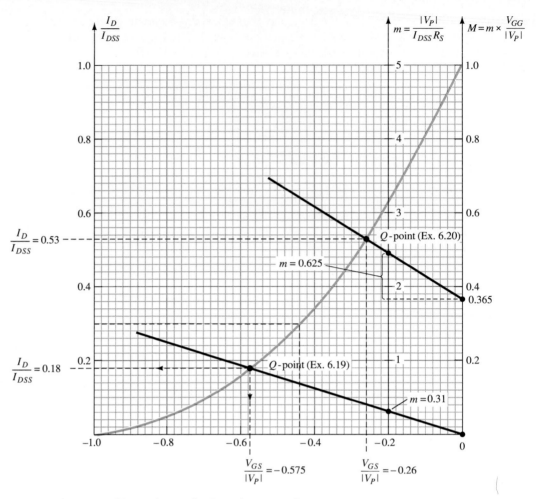

Figure 6.60 Universal curve for Examples 6.19 and 6.20.

The quiescent values of I_D and V_{GS} can then be determined as follows:

$$I_{D_Q} = 0.18I_{DSS} = 0.18(6 \text{ mA}) = \mathbf{1.08 \text{ mA}}$$

and

$$V_{GS_Q} = -0.575|V_P| = -0.575(3 \text{ V}) = \mathbf{-1.73 \text{ V}}$$

EXAMPLE 6.20   Determine the quiescent values of I_D and V_{GS} for the network of Fig. 6.61.

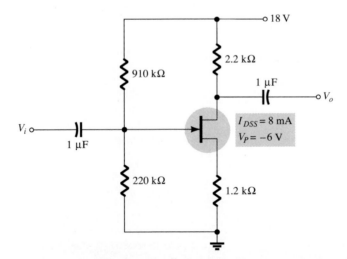

Figure 6.61 Example 6.20

Solution

Calculating m gives

$$m = \frac{|V_P|}{I_{DSS}R_S} = \frac{|-6 \text{ V}|}{(8 \text{ mA})(1.2 \text{ k}\Omega)} = 0.625$$

Determining V_G yields

$$V_G = \frac{R_2 V_{DD}}{R_1 + R_2} = \frac{(220 \text{ k}\Omega)(18 \text{ V})}{910 \text{ k}\Omega + 220 \text{ k}\Omega} = 3.5 \text{ V}$$

Finding M, we have

$$M = m \times \frac{V_G}{|V_P|} = 0.625\left(\frac{3.5 \text{ V}}{6 \text{ V}}\right) = 0.365$$

Now that m and M are known, the bias line can be drawn on Fig. 6.60. In particular, note that even though the levels of I_{DSS} and V_P are different for the two networks, the same universal curve can be employed. First find M on the M axis as shown in Fig. 6.60. Then draw a horizontal line over to the m axis and at the point of intersection add the magnitude of m as shown in the figure. Using the resulting point on the m axis and the M intersection, draw the straight line to intersect with the transfer curve and define the Q-point

That is,

$$\frac{I_D}{I_{DSS}} = 0.53 \quad \text{and} \quad \frac{V_{GS}}{|V_P|} = -0.26$$

and

$$I_{D_Q} = 0.53 I_{DSS} = 0.53(8 \text{ mA}) = \textbf{4.24 mA}$$

with

$$V_{GS_Q} = -0.26|V_P| = -0.26(6 \text{ V}) = \textbf{-1.56 V}$$

6.13 COMPUTER ANALYSIS

The computer analysis of a voltage-divider JFET configuration is performed in this section using both BASIC and PSpice. The PSpice approach will be quite similar to that employed for the BJT configuration of Chapter 4. The use of BASIC will require a mathematical approach that will include finding the solution to a quadratic equation.

PSpice

The voltage-divider configuration of Fig. 6.61 is redrawn in Fig. 6.62 with the defined nodes and device parameters as introduced in Chapter 5. The parameters are

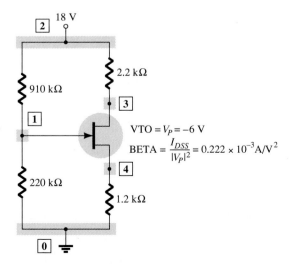

18 V

$VTO = V_P = -6 \text{ V}$

$BETA = \dfrac{I_{DSS}}{|V_P|^2} = 0.222 \times 10^{-3} \text{A/V}^2$

Figure 6.62 Network of Fig. 6.61 with defined nodes for a PSpice analysis.

```
**** 02/28/90 ****** Evaluation PSpice (January 1989) ****** 15:11:26 ****

 DC Bias of JFET configuration in Fig. 6.61

****      CIRCUIT DESCRIPTION

**********************************************************************

VDD 2 0 18V
R1 2 1 910K
R2 1 0 220K
RD 2 3 2.2K
RS 4 0 1.2K
J1 3 1 4 JN
.MODEL JN NJF(VTO=-6V BETA=.222E-3)
.DC VDD 18 18 1
.PRINT DC V(1,4) I(RD)
.OPTIONS NOPAGE
.END

****      Junction FET MODEL PARAMETERS
               JN
              NJF
        VTO    -6
        BETA   222.000000E-06


****      DC TRANSFER CURVES              TEMPERATURE =   27.000 DEG C

     VDD        V(1,4)       I(RD)
     1.800E+01  -1.565E+00   4.225E-03
```

Figure 6.63 PSpice analysis of the JFET configuration of Fig. 6.61.

entered as appearing in Fig. 6.63 in the same manner as described in previous chapters with the JFET introduced using the formats described in Chapter 5. The voltage requested as V(1,4) is V_{GS_Q} and the current I(RD) is I_{D_Q}. Note how closely the results match those of Example 6.20 with I_{D_Q} (Example 6.20) = 4.24 mA and I_{D_Q} (PSpice) = 4.23 mA, and V_{GS_Q} (Example 6.20) = -1.56 V and V_{GS_Q} (PSpice) = -1.57 V.

BASIC

If a language such as BASIC is employed, a common solution for the equations defined by the network and device must be found using mathematical techniques. For the network of Fig. 6.64a, we recognize that the device is described by Shockley's equation (6.64b):

$$I_D = I_{DSS}\left(1 - \frac{V_{GS}}{V_P}\right)^2 \tag{6.37}$$

while the network is defined by (Fig. 6.64b)

$$V_{GS} = V_G - I_D R_S \tag{6.38}$$

with

$$V_G = \frac{R_2 V_{DD}}{R_1 + R_2} \tag{6.39}$$

If we insert the equation for I_D [Eq. (6.37)] into Eq. (6.38), we obtain

$$V_{GS} = V_G - I_{DSS} R_S\left(1 - \frac{V_{GS}}{V_P}\right)^2$$

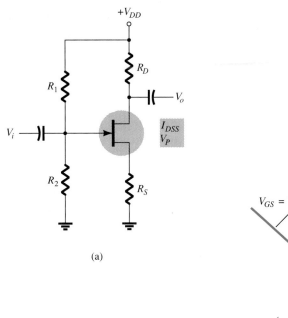

(a)

$$V_{GS} = V_G - I_D R_S$$

$$I_D = I_{DSS}\left(1 - \frac{V_{GS}}{V_P}\right)^2$$

Figure 6.64 Voltage-divider configuration to be analyzed using BASIC.

(b)

which, when expanded, results in the quadratic equation

$$\underbrace{\frac{I_{DSS}R_S}{V_P^2}}_{A}V_{GS}^2 + \underbrace{\left(1 - \frac{2I_{DSS}R_S}{V_P}\right)}_{B}V_{GS} + \underbrace{(I_{DSS}R_S - V_G)}_{C}=0$$

The solutions to the quadratic equation are then determined by

$$V_{GS_{1,2}} = \frac{-B \pm \sqrt{B^2 - 4AC}}{2A}$$

with the actual solution being that value of V_{GS} which falls within the range 0 to V_P. The program will, of course, test the value of $B^2 - 4AC$, indicating no solution for the case of a negative value. The drain and source voltages are then

$$V_D = V_{DD} - I_D R_D \tag{6.40}$$

$$V_S = I_D R_S \tag{6.41}$$

and

$$V_{DS} = V_D - V_S \tag{6.42}$$

A summary of the variables and equations used in module 11000 are provided in Tables 6.2 and 6.3. The program listing appears in Fig. 6.65 with a run for the same values employed in the PSpice analysis. Again note the close correspondence in solutions.

TABLE 6.2 Equations and Computer Statements for Module 11000

Equation	Computer Statement
$V_G = \dfrac{R_2}{R_1 + R_2}V_{DD}$	GG = (R2/(R1 + R2)) * DD
$V_S = I_D R_S$	VS = ID * RS
$V_{GS} = V_G - V_S$	GS = VG − VS
$I_D = I_{DSS}\left(1 - \dfrac{V_{GS}}{V_P}\right)^2$	ID = SS * (1 − GS/VP) ↑ 2
$A = \dfrac{I_{DSS}R_S}{V_P^2}$	A = SS * RS/VP ↑ 2
$B = 1 - \dfrac{2I_{DSS}R_S}{V_P}$	B = 1 − 2 * SS * RS/VP
$C = I_{DSS}R_S - V_G$	C = SS * RS − GG
$D = B^2 - 4AC$	D = B ↑ 2 − 4 * A * C
$V_1 = \dfrac{-B + \sqrt{D}}{2A}$	V1 = (−B + SQR(D))/(2 * A)
$V_2 = \dfrac{-B - \sqrt{D}}{2A}$	V2 = (−B − SQR(D))/(2 * A)
$V_D = V_{DD} - I_D R_D$	VD = DD − ID * RD
$V_S = I_D R_S$	VS = ID * RS
$V_{DS} = V_D - V_S$	DS = VD − VS

TABLE 6.3 Equation and Program Variables for Module 11000

Equation Variable	Program Variable
V_G	VG
V_S	VS
V_D	VD
V_G	GG
V_{DD}	DD
V_{GS}	GS
V_{DS}	DS
V_P	VP
I_D	ID
I_{DSS}	SS
R_1	R1
R_2	R2
R_S	RS
R_D	RD

```
10 REM **********************************************
20 REM
30 REM Module for FET dc Bias Calculations
40 REM
50 REM **********************************************
60 REM
100 PRINT "This program provides the dc bias calculations"
110 PRINT "for a JFET or depletion MOSFET"
120 PRINT "voltage-divider configuration."
130 PRINT
140 PRINT "Enter the following circuit data:"
150 PRINT
160 INPUT "R1 (use 1E30 if open)=";R1
170 INPUT "R2               =";R2
180 INPUT "RS=";RS
190 INPUT "RD=";RD
200 PRINT
210 INPUT "Supply voltage, VDD=";DD
220 PRINT
230 PRINT "Enter the following device data:"
240 INPUT "Drain-source saturation current, IDSS=";SS
250 INPUT "Gate-source pinchoff voltage, VP=";VP
260 PRINT :PRINT
270 REM Now do bias calculations
280 GOSUB 11000
290 PRINT "Bias current is, ID=";ID*1000;"mA"
300 PRINT "Bias voltages are:"
310 PRINT "VGS=";GS;"volts"
320 PRINT "VD=";VD;"volts"
330 PRINT "VS=";VS;"volts"
340 PRINT "VDS=";DS;"volts"
350 END
```

Figure 6.65 BASIC program for the analysis of the network of Fig. 6.64.

```
11000 REM Module for FET dc bias calculations
11010 GG=(R2/(R1+R2))*DD
11020 A=SS*RS/VP^2
11030 B=1-2*SS*RS/VP
11040 C=SS*RS-GG
11050 D=B^2-4*A*C
11060 IF D<0 THEN PRINT "No Solution!!!" :STOP
11070 V1=(-B+SQR(D))/(2*A)
11080 V2=(-B-SQR(D))/(2*A)
11090 IF ABS(V1)>ABS(VP) THEN GS=V2
11100 IF ABS(V2)>ABS(VP) THEN GS=V1
11110 ID=SS*(1-GS/VP)^2
11120 VS=ID*RS
11130 VG=GG
11140 VD=DD-ID*RD
11150 DS=VD-VS
11160 RETURN


RUN
This program provides the dc bias calculations
for a JFET or depletion MOSFET
voltage-divider configuration.

Enter the following circuit data:

R1 (use 1E30 if open)=? 910E3
R2                    =? 220E3
RS=? 1.2E3
RD=? 2.2E3

Supply voltage, VDD=? 18

Enter the following device data:
Drain-source saturation current, IDSS=? 8E-3
Gate-source pinchoff voltage, VP=? -6


Bias current is, ID= 4.26821 mA
Bias voltages are:
VGS=-1.617427 volts
VD= 8.609939 volts
VS= 5.121852 volts
VDS= 3.488087 volts
```

Figure 6.65   (Continued.)

§ 6.2 Fixed-Bias Configuration

1. For the fixed-bias configuration of Fig. 6.66:
   (a) Sketch the transfer characteristics of the device.
   (b) Superimpose the network equation on the same graph.
   (c) Determine I_{D_Q} and V_{DS_Q}.
   (d) Using Shockley's equation, solve for I_{D_Q} and then find V_{DS_Q}. Compare with the solutions of part (c).

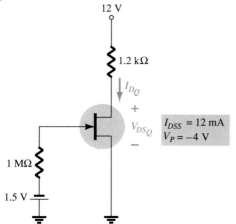

Figure 6.66   Problems 1, 35, 38

Problems

2. For the fixed-bias configuration of Fig. 6.67, determine:
  (a) I_{D_Q} and V_{GS_Q}.
  (b) V_{DS}, V_D, V_G, and V_S.

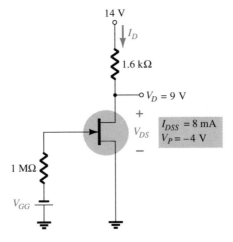

Figure 6.67 Problem 2

3. Given the measured value of V_D in Fig. 6.68, determine:
  (a) I_D.
  (b) V_{DS}.
  (c) V_{GG}.

Figure 6.68 Problem 3

4. Determine V_D for the fixed-bias configuration of Fig. 6.69.
5. Determine V_D for the fixed-bias configuration of Fig. 6.70.

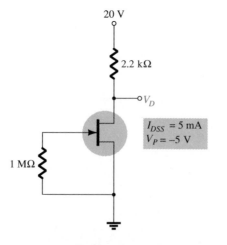

Figure 6.69 Problem 4

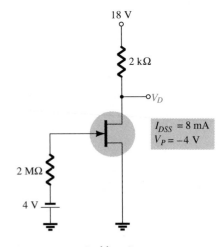

Figure 6.70 Problem 5

6. For the self-bias configuration of Fig. 6.71:
  (a) Sketch the transfer curve for the device.
  (b) Superimpose the network equation on the same graph.
  (c) Determine I_{D_Q} and V_{GS_Q}.
  (d) Calculate V_{DS}, V_D, V_G, and V_S.

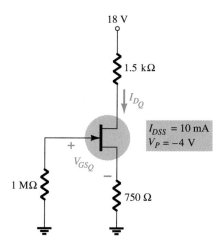

18 V

1.5 kΩ

I_{D_Q}

$I_{DSS} = 10$ mA
$V_P = -4$ V

V_{GS_Q}

1 MΩ

750 Ω

Figure 6.71 Problems 6, 7, 36, 39

* **7.** Determine I_{D_Q} for the network of Fig. 6.71 using a purely mathematical approach. That is, establish a quadratic equation for I_D and choose that solution compatible with the network characteristics. Compare to the solution obtained in Problem 6.

8. For the network of Fig. 6.72, determine:
  (a) V_{GS_Q} and I_{D_Q}.
  (b) V_{DS}, V_D, V_G, and V_S.

9. Given the measurement $V_S = 1.7$ V for the network of Fig. 6.73, determine:
  (a) I_{D_Q}.
  (b) V_{GS_Q}.
  (c) I_{DSS}.
  (d) V_D.
  (e) V_{DS}.

* **10.** For the network of Fig. 6.74, determine:
  (a) I_D.
  (b) V_{DS}.
  (c) V_D.
  (d) V_S.

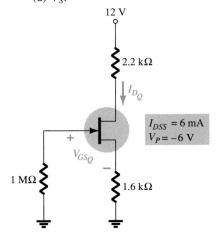

12 V

2.2 kΩ

I_{D_Q}

$I_{DSS} = 6$ mA
$V_P = -6$ V

V_{GS_Q}

1 MΩ

1.6 kΩ

Figure 6.72 Problem 8

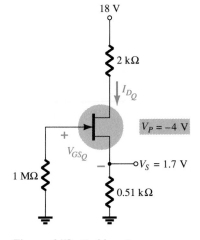

18 V

2 kΩ

I_{D_Q}

$V_P = -4$ V

V_{GS_Q}

1 MΩ

$V_S = 1.7$ V

0.51 kΩ

Figure 6.73 Problem 9

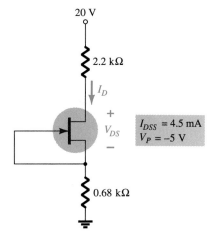

20 V

2.2 kΩ

I_D

$+$
V_{DS}
$-$

$I_{DSS} = 4.5$ mA
$V_P = -5$ V

0.68 kΩ

Figure 6.74 Problem 10

* **11.** Find V_S for the network of Fig. 6.75.

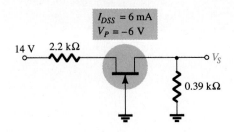

Figure 6.75 Problem 11

§ **6.4 Voltage-Divider Biasing**

12. For the network of Fig. 6.76, determine:
(a) V_G.
(b) I_{D_Q} and V_{GS_Q}.
(c) V_{DS_Q}.
(d) V_D and V_S.

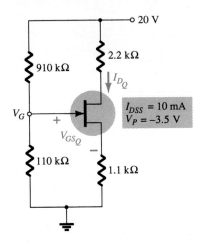

Figure 6.76 Problems 12, 13, 40

13. (a) Repeat Problem 12 with $R_S = 0.51 \text{ k}\Omega$ (about 50% of the value of 12). What is the effect of a smaller R_S on I_{D_Q} and V_{GS_Q}?
(b) What is the minimum possible value of R_S for the network of Fig. 6.76?

14. For the network of Fig. 6.77, $V_D = 9$ V. Determine:
(a) I_D.
(b) V_S, V_{DS}.
(c) V_G.
(d) V_P.

* **15.** For the network of Fig. 6.78, determine:
(a) I_{D_Q} and V_{GS_Q}.
(b) V_{DS} and V_S.

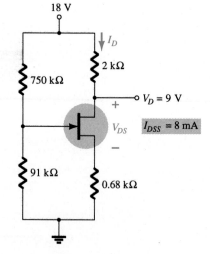

Figure 6.77 Problem 14

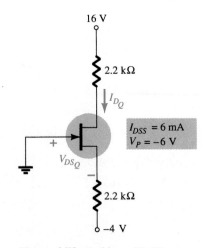

Figure 6.78 Problems 15, 37

Chapter 6 FET Biasing

* **16.** Given $V_{DS} = 8$ V for the network of Fig. 6.79, determine:
   (a) I_D.
   (b) V_D and V_S.
   (c) V_{GS}.

§ 6.5 Depletion-Type MOSFETs

17. For the self-bias configuration of Fig. 6.80, determine:
   (a) I_{D_Q} and V_{GS_Q}.
   (b) V_{DS} and V_D.

* **18.** For the network of Fig. 6.81, determine:
   (a) I_{D_Q} and V_{GS_Q}.
   (b) V_{DS} and V_S.

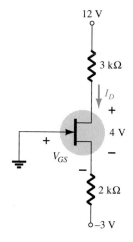

Figure 6.79 Problem 16

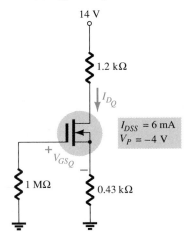

Figure 6.80 Problem 17

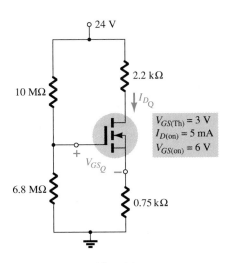

Figure 6.81 Problem 18

§ 6.6 Enhancement-Type MOSFETs

19. For the network of Fig. 6.82, determine:
   (a) I_{D_Q}.
   (b) V_{GS_Q} and V_{DS_Q}.
   (c) V_D and V_S.
   (d) V_{DS}.

20. For the voltage-divider configuration of Fig. 6.83, determine:
   (a) I_{D_Q} and V_{GS_Q}.
   (b) V_D and V_S.

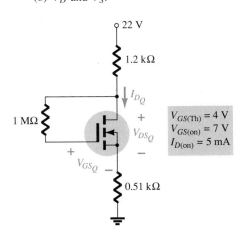

Figure 6.82 Problem 19

Figure 6.83 Problem 20

* **21.** For the network of Fig. 6.84, determine:

(a) V_G.

(b) V_{GS_Q} and I_{D_Q}.

(c) I_E.

(d) I_B.

(e) V_D.

(f) V_C.

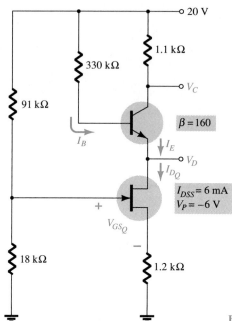

Figure 6.84 Problem 21

* **22.** For the combination network of Fig. 6.85, determine:

(a) V_B, V_G.

(b) V_E.

(c) I_E, I_C, I_D.

(d) I_B.

(e) V_C, V_S, V_D.

(f) V_{CE}.

(g) V_{DS}.

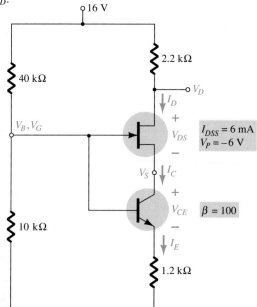

Figure 6.85 Problem 22

Chapter 6   FET Biasing

§ 6.9 Design

* **23.** Design a self-bias network using a JFET transistor with $I_{DSS} = 8$ mA and $V_P = -6$ V to have a Q-point at $I_{D_Q} = 4$ mA using a supply of 14 V. Assume that $R_D = 3R_S$ and use standard values.

* **24.** Design a voltage-divider bias network using a depletion-type MOSFET with $I_{DSS} = 10$ mA and $V_P = -4$ V to have a Q-point at $I_{D_Q} = 2.5$ mA using a supply of 24 V. In addition, set $V_G = 4$ V and use $R_D = 2.5R_S$ with $R_1 = 22$ MΩ. Use standard values.

25. Design a network such as appearing in Fig. 6.39 using an enhancement-type MOSFET with $V_{GS(Th)} = 4$ V, $k = 0.5 \times 10^{-3} A/V^2$ to have a Q-point of $I_{D_Q} = 6$ mA. Use a supply of 16 V and standard values.

§ 6.10 Troubleshooting

* **26.** What do the readings for each configuration of Fig. 6.86 suggest about the operation of the network?

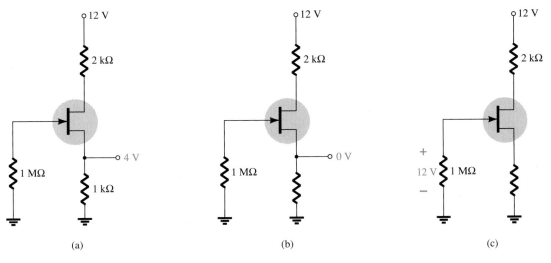

Figure 6.86 Problem 26

* **27.** Although the readings of Fig. 6.87 initially suggest that the network is behaving properly, determine a possible cause for the undesirable state of the network.

* **28.** The network of Fig. 6.88 is not operating properly. What is the specific cause for its failure?

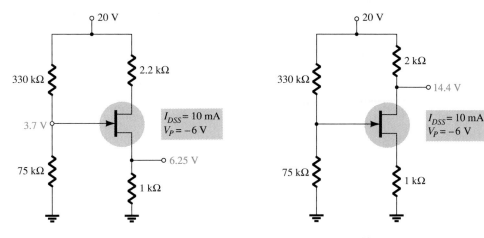

Figure 6.87 Problem 27   **Figure 6.88** Problem 28

§ 6.11 P-Channel FETs

29. For the network of Fig. 6.89, determine:
    (a) I_{D_Q} and V_{GS_Q}.
    (b) V_{DS}.
    (c) V_D.

30. For the network of Fig. 6.90, determine:
    (a) I_{D_Q} and V_{GS_Q}.
    (b) V_{DS}.
    (c) V_D.

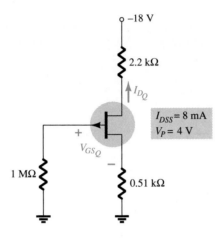

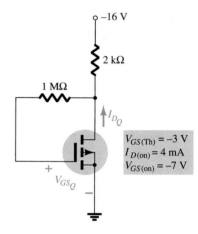

Figure 6.89 Problem 29   **Figure 6.90** Problem 30

§ 6.12 Universal JFET Bias Curve

31. Repeat Problem 1 using the universal JFET bias curve.

32. Repeat Problem 6 using the universal JFET bias curve.

33. Repeat Problem 12 using the universal JFET bias curve.

34. Repeat Problem 15 using the universal JFET bias curve.

§ 6.13 Computer Analysis

35. Perform a PSpice analysis of the network of Problem 1. Determine I_{D_Q} and V_{GS_Q}.

36. Perform a PSpice analysis of the network of Problem 6. Determine I_{D_Q} and V_{GS_Q}.

37. Perform a PSpice analysis of the network of Problem 15. Determine I_{D_Q}, V_{GS_Q}, and V_{DS_Q}.

38. Using BASIC, determine I_{D_Q} and V_{GS_Q} for the network of Problem 1.

39. Using BASIC, determine I_{D_Q} and V_{GS_Q} for the network of Problem 6.

40. Using BASIC, determine I_{D_Q}, V_{GS_Q}, and V_{DS_Q} for the network of Problem 12.

*Please Note: Asterisks indicate more difficult problems.

BJT Transistor Modeling

7.1 INTRODUCTION

The basic construction, appearance, and characteristics of the transistor were introduced in Chapter 3. The dc biasing of the device was then examined in detail in Chapter 4. We now begin to examine the *small-signal* ac response of the BJT amplifier by reviewing the *models* most frequently used to represent the transistor in the sinusoidal ac domain.

One of our first concerns in the sinusoidal ac analysis of transistor networks is the magnitude of the input signal. It will determine whether *small-signal* or *large-signal* techniques should be applied. There is no set dividing line between the two, but the application, and the magnitude of the variables of interest relative to the scales of the device characteristics, will usually make it quite clear which method is appropriate. The small-signal technique is introduced in this chapter and large-signal applications are examined in Chapter 16.

There are two models commonly used in the small-signal ac analysis of transistor networks: the r_e model and the *hybrid equivalent* model. This chapter not only introduces both models but defines the role of each and the relationship between the two.

7.2 AMPLIFICATION IN THE AC DOMAIN

It was demonstrated in Chapter 3 that the transistor can be employed as an amplifying device. That is, the output sinusoidal signal is greater than the input signal or, stated another way, the output ac power is greater than the input ac power. The question then arises as to how the ac power output can be greater than the input ac power? Conservation of energy dictates that over time the total power output, P_o, of a system cannot be greater than its power input, P_i, and that the efficiency defined by $\eta = P_o/P_i$ cannot be greater than 1. The factor missing from the discussion above that permits an ac power output greater than the ac input power is the applied dc power. It is a contributor to the total output power even though part of it is dissipated by the device and resistive elements. In other words, there is an "exchange" of dc power to the ac domain that permits establishing a higher output ac power. In fact, a *conversion efficiency* is defined by $\eta = P_{o(\text{ac})}/P_{i(\text{dc})}$, where $P_{o(\text{ac})}$ is the ac power to the load and $P_{i(\text{dc})}$ is the dc power supplied.

Perhaps the role of the dc supply can best be described by first considering the simple dc network of Fig. 7.1. The resulting direction of flow is indicated in the figure with a plot of the current i versus time. Let us now insert a control mechanism

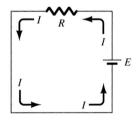

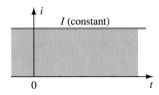

Figure 7.1 Steady current established by a dc supply.

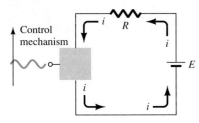

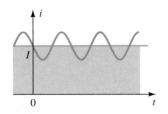

Figure 7.2 Effect of a control element on the steady-state flow of the electrical system of Fig. 7.1.

such as that shown in Fig. 7.2. The control mechanism is such that the application of a relatively small signal to the control mechanism can result in a much larger oscillation in the output circuit. For the system of Fig. 7.2 the peak value of the oscillation is controlled by the established dc level. Any attempt to exceed the limit set by the dc level will result in a "clipping" (flattening) of the peak region of the output signal. In total, therefore, proper amplifier design requires that the dc and ac components be sensitive to each others requirements and limitations.

However, it is indeed fortunate that transistor small-signal amplifiers can be considered linear for most applications, permitting the use of the superposition theorem to isolate the dc analysis from the ac analysis.

7.3 BJT TRANSISTOR MODELING

The key to transistor small-signal analysis is the use of equivalent circuits (models) to be introduced in this chapter.

A model is the combination of circuit elements, properly chosen, that best approximates the actual behavior of a semiconductor device under specific operating conditions.

Once the ac equivalent circuit has been determined, the graphic symbol of the device can be replaced in the schematic by this circuit and the basic methods of ac circuit analysis (mesh analysis, nodal analysis, and Thévenin's theorem) can be applied to determine the response of the circuit.

There are two schools of thought in prominence today regarding the equivalent circuit to be substituted for the transistor. For many years the industrial and educational institutions relied heavily on the *hybrid parameters* (to be introduced shortly). The hybrid-parameter equivalent circuit continues to be very popular, although it must now share the spotlight with an equivalent circuit derived directly from the operating conditions of the transistor—the r_e model. Manufacturers continue to specify the hybrid parameters for a particular operating region on their specification sheets. The parameters (or components) of the r_e model can be derived directly from the hybrid parameters in this region. However, the hybrid equivalent circuit suffers from being limited to a particular set of operating conditions if it is to be considered accurate. The parameters of the other equivalent circuit can be determined for any region of operation within the active region and are not limited by the single set of parameters provided by the specification sheet. In turn, however, the r_e model fails to account for the output impedance level of the device and the feedback effect from output to input.

Since both models are used extensively today, they are both examined in detail in this text. In some analysis and examples the hybrid model will be employed, while in others the r_e model will be used exclusively. The text will make every effort, however, to show how closely related the two models are and how a proficiency with one leads to a natural proficiency with the other.

In an effort to demonstrate the effect that the ac equivalent circuit will have on the analysis to follow, consider the circuit of Fig. 7.3. Let us assume for the moment that the small-signal ac equivalent circuit for the transistor has already been determined. Since we are interested only in the ac response of the circuit, all the dc supplies can be replaced by a zero-potential equivalent (short circuit) since they determine only the dc (quiescent level) of the output voltage and not the magnitude of the swing of the ac output. This is clearly demonstrated by Fig. 7.4. The dc levels were simply important for determining the proper Q-point of operation. Once determined, the dc levels can be ignored in the ac analysis of the network. In addition, the coupling capacitors C_1 and C_2 and bypass capacitor C_3 were chosen to have a very small reactance at the

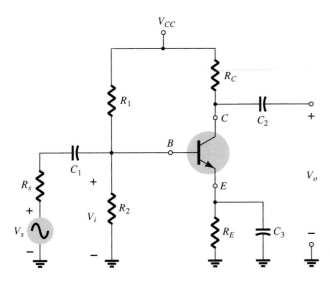

Figure 7.3 Transistor circuit under examination in this introductory discussion.

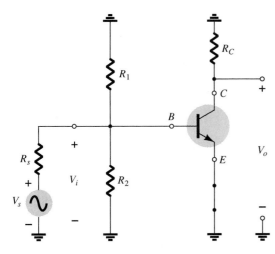

Figure 7.4 The network of Fig. 7.3 following the removal of the dc supply and inserting the short-circuit equivalent for the capacitors.

frequency of application. Therefore, they too may for all practical purposes be replaced by a low-resistance path or a short circuit. Note that this will result in the "shorting out" of the dc biasing resistor R_E. Recall that capacitors assume an "open-circuit" equivalent under dc steady-state conditions, permitting an isolation between stages for the dc levels and quiescent conditions.

If we establish a common ground and rearrange the elements of Fig. 7.4 R_1 and R_2 will be in parallel and R_C will appear from collector to emitter as shown in Fig. 7.5. Since the components of the transistor equivalent circuit appearing in Fig. 7.5 employ

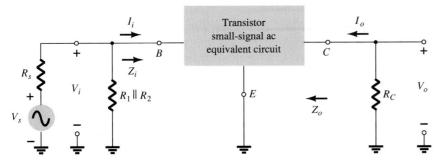

Figure 7.5 Circuit of Fig. 7.4 redrawn for small-signal ac analysis.

familiar components such as resistors and independent controlled sources, analysis techniques such as superposition, Thévenin's theorem, and so on, can be applied to determine the desired quantities.

Let us further examine Fig. 7.5 and identify the important quantities to be determined for the system. Since we know that the transistor is an amplifying device, we would expect some indication of how the output voltage V_o is related to the input voltage V_i—the *voltage gain*. Note in Fig. 7.5 for this configuration that $I_i = I_b$ and $I_o = I_c$, which define the *current gain* $A_i = I_o/I_i$. The input impedance Z_i, and output impedance Z_o will prove particularly important in the analysis to follow. A great deal more will be offered about these parameters in the sections to follow.

In summary, therefore, the ac equivalent of a network is obtained by:

1. *Setting all dc sources to zero and replacing them by a short-circuit equivalent*
2. *Replacing all capacitors by a short-circuit equivalent*
3. *Removing all elements bypassed by the short-circuit equivalents introduced by steps 1 and 2*
4. *Redrawing the network in a more convenient and logical form*

In the sections to follow the r_e and hybrid equivalent circuits will be introduced to complete the ac analysis of the network of Fig. 7.5.

7.4 THE IMPORTANT PARAMETERS: Z_i, Z_o, A_v, A_i

Before investigating the equivalent circuits for BJTs in some detail, let us concentrate on those parameters of a two-port system that are of paramount importance from an analysis and design viewpoint. For the two-port (two pairs of terminals) system of Fig. 7.6, the input side (the side to which the signal is normally applied) is to the left and the output side (where the load is connected) is to the right. In fact, for most electrical and electronic systems the general flow is usually from the left to the right. For both sets of terminals the impedance between each pair of terminals under normal operating conditions is quite important.

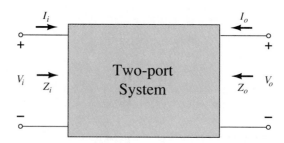

Figure 7.6 Two-port system.

Input Impedance, Z_i

For the input side, the input impedance Z_i is defined by Ohm's law as the following:

$$Z_i = \frac{V_i}{I_i} \tag{7.1}$$

If the input signal V_i is changed, the current I_i can be computed using the same level of input impedance. In other words:

For small-signal analysis, once the input impedance has been determined the same numerical value can be used for changing levels of applied signal.

In fact, we will find in the sections to follow that the input impedance of a transistor can be approximately determined by the dc biasing conditions—conditions that do not change simply because the magnitude of the applied ac signal has changed.

It is particularly noteworthy that for frequencies in the low to mid-range (typically ≤100 kHz):

The input impedance of a BJT transistor amplifier is purely resistive in nature, and depending on the manner in which the transistor is employed, can vary from a few ohms to megohms.

In addition:

An ohmmeter cannot be used to measure the small-signal ac input impedance since the ohmmeter operates in the dc mode.

Equation (7.1) is particularly useful in that it provides a method for measuring the input resistance in the ac domain. For instance, in Fig. 7.7 a sensing resistor has been added to the input side to permit a determination of I_i using Ohm's law. An oscilloscope or sensitive digital multimeter (DMM) can be used to measure the voltage V_s and V_i. Both voltages can be the peak-to-peak, peak, or rms values, as long as both levels use the same standard. The input impedance is then determined in the following manner:

$$I_i = \frac{V_s - V_i}{R_{\text{sense}}} \tag{7.2}$$

and

$$Z_i = \frac{V_i}{I_i} \tag{7.3}$$

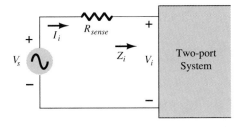

Figure 7.7 Determining Z_i.

The importance of the input impedance of a system can best be demonstrated by the network of Fig. 7.8. The signal source has an internal resistance of 600 Ω and the system (possibly a transistor amplifier) has an input resistance of 1.2 kΩ. If the source were ideal ($R_s = 0\ \Omega$), the full 10 mV would be applied to the system, but

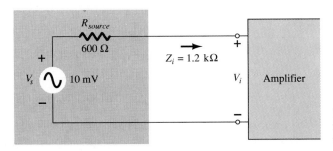

Figure 7.8 Demonstrating the impact of Z_i on an amplifier's response.

7.4 **The Important Parameters:** Z_i, Z_o, A_v, A_i

with a source impedance, the input voltage must be determined using the voltage-divider rule as follows:

$$V_i = \frac{Z_i V_s}{Z_i + R_{source}} = \frac{(1.2 \text{ k}\Omega)(10 \text{ mV})}{1.2 \text{ k}\Omega + 0.6 \text{ k}\Omega} = 6.67 \text{ mV}$$

Thus only 66.7% of the full-input signal is available at the input. If Z_i were only 600 Ω, then $V_i = \frac{1}{2}(10 \text{ mV}) = 5$ mV or 50% of the available signal. Of course, if $Z_i = 8.2$ kΩ, V_i will be 93.2% of the applied signal. The level of input impedance, therefore, can have a significant impact on the level of signal that reaches the system (or amplifier). In the sections and chapters to follow it will be demonstrated that the ac input resistance is dependent on whether the transistor is in the common-base, common-emitter, or common-collector configuration and the placement of the resistive elements.

EXAMPLE 7.1   For the system of Fig. 7.9, determine the level of input impedance.

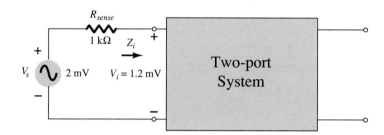

Figure 7.9 Example 7.1

Solution

$$I_i = \frac{V_s - V_i}{R_{sense}} = \frac{2 \text{ mV} - 1.2 \text{ mV}}{1 \text{ k}\Omega} = \frac{0.8 \text{ mV}}{1 \text{ k}\Omega} = 0.8 \ \mu\text{A}$$

and

$$Z_i = \frac{V_i}{I_i} = \frac{1.2 \text{ mV}}{0.8 \ \mu\text{A}} = \mathbf{1.5 \ k\Omega}$$

Output Impedance, Z_o

The output impedance is naturally defined at the output set of terminals, but the manner in which it is defined is quite different from that of the input impedance. That is:

> *The output impedance is determined at the output terminals looking back into the system with the applied signal set to zero.*

In Fig. 7.10, for example, the applied signal has been set to zero volts. To determine Z_o, a signal, V_s, is applied to the output terminals and the level of V_o is measured with an oscilloscope or sensitive DMM. The output impedance is then determined in the following manner:

$$I_o = \frac{V - V_o}{R_{sense}} \tag{7.4}$$

and

$$Z_o = \frac{V_o}{I_o} \tag{7.5}$$

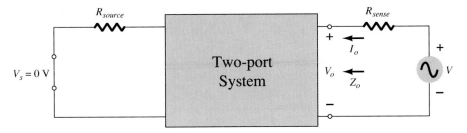

Figure 7.10 Determining Z_o.

In particular for frequencies in the low to mid-range (typically ≤ 100 kHz):

The output impedance of a BJT transistor amplifier is resistive in nature and depending on the configuration and the placement of the resistive elements, Z_o, can vary from a few ohms to a level that can exceed 2 MΩ.

In addition:

An ohmmeter cannot be used to measure the small-signal ac output impedance since the ohmmeter operates in the dc mode.

For amplifier configurations where significant gain in current is desired, the level of Z_o should be as large as possible. As demonstrated by Fig. 7.11, if $Z_o \gg R_L$, the majority of the amplifier output current will pass on to the load. It will be demonstrated in the sections and chapters to follow that Z_o is frequently so large compared to R_L that it can be replaced by an open-circuit equivalent.

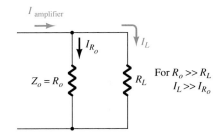

Figure 7.11 Effect of $Z_o = R_o$ on the load or output current I_L.

For the system of Fig. 7.12, determine the level of output impedance.

EXAMPLE 7.2

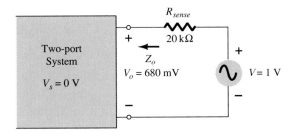

Figure 7.12 Example 7.2

Solution

$$I_o = \frac{V - V_o}{R_{\text{sense}}} = \frac{1\text{ V} - 680\text{ mV}}{20\text{ k}\Omega} = \frac{320\text{ mV}}{20\text{ k}\Omega} = 16\ \mu\text{A}$$

and

$$Z_o = \frac{V_o}{I_o} = \frac{680\text{ mV}}{16\ \mu\text{A}} = \textbf{42.5 k}\boldsymbol{\Omega}$$

Voltage Gain, A_v

One of the most important characteristics of an amplifier is the small-signal ac voltage gain as determined by

$$A_v = \frac{V_o}{V_i} \qquad (7.6)$$

7.4 **The Important Parameters: Z_i, Z_o, A_v, A_i**

For the system of Fig. 7.13, a load has not been connected to the output terminals and the level of gain determined by Eq. (7.6) is referred to as the no-load voltage gain. That is,

$$A_{v_{NL}} = \frac{V_o}{V_i}\bigg|_{R_L = \infty \ \Omega \ \text{(open-circuit)}}$$

(7.7)

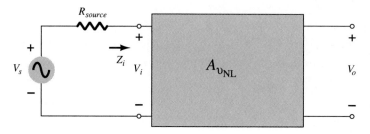

Figure 7.13 Determining the no-load voltage gain.

In Chapter 9 it will be demonstrated that:

For transistor amplifiers, the no-load voltage gain is greater than the loaded voltage gain.

For the system of Fig. 7.13, having a source resistance R_s, the level of V_i would first have to be determined using the voltage-divider rule before the gain V_o/V_s could be calculated. That is,

$$V_i = \frac{Z_i V_s}{Z_i + R_s}$$

with

$$\frac{V_i}{V_s} = \frac{Z_i}{Z_i + R_s}$$

and

$$A_{v_s} = \frac{V_o}{V_s} = \frac{V_i}{V_s} \cdot \frac{V_o}{V_i}$$

so that

$$A_{v_s} = \frac{V_o}{V_s} = \frac{Z_i}{Z_i + R_s} A_{v_{NL}}$$

(7.8)

Experimentally, the voltage gain A_{v_s} or $A_{v_{NL}}$ can be determined simply by measuring the appropriate voltage levels with an oscilloscope or sensitive DMM and substituting into the appropriate equation.

Depending on the configuration, the magnitude of the voltage gain for a loaded single-stage transistor amplifier typically ranges from just less than 1 to a few hundred. A multistage (multiunit) system, however, can have a voltage gain in the thousands.

EXAMPLE 7.3

For the BJT amplifier of Fig. 7.14, determine:
(a) V_i.
(b) I_i.
(c) Z_i.
(d) A_{v_s}.

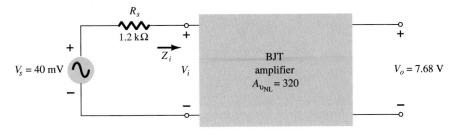

Figure 7.14 Example 7.3

Solution

(a) $A_{v_{NL}} = \dfrac{V_o}{V_i}$ and $V_i = \dfrac{V_o}{A_{v_{NL}}} = \dfrac{7.68\ \text{V}}{320} = \mathbf{24\ mV}$

(b) $I_i = \dfrac{V_s - V_i}{R_s} = \dfrac{40\ \text{mV} - 24\ \text{mV}}{1.2\ \text{k}\Omega} = \mathbf{13.33\ \mu A}$

(c) $Z_i = \dfrac{V_i}{I_i} = \dfrac{24\ \text{mV}}{13.33\ \mu\text{A}} = \mathbf{1.8\ k\Omega}$

(d) $A_{v_s} = \dfrac{Z_i}{Z_i + R_s} A_{v_{NL}}$

$= \dfrac{1.8\ \text{k}\Omega}{1.8\ \text{k}\Omega + 1.2\ \text{k}\Omega}\ (320)$

$= \mathbf{192}$

Current Gain, A_i

The last numerical characteristic to be discussed is the current gain defined by

$$A_i = \frac{I_o}{I_i} \tag{7.9}$$

Although typically the recipient of less attention than the voltage gain, it is, however, an important quantity that can have significant impact on the overall effectiveness of a design. In general:

For BJT amplifiers, the current gain typically ranges from a level just less than 1 to a level that may exceed 100.

For the loaded situation of Fig. 7.15,

$$I_i = \frac{V_i}{Z_i} \qquad \text{and} \qquad I_o = -\frac{V_o}{R_L}$$

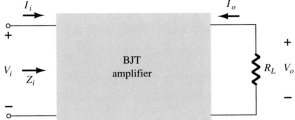

Figure 7.15 Determining the loaded current gain.

with

$$A_i = \frac{I_o}{I_i} = -\frac{V_o/R_L}{V_i/Z_i} = -\frac{V_o Z_i}{V_i R_L}$$

and

$$\boxed{A_i = -A_v \frac{Z_i}{R_L}}$$ (7.10)

Eq. (7.10) allows the determination of the current gain from the voltage gain and the impedance levels.

Phase Relationship

The phase relationship between input and output sinusoidal signals is important for a variety of practical reasons. Fortunately, however:

For the typical transistor amplifier at frequencies that permit ignoring the effects of the reactive elements, the input and output signals are either 180° out of phase or in phase.

The reason for the either–or situation will become quite clear in the chapters to follow.

Summary

The parameters of primary importance for an amplifier have now been introduced: the input impedance Z_i, the output impedance Z_o, the voltage gain A_v, the current gain A_i, and the resulting phase relationship. Other factors, such as the applied frequency at the low and high ends of the frequency spectrum, will affect some of these parameters, but this will be discussed in Chapter 11. In the sections and chapters to follow all the parameters will be determined for a variety of transistor networks to permit a comparison of the strengths and weaknesses for each configuration.

7.5 THE r_e TRANSISTOR MODEL

The r_e model employs a diode and controlled current source to duplicate the behavior of a transistor in the region of interest. Recall that a current-controlled current source is one where the parameters of the current source are controlled by a current elsewhere in the network. In fact: in general:

BJT transistor amplifiers are referred to as current-controlled devices.

Common Base Configuration

In Fig. 7.16a a common-base *pnp* transistor has been inserted within the two-port structure employed in our discussion of the last few sections. In Fig. 7.16b the r_e model for the transistor has been placed between the same four terminals. As noted in Section 7.3, the model (equivalent circuit) is chosen in such a way as to approximate the behavior of the device it is replacing in the operating region of interest. In other words, the results obtained with the model in place should be relatively close to those obtained with the actual transistor. You will recall from Chapter 3 that one junction of an operating transistor is forward-biased while the other is reverse-biased. The forward-biased junction will behave much like a diode (ignoring the effects of changing levels of V_{CE}) as verified by the curves of Fig. 3.7. For the base-to-emitter junction of the transistor of Fig. 7.16a the diode equivalence of Fig. 7.16b between the same two terminals seems to be quite appropriate. For the output side recall that the horizontal curves of Fig. 3.8 revealed that $I_c \cong I_e$ (as derived from $I_c = \alpha I_e$) for the range of values of V_{CE}. The current source of Fig. 7.16b establishes the fact that

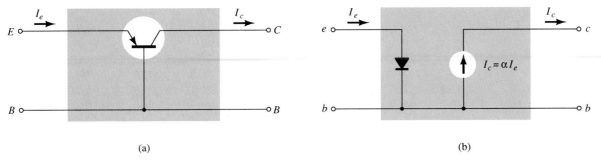

(a)   (b)

Figure 7.16 (a) Common-base BJT transistor; (b) r_e model for the configuration of Fig. 7.16a.

$I_c = \alpha I_e$ with the controlling current I_e appearing in the input side of the equivalent circuit as dictated by Fig. 7.16a. We have therefore established an equivalence at the input and output terminals with the current-controlled source, providing a link between the two—an initial review would suggest that the model of Fig. 7.16b is a valid model of the actual device.

Recall from Chapter 1 that the ac resistance of a diode can be determined by the equation $r_{ac} = 26$ mV/I_D, where I_D is the dc current through the diode at the Q (quiescent) point. This same equation can be used to find the ac resistance of the diode of Fig. 7.16b if we simply substitute the emitter current as follows:

$$r_e = \frac{26 \text{ mV}}{I_E} \qquad (7.11)$$

The subscript e of r_e was chosen to emphasize that it is the dc level of emitter current that determines the ac level of the resistance of the diode of Fig. 7.16b. Substituting the resulting value of r_e in Fig. 7.16b will result in the very useful model of Fig. 7.17.

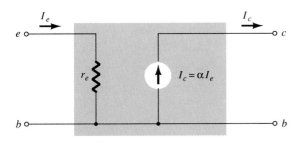

Figure 7.17 Common-base r_e equivalent circuit.

Due to the isolation that exists between input and output circuits of Fig. 7.17, it should be fairly obvious that the input impedance Z_i for the common-base configuration of a transistor is simply r_e. That is,

$$Z_i = r_e \Big|_{CB} \qquad (7.12)$$

For the common-base configuration, typical values of Z_i range from a few ohms to a maximum of about 50 Ω.

For the output impedance, if we set the signal to zero, then $I_e = 0$ A and $I_c = \alpha I_e = \alpha(0$ A$) = 0$ A, resulting in an open-circuit equivalence at the output terminals. The result is that for the model of Fig. 7.17,

$$Z_o \cong \infty \ \Omega \Big|_{CB} \qquad (7.13)$$

7.5 The r_e Transistor Model   311

In actuality:

For the common-base configuration, typical values of Z_o are in the megaohm range.

The output resistance of the common-base configuration is determined by the slope of the characteristic lines of the output characteristics as shown in Fig. 7.18. Assuming the lines to be perfectly horizontal (an excellent approximation) would result in the conclusion of Eq. (7.13). If care were taken to measure Z_o graphically or experimentally, levels typically in the range 1- to 2-MΩ would be obtained.

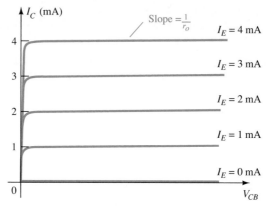

Figure 7.18 Defining Z_o.

In general, for the common-base configuration the input impedance is relatively small and the output impedance quite high.

The voltage gain will now be determined for the network of Fig. 7.19.

$$V_o = -I_o R_L = -(-I_c)R_L = \alpha I_e R_L$$

and

$$V_i = I_e Z_i = I_e r_e$$

so that

$$A_v = \frac{V_o}{V_i} = \frac{\alpha I_e R_L}{I_e r_e}$$

and

$$\boxed{A_v = \frac{\alpha R_L}{r_e} \cong \frac{R_L}{r_e}}_{CB} \tag{7.14}$$

For the current gain,

$$A_i = \frac{I_o}{I_i} = \frac{-I_c}{I_e} = -\frac{\alpha I_e}{I_e}$$

and

$$\boxed{A_i = -\alpha \cong -1}_{CB} \tag{7.15}$$

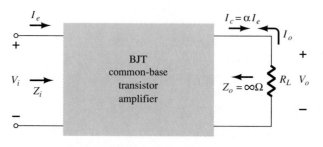

Figure 7.19 Defining $A_v = V_o/V_i$ for the common-base configuration.

The fact that the polarity of the voltage V_o as determined by the current I_c is the same as defined by Fig. 7.19 (i.e., the negative side is at ground potential) reveals that V_o and V_i are *in phase* for the common-base configuration. For an *NPN* transistor in the common-base configuration the equivalence would appear as shown in Fig. 7.20.

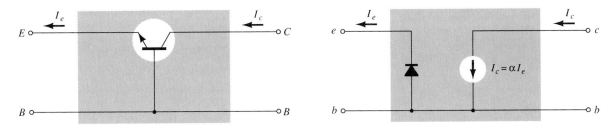

Figure 7.20 Approximate model for a common-base NPN transistor configuration.

For a common-base configuration of Fig. 7.17 with $I_E = 4$ mA, $\alpha = 0.98$, and an ac signal of 2 mV applied between the base and emitter terminals:
(a) Determine the input impedance.
(b) Calculate the voltage gain if a load of 0.56 kΩ is connected to the output terminals.
(c) Find the output impedance and current gain.

EXAMPLE 7.4

Solution

(a) $r_e = \dfrac{26 \text{ mV}}{I_E} = \dfrac{26 \text{ mV}}{4 \text{ mA}} = \mathbf{6.5\ \Omega}$

(b) $I_i = I_e = \dfrac{V_i}{Z_i} = \dfrac{2 \text{ mV}}{6.5 \text{ }\Omega} = 307.69 \text{ } \mu\text{A}$

$V_o = I_c R_L = \alpha I_e R_L = (0.98)(307.69 \text{ } \mu\text{A})(0.56 \text{ k}\Omega)$
$\quad = 168.86 \text{ mV}$

and $A_v = \dfrac{V_o}{V_i} = \dfrac{168.86 \text{ mV}}{2 \text{ mV}} = \mathbf{84.43}$

or from Eq. (7.14),

$$A_v = \frac{\alpha R_L}{r_e} = \frac{(0.98)(0.56 \text{ k}\Omega)}{6.5 \text{ }\Omega} = \mathbf{84.43}$$

(c) $Z_o \cong \infty \text{ }\Omega$

$A_i = \dfrac{I_o}{I_i} = -\alpha = \mathbf{-0.98}$ \qquad as defined by Eq. (7.15)

Common Emitter Configuration

For the common-emitter configuration of Fig. 7.21a, the input terminals are the base and emitter terminals, but the output set is now the collector and emitter terminals. In addition, the emitter terminal is now common between the input and output ports of the amplifier. Substituting the r_e equivalent circuit for the *npn* transistor will result in the configuration of Fig. 7.21b. Note that the controlled-current source is still connected between the collector and base terminals and the diode between the base and

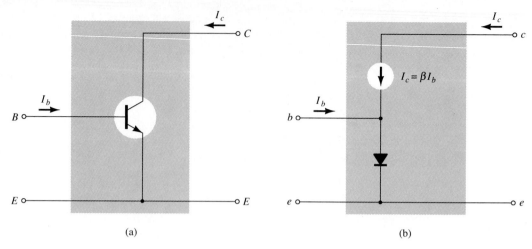

Figure 7.21 (a) Common-emitter BJT transistor; (b) approximate model for the configuration of Fig. 7.21a.

emitter terminals. In this configuration the base current is the input current, while the output current is still I_c. Recall from Chapter 3 that the base and collector currents are related by the following equation:

$$I_c = \beta I_b \qquad (7.16)$$

The current through the diode is therefore determined by

$$I_e = I_c + I_b = \beta I_b + I_b$$

and

$$I_e = (\beta + 1)I_b \qquad (7.17)$$

However, since the ac beta is typically much greater than 1, we will use the following approximation for the current analysis:

$$I_e \cong \beta I_b \qquad (7.18)$$

The input impedance is determined by the following ratio:

$$Z_i = \frac{V_i}{I_i} = \frac{V_{be}}{I_b}$$

The voltage V_{be} is across the diode resistance as shown in Fig. 7.22. The level of r_e is still determined by the dc current I_E. Using Ohm's law gives

$$V_i = V_{be} = I_e r_e \cong \beta I_b r_e$$

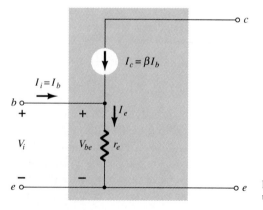

Figure 7.22 Determining Z_i using the approximate model.

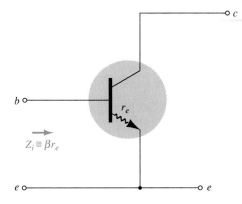

Substituting yields

$$Z_i = \frac{V_{be}}{I_b} \cong \frac{\beta I_b r_e}{I_b}$$

and

$$\boxed{Z_i \cong \beta r_e}_{CE} \qquad (7.19)$$

In essence, Eq. (7.19) states that the input impedance for a situation such as shown in Fig. 7.23 is beta times the value of r_e. In other words, a resistive element in the emitter leg is reflected into the input circuit by a multiplying factor β. For instance, if $r_e = 6.5\ \Omega$ as in Example 7.4 and $\beta = 160$ (quite typical), the input impedance has increased to a level of

$$Z_i \cong \beta r_e = (160)(6.5\ \Omega) = \textbf{1.04 k}\boldsymbol{\Omega}$$

For the common-emitter configuration, typical values of Z_i defined by βr_e range from a few hundred ohms to the kilohm range with maximums of about 6–7 kilohms

For the output impedance the characteristics of interest are the output set of Fig. 7.24. Note that the slope of the curves increases with increase in collector current. The steeper the slope, the less the level of output impedance (Z_o) The r_e model of Fig. 7.21 does not include an output impedance, but if available from a graphical analysis or from data sheets it can be included as shown in Fig. 7.25.

Figure 7.23 Impact of r_e on input impedance.

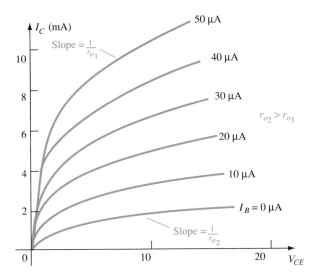

Figure 7.24 Defining r_o for the common-emitter configuration.

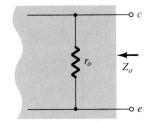

Figure 7.25 Including r_o in the transistor equivalent circuit.

For the common-emitter configuration, typical values of Z_o are in the range of 40 to 50 kΩ

For the model of Fig. 7.25, if the applied signal is set to zero, the current I_c is 0 A and the output impedance is

$$\boxed{Z_o = r_o}_{CE} \qquad (7.20)$$

Of course, if the contribution due to r_o is ignored as in the r_e model the output impedance is defined by $Z_o = \infty\ \Omega$.

The voltage gain for the common-emitter configuration will now be determined for the configuration of Fig. 7.26 using the assumption that $Z_o = \infty\ \Omega$. The effect of including r_o will be considered in Chapter 8. For the defined direction of I_o and polarity of V_o,

$$V_o = -I_o R_L$$

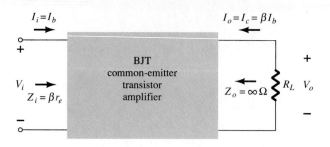

Figure 7.26 Determining the voltage and current gain for the common-emitter transistor amplifier.

The minus sign simply reflects the fact that the direction of I_o in Fig. 7.26 would establish a voltage V_o with the opposite polarity. Continuing gives

$$V_o = -I_o R_L = -I_c R_L = -\beta I_b R_L$$

and

$$V_i = I_i Z_i = I_b \beta r_e$$

so that

$$A_v = \frac{V_o}{V_i} = -\frac{\beta I_b R_L}{I_b \beta r_e}$$

and

$$\boxed{A_v = -\frac{R_L}{r_e}}_{CE, r_o = \infty \Omega}$$

(7.21)

The resulting minus sign for the voltage gain reveals that the output and input voltages are 180° out of phase.

The current gain for the configuration of Fig. 7.26:

$$A_i = \frac{I_o}{I_i} = \frac{I_c}{I_b} = \frac{\beta I_b}{I_b}$$

and

$$\boxed{A_i = \beta}_{CE, r_o = \infty \Omega}$$

(7.22)

Using the fact that the input impedance is βr_e, that the collector current is βI_b, and that the output impedance is r_o, the equivalent model of Fig. 7.27 can be an effective tool in the analysis to follow. For typical parameter values the common-emitter configuration can be considered one that has a moderate level of input impedance, a high voltage and current gain, and an output impedance that may have to be included in the network analysis.

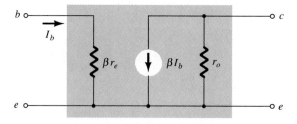

Figure 7.27 r_e model for the common-emitter transistor configuration.

EXAMPLE 7.5

Given $\beta = 120$ and $I_E = 3.2$ mA for a common-emitter configuration with $r_o = \infty \ \Omega$, determine:
(a) Z_i.
(b) A_i.
(c) A_v if a load of 2 kΩ is applied.

Solution

(a) $r_e = \dfrac{26 \text{ mV}}{I_E} = \dfrac{26 \text{ mV}}{3.2 \text{ mA}} = 8.125 \; \Omega$

and $Z_i = \beta r_e = (120)(8.125 \; \Omega) = \mathbf{975 \; \Omega}$

(b) $A_i = \dfrac{I_o}{I_i} = \beta = \mathbf{120}$

(c) Eq. (7.21): $A_v = -\dfrac{R_L}{r_e} = -\dfrac{2 \text{ k}\Omega}{8.125 \; \Omega} = \mathbf{-246.15}$

Common Collector Configuration

For the common-collector configuration the model defined for the common-emitter configuration of Fig. 7.21 is normally applied rather than defining a model for the common-collector configuration. In subsequent chapters a number of common-collector configurations will be investigated and the impact of using the same model will become quite apparent.

7.6   THE HYBRID EQUIVALENT MODEL

It was pointed out in Section 7.5 that the r_e model for a transistor is sensitive to the dc level of operation of the amplifier. The result is an input resistance that will vary with the dc operating point. For the hybrid equivalent model to be described in this section the parameters are defined at an operating point that may or may not reflect the actual operating conditions of the amplifier. This is due to the fact that specification sheets cannot provide parameters for an equivalent circuit at every possible operating point. They must choose operating conditions that they believe reflect the general characteristics of the device. The hybrid parameters as shown in Fig. 7.28 are drawn from the specification sheet for the 2N4400 transistor described in Chapter 3. The values are provided at a dc collector current of 1 mA and a collector-to-emitter voltage of 10 V. In addition, a range of values is provided for each parameter for guidance in the initial design or analysis of a system. One obvious advantage of the specification sheet listing is the immediate knowledge of typical levels for the parameters of the device as compared to other transistors.

The quantities h_{ie}, h_{re}, h_{fe}, and h_{oe} of Fig. 7.28 are called the hybrid parameters and are the components of a small-signal equivalent circuit to be described shortly. For years, the hybrid model with all its parameters was the chosen model for the educational and industrial communities. Presently, however, the r_e model is applied more frequently, but often with the h_{oe} parameter of the hybrid equivalent model to

		Min.	Max.	
Input impedance ($I_C = 1$ mA dc, $V_{CE} = 10$ V dc, $f = 1$ kHz) 2N4400	h_{ie}	0.5	7.5	k Ω
Voltage feedback ratio ($I_C = 1$ mA dc, $V_{CE} = 10$ V dc, $f = 1$ kHz)	h_{re}	0.1	8.0	$\times 10^{-4}$
Small-signal current gain ($I_C = 1$ mA dc, $V_{CE} = 10$ V dc, $f = 1$ kHz) 2N4400	h_{fe}	20	250	—
Output admittance ($I_C = 1$ mA dc, $V_{CE} = 10$ V dc, $f = 1$ kHz)	h_{oe}	1.0	30	1 μS

Figure 7.28 Hybrid parameters for the 2N4400 transistor.

provide some measure for the output impedance. Since specification sheets do provide the hybrid parameters and the hybrid model continues to receive a good measure of attention, it is quite important that the hybrid model be covered in some detail in this book. Once developed, the similarities between the r_e and hybrid models will be quite apparent. In fact, once the components of one are defined for a particular operating point, the parameters of the other model are immediately available.

Our description of the hybrid equivalent model will begin with the general two-port system of Fig. 7.29. The following set of equations (7.23) is only one of a number of ways in which the four variables of Fig. 7.29 can be related. It is the most frequently employed in transistor circuit analysis, however, and therefore is discussed in detail in this chapter.

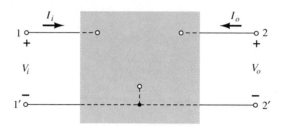

Figure 7.29 Two-port system.

$$V_i = h_{11}I_i + h_{12}V_o \qquad (7.23a)$$

$$I_o = h_{21}I_i + h_{22}V_o \qquad (7.23b)$$

The parameters relating the four variables are called *h-parameters* from the word "hybrid." The term "hybrid" was chosen because the mixture of variables (V and I) in each equation results in a "hybrid" set of units of measurement for the *h*-parameters. A clearer understanding of what the various *h*-parameters represent and how we can determine their magnitude can be developed by isolating each and examining the resulting relationship.

If we arbitrarily set $V_o = 0$ (short circuit the output terminals), and solve for h_{11} in Eq. (7.23a), the following will result:

$$h_{11} = \left. \frac{V_i}{I_i} \right|_{V_o=0} \qquad \text{ohms} \qquad (7.24)$$

The ratio indicates that the parameter h_{11} is an impedance parameter with the units of ohms. Since it is the ratio of the *input* voltage to the *input* current with the output terminals *shorted*, it is called the *short-circuit input-impedance parameter*. The subscript 11 of h_{11} defines the facts that the parameter is determined by a ratio of quantities measured at the input terminals.

If I_i is set equal to zero by opening the input leads, the following will result for h_{12}:

$$h_{12} = \left. \frac{V_i}{V_o} \right|_{I_i=0} \qquad \text{unitless} \qquad (7.25)$$

The parameter h_{12}, therefore, is the ratio of the input voltage to the output voltage with the input current equal to zero. It has no units since it is a ratio of voltage levels and is called the *open-circuit reverse transfer voltage ratio parameter*. The subscript 12 of h_{12} reveals that the parameter is a transfer quantity determined by a ratio of input to output measurements. The first integer of the subscript defines the measured quantity to appear in the numerator; the second integer defines the source of the quantity to appear in the denominator. The term *reverse* is included because the ratio is an input voltage over an output voltage rather than the reverse ratio typically of interest.

If in Eq. (7.23b), V_o is equal to zero by again shorting the output terminals, the following will result for h_{21}:

$$h_{21} = \frac{I_o}{I_i}\bigg|_{V_o=0} \quad \text{unitless} \tag{7.26}$$

Note that we now have the ratio of an output quantity to an input quantity. The term *forward* will now be used rather than *reverse* as indicated for h_{12}. The parameter h_{21} is the ratio of the output current to the input current with the output terminals shorted. This parameter, like h_{12}, has no units since it is the ratio of current levels. It is formally called the *short-circuit forward transfer current ratio parameter*. The subscript 21 again indicates that it is a transfer parameter with the output quantity in the numerator and the input quantity in the denominator.

The last parameter, h_{22}, can be found by again opening the input leads to set $I_1 = 0$ and solving for h_{22} in Eq. (7.23b):

$$h_{22} = \frac{I_o}{V_o}\bigg|_{I_i=0} \quad \text{siemens} \tag{7.27}$$

Since it is the ratio of the output current to the output voltage, it is the output conductance parameter and is measured in siemens (S). It is called the *open-circuit output admittance parameter*. The subscript 22 reveals that it is determined by a ratio of output quantities.

Since each term of Eq. (7.23a) has the unit volt, let us apply Kirchhoff's voltage law "in reverse" to find a circuit that "fits" the equation. Performing this operation will result in the circuit of Fig. 7.30. Since the parameter h_{11} has the unit ohm, it is represented by a resistor in Fig. 7.30. The quantity h_{12} is dimensionless and therefore simply appears as a multiplying factor of the "feedback" term in the input circuit.

Since each term of Eq. (7.23b) has the units of current, let us now apply Kirchhoff's current law "in reverse" to obtain the circuit of Fig. 7.31. Since h_{22} has the units of admittance, which for the transistor model is conductance, it is represented by the resistor symbol. Keep in mind, however, that the resistance in ohms of this resistor is equal to the reciprocal of conductance ($1/h_{22}$).

The complete "ac" equivalent circuit for the basic three-terminal linear device is indicated in Fig. 7.32 with a new set of subscripts for the h-parameters. The notation of Fig. 7.32 is of a more practical nature since it relates the h-parameters to the resulting ratio obtained in the last few paragraphs. The choice of letters is obvious from the following listing:

$$h_{11} \rightarrow \text{input resistance} \rightarrow h_i$$

$$h_{12} \rightarrow \text{reverse transfer voltage ratio} \rightarrow h_r$$

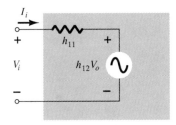

Figure 7.30 Hybrid input equivalent circuit.

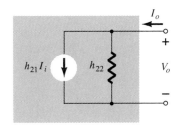

Figure 7.31 Hybrid output equivalent circuit.

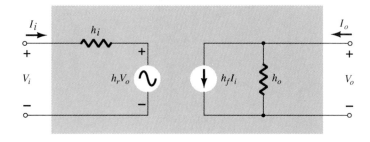

Figure 7.32 Complete hybrid equivalent circuit.

$$h_{21} \rightarrow \textit{forward transfer current ratio} \rightarrow h_f$$

$$h_{22} \rightarrow \textit{output conductance} \rightarrow h_o$$

The circuit of Fig. 7.32 is applicable to any linear three-terminal electronic device or system with no internal independent sources. For the transistor, therefore, even though it has three basic configurations, *they are all three-terminal configurations,* so that the resulting equivalent circuit will have the same format as shown in Fig. 7.32. In each case the bottom of the input and output sections of the network of Fig. 7.32 can be connected as shown in Fig. 7.33, since the potential level is the same. Essentially, therefore, the transistor model is a three-terminal two-port system. The *h*-parameters, however, will change with each configuration. To distinguish which parameter has been used or which is available, a second subscript has been added to the *h*-parameter notation. For the common-base configuration the lowercase letter *b* was added, while for the common-emitter and common-collector configurations the letters *e* and *c* were added, respectively. The hybrid equivalent network for the common-emitter configuration appears with the standard notation in Fig. 7.33. Note that $I_i = I_b$, $I_o = I_c$, and through an application of Kirchhoff's current law, $I_e = I_b + I_c$. The input voltage is now V_{be} with the output voltage V_{ce}. For the common-base configuration of Fig. 7.34, $I_i = I_e$, $I_o = I_c$ with $V_{eb} = V_i$ and $V_{cb} = V_o$. The networks of Figs. 7.33 and 7.34 are applicable for *pnp* or *npn* transistors.

The fact that both a Thévenin and Norton circuit appear in the circuit of Fig. 7.32 was further impetus for calling the resultant circuit a *hybrid* equivalent circuit. Two additional transistor equivalent circuits, not to be discussed in this text, called the

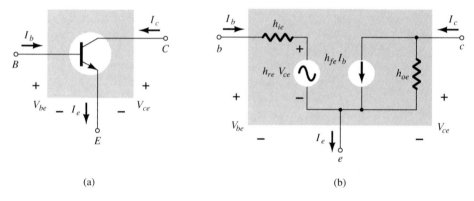

(a) (b)

Figure 7.33 Common-emitter configuration: (a) graphic symbol; (b) hybrid equivalent circuit.

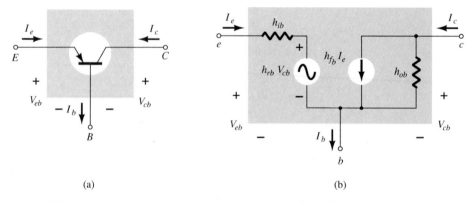

(a) (b)

Figure 7.34 Common-base configuration: (a) graphic symbol; (b) hybrid equivalent circuit.

Chapter 7 BJT Transistor Modeling

z-parameter and y-parameter equivalent circuits, use either the voltage source or the current source but not both in the same equivalent circuit. In Section 7.7 the magnitude of the various parameters will be found from the transistor characteristics in the region of operation resulting in the desired *small-signal equivalent network* for the transistor.

For the common-emitter and common-base configurations the magnitude of h_r and h_o is often such that the results obtained for the important parameters such as Z_i, Z_o, A_v and A_i are only slightly affected if they (h_r and h_o) are not included in the model.

Since h_r is normally a relatively small quantity, its removal is approximated by $h_r \cong 0$ and $h_r V_o = 0$, resulting in a short-circuit equivalent for the feedback element as shown in Fig. 7.35. The resistance determined by $1/h_o$ is often large enough to be ignored in comparison to a parallel load permitting its replacement by an open-circuit equivalent for the CE and CB models, as shown in Fig. 7.35.

The resulting equivalent of Fig. 7.36 is quite similar to the general structure of the common-base and common-emitter equivalent circuits obtained with the r_e model. In fact, the hybrid equivalent and the r_e models for each configuration have been repeated in Fig. 7.37 for comparison. It should be reasonably clear from Fig. 7.37a that

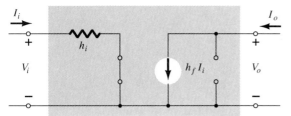

Figure 7.35 Effect of removing h_{rc} and h_{oe} from the hybrid equivalent circuit.

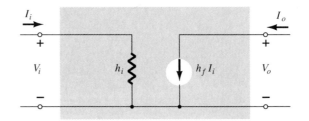

Figure 7.36 Approximate hybrid equivalent model.

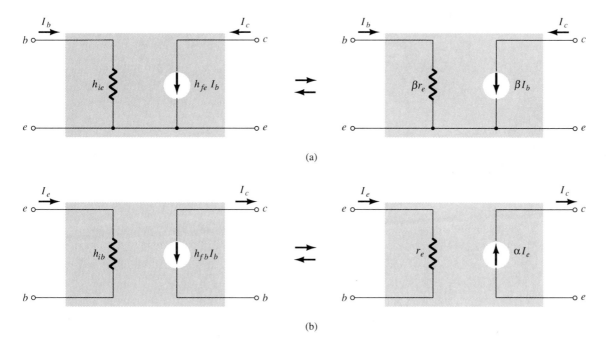

(a)

(b)

Figure 7.37 Hybrid versus r_e model: (a) common-emitter configuration; (b) common-base configuration.

7.6 **The Hybrid Equivalent Model**

$$\boxed{h_{ie} = \beta r_e} \qquad (7.28)$$

and
$$\boxed{h_{fe} = \beta_{ac}} \qquad (7.29)$$

From Fig. 7.37b,

$$\boxed{h_{ib} = r_e} \qquad (7.30)$$

and
$$\boxed{h_{fb} = -\alpha \cong -1} \qquad (7.31)$$

In particular note that the minus sign in Eq. (7.31) accounts for the fact that the current source of the standard hybrid equivalent circuit is pointing down rather than in the actual direction as shown in the r_e model of Fig. 7.37b.

EXAMPLE 7.6

Given $I_E = 2.5$ mA, $h_{fe} = 140$, $h_{oe} = 20$ μS (μmho) and $h_{ob} = 0.5$ μS, determine:
(a) The common-emitter hybrid equivalent circuit.
(b) The common-base r_e model.

Solution

(a) $r_e = \dfrac{26 \, mV}{I_E} = \dfrac{26 \text{ mV}}{2.5 \text{ mA}} = 10.4 \, \Omega$

$h_{ie} = \beta r_e = (140)(10.4 \, \Omega) = \mathbf{1.456 \ k\Omega}$

$r_o = \dfrac{1}{h_{oe}} = \dfrac{1}{20 \, \mu S} = 50 \text{ k}\Omega$

Note Fig. 7.38.

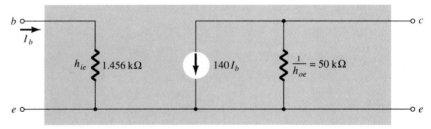

Figure 7.38 Common-emitter hybrid equivalent circuit for the parameters of example 7.6.

(b) $r_e = \mathbf{10.4 \ \Omega}$

$\alpha \cong 1, \qquad r_o = \dfrac{1}{h_{ob}} = \dfrac{1}{0.5 \, \mu S} = \mathbf{2 \ M\Omega}$

Note Fig. 7.39.

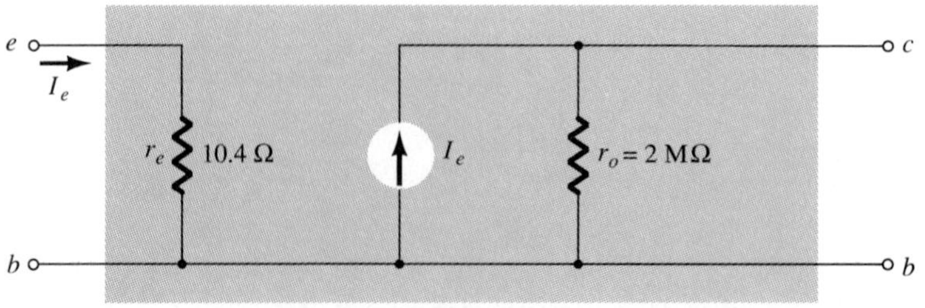

Figure 7.39 Common-base r_e model for the parameters of example 7.6.

Chapter 7   BJT Transistor Modeling

A series of equations relating the parameters of each configuration for the hybrid equivalent circuit is provided in Appendix A. In Section 7.8 we demonstrate that the hybrid parameter h_{fe} (β_{ac}) is the least sensitive of the hybrid parameters to a change in collector current. Assuming, therefore, that $h_{fe} = \beta$ is a constant for the range of interest is a fairly good approximation. It is $h_{ie} = \beta r_e$ that will vary significantly with I_C and should be determined at operating levels, since it can have a real impact on the gain levels of a transistor amplifier.

7.7 GRAPHICAL DETERMINATION OF THE h-PARAMETERS

Using partial derivatives (calculus), it can be shown that the magnitude of the h-parameters for the small-signal transistor equivalent circuit in the region of operation for the common-emitter configuration can be found using the following equations:*

$$h_{ie} = \frac{\partial v_i}{\partial i_i} = \frac{\partial v_{be}}{\partial i_b} \cong \frac{\Delta v_{be}}{\Delta i_b}\bigg|_{V_{CE}=\text{constant}} \qquad \text{(ohms)} \qquad (7.32)$$

$$h_{re} = \frac{\partial v_i}{\partial v_o} = \frac{\partial v_{be}}{\partial v_{ce}} \cong \frac{\Delta v_{be}}{\Delta v_{ce}}\bigg|_{I_B=\text{constant}} \qquad \text{(unitless)} \qquad (7.33)$$

$$h_{fe} = \frac{\partial i_o}{\partial i_i} = \frac{\partial i_c}{\partial i_b} \cong \frac{\Delta i_c}{\Delta i_b}\bigg|_{V_{CE}=\text{constant}} \qquad \text{(unitless)} \qquad (7.34)$$

$$h_{oe} = \frac{\partial i_o}{\partial v_o} = \frac{\partial i_c}{\partial v_{ce}} \cong \frac{\Delta i_c}{\Delta v_{ce}}\bigg|_{I_B=\text{constant}} \qquad \text{(siemens)} \qquad (7.35)$$

In each case the symbol Δ refers to a small change in that quantity around the quiescent point of operation. In other words, the h-parameters are determined in the region of operation for the applied signal, so that the equivalent circuit will be the most accurate available. The constant values of V_{CE} and I_B in each case refer to a condition that must be met when the various parameters are determined from the characteristics of the transistor. For the common-base and common-collector configurations the proper equation can be obtained by simply substituting the proper values of v_i, v_o, i_i, and i_o.

The parameters h_{ie} and h_{re} are determined from the input or base characteristics, while the parameters h_{fe} and h_{oe} are obtained from the output or collector characteristics. Since h_{fe} is usually the parameter of greatest interest, we shall discuss the operations involved with equations, such as Eqs. (7.32) through (7.35), for this parameter first. The first step in determining any of the four hybrid parameters is to find the quiescent point of operation as indicated in Fig. 7.40. In Eq. (7.34) the condition $V_{CE} = $ constant requires that the changes in base current and collector current be taken along a vertical straight line drawn through the Q-point representing a fixed collector-to-emitter voltage. Equation (7.34) then requires that a small change in collector current be divided by the corresponding change in base current. For the greatest accuracy these changes should be made as small as possible.

*The partial derivative $\partial v_i/\partial i_i$ provides a measure of the instantaneous change in v_i due to an instantaneous change in i_i.

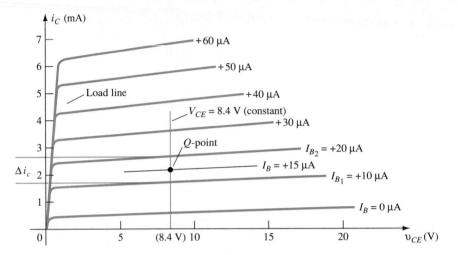

Figure 7.40 h_{fe} determination.

In Fig. 7.40 the change in i_b was chosen to extend from I_{B_1} to I_{B_2} along the perpendicular straight line at V_{CE}. The corresponding change in i_c is then found by drawing the horizontal lines from the intersections of I_{B_1} and I_{B_2} with V_{CE} = constant to the vertical axis. All that remains is to substitute the resultant changes of i_b and i_c into Eq. (7.34); that is,

$$|h_{fe}| = \left.\frac{\Delta i_c}{\Delta i_b}\right|_{V_{CE}=\text{constant}} = \left.\frac{(2.7 - 1.7)\text{ mA}}{(20 - 10)\ \mu A}\right|_{V_{CE}=8.4\text{ V}}$$

$$= \frac{10^{-3}}{10 \times 10^{-6}} = \mathbf{100}$$

In Fig. 7.41 a straight line is drawn tangent to the curve I_B through the Q-point to establish a line I_B = constant as required by Eq. (7.35) for h_{oe}. A change in v_{CE} was then chosen and the corresponding change in i_C is determined by drawing the horizontal lines to the vertical axis at the intersections on the I_B = constant line. Substituting into Eq. (7.35), we get

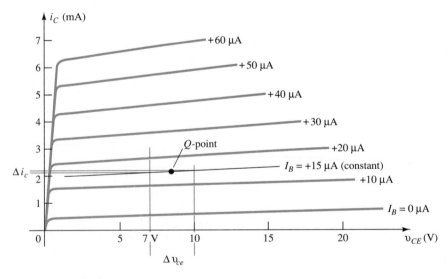

Figure 7.41 h_{oe} determination.

$$|h_{oe}| = \frac{\Delta i_c}{\Delta v_{ce}}\bigg|_{I_B=\text{constant}} = \frac{(2.2 - 2.1)\ \text{mA}}{(10 - 7)\ \text{V}}\bigg|_{I_B=+15\ \mu A}$$

$$= \frac{0.1 \times 10^{-3}}{3} = \mathbf{33\ \mu A/V = 33 \times 10^{-6}\ S = 33\ \mu S}$$

To determine the parameters h_{ie} and h_{re} the Q-point must first be found on the input or base characteristics as indicated in Fig. 7.42. For h_{ie}, a line is drawn tangent to the curve $V_{CE} = 8.4$ V through the Q-point, to establish a line $V_{CE} =$ constant as required by Eq. (7.32). A small change in v_{be} was then chosen, resulting in a corresponding change in i_b. Substituting into Eq. (7.32), we get

$$|h_{ie}| = \frac{\Delta v_{be}}{\Delta i_b}\bigg|_{V_{CE}=\text{constant}} = \frac{(733 - 718)\ \text{mV}}{(20 - 10)\ \mu A}\bigg|_{V_{CE}=8.4\ V}$$

$$= \frac{15 \times 10^{-3}}{10 \times 10^{-6}} = \mathbf{1.5\ k\Omega}$$

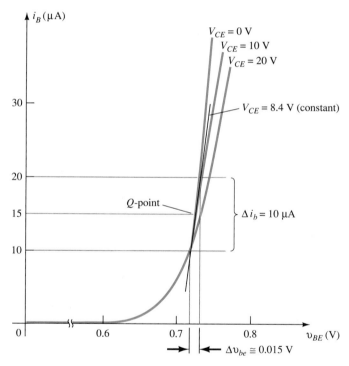

Figure 7.42 h_{ie} determination.

The last parameter, h_{re}, can be found by first drawing a horizontal line through the Q-point at $I_B = 15\ \mu A$. The natural choice then is to pick a change in v_{CE} and find the resulting change in v_{BE} as shown in Fig. 7.43.

Substituting into Eq. (7.33), we get

$$|h_{re}| = \frac{\Delta v_{be}}{\Delta v_{ce}}\bigg|_{I_B=\text{constant}} = \frac{(733 - 725)\ \text{mV}}{(20 - 0)\ \text{V}} = \frac{8 \times 10^{-3}}{20} = \mathbf{4 \times 10^{-4}}$$

For the transistor whose characteristics have appeared in Figs. 7.40 through 7.43 the resulting hybrid small-signal equivalent circuit is shown in Fig. 7.44.

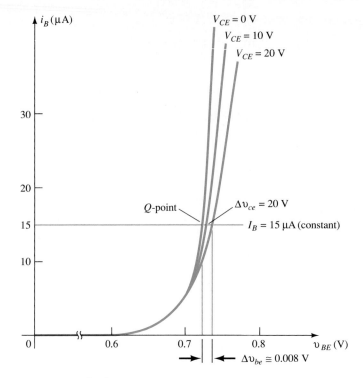

Figure 7.43 h_{re} determination.

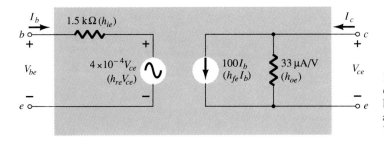

Figure 7.44 Complete hybrid equivalent circuit for a transistor having the characteristics that appear in Figs. 7.40 through 7.43.

As mentioned earlier, the hybrid parameters for the common-base and common-collector configurations can be found using the same basic equations with the proper variables and characteristics.

Typical values for each parameter for the broad range of transistors available today in each of its three configurations are provided in Table 7.1. The minus sign indicates that in Eq. (7.34) as one quantity increased in magnitude, within the change chosen, the other decreased in magnitude.

TABLE 7.1 Typical Parameter Values for the CE, CC, and CB Transistor Configurations

Parameter	CE	CC	CB
h_i	1 kΩ	1 kΩ	20 Ω
h_r	2.5×10^{-4}	$\cong 1$	3.0×10^{-4}
h_f	50	−50	−0.98
h_o	25 μA/V	25 μA/V	0.5 μA/V
$1/h_o$	40 kΩ	40 kΩ	2 MΩ

Note in retrospect (Section 3.5: Transistor Amplifying Action) that the input resistance of the common-base configuration is low, while the output resistance is high. Consider also that the short-circuit current gain is very close to 1. For the common-emitter and common-collector configurations note that the input resistance is much higher than that of the common-base configuration and that the ratio of output to input resistance is about 40:1. Consider also for the common-emitter and common-base configuration that h_r is very small in magnitude. Transistors are available today with values of h_{fe} that vary from 20 to 600. For any transistor the region of operation and conditions under which it is being used will have an effect on the various h-parameters. The effect of temperature and collector current and voltage on the h-parameters is discussed in Section 7.8.

7.8 VARIATIONS OF TRANSISTOR PARAMETERS

There are a large number of curves that can be drawn to show the variations of the h-parameters with temperature, frequency, voltage, and current. The most interesting and useful at this stage of the development include the h-parameter variations with junction temperature and collector voltage and current.

In Fig. 7.45 the effect of the collector current on the h-parameter has been indicated. Take careful note of the logarithmic scale on the vertical and horizontal axes. Logarithmic scales will be examined in Chapter 11. The parameters have all been normalized to unity so that the relative change in magnitude with collector current can easily be determined. On every set of curves, such as in Fig. 7.46, the operating point at which the parameters were found is always indicated. For this particular situation the quiescent point is at the intersection of $V_{CE} = 5.0$ V and $I_C = 1.0$ mA. Since the frequency and temperature of operation will also affect the h-parameters, these quantities are also indicated on the curves. At 0.1 mA, h_{fe} is about 0.5 or 50% of its value at 1.0 mA, while at 3 mA, it is 1.5 or 150% of that value. In other words, if $h_{fe} = 50$ at $I_C = 1.0$ mA, h_{fe} has changed from a value of 0.5(50) = 25 to 1.5(50) = 75 with a

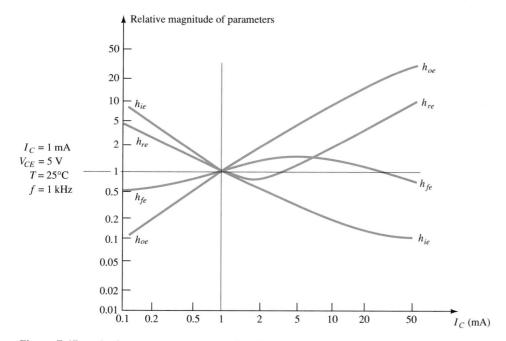

Figure 7.45 Hybrid parameter variations with collector current.

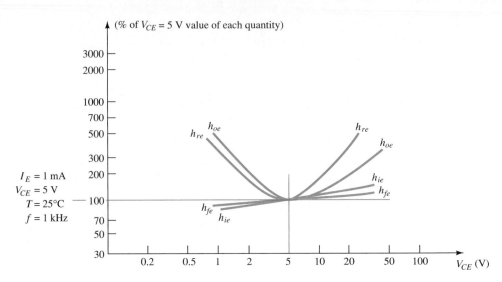

Figure 7.46 Hybrid parameter variations with collector–emitter potential.

change of I_C from 0.1 mA to 3 mA. Consider, however, the point of operation at $I_C = 50$ mA. The magnitude of h_{re} is now approximately 11 times that at the defined Q-point, a magnitude that may not permit eliminating this parameter from the equivalent circuit. The parameter h_{oe} is approximately 35 times the normalized value. This increase in h_{oe} will decrease the magnitude of the output resistance of the transistor to a point where it may approach the magnitude of the load resistor. There would then be no justification in eliminating h_{oe} from the equivalent circuit on an approximate basis.

In Fig. 7.46 the variation in magnitude of the h-parameters on a normalized basis has been indicated with change in collector voltage. This set of curves was normalized at the same operating point of the transistor discussed in Fig. 7.45 so that a comparison between the two sets of curves can be made. Note that h_{ie} and h_{fe} are relatively steady in magnitude, while h_{oe} and h_{re} are much larger to the left and right of the chosen operating point. In other words, h_{oe} and h_{re} are much more sensitive to changes in collector voltage than are h_{ie} and h_{fe}.

It is interesting to note from Figs. 7.45 and 7.46 that the value of h_{fe} appears to change the least. Therefore, the specific value of current gain, whether h_{fe} or β, can, on an approximate and relative basis, be considered fairly constant for the range of collector current and voltage.

The value of $h_{ie} = \beta r_e$ does vary considerably with collector current as one might expect due to the sensitivity of r_e to emitter ($I_E \cong I_C$) current. It is therefore a quantity that should be determined as close to operating conditions as possible. For values below the specified V_{CE}, h_{re} is fairly constant, but it does increase measurably for higher values. It is indeed fortunate that for most applications the magnitude of h_{re} and h_{oe} are such that they can usually be ignored. They are quite sensitive to collector current and collector-to-emitter voltage.

In Fig. 7.47 the variation in h-parameters has been plotted for changes in junction temperature. The normalization value is taken to be room temperature: $T = 25°C$. The horizontal scale is a linear scale rather than a logarithmic scale as was employed for Figs. 7.45 and 7.46. In general, all the parameters increase in magnitude with temperature. The parameter least affected, however, is h_{oe}, while the input impedance h_{ie} changes at the greatest rate. The fact that h_{fe} will change from 50% of its normalized value at $-50°C$ to 150% of its normalized value at $+150°C$ indicates clearly that the operating temperature must be carefully considered in the design of transistor circuits.

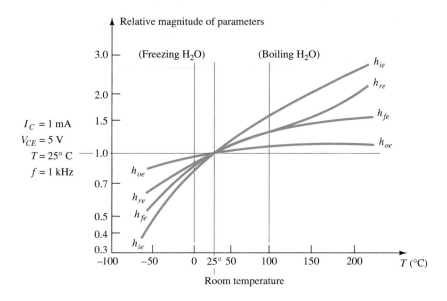

Figure 7.47 Hybrid parameter variations with temperature.

7.9 COMPUTER ANALYSIS

The appearance of a current-controlled current source (CCCS) in the equivalent models of a transistor requires that the PSpice format for such a source be introduced. The format is initialized by the letter F as shown below:

$$\underbrace{\text{FBJT}}_{\text{name}} \quad \underbrace{\underset{(+N)}{3} \quad \underset{(-N)}{2}}_{\substack{\text{current-controlled} \\ \text{source}}} \quad \underbrace{\text{VSENSE}}_{\substack{\text{name of} \\ \text{controlling} \\ \text{voltage source}}} \quad \underbrace{0.98}_{\substack{\text{magnitude of} \\ \text{multiplier for} \\ \text{controlled source}}}$$

The name (up to seven characters) assigned to the controlled source is followed by the positive and negative nodes for the source. The letter V must appear before the name of the dc voltage source establishing the direction of the controlling current. The voltage source must be in the same series circuit as the controlling current and polarized such as to establish a current in the *opposite* direction as the controlling current. The opposite direction is required because in PSpice the current of an independent voltage source is defined to have a direction opposite to the applied pressure of the source. Its magnitude is 0 V if its sole purpose is to establish the direction of the controlling current. The last factor of the format is the multiplying factor for the controlled-current source. Since the defining voltage source must be part of the network to appear in the input file, a separate line must define the name, polarity, and magnitude of the dc source.

For the common-base transistor configuration the model of Fig. 7.48 will be employed. For the common-emitter transistor configuration the model of Fig. 7.49 will be employed.

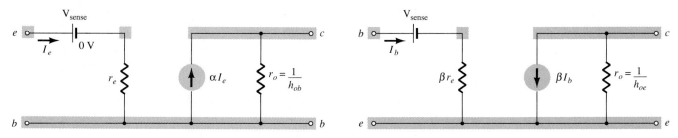

Figure 7.48 PSpice common-base model.

Figure 7.49 PSpice common-emitter model.

EXAMPLE 7.7

Write the input file for the common-emitter amplifier of Fig. 7.50 and request the magnitude and phase angle of the output voltage $\mathbf{V}_o$.

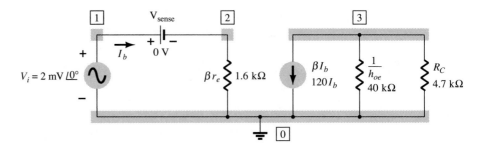

Figure 7.50 Example 7.7

Solution

The input file for the network of Fig. 7.50 appears in Fig. 7.51. The first two lines describe the two sources of the network with the angle 0° not included in the description of the ac source since it is the default value if not specified. The input impedance βr_e is defined on the third line and the controlled-current source on the next line. Compare the description of the controlled-current source to the description of CCCS sources above. The output impedance is 40 kΩ between terminals 3 and 0. The resistor R_C is the collector resistance of the design. The chosen frequency for the ac analysis (a frequency must be specified) is 1 kHz and the next line requests the magnitude and phase angle of the output voltage $\mathbf{V}_o$. Recall that the .OPTIONS NOPAGE command eliminates some of the superfluous material in the output file.

```
**** 06/13/89 ******* Evaluation PSpice (January 1989) ******* 19:49:57 ****

  Common-emitter amplifier of Fig. 7.50

****        CIRCUIT DESCRIPTION

*******************************************************************************

VI 1 0 AC 2MV
VSENSE 1 2 0
RBRE 2 0 1.6K
FBETA 3 0 VSENSE 120
RO 3 0 40K
RC 3 0 4.7K
.AC LIN 1 1K 1K
.PRINT AC VM(3,0) VP(3,0)
.OPTIONS NOPAGE
.END


****        SMALL SIGNAL BIAS SOLUTION        TEMPERATURE =   27.000 DEG C
  NODE    VOLTAGE        NODE    VOLTAGE       NODE    VOLTAGE      NODE    VOLTAGE
(    1)    0.0000  (     2)     0.0000   (    3)      0.0000


      VOLTAGE SOURCE CURRENTS
      NAME            CURRENT
      VI            0.000E+00
      VSENSE        0.000E+00
      TOTAL POWER DISSIPATION      0.00E+00   WATTS


****        AC ANALYSIS                        TEMPERATURE =   27.000 DEG C
   FREQ          VM(3,0)        VP(3,0)
   1.000E+03     6.309E-01      1.800E+02
```

Figure 7.51 PSpice analysis of the CE network of Fig. 7.50.

The results indicate that the magnitude of the output voltage is 630.9 mV, resulting in a no-load gain of

$$A_{v_{NL}} = \left| \frac{V_o}{V_i} \right| = \frac{630.9 \text{ mV}}{2 \text{ mV}} = 315.45$$

a level that will drop with an applied load. The results also indicate a 180° phase shift between $\mathbf{V}_o$ and $\mathbf{V}_i$ as expected for the common-emitter configuration.

PROBLEMS

§ 7.2 Amplification in the AC Domain

1. (a) What is the expected amplification of a BJT transistor amplifier if the dc supply is set to zero volts?
   (b) What will happen to the output ac signal if the dc level is insufficient? Sketch the effect on the waveform.
   (c) What is the conversion efficiency of an amplifier in which the effective value of the current through a 2.2-kΩ load is 5 mA and the drain on the 18-V dc supply is 3.8 mA?

2. Can you think of an analogy that would explain the importance of the dc level on the resulting ac gain?

§ 7.3 BJT Transistor Modeling

3. What is the reactance of a 10-μF capacitor at a frequency of 1 kHz? For networks in which the resistor levels are typically in the kilohm range, is it a good assumption to use the short-circuit equivalence for the conditions just described? How about at 100 kHz?

4. Given the common-base configuration of Fig. 7.52, sketch the ac equivalent using the notation for the transistor model appearing in Fig. 7.5.

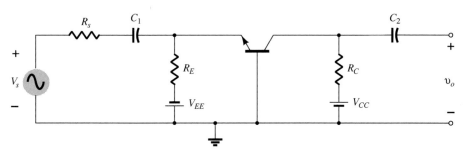

Figure 7.52 Problem 4

5. (a) Describe the differences between the r_e and hybrid equivalent models for a BJT transistor.
   (b) For each model, list the conditions under which it should be applied.

§ 7.4 The Important Parameters: Z_i, Z_o, A_v, A_i

6. (a) For the configuration of Fig. 7.7, determine Z_i if $V_s = 40$ mV, $R_{sense} = 0.5$ kΩ, and $I_i = 20$ μA.
   (b) Using the results of part (a), determine V_i if the applied source is changed to 12 mV with an internal resistance of 0.4 kΩ.

7. (a) For the network of Fig. 7.10, determine Z_o if $V = 600$ mV, $R_{sense} = 10$ kΩ, and $I_o = 10$ μA.
   (b) Using the Z_o obtained in part (a), determine I_L for the configuration of Fig. 7.11 if $R_L = 2.2$ kΩ and $I_{amplifier} = 6$ mA.

8. Given the BJT configuration of Fig. 7.53, determine:
   (a) V_i.
   (b) Z_i.
   (c) $A_{v_{NL}}$.
   (d) A_{v_s}.

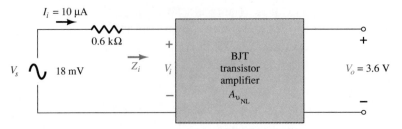

Figure 7.53 Problem 8

9. For the BJT amplifier of Fig. 7.54, determine:
   (a) I_i.
   (b) Z_i.
   (c) V_o.
   (d) I_o.
   (e) A_i using the results of parts (a) and (d).
   (f) A_i using Eq. (7.10).

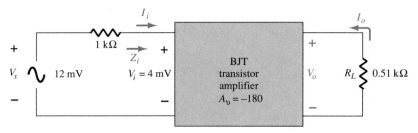

Figure 7.54 Problem 9

§ **7.5 The r_e Transistor Model**

10. For the common-base configuration of Fig. 7.17 an ac signal of 10 mV is applied, resulting in an emitter current of 0.5 mA. If $\alpha = 0.980$, determine:
    (a) Z_i.
    (b) V_o if $R_L = 1.2$ kΩ.
    (c) $A_v = V_o/V_i$.
    (d) Z_o with $r_o = \infty\Omega$.
    (e) $A_i = I_o/I_i$.
    (f) I_b.

11. For the common-base configuration of Fig. 7.17, the emitter current is 3.2 mA and α is 0.99. Determine the following if the applied voltage is 48 mV and the load is 2.2 kΩ.
    (a) r_e.
    (b) Z_i.
    (c) I_c.
    (d) V_o.
    (e) A_v.
    (f) I_b.

12. Using the model of Fig. 7.27, determine the following for a common-emitter amplifier if $\beta = 80$, $I_E(dc) = 2$ mA, and $r_o = 40$ kΩ.
    (a) Z_i.
    (b) I_b.
    (c) $A_i = I_o/I_i = I_L/I_b$ if $R_L = 1.2$ kΩ.
    (d) A_v if $R_L = 1.2$ kΩ.

13. The input impedance to a common-emitter transistor amplifier is 1.2 kΩ with $\beta = 140$, $r_o = 50$ kΩ, and $R_L = 2.7$ kΩ. Determine:
   (a) r_e.
   (b) I_b if $V_i = 30$ mV.
   (c) I_c.
   (d) $A_i = I_o/I_i = I_L/I_b$.
   (e) $A_v = V_o/V_i$.

§ 7.6 The Hybrid Equivalent Model

14. Given $I_E(\text{dc}) = 1.2$ mA, $\beta = 120$, and $r_o = 40$ kΩ, sketch the:
   (a) Common-emitter hybrid equivalent model
   (b) Common-emitter r_e equivalent model.
   (c) Common-base hybrid equivalent model.
   (d) Common-base r_e equivalent model.

15. Given $h_{ie} = 2.4$ kΩ, $h_{fe} = 100$, $h_{re} = 4 \times 10^{-4}$, and $h_{oe} = 25$ μS, sketch the:
   (a) Common-emitter hybrid equivalent model.
   (b) Common-emitter r_e equivalent model.
   (c) Common-base hybrid equivalent model.
   (d) Common-base r_e equivalent model.

16. Redraw the common-emitter network of Fig. 7.3 for the ac response with the approximate hybrid equivalent model substituted between the appropriate terminals.

17. Redraw the network of Fig. 7.55 for the ac response with the r_e model inserted between the appropriate terminals. Include r_o.

18. Redraw the network of Fig. 7.56 for the ac response with the r_e model inserted between the appropriate terminals. Include r_o.

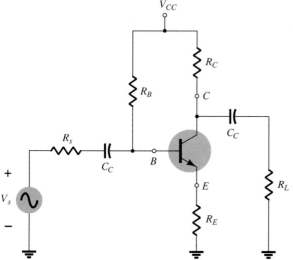

Figure 7.55 Problem 17

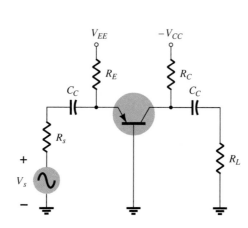

Figure 7.56 Problem 18

19. Given the typical values of $h_{ie} = 1$ kΩ, $h_{re} = 2 \times 10^{-4}$, and $A_v = -160$ for the input configuration of Fig. 7.57:
   (a) Determine V_o in terms of V_i.
   (b) Calculate I_b in terms of V_i.
   (c) Calculate I_b if $h_{re}V_o$ is ignored.
   (d) Determine the percent difference in I_b using the following equation:

$$\% \text{ difference in } I_b = \frac{I_b(\text{without } h_{re}) - I_b(\text{with } h_{re})}{I_b(\text{without } h_{re})} \times 100\%$$

   (e) Is it a valid approach to ignore the effects of $h_{re}V_o$ for the typical values employed in this example?

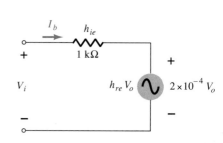

Figure 7.57 Problems 19, 21

20. Given the typical values of $R_L = 2.2$ kΩ and $h_{oe} = 20$ μS, is it a good approximation to ignore the effects of $1/h_{oe}$ on the total load impedance? What is the percent difference in total loading on the transistor using the following equation?

$$\% \text{ difference in total load} = \frac{R_L - R_L \| (1/h_{oe})}{R_L} \times 100\%$$

21. Repeat Problem 19 using the average values of the parameters of Fig. 7.28 with $A_v = -180$.

22. Repeat Problem 20 for $R_L = 3.3$ kΩ and the average value of h_{oe} in Fig. 7.28.

§ 7.7 Graphical Determination of the h-Parameters

23. (a) Using the characteristics of Fig. 7.40, determine h_{fe} at $I_C = 6$ mA and $V_{CE} = 5$ V.
(b) Repeat part (a) at $I_C = 1$ mA and $V_{CE} = 15$ V.

24. (a) Using the characteristics of Fig. 7.41, determine h_{oe} at $I_C = 6$ mA and $V_{CE} = 5$ V.
(b) Repeat part (a) at $I_C = 1$ mA and $V_{CE} = 15$ V.

25. (a) Using the characteristics of Fig. 7.42, determine h_{ie} at $I_B = 20$ μA and $V_{CE} = 20$ V.
(b) Repeat part (a) at $I_B = 5$ μA and $V_{CE} = 10$ V.

26. (a) Using the characteristics of Fig. 7.43, determine h_{re} at $I_B = 20$ μA.
(b) Repeat part (a) at $I_B = 30$ μA.

* **27.** Using the characteristics of Figs. 7.40 and 7.42, determine the approximate CE hybrid equivalent model at $I_B = 25$ μA and $V_{CE} = 12.5$ V.

* **28.** Determine the CE r_e model at $I_B = 25$ μA and $V_{CE} = 12.5$ V using the characteristics of Figs. 7.40 and 7.42.

29. Using the results of Fig. 7.44, sketch the r_e equivalent model for the transistor having the characteristics appearing in Figs. 7.40 through 7.43. Include r_o.

§ 7.8 Variations of Transistor Parameters

For Problems 30 through 34, use Figs. 7.45 through 7.47.

30. (a) Using Fig. 7.45 determine the magnitude of the percent change in h_{fe} for an I_C change from 0.2 mA to 1 mA using the equation

$$\% \text{ change} = \left| \frac{h_{fe}(0.2 \text{ mA}) - h_{fe}(1 \text{ mA})}{h_{fe}(0.2 \text{ mA})} \right| \times 100\%$$

(b) Repeat part (a) for an I_C change from 1 mA to 5 mA.

31. Repeat Problem 30 for h_{ie} (same changes in I_C).

32. (a) If $h_{oe} = 20$ μS at $I_C = 1$ mA on Fig. 7.45, what is the approximate value of h_{oe} at $I_C = 0.2$ mA?
(b) Determine its resistive value at 0.2 mA and compare to a resistive load of 6.8 kΩ. Is it a good approximation to ignore the effects of $1/h_{oe}$ in this case?

33. (a) If $h_{oe} = 20$ μS at $I_C = 1$ mA on Fig. 7.45, what is the approximate value of h_{oe} at $I_C = 10$ mA?
(b) Determine its resistive value at 10 mA and compare to a resistive load of 6.8 kΩ. Is it a good approximation to ignore the effects of $1/h_{oe}$ in this case?

34. (a) If $h_{re} = 2 \times 10^{-4}$ at $I_C = 1$ mA on Fig. 7.45, determine the approximate value of h_{re} at 0.1 mA.
(b) Using the value of h_{re} determined in part (a), can h_{re} be ignored as a good approximation if $A_v = 210$?

* **35.** (a) Reviewing the characteristics of Fig. 7.45, which parameter changed the least for the full range of collector current?
(b) Which parameter changed the most?
(c) What are the maximum and minimum values of $1/h_{oe}$? Is the approximation $1/h_{oe} \| R_L \cong R_L$ more appropriate at high or low levels of collector current?
(d) In which region of current spectrum is the approximation $h_{re}V_{ce} \cong 0$ the most appropriate?

* **36.** (a) Reviewing the characteristics of Fig. 7.47, which parameter changed the most with increase in temperature?
   (b) Which changed the least?
   (c) What are the maximum and minimum values of h_{fe}? Is the change in magnitude significant? Was it expected?
   (d) How does r_e vary with increase in temperature? Simply calculate its level at three or four points and compare their magnitudes.
   (e) In which temperature range do the parameters change the least?

§ 7.9 **Computer Analysis**

PSpice

37. Write the input file for the common-base network of Fig. 7.58 and request:
   (a) The magnitude and phase of V_o.
   (b) The magnitude of the output current I_o.
   (c) The magnitude of the current I_{r_o} (and compare with I_o).
   (d) The magnitude of the current I_e.

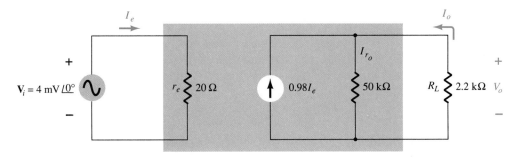

Figure 7.58 Problems 37, 39

38. Write the input file for the common-emitter network of Fig. 7.59 and request:
   (a) The magnitude and phase of V_o.
   (b) The magnitude of I_o.
   (c) The magnitude of I_{r_o} (and compare to I_o).
   (d) The magnitude of the input current I_b.

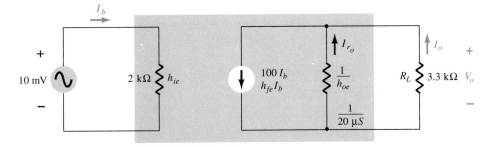

Figure 7.59 Problems 38, 40

BASIC

39. Repeat Problem 37 using BASIC.

40. Repeat Problem 38 using BASIC.

---

*Please Note: Asterisks indicate more difficult problems.

8 BJT Small-Signal Analysis

8.1 INTRODUCTION

The transistor models introduced in Chapter 7 will now be used to perform a small-signal ac analysis of a number of standard transistor network configurations. The networks analyzed represent the majority of those appearing in practice today. Modifications of the standard configurations will be relatively easy to examine once the content of this chapter is reviewed and understood.

Since the r_e model is sensitive to the actual point of operation it will be our primary model for the analysis to be performed. For each configuration, however, the effect of an output impedance is examined as provided by the h_{oe} parameter of the hybrid equivalant model. To demonstrate the similarities in analysis that exist between models a section is devoted to the small-signal analysis of BJT networks using solely the hybrid equivalent model. The analysis of this chapter does not include a load resistance R_L or source resistance R_s. The effect of both parameters is reserved for a systems approach in Chapter 10.

The computer analysis section includes a brief description of the transistor model employed in the PSpice software package. It demonstrates the range and depth of the computer analysis systems available today and how relatively easy it is to enter a complex network and print out the desired results. A program in BASIC is included to permit a comparison between the use of a software package and a computer language.

8.2 COMMON-EMITTER FIXED-BIAS CONFIGURATION

The first configuration to be analyzed in detail is the common-emitter *fixed-bias* network of Fig. 8.1. Note that the input signal V_i is applied to the base of the transistor while the output V_o is off the collector. In addition, recognize that the input current I_i is not the base current but the source current, while the output current I_o is the collector current. The small-signal ac analysis begins by removing the dc effects of V_{CC} and replacing the dc blocking capacitors C_1 and C_2 by short-circuit equivalents, resulting in the network of Fig. 8.2.

Note in Fig. 8.2 that the common ground of the dc supply and the transistor emitter terminal permits the relocation of R_B and R_C in parallel with the input and output sections of the transistor, respectively. In addition, note the placement of the important network parameters Z_i, Z_o, I_i, and I_o on the redrawn network. Substituting the r_e model for the common-emitter configuration of Fig. 8.2 will result in the network of Fig. 8.3.

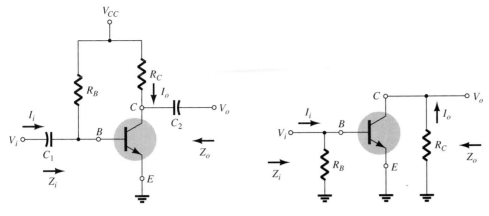

Figure 8.1 Common-emitter fixed-bias configuration.

Figure 8.2 Network of Figure 8.1 following the removal of the effects of V_{CC}, C_1 and C_2.

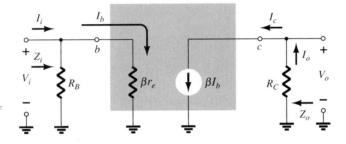

Figure 8.3 Substituting the r_e model into the network of Fig. 8.2.

The next step is find β and r_e. The magnitude of β is typically obtained from a specification sheet or by direct measurement using a curve tracer or transistor testing instrument. The value of r_e must be determined from a dc analysis of the system. Assuming that both β and r_e have been determined will result in the following equations for the important two-port characteristics of the system.

Z_i: Figure 8.3 clearly reveals that

$$\boxed{Z_i = R_B \| \beta r_e}$$ ohms (8.1)

For the majority of situations R_B is greater than βr_e by more than a factor of 10 (recall from the analysis of parallel elements that the total resistance of two parallel resistors is always less than the smallest and very close to the smallest if one is much larger than the other), permitting the following approximation:

$$\boxed{Zi \cong \beta r_e}$$ ohms (8.2)

Z_o: Recall that the output impedance of any system is defined as the impedance Z_o determined when $V_i = 0$. For Fig. 8.3, when $V_i = 0$, $I_i = I_b = 0$, resulting in an open-circuit equivalence for the current source. The result is

$$\boxed{Z_o = R_C}$$ ohms (8.3)

A_v: Solving for V_o gives

$$V_o = -I_o R_C$$

The minus sign specifies that the polarity of V_o is opposite to that defined by the indicated direction of I_o. Substituting $I_o = \beta I_b$ yields

$$V_o = -\beta I_b R_C$$

8.2 Common-Emitter Fixed-Bias Configuration

Assuming that $R_B \gg \beta r_e$ permits the approximation $I_b = I_i$, resulting in

$$V_o = -\beta I_i R_C$$

but

$$I_i = \frac{V_i}{Z_i} = \frac{V_i}{\beta r_e}$$

and

$$V_o = -\beta \left(\frac{V_i}{\beta r_e}\right) R_C$$

or

$$A_v = \frac{V_o}{V_i} = -\frac{R_C}{r_e} \qquad (8.4)$$

Note the explicit absence of β in Eq. (8.4), although we recognize that β must be utilized to determine r_e.

A_i: The current gain is determined in the following manner:

$$I_o = \beta I_b \cong \beta I_i$$

and

$$A_i = \frac{I_o}{I_i} \cong \beta \qquad (8.5)$$

The approximation sign of Eq. (8.5) is present due to the assumption that $R_B \gg \beta r_e$.

Phase Relationship: The negative sign in the resulting equation for A_v reveals that a 180° phase shift occurs between the input and output signals, as shown in Fig. 8.4.

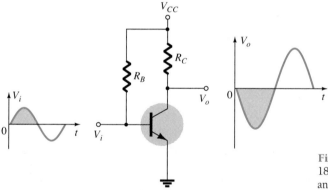

Figure 8.4 Demonstrating the 180° phase shift between input and output waveforms.

Effect of r_o: Including r_o in the output circuit of Fig. 8.3 will result in the output network of Fig. 8.5. Z_i will remain the same but Z_o is now defined by

$$Z_o = r_o \| R_C \qquad (8.6)$$

However, r_o is typically so much larger than R_C that Eq. (8.3) can often be applied with a high degree of accuracy. In fact, if we assume that $r_o \| R_C \cong R_C$, the equations for A_v and A_i are unaffected. If r_o must be included, the voltage gain is now

$$A_v = -\frac{R_C \| r_o}{r_e} \qquad (8.7)$$

To determine A_i, the current gain, I_o of Fig. 8.5, is first determined by the current-divider rule:

$$I_o = \frac{r_o \beta I_b}{r_o + R_C} \qquad \text{or} \qquad \frac{I_o}{I_b} = \frac{r_o \beta}{r_o + R_C}$$

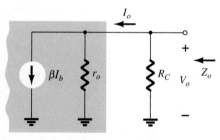

Figure 8.5 Including r_o in the output circuit of the common-emitter equivalent circuit.

Chapter 8   BJT Small-Signal Analysis

and the current gain

$$A_i = \frac{I_o}{I_i} = \frac{I_o}{I_b}\frac{I_b}{I_i} = \left(\frac{r_o\beta}{r_o + R_C}\right)(\cong 1)$$

and

$$\boxed{A_i = \frac{r_o\beta}{r_o + R_C}}$$

(8.8)

EXAMPLE 8.1

For the network of Fig. 8.6:
(a) Determine r_e.
(b) Find Z_i (with $r_o = \infty \ \Omega$).
(c) Calculate Z_o (with $r_o = \infty \ \Omega$).
(d) Determine A_v (with $r_o = \infty \ \Omega$).
(e) Find A_i (with $r_o = \infty \ \Omega$).
(f) Repeat parts (c) through (e) including $r_o = 50 \ k\Omega$ in all calculations and compare results.

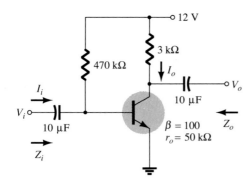

Figure 8.6 Example 8.1

Solution

(a) Dc analysis:

$$I_B = \frac{V_{CC} - V_{BE}}{R_B} = \frac{12 \text{ V} - 0.7 \text{ V}}{470 \text{ k}\Omega} = 24.04 \ \mu A$$

$$I_E = (\beta + 1)I_B = (101)(24.04 \ \mu A) = 2.428 \text{ mA}$$

$$r_e = \frac{26 \text{ mV}}{I_E} = \frac{26 \text{ mV}}{2.428 \text{ mA}} = \mathbf{10.71 \ \Omega}$$

(b) $\beta r_e = (100)(10.71 \ \Omega) = 1.071 \text{ k}\Omega$

$Z_i = R_B\|\beta r_e = 470 \text{ k}\Omega\|1.071 \text{ k}\Omega = \mathbf{1.069 \ k\Omega}$

(c) $Z_o = R_C = \mathbf{3 \ k\Omega}$

(d) $A_v = -\dfrac{R_C}{r_e} = -\dfrac{3 \text{ k}\Omega}{10.71 \ \Omega} = \mathbf{-280.11}$

(e) $A_i \cong \beta = \mathbf{100}$

(f) $Z_o = r_o\|R_C = 50 \text{ k}\Omega\|3 \text{ k}\Omega = \mathbf{2.83 \ k\Omega}$ vs. 3 kΩ

$$A_v = -\frac{r_o\|R_C}{r_e} = \frac{2.83 \text{ k}\Omega}{10.71 \ \Omega} = \mathbf{-264.24} \text{ vs. } -280.11$$

$$A_i = \frac{r_o\beta}{r_o + R_C} = \frac{(50 \text{ k}\Omega)(100)}{50 \text{ k}\Omega + 3 \text{ k}\Omega} = \mathbf{94.34} \text{ vs. } 100$$

8.3 VOLTAGE-DIVIDER BIAS

The next configuration to be analyzed is the *voltage-divider* bias network of Fig. 8.7. Recall that the name of the configuration is a result of the voltage-divider bias at the input side to determine the dc level of V_B.

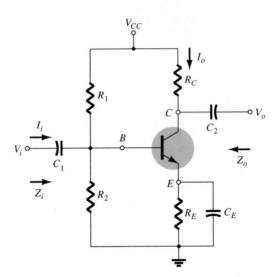

Figure 8.7 Voltage-divider bias configuration.

Substituting the r_e equivalent circuit will result in the network of Fig. 8.8. Note the absence of R_E due to the low-impedance shorting effect of the bypass capacitor, C_E. That is, at the frequency (or frequencies) of operation, the reactance of the capacitor is so small compared to R_E that it is treated as a short circuit across R_E. When V_{CC} is set to zero it places one end of R_1 and R_C at ground potential as shown in Fig. 8.8. In addition, note that R_1 and R_2 remain part of the input circuit while R_C is part of the output circuit. The parallel combination of R_1 and R_2 is defined by

$$R' = R_1 \| R_2 = \frac{R_1 R_2}{R_1 + R_2} \qquad (8.9)$$

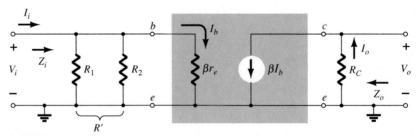

Figure 8.8 Substituting the r_e equivalent circuit into the ac equivalent network of Fig. 8.7.

Z_i: From Fig. 8.8,

$$Z_i = R' \| \beta r_e \qquad (8.10)$$

Z_o: From Fig. 8.8,

$$Z_o = R_C \qquad (8.11)$$

A_v:

$$V_o = -I_o R_C = -\beta I_b R_C$$

$$= -\beta \left(\frac{V_i}{\beta r_e} \right) R_C$$

$$= -\frac{R_C}{r_e} V_i$$

and

$$\boxed{A_v = \frac{V_o}{V_i} = -\frac{R_C}{r_e}} \qquad (8.12)$$

A_i: Since the resistance R' is often too close in magnitude to βr_e to be ignored, the effect of R' should be included in the current gain equation. Referencing Fig. 8.8, we have

$$I_b = \frac{R' I_i}{R' + \beta r_e} \qquad \text{(current-divider rule)}$$

or

$$\frac{I_b}{I_i} = \frac{R'}{R' + \beta r_e}$$

For the output side,

$$I_o = \beta I_b$$

or

$$\frac{I_o}{I_b} = \beta$$

The current gain

$$A_i = \frac{I_o}{I_i} = \frac{I_o}{I_b} \frac{I_b}{I_i}$$

$$= \beta \frac{R'}{R' + \beta r_e}$$

and

$$\boxed{A_i = \frac{\beta R'}{R' + \beta r_e}} \qquad (8.13)$$

Phase relationship: The negative sign of Eq. (8.12) reveals a 180° phase shift between V_o and V_i.

Effect of r_o: If r_o is included, the output circuit will have the same appearance as Fig. 8.5. The *input impedance is unaffected* and

$$\boxed{Z_o = r_o \| R_C} \qquad (8.14)$$

For the voltage gain,

$$\boxed{A_v = -\frac{R_C \| r_o}{r_e}} \qquad (8.15)$$

For the current gain the following equation was derived from Fig. 8.5:

$$\frac{I_o}{I_b} = \frac{r_o \beta}{r_o + R_C}$$

while the following was derived for this configuration:

$$\frac{I_b}{I_i} = \frac{R'}{R' + \beta r_e}$$

Using the fact that

$$A_i = \frac{I_o}{I_i} = \frac{I_o}{I_b}\frac{I_b}{I_i}$$

we have

$$A_i = \frac{r_o\beta}{r_o + R_C}\frac{R'}{R' + \beta r_e} \qquad (8.16)$$

The complexity of Eq. (8.16) suggests that we return to an equation such as Eq. (7.10) which utilizes the above results. That is,

$$A_i = -A_v\frac{Z_i}{R_C} \qquad (8.17)$$

EXAMPLE 8.2

For the network of Fig. 8.9, determine:
(a) r_e.
(b) Z_i.
(c) Z_o.
(d) A_v.
(e) A_i.
(f) The parameters of parts (b) through (e) if $r_o = 1/h_{oe} = 50$ kΩ and compare results.

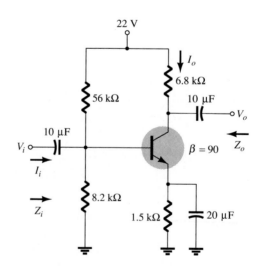

Figure 8.9 Example 8.2.

Solution

(a) DC: Testing $\beta R_E > 10 R_2$

$$(90)(1.5 \text{ k}\Omega) > 10(8.2 \text{ k}\Omega)$$

$$135 \text{ k}\Omega > 82 \text{ k}\Omega \; \textit{satisfied}$$

Chapter 8   BJT Small-Signal Analysis

Using the approximate approach

$$V_B = \frac{R_2}{R_1 + R_2} V_{CC} = \frac{8.2 \text{ k}\Omega \ (22 \text{ V})}{56 \text{ k}\Omega + 8.2 \text{ k}\Omega} = 2.81 \text{ V}$$

$$V_E = V_B - V_{BE} = 2.81 \text{ V} - 0.7 \text{ V} = 2.11 \text{ V}$$

$$I_E = \frac{V_E}{R_E} = \frac{2.11 \text{ V}}{1.5 \text{ k}\Omega} = 1.41 \text{ mA}$$

$$r_e = \frac{26 \text{ mV}}{I_E} = \frac{26 \text{ mV}}{1.41 \text{ mA}} = \mathbf{18.44 \ \Omega}$$

(b) $R' = R_1 \| R_2 = (56 \text{ k}\Omega) \| (8.2 \text{ k}\Omega) = 7.15 \text{ k}\Omega$

$Z_i = R' \| \beta r_e = 7.15 \text{ k}\Omega \| (90)(18.44 \ \Omega) = 7.15 \text{ k}\Omega \| 1.66 \text{ k}\Omega$

$\quad = \mathbf{1.35 \text{ k}\Omega}$

(c) $Z_o = R_C = \mathbf{6.8 \text{ k}\Omega}$

(d) $A_v = -\dfrac{R_C}{r_e} = -\dfrac{6.8 \text{ k}\Omega}{18.44 \ \Omega} = \mathbf{-368.76}$

(e) $A_i = \dfrac{R'\beta}{R' + \beta r_e} = \dfrac{(7.15 \text{ k}\Omega)(90)}{7.15 \text{ k}\Omega + (90)(18.44 \ \Omega)} = \mathbf{73.0}$

(f) $Z_i = \mathbf{1.35 \text{ k}\Omega}$

$Z_o = R_C \| r_o = 6.8 \text{ k}\Omega \| 50 \text{ k}\Omega = \mathbf{5.98 \text{ k}\Omega}$ vs. $6.8 \text{ k}\Omega$

$A_v = -\dfrac{R_C \| r_o}{r_e} = -\dfrac{5.98 \text{ k}\Omega}{18.44 \ \Omega} = \mathbf{-324.3}$ vs. -368.76

$A_i = \dfrac{r_o \beta}{r_o + R_C} \dfrac{R'}{R' + \beta r_e} = \dfrac{(50 \text{ k}\Omega)(90)}{50 \text{ k}\Omega + 6.8 \text{ k}\Omega} \left[\dfrac{7.15 \text{ k}\Omega}{7.15 \text{ k}\Omega + (90)(18.44 \ \Omega)} \right]$

$\quad = (79.225)(0.812) = \mathbf{64.33}$ vs. 73.0

or using Eq. (8.17):

$$A_i = -A_v \frac{Z_i}{R_C} = -(-324.3)\left(\frac{1.35 \text{ k}\Omega}{6.8 \text{ k}\Omega}\right) = \mathbf{64.38}$$

with the slight difference due to the level of accuracy carried through the calculations.

8.4 CE EMITTER-BIAS CONFIGURATION

The networks examined in this section include an emitter resistor that may or may not be bypassed in the ac domain. We will first consider the unbypassed situation and then modify the resulting equations for the bypassed configuration.

Unbypassed

The most fundamental of unbypassed configurations appears in Fig. 8.10. Substituting the r_e model will result in the configuration of Fig. 8.11. Applying Kirchhoff's voltage law to the input side of Eq. (8.11) will result in

$$V_i = I_b \beta r_e + I_e R_E$$

or

$$V_i = I_b \beta r_e + (\beta + 1)I_b R_E$$

and the input impedance looking into the network to the right of R_B is

$$Z_b = \frac{V_i}{I_b} = \beta r_e + (\beta + 1)R_E$$

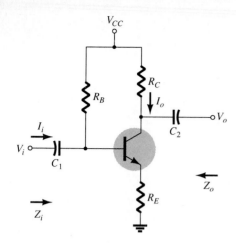

Figure 8.10 CE emitter-bias configuration.

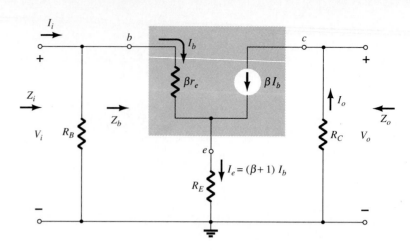

Figure 8.11 Substituting the r_e equivalent circuit into the ac equivalent network of Fig. 8.10.

The result as displayed in Fig. 8.12 reveals that the input impedance of a transistor with an unbypassed resistor R_E is determined by

$$Z_b = \beta r_e + (\beta + 1)R_E \qquad (8.18a)$$

Since β is normally much greater than 1, the approximate equation is the following:

$$Z_b \cong \beta r_e + \beta R_E$$

and

$$Z_b \cong \beta(r_e + R_E) \qquad (8.18b)$$

Since R_E is normally much greater than r_e, Eq. (8.19) can be further reduced to

$$Z_b \cong \beta R_E \qquad (8.19)$$

Z_i: Returning to Fig. 8.11, we have

$$Z_i = R_B \| Z_b \qquad (8.20)$$

Z_o: With V_i set to zero, $I_b = 0$ and βI_b can be replaced by an open-circuit equivalent. The result is

$$Z_o = R_C \qquad (8.21)$$

A_v:

$$I_b = \frac{V_i}{Z_b}$$

and

$$V_o = -I_o R_C = -\beta I_b R_C$$

$$= -\beta \left(\frac{V_i}{Z_b} \right) R_C$$

with

$$A_v = \frac{V_o}{V_i} = -\frac{\beta R_C}{Z_b} \qquad (8.22)$$

Substituting $Z_b = \beta(r_e + R_E)$ gives

$$A_v = \frac{V_o}{V_i} = -\frac{R_C}{r_e + R_E} \qquad (8.23)$$

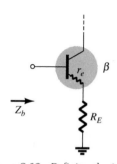

Figure 8.12 Defining the input impedance of a transistor with an unbypassed emitter resistor.

and for the approximation $Z_b \cong \beta R_E$

$$\boxed{A_v = \frac{V_o}{V_i} \cong -\frac{R_C}{R_E}}$$ (8.24)

Note again the absence of β from the equation for A_v.

A_i: The magnitude of R_B is often too close to Z_b to permit the approximation $I_b = I_i$. Applying the current-divider rule to the input circuit will result in

$$I_b = \frac{R_B I_i}{R_B + Z_b}$$

and

$$\frac{I_b}{I_i} = \frac{R_B}{R_B + Z_b}$$

In addition,

$$I_o = \beta I_b$$

and

$$\frac{I_o}{I_b} = \beta$$

so that

$$A_i = \frac{I_o}{I_i} = \frac{I_o}{I_b} \frac{I_b}{I_i}$$

$$= \beta \frac{R_B}{R_B + Z_b}$$

and

$$\boxed{A_i = \frac{I_o}{I_i} = \frac{\beta R_B}{R_B + Z_b}}$$ (8.25)

or

$$\boxed{A_i = -A_v \frac{Z_i}{R_C}}$$ (8.26)

Phase relationship: The negative sign in Eq. (8.22) through (8.24) again reveals a 180° phase shift between V_o and V_i.

Effect of r_o: The placement of r_o for this configuration is such that for typical parameter values the effect of r_o on the output impedance and voltage gain can be ignored.

The result is

$$\boxed{Z_o \cong R_C}$$ (8.27)

with

$$\boxed{A_v = -\frac{\beta R_C}{Z_b} \cong -\frac{R_C}{R_E}}$$ (8.28)

and

$$\boxed{A_i = -A_v \frac{Z_i}{R_C}}$$ (8.29)

Bypassed

If R_E of Fig. 8.10 is bypassed by an emitter capacitor C_E, the equations developed will be modified as follows:

$$Z_b = \beta r_e \quad \text{and} \quad A_v = -\frac{R_C}{r_e}$$

EXAMPLE 8.3

For the network of Fig. 8.13, without C_E (unbypassed), determine:

(a) r_e.

(b) Z_i.

(c) Z_o.

(d) A_v.

(e) A_i.

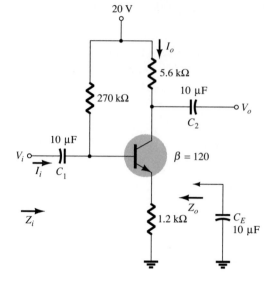

Figure 8.13 Example 8.3

Solution

(a) DC: $I_B = \dfrac{V_{CC} - V_{BE}}{R_B + (\beta + 1)R_E} = \dfrac{20\ V - 0.7\ V}{270\ k\Omega + (121)1.2\ k\Omega} = 46.5\ \mu A$

$I_E = (\beta + 1)I_B = (121)(46.5\ \mu A) = 5.63\ mA$

and $\quad r_e = \dfrac{26\ mV}{I_E} = \dfrac{26\ mV}{5.63\ mA} = \mathbf{4.62\ \Omega}$

(b) $Z_b = \beta r_e + (\beta + 1)R_E$

$\quad = (120)(4.62\ \Omega) + (121)1.2\ k\Omega$

$\quad = 553.2\ \Omega + 145.2\ k\Omega$

$\quad = 145.75\ k\Omega$

versus 144 kΩ using $Z_b \cong \beta R_E$.

$Z_i = R_B \| Z_b$

$\quad = 270\ k\Omega \| 145.75\ k\Omega$

$\quad = \mathbf{94.65\ k\Omega}$

versus 93.91 kΩ using $Z_i = R_B \| \beta R_E$.

(c) $Z_o = R_C = \mathbf{5.6\ k\Omega}$

(d) $A_v = -\dfrac{\beta R_C}{Z_b} = -\dfrac{(120)(5.6\ k\Omega)}{145.75\ k\Omega}$

$\quad = \mathbf{-4.61}$

versus -4.67 using $A_v \cong -R_C/R_E$.

(e) $A_i = \dfrac{\beta R_B}{R_B + Z_b} = \dfrac{(120)(270\ k\Omega)}{270\ k\Omega + 145.75\ k\Omega}$

$\quad = \mathbf{77.93}$

or $\quad A_i = -A_v \dfrac{Z_i}{R_C} = -(-4.61)\left(\dfrac{94.65\ k\Omega}{5.6\ k\Omega}\right)$

$\quad = \mathbf{77.92}$

EXAMPLE 8.4

Repeat the analysis of Example 8.3 with C_E in place.

Solution

(a) The dc analysis is the same and $r_e = 4.62 \ \Omega$.

(b) R_E is "shorted out" by C_E for the ac analysis. Therefore,

$$Z_i = R_B \| Z_b = R_B \| \beta r_e = 270 \ \text{k}\Omega \| (120)(4.62 \ \Omega)$$

$$= 270 \ \text{k}\Omega \| 554.4 \ \Omega \cong \mathbf{553.26 \ \Omega}$$

(c) $Z_o = R_C = \mathbf{5.6 \ k\Omega}$

(d) $A_v = -\dfrac{R_C}{r_e}$

$$= -\dfrac{5.6 \ \text{k}\Omega}{4.62 \ \Omega} = \mathbf{-1212.12} \quad \text{(a significant increase)}$$

(e) $A_i = \dfrac{\beta R_B}{R_B + Z_b} = \dfrac{(120)(270 \ \text{k}\Omega)}{270 \ \text{k}\Omega + 554.4 \ \Omega}$

$$= \mathbf{119.75}$$

For the network of Fig. 8.14, determine (using appropriate approximations):
(a) r_e.
(b) Z_i.
(c) Z_o.
(d) A_v.
(e) A_i.

EXAMPLE 8.5

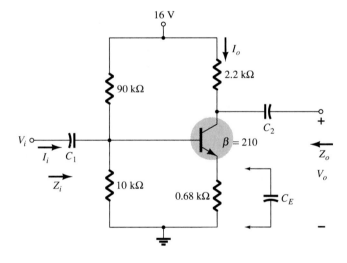

Figure 8.14 Example 8.5

Solution

(a) Testing $\beta R_E > 10 \ R_2$

$$(210)(0.68 \ \text{k}\Omega) > 10(10 \ \text{k}\Omega)$$

$$142.8 \ \text{k}\Omega > 100 \ \text{k}\Omega \ \textit{satisfied}$$

$$V_B = \frac{R_2}{R_1 + R_2} V_{CC} = \frac{10 \ \text{k}\Omega}{90 \ \text{k}\Omega + 10 \ \text{k}\Omega} (16 \ \text{V}) = 1.6 \ \text{V}$$

$$V_E = V_B - V_{BE} = 1.6 \ \text{V} - 0.7 \ \text{V} = 0.9 \ \text{V}$$

$$I_E = \frac{V_E}{R_E} = \frac{0.9 \ \text{V}}{0.68 \ \text{k}\Omega} = 1.324 \ \text{mA}$$

$$r_e = \frac{26 \ \text{mV}}{I_E} = \frac{26 \ \text{mV}}{1.324 \ \text{mA}} = \mathbf{19.64 \ \Omega}$$

(b) The ac equivalent circuit is provided in Fig. 8.15. The resulting configuration is now different from Fig. 8.11 only by the fact that now

$$R' = R_1\|R_2 = 9 \text{ k}\Omega$$

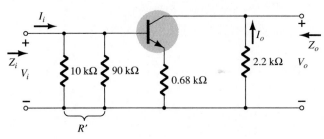

Figure 8.15 The ac equivalent circuit of Fig. 8.14.

Using the appropriate approximations yields

$$Z_b \cong \beta R_E = 142.8 \text{ k}\Omega$$

$$Z_i = R_B\|Z_b = 9 \text{ k}\Omega\|142.8 \text{ k}\Omega$$

$$= \mathbf{8.47 \text{ k}\Omega}$$

(c) $Z_o = R_C = \mathbf{2.2 \text{ k}\Omega}$

(d) $A_v = -\dfrac{R_C}{R_E} = -\dfrac{2.2 \text{ k}\Omega}{0.68 \text{ k}\Omega} = \mathbf{-3.24}$

(e) $A_i = -A_v\dfrac{Z_i}{R_C} = -(-3.24)\left(\dfrac{8.47 \text{ k}\Omega}{2.2 \text{ k}\Omega}\right)$

$$= \mathbf{12.47}$$

EXAMPLE 8.6   Repeat Example 8.5 with C_E in place.

Solution

(a) The dc analysis is the same and $r_e = \mathbf{19.64 \ \Omega}$.
(b) $Z_b = \beta r_e = (210)(19.64 \ \Omega) \cong 4.12 \text{ k}\Omega$

$\quad Z_i = R_B\|Z_b = 9 \text{ k}\Omega\|4.12 \text{ k}\Omega$

$\quad\quad = \mathbf{2.83 \text{ k}\Omega}$

(c) $Z_o = R_C = \mathbf{2.2 \text{ k}\Omega}$

(d) $A_v = -\dfrac{R_C}{r_e} = -\dfrac{2.2 \text{ k}\Omega}{19.64 \text{ k}\Omega} = \mathbf{-112.02}$ (a significant increase)

(e) $A_i = -A_v\dfrac{Z_i}{R_L} = -(-112.02)\left(\dfrac{2.83 \text{ k}\Omega}{2.2 \text{ k}\Omega}\right)$

$$= \mathbf{144.1}$$

Another variation of an emitter-bias configuration appears in Fig. 8.16. For the dc analysis the emitter resistance is $R_{E_1} + R_{E_2}$, while for the ac analysis the resistor R_E in the equations above is simply R_{E_1} and R_{E_2} is bypassed by C_E.

Chapter 8   BJT Small-Signal Analysis

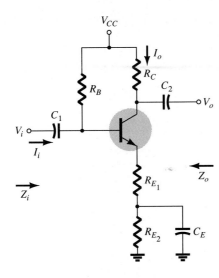

Figure 8.16 An emitter-bias configuration with a portion of the emitter-bias resistance by-passed in the ac domain.

8.5 EMITTER-FOLLOWER CONFIGURATION

When the output is taken from the emitter terminal of the transistor as shown in Fig. 8.17 the network is referred to as an *emitter-follower*. The output voltage is always slightly less than the input signal, due to the drop from base to emitter, but the approximation $A_v \cong 1$ is usually a good one. Unlike the collector voltage, the emitter voltage is in phase with the signal V_i. That is, both V_o and V_i will attain their positive and negative peak values at the same time. The fact that V_o "follows" the magnitude of V_i with an in-phase relationship accounts for the terminology emitter-follower.

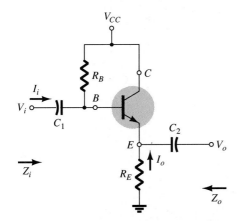

Figure 8.17 Emitter-follower configuration.

The most common emitter-follower configuration appears in Fig. 8.17. In fact, because the collector is grounded for ac analysis, it is actually a *common-collector* configuration. Other variations of Fig. 8.17 that draw the output off the emitter with $V_o \cong V_i$ will appear later in this section.

The emitter-follower configuration is frequently used for impedance-matching purposes. It presents a high impedance at the input and a low impedance at the output, which is the direct opposite of the standard fixed-bias configuration. The resulting effect is much the same as that obtained with a transformer, where a load is matched to the source impedance for maximum power transfer through the system.

Substituting the r_e equivalent circuit into the network of Fig. 8.17 will result in the network of Fig. 8.18.

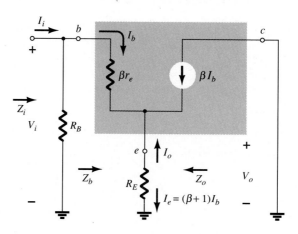

Figure 8.18 Substituting the r_e equivalent circuit into the ac equivalent network of Fig. 8.17.

Z_i: The input impedance is determined in the same manner as described in the preceding section:

$$Z_i = R_B \| Z_b \tag{8.30}$$

with

$$Z_b = \beta r_e + (\beta + 1)R_E \tag{8.31}$$

or

$$Z_b \cong \beta(r_e + R_E) \tag{8.32}$$

and

$$Z_b \cong \beta R_E \tag{8.33}$$

Z_o: The output impedance is best described by first writing the equation for the current I_b:

$$I_b = \frac{V_i}{Z_b}$$

and then multiplying by $(\beta + 1)$ to establish I_e. That is,

$$I_e = (\beta + 1)I_b = (\beta + 1)\frac{V_i}{Z_b}$$

Substituting for Z_b gives

$$I_e = \frac{(\beta + 1)V_i}{\beta r_e + (\beta + 1)R_E}$$

or

$$I_e = \frac{V_i}{[\beta r_e/(\beta + 1)] + R_E}$$

but

$$\beta + 1 \cong \beta$$

and

$$\frac{\beta r_e}{\beta + 1} \cong \frac{\beta r_e}{\beta} = r_e$$

so that

$$I_e \cong \frac{V_i}{r_e + R_E} \tag{8.34}$$

If we now construct the network defined by Eq. (8.34), the configuration of Fig. 8.19 will result.

To determine Z_o, V_i is set to zero and

$$Z_o = R_E \| r_e \tag{8.35}$$

Since R_E is typically much greater than r_e the following approximation is often applied:

$$Z_o \cong r_e \tag{8.36}$$

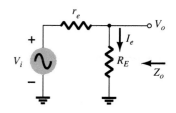

Figure 8.19 Defining the output impedance for the emitter-follower configuration.

A_v: Figure 8.19 can be utilized to determine the voltage gain through an application of the voltage-divider rule:

$$V_o = \frac{R_E V_i}{R_E + r_e}$$

and

$$A_v = \frac{V_o}{V_i} = \frac{R_E}{R_E + r_e} \tag{8.37}$$

Since R_E is usually much greater than r_e, $R_E + r_e \cong R_E$ and

$$A_v = \frac{V_o}{V_i} \cong 1 \tag{8.38}$$

A_i: From Fig. 8.18,

$$I_b = \frac{R_B I_i}{R_B + Z_b}$$

or

$$\frac{I_b}{I_i} = \frac{R_B}{R_B + Z_b}$$

and

$$I_o = -I_e = -(\beta + 1)I_b$$

or

$$\frac{I_o}{I_b} = -(\beta + 1)$$

so that

$$A_i = \frac{I_o}{I_i} = \frac{I_o}{I_b} \frac{I_b}{I_i}$$

$$= -(\beta + 1)\frac{R_B}{R_B + Z_b}$$

and since

$$\beta + 1 \cong \beta,$$

$$A_i \cong -\frac{\beta R_B}{R_B + Z_b} \tag{8.39}$$

or

$$A_i = -A_v \frac{Z_i}{R_E} \tag{8.40}$$

Phase relationship: As revealed by Eq. (8.37) and earlier discussions of this section, V_o and V_i are in phase for the emitter-follower configuration.

Effect of r_o: For this configuration r_o appears in parallel with the emitter resistor R_E, which can affect both the input and output impedances.

8.5 **Emitter-Follower Configuration**

351

EXAMPLE 8.7

For the emitter-follower network of Fig. 8.20, determine:
(a) r_e.
(b) Z_i.
(c) Z_o.
(d) A_v.
(e) A_i.

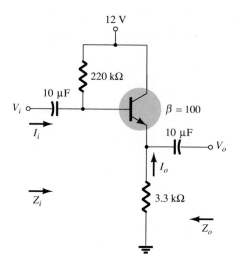

Figure 8.20 Example 8.7

Solution

(a) $$I_B = \frac{V_{CC} - V_{BE}}{R_B + (\beta + 1)R_E}$$

$$= \frac{12\ V - 0.7\ V}{220\ k\Omega + (101)3.3\ k\Omega} = 20.42\ \mu A$$

$$I_E = (\beta + 1)I_B$$

$$= (101)(20.42\ \mu A) = 2.062\ mA$$

$$r_e = \frac{26\ mV}{I_E} = \frac{26\ mV}{2.062\ mA} = \textbf{12.61}\ \boldsymbol{\Omega}$$

(b) $$Z_b = \beta r_e + (\beta + 1)R_E$$

$$= (100)(12.61\ \Omega) + (101)(3.3\ k\Omega)$$

$$= 1.261\ k\Omega + 333.3\ k\Omega$$

$$= 334.56\ k\Omega \cong \beta R_E$$

$$Z_i = R_B \| Z_b = 220\ k\Omega \| 334.56\ k\Omega$$

$$= \textbf{132.72}\ \textbf{k}\boldsymbol{\Omega}$$

(c) $$Z_o = R_E \| r_e = 3.3\ k\Omega \| 12.61\ \Omega$$

$$= \textbf{12.56}\ \boldsymbol{\Omega} \cong r_e$$

(d) $$A_v = \frac{V_o}{V_i} = \frac{R_E}{R_E + r_e} = \frac{3.3\ k\Omega}{3.3\ k\Omega + 12.61\ \Omega}$$

$$= \textbf{0.996} \cong 1$$

(e) $$A_i \cong -\frac{\beta R_B}{R_B + Z_b} = -\frac{(100)(220\ k\Omega)}{220\ k\Omega + 334.56\ k\Omega} = \textbf{--39.67}$$

versus

$$A_i = -A_v \frac{Z_i}{R_E} = -(0.996)\left(\frac{132.72\ k\Omega}{3.3\ k\Omega}\right) = \textbf{--40.06}$$

The network of Fig. 8.21 is a variation of the network of Fig. 8.17 which employs a voltage-divider input section to set the bias conditions. Equations (8.30) through (8.40) are changed only by replacing R_B by $R' = R_1 \| R_2$.

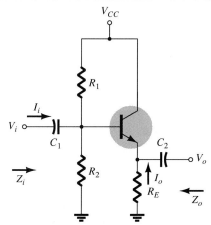

Figure 8.21 Emitter-follower configuration with a voltage-divider biasing arrangement.

The network of Fig. 8.22 will also provide the input/output characteristics of an emitter-follower but includes a collector resistor R_C. In this case R_B is again replaced by the parallel combination of R_1 and R_2. The input impedance Z_i and output impedance Z_o are unaffected by R_C since it is not reflected into the base or emitter equivalent networks. In fact, the only effect of R_C will be to determine the Q-point of operation.

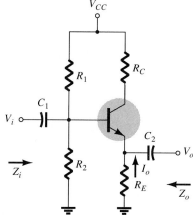

Figure 8.22 Emitter-follower configuration with a collector resistor R_C.

8.6 COMMON-BASE CONFIGURATION

The common-base configuration is characterized as having a relatively low input and a high output impedance and a current gain less than 1. The voltage gain, however, can be quite large. The standard configuration appears in Fig. 8.23 with the common-base r_e equivalent model substituted in Fig. 8.24.

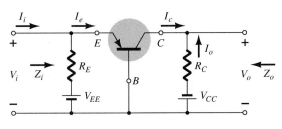

Figure 8.23 Common-base configuration.

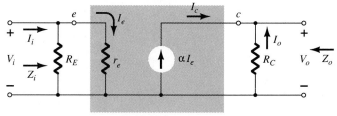

Figure 8.24 Substituting the r_e equivalent circuit into the ac equivalent network of Fig. 8.23.

Z_i:

$$Z_i = R_E \| r_e \qquad (8.41)$$

Z_o:

$$Z_o = R_C \qquad (8.42)$$

A_v:

$$V_o = -I_o R_C = -(-I_c)R_C = \alpha I_e R_C$$

with

$$I_e = \frac{V_i}{r_e}$$

so that

$$V_o = \alpha \left(\frac{V_i}{r_e} \right) R_C$$

and

$$A_v = \frac{V_o}{V_i} = \frac{\alpha R_C}{r_e} \cong \frac{R_C}{r_e} \qquad (8.43)$$

A_i: Assuming that $R_E \gg r_e$ yields

$$I_e = I_i$$

and

$$I_o = -\alpha I_e = -\alpha I_i$$

with

$$A_i = \frac{I_o}{I_i} = -\alpha \cong -1 \qquad (8.44)$$

Phase relationship: The fact that A_v is a positive number reveals that V_o and V_i are in phase for the common-base configuration.

Effect of r_o: For the common-base configuration, $r_o = 1/h_{ob}$ is typically in the megaohm range and sufficiently larger than the parallel resistance R_C to permit the approximation $r_o \| R_C \cong R_C$.

EXAMPLE 8.8

For the network of Fig. 8.25, determine:
(a) r_e.
(b) Z_i.
(c) Z_o.
(d) A_v.
(e) A_i.

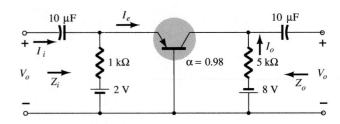

Figure 8.25 Example 8.8

Solution

(a) $I_E = \dfrac{V_{EE} - V_{BE}}{R_E} = \dfrac{2 \text{ V} - 0.7 \text{ V}}{1 \text{ k}\Omega} = \dfrac{1.3 \text{ V}}{1 \text{ k}\Omega} = 1.3 \text{ mA}$

$r_e = \dfrac{26 \text{ mV}}{I_E} = \dfrac{26 \text{ mV}}{1.3 \text{ mA}} = \mathbf{20 \ \Omega}$

(b) $Z_i = R_E \| r_e = 1 \text{ k}\Omega \| 20 \ \Omega$

$\quad\quad = \mathbf{19.61 \ \Omega} \cong r_e$

(c) $Z_o = R_C = \mathbf{5 \ k\Omega}$

(d) $A_v \cong \dfrac{R_C}{r_e} = \dfrac{5 \text{ k}\Omega}{20 \ \Omega} = \mathbf{250}$

(e) $A_i = \mathbf{-0.98} \cong -1$

8.7 COLLECTOR FEEDBACK CONFIGURATION

The collector feedback network of Fig. 8.26 employs a feedback path from collector to base to increase the stability of the system as discussed in Section 4.12. However, the simple maneuver of connecting a resistor from base to collector rather than base to dc supply has a significant impact on the level of difficulty encountered when analyzing the network.

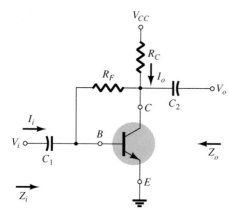

Figure 8.26 Collector feedback configuration.

Some of the steps to be performed below are the result of experience working with such configurations. It is not expected that a new student of the subject would choose the sequence of steps described below without taking a wrong step or two. Substituting the equivalent circuit and redrawing the network will result in the configuration of Fig. 8.27.

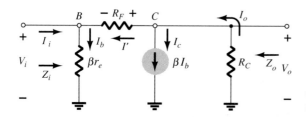

Figure 8.27 Substituting the r_e equivalent circuit into the ac equivalent network of Fig. 8.26.

The voltage gain will be determined first, followed by the current gain and the impedance levels.

A_v: At node C:

$$I_o = \beta I_b + I'$$

For typical values, $\beta I_b \gg I'$ and $I_o \cong \beta I_b$.

$$V_o = -I_o R_C = -(\beta I_b) R_C$$

Substituting $I_b = V_i/\beta r_e$ gives us

$$V_o = -\beta\frac{V_i}{\beta r_e}R_C$$

and

$$\boxed{A_v = \frac{V_o}{V_i} = -\frac{R_C}{r_e}} \tag{8.45}$$

A_i: Applying Kirchhoff's voltage law around the outside network loop yields

$$V_i + V_{R_F} - V_o = 0$$

and

$$I_b\beta r_e + (I_b - I_i)R_F + I_o R_C = 0$$

Using $I_o \cong \beta I_b$, we have

$$I_b\beta r_e + I_b R_F - I_i R_F + \beta I_b R_C = 0$$

and

$$I_b(\beta r_e + R_F + \beta R_C) = I_i R_F$$

Substituting $I_b = I_o/\beta$ from $I_o \cong \beta I_b$ yields

$$\frac{I_o}{\beta}(\beta r_e + R_F + \beta R_C) = I_i R_F$$

and

$$I_o = \frac{\beta R_F I_i}{\beta r_e + R_F + \beta R_C}$$

Ignoring βr_e compared to R_F and βR_C gives us

$$\boxed{A_i = \frac{I_o}{I_i} = \frac{\beta R_F}{R_F + \beta R_C}} \tag{8.46}$$

For $\beta R_C \gg R_F$,

$$A_i = \frac{I_o}{I_i} = \frac{\beta R_F}{\beta R_C}$$

and

$$\boxed{A_i = \frac{I_o}{I_i} \cong \frac{R_F}{R_C}} \tag{8.47}$$

Z_i: From Fig. 8.27,

$$I_b = I_i + \frac{V_o - V_i}{R_F}$$

Since $V_o \gg V_i$,

$$I_b \cong I_i + \frac{V_o}{R_F}$$

and

$$V_i = I_b\beta r_e$$

$$= \left(I_i + \frac{V_o}{R_F}\right)\beta r_e$$

$$= I_i\beta r_e + \frac{\beta r_e}{R_F}V_o$$

Substituting $V_o = A_v V_i$ from $A_v = V_o/V_i$, we have

$$V_i = I_i\beta r_e + \frac{\beta r_e A_v V_i}{R_F}$$

Then
$$V_i\left(1 - \frac{\beta r_e A_v}{R_F}\right) = I_i\beta r_e$$

with
$$\frac{V_i}{I_i} = \frac{\beta r_e}{1 - \beta r_e(A_v/R_F)} = \frac{\dfrac{\beta r_e}{1 + \beta r_e}}{\dfrac{R_F}{|A_v|}}$$

For parallel elements,
$$x\|y = \frac{xy}{x + y} = \frac{x}{1 + x/y}$$

which has the same format as the equation above with $x = \beta r_e$ and $y = -R_F/A_v$. However, since A_v is a negative quantity for this configuration the magnitude of Z_i is given by

$$\boxed{Z_i = \frac{V_i}{I_i} = \beta r_e \left\| \frac{R_F}{|A_v|} \right.}$$ (8.48)

and we note that the magnitude of the voltage gain must be determined before the input impedance can be calculated.

Z_o: If we set V_i to zero as required to define Z_o, the network will appear as shown in Fig. 8.28. The effect of βr_e is removed and R_F appears in parallel with R_C and

$$\boxed{Z_o \cong R_C \| R_F}$$ (8.49)

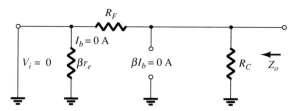

Figure 8.28 Defining Z_o for the collector feedback configuration.

Phase relationship: The negative sign of Eq. (8.45) reveals a 180° phase shift between V_o and V_i.

Effect of r_o: As demonstrated by Fig. 8.27, the output resistance of the transistor will appear in parallel with the resistor R_C throughout the analysis. Equations including the effects of r_o can therefore be obtained for A_v and Z_o simply by replacing R_C by the parallel combination, $r_o \| R_C$.

For the network of Fig. 8.29, determine:

(a) r_e.
(b) A_v.
(c) A_i.
(d) Z_i.
(e) Z_o.

EXAMPLE 8.9

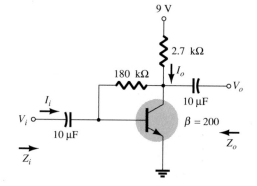

Figure 8.29 Example 8.9

Solution

(a) $I_B = \dfrac{V_{CC} - V_{BE}}{R_F + \beta R_C} = \dfrac{9\ \text{V} - 0.7\ \text{V}}{180\ \text{k}\Omega + (200)2.7\ \text{k}\Omega}$

$\qquad = 11.53\ \mu\text{A}$

$\quad I_E = (\beta + 1)I_B = (201)(11.53\ \mu\text{A}) = 2.32\ \text{mA}$

$\quad r_e = \dfrac{26\ \text{mV}}{I_E} = \dfrac{26\ \text{mV}}{2.32\ \text{mA}} = \mathbf{11.21\ \Omega}$

(b) $A_v = -\dfrac{R_C}{r_e} = -\dfrac{2.7\ \text{k}\Omega}{11.21\ \Omega} = \mathbf{-240.86}$

(c) $A_i = \dfrac{\beta R_F}{R_F + \beta R_C} = \dfrac{(200)(180\ \text{k}\Omega)}{180\ \text{k}\Omega + (200)(2.7\ \text{k}\Omega)}$

$\qquad = \mathbf{50}$

(d) $Z_i = \beta r_e \| \dfrac{R_F}{|A_v|} = (200)(11.21\ \Omega) \| \dfrac{180\ \text{k}\Omega}{240.86}$

$\qquad = 2.24\ \text{k}\Omega \| 0.747\ \text{k}\Omega = \mathbf{0.56\ \text{k}\Omega}$

(e) $Z_o = R_C \| R_F = 2.7\ \text{k}\Omega \| 180\ \text{k}\Omega$

$\qquad = \mathbf{2.66\ \text{k}\Omega}$

For the configuration of Fig. 8.30, Eqs. (8.50) through (8.53) will determine the variables of interest. The derivations are left as an exercise at the end of the chapter.

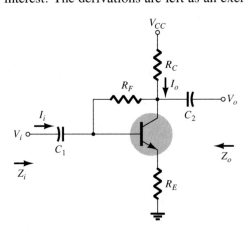

Figure 8.30 Collector feedback configuration with an emitter resistor R_E.

A_v:

$$\boxed{A_v \cong -\frac{R_C}{R_E}} \qquad (8.50)$$

A_i:

and

$$\boxed{A_i \cong \frac{R_F}{R_E + R_C + R_F/\beta}} \qquad (8.51)$$

Z_i:

$$\boxed{Z_i \cong \frac{\beta R_E R_F}{R_F + \beta R_E(1 + |A_v|)}} \qquad (8.52)$$

Z_o:

$$\boxed{Z_o \cong R_C \| R_F} \qquad (8.53)$$

8.8 COLLECTOR DC FEEDBACK CONFIGURATION

The network of Fig. 8.31 has a dc feedback resistor for increased stability, yet the capacitor C_3 will shift portions of the feedback resistance to the input and output sections of the network in the ac domain. The portion of R_F shifted to the input or output side will be determined by the desired ac input and output resistance levels.

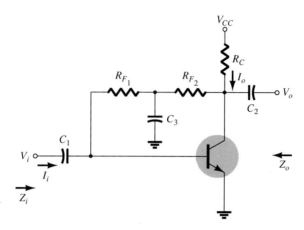

Figure 8.31 Collector DC feedback configuration.

At the frequency or frequencies of operation, the capacitor will assume a short-circuit equivalent to ground due to its low impedance level compared to the other elements of the network. The small-signal ac equivalent circuit will then appear as shown in Fig. 8.32.

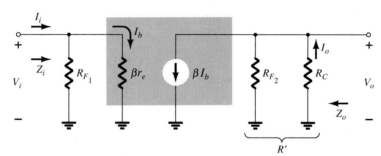

Figure 8.32 Substituting the r_e equivalent circuit into the ac equivalent network of Fig. 8.31.

Z_i:

$$\boxed{Z_i = R_{F_1}\|\beta r_e} \qquad (8.54)$$

Z_o:

$$\boxed{Z_o = R_C\|R_{F_2}} \qquad (8.55)$$

A_v:

$$R' = R_{F_2}\|R_C$$

and

$$V_o = -\beta I_b R'$$

but

$$I_b = \frac{V_i}{\beta r_e}$$

and

$$V_o = -\beta \frac{V_i}{\beta r_e} R'$$

so that

$$A_v = \frac{V_o}{V_i} = -\frac{R_{F_2} \| R_C}{r_e} \qquad (8.56)$$

A_i: For the input section,

$$I_b = \frac{R_{F_1} I_i}{R_{F_1} + \beta r_e} \quad \text{or} \quad \frac{I_b}{I_i} = \frac{R_{F_1}}{R_{F_1} + \beta r_e}$$

and for the output section,

$$I_o = \frac{R_{F_2} \beta I_b}{R_{F_2} + R_C} \quad \text{or} \quad \frac{I_o}{I_b} = \frac{R_{F_2} \beta}{R_{F_2} + R_C}$$

The current gain

$$A_i = \frac{I_o}{I_i} = \frac{I_o}{I_b} \frac{I_b}{I_i}$$

$$= \frac{R_{F_2} \beta}{R_{F_2} + R_C} \frac{R_{F_1}}{R_{F_1} + \beta r_e}$$

and

$$A_i = \frac{\beta R_{F_1} R_{F_2}}{(R_{F_1} + \beta r_e)(R_{F_2} + R_C)} \qquad (8.57)$$

or

$$A_i = -A_v \frac{Z_i}{R_C} \qquad (8.58)$$

Phase relationship: The negative sign in Eq. (8.56) clearly reveals a 180° phase shift between input and output voltages.

Effect of r_o: Since r_o is in parallel with R_C the output impedance will be changed to

$$Z_o = r_o \| R_C \| R_{F_2} \qquad (8.59)$$

and the voltage gain to

$$A_v = -\frac{r_o \| R_C \| R_{F_2}}{r_e} \qquad (8.60)$$

EXAMPLE 8.10

For the network of Fig. 8.33, determine:
(a) r_e.
(b) Z_i.
(c) Z_o.
(d) A_v.
(e) A_i.

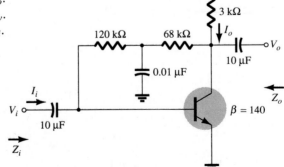

Figure 8.33 Example 8.10

Chapter 8 BJT Small-Signal Analysis

Solution

(a) DC: $I_B = \dfrac{V_{CC} - V_{BE}}{R_F + \beta R_C}$

$= \dfrac{12\ \text{V} - 0.7\ \text{V}}{(120\ \text{k}\Omega + 68\ \text{k}\Omega) + (140)3\ \text{k}\Omega}$

$= \dfrac{11.3\ \text{V}}{608\ \text{k}\Omega} = 18.6\ \mu\text{A}$

$I_E = (\beta + 1)I_B = (141)(18.6\ \mu\text{A})$

$= 2.62\ \text{mA}$

$r_e = \dfrac{26\ \text{mV}}{I_E} = \dfrac{26\ \text{mV}}{2.62\ \text{mA}} = \textbf{9.92}\ \boldsymbol{\Omega}$

(b) $\beta r_e = (140)(9.92\ \Omega) = 1.39\ \text{k}\Omega$

The ac equivalent network appears in Fig. 8.34.

$Z_i = R_{F_1} \| \beta r_e = 120\ \text{k}\Omega \| 1.39\ \text{k}\Omega$

$\cong \textbf{1.37}\ \textbf{k}\boldsymbol{\Omega}$

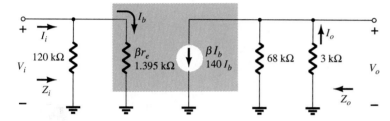

Figure 8.34 Substituting the r_e equivalent circuit into the ac equivalent network of Fig. 8.33.

(c) $Z_o = R_C \| R_{F_2} = 3\ \text{k}\Omega \| 68\ \text{k}\Omega$

$\cong \textbf{2.87}\ \textbf{k}\boldsymbol{\Omega}$

(d) $R' = R_C \| R_{F_2} = 2.87\ \text{k}\Omega$

$A_v = -\dfrac{R_C \| R_{F_2}}{r_e}$

$= -\dfrac{2.87\ \text{k}\Omega}{9.92\ \Omega}$

$= \textbf{-289.3}$

(e) $A_i = \dfrac{\beta R_{F_1} R_{F_2}}{(R_{F_1} + \beta r_e)(R_{F_2} + R_C)}$

$= \dfrac{(140)(120\ \text{k}\Omega)(68\ \text{k}\Omega)}{(120\ \text{k}\Omega + 1.39\ \text{k}\Omega)(68\ \text{k}\Omega + 3\ \text{k}\Omega)}$

$= \textbf{132.54}$

or

$A_i = -A_v \dfrac{Z_i}{R_C}$

$= -(-289.3)\left(\dfrac{1.37\ \text{k}\Omega}{3\ \text{k}\Omega}\right)$

$= \textbf{132.11}$

8.9 APPROXIMATE HYBRID EQUIVALENT CIRCUIT

The analysis using the approximate hybrid equivalent circuit of Fig. 8.35 for the common-emitter configuration and of Fig. 8.36 for the common-base configuration is very similar to that just performed using the r_e model. Although time and priorities do not permit a detailed analysis of all the configurations discussed thus far, a brief overview of some of the most important will be included in this section to demonstrate the similarities in approach and the resulting equations.

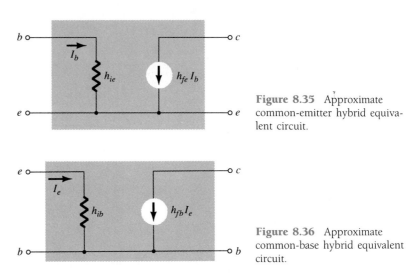

Figure 8.35 Approximate common-emitter hybrid equivalent circuit.

Figure 8.36 Approximate common-base hybrid equivalent circuit.

Since the various parameters of the hybrid model are specified by a data sheet or experimental analysis, the dc analysis associated with the use of the r_e model is not an integral part of the use of the hybrid parameters. In other words, when the problem is presented, the parameters such as h_{ie}, h_{fe}, h_{ib}, and so on, are specified. Keep in mind, however, that the hybrid parameters and components of the r_e model are related by the following equations as discussed in detail in chapter 7: $h_{ie} = \beta r_e$, $h_{fe} = \beta$, $h_{oe} = 1/r_o$, $h_{fb} = -\alpha$, and $h_{ib} = r_e$ (note Appendix A).

Fixed-Bias Configuration

For the fixed-bias configuration of Fig. 8.37 the small-signal ac equivalent network will appear as shown in Fig. 8.38 using the approximate common-emitter hybrid equivalent model. Compare the similarities in appearance with Fig. 8.3 and the r_e model analysis. The similarities suggest that the analysis will be quite similar and the results of one can be directly related to the other.

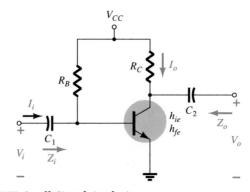

Figure 8.37 Fixed-bias configuration.

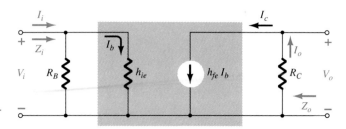

Figure 8.38 Substituting the approximate hybrid equivalent circuit into the ac equivalent network of Fig. 8.37.

Z_i: From Fig. 8.38,

$$Z_i = R_B \| h_{ie} \qquad (8.61)$$

Z_o: From Fig. 8.38,

$$Z_o = R_C \qquad (8.62)$$

A_v:

$$V_o = -I_o R_C = -I_C R_C$$
$$= -h_{fe} I_b R_C$$

and

$$I_b = \frac{V_i}{h_{ie}}$$

with

$$V_o = -h_{fe}\frac{V_i}{h_{ie}} R_C$$

so that

$$A_v = \frac{V_o}{V_i} = -\frac{h_{fe} R_C}{h_{ie}} \qquad (8.63)$$

A_i: Assuming that $R_B \gg h_{ie}$, then $I_b \cong I_i$ and $I_o = I_c = h_{fe}I_b = h_{fe}I_i$ with

$$A_i = \frac{I_o}{I_i} \cong h_{fe} \qquad (8.64)$$

Impact of h_{oe}: Since $r_o = 1/h_{oe}$ is directly in parallel with R_C,

$$Z_o = \frac{1}{h_{oe}} \| R_C \qquad (8.65)$$

and

$$A_v = \frac{-h_{fe}\left(\dfrac{1}{h_{oe}} \| R_C\right)}{h_{ie}} \qquad (8.66)$$

For the network of Fig. 8.39, determine:

(a) Z_i.
(b) Z_o.
(c) A_v.
(d) A_i.

EXAMPLE 8.11

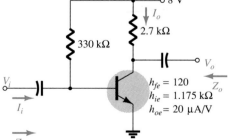

Figure 8.39 Example 8.11

8.9 **Approximate Hybrid Equivalent Circuit**

Solution

(a) $Z_i = R_B \| h_{ie} = 330 \text{ k}\Omega \| 1.175 \text{ k}\Omega$

$\quad\quad \cong h_{ie} = \mathbf{1.171 \text{ k}\Omega}$

(b) $r_o = \dfrac{1}{h_{oe}} = \dfrac{1}{20 \text{ }\mu\text{A/V}} = 50 \text{ k}\Omega$

$\quad\quad Z_o = \dfrac{1}{h_{oe}} \| R_C = 50 \text{ k}\Omega \| 2.7 \text{ k}\Omega = \mathbf{2.56 \text{ k}\Omega} \cong R_C$

(c) $A_v = -\dfrac{h_{fe} R_C}{h_{ie}} = -\dfrac{(120)(2.7 \text{ k}\Omega)}{1.171 \text{ k}\Omega} = \mathbf{-276.69}$

(d) $A_i \cong h_{fe} = \mathbf{120}$

Voltage-Divider Configuration

For the voltage-divider bias configuration of Fig. 8.40 the resulting small-signal ac equivalent network will have the same appearance as Fig. 8.38 with R_B replaced by $R' = R_1 \| R_2$.

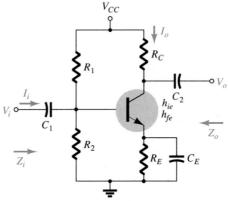

Figure 8.40 Voltage-divider bias configuration.

Z_i: From Fig. 8.38 with $R_B = R'$,

$$\boxed{Z_i = R' \| h_{ie}} \tag{8.67}$$

Z_o: From Fig. 8.38,

$$\boxed{Z_o = R_C} \tag{8.68}$$

A_v:

$$\boxed{A_v = -\dfrac{h_{fe} R_C}{h_{ie}}} \tag{8.69}$$

A_i:

$$\boxed{A_i = -\dfrac{h_{fe} R_{BB}}{R_{BB} + h_{ie}}} \tag{8.70}$$

The effect of $r_o = 1/h_{oe}$ is the same as encountered for the fixed-bias configuration.

Unbypassed Emitter-Bias Configuration

For the *CE* unbypassed emitter-bias configuration of Fig. 8.41 the small-signal ac model will be the same as Fig. 8.11, with βr_e replaced by h_{ie} and βI_b by $h_{fe} I_b$. The analysis will proceed in exactly the same manner with

Chapter 8 BJT Small-Signal Analysis

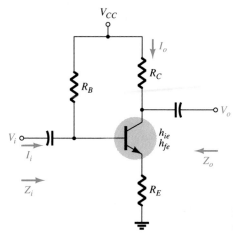

Figure 8.41 CE unbypassed emitter-bias configuration.

Z_i:

$$\boxed{Z_b \cong h_{fe}R_E} \tag{8.71}$$

and

$$\boxed{Z_i = R_B \| Z_b} \tag{8.72}$$

Z_o:

$$\boxed{Z_o = R_C} \tag{8.73}$$

A_v:

$$A_v = -\frac{h_{fe}R_C}{Z_b} \cong -\frac{h_{fe}R_C}{h_{fe}R_E}$$

and

$$\boxed{A_v \cong -\frac{R_C}{R_E}} \tag{8.74}$$

A_i:

$$\boxed{A_i = \frac{h_{fe}R_B}{R_B + Z_b}} \tag{8.75}$$

or

$$\boxed{A_i = -A_v\frac{Z_i}{R_C}} \tag{8.76}$$

Emitter-Follower Configuration

For the emitter-follower of Fig. 8.42 the small-signal ac model will match Fig. 8.18 with $\beta r_e = h_{ie}$ and $\beta = h_{fe}$. The resulting equations will therefore be quite similar.

Z_i:

$$\boxed{Z_b \cong h_{fe}R_E} \tag{8.77}$$

$$\boxed{Z_i = R_B \| Z_b} \tag{8.78}$$

8.9 **Approximate Hybrid Equivalent Circuit**   365

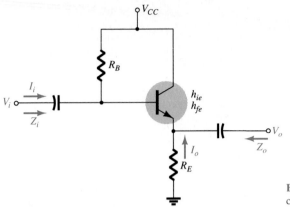

Figure 8.42 Emitter-follower configuration.

Z_o: For Z_o the output network defined by the resulting equations will appear as shown in Fig. 8.43. Review the development of the equations in Section 8.5 and

$$Z_o = R_E \| \frac{h_{ie}}{1 + h_{fe}}$$

or since $1 + h_{fe} \cong h_{fe}$,

$$\boxed{Z_o \cong R_E \| \frac{h_{ie}}{h_{fe}}} \tag{8.79}$$

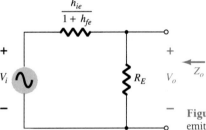

Figure 8.43 Defining Z_o for the emitter-follower configuration.

A_v: For the voltage gain the voltage-divider rule can be applied to Fig. 8.43 as follows:

$$V_o = \frac{R_E(V_i)}{R_E + h_{ie}/(1 + h_{fe})}$$

but since $1 + h_{fe} \cong h_{fe}$,

$$\boxed{A_v = \frac{V_o}{V_i} \cong \frac{R_E}{R_E + h_{ie}/h_{fe}}} \tag{8.80}$$

A_i:

$$\boxed{A_i = \frac{h_{fe}R_B}{R_B + Z_b}} \tag{8.81}$$

or

$$\boxed{A_i = -A_v\frac{Z_i}{R_E}} \tag{8.82}$$

Common-Base Configuration

The last configuration to be examined with the approximate hybrid equivalent circuit will be the common-base amplifier of Fig. 8.44. Substituting the approximate common-base hybrid equivalent model will result in the network of Fig. 8.45, which is very similar to Fig. 8.24. From Fig. 8.45,

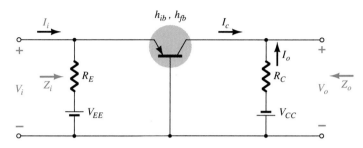

Figure 8.44 Common-base configuration.

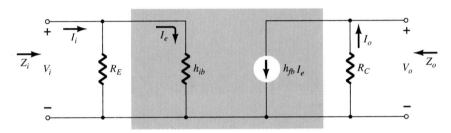

Figure 8.45 Substituting the approximate hybrid equivalent circuit into the ac equivalent network of Fig. 8.44.

Z_i:

$$Z_i = R_E \| h_{ib} \qquad (8.83)$$

Z_o:

$$Z_o = R_C \qquad (8.84)$$

A_v:

$$V_o = -I_o R_C = -(h_{fb} I_e) R_C$$

with $\qquad I_e = \dfrac{V_i}{h_{ib}} \qquad$ and $\qquad V_o = -h_{fb} \dfrac{V_i}{h_{ib}} R_C$

so that $\qquad A_v = \dfrac{V_o}{V_i} = -\dfrac{h_{fb} R_C}{h_{ib}} \qquad (8.85)$

A_i:

$$A_i = \dfrac{I_o}{I_i} = h_{fb} \cong -1 \qquad (8.86)$$

EXAMPLE 8.12

For the network of Fig. 8.46, determine:

(a) Z_i.

(b) Z_o.

(c) A_v.

(d) A_i.

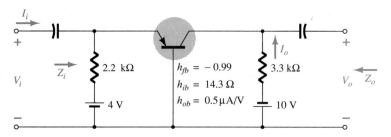

Figure 8.46 Example 8.12

Solution

(a) $Z_i = R_E \| h_{ib} = 2.2 \text{ k}\Omega \| 14.3 \ \Omega = \textbf{14.21 } \boldsymbol{\Omega} \cong h_{ib}$

(b) $r_o = \dfrac{1}{h_{ob}} = \dfrac{1}{0.5 \ \mu\text{A/v}} = 2 \text{ M}\Omega$

$Z_o = \dfrac{1}{h_{ob}} \| R_C \cong R_C = \textbf{3.3 k}\boldsymbol{\Omega}$

(c) $A_v = -\dfrac{h_{fb} R_C}{h_{ib}} = -\dfrac{(-0.99)(3.3 \text{ k}\Omega)}{14.21} = \textbf{229.91}$

(d) $A_i \cong h_{fb} = \textbf{-1}$

The remaining configurations of Sections 8.1 through 8.8 that were not analyzed in this section are left as an exercise in the problem section of this chapter. It is assumed that the analysis above clearly reveals the similarities in approach using the r_e or approximate hybrid equivalent models, thereby removing any real difficulty with analyzing the remaining networks of the earlier sections.

8.10 COMPLETE HYBRID EQUIVALENT MODEL

The analysis of Section 8.9 was limited to the approximate hybrid equivalent circuit with some discussion about the output impedance. In this section we employ the complete equivalent circuit to show the impact of h_r and define in more specific terms the impact of h_o. It is important to realize that since the hybrid equivalent model has the same appearance for the common-base, common-emitter, and common-collector configurations, the equations developed in this section can be applied to each configuration. It is only necessary to insert the parameters defined for each configuration. That is, for a common-base configuration, h_{fb}, h_{ib}, and so on, are employed, while for a common-emitter configuration, h_{fe}, h_{ie}, and so on, are utilized. Recall that Appendix A permits a conversion from one set to the other if one set is provided and the other is required.

Consider the general configuration of Fig. 8.47 with the two-port parameters of particular interest. The complete hybrid equivalent model is then substituted in Fig. 8.48 using parameters that do not specify the type of configuration. In other words,

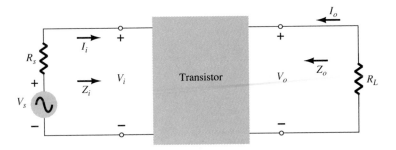

Figure 8.47 Two-port system

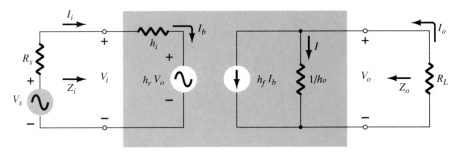

Figure 8.48 Substituting the complete hybrid equivalent circuit into the two-port system of Fig. 8.47.

the solutions will be in terms of h_i, h_r, h_f, and h_o. Unlike the analysis of previous sections of this chapter, the current gain A_i will be determined first since the equations developed will prove useful in the determination of the other parameters.

Current Gain, $A_i = I_o/I_i$

Applying Kirchhoff's current law to the output circuit yields

$$I_o = h_f I_b + I = h_f I_i + \frac{V_o}{1/h_o} = h_f I_i + h_o V_o$$

Substituting $V_o = -I_o R_L$ gives us

$$I_o = h_f I_i - h_o R_L I_o$$

Rewriting the equation above, we have

$$I_o + h_o R_L I_o = h_f I_i$$

and

$$I_o(1 + h_o R_L) = h_f I_i$$

so that

$$\boxed{A_i = \frac{I_o}{I_i} = \frac{h_f}{1 + h_o R_L}} \qquad (8.87)$$

Note that the current gain will reduce to the familiar result of $A_i = h_f$ if the factor $h_o R_L$ is sufficiently small compared to 1.

Voltage Gain, $A_v = V_o/V_i$

Applying Kirchhoff's voltage law to the input circuit results in

$$V_i = I_i h_i + h_r V_o$$

Substituting $I_i = (1 + h_o R_L)I_o/h_f$ from Eq. (8.87) and $I_o = -V_o/R_L$ from above results in

$$V_i = \frac{-(1 + h_o R_L)h_i}{h_f R_L}V_o + h_r V_o$$

Solving for the ratio V_o/V_i yields

$$\boxed{A_v = \frac{V_o}{V_i} = \frac{-h_f R_L}{h_i + (h_i h_o - h_f h_r)R_L}} \tag{8.88}$$

In this case the familiar form of $A_v = -h_f R_L/h_i$ will return if the factor $(h_i h_o - h_f h_r)R_L$ is sufficiently small compared to h_i.

Input Impedance, $Z_i = V_i/I_i$

For the input circuit,

$$V_i = h_i I_i + h_r V_o$$

Substituting

$$V_o = -I_o R_L$$

we have

$$V_i = h_i I_i - h_r R_L I_o$$

Since

$$A_i = \frac{I_o}{I_i}$$

$$I_o = A_i I_i$$

so that the equation above becomes

$$V_i = h_i I_i - h_r R_L A_i I_i$$

Solving for the ratio V_i/I_i, we obtain

$$Z_i = \frac{V_i}{I_i} = h_i - h_r R_L A_i$$

and substituting

$$A_i = \frac{h_f}{1 + h_o R_L}$$

yields

$$\boxed{Z_i = \frac{V_i}{I_i} = h_i - \frac{h_f h_r R_L}{1 + h_o R_L}} \tag{8.89}$$

The familiar form of $Z_i = h_i$ will be obtained if the second factor is sufficiently smaller than the first.

Output Impedance, $Z_o = V_o/I_o$

The output impedance of an amplifier is defined to be the ratio of the output voltage to the output current with the signal V_s set to zero. For the input circuit with $V_s = 0$,

$$I_i = \frac{-h_r V_o}{R_s + h_i}$$

Substituting this relationship into the following equation obtained from the output circuit yields

$$I_o = h_f I_i + h_o V_o$$

$$= \frac{-h_f h_r V_o}{R_s + h_i} + h_o V_o$$

and

$$Z_o = \frac{V_o}{I_o} = \frac{1}{h_o - [h_f h_r / (h_i + R_s)]} \qquad (8.90)$$

In this case the output impedance will reduce to the familiar form $Z_o = 1/h_o$ for the transistor when the second factor in the denominator is sufficiently less than the first.

EXAMPLE 8.13

For the network of Fig. 8.49 determine the following parameters using the complete hybrid equivalent model and compare to the results obtained using the approximate model.

(a) Z_i and Z_i'.

(b) A_v.

(c) $A_i = I_o/I_i$ and $A_i' = I_o/I_i'$.

(d) Z_o (within R_C) and Z_o' (including R_C).

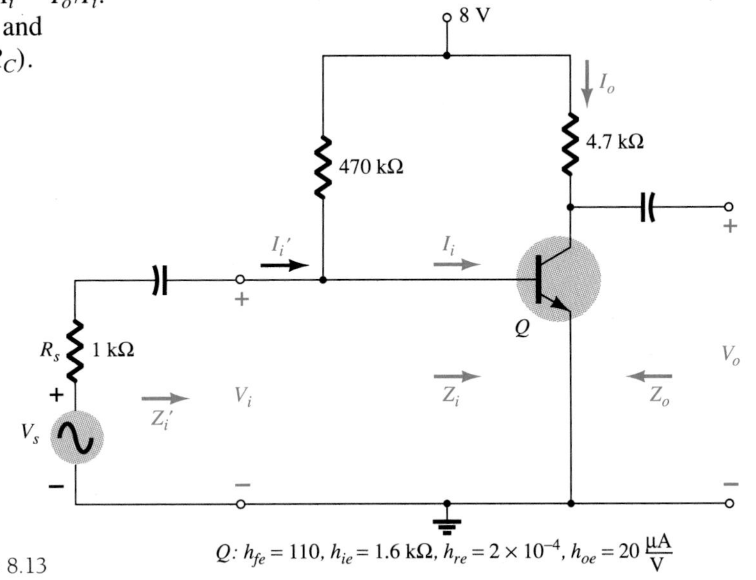

$Q: h_{fe} = 110, h_{ie} = 1.6 \text{ k}\Omega, h_{re} = 2 \times 10^{-4}, h_{oe} = 20 \frac{\mu A}{V}$

Figure 8.49 Example 8.13

Solution

Now that the basic equations for each quantity have been derived the order in which they are calculated is arbitrary. However, the input impedance is often a useful quantity to know and therefore will be calculated first. The complete common-emitter hybrid equivalent circuit has been substituted and the network redrawn as shown in Fig. 8.50. A Thévenin equivalent circuit for the input section of Fig. 8.50 will result in the input equivalent of Fig. 8.51 since $E_{Th} \cong V_s$ and $R_{Th} \cong R_s = 1 \text{ k}\Omega$ (a result of

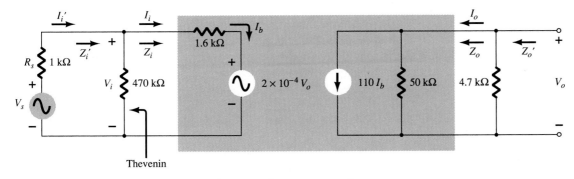

Figure 8.50 Substituting the complete hybrid equivalent circuit into the ac equivalent network of Fig. 8.49.

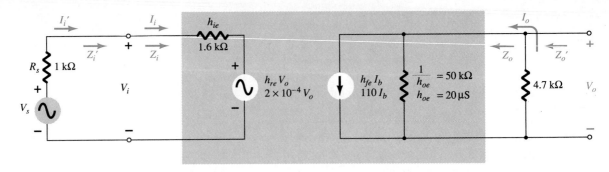

Figure 8.51 Replacing the input section of Fig. 8.50 with a Thévenin equivalent circuit.

$R_B = 470$ kΩ being much greater than $R_s = 1$ kΩ). In this example $R_L = R_C$ and I_o is defined as the current through R_C as in previous examples of this chapter. The output impedance Z_o as defined by Eq. (8.90) is for the output transistor terminals only. It does not include the effects of R_C. Z_o' is simply the parallel combination of Z_o and R_L. The resulting configuration of Fig. 8.51 is then an exact duplicate of the defining network of Fig. 8.48 and the equations derived above can be applied.

(a) Eq. (8.89): $Z_i = \dfrac{V_i}{I_i} = h_{ie} - \dfrac{h_{fe}h_{re}R_L}{1 + h_{oe}R_L}$

$\qquad\qquad = 1.6 \text{ k}\Omega - \dfrac{(110)(2 \times 10^{-4})(4.7 \text{ k}\Omega)}{1 + (20 \text{ } \mu\text{S})(4.7 \text{ k}\Omega)}$

$\qquad\qquad = 1.6 \text{ k}\Omega - 94.52 \text{ } \Omega$

$\qquad\qquad = \mathbf{1.51 \text{ k}\Omega}$

versus 1.6 kΩ using simply h_{ie}.

$$Z_i' = 470 \text{ k}\Omega \| Z_i \cong Z_i = \mathbf{1.51 \text{ k}\Omega}$$

(b) Eq. (8.88): $A_v = \dfrac{V_o}{V_i} = \dfrac{-h_{fe}R_L}{h_{ie} + (h_{ie}h_{oe} - h_{fe}h_{re})R_L}$

$\qquad\qquad = \dfrac{-(110)(4.7 \text{ k}\Omega)}{1.6 \text{ k}\Omega + [(1.6 \text{ k}\Omega)(20 \text{ } \mu\text{S}) - (110)(2 \times 10^{-4})]4.7 \text{ k}\Omega}$

$\qquad\qquad = \dfrac{-517 \times 10^3 \text{ } \Omega}{1.6 \text{ k}\Omega + (0.032 - 0.022)4.7 \text{ k}\Omega}$

$\qquad\qquad = \dfrac{-517 \times 10^3 \text{ } \Omega}{1.6 \text{ k}\Omega + 47 \text{ } \Omega}$

$\qquad\qquad = \mathbf{-313.9}$

versus − 323.125 using $A_v \cong -h_{fe}R_L/h_{ie}$.

(c) Eq. (8.87): $A_i = \dfrac{I_o}{I_i} = \dfrac{h_{fe}}{1 + h_{oe}R_L} = \dfrac{110}{1 + (20 \text{ } \mu\text{S})(4.7 \text{ k}\Omega)}$

$\qquad\qquad = \dfrac{110}{1 + 0.094} = \mathbf{100.55}$

versus 110 using simply h_{fe}. Since 470 kΩ ≫ Z_i, $I_i' \cong I_i$ and $A_i' \cong \mathbf{100.55}$ also.

(d) Eq. (8.90):

$$Z_o = \frac{V_o}{I_o} = \frac{1}{h_{oe} - [h_{fe}h_{re}/(h_{ie} + R_s)]}$$

$$= \frac{1}{20\ \mu S - [(110)(2 \times 10^{-4})/(1.6\ k\Omega + 1\ k\Omega)]}$$

$$= \frac{1}{20\ \mu S - 8.46\ \mu S}$$

$$= \frac{1}{11.54\ \mu S}$$

$$= \mathbf{86.66\ k\Omega}$$

which is greater than the value determined from $1/h_{oe} = 50\ k\Omega$.

$$Z'_o = R_C\|Z_o = 4.7\ k\Omega\|86.66\ k\Omega = \mathbf{4.46\ k\Omega}$$

versus 4.7 kΩ using only R_C.

Note from the results above that the approximate solutions for A_v and Z_i were very close to those calculated with the complete equivalent model. In fact, even A_i was off by less than 10%. The higher value of Z_o only contributed to our earlier conclusion that Z_o is often so high that it can be ignored compared to the applied load. However, keep in mind that when there is a need to determine the impact of h_{re} and h_{oe} the complete hybrid equivalent model, must be used, as described above.

The specification sheet for a particular transistor typically provides the common-emitter parameters as noted in Fig. 7.28. The next example will employ the same transistor parameters appearing in Fig. 8.49 in a *pnp* common-base configuration to introduce the parameter conversion procedure and emphasize the fact that the hybrid equivalent model maintains the same layout.

For the common-base amplifier of Fig. 8.52, determine the following parameters using the complete hybrid equivalent model and compare the results to those obtained using the approximate model.

(a) Z_i and Z'_i.
(b) A_i and A'_i.
(c) A_v.
(d) Z_o and Z'_o.

EXAMPLE 8.14

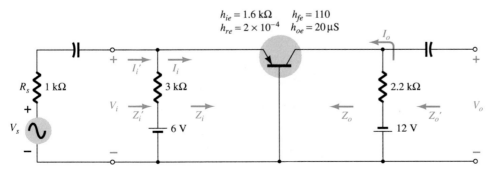

Figure 8.52 Example 8.14

Solution

The common-base hybrid parameters are derived from the common-emitter parameters using the approximate equations of Appendix A:

$$h_{ib} \cong \frac{h_{ie}}{1 + h_{fe}} = \frac{1.6 \text{ k}\Omega}{1 + 110} = \mathbf{14.41 \ \Omega}$$

Note how closely the magnitude compares with the value determined from

$$h_{ib} = r_e = \frac{h_{ie}}{\beta} = \frac{1.6 \text{ k}\Omega}{110} = 14.55 \ \Omega$$

$$h_{rb} \cong \frac{h_{ie}h_{oe}}{1 + h_{fe}} - h_{re} = \frac{(1.6 \text{ k}\Omega)(20 \ \mu\text{S})}{1 + 110} - 2 \times 10^{-4}$$

$$= \mathbf{0.883 \times 10^{-4}}$$

$$h_{fb} \cong \frac{-h_{fe}}{1 + h_{fe}} = \frac{-110}{1 + 110} = \mathbf{-0.991}$$

$$h_{ob} \cong \frac{h_{oe}}{1 + h_{fe}} = \frac{20 \ \mu\text{S}}{1 + 110} = \mathbf{0.18 \ \mu\text{S}}$$

Substituting the common-base hybrid equivalent circuit into the network of Fig. 8.52 will then result in the small-signal equivalent network of Fig. 8.53. The Thévenin network for the input circuit will result in $R_{\text{Th}} = 3 \text{ k}\Omega \| 1 \text{ k}\Omega = 0.75 \text{ k}\Omega$ for R_s in the equation for Z_o.

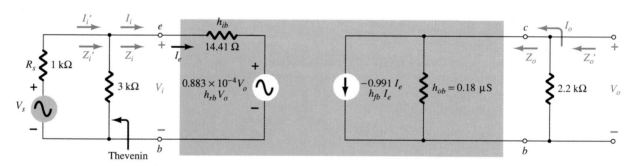

Figure 8.53 Small-signal equivalent for the network of Fig. 8.52.

(a) Eq. (8.89): $Z_i = \dfrac{V_i}{I_i} = h_{ib} - \dfrac{h_{fb}h_{rb}R_L}{1 + h_{ob}R_L}$

$$= 14.41 \ \Omega - \frac{(-0.991)(0.883 \times 10^{-4})(2.2 \text{ k}\Omega)}{1 + (0.18 \ \mu\text{S})(2.2 \text{ k}\Omega)}$$

$$= 14.41 \ \Omega + 0.19 \ \Omega$$

$$= \mathbf{14.60 \ \Omega}$$

versus $14.41 \ \Omega$ using $Z_i \cong h_{ib}$.

$$Z_i' = 3 \text{ k}\Omega \| Z_i \cong Z_i = \mathbf{14.60 \ \Omega}$$

(b) Eq. (8.87): $A_i = \dfrac{I_o}{I_i} = \dfrac{h_{fb}}{1 + h_{ob}R_L}$

$$= \dfrac{-0.991}{1 + (0.18\ \mu S)(2.2\ k\Omega)}$$

$$= -0.991 = h_{fb}$$

Since $3\ k\Omega \gg Z_i$, $I'_i \cong I_i$ and $A'_i = I_o/I'_i \cong -1$ also.

(c) Eq. (8.88): $A_v = \dfrac{V_o}{V_i} = \dfrac{-h_{fb}R_L}{h_{ib} + (h_{ib}h_{ob} - h_{fb}h_{rb})R_L}$

$$= \dfrac{-(-0.991)(2.2\ k\Omega)}{14.41\ \Omega + [(14.41\ \Omega)(0.18\ \mu S) - (-0.991)(0.883 \times 10^{-4})]2.2\ k\Omega}$$

$$= 149.25$$

versus 151.3 using $A_v \cong -h_{fb}R_L/h_{ib}$.

(d) Eq. (8.90): $Z_o = \dfrac{1}{h_{ob} - [h_{fb}h_{rb}/(h_{ib} + R_s)]}$

$$= \dfrac{1}{0.18\ \mu S - [(-0.991)(0.883 \times 10^{-4})/(14.41\ \Omega + 0.75\ k\Omega)]}$$

$$= \dfrac{1}{0.295\ \mu S}$$

$$= 3.39\ M\Omega$$

versus 5.56 MΩ using $Z_o \cong 1/h_{ob}$. For Z'_o as defined by Fig. 8.53:

$$Z'_o = R_C \| Z_o = 2.2\ k\Omega \| 3.39\ M\Omega = 2.199\ k\Omega$$

versus 2.2 kΩ using $Z'_o \cong R_C$.

8.11 SUMMARY TABLE

Now that the most familiar configurations of the small-signal transistor amplifiers have been introduced, Table 8.1 is presented to review the general characteristics of each for immediate recall. It must be absolutely clear that the values listed are simply typical values to establish a basis for comparison. The levels obtained in an actual analysis will most likely be different and certainly different from one configuration to another. Being able to repeat most of the information in the table is an important first step in developing a general familiarity with the subject matter. For instance, one should now be able to state with some assurance that the emitter-follower configuration typically has a high input impedance, low output impedance, and a voltage gain slightly less than 1. There should be no need to perform a variety of calculations to recall salient facts such as those above. For the future it will permit the study of a network or system without becoming mathematically involved. The function of each component of a design will become increasingly familiar as general facts such as those above become part of your background.

TABLE 8.1 Relative Levels for the Important Parameters of the CE, CB, and CC Transistor Amplifiers

Configuration		Z_i	A_v	A_i	Z_o
Fixed-bias:		Medium (1 kΩ) $\cong \beta r_e$	High (−200) $\cong -\dfrac{R_C}{r_e}$	High (100) $\cong \beta$	Medium $\cong R_C$ ($r_o = 40$ kΩ)
Voltage-divider bias:		Medium (1 kΩ) $\cong R'\|\beta r_e$ ($R' = R_1\|R_2$)	High (−200) $\cong -\dfrac{R_C}{r_e}$	High (50) $= \dfrac{\beta R'}{R' + \beta r_e}$	Medium $\cong R_C$ ($r_o = 40$ kΩ)
Unbypassed emitter bias:		High (100 kΩ) $\cong \beta R_E$	Low (−5) $\cong -\dfrac{R_C}{R_E}$	High (50) $= \dfrac{\beta R_B}{R_B + \beta R_E}$	Medium $\cong R_C$ ($r_o = 40$ kΩ)
Emitter-follower		High (100 kΩ) $\cong R_B\|\beta R_E$	Low $\cong 1$	High (−50) $\cong \dfrac{-\beta R_B}{R_B + \beta R_E}$	Low (20 Ω) $\cong r_e$ ($r_o = 40$ kΩ)
Common-base		Low (20 Ω) $\cong r_e$	High (200) $\cong \dfrac{R_C}{r_e}$	Low $\cong -1$	Medium $\cong R_C$ ($r_o = 2$ MΩ)
Collector feedback		Medium (1 kΩ) $= \beta r_e\|\dfrac{R_F}{|A_v|}$	High (−200) $\cong -\dfrac{R_C}{r_e}$	High (50) $= \dfrac{\beta R_F}{R_F + \beta R_C}$	Medium $\cong R_C\|R_F$ ($r_o = 40$ kΩ)

One obvious advantage of being able to recall general facts like the above is an ability to check the results of a mathematical analysis. If the input impedance of a common-base configuration is in the kilohm range there is good reason to recheck the analysis. However, on the other side of the coin, a result of 22 Ω suggests that the analysis may be correct.

8.12 TROUBLESHOOTING

Although the terminology *troubleshooting* suggests that the procedures to be described are designed simply to isolate a malfunction, it is important to realize that the same techniques can be applied to ensure that a system is operating properly. In any case, the testing, checking, or isolating procedures require an understanding of what to expect at various points in the network in both the dc and ac domains. In most cases, a network operating correctly in the dc mode will also behave properly in the ac domain. In addition, a network providing the expected ac response is most likely biased as planned. In a typical laboratory setting both the dc and ac supplies are applied and the ac response at various points in the network is checked with an oscilloscope as shown in Fig. 8.54. Note that the black (gnd) lead of the oscilloscope is connected directly to ground and the red lead is moved from point to point in the network, providing the patterns appearing in Fig. 8.54. The vertical channels are set in the ac mode to remove any dc component associated with the voltage at a particular point. The small ac signal applied to the base is amplified to the level appearing from collector to ground. Note the difference in vertical scales for the two voltages. There is no ac response at the emitter terminal due to the short-circuit characteristics of the capacitor at the applied frequency. The fact that v_o is measured in volts and v_i in millivolts suggests a sizable gain for the amplifier. In general, the network appears to be operating properly. If desired, the dc mode of the multimeter could be used to check V_{BE} and the levels of V_B, V_{CE}, and V_E to review whether they lie in the expected range. Of course, the oscilloscope can also be used to compare dc levels simply by switching to the dc mode for each channel.

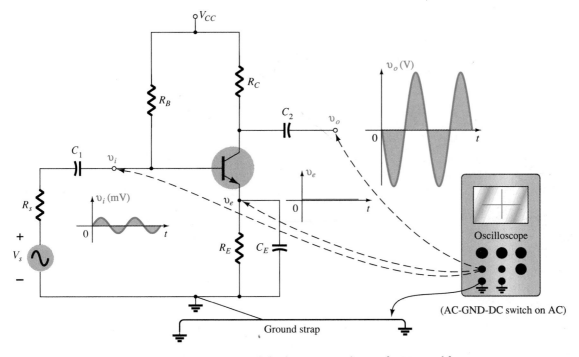

Figure 8.54 Using the oscilloscope to measure and display various voltages of a BJT amplifier.

Needless to say, a poor ac response can be due to a variety of reasons. In fact, there may be more than one problem area in the same system. Fortunately, however, with time and experience the probability of malfunctions in some areas can be predicted and an experienced person can isolate problem areas fairly quickly.

In general, there is nothing mysterious about the general troubleshooting process. If you decide to follow the ac response, it is good procedure to start with the applied signal and progress through the system toward the load checking critical points along the way. An unexpected response at some point suggests that the network is fine up to that area, thereby defining the region that must be investigated further. The waveform obtained on the oscilloscope will certainly help in defining the possible problems with the system.

If the response for the network of Fig. 8.54 is as appearing in Fig. 8.55, the network has a malfunction that is probably in the emitter area. An ac response across the emitter is unexpected and the gain of the system as revealed by v_o is much lower. Recall for this configuration that the gain is much greater if R_E is bypassed. The response obtained suggests that R_E is not bypassed by the capacitor and the terminal connections of the capacitor and the capacitor itself should be checked. In this case a checking of the dc levels will probably not isolate the problem area since the capacitor has an "open-circuit" equivalent for dc. In general, a prior knowledge of what to expect, a familiarity with the instrumentation, and most important, experience are all factors that contribute to the development of an effective approach to the art of trouble-shooting.

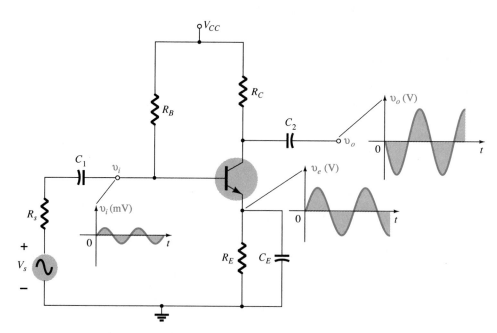

Figure 8.55 The waveforms resulting from a malfunction in the emitter area.

8.13 COMPUTER ANALYSIS

The analysis of a small-signal BJT amplifier can be performed using a software package such as PSpice or a language such as BASIC. Both will be employed in the analysis of the same voltage-divider bias configuration to permit a comparison of methods. PSpice is well equipped to analyze transistor networks using an enhanced Gummel–Poon model described in detail in the PSpice manual. If specific parameters

Chapter 8 BJT Small-Signal Analysis

of this model are left unspecified, the model takes on the appearance of the r_e model employed in this book. The use of a language such as BASIC requires that the various equations developed in the book be applied in a specific order to obtain the desired unknowns. In reality the same sequence of steps used to analyze the network in the longhand fashion (with calculator in hand) would be the general direction of a BASIC program. Of course, the use of BASIC provides the opportunity for the user to define the scope and type of output for an analysis, while PSpice is limited to a specific list of output quantities. In general, however, the PSpice list is extensive enough for most investigations. The analysis will first be described using PSpice followed by the use of the BASIC language.

PSpice

The list of parameters that can be specified for the PSpice model is so extensive (40 in total) that we will limit our attention to those parameters required to perform the type of analysis covered in this chapter. As additional parameters are required in the chapters to follow, they will be defined in the same detail. One does not have to be concerned about not specifying all the parameters. If a particular parameter is required to perform the PSpice analysis but is not specified, the software package will employ a *default* value that is typical for the device under investigation. Some of the parameters need to be specified only if required by the depth of the analysis or design. The primary intent of this section is to provide an introduction to the use of models that is as clear and uncluttered as possible. As your expertise grows, the PSpice manuals and a growing list of publications are available for extensive detail and additional instruction.

In general, once the network nodes have been defined and the basic structure (resistors, capacitors, sources, etc.) entered into the input file, a minimum of two lines is required to describe a transistor. The first is the element line that has the following format:

QXISTOR	9	8	7	QMODEL
required ——⌐ name	collector node	base node	emitter node	name of transistor model to be defined by model line below

Other parameters can be defined on this line but they are beyond the needs of this book and can be referenced in the PSpice manual.

The next required line to define the transistor is the model line, which has the following basic format:

.MODEL	QMODEL	NPN	(BF = 90, IS = 5E − 15)
required	model name specified in element line above	type of transistor (required)	specifying parameters of the model

The last grouping of the line above permits specification of particular parameters of the model (a list that can include 40 parameters). BF stands for the ideal maximum forward beta (in this case, $\beta = 90$). Its default value is 100, revealing that if the parameter is unspecified as above, the software package will use a value of 100. In the model the reverse saturation current has an important impact on the general characteristics of the model. Its default value is 1E-16 or 0.0001 pA. Changing the level of I_s will change the level of important design voltages and currents such as V_{BE} for the dc analysis and I_C for the ac analysis. In fact, since V_{BE} is set at 0.7 V for the dc analysis of this book, a level of 5×10^{-15} A was chosen for I_s since the resulting level of V_{BE} is usually very close to 0.7 V for the range of current levels expected for BJT small-signal analysis. In other words, PSpice does not permit specifying the level of V_{BE} for dc analysis but simply uses the saturation current and a series of important equations to determine the resulting level of V_{BE}. For this reason V_{BE} will seldom be exactly 0.7 V, but it will be just above or below this value. Consider the 0.7 V level to be an average of expected levels using PSpice if I_s is specified as 5×10^{-15} A.

We are now ready to apply PSpice to the voltage-divider network of Fig. 8.9 (Example 8.2). The network is redrawn in Fig. 8.56 with the defined nodes for the analysis. Since specific characteristics such as A_v and A_i are not part of the list of output options in PSpice, we will apply a signal of 1 mV and calculate the gain using the output level.

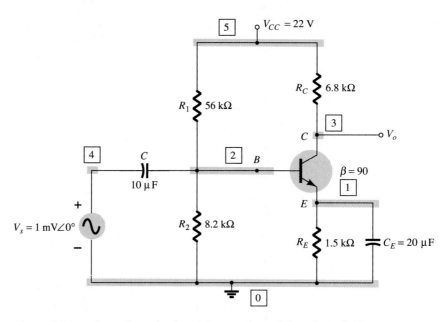

Figure 8.56 Defining the nodes for a PSpice analysis of the voltage-divider configuration.

By now the first eight lines of the input file of Fig. 8.57 should be fairly familiar and understandable. The transistor is then defined on the next two lines, with QMODEL being the name of the model for the transistor. Note on the model line that beta is specified as 90. No value of IS was specified to demonstrate the impact on the results obtained. The second run will include the suggested level of IS for comparison purposes. The .PRINT command requests both the magnitude and phase angle for the output voltage from collector to ground. As required by the ac source, a frequency of 10 kHz was chosen for the run. The only real impact of the applied frequency will be on the capacitive elements and their effectiveness as equivalent short circuits for the ac analysis.

Once the input file is entered, PSpice is run and a list of BJT model parameters is listed. Note that β (BF) is 90 and I_s (IS) is the default value of 1×10^{-4} pA. NF (the forward current emission coefficient), BR (the ideal maximum reverse beta), and NR (the reverse current emission coefficient) take on the default value of 1. The last three quantities define the behavior of the model in a way that is beyond the needs of this book and will have minimal impact on the current small-signal ac analysis.

```
**** 05/03/90 ******* Evaluation PSpice (January 1989) ******* 17:32:07 ****

 Voltage-Divider Bias - Configuration of Fig. 8.56(IS = default value)

****        CIRCUIT DESCRIPTION

************************************************************************

VCC 5 0 DC 22V
RB1 5 2 56K
RB2 2 0 8.2K
RE 1 0 1.5K
RC 5 3 6.8K
C1 4 2 10UF
CE 1 0 20UF
VS 4 0 AC 1MV 0
Q1 3 2 1 QMODEL
.MODEL QMODEL NPN(BF=90)
.OP
.AC LIN 1 10KH 10KH
.PRINT AC VM(3,0) VP(3,0)
.OPTIONS NOPAGE
.END

****        BJT MODEL PARAMETERS

             QMODEL
             NPN
      IS     100.000000E-18
      BF      90
      NF      1
      BR      1
      NR      1


****        SMALL SIGNAL BIAS SOLUTION        TEMPERATURE =   27.000 DEG C

 NODE    VOLTAGE       NODE    VOLTAGE      NODE    VOLTAGE      NODE    VOLTAGE
(   1)    1.9285    (    2)    2.7089    (    3)   13.3540    (    4)    0.0000
(   5)   22.0000


    VOLTAGE SOURCE CURRENTS
    NAME         CURRENT
    VCC         -1.616E-03
    VS           0.000E+00

    TOTAL POWER DISSIPATION    3.56E-02   WATTS


****        OPERATING POINT INFORMATION       TEMPERATURE =   27.000 DEG C

**** BIPOLAR JUNCTION TRANSISTORS

NAME         Q1
MODEL        QMODEL
IB           1.41E-05
IC           1.27E-03
VBE          7.80E-01
VBC         -1.06E+01
VCE          1.14E+01
BETADC       9.00E+01
GM           4.92E-02
RPI          1.83E+03
RX           0.00E+00
RO           1.00E+12
CBE          0.00E+00
CBC          0.00E+00
CBX          0.00E+00
CJS          0.00E+00
BETAAC       9.00E+01
FT           7.82E+17


****        AC ANALYSIS                       TEMPERATURE =   27.000 DEG C

  FREQ         VM(3,0)       VP(3,0)
  1.000E+04    3.340E-01   -1.777E+02
```

Figure 8.57 PSpice analysis of the voltage-divider configuration of Figure 8.56 with IS = default value.

PSpice is then designed to perform a dc analysis of the network automatically. The results are

$$V_1 = V_E = 1.9285 \text{ V}$$

$$V_2 = V_B = 2.7089 \text{ V}$$

$$V_3 = V_C = 13.354 \text{ V}$$

$$V_4 = V_{\text{gnd(for dc)}} = 0 \text{ V}$$

$$V_5 = V_{CC} = 22 \text{ V}$$

The output file then provides the source current for V_{CC} with the dc level of the ac source V_s at 0.0 A. The total dc power dissipated by the resistors and transistor is 35.6 mW.

Various other dc levels are then provided for the network such as $I_B = 14.1$ μA, $I_C = 1.27$ mA (compared to 1.41 mA in Example 8.2), and $V_{BE} = 0.78$ V (exceeding the level of 0.7 V used in Example 8.2). Keep the level of V_{BE} in mind when we review the results with I_s set to 5×10^{-15} A in the next run. The dc values of V_{BC} and V_{CE} are then specified as -10.6 V and 11.4 V, respectively, and the dc beta matches the ac beta of 90. The transconductance $g_m = 1/r_e$ and $r_e = 20.3$ Ω. The input impedance is then $\beta r_e = (90)(20.3 \ \Omega) = 1.827$ kΩ or 1.83 kΩ as specified by RPI. The output resistance is listed as 1×10^{12} Ω and the ac beta is 90 with FT (the ideal forward transit time) equal to 7.82×10^{17} s. Again, some of the parameters are probably meaningless at this point, but a number are quite recognizable and can be helpful in checking a design or analysis.

The ac analysis to follow reveals that the magnitude of V_o is 334 mV for a voltage gain of 334 compared to a gain of 368.76 determined in Example 8.2. The phase shift is 177.7° rather than 180° due to the capacitive elements of the network. Choosing a higher frequency or increasing the capacitance level would bring the phase shift closer to 180°.

The effect of changing I_s to 5×10^{-15} A will be clearly demonstrated by the run of Fig. 8.58. The level of V_E is now 2.0235 V compared to 2.11 V for Example 8.2. The level of I_C is 1.33 mA compared to 1.41 mA, and the ac voltage gain is now 350.4 compared to 368.76 in Example 8.2. In general, a definite improvement in the match between hand-calculated and PSpice results. A further improvement will result

```
**** 05/03/90 ******* Evaluation PSpice (January 1989) ******* 17:36:27 ****

   Voltage-Divider Bias - Configuration of Fig. 8.56(specified IS)

****       CIRCUIT DESCRIPTION

***********************************************************************************
VCC 5 0 DC 22V
RB1 5 2 56K
RB2 2 0 8.2K
RE 1 0 1.5K
RC 5 3 6.8K
C1 4 2 10UF
CE 1 0 20UF
VS 4 0 AC 1MV 0
Q1 3 2 1 QMODEL
.MODEL QMODEL NPN(BF=90 IS=5E-15)
.OP
.AC LIN 1 10KH 10KH
.PRINT AC VM(3,0) VP(3,0)
.OPTIONS NOPAGE
.END
```

Figure 8.58 PSpice analysis of the voltage-divider configuration of Fig. 8.56 with IS = 5×10^{-15} A.

```
****        BJT MODEL PARAMETERS

            QMODEL
            NPN
      IS      5.000000E-15
      BF     90
      NF      1
      BR      1
      NR      1



****        SMALL SIGNAL BIAS SOLUTION          TEMPERATURE =    27.000 DEG C

  NODE    VOLTAGE     NODE    VOLTAGE     NODE    VOLTAGE     NODE    VOLTAGE
(    1)    2.0235   (    2)    2.7039   (    3)   12.9280   (    4)    0.0000
(    5)   22.0000


      VOLTAGE SOURCE CURRENTS
      NAME           CURRENT
      VCC          -1.679E-03
      VS            0.000E+00


    TOTAL POWER DISSIPATION    3.69E-02   WATTS


****        OPERATING POINT INFORMATION          TEMPERATURE =    27.000 DEG C

**** BIPOLAR JUNCTION TRANSISTORS

NAME          Q1
MODEL         QMODEL
IB            1.48E-05
IC            1.33E-03
VBE           6.80E-01
VBC          -1.02E+01
VCE           1.09E+01
BETADC        9.00E+01
GM            5.16E-02
RPI           1.74E+03
RX            0.00E+00
RO            1.00E+12
CBE           0.00E+00
CBC           0.00E+00
CBX           0.00E+00
CJS           0.00E+00
BETAAC        9.00E+01
FT            8.21E+17



****        AC ANALYSIS                          TEMPERATURE =    27.000 DEG C

    FREQ        VM(3,0)       VP(3,0)
   1.000E+04    3.504E-01    -1.776E+02
```

Figure 8.58 Continued.

if the exact rather than approximate solution were obtained in Ex. 8.2. In particular, note that V_{BE} is now 0.68 V, which compares very favorably with the fixed approximate value of 0.7 V. In general, therefore, for the small-signal analysis performed in this book using PSpice, IS will be specified as 5×10^{-15} A.

The introduction above was relatively brief due to space and priority constraints, but its purpose was served if the relative simplicity of applying PSpice to determine the small-signal response is now evident. When time permits, the manuals should be read carefully to fully understand the effect of the various parameters and the equations involved with the PSpice model. Be aware that a commercial version of PSpice is available that has an entire catalog of specific transistors in memory for use when called up by the PSpice software package. In other words, the input file can include reference to a particular transistor and the package will automatically insert the parameters that best describe that transistor for the analysis to be performed. Additional information on the commercial version can be obtained by writing directly to the Microsim Corp. Let us now compare the analysis above with the analysis of the same circuit using the BASIC language.

BASIC

The BASIC program of Fig. 8.59 will analyze the voltage-divider bias configuration of Fig. 8.56 with the added features that it can also provide a solution if a portion of the emitter resistor is unbypassed and can include the effects of a source and load resistance. The emitter resistance will be designated R_{E_1} if unbypassed and R_{E_2} if bypassed.

```
10 REM ********************************************************
20 REM                     PROGRAM 8.1
30 REM ********************************************************
40 REM                    BJT AC ANALYSIS
50 REM             USING re AND BETA PARAMETERS
60 REM ********************************************************
70 REM
100 CLS
110 PRINT "This program performs the ac calculations"
120 PRINT "for a BJT voltage-divider using the re and beta parameters."
130 PRINT
140 PRINT "Enter the following circuit data:"
150 PRINT
160 INPUT "RB1=";R1
170 INPUT "RB2=";R2
180 INPUT "RC=";RC
190 INPUT "Unbypassed emitter resistance, RE1=";E1
200 INPUT "Bypassed emitter resistance, RE2=";E2
210 PRINT
220 INPUT "Beta=";BETA
230 INPUT "Supply voltage, VCC=";CC
240 INPUT "Load resistance, RL=";RL
250 INPUT "Source resistance, RS=";RS
260 INPUT "Source voltage, VS=";VS
270 PRINT:PRINT
280 GOSUB 11200:REM Perform ac analysis
290 PRINT "The results of the ac analysis are:"
300 PRINT
310 PRINT "Transistor dynamic resistance, re=";RE;"ohms"
320 PRINT
330 IF CC-IE*(RC+E1+E2)<=0 THEN PRINT "Circuit in saturation." :GOTO 420
340 PRINT "Input impedance, Ri=";RI;"ohms"
350 PRINT "Output impedance, Ro=";RO;"ohms"
360 PRINT "Voltage-gain(no-load), Av=";AV
370 PRINT "Current gain, Ai=";AI
380 PRINT
390 PRINT "Output voltage(no load), Vo=";VO;"volts"
400 PRINT
410 PRINT "Output voltage(under load), VL=";VL;"volts"
420 PRINT
430 VM=CC-IE*(BETA/(BETA+1))*(RC+E1+E2) :REM Maximum signal swing
440 IF ABS(VL)>VM THEN PRINT "but maximum undistorted output is";VM;"volts"
450 END
11200 REM Module to perform BJT ac analysis using re model
11210 RB=R1*(R2/(R1+R2))
11220 RP=RC*(RL/(RC+RL))
11230 BB=R2*CC/(R1+R2)
11240 IE=(BB-.7)*(BETA+1)/(RB+BETA*(E1+E2))
11250 RE=.026/IE
11260 R3=BETA*(RE+E1)
11270 RI=RB*(R3/(RB+R3))
11280 RO=RC
11290 AI=(RC/(RC+RL))*BETA*(RB/(RB+R3))
11300 AV=-RC/(E1+RE)
11310 VI=VS*(RI/(RI+RS))
11320 VO=AV*VI
11330 VL=VO*(RL/(RO+RL))
11340 RETURN
```

Figure 8.59 BASIC program for the ac analysis of a BJT configuration.

```
RUN
This program performs the ac calculations
for a BJT voltage-divider using the re and beta parameters.

Enter the following circuit data:

RB1=? 56E3
RB2=? 8.2E3
RC=? 6.8E3
Unbypassed emitter resistance, RE1=? 0
Bypassed emitter resistance, RE2=? 1.5E3

Beta=? 90
Supply voltage, VCC=? 22
Load resistance, RL=? 10E3
Source resistance, RS=? 600
Source voltage, VS=? 1E-3

The results of the ac analysis are:

Transistor dynamic resistance, re= 19.24912 ohms

Input impedance, Ri= 1394.631 ohms
Output impedance, Ro= 6800 ohms
Voltage-gain(no-load), Av=-353.263
Current gain, Ai= 29.32569

Output voltage(no load), Vo=-.2469988 volts

Output voltage(under load), VL=-.1470231 volts
```

Figure 8.59 Continued.

The module of lines 11210 through 11260 will determine the important parameters for the transistor model for Fig. 8.60 and perform the required analysis. The sequential steps of the module should be carefully reviewed and compared to calculations performed in the longhand (calculator) manner.

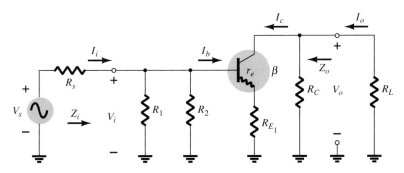

Figure 8.60 Network analyzed by the module extending from line 11210 through 11260 of the BASIC program of Fig. 8.59.

A run of the program with the values of Fig. 8.56 will provide the results appearing at the end of Fig. 8.59. In particular, note how the BASIC program can be written to provide information about the system in a clear, concise, tabulated manner. The level of $R_i = R' \parallel \beta r_e = 1{,}394.63 \ \Omega$, which is different from RI in the PSpice run since RI includes only the input impedance of the transistor configuration (βr_e). The no-load gain is 353.26, which compares nicely with the 368.76 obtained using PSpice. The current gain of $4.9 \times 10^{-25} \ A \cong 0 \ A$, due to the absence of a load to define an output current. The absence of a load also results in $A_v = A_{v_{NL}}$.

§ 8.2 Common-Emitter Fixed-Bias Configuration

1. For the network of Fig. 8.61:
  (a) Determine Z_i and Z_o.
  (b) Find A_v and A_i.
  (c) Repeat part (a) with $r_o = 40$ kΩ.
  (d) Repeat part (b) with $r_o = 40$ kΩ.

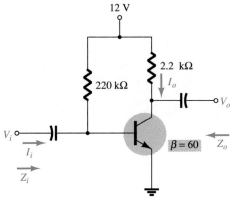

Figure 8.61 Problems 1, 21

2. For the network of Fig. 8.62, determine V_{CC} for a voltage gain of $A_v = -200$.

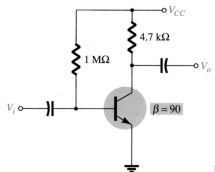

Figure 8.62 Problem 2

* **3.** For the network of Fig. 8.63:
  (a) Calculate I_B, I_C, and r_e.
  (b) Determine Z_i and Z_o.
  (c) Calculate A_v and A_i.
  (d) Determine the effect of $r_o = 40$ kΩ on A_v and A_i.

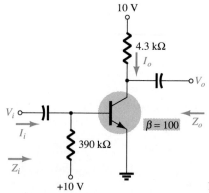

Figure 8.63 Problem 3

4. For the network of Fig. 8.64:
   (a) Determine r_e.
   (b) Calculate Z_i and Z_o.
   (c) Find A_v and A_i.
   (d) Repeat parts (b) and (c) with $r_o = 50\ k\Omega$.

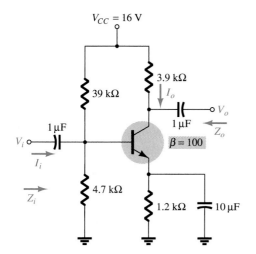

Figure 8.64 Problem 4

Figure 8.65 Problem 5

5. Determine V_{CC} for the network of Fig. 8.65 if $A_V = -160$ and $r_o = \infty\Omega$.

6. For the network of Fig. 8.66:
   (a) Determine r_e.
   (b) Calculate V_B and V_C.
   (c) Determine Z_i and $A_v = V_o/V_i$.

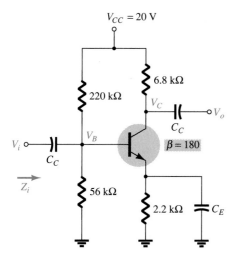

Figure 8.66 Problem 6

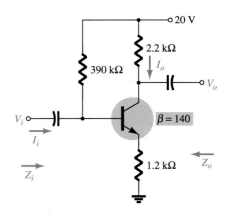

Figure 8.67 Problems 7, 9

§ **8.4 CE Emitter-Bias Configuration**

7. For the network of Fig. 8.67:
   (a) Determine r_e.
   (b) Find Z_i and Z_o.
   (c) Calculate A_v and A_i.
   (d) Repeat parts (b) and (c) with $r_o = 50\ k\Omega$.

8. For the network of Fig. 8.68, determine R_E and R_B if $A_v = -10$ and $r_e = 3.8 \ \Omega$. Assume that $Z_b = \beta R_E$.

9. Repeat Problem 7 with R_E bypassed. Compare results.

* **10.** For the network of Fig. 8.69:
  (a) Determine r_e.
  (b) Find Z_i and A_v.
  (c) Calculate A_i.
  (d) Repeat parts (b) and (c) with $r_o = 40 \ k\Omega$.

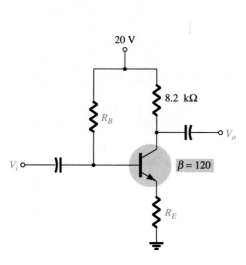

Figure 8.68 Problem 8

Figure 8.69 Problem 10

§ 8.5 Emitter-Follower Configuration

11. For the network of Fig. 8.70:
  (a) Determine r_e and βr_e.
  (b) Find Z_i and Z_o.
  (c) Calculate A_v and A_i.
  (d) Repeat parts (b) and (c) with $r_o = 40 \ k\Omega$.

* **12.** For the network of Fig. 8.71:
  (a) Determine Z_i and Z_o.
  (b) Find A_v.
  (c) Calculate V_o if $V_i = 1 \ mV$.
  (d) Repeat parts (a) through (c) using $r_o = 25 \ k\Omega$ and comment on the effect of this relatively low level of r_o.

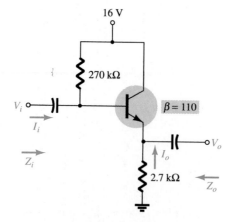

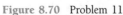

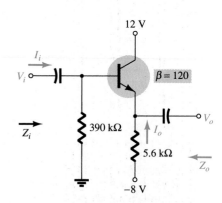

Figure 8.70 Problem 11

Figure 8.71 Problem 12

* **13.** For the network of Fig. 8.72:
    (a) Calculate I_B and I_C.
    (b) Determine r_e.
    (c) Determine Z_i and Z_o.
    (d) Find A_v and A_i.

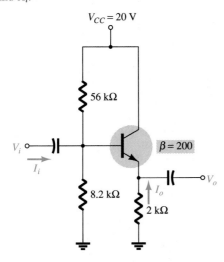

Figure 8.72 Problem 13

§ 8.6 **Common-Base Configuration**

14. For the common-base configuration of Fig. 8.73:
    (a) Determine r_e.
    (b) Find Z_i and Z_o.
    (c) Calculate A_v and A_i.

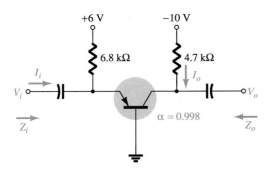

Figure 8.73 Problem 14

* **15.** For the network of Fig. 8.74, determine A_v and A_i.

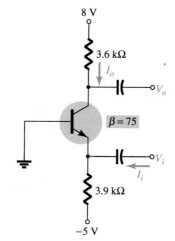

Figure 8.74 Problem 15

§ 8.7 Collector Feedback Configuration

16. For the collector FB configuration of Fig. 8.75:
  (a) Determine r_e.
  (b) Find Z_i and Z_o.
  (c) Calculate A_v and A_i.

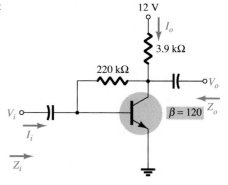

Figure 8.75 Problem 16

* **17.** Given $r_e = 10\ \Omega$, $\beta = 200$, $A_v = -160$, and $A_i = 19$ for the network of Fig. 8.76, determine R_C, R_F, and V_{CC}.

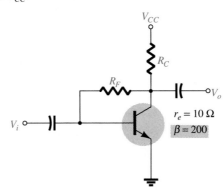

Figure 8.76 Problem 17

* **18.** For the network of Fig. 8.30:
  (a) Derive the approximate equation for A_v.
  (b) Derive the approximate equation for A_i.
  (c) Derive the approximate equations for Z_i and Z_o.
  (d) Given $R_C = 2.2\ \text{k}\Omega$, $R_F = 120\ \text{k}\Omega$, $R_E = 1.2\ \text{k}\Omega$, $\beta = 90$, and $V_{CC} = 10$ V, calculate the magnitude of A_v, A_i, Z_i, and Z_o using the equations of parts (a) through (c).

§ 8.8 Collector DC Feedback Configuration

19. For the network of Fig. 8.77:
  (a) Determine Z_i and Z_o.
  (b) Find A_v and A_i.
  (c) Repeat parts (a) and (b) if $r_o = 45\ \text{k}\Omega$.

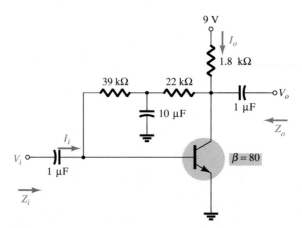

Figure 8.77 Problem 19

20. (a) Given $\beta = 120$, $r_e = 4.5\ \Omega$, and $r_o = 40\ \text{k}\Omega$, sketch the approximate hybrid equivalent circuit.

(b) Given $h_{ie} = 1\ \text{k}\Omega$, $h_{re} = 2 \times 10^{-4}$, $h_{fe} = 90$, and $h_{oe} = 20\ \mu\text{S}$, sketch the r_e model.

21. For the network of Problem 1:

(a) Determine r_e.

(b) Find h_{fe} and h_{ie}.

(c) Find Z_i and Z_o using the hybrid parameters.

(d) Calculate A_v and A_i using the hybrid parameters.

(e) Determine Z_i and Z_o if $h_{oe} = 25\ \mu\text{S}$.

(f) Determine A_v and A_i if $h_{oe} = 25\ \mu\text{S}$.

(g) Compare the solutions above with those of Problem 1. (Note: The solutions are available in Appendix E if Problem 1 was not performed.)

22. For the network of Fig. 8.78:

(a) Determine Z_i and Z_o.

(b) Calculate A_v and A_i.

(c) Determine r_e and compare βr_e to h_{ie}.

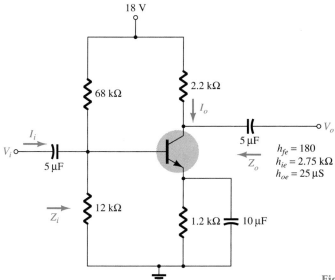

Figure 8.78 Problems 22, 24

* **23.** For the common-base network of Fig. 8.79:

(a) Determine Z_i and Z_o.

(b) Calculate A_v and A_i.

(c) Determine α, β, r_e, and r_o.

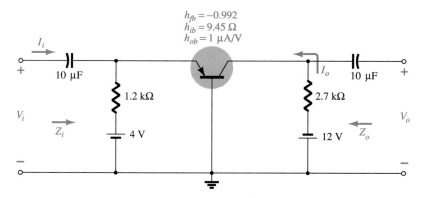

Figure 8.79 Problem 23

§ 8.10 Complete Hybrid Equivalent Model

* **24.** Repeat parts (a) and (b) of Problem 22 with $h_{re} = 2 \times 10^{-4}$ and compare results.

* **25.** For the network of Fig. 8.80, determine:
  - (a) Z_i.
  - (b) A_v.
  - (c) $A_i = I_o / I_i$.
  - (d) Z_o.

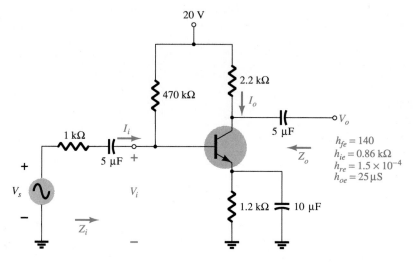

Figure 8.80 Problem 25

* **26.** For the common-base amplifier of Fig. 8.81, determine:
  - (a) Z_i.
  - (b) A_i.
  - (c) A_v.
  - (d) Z_o.

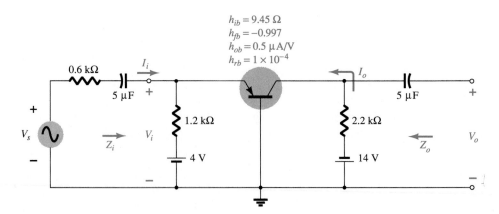

Figure 8.81 Problem 26

§ 8.12 Troubleshooting

* **27.** Given the network of Fig. 8.82:
  - (a) Determine if the system is operating properly based on the voltage-divider bias levels and expected waveforms for v_o and v_E.
  - (b) Determine the reason for the dc levels obtained and why the waveform for v_o was obtained.

Chapter 8 BJT Small-Signal Analysis

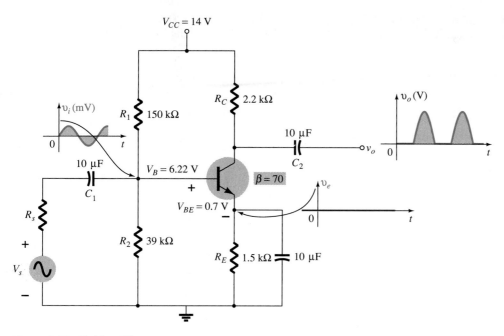

Figure 8.82 Problem 27

§ 8.13 Computer Analysis

28. (a) Write the PSpice input file for the network of Fig. 8.6 (Example 8.1) and request the level of V_o for $V_i = 1$ mV.
    (b) Perform the PSpice analysis and compare the result for V_o with the results of Example 8.1.

29. (a) Write the input file for the network of Fig. 8.13 (Example 8.3) and request the level of V_o for $V_i = 1$ mV.
    (b) Perform the PSpice analysis and compare the result for V_o with the results of Example 8.3.

30. (a) Write the input file for the network of Fig. 8.25 (Example 8.8) and request the level of V_o for $V_i = 1$ mV.
    (b) Perform the PSpice analysis and compare the result for V_o with the results of Example 8.8.

31. (a) Write a BASIC program to determine Z_i, Z_o, A_v, and A_i for the network of Fig. 8.9 (Example 8.2).
    (b) Perform the analysis of part (a) and compare with the results of Example 8.2.

32. (a) Write a BASIC program to determine Z_i, Z_o, A_v, and A_i for the network of Fig. 8.13 (Example 8.3).
    (b) Perform the analysis of part (a) and compare with the results of Example 8.3.

33. (a) Write a BASIC program to determine Z_i, Z_o, A_v, and A_i for the network of Fig. 8.25 (Example 8.8).
    (b) Perform the analysis of part (a) and compare with the results of Example 8.8.

---

*Please Note: Asterisks indicate more difficult problems.

FET Small-Signal Analysis

g_m

9.1 INTRODUCTION

FET devices can be used to build small-signal amplifier circuits providing voltage gain at very high input impedance. Both JFET and depletion MOSFET devices can be used to provide amplifiers having similar ac voltage gains. The depletion MOSFET circuit, however, has much higher input impedance than a similar JFET circuit.

While a BJT device controls a large output (collector) current by means of a smaller input (base) current, the FET device controls an output (drain) current by means of a small input (gate–source) voltage. The FET ac equivalent circuit is basically simpler than that for a BJT. While a BJT has current gain (beta) the FET has a transconductance, g_m.

The FET can be used as a linear device in amplifier circuits or as a digital device in logic circuits. The enhancement MOSFET is quite popular in digital circuitry, especially in CMOS circuits that require very low power consumption. FET devices are also widely used in high-frequency applications and in buffering (interfacing) applications. Table 9.1, located at the end of the chapter, provides a summary of FET small-signal amplifier circuits and circuit calculating formula.

While the common-source circuit is most popular providing an inverted, amplified signal, one also finds common-drain (source-follower) circuits providing unity gain with no inversion and common-gate circuits providing gain with no inversion. As with BJT amplifiers, the important circuit features described in this chapter include voltage gain, input impedance, and output impedance. While the voltage gain of a FET amplifier is generally less than that using a BJT amplifier, the FET amplifier provides a much higher input impedance than that of a BJT circuit. Output impedance values are comparable for both BJT and FET circuits.

FET ac amplifier circuits can also be analyzed using computer software. Using PSpice, one can do dc analysis to obtain the circuit bias conditions and ac analysis to determine voltage gain. Using PSpice transistor models, one can analyze the circuit using specific transistor models. On the other hand, one can use individually written programs, as in BASIC, to determine the dc bias conditions, voltage gain, and input and output impedances of a particular circuit. When writing programs in a language such as BASIC, one can provide the input data and output results can be listed in any specific form desired.

9.2 FET SMALL-SIGNAL MODEL

To perform ac analysis of a circuit using FET devices, we first need to obtain an ac equivalent circuit for the device. The ac circuit model to be used is similar for various types of FET devices. A major feature of FET ac operation is that an ac voltage applied to the device gate–source terminals controls the current between the drain–source terminals.

A gate-to-source voltage controls the drain–source (channel) current in a FET.

This action of a voltage signal at one set of terminals affecting a current at another set of terminals is described by a transfer (from one circuit place to another) conductance (current divided by voltage), or transconductance parameter g_m, where g_m is generally defined as

$$g_m = \frac{\text{drain–source current}}{\text{gate–source voltage}}$$

For a FET device the value of g_m can be obtained from Shockley's equation,[*]

$$g_m = g_{mo}\left(1 - \frac{V_{GS}}{V_P}\right) \tag{9.1}$$

where

$$g_{mo} = \frac{2I_{DSS}}{|V_P|} \tag{9.2}$$

The value g_{mo} is the FET transconductance at the $V_{GS} = 0$ V bias point and represents a fixed value of the maximum transconductance of the FET device. The value of g_{mo} is a constant for a particular FET and is not affected by the choice of dc bias point. At any bias point in the operating region, a value of g_m smaller than the value g_{mo} is obtained.

The largest value of g_m is defined by Eq. (9.2) as the value g_{mo}, which occurs at a dc bias point of $V_{GS} = 0$ V.

Graphical Description of g_m

The value of g_m can be obtained at the dc bias point using Eq. (9.1), or graphically using the slope of the device transfer characteristic at the particular dc bias point. Figure 9.1 shows that g_m represents the slope of the transfer curve at the dc bias point (Q-point). Recall that g_m represents the amount I_D changes due to the change in the voltage V_{GS}—which is the slope of the transfer curve.

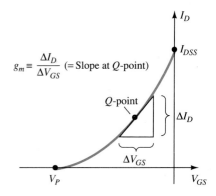

$$g_m \equiv \frac{\Delta I_D}{\Delta V_{GS}} \ (= \text{Slope at } Q\text{-point})$$

Figure 9.1 Definition of g_m using transfer characteristic.

---

[*]Using Shockley's equation,

$$I_D = I_{DSS}\left(1 - \frac{V_{GS}}{V_P}\right)^2$$

and the basic definition of g_m,

$$g_m = \frac{\Delta I_D}{\Delta V_{GS}}\bigg|_{V_{DS}=\text{constant}}$$

we obtain the equation for g_m in terms of the device parameters:

$$g_m = \frac{2I_{DSS}}{|V_P|}\left(1 - \frac{V_{GS}}{V_P}\right) = g_{mo}\left(1 - \frac{V_{GS}}{V_P}\right)$$

where

$$g_{mo} = \frac{2I_{DSS}}{|V_P|}$$

EXAMPLE 9.1

Determine the value of g_m for a JFET having parameters $I_{DSS} = 8$ mA and $V_P = -4$ V at the following dc bias points.

(a) $V_{GS_Q} = -0.5$ V
(b) $V_{GS_Q} = -1.5$ V.
(c) $V_{GS_Q} = -2.5$ V.

Solution

Using the transfer curve of the device as shown in Fig. 9.2, the value of g_m is determined at each of the dc bias points to be

(a) At $V_{GS_Q} = -0.5$ V

$$g_m = \frac{\Delta I_D}{\Delta V_{GS}} = \frac{2 \text{ mA}}{0.6 \text{ V}} = \textbf{3.3 mS}$$

(b) At $V_{GS_Q} = -1.5$ V,

$$g_m = \frac{\Delta I_D}{\Delta V_{GS}} = \frac{1.8 \text{ mA}}{0.75 \text{ V}} = \textbf{2.4 mS}$$

(c) At $V_{GS_Q} = -2.5$ V,

$$g_m = \frac{\Delta I_D}{\Delta V_{GS}} = \frac{1.5 \text{ mA}}{1.1 \text{ V}} = \textbf{1.4 mS}$$

Notice that the value of g_m becomes smaller at more negative dc bias points (as the slope of the transfer curve gets less steep).

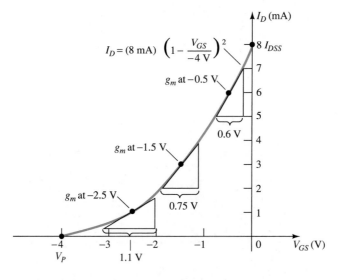

Figure 9.2 Calculating g_m at various bias points.

From Example 9.1 we can get a feeling for how different dc bias points will affect the value of g_m. The value of g_m is greatest at $V_{GS} = 0$ V, the maximum value being that defined as g_{mo}. As the dc bias point is made more negative, the value of g_m is smaller.

Plot of g_m versus V_{GS}

A helpful picture showing how g_m varies with the choice of dc bias point is provided by plotting Eq. (9.1). The resulting plot is shown in Fig. 9.3 and is seen to be a straight line, going from $g_m = 0$ at the pinch-off voltage, V_P, up to g_{mo} at $V_{GS} = 0$ V.

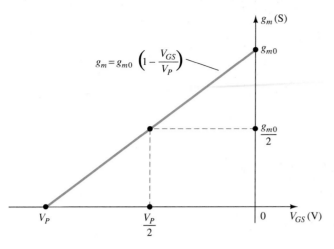

Figure 9.3 Plot of g_m versus V_{GS}.

Mathematical Calculations of g_m

While a graphical description of how g_m is affected by the choice of dc bias point is helpful, using an equation to calculate the value is more appropriate. To show that one could easily obtain the value of g_m at a particular dc bias point using Eqs. (9.1) and (9.2), consider the following example.

EXAMPLE 9.2

Calculate the value of g_m for a JFET having parameters of $I_{DSS} = 8$ mA and $V_P = -4$ V, for dc bias points of:
(a) -0.5 V.
(b) -1.5 V.
(c) -2.5 V.

Solution

Using Eq. (9.2) to calculate g_{mo}, we obtain

$$g_{mo} = \frac{2I_{DSS}}{|V_P|} = \frac{2(8 \text{ mA})}{|-4 \text{ V}|} = 4 \text{ mS}$$

(a) At $V_{GS} = -0.5$ V,

$$g_m = g_{mo}\left(1 - \frac{V_{GS_Q}}{V_P}\right) = 4 \text{ mS}\left(1 - \frac{-0.5 \text{ V}}{-4 \text{ V}}\right) = \textbf{3.5 mS}$$

(b) At $V_{GS} = -1.5$ V

$$g_m = g_{mo}\left(1 - \frac{V_{GS_Q}}{V_P}\right) = 4 \text{ mS}\left(1 - \frac{-1.5 \text{ V}}{-4 \text{ V}}\right) = \textbf{2.5 mS}$$

(c) At $V_{GS} = -2.5$ V,

$$g_m = g_{mo}\left(1 - \frac{V_{GS_Q}}{V_P}\right) = 4 \text{ mS}\left(1 - \frac{-2.5 \text{ V}}{-4 \text{ V}}\right) = \textbf{1.5 mS}$$

Compare the values calculated mathematically with those obtained graphically in Example 9.1 to see that the two methods agree reasonably well.

g_m Relationship to I_D

A mathematical relation between the value of g_m and the dc bias value of I_D could also be used. Using Shockley's equation, we can obtain the expression

$$I_D = I_{DSS}\left(1 - \frac{V_{GS}}{V_P}\right)^2$$

$$1 - \frac{V_{GS}}{V_P} = \sqrt{\frac{I_D}{I_{DSS}}} \tag{9.3}$$

Using Eq. (9.3) in Eq. (9.1) yields

$$g_m = g_{mo}\left(1 - \frac{V_{GS}}{V_P}\right) = g_{mo}\sqrt{\frac{I_D}{I_{DSS}}} \tag{9.4}$$

Using Eq. (9.4) to determine g_m for a few specific values of I_D:

(a) If $I_D = I_{DSS}$,

$$g_m = g_{mo}\sqrt{\frac{I_{DSS}}{I_{DSS}}} = g_{mo} \tag{9.5a}$$

(b) If $I_D = I_{DSS}/2$,

$$g_m = g_{mo}\sqrt{\frac{I_{DSS}/2}{I_{DSS}}} = 0.707 g_{mo} \tag{9.5b}$$

(c) If $I_D = I_{DSS}/4$,

$$g_m = g_{mo}\sqrt{\frac{I_{DSS}/4}{I_{DSS}}} = \frac{g_{mo}}{2} = 0.5 g_{mo} \tag{9.5c}$$

FET AC Equivalent Circuit

The ac equivalent circuit of a FET is shown in Fig. 9.4. Figure 9.4 shows only the FET device (the rest of the amplifier circuit will be considered later) with an ac input voltage, V_{gs}. The ac model, or ac equivalent circuit for the FET device only, is seen to consist of a voltage-controlled current source between drain and source terminals which is dependent on the device value g_m and the input ac voltage V_{gs}, and a device ac resistance between drain-to-source terminals of value r_d (ac output resistance). In particular, note that the input resistance from gate to source is shown in this model to be an open circuit or infinite resistance. For a practical JFET device the measured input resistance is typically over 10^9 Ω, while for a MOSFET it is typically 10^{12} to 10^{15} Ω.[*]

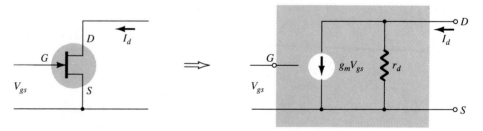

Figure 9.4 FET ac equivalent model.

---

[*]In comparison, the BJT model provides an input resistance, $h_{ie} = \beta r_e$, typically a few kilohms.

Chapter 9   FET Small-Signal Analysis

Calculate the value of g_m for a JFET device having $I_{DSS} = 12$ mA and $V_P = -3$ V when operated at a dc bias point of $V_{GSQ} = -1$ V.   *EXAMPLE 9.3*

Solution

Eq. (9.1): $g_{mo} = \dfrac{2I_{DSS}}{|V_P|} = \dfrac{2(12 \text{ mA})}{|-3 \text{ V}|} = 8 \times 10^{-3} = 8 \text{ mS} = 8000 \text{ } \mu\text{S}$

Eq. (9.2): $g_m = g_{mo}\left(1 - \dfrac{V_{GSQ}}{V_P}\right) = 8 \text{ mS}\left(1 - \dfrac{-1 \text{ V}}{-3 \text{ V}}\right) = \mathbf{5.3 \text{ mS}}$

Calculate the value of g_m for a JFET ($I_{DSS} = 8$ mA, $V_P = -4$ V) biased to operate at $V_{GS} = -1.8$ V.   *EXAMPLE 9.4*

Solution

Using Eq. (9.2), the device value of g_{mo} is

$$g_{mo} = \frac{2I_{DSS}}{|V_P|} = \frac{2(8 \text{ mA})}{|-4 \text{ V})|} = 4 \times 10^{-3} = 4 \text{ mS} = 4000 \text{ } \mu\text{S}$$

At the bias point, g_m is calculated using Eq. (9.1) to be

$$g_m = g_{mo}\left(1 - \frac{V_{GS}}{V_P}\right) = 4 \text{ mS}\left(1 - \frac{-1.8 \text{ V}}{-4 \text{ V}}\right) = \mathbf{2.2 \text{ mS}}$$

FET Output Resistance, r_d

The ac equivalent model of the FET also includes an output resistance r_d, which is in the range of tens of kilohms, 30 to 100 kΩ being typical. This value is similar to that of a BJT device. The value of r_d can be seen on the FET device's drain characteristic. Figure 9.5 shows a FET drain characteristic, the value of r_d being defined as

$$r_d = \frac{\Delta V_{DS}}{\Delta I_D}\bigg|_{V_{GS}=\text{constant}} \tag{9.6}$$

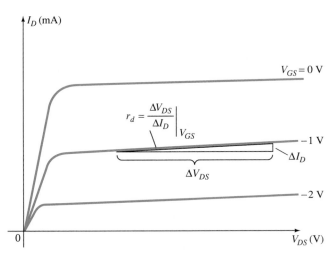

Figure 9.5 Definition of r_d using FET drain characteristic.

9.2 FET Small-Signal Model

EXAMPLE 9.5   Graphically determine the value of r_d for the FET drain characteristic of Fig. 9.6 at $V_{GS} = 0$ V.

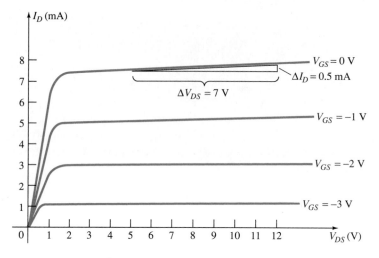

Figure 9.6   Drain characteristic used to calculate r_d in Example 9.5.

Solution

From the device drain characteristic, using Eq. (9.6) for the $V_{GS} = 0$ V curve, we have

$$r_d = \frac{\Delta V_{DS}}{\Delta I_D} = \frac{7 \text{ V}}{0.5 \text{ mA}} = \mathbf{14 \ k\Omega}$$

The value of r_d depicted in Example 9.5 was relatively low to allow reasonable graphical values to be shown. Actual FET drain characteristics will show flatter curve lines. Recall that a perfectly flat line would represent an infinite value of drain resistance

$$r_d = \frac{\Delta V_{DS}}{\Delta I_D} = \frac{\Delta V_{DS}}{0} = \text{infinite resistance}$$

Typically, the device characteristic provides a reasonably flat line with resulting drain resistance, r_d, of tens of kilohms. One commonly gets a typical output resistance value of r_d from the device's specifications sheet data. The data sheet provides a value of output conductance, y_{os}, whose reciprocal value is the output resistance, r_d. That is,

$$r_d = \frac{1}{y_{os}} \tag{9.7}$$

Circuit Features of FET Device

We can first consider the various circuit features due only to the FET device before considering the overall circuit operation. Three main features of any amplifier circuit are input impedance, output impedance, and voltage (or current) gain.

INPUT IMPEDANCE, Z_i

The input impedance of the FET device is ideally an open circuit. This is expressed as

$$\boxed{Z_i(\text{FET}) = \infty} \qquad (9.8)$$

For a JFET a practical value of 10^9 Ω is typical, while values from 10^{12} to 10^{15} Ω are typical for a MOSFET.

OUTPUT IMPEDANCE, Z_o

The output impedance of the FET device is shown in Fig. 9.4 as a resistance, r_d, across the drain–source terminals. The value of this resistance can be found in FET specification sheet data as output admittance value, y_{os}. The output resistance is then calculated from the reciprocal value:

Small-Signal Common-Source	Parameter	Units
Output admittance	y_{os}	μS

The output admittance value listed can be used to obtain the output resistance, r_d, of the FET device. While the value of admittance is listed in siemens, its reciprocal is the output resistance in ohms.

$$\boxed{Z_o(\text{FET}) = r_d = \frac{1}{y_{os}}} \qquad (9.9)$$

Determine the output impedance of a 2N4421 JFET.

EXAMPLE 9.6

Solution

The 2N4421 data sheet provides the following value:

$$y_{os} = 20 \ \mu\text{S}$$

The output impedance is then

$$Z_o = r_d = \frac{1}{y_{os}} = \frac{1}{20 \times 10^{-6}} = 50 \times 10^3 = \mathbf{50 \ k\Omega}$$

VOLTAGE GAIN, A_V

Referring to Fig. 9.4, the voltage gain provided by the FET can be determined as the ratio of the output voltage, V_{ds}, to the input voltage, V_{gs}. Since the drain–source voltage can be calculated using Ohm's law as a voltage drop due to current source $g_m V_{gs}$ through device resistance, r_d, the voltage gain is calculated to be

$$\boxed{A_v(\text{FET}) = \frac{V_o}{V_i} = \frac{V_{ds}}{V_{gs}} = \frac{-(g_m V_{gs})r_d}{V_{gs}} = -g_m r_d} \qquad (9.10)$$

The minus sign occurs because the voltage drop due to the current is negative to positive from drain to source and reflects the phase inversion that occurs between input at gate–source and output at drain–source. Specification sheet data list the value of g_m as:

Small-Signal Common-Source	Parameter	Units
Forward transfer admittance	y_{fs}	mS

That is,

$$\boxed{g_m = y_{fs}} \tag{9.11}$$

EXAMPLE 9.7

What maximum voltage gain can be achieved from a 2N4421 FET having $y_{fs} = 5$ mS and $y_{os} = 20$ μS?

Solution

From the specification sheet data

$$g_m = y_{fs} = 5 \text{ mS}$$

$$r_d = \frac{1}{y_{os}} = \frac{1}{20 \times 10^{-6}} = 50 \times 10^3 = 50 \text{ k}\Omega$$

Using Eq. (9.10) gives

$$A_v = -g_m r_d = -5 \text{ mS}(50 \text{ k}\Omega) = \mathbf{-250}$$

9.3   AC EQUIVALENT CIRCUIT: BASIC JFET CIRCUITS

In this section we will analyze FET circuits by replacing components by their ac equivalent in various FET circuits. Consider first the FET amplifier circuit of Fig. 9.7a. The dc bias is provided by a fixed battery voltage, V_{GG}. The ac input signal, V_i, is applied through capacitor C_1, and the output, V_o, taken from the amplifier stage via capacitor C_2. The ac equivalent circuit is then drawn in Fig. 9.7b. The battery is replaced by a short circuit (battery ac resistance is ideally 0 Ω), the capacitors are also replaced by short circuits (at mid-frequency the capacitive impedances are both very low—approximately 0 Ω). Resistors R_G and R_D are connected as placed in the actual circuit with the supply batteries both providing connection to ground. Finally, the FET is replaced by its ac equivalent circuit. In this case the FET has an applied voltage across gate–source, V_{gs}, which results in a current, $g_m V_{gs}$, from drain to source. The FET model also includes a drain–source resistance, r_d. The ac equivalent circuit is then redrawn in Fig. 9.7c.

Figure 9.8a next shows an amplifier circuit with self-bias set by resistor, R_S. The ac equivalent circuit is redrawn in Fig. 9.8b and then simplified in Fig. 9.8c. Figure 9.9a shows a FET amplifier with voltage-divider dc bias, and the resulting ac equivalent circuit in Fig. 9.9b and then redrawn in Fig. 9.9c. The ac equivalent circuit of a depletion MOSFET is the same as that for a JFET.

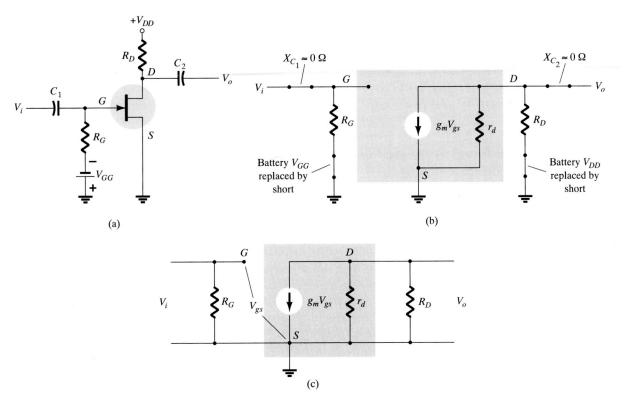

Figure 9.7 JFET amplifier with fixed dc bias: (a) circuit; (b) ac equivalent circuit; (c) redrawn ac equivalent circuit.

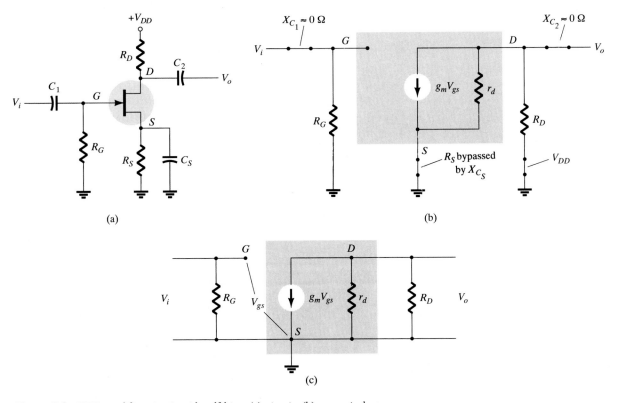

Figure 9.8 JFET amplifier circuit with self-bias: (a) circuit; (b) ac equivalent circuit; (c) redrawn ac equivalent circuit.

9.3   AC Equivalent Circuit; Basic JFET Circuits   **403**

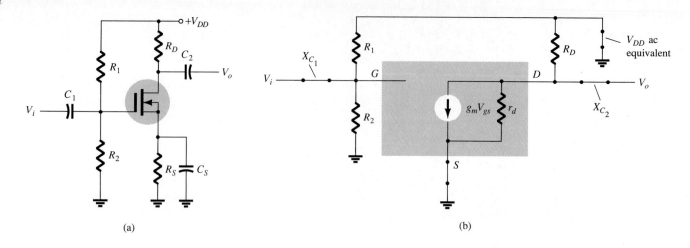

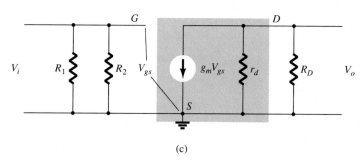

(c)

Figure 9.9 FET amplifier with voltage-divider bias: (a) circuit; (b) ac circuit; (c) redrawn ac circuit.

9.4 VOLTAGE GAIN

The voltage gain of a FET amplifier can be obtained from the ac equivalent circuit. The ac equivalent circuits of Figs. 9.7, 9.8, and 9.9 are all quite similar, providing an output voltage, V_o, due to the current $g_m V_{gs}$ through a parallel combination of r_d and R_D. The output voltage is then

$$V_o = -(g_m V_{gs})(R_D \| r_d)$$

$$A_v = \frac{V_o}{V_i} = \frac{-g_m V_{gs}(R_D \| r_d)}{V_{gs}}$$

$$\boxed{A_v = -g_m(R_D \| r_d)} \qquad (9.12)$$

If the value of device resistance, r_d, is much larger than the circuit resistance, R_D, the voltage gain equation is approximately equal to

$$\boxed{A_v = -g_m R_D} \qquad (9.13)$$

Calculate the voltage gain of the FET amplifier circuit in Fig. 9.10.

EXAMPLE 9.8

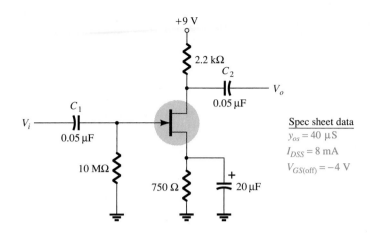

Figure 9.10 Ac FET amplifier for Examples 9.8, 9.12, and 9.13.

Solution

Using Eq. (9.2):

$$g_{mo} = \frac{2I_{DSS}}{|V_P|} = \frac{2(8 \times 10^{-3})}{|-4 \text{ V}|} = 4 \times 10^{-3} = 4 \text{ mS}$$

Dc bias calculations (if necessary, refer to Chapter 6) result in $I_{DQ} = 2.4$ mA, $V_{GS_Q} = -1.8$ V. At the dc bias point, Eq. (9.1) then provides

$$g_m = g_{mo}\left(1 - \frac{V_{GS_Q}}{V_P}\right) = 4 \text{ mS}\left(1 - \frac{-1.8 \text{ V}}{-4 \text{ V}}\right) = 2.2 \text{ mS}$$

From Eq. (9.7),

$$r_d = \frac{1}{y_{os}} = \frac{1}{40 \times 10^{-6}} = 25 \times 10^3 = 25 \text{ k}\Omega$$

The circuit voltage gain is then calculated using Eq. (9.12).

$$A_v = -g_m(R_D \| r_d)$$
$$= -(2.2 \times 10^{-3})(2.2 \text{ k}\Omega \| 25 \text{ k}\Omega) = \mathbf{-4.45}$$

Determine the output voltage, V_o, for the amplifier shown in Fig. 9.11.

EXAMPLE 9.9

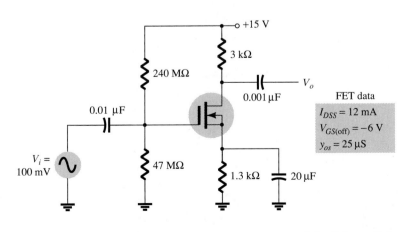

Figure 9.11 FET amplifier circuit for Examples 9.9, 9.12, and 9.13.

Solution

The device maximum transconductance is:

Eq. (9.2): $g_{mo} = \dfrac{2I_{DSS}}{|V_P|} = \dfrac{2(12\text{ mA})}{|-6\text{ V}|} = 4\text{ mS}$

The dc bias condition is determined to be

$$V_{GS_Q} = -3\text{ V}, \qquad I_{DQ} = 3\text{ mA}$$

At the dc bias point the device transconductance is calculated to be

$$g_m = g_{mo}\left(1 - \dfrac{V_{GS_Q}}{V_P}\right) = 4\text{ mS}\left(1 - \dfrac{-3\text{ V}}{-6\text{ V}}\right) = 2\text{ mS}$$

With the device output resistance,

$$r_d = \dfrac{1}{y_{os}} = \dfrac{1}{25 \times 10^{-6}} = 40\text{ k}\Omega$$

the voltage gain, A_v, is then calculated to be

$$A_v = \dfrac{V_o}{V_i} = -g_m(R_D \| r_d) = -2\text{ mS}(3\text{ k}\Omega \| 40\text{ k}\Omega) = -5.58$$

The output voltage is

$$V_o = A_v V_i = (-5.58)(100\text{ mV}) = -558\text{ mV} = \mathbf{-0.558\text{ V}}$$

Voltage Gain of FET Amplifier with Source Resistance (Assume that $r_d \approx \infty$)

If the FET amplifier has an unbypassed source resistance, R_S, the ac voltage gain is reduced. Figure 9.12a shows a depletion MOSFET amplifier with an unbypassed source resistance, R_S. The ac equivalent circuit shown in Fig. 9.12b is a simplified FET model with device resistance r_d assumed to be much larger than R_D (r_d being neglected in this case).

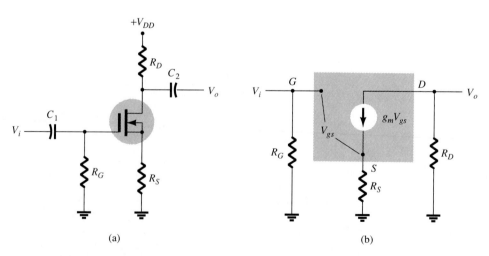

Figure 9.12 FET amplifier with source resistance: (a) amplifier circuit; (b) ac equivalent circuit.

The voltage measured from gate to source, V_{gs}, is

$$V_{gs} = V_i - V_s$$

The voltage across the source resistor, R_S, is

$$V_s = g_m V_{gs} R_S$$

Solving for V_{gs} using the two equations above,

$$V_{gs} = V_i - g_m V_{gs} R_S$$

we can solve for the input voltage

$$V_i = V_{gs}(1 + g_m R_S)$$

Since the output voltage due to the current $g_m V_{gs}$ through the drain resistor, R_D, is

$$V_o = -(g_m V_{gs})R_D$$

The voltage gain is then

$$A_v = \frac{V_o}{V_i} = \frac{-(g_m V_{gs})R_D}{V_{gs}(1 + g_m R_S)}$$

$$\boxed{A_v = \frac{-g_m R_D}{1 + g_m R_S}} \tag{9.14}$$

Notice that the gain would reduce to $-g_m R_D$ [see Eq. (9.13)] if R_S is bypassed—the resistance from source to ground nears zero ohms.

For the depletion MOSFET amplifier circuit of Fig. 9.13, calculate the voltage gain with:
(a) Resistor R_S bypassed by a capacitor.
(b) Resistor R_S unbypassed.

EXAMPLE 9.10

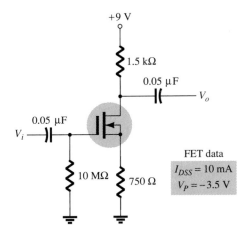

Figure 9.13 Amplifier circuit for Examples 9.10 and 9.11.

Solution

From the dc bias analysis we obtain

$$I_{DQ} = 2.3 \text{ mA}, \qquad V_{GS_Q} = -1.8 \text{ V}$$

The value of g_m at the dc bias condition is then

$$g_m = \frac{2I_{DSS}}{|V_P|}\left(1 - \frac{V_{GS_Q}}{V_P}\right) = (5.7 \text{ mS})(0.486) = 2.77 \text{ mS}$$

(a) The voltage gain with resistor R_S bypassed is

Eq. (9.13): $A_v = -g_m R_D = -(2.77 \text{ mS})(1.5 \text{ k}\Omega) = \mathbf{-4.155}$

(b) The voltage gain with resistor R_S present is

Eq. (9.14): $A_v = \dfrac{-g_m R_D}{1 + g_m R_S} = \dfrac{-4.155}{1 + (2.77 \text{ mS})(0.75 \text{ k}\Omega)} = \mathbf{-1.35}$

Notice that the effect of an unbypassed source resistance is to reduce the gain considerably.

Voltage Gain of FET Amplifier with Source Resistance (r_d Present)

For a depletion MOSFET amplifier circuit, such as that of Fig. 9.12a, the ac equivalent circuit having a FET model that includes r_d is shown in Fig. 9.14. The output current through resistor R_D is the sum of the current of the source and that through r_d:

$$I_o = g_m V_{gs} + \frac{V_o - V_s}{r_d} = g_m V_{gs} + \frac{-I_o R_D - I_o R_S}{r_d}$$

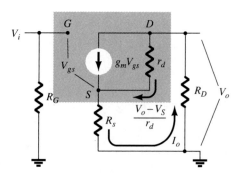

Figure 9.14 Ac equivalent circuit for amplifier of Fig. 9.12a, including r_d.

Since $V_{gs} = V_i - I_o R_S$

The current I_o can be expressed as

$$I_o = g_m(V_i - I_o R_S) + \frac{-I_o R_D - I_o R_S}{r_d}$$

so that $I_o\left(1 + g_m R_S + \dfrac{R_D + R_S}{r_d}\right) = g_m V_i$

which results in

$$I_o = \frac{g_m V_i}{1 + g_m R_S + (R_D + R_S)/r_d}$$

The output voltage gain is then calculated to be

$$V_o = -I_o R_D = \frac{-g_m R_D V_i}{1 + g_m R_S + (R_D + R_S)/r_d}$$

resulting in a voltage gain of

$$\boxed{A_v = \frac{V_o}{V_i} = \frac{-g_m R_D}{1 + g_m R_S + (R_D + R_S)/r_d}}$$

(9.15)

Calculate the ac voltage gain of the circuit of Fig. 9.13 using a FET model having $y_{os} = 25 \ \mu S$.

EXAMPLE 9.11

Solution

The FET ac output impedance is

$$r_d = \frac{1}{y_{os}} = \frac{1}{25 \times 10^{-6}} = 40 \times 10^3$$

From Example 9.10, $g_m = 2.77$ mS. Using Eq. (9.15) for the ac gain using the FET model with r_d,

$$A_v = \frac{V_o}{V_i} = \frac{-g_m R_D}{1 + g_m R_S + (R_D + R_S)/r_d}$$

$$= \frac{-(2.77 \times 10^{-3})(1.5 \times 10^3)}{1 + (2.77 \times 10^{-3})(750) + (1.5 \times 10^3 + 750)/25 \times 10^3}$$

$$= \frac{-4.155}{1 + 2.078 + 0.090} = \mathbf{-1.3}$$

Compare the resulting gain (-1.3) with that calculated in Example 9.10(b) (-1.35). Notice that the additional denominator term (0.090) was negligible in this example compared to the other values in the denominator. This will generally be true and the simpler calculation using Eq. (9.14) can be used in most cases.

Input Impedance

The input impedance of a FET circuit is due primarily to the resistance from gate terminal to ground. While the ac impedance of the FET device is nearly open-circuit, the impedance seen at the input is that from the gate to ground. For the circuit of Fig. 9.15, the input impedance is

$$\boxed{Z_i = R_G} \tag{9.16}$$

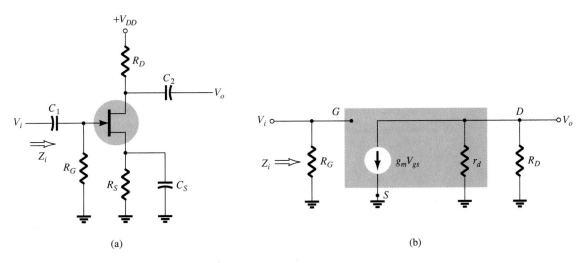

(a)   (b)

Figure 9.15 Common-source JFET amplifier.

For the voltage-divider circuit of Fig. 9.16, it is

$$Z_i = R_1 \| R_2 = \frac{R_1 R_2}{R_1 + R_2}$$

(9.17)

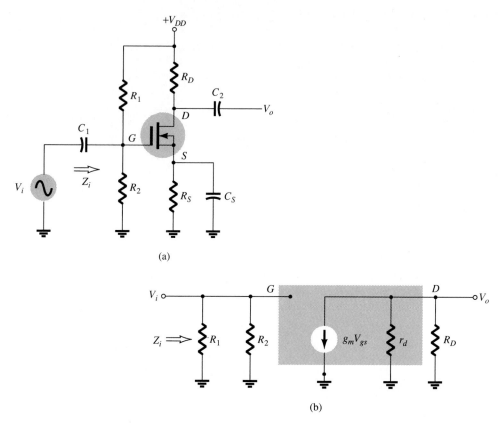

(a)

(b)

Figure 9.16 FET amplifier showing ac input impedance, Z_i.

EXAMPLE 9.12

Calculate the ac input impedance of the circuits in Figs. 9.10 and 9.11.

Solution

For the circuit of Fig. 9.10,

$$Z_i = R_G = \mathbf{10\ M\Omega}$$

For the circuit of Fig. 9.11,

$$Z_i = R_1 \| R_2 = 47\ M\Omega \| 240\ M\Omega = \mathbf{39.3\ M\Omega}$$

AC Output Impedance

The impedance seen looking from the output into the amplifier circuit is basically that due to the device resistance from drain to source, and the bias resistor, R_D. For the amplifier of Fig. 9.17, the output impedance is

$$Z_o = r_d \| R_D = \frac{r_d R_D}{r_d + R_D}$$

(9.18)

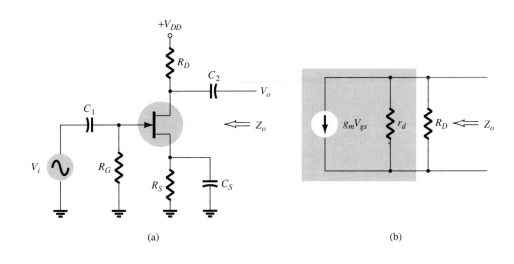

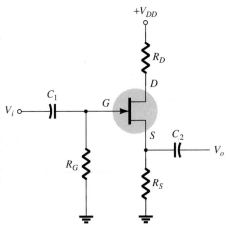

Figure 9.17 FET amplifier circuit showing output impedance, Z_o.

(a)　　　　　　　　　　　　　(b)

Calculate the output impedance of the circuits in Figs. 9.10 and 9.11.

EXAMPLE 9.13

Solution

For the given device specifications in Fig. 9.5, the value of r_d is

$$r_d = \frac{1}{y_{os}} = \frac{1}{40 \ \mu\text{S}} = 25 \ \text{k}\Omega$$

The ac output impedance of the circuit is then calculated using Eq. (9.18) to be

$$Z_o = r_d \| R_D = 25 \ \text{k}\Omega \| 2.2 \ \text{k}\Omega = \mathbf{2.0 \ k\Omega}$$

For the circuit of Fig. 9.11,

$$r_d = \frac{1}{y_{os}} = \frac{1}{25 \ \mu\text{S}} = 40 \ \text{k}\Omega$$

and the output impedance is

$$Z_o = r_d \| R_D = 40 \ \text{k}\Omega \| 3 \ \text{k}\Omega = \mathbf{2.8 \ k\Omega}$$

9.5 SOURCE-FOLLOWER (COMMON-DRAIN) CIRCUIT

A second ac circuit configuration is the common-drain or source-follower circuit shown in Fig. 9.18. The circuit is seen to be a JFET version similar to a bipolar emitter-follower configuration. As with the bipolar circuit, the JFET source follower provides voltage gain less than 1 with no polarity inversion. In addition, the JFET circuit provides very high input impedance and low output impedance.

Figure 9.18 Common-drain (source-follower) circuit.

Voltage Gain

The ac signal output is taken from the source terminal. Figure 9.19a shows the ac equivalent circuit. Redrawing that circuit results in that shown in Fig. 9.19b, the output voltage then determined to be

$$V_o = (g_m V_{gs})(R_S \| r_d)$$

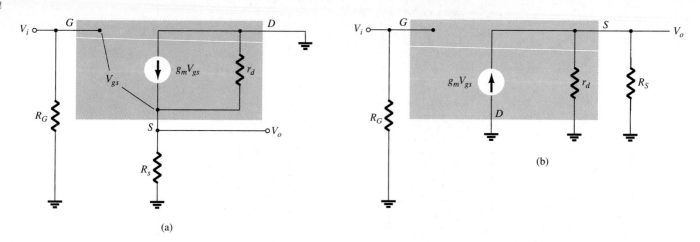

Figure 9.19 (a) Ac equivalent circuit of Fig. 9.18 source follower; (b) redrawn ac equivalent circuit.

Since the voltage V_{gs} is

$$V_{gs} = V_g - V_s = V_i - V_o$$

the output voltage is then

$$V_o = g_m(V_i - V_o)(R_s\|r_d) = g_m V_i(R_s\|r_d) - g_m V_o(R_s\|r_d)$$

which can be rewritten as

$$V_o[1 + g_m(R_s\|r_d)] = g_m(R_s\|r_d)V_i$$

We can then solve for the voltage gain

$$\boxed{A_v = \frac{V_o}{V_i} = \frac{g_m(R_s\|r_d)}{1 + g_m(R_s\|r_d)}} \tag{9.19}$$

The voltage gain is seen in Eq. (9.19) to be noninverting and have a value of less than 1. The gain approaches a value of 1 as the value of $g_m(R_s\|r_d)$ becomes much larger than 1.

Input Impedance

The input impedance of a FET source-follower circuit, as that of Fig. 9.18 is the value of gate resistance.

$$\boxed{Z_i = R_G} \tag{9.20a}$$

If the circuit has a voltage divider at the gate, the input resistance is

$$\boxed{Z_i = R_1\|R_2} \tag{9.20b}$$

Output Impedance

The output impedance of a FET source-follower circuit is the impedance seen looking back into the source. Using the circuit shown in Fig. 9.20, we see source resistor R_S in parallel with device output resistance, r_d, in parallel with the current source, $g_m V_{gs}$. If we replace the current source by a voltage source in series with resistance $1/g_m$ and

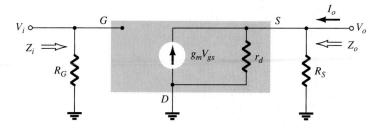

Figure 9.20 Source-follower ac equivalent circuit showing Z_i and Z_o.

set the voltage source to zero in order to determine Z_o, we can express the output impedance as

$$Z_o = R_s \| r_d \| \frac{1}{g_m} \qquad (9.21)$$

Calculate the voltage gain and the input and output impedances for the circuit of Fig. 9.21.

EXAMPLE 9.14

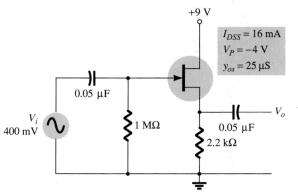

Figure 9.21 Source-follower amplifier circuit for Example 9.14.

Solution

From dc calculations, $V_{GSQ} = -2.86$ V. The device transconductance is

$$g_{mo} = \frac{2I_{DSS}}{|V_P|} = \frac{2(16 \text{ mA})}{|-4 \text{ V}|} = 8 \text{ mS}$$

At the dc bias point the value of g_m is

$$g_m = g_{mo}\left(1 - \frac{V_{GSQ}}{V_P}\right) = 8 \text{ mS}\left(1 - \frac{-2.86 \text{ V}}{-4 \text{ V}}\right) = 2.28 \text{ mS}$$

The device ac output resistance is

$$r_d = \frac{1}{y_{os}} = \frac{1}{25 \text{ } \mu\text{S}} = 40 \text{ k}\Omega$$

The ac voltage gain is then calculated to be

$$A_v = \frac{g_m(R_S \| r_d)}{1 + g_m(R_S \| r_d)} = \frac{(2.28 \text{ mS})(2.2 \text{ k}\Omega \| 40 \text{ k}\Omega)}{1 + (2.28 \text{ mS})(2.2 \text{ k}\Omega \| 40 \text{ k}\Omega)}$$

$$= \mathbf{0.83}$$

The amplifier input impedance is

$$Z_i = R_G = \mathbf{1 \ M\Omega}$$

The amplifier output impedance is

$$Z_o = R_S \| r_d \| \frac{1}{g_m} = 2.2 \ \text{k}\Omega \| 40 \ \text{k}\Omega \| \frac{1}{2.28 \times 10^{-3}}$$

$$= 2.2 \ \text{k}\Omega \| 40 \ \text{k}\Omega \| 438.6 \ \Omega \approx \mathbf{362 \ \Omega}$$

9.6 COMMON-GATE CIRCUIT

A third FET circuit configuration is the common-gate shown in Fig. 9.22(a). As will be shown, this circuit has a low input impedance, output impedance similar to the common-source, and a noninverting gain the same magnitude as that of the common-source.

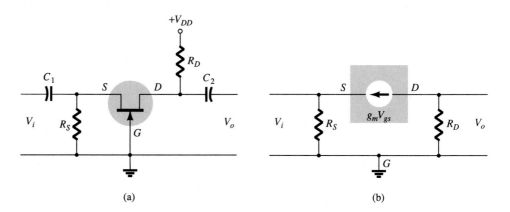

Figure 9.22 Common-gate amplifier circuit.

(a) (b)

The ac equivalent circuit is shown in Fig. 9.22(b). With the gate–source voltage, $V_{gs} = -V_i$, the output voltage is

$$V_o = -(g_m V_{gs})R_D = g_m V_i R_D$$

so that the amplifier voltage gain is

$$\boxed{A_v = \frac{V_o}{V_i} = g_m R_D} \tag{9.22a}$$

If the FET ac model including r_d is used, the amplifier gain equation can be approximately expressed by

$$\boxed{A_v \approx g_m(R_D \| r_d)} \tag{9.22b}$$

For large r_d the input resistance is approximately

$$\boxed{Z_i \approx R_S \| \frac{1}{g_m}} \tag{9.23}$$

The output impedance is

$$\boxed{Z_o = R_D \| r_d} \tag{9.24}$$

Calculate V_o for the circuit of Fig. 9.23.

EXAMPLE 9.15

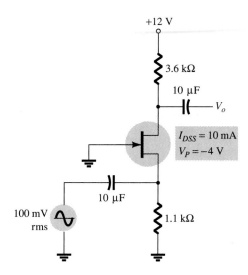

Figure 9.23 Circuit for Examples 9.15 and 9.16.

Solution

Dc bias calculations result in $V_{GS_Q} = -2.2$ V. With $g_{mo} = 5$ mS, the value of g_m at the bias condition is

$$g_m = g_{mo}\left(1 - \frac{V_{GS_Q}}{V_P}\right) = 5 \text{ mS}\left(1 - \frac{-2.2 \text{ V}}{-4 \text{ V}}\right) = 2.25 \text{ mS}$$

The amplifier gain is (r_d very large)

$$A_v = g_m R_D = 2.25 \text{ mS}(3.6 \text{ k}\Omega) = 8.1$$

and the output voltage is

$$V_o = A_v V_i = 8.1(100 \text{ mV}) = \textbf{0.81 V}$$

Calculate the voltage gain of the circuit in Fig. 9.23 if the FET model included a device parameter of $y_{os} = 25$ μS.

EXAMPLE 9.16

Solution

With $r_d = 1/y_{os} = 1/(25 \ \mu\text{S}) = 40$ kΩ, the amplifier voltage gain is calculated to be

$$A_v \approx g_m(R_D \| r_d) = 2.25 \text{ mS}(3.6 \text{ k}\Omega \| 40 \text{ k}\Omega) = \textbf{7.4}$$

9.7 ENHANCEMENT MOSFET AMPLIFIER

The enhancement MOSFET can be either a *p*-channel (*p*MOS) or *n*-channel (*n*MOS) device, as shown in Fig. 9.24. The ac small-signal equivalent circuit of either device is shown in Fig. 9.24 to provide an open circuit between gate and the drain–source channel and a current source from drain to source, which is dependent on the voltage applied to the gate–source terminals. The device also has an output resistance from drain to source, r_d, which is specified in spec sheet data by the value of output admittance, y_{os}. The device transconductance, g_m is given in the spec sheet data as forward transfer admittance, y_{fs}.

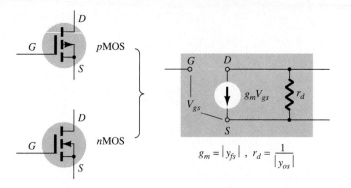

Figure 9.24 Enhancement MOSFET ac small-signal model.

An enhancement MOSFET amplifier circuit can be constructed as shown in Fig. 9.25a. Dc bias is provided by the voltage supply, V_{DD}, and the resistors R_G and R_D. At the established dc bias condition, the ac small-signal operation of the circuit can be determined using an ac small-signal equivalent as shown in Fig. 9.25b.

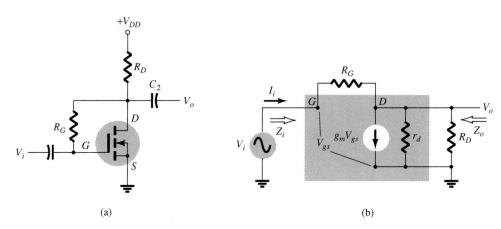

(a)   (b)

Figure 9.25 Enhancement MOSFET amplifier; (a) circuit; (b) small-signal equivalent circuit.

Voltage Gain

For the ac equivalent circuit of Fig. 9.25b, the output voltage is seen to be the result of two currents through the parallel resistance of R_D and r_d:

$$V_o = -\left(\frac{V_o - V_i}{R_G} + g_m V_{gs}\right)(R_D \| r_d)$$

Since $V_{gs} = V_i$, the voltage gain can be expressed as

$$A_v = \frac{V_o}{V_i} = \frac{(1 - g_m R_G)(R_D \| r_d)}{R_G + R_D \| r_d}$$

For $g_m R_G \gg 1$, as is usually true, the expression for the voltage gain reduces to

$$A_v = \frac{V_o}{V_i} = -\frac{g_m R_G(R_D \| r_d)}{R_G + R_D \| r_d} = -g_m(R_G \| R_D \| r_d)$$

Since R_G is typically much larger than either R_D or r_d, the equation further simplifies to

$$A_v = \frac{V_o}{V_i} = -g_m(R_D \| r_d) \qquad (9.25)$$

DETERMINING THE VALUE OF g_m

The value of g_m to use may be that provided on a specification sheet for a particular bias condition, or it may be a value calculated at the bias point for the given circuit. In the later case the value of g_m is determined to be[*]

$$g_m = 2k(V_{GS_Q} - V_T) \qquad (9.26)$$

where k is typically 0.3 mA/V^2.

Input Impedance

The input impedance seen by the source can be obtained using

$$Z_i = \frac{V_i}{I_i}$$

The input current, I_i, can be expressed as

$$I_i = \frac{V_i - V_o}{R_G} = \frac{V_i(1 - V_o/V_i)}{R_G}$$

We can then solve for Z_i as

$$Z_i = \frac{V_i}{I_i} = \frac{V_i}{V_i(1 - V_o/V_i)/R_G} = \frac{R_G}{1 - V_o/V_i}$$

Using Eq. (9.25), we have

$$Z_i = \frac{R_G}{1 - [-g_m(R_D \| r_d)]} = \frac{R_G}{1 + g_m(R_D \| r_d)}$$

$$Z_i = \frac{R_G}{1 + g_m(R_D \| r_d)} \qquad (9.27)$$

Output Impedance

The output impedance seen by a load connected to V_o, as in Fig. 9.25, is due primarily to the parallel resistance of R_D, r_d, and R_G:

$$Z_o = R_D \| r_d \| R_G$$

---

[*]The value of g_m is obtained by differentiating the expression

$$I_D = k(V_{GS_Q} - V_T)^2$$

resulting in

$$g_m = \frac{\Delta I_D}{\Delta V_{GS}} = 2k(V_{GS_Q} - V_T)$$

Since R_G is typically much larger than either R_D or r_d, we can express the output impedance as

$$\boxed{Z_o = R_D \| r_d} \tag{9.28}$$

EXAMPLE 9.17

Calculate the output voltage for the circuit of Fig. 9.26.

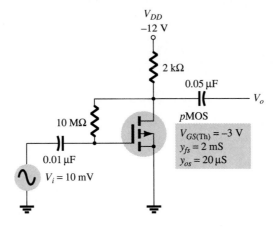

V_{DD}
–12 V

2 kΩ

0.05 μF

V_o

10 MΩ

pMOS
$V_{GS(Th)} = -3$ V
$y_{fs} = 2$ mS
$y_{os} = 20$ μS

0.01 μF

$V_i = 10$ mV

Figure 9.26 Enhancement small-signal amplifier circuit for Example 9.17.

Solution

Dc bias calculations result in $V_{GS_Q} = -6.1$ V and $I_D = 2.9$ mA. From specification sheet data,

$$r_d = \frac{1}{y_{os}} = \frac{1}{20 \times 10^{-6}} = 50 \text{ k}\Omega$$

Using Eq. (9.26) gives

$$g_m = 2k(V_{GS_Q} - V_T) = 2(0.3 \times 10^{-3})(6.1 \text{ V} - 3 \text{ V})$$
$$= 1.86 \text{ mS}$$

so that the voltage gain can be calculated to be

$$A_v = \frac{V_o}{V_i} = -g_m(R_D \| r_d) = -1.86 \times 10^{-3}(2 \text{ k}\Omega \| 50 \text{ k}\Omega) = -3.6$$

The output voltage is then

$$V_o = A_v V_i = -3.6(10 \text{ mV}) = \mathbf{-36 \text{ mV}}$$

9.8 DESIGN OF JFET AMPLIFIER CIRCUITS

Design problems at this stage are limited to obtaining a desired dc bias condition or a desired ac voltage gain. The various relations, such as Shockley's equation or the ac voltage gain equation, are used to calculate a desired circuit component value. A few examples will help show the type of approach one can use to obtain some desired circuit operation.

Design for Specified Voltage Gain

To obtain a desired voltage gain involves selecting a value of R_D and a value of g_m (at the dc bias point). Recall that the ac voltage gain for a circuit such as that in Fig. 9.26 is obtained from the equation

$$A_v = -g_m(R_D\|r_d)$$

For example, consider designing a circuit such as that in Fig. 9.27 to provide a voltage gain of magnitude 10.

EXAMPLE 9.18

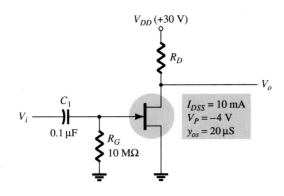

V_{DD} (+30 V)

R_D

V_o

C_1

V_i

0.1 μF

R_G
10 MΩ

$I_{DSS} = 10$ mA
$V_P = -4$ V
$y_{os} = 20$ μS

Figure 9.27 Circuit for desired voltage gain in Example 9.18.

Solution

Since the dc bias provides $g_m = g_{mo}$ (dc bias at $V_{GS_Q} = 0$ V), the ac voltage gain equation provides

$$A_v = -g_m(R_D\|r_d) = -g_{mo}(R_D\|r_d)$$

$$-10 = -\frac{2I_{DSS}}{|V_P|}(R_D\|r_d)$$

$$-10 = -\frac{2(10 \text{ mA})}{|-4 \text{ V}|}(R_D\|r_d)$$

$$R_D\|r_d = 2 \text{ k}\Omega$$

From the device specifications

$$r_d = \frac{1}{y_{os}} = \frac{1}{20 \times 10^{-6}} = 50 \text{ k}\Omega$$

The value of circuit resistor, R_D, should then be

$$R_D\|r_d = R_D\|(50 \text{ k}\Omega) = 2 \text{ k}\Omega$$

so that $\qquad\qquad R_D = 2.08 \text{ k}\Omega \approx 2.1 \text{ k}\Omega$

The dc bias point resulting for $R_D = 2.1$ kΩ would then be

$$V_{DS_Q} = V_{DD} - I_{DQ}R_D = 30 \text{ V} - (10 \text{ mA})(2.1 \text{ k}\Omega) = 9 \text{ V}$$

As another example, consider designing a circuit such as that in Fig. 9.28 to obtain a voltage gain of magnitude 8. To limit the design steps, consider selecting V_{GS_Q} near the maximum g_m value, say at $V_P/4$.

EXAMPLE 9.19

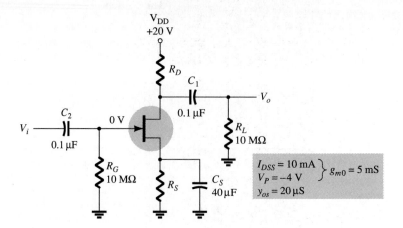

Figure 9.28 Circuit for ac circuit design in Example 9.19.

Solution

$$V_{GS_Q} = \tfrac{1}{4}V_P = \tfrac{1}{4}(-4 \text{ V}) = -1 \text{ V}$$

The drain current calculated using Shockley's equation is then

$$I_{DQ} = I_{DSS}\left(1 - \frac{V_{GS_Q}}{V_P}\right)^2$$

$$= 10 \text{ mA}\left(1 - \frac{-1 \text{ V}}{-4 \text{ V}}\right)^2 = 5.625 \text{ mA}$$

At this dc bias point we calculate g_m to be

$$g_m = g_{mo}\left(1 - \frac{V_{GS_Q}}{V_P}\right)^2$$

$$= \frac{2I_{DSS}}{|V_P|}\left(1 - \frac{V_{GS_Q}}{V_P}\right)^2$$

$$= \frac{2 \times 10 \text{ mA}}{|-4 \text{ V}|}\left(1 - \frac{-1 \text{ V}}{-4 \text{ V}}\right)^2 = 3.75 \text{ mS}$$

For the voltage gain of magnitude 8:

$$A_v = g_m(R_D\|r_d)$$

$$8 = 3.75 \times 10^{-3}(R_D\|r_d)$$

so that

$$R_D\|r_d = \frac{8}{3.75 \times 10^{-3}} = 2.13 \text{ k}\Omega$$

Since r_d is

$$r_d = \frac{1}{y_{os}} = \frac{1}{20 \text{ } \mu\text{S}} = 50 \text{ k}\Omega$$

the value of R_D would be

$$R_D\|50 \text{ k}\Omega = 2.13 \text{ k}\Omega$$

and

$$R_D = 2.2 \text{ k}\Omega$$

The value of R_S for the desired value of V_{GS_Q} is

$$V_{GS_Q} = -I_D R_S$$

$$-1 \text{ V} = -(5.625 \text{ mA})R_S$$

$$R_S = \frac{-1 \text{ V}}{-5.625 \text{ mA}} = 177.8 \text{ } \Omega$$

Complete the design of a JFET amplifier circuit such as that in Fig. 9.29 using the given circuit values and device parameters. The desired amplifier circuit should provide an ac voltage gain of at least 5. (Choose V_{GS_Q} at $V_P/4$.)

EXAMPLE 9.20

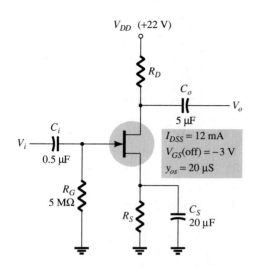

Figure 9.29 Circuit for Example 9.20.

Solution

At $V_{GS_Q} = V_P/4 = -3 \text{ V}/4 = -0.75 \text{ V}$,

$$I_{D_Q} = I_{DSS}\left(1 - \frac{V_{GS_Q}}{V_P}\right) = 12 \text{ mA}\left(1 - \frac{-0.75 \text{ V}}{-3 \text{ V}}\right) = 6.75 \text{ mA}$$

At this bias point the circuit transconductance is

$$g_m = g_{mo}\left(1 - \frac{V_{GS_Q}}{V_P}\right) = \frac{2(12 \text{ mA})}{|-3 \text{ V}|}\left(1 - \frac{-0.75 \text{ V}}{-3 \text{ V}}\right) = 6 \text{ mS}$$

For a voltage gain of magnitude $A_v = 5$,

$$A_v = g_m(R_D\|r_d)$$

so that we calculate

$$R_D\|r_d = \frac{A_v}{g_m} = \frac{5}{6 \text{ mS}} = 833.3 \text{ }\Omega$$

Since r_d is

$$r_d = \frac{1}{y_{os}} = \frac{1}{20 \text{ }\mu\text{S}} = 50 \text{ k}\Omega$$

$$R_D = 847 \text{ }\Omega \quad (\text{use } 820 \text{ }\Omega)$$

At the bias point

$$V_{GS_Q} = I_{D_Q}R_S$$

we calculate

$$R_S = \frac{V_{GS_Q}}{I_{D_Q}} = \frac{0.75 \text{ V}}{6.75 \text{ mA}} = 111.1\Omega \quad (\text{use } 110 \text{ }\Omega)$$

9.9   TROUBLESHOOTING

As mentioned before, troubleshooting a circuit is a combination of knowing the theory and having experience using meters and an oscilloscope to check the operation of the circuit. A good troubleshooter has a "nose" for finding the trouble in a circuit—this ability to "see" what is happening being greatly developed through building, testing, and repairing many different circuits. For a FET small-signal amplifier one could go about troubleshooting a circuit by performing a number of basic steps.

1. Look at the circuit board to see if any obvious problems can be seen: an area charred by excess heating of a component; a component that feels or seems too hot to touch; what appears to be a poor solder joint; any connection that appears to have come loose.

2. Using a dc meter: make some measurements as marked in a repair manual containing the circuit schematic diagram and a listing of test dc voltages.

3. Apply a test ac signal: measure the ac voltages starting at the input and working along toward the output.

4. If the problem is identified at a particular stage, the ac signal at various points should be checked using an oscilloscope to see the waveform, its polarity, amplitude, and frequency, as well as any unusual waveform "glitches" that may be present. In particular, observe that the signal is present for the full signal cycle.

Possible Symptoms and Actions

If there is no output ac voltage:

1. Check if the supply voltage is present.
2. Check if the output voltage at V_D is between 0 V and V_{DD}.
3. Check if there is any input ac signal at the gate terminal.
4. Check the ac voltage at each side of the coupling capacitor terminals.

When building and testing a FET amplifier circuit in the laboratory:

1. Check the color code of resistor values to be sure that they are correct. Even better, measure the resistor value, as components used repeatedly may get overheated when used incorrectly, causing the nominal value to change.

2. Check that all dc voltages are present at the component terminals. Be sure that all ground connections are made common.

3. Measure the ac input signal to be sure the expected value is provided to the circuit.

9.10   COMPUTER ANALYSIS

Since ac calculations of voltage gain, input impedance, and output impedance for various FET circuits requires calculating dc bias values to use in determining device parameters at the Q-point, a computer analysis can be quite helpful. The SPICE analysis provides models of JFET, depletion MOSFET, and enhancement MOSFET devices. A few examples will demonstrate how a program circuit description is written and how one can get desired output results for the ac operation of the circuit.

JFET Description

JFET ELEMENT LINE

The general form for an element line for a junction field-effect transistor is

$$\text{JXXXX} \quad \text{ND} \quad \text{NG} \quad \text{NS} \quad \text{MODNAME}$$

where JXXXX is the name of the transistor; ND, NG, and NS are the node numbers for the drain, gate, and source, respectively; and MODNAME is the model name used in the .MODEL line described next.

JFET MODEL LINE

The general form for a model line for a JFET is

$$\text{.MODEL MODNAME NJF VTO} = \quad \text{BETA} =$$
$$\text{.MODEL MODNAME PJF VTO} = \quad \text{BETA} =$$

where MODNAME is the model name given in the element line, NJF identifies an *n*-channel device, and PJF identifies a *p*-channel device. Of the numerous JFET model parameters, two of the more important are

VTO = V_P: gate–source pinch-off voltage

BETA = I_{DSS/V_P^2}: parameter combining two JFET device parameters, along with VTO used to specify I_{DSS}

Write the SPICE circuit lines to describe the following JFET devices.

EXAMPLE 9.21

(a) An *n*-channel JFET having I_{DSS} = 12 mA and V_P = −4 V.
(b) An *n*-channel JFET having I_{DSS} = 8 mA and V_P = −3 V.
Assume that each device is connected at nodes: drain = 5, source = 4, and gate = 2.

Solution

(a) JUP 5 2 4 JN
　.MODEL JN NJF VTO = −4　BETA = 750E−6
(b) JDOWN 5 2 4 JJ
　.MODEL JJ NJF VTO = −3　BETA = 889E−6

Program 9.1: JFET Amplifier Circuit

A JFET amplifier circuit is shown in Fig. 9.30. Dc bias of the JFET is provided by the V_{GG} supply voltage, the V_{DD} supply voltage, and the drain resistor, R_D. An input ac voltage is applied through capacitor C_1, the resulting amplified output being provided through capacitor C_2. PSpice requires that the output path be connected to ground so that a very high impedance load resistance, R_L, is specified. With a value of 10 MΩ the output is essentially open-circuit.

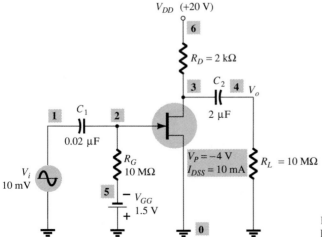

Figure 9.30 JFET amplifier for PSpice analysis.

The circuit description file is listed in Fig. 9.31 for the circuit being analyzed (Fig. 9.30) showing all nodes marked and the resulting output data. Some comments about the PSpice program are:

Form of JFET component line:

$$\text{J1} \quad 3 \quad 2 \quad 0 \qquad \text{JFET}$$

Form of JFET MODEL line:

$$.\text{MODEL JFET NJF VTO} = -4\text{V BETA} = 6.25\text{E}{-}4$$

Notice also:

1. The megohm units are marked MEGOHM (MEG would also be OK).
2. The polarity of the battery, V_{GG}, is provided by identifying a 1.5-V supply from node 0 (positive) to node 5 (negative).
3. The .AC LIN provides one frequency of 10 kHz, so that the .PRINT line can be used to provide ac voltages at nodes 1, 2, 3, and 4.

The circuit has an ac voltage gain, $V(4)/V(1) = 6.249$.

```
*** 01/25/90 ******* Evaluation PSpice (January 1989) ******* 19:29:50 ****

JFET Amplifier - Fixed bias

***        CIRCUIT DESCRIPTION

*****************************************************************************

VDD 6 0 DC 20VOLTS
VGG 0 5 DC 1.5VOLTS
J1  3 2 0 JFET
RG 2 5 10MEGOHM
RD 6 3 2KOHM
RL 4 0 10MEGOHM
C1 1 2 0.02UF
C2 3 4 2UF
VI 1 0 AC 10MV
.MODEL JFET NJF VTO=-4V BETA=6.25E-4
.AC LIN 1 10KH 10KH
.PRINT AC V(1) V(2) V(3) V(4)
.OPTIONS NOPAGE
.END

****      Junction FET MODEL PARAMETERS
          JFET
          NJF
     VTO    -4
     BETA   625.000000E-06

****      SMALL SIGNAL BIAS SOLUTION       TEMPERATURE =   27.000 DEG C
   NODE   VOLTAGE      NODE   VOLTAGE    NODE   VOLTAGE      NODE    VOLTAGE
 (    1)   0.0000    (    2)  -1.4998  (    3)  12.1870   (    4)    0.0000
 (    5)  -1.5000    (    6)  20.0000


     VOLTAGE SOURCE CURRENTS
     NAME          CURRENT
     VDD          -3.907E-03
     VGG          -1.521E-11
     TOTAL POWER DISSIPATION   7.81E-02   WATTS


****      AC ANALYSIS                       TEMPERATURE =   27.000 DEG C
    FREQ       V(1)        V(2)       V(3)        V(4)
   1.000E+04   1.000E-02   1.000E-02  6.249E-02   6.249E-02
```

Figure 9.31 PSpice output for circuit of Fig. 9.30.

Program 9.2: JFET Amplifier with DC Self-Bias

Figure 9.32 is an amplifier having dc self-bias. The bias resistor, R_S, is bypassed by capacitor C_S. Figure 9.33 provides the PSpice circuit description and the output results of the dc bias and ac operation. The voltage gain is seen to be V(5)/V(1) =

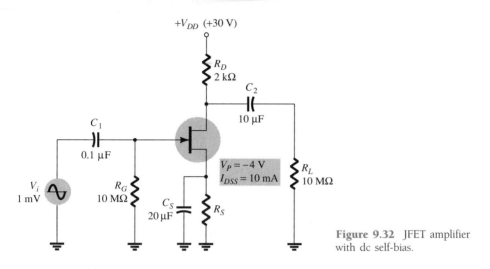

Figure 9.32 JFET amplifier with dc self-bias.

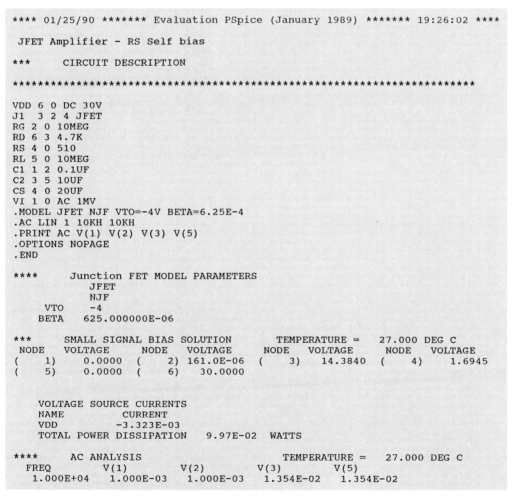

```
**** 01/25/90 ******* Evaluation PSpice (January 1989) ******* 19:26:02 ****

JFET Amplifier - RS Self bias

***      CIRCUIT DESCRIPTION

*********************************************************************

VDD 6 0 DC 30V
J1   3 2 4 JFET
RG 2 0 10MEG
RD 6 3 4.7K
RS 4 0 510
RL 5 0 10MEG
C1 1 2 0.1UF
C2 3 5 10UF
CS 4 0 20UF
VI 1 0 AC 1MV
.MODEL JFET NJF VTO=-4V BETA=6.25E-4
.AC LIN 1 10KH 10KH
.PRINT AC V(1) V(2) V(3) V(5)
.OPTIONS NOPAGE
.END

****     Junction FET MODEL PARAMETERS
           JFET
           NJF
     VTO    -4
     BETA   625.000000E-06

***      SMALL SIGNAL BIAS SOLUTION     TEMPERATURE =   27.000 DEG C
  NODE   VOLTAGE     NODE   VOLTAGE    NODE   VOLTAGE     NODE   VOLTAGE
(    1)   0.0000   (    2) 161.0E-06  (   3)  14.3840   (   4)   1.6945
(    5)   0.0000   (    6)  30.0000


     VOLTAGE SOURCE CURRENTS
     NAME           CURRENT
     VDD            -3.323E-03
     TOTAL POWER DISSIPATION   9.97E-02  WATTS

****     AC ANALYSIS                    TEMPERATURE =   27.000 DEG C
  FREQ       V(1)        V(2)        V(3)        V(5)
  1.000E+04  1.000E-03   1.000E-03   1.354E-02   1.354E-02
```

Figure 9.33 PSpice output for circuit of Fig. 9.32.

13.54. Dc bias results in $V_D = V(3) = 14.384$ V, while the gate–source voltage, $V_{GS} = V(2) - V(4) = -1.69$ V. The JFET model line is seen to be the same as in the preceding circuit in Fig. 9.30 and 9.31.

Program 9.3: JFET Amplifier with Voltage-Divider DC Bias

Figure 9.34 provides voltage-divider dc bias and an amplification from V_i to V_o. The circuit description in Fig. 9.35 includes the same transistor model as in the previous two circuits, with the R_S bias resistor bypassed by capacitor C_S. The ac voltage gain is seen to be $V(5)/V(1) = 5.499$.

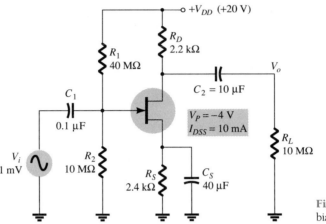

Figure 9.34 Voltage-divider bias ac amplifier.

Figure 9.35 PSpice output for circuit of Fig. 9.34.

```
**** 01/25/90 ******* Evaluation PSpice (January 1989) ******* 19:22:58 ****

JFET Amplifier - Voltage divider, self-bias

**      CIRCUIT DESCRIPTION

*********************************************************************************

VDD 6 0 DC 20V
J1 3 2 4 JFET
R1 6 2 40MEG
R2 2 0 10MEG
RD 3 6 2.2K
RS 4 0 2.4K
C1 1 2 0.1UF
CS 4 0 40UF
C2 3 5 10UF
RL 5 0 10MEG
.MODEL JFET NJF VTO=-4V BETA=6.25E-4
VI 1 0 AC 1MV
.AC LIN 1 10KH 10KH
.PRINT AC V(1) V(2) V(4) V(3) V(5)
.OPTIONS NOPAGE
.END


****      Junction FET MODEL PARAMETERS
          JFET
          NJF
    VTO    -4
    BETA   625.000000E-06


****      SMALL SIGNAL BIAS SOLUTION        TEMPERATURE =   27.000 DEG C
   NODE   VOLTAGE     NODE   VOLTAGE    NODE   VOLTAGE    NODE   VOLTAGE
(    1)    0.0000  (    2)    4.0001  (    3)   14.5000  (    4)    6.0001
(    5)    0.0000  (    6)   20.0000


     VOLTAGE SOURCE CURRENTS
     NAME         CURRENT
     VDD         -2.500E-03
     TOTAL POWER DISSIPATION    5.00E-02  WATTS


****      AC ANALYSIS                       TEMPERATURE =   27.000 DEG C
    FREQ        V(1)        V(2)        V(4)        V(3)        V(5)
   1.000E+04   1.000E-03   1.000E-03   9.947E-07   5.499E-03   5.499E-03
```

Program 9.4: Enhancement MOSFET Amplifier

Figure 9.36 is an enhancement amplifier with ac input V_i and resulting output V_o. The PSpice circuit description is provided in Fig. 9.37. The output listing shows dc bias at V_D = V(3) = 9.529 V and a voltage gain of V(5)/V(1) = 3.296. Notice the MOSFET device line

$$M1 \quad 3 \quad 2 \quad 0 \quad 4 \quad NFET$$

which identifies the element as a MOSFET device (M1), connected from drain (node 3), gate (node 2), source (node 0), and substrate (node 4), with the device an

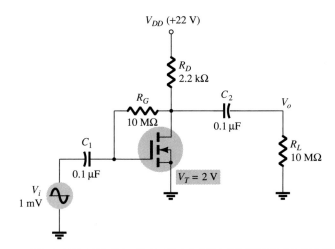

Figure 9.36 Enhancement MOSFET amplifier.

Figure 9.37 PSpice output for circuit of Fig. 9.36.

```
**** 01/25/90 ******* Evaluation PSpice (January 1989) ******* 18:13:59 ****

JFET AC Amplifier

**      CIRCUIT DESCRIPTION

***********************************************************************

VDD 6 0 DC 22V
MFET 3 2 0 4 NFET
RG 2 3 10MEG
RD 3 6 22K
C1 1 2 0.1UF
C2 3 5 0.1UF
RL 5 0 10MEG
.MODEL NFET NMOS(VTO=2V)
VI 1 0 AC 1MV
.AC LIN 1 10KH 10KH
.PRINT AC V(1) V(5)
.OPTIONS NOPAGE
.END

****      MOSFET MODEL PARAMETERS
                NFET
                NMOS
      LEVEL     1
        VTO     2
         KP     20.000000E-06


****      SMALL SIGNAL BIAS SOLUTION      TEMPERATURE =   27.000 DEG C
   NODE    VOLTAGE      NODE    VOLTAGE      NODE    VOLTAGE      NODE    VOLTAGE
   (   1)   0.0000   (    2)    9.5290   (    3)    9.5290   (    4)    .1765
   (   5)   0.0000   (    6)   22.0000

      VOLTAGE SOURCE CURRENTS
      NAME       CURRENT
      VDD        -5.669E-04
      TOTAL POWER DISSIPATION    1.25E-02  WATTS


****      AC ANALYSIS                      TEMPERATURE =   27.000 DEG C
   FREQ        V(1)         V(5)
   1.000E+04   1.000E-03    3.296E-03
```

427

n-channel MOSFET (NFET). The device model line

$$.\text{MODEL} \quad \text{NFET} \quad \text{NMOS (VTO} = 2\text{V)}$$

provides the specification that the enhancement MOSFET has a threshold voltage of VTO = V_T = 2 V.

PROBLEMS

§ 9.2 FET Small-Signal Model

1. Calculate g_{mo} for a JFET having device parameters I_{DSS} = 15 mA, V_P = −5 V.

2. Determine the pinch-off voltage of a JFET with g_{mo} = 10 mS and I_{DSS} = 12 mA.

3. For a JFET having device parameter g_{mo} = 5 mS and V_P = −3.5 V, what is the device current at V_{GS} = 0 V?

4. Calculate the value of g_m for a JFET (I_{DSS} = 12 mA, V_P = −3 V) at a bias point of V_{GS} = −1 V.

5. For a JFET having g_m = 6 mS at V_{GS_Q} = −1 V, what is the value of I_{DSS} if V_P = −2.5 V?

6. A JFET (I_{DSS} = 10 mA, V_P = −5 V) is biased at I_D = $I_{DSS}/4$. What is the value of g_m at that bias point?

7. Determine the value of g_m for a JFET (I_{DSS} = 8 mA, V_P = −5 V) when biased at V_{GS_Q} = $V_P/4$.

8. A specification sheet provides the following data (at a listed drain–source current)

$$y_{fs} = 4.5 \text{ mS}, \qquad y_{os} = 25 \text{ } \mu\text{S}$$

At the listed drain–source current, determine:
(a) g_m.   (b) r_d.

9. For a JFET having specified values of y_{fs} = 4.5 mS and y_{os} = 25 μS, determine the device output impedance, Z_o(FET), and device ideal voltage gain, A_v(FET).

10. If a JFET having a specified value of r_d = 100 kΩ has an ideal voltage gain of A_v(FET) = −200, what is the value of g_m?

11. Using the transfer characteristic of Fig. 9.38:
(a) What is the value of g_{mo}?
(b) What is the value of g_m at V_{GS_Q} = −1.5 V?
(c) What is the value of g_m at V_{GS_Q} = −2.5 V?

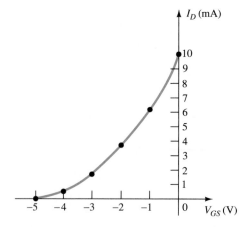

Figure 9.38 JFET transfer characteristic for problem 11.

12. Using the drain characteristic of Fig. 9.39:
(a) What is the value of r_d for V_{GS} = 0 V?
(b) What is the value of g_{mo} at V_{DS} = 10 V?

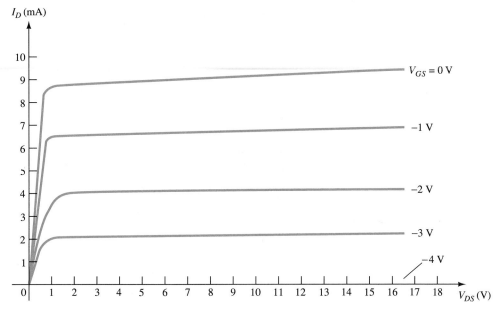

Figure 9.39 JFET drain characteristic for problem 12.

13. For a 2N4220 n-channel JFET (y_{fs}(minimum) = 750 μS, y_{os}(maximum) = 10 μS):
(a) What is the value of g_m?
(b) What is the value of r_d?

§ 9.4 Voltage Gain

14. Calculate the voltage gain of the amplifier in Fig. 9.40 for device parameters I_{DSS} = 10 mA and $V_P = -4$ V. (Assume that $r_d = \infty$.)

15. Calculate the voltage gain of the amplifier in Fig. 9.40 for device parameters I_{DSS} = 12 mA, $V_P = -6$ V, and y_{os} = 25 μS.

16. Calculate the voltage gain of the amplifier in Fig. 9.41 for specification parameters of y_{fs} = 3000 μS and y_{os} = 40 μS.

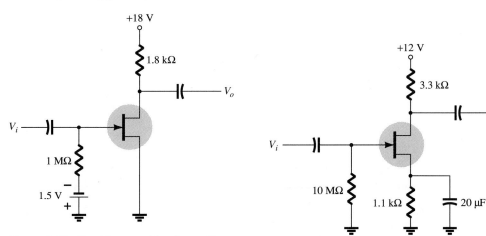

Figure 9.40 Fixed bias amplifier for problems 14, 15, and 36.

Figure 9.41 Problems 16, 17, 18, and 37

17. For the circuit of Fig. 9.41, calculate (a) Z_i; (b) Z_o.

18. For the circuit of Fig. 9.41, calculate A_V for a JFET with I_{DSS} = 8 mA, $V_P = -3.5$ V.

g_m

19. For the circuit of Fig. 9.42, calculate the output voltage V_o if:
  (a) R_S is bypassed by capacitor $C_S = 40\ \mu F$.
  (b) No bypass capacitor is present.

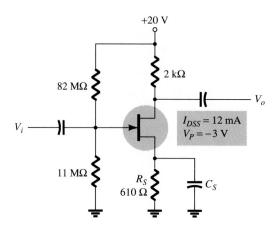

Figure 9.42 Problems 19, 20, and 38

20. For the circuit of Fig. 9.42, calculate the circuit voltage gain, A_v, if $r_d = 35\ k\Omega$.

21. Calculate V_o for the circuit of Fig. 9.43.

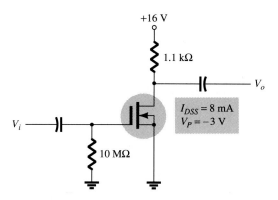

Figure 9.43 Problems 21, 22

22. Calculate V_o for the circuit of Fig. 9.43 for $r_d = 30\ k\Omega$.

23. For the circuit of Fig. 9.44, calculate the output voltage V_o assuming that r_d is infinite.

* **24.** Calculate V_o for the circuit of Fig. 9.44 for $r_d = 40\ k\Omega$.

* **25.** Calculate A_v for the circuit of Fig. 9.45.

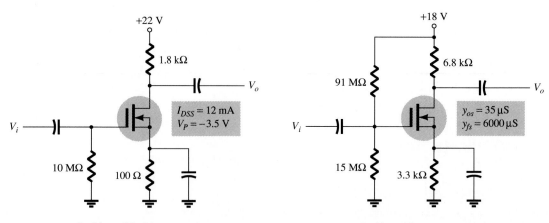

Figure 9.44 Problems 23, 24

Figure 9.45 Problem 25

§ 9.5 Source-Follower (Common-Drain) Circuit

26. Calculate V_o, Z_i, and Z_o for the circuit of Fig. 9.46 ($I_{DSS} = 9$ mA, $V_P = -4.5$ V), when $V_i = 0.2$ V, rms.

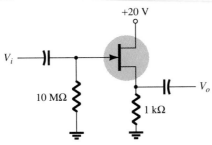

+20 V

V_i

10 MΩ

1 kΩ

V_o

Figure 9.46 Problems 26–28, 39

27. What is the value of g_m in the circuit of Fig. 9.46 if $V_o = 0.86$ mV when $V_i = 1$ mV (assuming that $r_d = \infty$).

28. What value of g_m is needed to provide a voltage gain of 0.9 in the circuit of Fig. 9.46 if $r_d = 50$ kΩ?

* **29.** Calculate Z_i and Z_o for the circuit of Fig. 9.47.

* **30.** Calculate A_v for the circuit of Fig. 9.47.

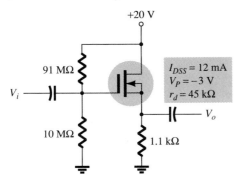

+20 V

91 MΩ

V_i

10 MΩ

1.1 kΩ

$I_{DSS} = 12$ mA
$V_P = -3$ V
$r_d = 45$ kΩ

V_o

Figure 9.47 Problems 29, 30

§ 9.6 Common-Gate Circuit

31. For the circuit in Fig. 9.48, calculate A_v, Z_i, and Z_o.

32. Calculate A_v, Z_i, and Z_o for the circuit of Fig. 9.48 with device $r_d = 33$ kΩ.

* **33.** Calculate Z_i, Z_o, and A_v for the circuit of Fig. 9.49.

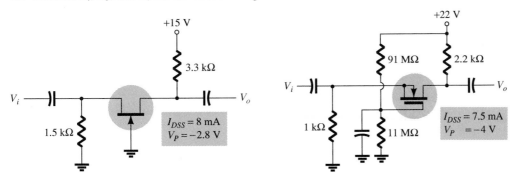

+15 V

3.3 kΩ

V_i

1.5 kΩ

$I_{DSS} = 8$ mA
$V_P = -2.8$ V

V_o

+22 V

91 MΩ

2.2 kΩ

V_i

1 kΩ

11 MΩ

$I_{DSS} = 7.5$ mA
$V_P = -4$ V

V_o

Figure 9.48 Problems 31, 32, 40

Figure 9.49 Problem 33

§ 9.7 Enhancement MOSFET Amplifier

34. Calculate g_m for a MOSFET having $V_T = 3$ V when biased at $V_{GS_Q} = 8$ V. (Assume that $k = 0.3 \times 10^{-3}$.)

35. Calculate the voltage gain of the amplifier in Fig. 9.50. (Use $k = 0.3 \times 10^{-3}$.)

36. Calculate the voltage gain of the amplifier in Fig. 9.50. (Use $k = 0.25 \times 10^{-3}$.)

* **37.** Calculate Z_i, Z_o, and A_v for the circuit of Fig. 9.51.

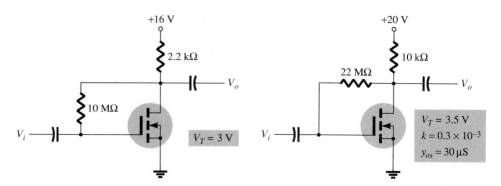

Figure 9.50 Problems 35, 36, 41   **Figure 9.51** Problem 37

§ 9.8 Design of JFET Amplifier Circuits

* **38.** Determine a value of R_D to provide dc bias at $V_{D_Q} = \frac{1}{2}V_{DD}$ in the circuit of Fig. 9.52.

* **39.** Determine values of R_D and R_S for dc bias of $V_{GS_Q} = \frac{1}{2}V_P$ and $V_{DS_Q} = \frac{1}{2}V_{DD}$ in the circuit of Fig. 9.53. Calculate A_v for the values selected.

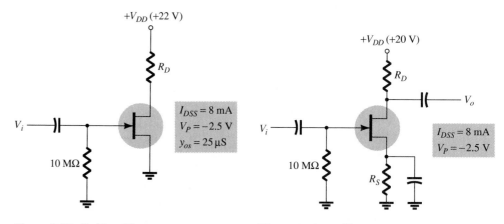

Figure 9.52 Problem 38   **Figure 9.53** Problem 39

§ 9.10 Computer Analysis

* **40.** Use PSpice to calculate the dc bias conditions and the voltage gain of the circuit in Fig. 9.40 for $I_{DSS} = 10$ mA, $V_P = -4$ V.

* **41.** Use PSpice to calculate the dc bias conditions and voltage gain for the circuit of Fig. 9.41 for $I_{DSS} = 8$ mA, $V_P = -3.5$ V.

* **42.** Use PSpice to calculate the dc bias conditions and voltage gain for the circuit of Fig. 9.42.

* **43.** Use PSpice to calculate the dc bias conditions and voltage gain for the circuit of Fig. 9.46 for $I_{DSS} = 9$mA, $V_P = -4.5$ V.

* **44.** Use PSpice to calculate the dc bias conditions and voltage gain for the circuit of Fig. 9.48.

* **45.** Use PSpice to calculate the dc bias conditions and voltage gain for the circuit of Fig. 9.50.

---

*Please Note: Asterisks indicate more difficult problems.

TABLE 9.1

Configuration	Z_i	A_v	Z_o
Fixed-bias	High (10 MΩ) $= R_G$	Medium (−20) $= -g_m(R_D\|r_d)$	Medium (2 kΩ) $= (R_D\|r_d)$
Voltage-divider	Medium (1 MΩ) $= R_1\|R_2$	Medium (−20) $= -g_m(R_D\|r_d)$	Medium (2 kΩ) $= (R_D\|r_d)$
Source-follower	High (10 MΩ) $= R_G$	Low $\cong 1$	Low (100 Ω) $= \left(R_s\|r_d\|\dfrac{1}{gm}\right)$
Unbypassed source	High (10 MΩ) $= R_G$	Low (−2) $= -\dfrac{g_m R_D}{1 + g_m R_s}$	Medium (2 kΩ) $= (R_D\|r_d)$
Common-gate	Low (1 kΩ) $= R_s$	Medium (+20) $= g_m(R_D\|r_d)$	Medium (2 kΩ) $\cong (R_D\|r_d)$
MOSFET	Medium (1 MΩ) $= \dfrac{R_G}{1 + g_m(R_D\|r_d)}$	Medium (+20) $= -g_m(R_D\|r_d)$	Medium (2 kΩ) $= (R_D\|r_d)$

10

Systems Approach— Effects of R_s and R_L

R_S/R_L

10.1 INTRODUCTION

In recent years the introduction of a wide variety of packaged networks and systems has generated an increasing interest in the systems approach to design and analysis. Fundamentally, this approach concentrates on the terminal characteristics of a package and treats each as a building block in the formation of the total package. The content of this chapter is a first step in developing some familiarity with this approach. The techniques introduced will be used in the remaining chapters and broadened as the need arises. The trend to packaged systems is quite understandable when you consider the enormous advances in the design and manufacturing of integrated circuits (ICs). The small IC packages contain stable, reliable, self-testing, sophisticated designs that would be quite bulky if built with discrete (individual) components. The systems approach is not a difficult one to apply once the basic definitions of the various parameters are correctly understood and the manner in which they are utilized is clearly demonstrated. In the next few sections we develop the systems approach in a slow deliberate manner that will include numerous examples to make each salient point. If the content of this chapter is clearly and correctly understood, a first plateau in the understanding of system analysis will be accomplished.

10.2 TWO-PORT SYSTEMS

The description to follow can be applied to any two-port system—not only those containing BJTs and FETs—although the emphasis in this chapter is on these active devices. The emphasis in previous chapters on determining the two-port parameters for various configurations will be quite helpful in the analysis to follow. In fact, many of the results obtained in the last two chapters are utilized in the analysis to follow.

In Fig. 10.1 the important parameters of a two-port system have been identified. Note, in particular, the absence of a load and a source resistance. The impact of these important elements is considered in detail in a later section. For the moment recognize that the impedance levels and the gains of Fig. 10.1 are determined for no-load (absence of R_L) and no source resistance (R_s) conditions.

If we take a "Thévenin look" at the output terminals we find with V_i set to zero that

$$\boxed{Z_{\text{Th}} = Z_o = R_o}$$

(10.1)

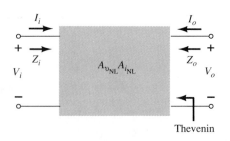

Figure 10.1 Two-port system.

E_{Th} is the open-circuit voltage between the output terminals identified as V_o. However,

$$A_{v_{NL}} = \frac{V_o}{V_i}$$

and

$$V_o = A_{v_{NL}} V_i$$

so that

$$\boxed{E_{Th} = A_{v_{NL}} V_i}$$ (10.2)

Note the use of the additional subscript notation NL to identify a no-load voltage gain.

Substituting the Thévenin equivalent circuit between the output terminals will result in the output configuration of Fig. 10.2. For the input circuit the parameters V_i and I_i are related by $Z_i = R_i$ permitting the use of R_i to represent the input circuit. Since our present interest is in BJT and FET amplifiers both Z_o and Z_i can be represented by resistive elements.

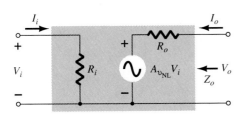

Figure 10.2 Substituting the internal elements for the two-port system of Fig. 10.1.

Before continuing let us check the results of Fig. 10.2 by finding Z_o and $A_{v_{NL}}$ in the usual manner. To find Z_o, V_i is set to zero, resulting in $A_{v_{NL}} V_i = 0$, permitting a short-circuit equivalent for the source. The result is an output impedance equal to R_o as originally defined. The absence of a load will result in $I_o = 0$, and the voltage drop across the impedance R_o will be 0 V. The open-circuit output voltage is therefore $A_{v_{NL}} V_i$, as it should be. Before looking at an example, take note of the fact that A_i does not appear in the two-port model of Fig. 10.2 and in fact is seldom part of the two-port system analysis of active devices. This is not to say that the quantity is seldom calculated, but it is most frequently calculated from the expression $A_i = -A_v(Z_i/R_L)$, where R_L is the defined load for the analysis of interest.

For the fixed-bias transistor network of Fig. 10.3 (Example 8.1), sketch the two-port equivalent of Fig. 10.2.

EXAMPLE 10.1

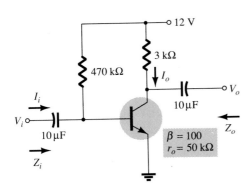

Figure 10.3 Example 10.1

Solution

From Example 8.1,

$$Z_i = 1.071 \text{ k}\Omega$$

$$Z_o = 3 \text{ k}\Omega$$

$$A_{v_{\text{NL}}} = -280.11$$

Using the information above, the two-port equivalent of Fig. 10.4 can be drawn. Note, in particular, the negative sign associated with the controlled voltage source, revealing an opposite polarity for the controlled source than that indicated in the figure. It also reveals a 180° phase-shift between the input and output voltages.

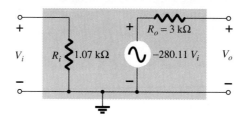

Figure 10.4 Two-port equivalent for the parameters specified in Example 10.1.

In Example 10.1, $R_C = 3 \text{ k}\Omega$ was included in defining the no-load voltage gain. Although this need not be the case (R_C could be defined as the load resistor in Chapter 8), the analysis of this chapter will assume that all biasing resistors are part of the no-load gain and that a loaded system requires an additional load R_L connected to the output terminals.

A second format for Fig. 10.2, particularly popular with op-amps (operational amplifiers), appears in Fig. 10.5. The only change is the general appearance of the model.

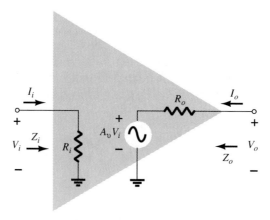

Figure 10.5 Operational amplifier (Op-amp) notation.

10.3 EFFECT OF A LOAD IMPEDANCE (R_L)

In this section the effect of an applied load is investigated using the two-port model of Fig. 10.2. The model can be applied to any current- or voltage-controlled amplifier. $A_{v_{\text{NL}}}$ is, as defined earlier, the gain of the system without an applied load. R_i and R_o are the input and output impedances of the amplifier as defined by the configuration. Ideally, all the parameters of the model are unaffected by changing loads or source resistances (as normally encountered for op-amps to be described in Chapter 14). However, for some transistor amplifier configurations R_i can be quite sensitive to the applied load, while for others R_o can be sensitive to the source resistance. In any case, once $A_{v_{NL}}$, R_i, and R_o are defined for a particular configuration, the equations about to be derived can be employed.

Applying a load to the two-port system of Fig. 10.2 will result in the configuration of Fig. 10.6. Applying the voltage-divider rule to the output circuit will result in

$$V_o = \frac{R_L A_{v_{NL}} V_i}{R_L + R_o}$$

and

$$\boxed{A_v = \frac{V_o}{V_i} = \frac{R_L}{R_L + R_o} A_{v_{NL}}}$$

(10.3)

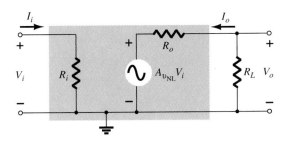

Figure 10.6 Applying a load to the two-port system of Fig. 10.2.

Since the ratio $R_L/(R_L + R_o)$ will always be less than 1:

The loaded voltage gain of an amplifier is always less than the no-load level.

Note also that the formula for the voltage gain does not include the input impedance or current gain.

Although the level of R_i may change with the configuration, the applied voltage and input current will always be related by

$$\boxed{I_i = \frac{V_i}{Z_i} = \frac{V_i}{R_i}}$$

(10.4)

Defining the output current as the current through the load will result in

$$\boxed{I_o = -\frac{V_o}{R_L}}$$

(10.5)

with the minus sign occurring due to the defined direction for I_o in Fig. 10.6.

The current gain is then determined by

$$A_i = \frac{I_o}{I_i} = \frac{-V_o/R_L}{V_i/Z_i} = -\frac{V_o}{V_i}\frac{Z_i}{R_L}$$

and

$$\boxed{A_i = -A_v\frac{Z_i}{R_L}}$$

(10.6)

for the unloaded situation. In general, therefore, the current gain can be obtained from the voltage gain and impedance parameters Z_i and R_L. The next example will demonstrate the usefulness and validity of Eqs. (10.3) through (10.6).

EXAMPLE 10.2

In Fig. 10.7 a load has been applied to the fixed-bias transistor amplifier of Example 10.1 (Fig. 10.3).
(a) Determine the voltage and current gain using the two-port systems approach defined by the model of Fig. 10.4.
(b) Determine the voltage and current gain using the r_e model and compare results.

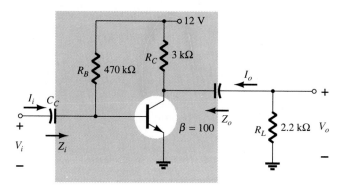

Figure 10.7 Example 10.2

Solution

(a) Recall from Example 10.1 that

$$Z_i = 1.071 \text{ k}\Omega \quad (\text{with } r_e = 10.71 \text{ }\Omega \text{ and } \beta = 100)$$

$$Z_o = 3 \text{ k}\Omega$$

$$A_{v_{NL}} = -280.11$$

Applying Eq. (10.3) yields

$$A_v = \frac{R_L}{R_L + R_o} A_{v_{NL}}$$

$$= \frac{2.2 \text{ k}\Omega}{2.2 \text{ k}\Omega + 3 \text{ k}\Omega} (-280.11)$$

$$= (0.423)(-280.11)$$

$$= \mathbf{-118.5}$$

For the current gain,

$$A_i = -A_v \frac{Z_i}{R_L}$$

In this case, Z_i is unaffected by the applied load and

$$A_i = -(-118.51) \frac{1.071 \text{ k}\Omega}{2.2 \text{ k}\Omega} = \mathbf{57.69}$$

(b) Substituting the r_e model will result in the network of Fig. 10.8. Note, in particular, that the applied load is in parallel with the collector resistor R_C defining a net parallel resistance

$$R_L' = R_C \| R_L = 3 \text{ k}\Omega \| 2.2 \text{ k}\Omega = 1.269 \text{ k}\Omega$$

The output voltage

$$V_o = -\beta I_B R_L'$$

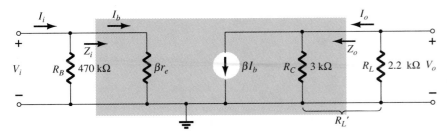

Figure 10.8 Substituting the r_e model in the ac equivalent network of Fig. 10.7.

with
$$I_b = \frac{V_i}{\beta r_e}$$

and
$$V_o = -\beta \frac{V_i}{\beta r_e} R_L'$$

so that
$$\boxed{A_v = \frac{V_o}{V_i} = -\frac{R_L'}{r_e} = -\frac{R_C \| R_L}{r_e}}$$ (10.7)

Substituting values gives
$$A_v = -\frac{1.269 \text{ k}\Omega}{10.71 \text{ }\Omega} = -\mathbf{118.5}$$

as obtained above. For the current gain, by the current-divider rule,
$$I_b = \frac{(470 \text{ k}\Omega)I_i}{470 \text{ k}\Omega + 1.071 \text{ k}\Omega} = 0.9977I_i \cong I_i$$

and
$$I_o = \frac{3 \text{ k}\Omega(\beta I_b)}{3 \text{ k}\Omega + 2.2 \text{ k}\Omega}$$
$$= 0.5769 \, \beta I_b$$

so that
$$A_i = \frac{I_o}{I_i} = \frac{0.5769 \beta I_b}{I_i} = \frac{0.5769 \beta I_i}{I_i}$$
$$= 0.5769(100) = \mathbf{57.69}$$

as obtained using Eq. (10.6).

Example 10.2 demonstrated two techniques to solve the same problem. Although any network can be solved using the r_e model approach, the advantage of the systems approach is that once the two-port parameters of a system are known, the effect of changing the load can be determined directly from Eq. (10.3). No need to go back to the ac equivalent model and analyze the entire network. The advantages of the systems approach are similar to those associated with applying Thévenin's theorem. They permit concentrating on the effects of the load without having to re-examine the entire network. Of course, if the network of Fig. 10.7 were presented for analysis without the unloaded parameters, it would be a toss-up as to which approach would yield the desired results in the most direct, efficient manner. However, keep in mind that the "package" approach is the developing trend. When you purchase a "system" the two-port parameters are provided, and as with any trend, the user must be aware of how to utilize the given data.

The AC Load Line

For a system such as appearing in Fig. 10.9a, the dc load line was drawn on the output characteristics as shown in Fig. 10.9b. The load resistance did not contribute to the dc load line since it was isolated from the biasing network by the coupling capacitor (C_C). For the ac analysis the coupling capacitors are replaced by a short-circuit equivalence that will place the load and collector resistors in a parallel arrangement defined by

$$R_L' = R_C\|R_L$$

The effect on the load line is shown in Fig. 10.9b with the levels to determine the new axes intersections. Note of particular importance that the ac and dc load lines pass through the same Q-point—a condition that must be satisfied to ensure a common solution for the network under dc and/or ac conditions.

For the unloaded situation, the application of a relatively small sinusoidal signal to the base of the transistor could cause the base current to swing from a level of I_{B_2} to I_{B_4} as shown in Fig. 10.9b. The resulting output voltage v_{ce} would then have the swing appearing in the same figure. The application of the same signal for a loaded situation would result in the same swing in the I_B level, as shown in Fig. 10.9b. The result, however, of the steeper slope of the ac load line is a smaller output voltage

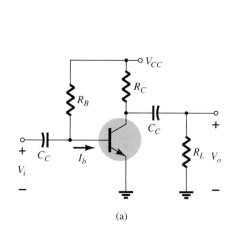

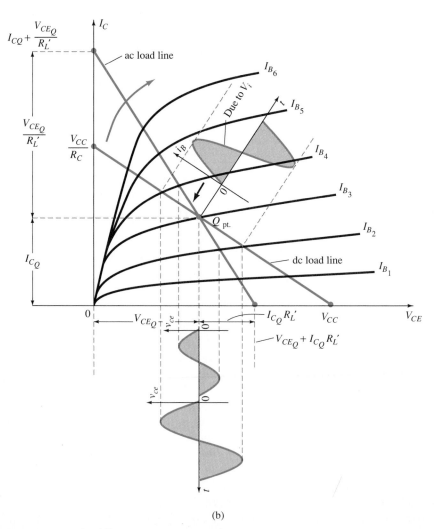

Figure 10.9 Demonstrating the differences between the dc and ac load lines.

Chapter 10 Systems Approach—Effects of R_S and R_L

swing (v_{ce}) and a drop in the gain of the system as demonstrated in the numerical analysis above. It should be obvious from the intersection of the ac load line on the vertical axis that the smaller the level of R_L', the steeper the slope and the smaller the ac voltage gain. Since R_L' is smaller for reduced levels of R_L, it should be fairly clear that:

For a particular design the smaller the level of R_L, the lower the level of ac voltage gain.

10.4 EFFECT OF THE SOURCE IMPEDANCE (R_s)

Our attention will now turn to the input side of the two-port system and the effect of an internal source resistance on the gain of an amplifier. In Fig. 10.10 a source with an internal resistance has been applied to the basic two-port system. The definitions of Z_i and $A_{v_{NL}}$ are such that:

The parameters Z_i and $A_{v_{NL}}$ of a two-port system are unaffected by the internal resistance of the applied source.

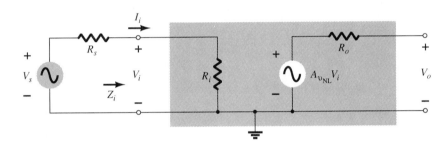

Figure 10.10 Including the effects of the source resistance R_s.

However:

The output impedance may be affected by the magnitude of R_s.

(Recall Eq. (8.8) for the complete hybrid equivalent model.) The fraction of the applied signal reaching the input terminals of the amplifier of Fig. 10.10 is determined by the voltage-divider rule. That is,

$$V_i = \frac{R_i V_s}{R_i + R_s} \qquad (10.8)$$

Equation (10.8) clearly shows that the larger the magnitude of R_s, the less the voltage at the input terminals of the amplifier. In general, therefore:

For a particular amplifier the larger the internal resistance of a signal source, the less the overall gain of the system.

For the two-port system of Fig. 10.10,

$$V_o = A_{v_{NL}} V_i$$

and

$$V_i = \frac{R_i V_s}{R_i + R_s}$$

so that

$$V_o = A_{v_{NL}} \frac{R_i}{R_i + R_s} V_s$$

and

$$A_{v_s} = \frac{V_o}{V_s} = \frac{R_i}{R_i + R_s} A_{v_{NL}} \qquad (10.9)$$

The result clearly supports the statement above regarding the reduction in gain with increase in R_s. Using Eq. (10.9), if $R_s = 0 \ \Omega$ (ideal voltage source), $A_{v_s} = A_{v_{NL}}$, which is the maximum possible value.

The input current is also altered by the presence of a source resistance as follows:

$$I_i = \frac{V_s}{R_s + R_i} \qquad (10.10)$$

EXAMPLE 10.3

In Fig. 10.11, a source with an internal resistance has been applied to the fixed-bias transistor amplifier of Example 10.1 (Fig. 10.4).
(a) Determine the voltage gain $A_{v_s} = V_o/V_s$. What percent of the applied signal appears at the input terminals of the amplifier?
(b) Determine the voltage gain $A_{v_s} = V_o/V_s$ using the r_e model.

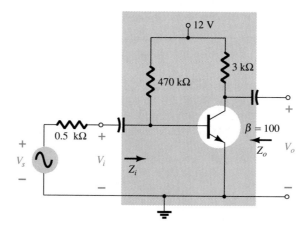

Figure 10.11 Example 10.3

Solution

(a) The two-port equivalent for the network appears in Fig. 10.12.

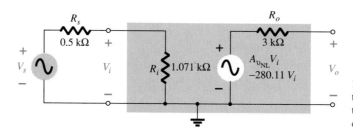

Figure 10.12 Substituting the two-port equivalent network for the fixed-bias transistor amplifier of Fig. 10.11.

$$\text{Eq. (10.9):} \quad A_{v_s} = \frac{V_o}{V_s} = \frac{R_i}{R_i + R_s} A_{v_{NL}} = \frac{1.071 \ \text{k}\Omega}{1.071 \ \text{k}\Omega + 0.5 \ \text{k}\Omega} (-280.11)$$

$$= (0.6817)(-280.11)$$

$$= \mathbf{-190.96}$$

$$\text{Eq. (10.8):} \quad V_i = \frac{R_i V_s}{R_i + R_s} = \frac{(1.071 \ \text{k}\Omega) V_s}{1.071 \ \text{k}\Omega + 0.5 \ \text{k}\Omega} = 0.6817 V_s$$

or **68.2%** of the available signal reached the amplifier and 31.8% was lost across the internal resistance of the source.

(b) Substituting the r_e model will result in the equivalent circuit of Fig. 10.13. Solving for V_o gives

$$V_o = -(100I_b)3 \text{ k}\Omega$$

with

$$Z_i \cong \beta r_e \quad \text{and} \quad I_b \cong I_i = \frac{V_s}{R_s + \beta r_e} = \frac{V_s}{1.571 \text{ k}\Omega}$$

and

$$V_o = -100\left(\frac{V_s}{1.571 \text{ k}\Omega}\right)3 \text{ k}\Omega$$

so that

$$A_{v_s} = \frac{V_o}{V_s} = -\frac{(100)(3 \text{ k}\Omega)}{1.57 \text{ k}\Omega}$$

$$= -190.96$$

as above.

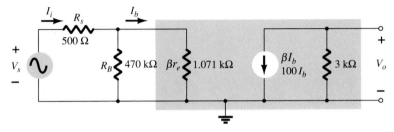

Figure 10.13 Substituting the r_e equivalent circuit for the fixed-bias transistor amplifier of Fig. 10.11.

Throughout the analysis above, note that R_s was not included in the definition of Z_i for the two-port system. Of course, the resistance "seen" by the source is now $R_s + Z_i$, but R_s remains a quantity associated only with the applied source.

Note again in Example 10.3 that the same results were obtained with the systems approach and using the r_e model. Certainly, if the two-port parameters are available, they should be applied. If not, the approach to the solution is simply a matter of preference.

10.5 COMBINED EFFECT OF R_s AND R_L

The effects of R_s and R_L has now been demonstrated on an individual basis. The next natural question is how the presence of both factors in the same network will affect the total gain. In Fig. 10.14 a source with an internal resistance R_s and a load R_L have been applied to a two-port system for which the parameters Z_i, $A_{v_{NL}}$, and Z_o have been specified. For the moment, let us assume that Z_i and Z_o are unaffected by R_L and R_s, respectively.

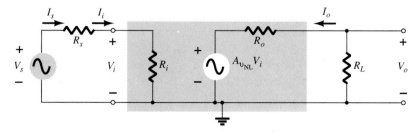

Figure 10.14 Considering the effects of R_s and R_L on the gain of an amplifier.

At the input side we find

$$\text{Eq. (10.8):} \quad V_i = \frac{R_i V_s}{R_i + R_s}$$

or

$$\boxed{\frac{V_i}{V_s} = \frac{R_i}{R_i + R_s}} \tag{10.11}$$

and at the output side,

$$V_o = \frac{R_L A_{v_{NL}} V_i}{R_L + R_o}$$

or

$$\boxed{A_v = \frac{V_o}{V_i} = \frac{R_L A_{v_{NL}}}{R_L + R_o}} \tag{10.12}$$

For the total gain $A_{v_s} = V_o/V_s$, the following mathematical steps can be performed:

$$A_{v_s} = \frac{V_o}{V_s} = \frac{V_o}{V_i}\frac{V_i}{V_s} \tag{10.13}$$

and substituting Eqs. (10.11) and (10.12) will result in

$$A_{v_s} = \frac{R_L A_{v_{NL}}}{R_L + R_o}\frac{R_i}{R_i + R_s}$$

and

$$\boxed{A_{v_s} = \frac{V_o}{V_s} = \frac{R_i}{R_i + R_s}\frac{R_L}{R_L + R_o}A_{v_{NL}}} \tag{10.14}$$

Since $I_i = V_i/R_i$, as before,

$$\boxed{A_i = -A_v \frac{R_i}{R_L}} \tag{10.15}$$

or using $I_s = V_s/(R_s + R_i)$,

$$\boxed{A_{i_s} = -A_{v_s}\frac{R_s + R_i}{R_L}} \tag{10.16}$$

However, $I_i = I_s$ so Eqs. 10.15 and 10.16 will generate the same result. Equation (10.14) clearly reveals that both the source and the load resistance will reduce the overall gain of the system. In fact:

The larger the source resistance and/or smaller the load resistance, the less the overall gain of an amplifier.

The two reduction factors of Eq. (10.14) form a product that has to be carefully considered in any design procedure. It is not sufficient to ensure that R_s is relatively small if the impact of the magnitude of R_L is ignored. For instance, in Eq. (10.14), if the first factor is 0.9 and the second factor is 0.2, the product of the two results in an overall reduction factor equal to $(0.9)(0.2) = 0.18$ which is close to the lower factor. The effect of the excellent 0.9 level was completely wiped out by the significantly lower second multiplier. If both were 0.9-level factors, the net result would be $(0.9)(0.9) = 0.81$, which is still quite high. Even if the first were 0.9 and the second 0.7, the net result of 0.63 would still be respectable. In general, therefore, for good overall gain the effect of both R_s and R_L must be evaluated individually and as a product.

Chapter 10 Systems Approach—Effects of R_S and R_L

For the single-stage amplifier of Fig. 10.15, with $R_L = 4.7$ kΩ and $R_s = 0.3$ kΩ, determine:

EXAMPLE 10.4

(a) A_{v_s}.
(b) $A_v = V_o/V_i$.
(c) A_i.

The two-port parameters for the fixed-bias configuration are $Z_i = 1.071$ kΩ, $Z_o = 3$ kΩ, and $A_{v_{NL}} = -280.11$.

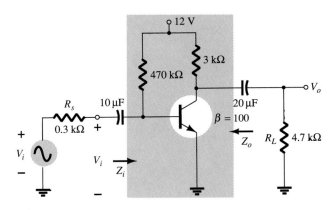

Figure 10.15 Example 10.4

Solution

(a) Eq. (10.14): $A_{v_s} = \dfrac{V_o}{V_s} = \dfrac{R_i}{R_i + R_s} \dfrac{R_L}{R_L + R_o} A_{v_{NL}}$

$$= \left(\dfrac{1.071 \text{ k}\Omega}{1.071 \text{ k}\Omega + 0.3 \text{ k}\Omega}\right)\left(\dfrac{4.7 \text{ k}\Omega}{4.7 \text{ k}\Omega + 3 \text{ k}\Omega}\right)(-280.11)$$

$$= (0.7812)(0.6104)(-280.11)$$

$$= (0.4768)(-280.11)$$

$$= \mathbf{-133.57}$$

(b) $A_v = \dfrac{V_o}{V_i} = \dfrac{R_L A_{v_{NL}}}{R_L + R_o} = \dfrac{(4.7 \text{ k}\Omega)(-280.11)}{4.7 \text{ k}\Omega + 3 \text{ k}\Omega}$

$$= (0.6104)(-280.11) = \mathbf{-170.98}$$

(c) $A_i = -A_v \dfrac{R_i}{R_L} = -(-170.98)\left(\dfrac{1.071 \text{ k}\Omega}{4.7 \text{ k}\Omega}\right)$

$$= \mathbf{38.96}$$

or $A_{i_s} = -A_{v_s} \dfrac{R_s + R_i}{R_L} = -(-133.57)\left(\dfrac{1.071 \text{ k}\Omega + 0.3 \text{ k}\Omega}{4.7 \text{ k}\Omega}\right)$

$$= \mathbf{38.96}$$

as above.

10.6 BJT CE NETWORKS

The fixed-bias configuration has been employed throughout the analysis of the early sections of this chapter to clearly show the effects of R_s and R_L. In this section various CE configurations are examined with a load and a source resistance. A detailed analysis will not be performed for each configuration since they follow a very similar path to that demonstrated in the last few sections.

Fixed Bias

For the fixed-bias configuration examined in detail in recent sections, the system model with a load and source resistance will appear as shown in Fig. 10.16. In general,

$$V_o = \frac{R_L A_{v_{NL}} V_i}{R_L + R_o}$$

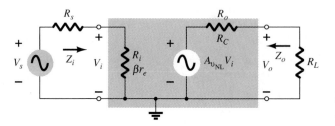

Figure 10.16 Fixed-bias configuration with R_s and R_L.

Substituting Eq. (8.4), $A_{v_{NL}} = -R_C/r_e$ and $R_o = R_C$,

$$V_o = \frac{R_L(-R_C/r_e)V_i}{R_L + R_C}$$

and

$$A_v = \frac{V_o}{V_i} = -\frac{R_L R_C}{R_L + R_C}\frac{1}{r_e}$$

but

$$R_L \| R_C = \frac{R_L R_C}{R_L + R_C}$$

and

$$\boxed{A_v = -\frac{R_L \| R_C}{r_e}}$$

(10.17)

If the r_e model were substituted for the transistor in the fixed-bias configuration, the network of Fig. 10.17 would result, clearly revealing that R_C and R_L are in parallel.

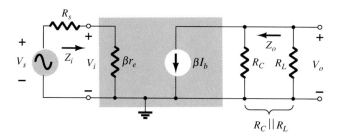

Figure 10.17 Fixed-bias configuration with the substitution of the r_e model.

For the voltage gain A_{v_s} of Fig. 10.16,

$$V_i = \frac{Z_i V_s}{Z_i + R_s}$$

and

$$\frac{V_i}{V_s} = \frac{Z_i}{Z_i + R_s}$$

with

$$A_{v_s} = \frac{V_o}{V_s} = \frac{V_i}{V_s}\frac{V_o}{V_i}$$

so that

$$A_{v_s} = \frac{Z_i}{Z_i + R_s} A_v \qquad (10.18)$$

Since the load is connected to the collector terminal of the common-emitter configuration,

$$Z_i = \beta r_e \qquad (10.19)$$

and

$$Z_o = R_C \qquad (10.20)$$

as obtained earlier.

Voltage-Divider Bias

For the loaded voltage-divider bias configuration of Fig. 10.18, the load is again connected to the collector terminal and Z_i remains

$$Z_i \cong R'\|\beta r_e \qquad (R' = R_1\|R_2) \qquad (10.21)$$

and for the system's output impedance

$$Z_o = R_C \qquad (10.22)$$

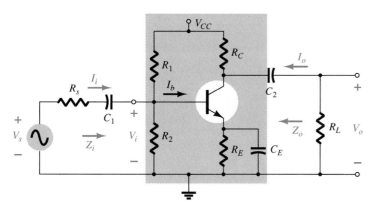

Figure 10.18 Voltage-divider bias configuration with R_s and R_L.

In the small-signal ac model R_C and R_L will again be in parallel

and

$$A_v = -\frac{R_C\|R_L}{r_e} \qquad (10.23)$$

with

$$A_{v_s} = \frac{Z_i}{Z_i + R_s} A_v \qquad (10.24)$$

CE Unbypassed Emitter Bias

For the common-emitter unbypassed emitter-bias configuration of Fig. 10.19, Z_i remains independent of the applied load and

$$Z_i \cong R_B\|\beta R_E \qquad (10.25)$$

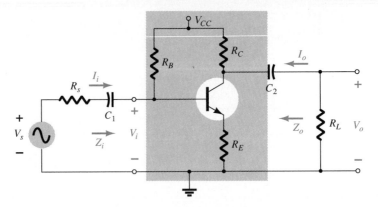

Figure 10.19 CE unbypassed emitter-bias configuration with R_s and R_L.

For the output impedance,

$$Z_o = R_C \tag{10.26}$$

For the voltage gain, the resistance R_C will again drop down in parallel with R_L and

$$A_v = \frac{V_o}{V_i} = -\frac{R_C \| R_L}{R_E} \tag{10.27}$$

with

$$A_{v_s} = \frac{V_o}{V_s} = \frac{Z_i}{Z_i + R_s} A_v \tag{10.28}$$

and

$$A_i = \frac{I_o}{I_i} = -A_v \frac{Z_i}{R_L} \tag{10.29}$$

but keep in mind that $I_i = I_s = V_s/(R_s + Z_i) = V_i/Z_i$.

Collector Feedback

To keep with our connection of the load to the collector terminal the next configuration to be examined is the collector feedback configuration of Fig. 10.20. In the small-signal ac model of the system R_C and R_L will again drop down in parallel and

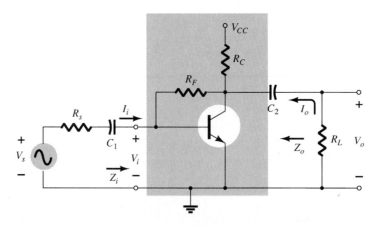

Figure 10.20 Collector feedback configuration with R_s and R_L.

Chapter 10 Systems Approach—Effects of R_S and R_L

$$A_v = -\frac{R_C \| R_L}{r_e} \qquad (10.30)$$

with

$$A_{v_s} = \frac{Z_i}{Z_i + R_s} A_v \qquad (10.31)$$

The output impedance

$$Z_o \cong R_C \| R_F \qquad (10.32)$$

and

$$Z_i = \beta r_e \| \frac{R_F}{|A_v|} \qquad (10.33)$$

The fact that A_v [Eq. (10.30)] is a function of R_L will alter the level of Z_i from the no-load value. Therefore, if the no-load model is available, the level of Z_i must be modified as demonstrated in the next example.

The collector feedback amplifier of Fig. 10.21 has the following no-load system parameters: $A_{v_{NL}} = -238.94$, $Z_o = R_C \| R_F = 2.66$ kΩ, and $Z_i = 0.553$ kΩ, with $r_e = 11.3$ Ω, and $\beta = 200$. Using the systems approach, determine:
(a) A_v.
(b) A_{v_s}.
(c) A_i.

EXAMPLE 10.5

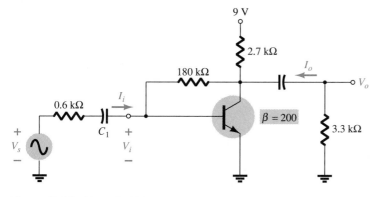

Figure 10.21 Example 10.5

Solution

(a) For the two-port system:

$$A_v = -\frac{R_C \| R_L}{r_e} = -\frac{2.7 \text{ k}\Omega \| 3.3 \text{ k}\Omega}{11.3 \text{ }\Omega}$$

$$= -\frac{1.485 \text{ k}\Omega}{11.3 \text{ }\Omega} = -\mathbf{131.42}$$

and $Z_i = \beta r_e \| \dfrac{R_F}{|A_v|} = (200)(11.3 \text{ }\Omega) \| \dfrac{180 \text{ k}\Omega}{131.42}$

$$= 2.26 \text{ k}\Omega \| 1.37 \text{ k}\Omega$$

$$= \mathbf{0.853 \text{ k}\Omega}$$

The system approach will result in the configuration of Fig. 10.22 with the value of Z_i as controlled by R_L and the voltage gain. Now the two-port gain equation can be applied (slight difference in A_v due to approximation $\beta I_b \gg I_{R_F}$ in section 8.7):

$$A_v = \frac{R_L A_{v_{NL}}}{R_L + R_o} = \frac{(3.3 \text{ k}\Omega)(-238.94)}{3.3 \text{ k}\Omega + 2.66 \text{ k}\Omega} = -132.3$$

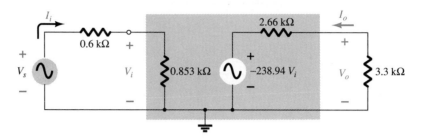

Figure 10.22 The ac equivalent circuit for the network of Fig. 10.21.

(b) $A_{v_s} = \dfrac{Z_i}{Z_i + R_s} A_v = \dfrac{0.853 \text{ k}\Omega}{0.853 \text{ k}\Omega + 0.6 \text{ k}\Omega}(-132.3)$

$= -77.67$

(c) $A_i = -A_v \dfrac{Z_i}{R_L} = -(-132.3)\left(\dfrac{0.853 \text{ k}\Omega}{3.3 \text{ k}\Omega}\right) = \dfrac{(132.3)(0.853 \text{ k}\Omega)}{3.3 \text{ k}\Omega}$

$= 34.2$

or $A_i = -A_{v_s} \dfrac{Z_i + R_s}{R_L} = -(-77.67)\left(\dfrac{0.853 \text{ k}\Omega + 0.6 \text{ k}\Omega}{3.3 \text{ k}\Omega}\right)$

$= 34.2$

10.7 BJT EMITTER-FOLLOWER NETWORKS

The input and output impedance parameters of the two-port model for the emitter-follower network are sensitive to the applied load and source resistance. For the emitter-follower configuration of Fig. 10.23 the small-signal ac model would appear as shown in Fig. 10.24. For the input section of Fig. 10.24 the resistance R_B is neglected because it is usually so much larger than the source resistance that a Théve-

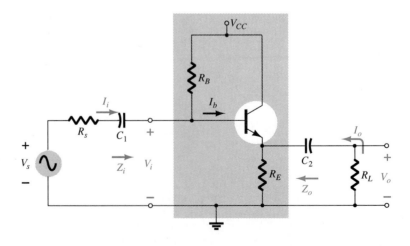

Figure 10.23 Emitter-follower configuraton with R_s and R_L.

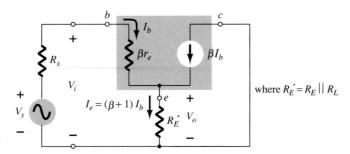

Figure 10.24 Emitter-follower configuration of Fig. 10.23 following the substitution of the r_e equivalent circuit.

where $R_E' = R_E \,\|\, R_L$

nin equivalent circuit for the configuration of Fig. 10.25 would result in simply R_s and V_s as shown in Fig. 10.24. Of course, if current levels are to be determined such as I_i in the original diagram, the effect of R_B must be included.

Applying Kirchhoff's voltage law to the input circuit of Fig. 10.24 will result in

$$V_s - I_b R_s - I_b \beta r_e - (\beta + 1)I_b R_E' = 0$$

and

$$V_s - I_b(R_s + \beta r_e + (\beta + 1)R_E') = 0$$

so that

$$I_b = \frac{V_s}{R_s + \beta r_e + (\beta + 1)R_E'}$$

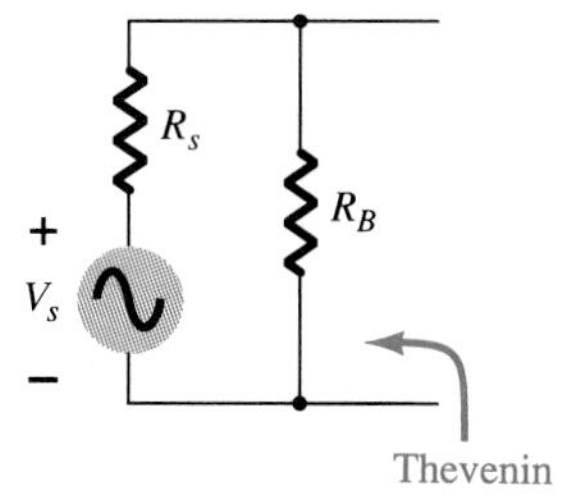

Establishing I_e, we have

$$I_e = (\beta + 1)I_b = \frac{(\beta + 1)V_s}{R_s + \beta r_e + (\beta + 1)R_E'}$$

and

$$I_e = \frac{V_s}{[(R_s + \beta r_e)/(\beta + 1)] + R_E'}$$

Using $\beta + 1 \cong \beta$ yields

$$\boxed{I_e = \frac{V_s}{(R_s/\beta + r_e) + R_E'}} \qquad (10.34)$$

Drawing the network to "fit" Eq. (10.34) will result in the configuration of Fig. 10.26a. In Fig. 10.26b, R_E and the load resistance R_L have been separated to permit a definition of Z_o and I_o.

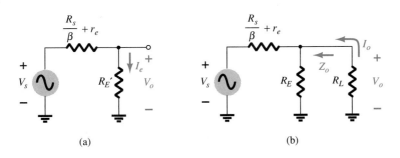

(a)

(b)

Figure 10.26 Networks resulting from the application of Kirchhoff's voltage law to the input circuit of Fig. 10.24.

The voltage gain can then be obtained directly from Fig. 10.26a using the voltage-divider rule.

$$V_o = \frac{R_E' V_s}{R_E' + (R_s/\beta + r_e)}$$

or

$$A_{v_s} = \frac{V_o}{V_s} = \frac{R_E'}{R_E' + (R_s/\beta + r_e)}$$

Figure 10.25 Determining the Thévenin equivalent circuit for the input circuit of Fig. 10.23.

Thevenin

10.7 **BJT Emitter-Follower Networks**

451

and

$$A_{v_s} = \frac{V_o}{V_s} = \frac{R_E\|R_L}{R_E\|R_L + R_s/\beta + r_e}$$

(10.35)

Setting $V_s = 0$ and solving for Z_o will result in

$$Z_o = R_E\left\|\left(\frac{R_s}{\beta} + r_e\right)\right.$$

(10.36)

For the input impedance,

$$Z_b = \beta(r_e + R_E')$$

and

$$Z_i = R_B\|Z_b$$

or

$$Z_i = R_B\|\beta(r_e + R_E\|R_L)$$

(10.37)

For no-load conditions the gain equation is

$$A_{v_{NL}} \cong \frac{R_E}{R_E + r_e}$$

while for loaded conditions,

$$A_v \cong \frac{V_o}{V_i} = \frac{R_E\|R_L}{R_E\|R_L + r_e}$$

(10.38)

EXAMPLE 10.6

For the loaded emitter-follower configuration of Fig. 10.27 with a source resistance and the following no-load two-port parameters: $Z_i = 157.54$ kΩ, $Z_o = 21.6$ Ω, and $A_{v_{NL}} = 0.993$ with $r_e = 21.74$ Ω and $\beta = 65$, determine:
(a) The new value of Z_i and Z_o as determined by the load and R_s, respectively.
(b) A_v using the systems approach.
(c) A_{v_s} using the systems approach.
(d) $A_i = I_o/I_i$.

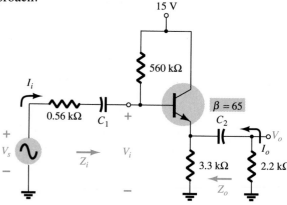

Figure 10.27 Example 10.6

Solution

Eq. (10.37): $Z_i = R_B\|\beta(r_e + R_E\|R_L)$

$= 560$ k$\Omega\|65(21.74$ $\Omega + \underbrace{3.3\ \text{k}\Omega\|2.2\ \text{k}\Omega}_{1.32\ \text{k}\Omega})$

$= 560$ k$\Omega\|87.21$ kΩ

$= \mathbf{75.46}$ **kΩ**

versus 157.54 kΩ (no-load).

$$Z_o = R_E \| \left(\frac{R_s}{\beta} + r_e \right)$$

$$= 3.3 \text{ k}\Omega \| \left(\frac{0.56 \text{ k}\Omega}{65} + 21.74 \text{ }\Omega \right)$$

$$= 3.3 \text{ k}\Omega \| 30.36 \text{ }\Omega$$

$$= \mathbf{30.08 \text{ }\Omega}$$

versus 21.6 Ω (no R_s).

(b) Substituting the two-port equivalent network will result in the small-signal ac equivalent network of Fig. 10.28.

$$V_o = \frac{R_L A_{v_{NL}} V_i}{R_L + R_o} = \frac{(2.2 \text{ k}\Omega)(0.993)V_i}{2.2 \text{ k}\Omega + 30.08 \text{ }\Omega}$$

$$\cong 0.98 \text{ } V_i$$

with $A_v = \dfrac{V_o}{V_i} \cong \mathbf{0.98}$

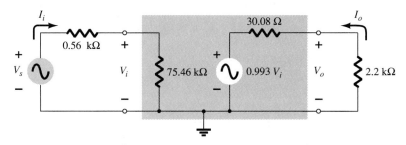

Figure 10.28 Small-signal ac equivalent circuit for the network of Fig. 10.27.

(c) $V_i = \dfrac{Z_i V_s}{Z_i + R_s} = \dfrac{(75.46 \text{ k}\Omega)V_s}{75.46 \text{ k}\Omega + 0.56 \text{ k}\Omega} = 0.993 \text{ } V_s$

so that $\quad A_{v_s} = \dfrac{V_o}{V_s} = \dfrac{V_o}{V_i} \dfrac{V_i}{V_s} = (0.98)(0.993) = \mathbf{0.973}$

(d) $A_i = \dfrac{I_o}{I_i} = -A_v \dfrac{Z_i}{R_L}$

$$= -(0.98)\left(\frac{75.46 \text{ k}\Omega}{2.2 \text{ k}\Omega} \right)$$

$$= \mathbf{-33.61}$$

10.8 BJT CB NETWORKS

A common-base amplifier with an applied load and source resistance appear in Fig. 10.29. The fact that the load is connected between the collector and base terminals isolates it from the input circuit and Z_i remains essentially the same for no-load or loaded conditions. The isolation that exists between input and output circuits also maintains Z_o at a fixed level even though the level of R_s may change. The voltage gain is now determined by

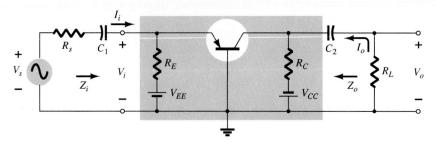

Figure 10.29 Common-base configuration with R_s and R_L.

$$A_v \cong \frac{R_C \| R_L}{r_e} \qquad (10.39)$$

and the current gain:

$$A_i \cong -1 \qquad (10.40)$$

EXAMPLE 10.7

For the common-base amplifier of Fig. 10.30, the no-load two-port parameters are (using $\alpha \cong 1$) $Z_i \cong r_e = 20\ \Omega$, $A_{v_{NL}} = 250$, and $Z_o = 5\ k\Omega$. Using the two-port equivalent model, determine:
(a) A_v.
(b) A_{v_s}.
(c) A_i.

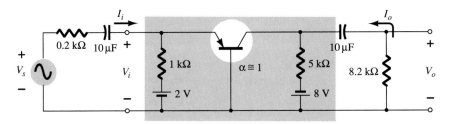

Figure 10.30 Example 10.7

Solution

(a) The small-signal ac equivalent network appears in Fig. 10.31.

$$V_o = \frac{R_L A_{v_{NL}} V_i}{R_L + R_o} = \frac{(8.2\ k\Omega)(250)V_i}{8.2\ k\Omega + 5\ k\Omega} = 155.3 V_i$$

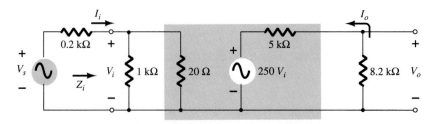

Figure 10.31 Small-signal ac equivalent circuit for the network of Fig. 10.30.

Chapter 10   Systems Approach—Effects of R_S and R_L

and $\qquad A_v = \dfrac{V_o}{V_i} = \mathbf{155.3}$

or $\qquad A_v \cong \dfrac{R_C\|R_L}{r_e} = \dfrac{5\ \text{k}\Omega\|8.2\ \text{k}\Omega}{20\ \Omega} = \dfrac{3.106\ \text{k}\Omega}{20\ \Omega}$

$\qquad\qquad = \mathbf{155.3}$

(b) $A_{v_s} = \dfrac{V_o}{V_s} = \dfrac{V_i}{V_s}\dfrac{V_o}{V_i}$

$\qquad\quad = \dfrac{R_i}{R_i + R_s}A_v = \left(\dfrac{20\ \Omega}{20\ \Omega + 200\ \Omega}\right)(155.3)$

$\qquad\quad = \mathbf{14.12}$

Note the relatively low gain due to a source impedance much larger than the input impedance of the amplifier.

(c) $A_i = -A_v\dfrac{Z_i}{R_L} = -(155.3)\left(\dfrac{20\ \Omega}{8.2\ \text{k}\Omega}\right)$

$\qquad\quad = \mathbf{-0.379}$

which is significantly less than 1 due to the division of output current between R_C and R_L.

10.9 FET NETWORKS

As noted in Chapter 9, the isolation that exists between gate and drain or source of a FET amplifier ensures that changes in R_L do not affect the level of Z_i and changes in R_{sig} do not affect R_o. In essence, therefore:

The no-load two-port model of Fig. 10.2 for a FET amplifier is unaffected by an applied load or source resistance.

Bypassed Source Resistance

For the FET amplifier of Fig. 10.32 the applied load will appear in parallel with R_D in the small-signal model, resulting in the following equation for the loaded gain:

$$A_v = -g_m(R_D\|R_L) \qquad\qquad (10.41)$$

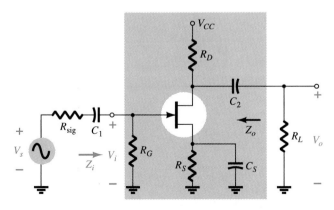

Figure 10.32 JFET amplifier with R_{sig} and R_L.

The impedance levels remain at

$$Z_i = R_G \qquad (10.42)$$

$$Z_o = R_D \qquad (10.43)$$

Unbypassed Source Resistance

For the FET amplifier of Fig. 10.33 the load will again appear in parallel with R_D and the loaded gain becomes

$$A_v = \frac{V_o}{V_i} = -\frac{g_m(R_D\|R_L)}{1 + g_mR_S} \qquad (10.44)$$

with

$$Z_i = R_G \qquad (10.45)$$

and

$$Z_o = R_D \qquad (10.46)$$

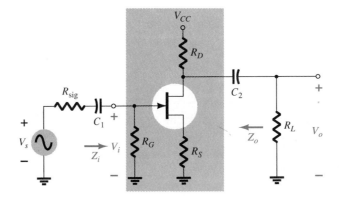

Figure 10.33 JFET amplifier with unbypassed R_s.

EXAMPLE 10.8

For the FET amplifier of Fig. 10.34, the no-load two-port parameters are $A_{v_{NL}} = -3.18$, $Z_i = R_1\|R_2 = 239$ kΩ, and $Z_o = 2.4$ kΩ, with $g_m = 2.2$ mS.
(a) Using the two-port parameters above, determine A_v and A_{v_s}.
(b) Using Eq. (10.44), calculate the loaded gain and compare to the result of part (a).

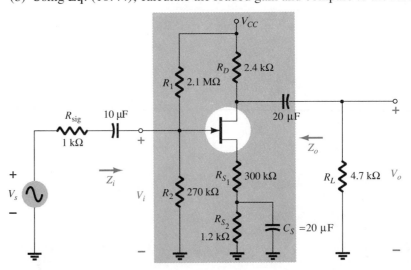

Figure 10.34 Example 10.8

Chapter 10 Systems Approach—Effects of R_S and R_L

Solution

(a) The small-signal ac equivalent network appears in Fig. 10.35 and

$$A_v = \frac{V_o}{V_i} = \frac{R_L A_{v_{NL}}}{R_L + R_o} = \frac{(4.7\ k\Omega)(-3.18)}{4.7\ k\Omega + 2.4\ k\Omega}$$

$$= -2.105$$

$$A_{v_s} = \frac{V_o}{V_s} = \frac{V_i}{V_s}\frac{V_o}{V_i} = \frac{R_i}{R_i + R_{sig}} A_v$$

$$= \frac{(239\ k\Omega)(-2.105)}{239\ k\Omega + 1\ k\Omega}$$

$$= -2.096 \cong A_v$$

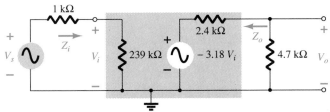

Figure 10.35 Small-signal ac equivalent circuit for the network of Fig. 10.34.

(b) Eq. (10.44): $A_v = \dfrac{-g_m(R_D\|R_L)}{1 + g_m R_{S_1}}$

$$= \frac{-(2.2\ mS)(4.7\ k\Omega\|2.4\ k\Omega)}{1 + (2.2\ mS)(0.3\ k\Omega)} = \frac{-3.498}{1.66}$$

$$= -2.105 \quad \text{as above}$$

Source Follower

For the source-follower configuration of Fig. 10.36 the level of Z_i is independent of the magnitude of R_L and determined by

$$\boxed{Z_i = R_G} \tag{10.47}$$

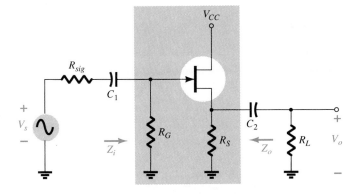

Figure 10.36 Source-follower configuration with R_s and R_L.

The loaded voltage gain has the same format as the unloaded gain with R_S replaced by the parallel combination of R_s and R_L.

$$A_v = \frac{V_o}{V_i} = \frac{g_m(R_S\|R_L)}{1 + g_m(R_S\|R_L)} \qquad (10.48)$$

The level of output impedance is as determined in Chapter 9:

$$Z_o = R_S\|\frac{1}{g_m} \qquad (10.49)$$

revealing an insensitivity to the magnitude of the source resistance R_s.

Common Gate

Even though the common-gate configuration of Fig. 10.37 is somewhat different from those described above with regard to the placement of R_L and R_s, the input and output circuits remain isolated and

$$Z_i = \frac{R_S}{1 + g_mR_S} \qquad (10.50)$$

$$Z_o = R_D \qquad (10.51)$$

The loaded voltage gain is given by

$$A_v = g_m(R_D\|R_L) \qquad (10.52)$$

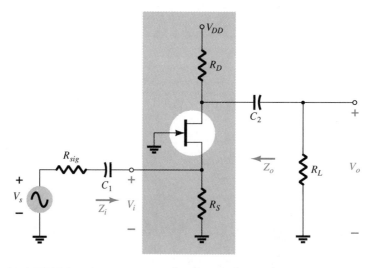

Figure 10.37 Common-gate configuration with R_s and R_L.

10.10 SUMMARY TABLE

Now that the loaded and unloaded (Chapters 8 and 9) BJT and JFET amplifiers have been examined in some detail, a review of the equations developed is provided by Table 10.1. Although all the equations are for the loaded situation, the removal of R_L will result in the equations for the unloaded amplifier. The same is true for the effect of R_s (for BJTs) and R_{sig} (for JFETs) on Z_o. In each case the phase relationship between the input and output voltages is also provided for quick reference. A review

Chapter 10 Systems Approach—Effects of R_S and R_L

of the equations will reveal that the isolation provided by the JFET between the gate and channel by the SiO_2 layer results in a series of less complex equations than those encountered for the BJT configurations. The linkage provided by I_b between input and output circuits of the BJT transistor amplifier adds a touch of complexity to some of the equations.

TABLE 10.1 Summary of Transistor Configurations (A_v, Z_i, Z_o)

Configuration	$A_v = V_o/V_i$	Z_i	Z_o
(circuit diagram: fixed-bias configuration with V_{CC}, R_C, R_B, R_s, V_i, V_s, R_L, Z_i, Z_o, V_o)	$\dfrac{-(R_L\|R_C)}{r_e}$	$R_B\|\beta r_e$	R_C
	$\dfrac{-h_{fe}}{h_{ie}}(R_L\|R_C)$	$R_B\|h_{ie}$	R_C
	Including r_o: $-\dfrac{(R_L\|R_C\|r_o)}{r_e}$	$R_B\|\beta r_e$	$R_C\|r_o$
(circuit diagram: voltage-divider bias with V_{CC}, R_C, R_1, R_2, R_E, C_E, R_s, V_i, V_s, R_L, Z_i, Z_o, V_o)	$\dfrac{-(R_L\|R_C)}{r_e}$	$R_1\|R_2\|\beta r_e$	R_C
	$\dfrac{-h_{fe}}{h_{ie}}(R_L\|R_C)$	$R_1\|R_2\|h_{ie}$	R_C
	Including r_o: $\dfrac{-(R_L\|R_C\|r_o)}{r_e}$	$R_1\|R_2\|\beta r_e$	$R_C\|r_o$
(circuit diagram: emitter-follower with V_{CC}, R_C, R_1, R_2, R_E, R_s, V_i, V_s, R_L, Z_i, Z_o, V_o)	$\cong 1$	$R_E{}' = R_L\|R_E$ $R_1\|R_2\|\beta(r_e + R_E')$	$R_s{}' = R_s\|R_1\|R_2$ $R_E\|\left(\dfrac{R_s'}{\beta} + r_e\right)$
	$\cong 1$	$R_1\|R_2\|(h_{ie} + h_{fe}R_E')$	$R_E\|\left(\dfrac{R_s' + h_{ie}}{h_{fe}}\right)$
	Including r_o: $\cong 1$	$R_1\|R_2\|\beta(r_e + R_E')$	$R_E\|\left(\dfrac{R_s'}{\beta} + r_e\right)$
(circuit diagram: common-base with R_s, V_i, R_E, R_C, R_L, V_{EE}, V_{CC}, V_s, Z_i, Z_o, V_o)	$\cong \dfrac{-(R_L\|R_C)}{r_e}$	$R_E\|r_e$	R_C
	$\cong \dfrac{-h_{fb}}{h_{ib}}(R_L\|R_C)$	$R_E\|h_{ib}$	R_C
	Including r_o: $\cong \dfrac{-(R_L\|R_C\|r_o)}{r_e}$	$R_E\|r_e$	$R_C\|r_o$

TABLE 10.1 Summary of Transistor Configurations (A_v, Z_i, Z_o) (Continued)

Configuration	$A_v = V_o/V_i$	Z_i	Z_o
	$\dfrac{-(R_L\|R_C)}{R_E}$	$R_1\|R_2\|\beta(r_e + R_E)$	R_C
	$\dfrac{-(R_L\|R_C)}{R_E}$	$R_1\|R_2\|(h_{ie} + h_{fe}R_E)$	R_C
	Including r_o: $\dfrac{-(R_L\|R_C)}{R_E}$	$R_1\|R_2\|\beta(r_e + R_E)$	$\cong R_C$
	$\dfrac{-(R_L\|R_C)}{R_{E_1}}$	$R_B\|\beta(r_e + R_{E_1})$	R_C
	$\dfrac{-(R_L\|R_C)}{R_{E_1}}$	$R_B\|(h_{ie} + h_{fe}R_{E_1})$	R_C
	Including r_o: $\dfrac{-R_L\|R_C}{R_{E_1}}$	$R_B\|\beta(r_e + R_{E_1})$	$\cong R_C$
	$\dfrac{-(R_L\|R_C)}{r_e}$	$\beta r_e\|\dfrac{R_F}{\|A_v\|}$	R_C
	$\dfrac{-h_{fe}}{h_{ie}}(R_L\|R_C)$	$h_{ie}\|\dfrac{R_F}{\|A_v\|}$	R_C
	Including r_o: $\dfrac{-(R_L\|R_C\|r_o)}{r_e}$	$\beta r_e\|\dfrac{R_F}{\|A_v\|}$	$R_C\|R_F\|r_o$
	$\dfrac{-(R_L\|R_C)}{R_E}$	$\beta R_E\|\dfrac{R_F}{\|A_v\|}$	$\cong R_C\|R_F$
	$\dfrac{-(R_L\|R_C)}{R_E}$	$h_{fe}R_E\|\dfrac{R_F}{\|A_v\|}$	$\cong R_C\|R_F$
	Including r_o: $\cong \dfrac{-(R_L\|R_C)}{R_E}$	$\cong \beta R_E\|\dfrac{R_F}{\|A_v\|}$	$\cong R_C\|R_F$

TABLE 10.1 (Continued)

Configuration	$A_v = V_o/V_i$	Z_i	Z_o
	$-g_m(R_D\|R_L)$ Including r_d: $-g_m(R_D\|R_L\|r_d)$	R_G R_G	R_D $R_D\|r_d$
	$\dfrac{-g_m(R_D\|R_L)}{1 + g_mR_S}$ Including r_d: $\dfrac{-g_m(R_D\|R_L)}{1 + g_mR_S + \dfrac{R_D + R_S}{r_d}}$	R_G R_G	$\dfrac{R_D}{1 + g_mR_S}$ $\cong \dfrac{R_D}{1 + g_mR_S}$
	$-g_m(R_D\|R_L)$ Including r_d: $-g_m(R_D\|R_L\|r_d)$	$R_1\|R_2$ $R_1\|R_2$	R_D $R_D\|r_d$
	$\dfrac{g_m(R_S\|R_L)}{1 + g_m(R_S\|R_L)}$ Including r_d: $= \dfrac{g_m r_d(R_S\|R_L)}{r_d + R_D + g_m r_d(R_S\|R_L)}$	R_G R_G	$R_S\|1/g_m$ $\dfrac{R_S}{1 + \dfrac{g_m r_d R_S}{r_d + R_D}}$
	$g_m(R_D\|R_L)$ Including r_d: $\cong g_m(R_D\|R_L)$	$\dfrac{R_S}{1 + g_mR_S}$ $Z_i = \dfrac{R_S}{1 + \dfrac{g_m r_d R_S}{r_d + R_D\|R_L}}$	R_D $R_D\|r_d$

10.11 CASCADED SYSTEMS

The two-port systems approach is particularly useful for cascaded systems such as that appearing in Fig. 10.38, where A_{v_1}, A_{v_2}, A_{v_3}, and so on, are the voltage gains of each stage *under loaded conditions*. That is, A_{v_1} is determined with the input impedance to A_{v_2} acting as the load on A_{v_1}. For A_{v_2}, A_{v_1} will determine the signal strength and source impedance at the input to A_{v_2}. The total gain of the system is then determined by the product of the individual gains as follows:

$$A_{v_T} = A_{v_1} \cdot A_{v_2} \cdot A_{v_3} \cdots$$

(10.53)

and the total current gain by

$$A_{i_T} = -A_{v_T}\frac{Z_{i_1}}{R_L}$$

(10.54)

No matter how perfect the system design the application of a load to a two-port system will affect the voltage gain. Therefore, there is no possibility of a situation where A_{v_1}, A_{v_2}, and so on, of Fig. 10.38 are simply the no-load values. The loading of each succeeding stage must be considered. The no-load parameters can be used to determine the loaded gains of Fig. 10.38, but Eq. (10.53) requires the loaded values.

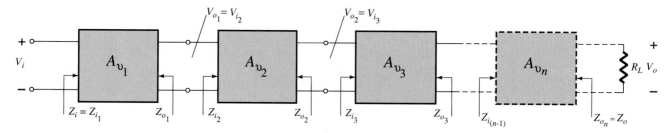

Figure 10.38 Cascaded system.

EXAMPLE 10.9

The two-stage system of Fig. 10.39 employed a transistor emitter-follower configuration prior to a common-base configuration to ensure that the maximum percent of the applied signal appears at the input terminals of the common-base amplifier. In Fig. 10.39 the no-load values are provided for each system, with the exception of Z_i and Z_o for the emitter-follower, which are the loaded values. For the configuration of Fig. 10.39, determine:
(a) The loaded gain for each stage.
(b) The total gain for the system, A_v and A_{v_s}.
(c) The total current gain for the system.
(d) The total gain for the system if the emitter-follower configuration were removed.

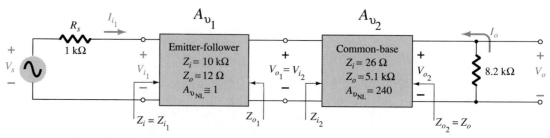

Figure 10.39 Example 10.9

Solution

(a) For the emitter-follower configuration the loaded gain is

$$V_{o_1} = \frac{Z_{i_2} A_{v_{NL}} V_{i_1}}{Z_{i_2} + Z_{o_1}} = \frac{(26\ \Omega)(1) V_{i_1}}{26\ \Omega + 12\ \Omega} = 0.684\ V_{i_1}$$

and $\qquad A_{v_1} = \dfrac{V_{o_1}}{V_{i_1}} = \mathbf{0.684}$

For the common-base configuration,

$$V_{o_2} = \frac{R_L A_{v_{NL}} V_{i_2}}{R_L + R_{o_2}} = \frac{(8.2\ k\Omega)(240) V_{i_2}}{8.2\ k\Omega + 5.1\ k\Omega} = 147.97\ V_{i_2}$$

and $\qquad A_{v_L} = \dfrac{V_{o_2}}{V_{i_2}} = \mathbf{147.97}$

(b) $A_{v_T} = A_{v_1} A_{v_2}$

$\qquad = (0.684)(147.97)$

$\qquad = \mathbf{101.20}$

$$A_{v_s} = \frac{Z_{i_1}}{Z_{i_1} + R_s} A_{v_T} = \frac{(10\ k\Omega)(101.20)}{10\ k\Omega + 1\ k\Omega}$$

$\qquad = \mathbf{92}$

(c) $A_{i_T} = -A_{v_T} \dfrac{Z_{i_1}}{R_L}$

$\qquad = -(101.20)\left(\dfrac{10\ k\Omega}{8.2\ k\Omega}\right)$

$\qquad = \mathbf{-123.41}$

(d) $V_{i_{CB}} = \dfrac{Z_{i_{CB}} V_s}{Z_{i_{CB}} + R_s} = \dfrac{(26\ \Omega) V_s}{26\ \Omega + 1\ k\Omega} = 0.025\ V_s$

and $\dfrac{V_i}{V_s} = 0.025 \qquad$ with $\qquad \dfrac{V_o}{V_i} = 147.97 \qquad$ from above

and $A_{v_s} = \dfrac{V_i}{V_s} \dfrac{V_o}{V_i} = (0.025)(147.97) = \mathbf{3.7}$

In total, therefore, the gain is about 25 times greater with the emitter-follower configuration to draw the signal to the amplifier stages. Take note, however, that it was also important that the output impedance of the first stage was relatively close to the input impedance of the second stage or the signal would have been "lost" again by the voltage-divider action.

10.12 COMPUTER ANALYSIS

The computer analysis of this section includes a PSpice evaluation of the response of a loaded BJT and FET amplifier with a source resistance. The BJT network of Fig. 10.40 employs the same unloaded configuration examined in the PSpice analysis of Chapter 8, where the unloaded gain was 350.4. The nodes are identified in Fig. 10.40 and appear in the description of the network in the input file of Fig. 10.41. Note in the transistor description that IS is our chosen value of 5×10^{-15} A, as discussed in Chapter 8. In addition, note the use of a very large resistance (essentially an open circuit) from node 4 to ground to ensure a dc path to ground for the capacitor (a

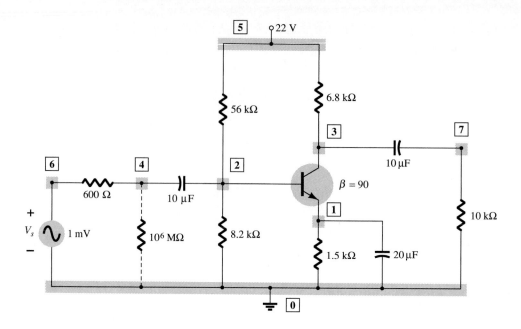

Figure 10.40 Defining the nodes of a voltage-divider configuration with R_s and R_L.

```
**** 06/12/89 ******* Evaluation PSpice (January 1989) ******* 13:22:58 ****

BJT Voltage-Divider Bias Configuration with Rs and RL - Fig. 10.40

****        CIRCUIT DESCRIPTION

*****************************************************************************

VCC 5 0 DC 22V
RB1 5 2 56K
RB2 2 0 8.2K
RE 1 0 1.5K
RC 5 3 6.8K
C1 4 2 10UF
CE 1 0 20UF
VS 6 0 AC 1MV 0
RS 6 4 600
RR 4 0 1E12
C2 3 7 10UF
RL 7 0 10K
Q1 3 2 1 QMODEL
.MODEL QMODEL NPN(BF=90 IS=5E-15)
.OP
.AC LIN 1 10KH 10KH
.PRINT AC VM(3) VM(7) VM(4)
.OPTIONS NOPAGE
.END


****        BJT MODEL PARAMETERS
            QMODEL
            NPN
       IS     5.000000E-15
       BF     90
       NF     1
       BR     1
       NR     1

****     SMALL SIGNAL BIAS SOLUTION       TEMPERATURE =   27.000 DEG C
   NODE    VOLTAGE       NODE    VOLTAGE       NODE    VOLTAGE       NODE    VOLTAGE
 (    1)     2.0235  (    2)     2.7039  (    3)    12.9280  (    4)     0.0000
 (    5)    22.0000  (    6)     0.0000  (    7)     0.0000

        VOLTAGE SOURCE CURRENTS
        NAME        CURRENT
        VCC        -1.679E-03
        VS          0.000E+00
        TOTAL POWER DISSIPATION    3.69E-02  WATTS
```

Figure 10.41 PSpice analysis of the BJT amplifier of Fig. 10.40.

```
****        OPERATING POINT INFORMATION           TEMPERATURE =     27.000 DEG C

**** BIPOLAR JUNCTION TRANSISTORS
NAME            Q1
MODEL           QMODEL
IB              1.48E-05
IC              1.33E-03
VBE             6.80E-01
VBC            -1.02E+01
VCE             1.09E+01
BETADC          9.00E+01
GM              5.16E-02
RPI             1.74E+03
RX              0.00E+00
RO              1.00E+12
CBE             0.00E+00
CBC             0.00E+00
CBX             0.00E+00
CJS             0.00E+00
BETAAC          9.00E+01
FT              8.21E+17


****      AC ANALYSIS                          TEMPERATURE =     27.000 DEG C
    FREQ          VM(3)         VM(7)         VM(4)
    1.000E+04    1.462E-01    1.462E-01    7.007E-04
```

Figure 10.41 Continued.

PSpice requirement). The PRINT statement includes a request for the magnitude of the voltage at nodes 3, 7, and 4 for an input signal of 1 mV.

Note in the bias solution that nodes 4, 6, and 7 have a response of 0 V due to the isolation offered by the capacitors. Node 5 is 22 V as it should be and $V_E = 2.0235$ V, $V_B = 2.7039$ V, and $V_C = 12.9280$ V compare well with the dc solution of Chapter 8.

The ac analysis reveals that V_3 and V_7 are at essentially the same level since the capacitors provide a direct link of minimal impedance from one node to the other at the applied frequency. Their magnitude reveals a gain of 146.2 compared with a no-load gain of 350.4. The magnitude of V_4 reveals that 30% (0.3 mV) of the applied signal is lost across the source resistance of 0.6 kΩ.

For interest's sake, let us now calculate the loaded voltage gain and compare to the PSpice solution of 146.2.

$$r_e = 18.44 \ \Omega$$

and

$$Z_i \cong R_1 \| R_2 \| \beta r_e$$

$$= 56 \ \text{k}\Omega \| 8.2 \ \text{k}\Omega \| (90)(18.44 \ \Omega)$$

$$\cong 1.35 \ \text{k}\Omega$$

$$V_i = \frac{Z_i V_s}{Z_i + R_s} = \frac{(1.35 \ \text{k}\Omega)V_s}{1.35 \ \text{k}\Omega + 0.6 \ \text{k}\Omega} = 0.69 V_s$$

and

$$\frac{V_i}{V_s} = 0.69$$

$$A_v = \frac{V_o}{V_i} = \frac{R_L A_{v_{\text{NL}}}}{R_L + R_o} = \frac{(10 \ \text{k}\Omega)(-350.4)}{10 \ \text{k}\Omega + 6.8 \ \text{k}\Omega}$$

$$= -208.57$$

with

$$A_{v_s} = \frac{V_i}{V_s} \frac{V_o}{V_i} = (0.69)(-208.57)$$

$$\cong \mathbf{-144}$$

which compares very favorably with the -146.2 obtained above using PSpice.

The loaded FET amplifier to be analyzed appears in Fig. 10.42. It is a system that appeared in Chapter 9 modified to show the effects of R_{sig} and R_L. The JFET description of Fig. 10.43 reveals that VTO = V_{gs}(off) = V_P = −4 V and beta defined by $I_{DSS}/|V_P|^2 = 6.25 \times 10^{-4}$ A/V². The isolation provided by the capacitors is again obvious from the bias solutions for V_1, V_2, and V_7. Certainly, $V_3 = 67.14 \times 10^{-6}$ V is approximately 0 V for all practical purposes. Node 6 is at 18 V as defined by the dc source and $V_D = 5.6862$ V and $V_E = 1.0075$ V as supported by the dc analysis.

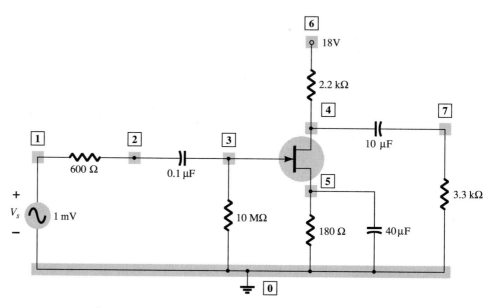

Figure 10.42 Defining the nodes of a JFET amplifier having a source resistance R_s and load resistance R_L.

```
****  05/07/90  *******  Evaluation PSpice (January 1989)  *******  18:00:01  ****

JFET ac Amplifier of Fig. 10.42

****        CIRCUIT DESCRIPTION

*********************************************************************************

VDD 6 0 DC 18V
J1 4 3 5 JFET
RG 3 0 10MEG
RD 6 4 2.2K
RS 5 0 180
CI 2 3 0.1UF
CS 5 0 40UF
CO 4 7 10UF
.MODEL JFET NJF VTO=-4V BETA=6.25E-4
VSIG 1 0 AC 1MV
RSIG 1 2 600
RL 7 0 3.3K
.OP
.AC LIN 1 10KH 10KH
.PRINT AC V(1) V(3) V(4) V(5) V(7)
.OPTIONS NOPAGE
.END
```

Figure 10.43 PSpice analysis of the JFET amplifier of Fig. 10.42.

```
****      Junction FET MODEL PARAMETERS


                JFET
                NJF
      VTO       -4
     BETA       625.000000E-06


  ***      SMALL SIGNAL BIAS SOLUTION      TEMPERATURE =    27.000 DEG C
  NODE    VOLTAGE      NODE    VOLTAGE      NODE    VOLTAGE      NODE    VOLTAGE
 (    1)   0.0000   (     2)    0.0000   (    3)  67.14E-06   (    4)     5.6862
 (    5)   1.0075   (     6)   18.0000   (    7)    0.0000


          VOLTAGE SOURCE CURRENTS
          NAME            CURRENT

          VDD             -5.597E-03
          VSIG             0.000E+00

          TOTAL POWER DISSIPATION   1.01E-01   WATTS

  ****      OPERATING POINT INFORMATION      TEMPERATURE =    27.000 DEG C

  **** JFETS

  NAME            J1
  MODEL           JFET
  ID              5.60E-03
  VGS             -1.01E+00
  VDS             4.68E+00
  GM              3.74E-03
  GDS             0.00E+00
  CGS             0.00E+00
  CGD             0.00E+00


  ****      AC ANALYSIS                    TEMPERATURE =    27.000 DEG C
    FREQ        V(1)         V(3)        V(4)         V(5)         V(7)
   1.000E+04   1.000E-03   9.999E-04   4.937E-03   1.488E-06   4.937E-03
```

Figure 10.43 Continued.

The ac solution reveals that $V_4 = V_7$ (capacitors in their short-circuit state equivalence) with a magnitude of 4.937 mV for a gain of 4.937 for A_{v_s} since the applied signal is 1 mV.

Let us again check the results using the equations developed in Chapters 8 and 9.

$$g_{mo} = \frac{2I_{DSS}}{V_P} = \frac{2(10 \text{ mA})}{-4 \text{ V}} = 5 \text{ mS}$$

$$g_m(\text{at} - 1 \text{ V}) = g_{mo}\left(1 - \frac{V_{GS}}{V_P}\right) = 5 \text{ mS}\left(1 - \frac{-1V}{-4V}\right)$$

$$= 3.75 \text{ mS}$$

to compare with 3.74 mS in the JFET description of the PSpice output.

$$A_v = -g_m(R_D\|R_L)$$
$$= -(3.75 \text{ mS})(2.2 \text{ k}\Omega\|3.3 \text{ k}\Omega)$$
$$= -(3.75 \text{ mS})(1.32 \text{ k}\Omega)$$
$$= \mathbf{-4.95}$$

to compare with -4.937 above.

§ 10.3 Effect of a Load Impedance (R_L)

1. For the fixed-bias configuration of Fig. 10.44:
   (a) Determine $A_{v_{NL}}$, Z_i, and Z_o.
   (b) Sketch the two-port model of Fig. 10.2 with the parameters determined in part (a) in place.
   (c) Calculate the gain A_v using the model of part (b) and Eq. (10.3).
   (d) Determine the current gain using Eq. (10.6).
   (e) Determine A_v, Z_i, and Z_o using the r_e model and compare with the solutions above.

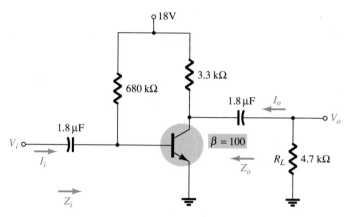

Figure 10.44 Problems 1, 2, 3

* 2. (a) Draw the dc and ac load lines for the network of Fig. 10.44 on the characteristics of Fig. 10.45.
   (b) Determine the peak-to-peak value of I_c and V_{ce} from the graph if V_i has a peak value of 10 mV. Determine the voltage gain $A_v = V_o/V_i$ and compare with the solution obtained in Problem 1.

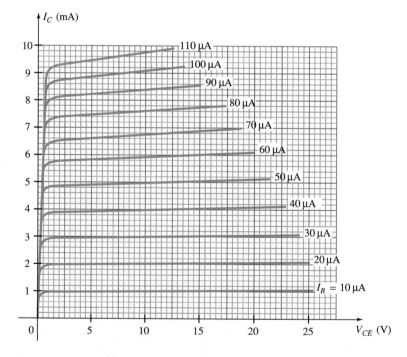

Figure 10.45 Problems 2, 7

Chapter 10 Systems Approach—Effects of R_S and R_L

3. (a) Determine the voltage gain A_v for the network of Fig. 10.44 for $R_L = 4.7$ kΩ, 2.2 kΩ, and 0.5 kΩ. What is the effect of decreasing levels of R_L on the voltage gain?
(b) How will Z_i, Z_o, and $A_{v_{NL}}$ change with decreasing values of R_L?

§ **10.4 Effect of a source impedance (R_s)**

* **4.** For the network of Fig. 10.46:
   (a) Determine $A_{v_{NL}}$, Z_i, and Z_o.
   (b) Sketch the two-port model of Fig. 10.2 with the parameters determined in part (a) in place.
   (c) Determine A_v using the results of part (b).
   (d) Determine A_{v_s}.
   (e) Determine A_{v_s} using the r_e model and compare the results to that obtained in part (d).
   (f) Change R_s to 1 kΩ and determine A_v. How does A_v change with the level of R_s?
   (g) Change R_s to 1 kΩ and determine A_{v_s}. How does A_{v_s} change with the level of R_s?
   (h) Change R_s to 1 kΩ and determine $A_{v_{NL}}$, Z_i, and Z_o. How do they change with change in R_s?

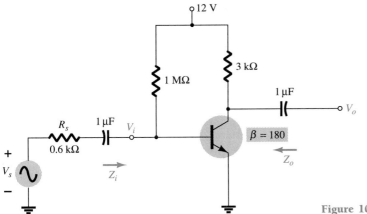

Figure 10.46 Problem 4

§ **10.5 Combined Effect of R_s and R_L.**

* **5.** For the network of Fig. 10.47:
   (a) Determine $A_{v_{NL}}$, Z_i, and Z_o.
   (b) Sketch the two-port model of Fig. 10.2 with the parameters determined in part (a) in place.
   (c) Determine A_v and A_{v_s}.
   (d) Calculate A_i.
   (e) Change R_L to 5.6 kΩ and calculate A_{v_s}. What is the effect of increasing levels of R_L on the gain?
   (f) Change R_s to 0.5 kΩ (with R_L at 2.7 kΩ) and comment on the effect of reducing R_s on A_{v_s}.
   (g) Change R_L to 5.6 kΩ and R_s to 0.5 kΩ and determine the new levels of Z_i and Z_o. How are the impedance parameters affected by changing levels of R_L and R_s?

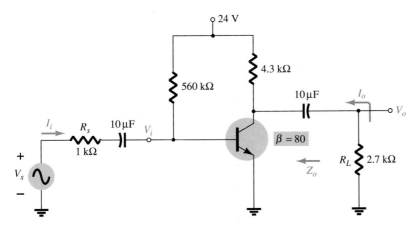

Figure 10.47 Problems 5, 17, 21

Problems   469

§ 10.6 BJT CE Networks

6. For the voltage-divider configuration of Fig. 10.48:
(a) Determine $A_{v_{NL}}$, Z_i, and Z_o.
(b) Sketch the two-port model of Fig. 10.2 with the parameters determined in part (a) in place.
(c) Calculate the gain A_v using the model of part (b).
(d) Determine the current gain A_i.
(e) Determine A_v, Z_i, and Z_o using the r_e model and compare solutions.

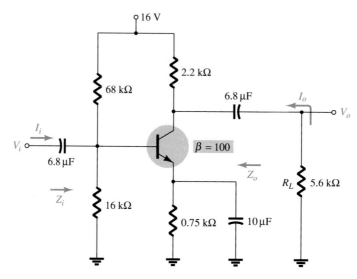

Figure 10.48 Problems 6, 7, 8

∗7. (a) Draw the dc and ac load lines for the network of Fig. 10.48 on the characteristics of Fig. 10.45.
(b) Determine the peak-to-peak value of I_c and V_{ce} from the graph if V_i has a peak value of 10 mV. Determine the voltage gain $A_v = V_o/V_i$ and compare the solution with that obtained in Problem 6.

8. (a) Determine the voltage gain A_v for the network of Fig. 10.48 with $R_L = 4.7$ kΩ, 2.2 kΩ, and 0.5 kΩ. What is the effect of decreasing levels of R_L on the voltage gain?
(b) How will Z_i, Z_o, and $A_{v_{NL}}$ change with decreasing levels of R_L?

9. For the emitter-stabilized network of Fig. 10.49:
(a) Determine $A_{v_{NL}}$, Z_i, and Z_o.
(b) Sketch the two-port model of Fig. 10.2 with the values determined in part (a).
(c) Determine A_v and A_{v_s}.
(d) Change R_s to 1 kΩ. What is the effect on $A_{v_{NL}}$, Z_i, and Z_o?
(e) Change R_s to 1 kΩ and determine A_v and A_{v_s}. What is the effect of increasing levels of R_s on A_v and A_{v_s}?

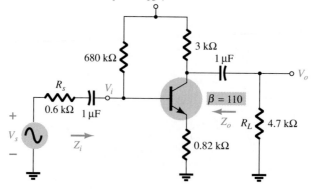

Figure 10.49 Problem 9

Chapter 10 Systems Approach—Effects of R_S and R_L

§ 10.7 BJT Emitter-Follower Networks

* **10.** For the network of Fig. 10.50:
  (a) Determine $A_{v_{NL}}$, Z_i, and Z_o.
  (b) Sketch the two-port model of Fig. 10.2 with the values determined in part (a).
  (c) Determine A_v and A_{v_s}.
  (d) Change R_s to 1 kΩ and determine A_v and A_{v_s}. What is the effect of increasing levels of R_s on the voltage gains?
  (e) Change R_s to 1 kΩ and determine $A_{v_{NL}}$, Z_i, and Z_o. What is the effect of increasing levels of R_s on the parameters?
  (f) Change R_L to 5.6 kΩ and determine A_v and A_{v_s}. What is the effect of increasing levels of R_L on the voltage gains? Maintain R_s at its original level of 0.6 kΩ.

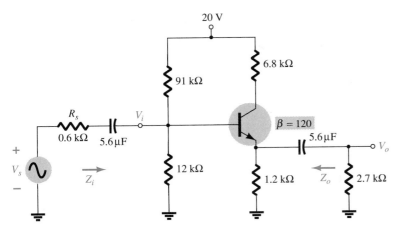

Figure 10.50 Problems 10, 18, 22

§ 10.8 BJT CB Networks

* **11.** For the common-base network of Fig. 10.51:
  (a) Determine Z_i, Z_o, and $A_{v_{NL}}$.
  (b) Sketch the two-port model of Fig. 10.2 with the parameters of part (a) in place.
  (c) Determine A_v and A_{v_s}.
  (d) Determine A_v and A_{v_s} using the r_e model and compare with the results of part (c).
  (e) Change R_s to 0.5 kΩ and R_L to 2.2 kΩ and calculate A_v and A_{v_s}. What is the effect of changing levels of R_s and R_L on the voltage gains?
  (f) Determine Z_o if R_s changed to 0.5 kΩ with all other parameters as appearing in Fig. 10.51. How is Z_o affected by changing levels of R_s?
  (g) Determine Z_i if R_L is reduced to 2.2 kΩ. What is the effect of changing levels of R_L on the input impedance?

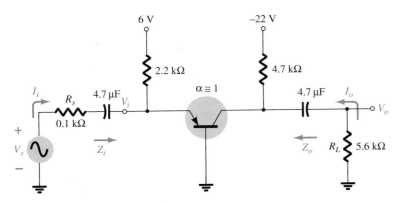

Figure 10.51 Problems 11, 19

§ 10.9 FET Networks

12. For the self-bias JFET network of Fig. 10.52:
    (a) Determine $A_{v_{NL}}$, Z_i, and Z_o.
    (b) Sketch the two-port model of Fig. 10.2 with the parameters determined in part (a) in place.
    (c) Determine A_v and A_{v_s}.
    (d) Change R_L to 6.8 kΩ and R_{sig} to 1 kΩ and calculate the new levels of A_v and A_{v_s}. How are the voltage gains affected by changes in R_{sig} and R_L?
    (e) For the same changes as part (d), determine Z_i and Z_o. What was the impact on both impedances?

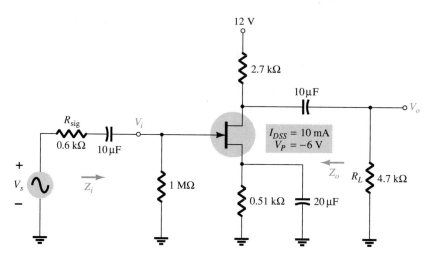

Figure 10.52 Problems 12, 20, 23

13. For the source-follower network of Fig. 10.53:
    (a) Determine $A_{v_{NL}}$, Z_i, and Z_o.
    (b) Sketch the two-port model of Fig. 10.2 with the parameters determined in part (a) in place.
    (c) Determine A_v and A_{v_s}.
    (d) Change R_L to 4.7 kΩ and calculate A_v and A_{v_s}. What was the effect of increasing levels of R_L on both voltage gains?
    (e) Change R_{sig} to 1 kΩ (with R_L at 2.2 kΩ) and calculate A_v and A_{v_s}. What was the effect of increasing levels of R_{sig} on both voltage gains?
    (f) Change R_L to 4.7 kΩ and R_{sig} to 1 kΩ and calculate Z_i and Z_o. What was the effect on both parameters?

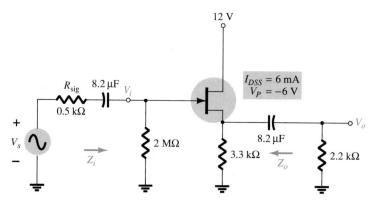

Figure 10.53 Problem 13

* **14.** For the common-gate configuration of Fig. 10.54:
   (a) Determine $A_{v_{NL}}$, Z_i, and Z_o.
   (b) Sketch the two-port model of Fig. 10.2 with the parameters determined in part (a) in place.
   (c) Determine A_v and A_{v_s}.
   (d) Change R_L to 2.2 kΩ and calculate A_v and A_{v_s}. What was the effect of changing R_L on the voltage gains?
   (e) Change R_{sig} to 0.5 kΩ (with R_L at 4.7 kΩ) and calculate A_v and A_{v_s}. What was the effect of changing R_{sig} on the voltage gains?
   (f) Change R_L to 2.2 kΩ and R_{sig} to 0.5 kΩ and calculate Z_i and Z_o. What was the effect on both parameters?

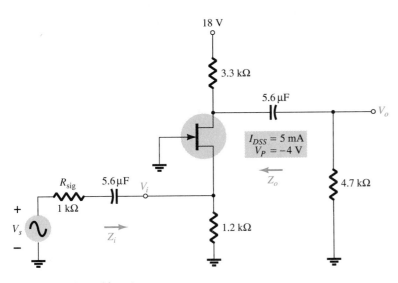

Figure 10.54 Problem 14

§ **10.11 Cascaded Systems**

* **15.** For the cascaded system of Fig. 10.55 with two identical stages, determine:
   (a) The loaded voltage gain of each stage.
   (b) The total gain of the system, A_v and A_{v_s}.
   (c) The loaded current gain of each stage.
   (d) The total current gain of the system.
   (e) How Z_i is affected by the second stage and R_L.
   (f) How Z_o is affected by the first stage and R_s.
   (g) The phase relationship between V_o and V_i.

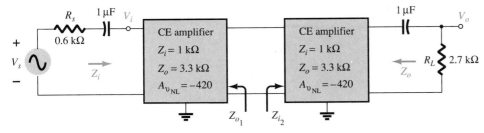

Figure 10.55 Problem 15

* **16.** For the cascaded system of Fig. 10.56, determine:
    (a) The loaded voltage gain of each stage.
    (b) The total gain of the system, A_v and A_{v_s}.
    (c) The loaded current gain of each stage.
    (d) The total current gain of the system.
    (e) How Z_i is affected by the second stage and R_L.
    (f) How Z_o is affected by the first stage and R_s.
    (g) The phase relationship between V_o and V_i.

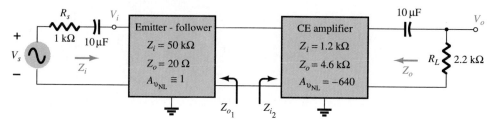

Figure 10.56 Problem 16

§ 10.12 Computer Analysis

17. (a) Write the PSpice input file for the network of Fig. 10.47 and request the level of V_o for $V_s = 1$ mV. For the capacitive elements assume a frequency of 1 kHz.
    (b) Perform the analysis and compare with the level of A_{v_s} for Problem 5.

18. Repeat Problem 17 for the network of Fig. 10.50 and compare the results with those of Problem 10.

19. Repeat Problem 17 for the network of Fig. 10.51 and compare with the results of Problem 11.

20. Repeat Problem 17 for the network of Fig. 10.52 and compare with the results of Problem 12.

21. Repeat Problem 17 using BASIC.

22. Repeat Problem 18 using BASIC.

23. Repeat Problem 20 using BASIC.

---

*Please Note: Asterisks indicate more difficult problems.

BJT and JFET Frequency Response

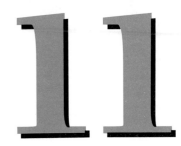

11.1 INTRODUCTION

The analysis thus far has been limited to a particular frequency. For the amplifier, it was a frequency that normally permitted ignoring the effects of the capacitive elements, reducing the analysis to one that included only resistive elements and sources of the independent and controlled variety. We will now investigate the frequency effects introduced by the larger capacitive elements of the network at the low-frequency end and the smaller capacitive elements of the active device at the high frequencies. Since the analysis will extend through a wide frequency range, the logarithmic scale will be defined and used throughout the analysis. In addition, since industry typically uses a decibel scale on its frequency plots, the concept of the decibel is introduced in some detail. The similarities between the frequency response analysis of both BJTs and FETs permit a coverage of each in the same chapter.

11.2 LOGARITHMS

There is no escaping the need to become comfortable with the logarithmic function. The plotting of a variable between wide limits, comparing levels without unwieldy numbers, and identifying levels of particular importance in the design, review, and analysis procedures are all positive features of using the logarithmic function.

As a first step in clarifying the relationship between the variables of a logarithmic function, consider the following mathematical equations:

$$a = b^x, \qquad x = \log_b a \tag{11.1}$$

The variables a, b, and x are the same in each equation. If a is determined by taking the base b to the x power, the same x will result if the log of a is taken to the base b. For instance, if $b = 10$ and $x = 2$,

$$a = b^x = (10)^2 = 100$$

but

$$x = \log_b a = \log_{10} 100 = 2$$

In other words, if you were asked to find the power of a number that would result in a particular level such as shown below:

$$10,000 = 10^x$$

the level of x could be determined using logarithms. That is,

$$x = \log_{10} 10{,}000 = 4$$

For the electrical/electronics industry and in fact for the vast majority of scientific research the base in the logarithmic equation is limited to 10 and the number $e = 2.71828. \ldots$

Logarithms taken to the base 10 are referred to as *common logarithms*, while logarithms taken to the base e are referred to as *natural logarithms*. In summary:

$$\boxed{\text{Common logarithm:} \quad x = \log_{10} a} \tag{11.2}$$

$$\boxed{\text{Natural logarithm:} \quad y = \log_e a} \tag{11.3}$$

The two are related by

$$\boxed{\log_e a = 2.3 \log_{10} a} \tag{11.4}$$

On today's scientific calculators the common logarithm is typically denoted by the $\boxed{\text{log}}$ key and the natural logarithm by the $\boxed{\text{ln}}$ key.

EXAMPLE 11.1

Using the calculator, determine the logarithm of the following numbers to the base indicated.
(a) $\log_{10} 10^6$.
(b) $\log_e e^3$.
(c) $\log_{10} 10^{-2}$.
(d) $\log_e e^{-1}$.

Solution

(a) **6**   (b) **3**   (c) **−2**   (d) **−1**

The results in Example 11.1 clearly reveal that the logarithm of a number taken to a power is simply the power of the number if the number matches the base of the logarithm. In the next example the base and the variable x are not related by an integer power of the base.

EXAMPLE 11.2

Using the calculator, determine the logarithm of the following numbers.
(a) $\log_{10} 64$.
(b) $\log_e 64$.
(c) $\log_{10} 1600$.
(d) $\log_{10} 8000$.

Solution

(a) **1.806**   (b) **4.159**   (c) **3.204**   (d) **3.903**

Note in parts (a) and (b) of Example 11.2 that the logarithms $\log_{10} a$ and $\log_e a$ are indeed related as defined by Eq. (11.4). In addition, note that the logarithm of a number does not increase in the same linear fashion as the number. That is, 8000 is 125 times larger than 64, but the logarithm of 8000 is only about 2.16 times larger than the magnitude of the logarithm of 64, revealing a very nonlinear relationship. In fact, Table 11.1 clearly shows how the logarithm of a number increases only as the exponent of the number. If the antilogarithm of a number is desired, the 10^x or e^x calculator functions are employed.

TABLE 11.1	
$\log_{10} 10^0$	$= 0$
$\log_{10} 10$	$= 1$
$\log_{10} 100$	$= 2$
$\log_{10} 1,000$	$= 3$
$\log_{10} 10,000$	$= 4$
$\log_{10} 100,000$	$= 5$
$\log_{10} 1,000,000$	$= 6$
$\log_{10} 10,000,000$	$= 7$
$\log_{10} 100,000,000$	$= 8$
and so on	

Using a calculator, determine the antilogarithm of the following expressions

EXAMPLE 11.3

(a) $1.6 = \log_{10} a$.
(b) $0.04 = \log_e a$.

Solution

(a) $a = 10^{1.6}$
Calculator keys: [1] [.] [6] [2nd F] [10ˣ]
and $a = \mathbf{39.81}$

(b) $a = e^{0.04}$
Calculator keys: [0] [.] [0] [4] [2nd F] [eˣ]
and $a = \mathbf{1.0408}$

Since the remaining analysis of this chapter employs the common logarithm, let us now review a few properties of logarithms using solely the common logarithm. In general, however, the same relationships hold true for logarithms to any base.

$$\log_{10} 1 = 0 \tag{11.5}$$

As clearly revealed by Table 11.1, since $10^0 = 1$,

$$\log_{10} \frac{a}{b} = \log_{10} a - \log_{10} b \tag{11.6}$$

which for the special case of $a = 1$ becomes

$$\log_{10} \frac{1}{b} = -\log_{10} b \tag{11.7}$$

revealing that for any b greater than 1 the logarithm of a number less than 1 is always negative.

$$\log_{10} ab = \log_{10} a + \log_{10} b \tag{11.8}$$

In each case the equations employing natural logarithms will have the same format.

EXAMPLE 11.4

Using a calculator determine the logarithm of the following numbers
(a) $\log_{10} 0.5$.

(b) $\log_{10} \dfrac{4000}{250}$.

(c) $\log_{10} (0.6 \times 30)$.

Solution

(a) **−0.3**

(b) $\log_{10} 4000 - \log_{10} 250 = 3.602 - 2.398 = \mathbf{1.204}$

   Check: $\log_{10} \dfrac{4000}{250} = \log_{10} 16 = \mathbf{1.204}$

(c) $\log_{10} 0.6 + \log_{10} 30 = -0.2218 + 1.477 = \mathbf{1.255}$

   Check: $\log_{10} (0.6 \times 30) = \log_{10} 18 = \mathbf{1.255}$

The use of log scales can significantly expand the range of variation of a particular variable on a graph. Most graph paper available is of the semilog or double-log (log-log) variety. The term *semi* (meaning one-half) indicates that only one of the two scales is a log scale, whereas double-log indicates that both scales are log scales. A semilog scale appears in Fig. 11.1. Note that the vertical scale is a linear scale with equal divisions. The spacing between the lines of the log plot is shown on the graph.

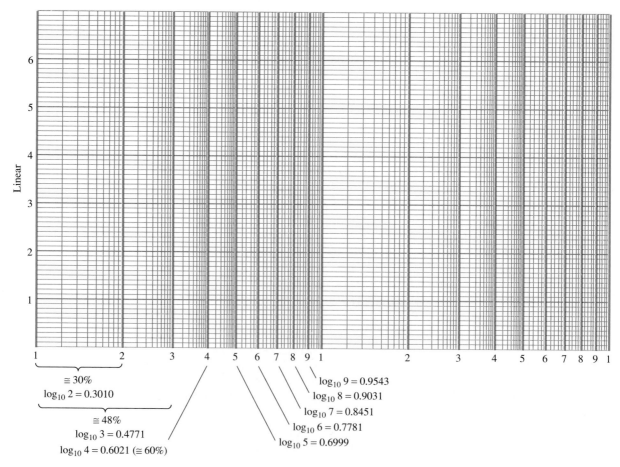

Figure 11.1 Semilog graph paper.

The log of 2 to the base 10 is approximately 0.3. The distance from $1(\log_{10} 1 = 0)$ to 2 is therefore 30% of the span. The log of 3 to the base 10 is 0.4771 or almost 48% of the span (very close to one-half the distance between power of 10 increments on the log scale). Since $\log_{10} 5 \cong 0.7$, it is marked off at a point 70% of the distance. Note that between any two digits the same compression of the lines appears as you progress from the left to the right. It is important to note the resulting numerical value and the spacing, since plots will typically only have the tic marks indicated in Fig. 11.2 due to a lack of space. You must realize that the longer bars for this figure have the numerical values of 0.3, 3, and 30 associated with them, whereas the next shorter bars have values of 0.5, 5, and 50, and the shortest bars 0.7, 7, and 70.

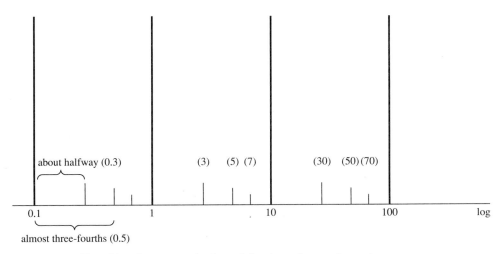

Figure 11.2 Identifying the numerical values of the tic marks on a log scale.

Be aware that plotting a function on a log scale can change the general appearance of the waveform as compared to a plot on a linear scale. A straight-line plot on a linear scale can develop a curve on a log scale and a nonlinear plot on a linear scale can take on the appearance of a straight line on a log plot. The important point is that the results extracted at each level be correctly labeled by developing a familiarity with the spacing of Figs. 11.1 and 11.2. This is particularly true for some of the log-log plots to appear later in the book.

11.3 DECIBELS

The concept of the *decibel* (dB) and the associated calculations will become increasingly important in the remaining sections of this chapter. The background surrounding the term *decibel* has its origin in the established fact that power and audio levels are related on a logarithmic basis. That is, an increase in power level, say 4 to 16 W, does not result in an audio level increase by a factor of $16/4 = 4$. It will increase by a factor of 2 as derived from the power of 4 in the following manner: $(4)^2 = 16$. For a change of 4 to 64 W the audio level will increase by a factor of 3 since $(4)^3 = 64$. In logarithmic form, the relationship can be written as $\log_4 64 = 3$.

For standardization, the *bel* (B) was defined by the following equation to relate power levels P_1 and P_2:

$$G = \log_{10} \frac{P_2}{P_1} \qquad \text{bel} \qquad (11.9)$$

The term *bel* was derived from the surname of Alexander Graham Bell.

It was found, however, that the bel was too large a unit of measurement for practical purposes, so the decibel (dB) was defined such that 10 decibels = 1 bel. Therefore,

$$G_{dB} = 10 \log_{10} \frac{P_2}{P_1} \quad \text{dB} \tag{11.10}$$

The terminal rating of electronic communication equipment (amplifiers, microphones, etc.) is commonly rated in decibels. Equation (11.10) indicates clearly, however, that the decibel rating is a measure of the difference in magnitude between *two* power levels. For a specified terminal (output) power (P_2) there must be a reference power level (P_1). The reference level is generally accepted to be 1 mW although on occasion the 6-mW standard of earlier years is applied. The resistance to be associated with the 1-mW power level is 600 Ω, chosen because it is the characteristic impedance of audio transmission lines. When the 1-mW level is employed as the reference level, the decibel symbol frequently appears as dBm. In equation form,

$$G_{dBm} = 10 \log_{10} \frac{P_2}{1 \text{ mW}} \bigg|_{600\Omega} \quad \text{dBm} \tag{11.11}$$

There exists a second equation for decibels that is applied frequently. It can be best described through the system of Fig. 11.3. For V_i equal to some value V_1, $P_1 = V_1^2/R_i$ where R_i is the input resistance of the system of Fig. 11.3. If V_i should be increased (or decreased) to some other level, V_2, then $P_2 = V_2^2/R_i$. If we substitute into Eq. (11.10) to determine the resulting difference in decibels between the power levels,

$$G_{dB} = 10 \log_{10} \frac{P_2}{P_1} = 10 \log_{10} \frac{V_2^2/R_i}{V_1^2/R_i} = 10 \log_{10} \left(\frac{V_2}{V_1}\right)^2$$

and

$$G_{dB} = 20 \log_{10} \frac{V_2}{V_1} \quad \text{dB} \tag{11.12}$$

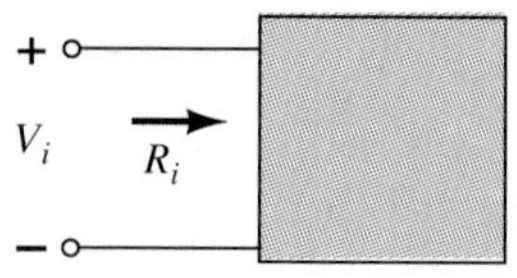

Figure 11.3 Configuration employed in the discussion of Eq. (11.12).

Frequently, the effect of different impedances ($R_1 \neq R_2$) is ignored and Eq. (11.12) applied simply to establish a basis of comparison between levels—voltage or current. For situations of this type the decibel gain should more correctly be referred to as the *voltage or current gain in decibels* to differentiate it from the common usage of decibel as applied to power levels.

One of the advantages of the logarithmic relationship is the manner in which it can be applied to cascaded stages. For example, the magnitude of the overall voltage gain of a cascaded system is given by

$$|A_{v_T}| = |A_{v_1}||A_{v_2}||A_{v_3}|\cdots|A_{v_n}| \tag{11.13}$$

Applying the proper logarithmic relationship results in

$$G_v = 20 \log_{10} |A_{v_T}| = 20 \log_{10} |A_{v_1}| + 20 \log_{10} |A_{v_2}|$$
$$+ 20 \log_{10} |A_{v_3}| + \cdots + 20 \log_{10} |A_{v_n}| \quad \text{(dB)} \qquad (11.14)$$

In words, the equation states that the decibel gain of a cascaded system is simply the sum of the decibel gains of each stage, that is,

$$\boxed{G_v = G_{v_1} + G_{v_2} + G_{v_3} + \cdots + G_{v_n}} \quad \text{dB} \qquad (11.15)$$

In an effort to develop some association between dB levels and voltage gains, Table 11.2 was developed. First note that a gain of 2 results in a dB level of +6 dB while a drop to $\frac{1}{2}$ results in a -6-dB level. A change in V_o/V_i from 1 to 10, 10 to 100, or 100 to 1000 results in the same 20-dB change in level. When $V_o = V_i$, $V_o/V_i = 1$ and the dB level is 0. At a very high gain of 1000 the dB level is 60, while at the much higher gain of 10,000 the dB level is 80 dB, an increase of only 20 dB—a result of the logarithmic relationship. Table 11.2 clearly reveals that voltage gains of 50 dB or higher should immediately be recognized as being quite high.

TABLE 11.2

Voltage Gain, V_o/V_i	dB Level
0.5	-6
0.707	-3
1	0
2	6
10	20
40	32
100	40
1000	60
10,000	80
etc.	

Find the magnitude gain corresponding to a decibel gain of 100.

EXAMPLE 11.5

Solution

By Eq. (11.10),

$$G_{dB} = 10 \log_{10} \frac{P_2}{P_1} = 100 \text{ dB} \rightarrow \log_{10} \frac{P_2}{P_1} = 10$$

so that

$$\frac{P_2}{P_1} = 10^{10} = \mathbf{10{,}000{,}000{,}000}$$

This example clearly demonstrates the range of decibel values to be expected from practical devices. Certainly, a future calculation giving a decibel result in the neighborhood of 100 should be questioned immediately.

The input power to a device is 10,000 W at a voltage of 1000 V. The output power is 500 W while the output impedance is 20 Ω.
(a) Find the power gain in decibels.
(b) Find the voltage gain in decibels.
(c) Explain why parts (a) and (b) agree or disagree.

EXAMPLE 11.6

Solution

(a) $G_{dB} = 10 \log_{10} \dfrac{P_o}{P_i} = 10 \log_{10} \dfrac{500 \text{ W}}{10 \text{ kW}} = 10 \log_{10} \dfrac{1}{20} = -10 \log_{10} 20$

$\quad\quad = -10(1.301) = \mathbf{-13.01\ dB}$

(b) $G_v = 20 \log_{10} \dfrac{V_o}{V_i} = 20 \log_{10} \dfrac{\sqrt{PR}}{1000} = 20 \log_{10} \dfrac{\sqrt{(500 \text{ W})(20\ \Omega)}}{1000 \text{V}}$

$\quad\quad = 20 \log_{10} \dfrac{100}{1000} = 20 \log_{10} \dfrac{1}{10} = -20 \log_{10} 10 = \mathbf{-20\ dB}$

(c) $R_i = \dfrac{V_i^2}{P_i} = \dfrac{(1 \text{ kV})^2}{10 \text{ kW}} = \dfrac{10^6}{10^4} = \mathbf{100\ \Omega} \neq R_o = \mathbf{20\ \Omega}$

EXAMPLE 11.7

An amplifier rated at 40-W output is connected to a 10-Ω speaker.
(a) Calculate the input power required for full power output if the power gain is 25 dB.
(b) Calculate the input voltage for rated output if the amplifier voltage gain is 40 dB.

Solution

(a) Eq. (11.10): $25 = 10 \log_{10} \dfrac{40 \text{ W}}{P_i} \Rightarrow P_i = \dfrac{40 \text{ W}}{\text{antilog } (2.5)} = \dfrac{40 \text{ W}}{3.16 \times 10^2}$

$$= \frac{40 \text{ W}}{316} \cong \mathbf{126.5 \text{ mW}}$$

(b) $G_v = 20 \log_{10} \dfrac{V_o}{V_i} \Rightarrow 40 = 20 \log_{10} \dfrac{V_o}{V_i}$

$\dfrac{V_o}{V_i} = \text{antilog } 2 = 100$

$V_o = \sqrt{PR} = \sqrt{(40 \text{ W})(10 \text{ }\Omega)} = 20 \text{ V}$

$V_i = \dfrac{V_o}{100} = \dfrac{20 \text{ V}}{100} = 0.2 \text{ V} = \mathbf{200 \text{ mV}}$

11.4 GENERAL FREQUENCY CONSIDERATIONS

The frequency of the applied signal can have a pronounced effect on the response of a single or multistage network. The analysis thus far has been for the midfrequency spectrum. At low frequencies we shall find that the coupling and bypass capacitors can no longer be replaced by the short-circuit approximation because of the increase in reactance of these elements. The frequency-dependent parameters of the small-signal equivalent circuits and the stray capacitive elements associated with the active device and the network will limit the high-frequency response of the system. An increase in the number of stages of a cascaded system will also limit both the high-and low-frequency response.

The magnitude of the gain response curves of an *RC*-coupled, direct-coupled, and transformer-coupled amplifier system are provided in Fig. 11.4. Note that the horizontal scale is a logarithmic scale to permit a plot extending from the low- to the high-frequency regions. For each plot, a low-, high-, and midfrequency region has been defined. In addition, the primary reasons for the drop in gain at low and high frequencies have also been indicated within the parentheses. For the *RC*-coupled amplifier the drop at low frequencies is due to the increasing reactance of C_C, C_s, or C_E, while its upper frequency limit is determined by either the parasitic capacitive elements of the network and frequency dependence of the gain of the active device. An explanation of the drop in gain for the transformer-coupled system requires a basic understanding of "transformer action" and the transformer equivalent circuit. For the moment let us say that it is simply due to the "shorting effect" (across the input terminals of the transformer) of the magnetizing inductive reactance at low frequencies ($X_L = 2\pi f L$). The gain must obviously be zero at $f = 0$ since at this point there is no longer a changing flux established through the core to induce a secondary or output voltage. As indicated in Fig. 11.4, the high-frequency response is controlled primarily by the stray capacitance between the turns of the primary and secondary windings. For the direct-coupled amplifier, there are no coupling or bypass

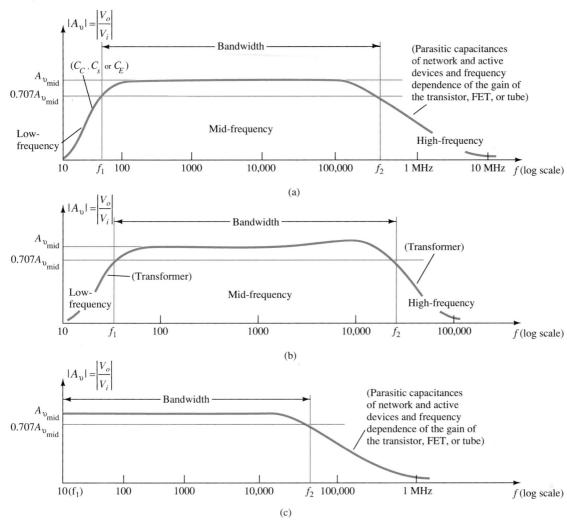

Figure 11.4 Gain versus frequency for (a) *RC*-coupled amplifiers; (b) transformer-coupled amplifiers; (c) direct-coupled amplifiers.

capacitors to cause a drop in gain at low frequencies. As the figure indicates, it is a flat response to the upper cutoff frequency which is determined by either the parasitic capacitances of the circuit or the frequency dependence of the gain of the active device.

For each system of Fig. 11.4 there is a band of frequencies in which the magnitude of the gain is either equal or relatively close to the midband value. To fix the frequency boundaries of relatively high gain, $0.707 A_{v_{mid}}$ was chosen to be the gain at the cutoff levels. The corresponding frequencies f_1 and f_2 are generally called the *corner, cutoff, band, break, or half-power frequencies.* The multiplier 0.707 was chosen because at this level the output power is half the midband power output, that is, at midfrequencies,

$$P_{o_{mid}} = \frac{|V_o^2|}{R_o} = \frac{|A_{v_{mid}} V_i|^2}{R_o}$$

and at the half-power frequencies,

$$P_{o_{HPF}} = \frac{|0.707 A_{v_{mid}} V_i|^2}{R_o} = 0.5 \frac{|A_{v_{mid}} V_i|^2}{R_o}$$

11.4 General Frequency Considerations 483

and

$$P_{O_{HPF}} = 0.5P_{O_{mid}}$$ (11.16)

The bandwidth (or passband) of each system is determined by f_1 and f_2, that is,

$$\text{bandwidth (BW)} = f_2 - f_1$$ (11.17)

For applications of a communications nature (audio, video), a decibel plot of the voltage gain versus frequency is more useful than that appearing in Fig. 11.4. Before obtaining the logarithmic plot, however, the curve is generally normalized as shown in Fig. 11.5. In this figure the gain at each frequency is divided by the midband value. Obviously, the midband value is then 1 as indicated. At the half-power frequencies the resulting level is $0.707 = 1/\sqrt{2}$. A decibel plot can now be obtained by applying Eq. (11.12) in the following manner:

$$\left.\frac{A_v}{A_{v_{mid}}}\right|_{dB} = 20 \log_{10} \frac{A_v}{A_{v_{mid}}}$$ (11.18)

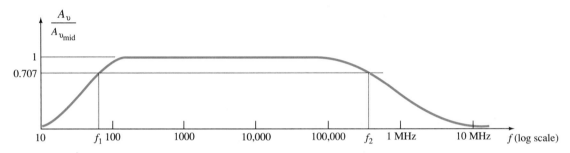

Figure 11.5 Normalized gain versus frequency plot.

At midband frequencies, $20 \log_{10} 1 = 0$, and at the cutoff frequencies, $20 \log_{10} 1/\sqrt{2} = -3$ dB. Both values are clearly indicated in the resulting decibel plot of Fig. 11.6. The smaller the fraction ratio, the more negative the decibel level.

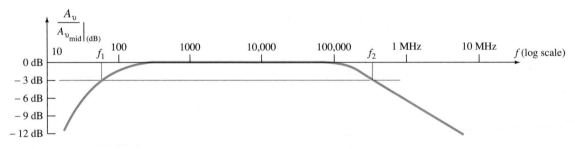

Figure 11.6 Decibel plot of the normalized gain versus frequency plot of Fig. 11.5.

For the greater part of the discussion to follow, a decibel plot will be made only for the low- and high-frequency regions. Keep Fig. 11.6 in mind, therefore, to permit a visualization of the broad system response.

It should be understood that most amplifiers introduce a 180° phase-shift between input and output signals. This fact must now be expanded to indicate that this is only the case in the midband region. At low frequencies there is a phase shift such that V_o lags V_i by an increased angle. At high frequencies the phase shift will drop below 180°. Figure 11.7 is a standard phase plot for an RC-coupled amplifier.

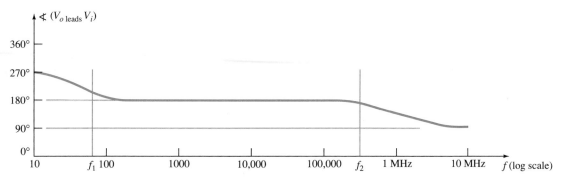

Figure 11.7 Phase plot for an *RC*-coupled amplifier system.

11.5 LOW-FREQUENCY ANALYSIS— BODE PLOT

In the low-frequency region of the single-stage BJT or FET amplifier it is the *R-C* combinations formed by the network capacitors C_C, C_E, and C_s and the network resistive parameters that determine the cutoff frequencies. In fact, an *R-C* network similar to Fig. 11.8 can be established for each capacitive element and the frequency at which the output voltage drops to 0.707 of its maximum value determined. Once the cutoff frequencies due to each capacitor are determined, they can be compared to establish which will determine the low-cutoff frequency for the system.

Our analysis, therefore, will begin with the series *R-C* combination of Fig. 11.8 and the development of a procedure that will result in a plot of the frequency response with a minimum of time and effort. At very high frequencies

$$X_C = \frac{1}{2\pi f C} \cong 0\ \Omega$$

and the short-circuit equivalent can be substituted for the capacitor as shown in Fig. 11.9. The result is that $V_o \cong V_i$ at high frequencies. At $f = 0$ Hz,

$$X_C = \frac{1}{2\pi f C} = \frac{1}{2\pi (0)C} = \infty\ \Omega$$

and the open-circuit approximation can be applied as shown in Fig. 11.10 with the result that $V_o = 0$ V.

Between the two extremes the ratio $A_v = V_o/V_i$ will vary as shown in Fig. 11.11. As the frequency increases the capacitive reactance decreases and more of the input voltage appears across the output terminals.

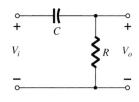

Figure 11.8 R-C combination that will define a low cut-off frequency.

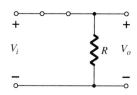

Figure 11.9 R-C circuit of Figure 11.8 at very high frequencies.

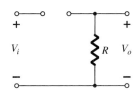

Figure 11.10 R-C circuit of Figure 11.8 at $f = 0$ Hz.

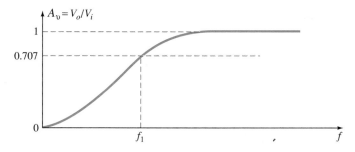

Figure 11.11 Low frequency response for the R-C circuit of Figure 11.8.

11.5 Low-Frequency Analysis—Bode Plot

485

The output and input voltages are related by the voltage-divider rule in the following manner:

$$\mathbf{V}_o = \frac{\mathbf{R}\mathbf{V}_i}{\mathbf{R} + \mathbf{X}_C}$$

with the magnitude of V_o determined by

$$V_o = \frac{RV_i}{\sqrt{R^2 + X_C^2}}$$

For the special case where $X_C = R$,

$$V_o = \frac{RV_i}{\sqrt{R^2 + X_C^2}} = \frac{RV_i}{\sqrt{R^2 + R^2}} = \frac{RV_i}{\sqrt{2R^2}} = \frac{RV_i}{\sqrt{2}R} = \frac{1}{\sqrt{2}}V_i$$

and

$$\boxed{|A_v| = \frac{V_o}{V_i} = \frac{1}{\sqrt{2}} = 0.707|_{X_C=R}}$$ (11.19)

the level of which is indicated on Fig. 11.11. In other words, at the frequency of which $X_C = R$, the output will be 70.7% of the input for the network of Fig. 11.8.

The frequency at which this occurs is determined from

$$X_C = \frac{1}{2\pi f_1 C} = R$$

and

$$\boxed{f_1 = \frac{1}{2\pi RC}}$$ (11.20)

In terms of logs,

$$G_v = 20 \log_{10} A_v = 20 \log_{10} \frac{1}{\sqrt{2}} = -3 \text{ dB}$$

while at $A_v = V_o/V_i = 1$ or $V_o = V_i$ (the maximum value),

$$G_v = 20 \log_{10} 1 = 20(0) = 0 \text{ dB}$$

In Fig. 11.6 we recognize that there is a 3-dB drop in gain from the midband level when $f = f_1$. In a moment we will find that an RC network will determine the low-frequency cutoff frequency for a BJT transistor and f_1 will be determined by Eq. (11.20).

If the gain equation is written as

$$A_v = \frac{V_o}{V_i} = \frac{R}{R - jX_C} = \frac{1}{1 - j(X_C/R)} = \frac{1}{1 - j(1/\omega CR)} = \frac{1}{1 - j(1/2\pi fCR)}$$

and using the frequency defined above,

$$\boxed{A_v = \frac{1}{1 - j(f_1/f)}}$$ (11.21)

In the magnitude and phase form,

$$A_v = \frac{V_o}{V_i} = \frac{1}{\underbrace{\sqrt{1 + (f_1/f)^2}}_{\text{magnitude of } A_v}} \underbrace{/\tan^{-1}(f_1/f)}_{\substack{\text{phase } \sphericalangle \text{ by which} \\ V_o \text{ leads } V_i}}$$ (11.22)

For the magnitude, when $f = f_1$,

$$|A_v| = \frac{1}{\sqrt{1 + (1)^2}} = \frac{1}{\sqrt{2}} = 0.707 \rightarrow -3 \text{ dB}$$

In the logarithmic form; the gain in dB is

$$A_{v(\text{dB})} = 20 \log_{10} \frac{1}{\sqrt{1 + (f_1/f)^2}} = -20 \log_{10} \left[1 + \left(\frac{f_1}{f} \right)^2 \right]^{1/2}$$

$$= -(\tfrac{1}{2})(20) \log_{10} \left[1 + \left(\frac{f_1}{f} \right)^2 \right]$$

$$= -10 \log_{10} \left[1 + \left(\frac{f_1}{f} \right)^2 \right]$$

For frequencies where $f \ll f_1$ or $(f_1/f)^2 \gg 1$, the equation above can be approximated by

$$= -10 \log_{10} \left(\frac{f_1}{f} \right)^2$$

and finally,

$$\boxed{A_{v(\text{dB})} = -20 \log_{10} \frac{f_1}{f}} \bigg|_{f \ll f_1} \tag{11.23}$$

Ignoring the condition $f \ll f_1$ for a moment, a plot of Eq. (11.23) on a frequency log scale will yield a result of a very useful nature for future decibel plots.

$$\text{At } f = f_1: \quad \frac{f_1}{f} = 1 \text{ and } -20 \log_{10} 1 = 0 \text{ dB}$$

$$\text{At } f = \tfrac{1}{2}f_1: \quad \frac{f_1}{f} = 2 \text{ and } -20 \log_{10} 2 \cong -6 \text{ dB}$$

$$\text{At } f = \tfrac{1}{4}f_1: \quad \frac{f_1}{f} = 4 \text{ and } -20 \log_{10} 4 \cong -12 \text{ dB}$$

$$\text{At } f = \tfrac{1}{10}f_1: \quad \frac{f_1}{f} = 10 \text{ and } -20 \log_{10} 10 = -20 \text{ dB}$$

A plot of these points is indicated in Fig. 11.12 from $0.1 f_1$ to f_1. Note that this results in a straight line when plotted against a log scale. In the same figure a straight

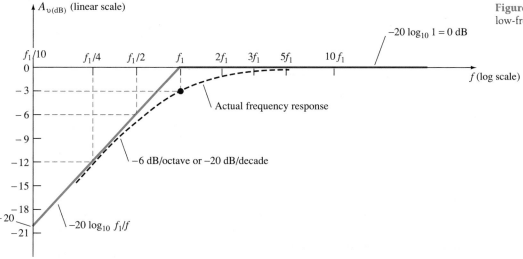

Figure 11.12 Bode plot for the low-frequency region.

line is also drawn for the condition of 0 dB for $f \gg f_1$. As stated earlier, the straight-line segments (asymptotes) are only accurate for 0 dB when $f \gg f_1$, and the sloped line when $f_1 \gg f$. We know, however, that when $f = f_1$, there is a 3-dB drop from the mid-band level. Employing this information in association with the straight-line segments permits a fairly accurate plot of the frequency response as indicated in the same figure. The piecewise linear plot of the asymptotes and associated breakpoints is called a *Bode plot* of the magnitude versus frequency.

The calculations above and the curve itself demonstrate clearly that:

A change in frequency by a factor of 2, equivalent to 1 octave, *results in a 6-dB change in the ratio as noted by the change in gain from $f_1/2$ to f_1.*

As noted by the change in gain from $f_1/2$ to f_1.

For a 10:1 change in frequency, equivalent to 1 decade, *there is a 20-dB change in the ratio as demonstrated between the frequencies of $f_1/10$ and f_1.*

In the future, therefore, a decibel plot can easily be obtained for a function having the format of Eq. (11.23). First simply find f_1 from the circuit parameters and then sketch two asymptotes—one along the 0-dB line and the other drawn through f_1 sloped at 6 dB/octave or 20 dB/decade. Then find the 3-dB point corresponding to f_1 and sketch the curve.

EXAMPLE 11.8

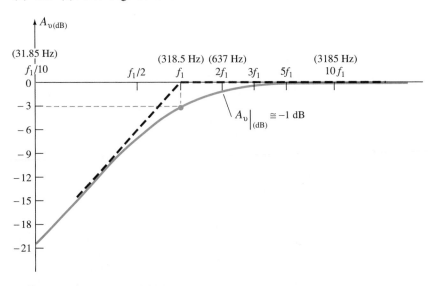

Figure 11.13 Example 11.8

For the network of Fig. 11.13:
(a) Determine the break frequency.
(b) Sketch the asymptotes and locate the −3-dB point.
(c) Sketch the frequency response curve.

Solution

(a) $f_1 = \dfrac{1}{2\pi RC} = \dfrac{1}{(6.28)(5 \times 10^3 \Omega)(0.1 \times 10^{-6}\text{F})}$

$\cong \mathbf{318.5\ Hz}$

(b) and (c). See Fig. 11.14.

Figure 11.14 Frequency response for the R-C circuit of Figure 11.13.

The gain at any frequency can then be determined from the frequency plot in the following manner:

$$A_{v(dB)} = 20 \log_{10}\frac{V_o}{V_i}$$

but

$$\frac{A_{v(dB)}}{20} = \log_{10}\frac{V_o}{V_i}$$

and

$$A_v = \frac{V_o}{V_i} = 10^{\left(\frac{A_{v(dB)}}{20}\right)}$$

(11.24)

For example, if $A_{v(dB)} = -3$ dB,

$$A_v = \frac{V_o}{V_i} = 10^{(-3/20)} = 10^{(-0.15)} \cong 0.707 \quad \text{as expected}$$

The quantity $10^{-0.15}$ is determined using the 10^x function found on most scientific calculators.

From Fig. 11.14, $A_{v(dB)} \cong -1$ dB at $f = 2 f_1 = 637$ Hz. The gain at this point is

$$A_v = \frac{V_o}{V_i} = 10^{\left(\frac{A_{v(dB)}}{20}\right)} = 10^{(-1/20)} = 10^{(-0.05)} = 0.891$$

and

$$V_o = 0.891 V_i$$

or V_o is 89.1% of V_i at $f = 637$ Hz.

The phase angle of θ is determined from

$$\theta = \tan^{-1}\frac{f_1}{f}$$

(11.25)

from Eq. (11.22).

For frequencies $f \ll f_1$,

$$\theta = \tan^{-1}\frac{f_1}{f} \rightarrow 90°$$

For instance, if $f_1 = 100\, f$,

$$\theta = \tan^{-1}\frac{f_1}{f} = \tan^{-1}(100) = 89.4°$$

For $f = f_1$,

$$\theta = \tan^{-1}\frac{f_1}{f} = \tan^{-1}1 = 45°$$

For $f \gg f_1$:

$$\theta = \tan^{-1}\frac{f_1}{f} \rightarrow 0°$$

For instance, if $f = 100 f_1$,

$$\theta = \tan^{-1}\frac{f_1}{f} = \tan^{-1}0.01 = 0.573°$$

11.5   Low-Frequency Analysis—Bode Plot

A plot of $\theta = \tan^{-1}(f_1/f)$ is provided in Fig. 11.15. If we add the additional 180° phase shift introduced by an amplifier, the phase plot of Fig. 11.7 will be obtained. The magnitude and phase response for an *R-C* combination has now been established. In Section 11.6 each capacitor of importance in the low-frequency region will be redrawn in an *R-C* format and the cutoff frequency for each determined to establish the low-frequency response for the BJT amplifier.

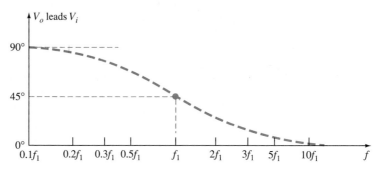

Figure 11.15 Phase response for the R-C circuit of Figure 11.8.

11.6 LOW-FREQUENCY RESPONSE— BJT AMPLIFIER

The analysis of this section will employ the loaded voltage-divider BJT bias configuration, but the results can be applied to any BJT configuration. It will simply be necessary to find the appropriate equivalent resistance for the *R-C* combination. For the network of Fig. 11.16, the capacitors C_s, C_C, and C_E will determine the low-frequency response. We will now examine the impact of each independently in the order listed.

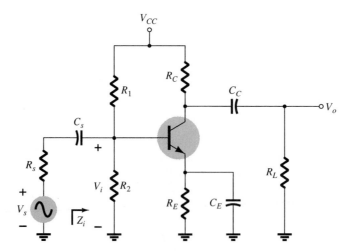

Figure 11.16 Loaded BJT amplifier with capacitors that affect the low-frequency response.

C_s

Since C_s is normally connected between the applied source and the active device, the general form of the *R-C* configuration is established by the network of Fig. 11.17. The total resistance is now $R_s + R_i$ and the cutoff frequency as established in Section 11.5 is

$$f_{L_S} = \frac{1}{2\pi(R_s + R_i)C_S} \quad \checkmark \quad (11.26)$$

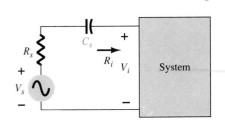

Figure 11.17 Determining the effect of C_S on the low frequency response.

At mid- or high frequencies the reactance of the capacitor will be sufficiently small to permit a short-circuit approximation for the element. The voltage V_i will then be related to V_s by

$$V_i|_{\text{mid}} = \frac{R_i V_s}{R_i + R_s} \quad (11.27)$$

At f_{L_s} the voltage V_i will be 70.7% of the value determined by Eq. (11.27), assuming that C_s is the only capacitive element controlling the low-frequency response.

For the network of Fig. 11.16 when we analyze the effects of C_s, we must make the assumption that C_E and C_C are performing their designed function or the analysis becomes too unwieldy, that is, that the magnitude of the reactances of C_E and C_C permits employing a short-circuit equivalent in comparison to the magnitude of the other series impedances. Using this hypothesis, the ac equivalent network for the input section of Fig. 11.16 will appear as shown in Fig. 11.18.

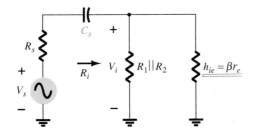

Figure 11.18 Localized ac equivalent for C_s.

The value of R_i for Eq. (11.26) is determined by

$$R_i = R_1 \| R_2 \| \beta r_e \quad (11.28)$$

The voltage $\mathbf{V}_i$ applied to the input of the active device can be calculated using the voltage-divider rule:

$$\mathbf{V}_i = \frac{R_i \mathbf{V}_s}{R_s + R_i - jX_{C_s}} \quad (11.29)$$

C_C

Since the coupling capacitor is normally connected between the output of the active device and the applied load, the R-C configuration that determines the low cutoff frequency due to C_C appears in Fig. 11.19. From Fig. 11.19 the total series resistance is now $R_o + R_L$ and the cutoff frequency due to C_C is determined by

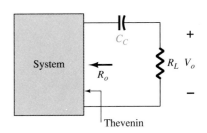

Figure 11.19 Determining the effect of C_C on the low-frequency response.

$$f_{L_C} = \frac{1}{2\pi(R_o + R_L)C_C} \quad \checkmark \quad (11.30)$$

Ignoring the effects of C_s and C_E, the output voltage V_o will be 70.7% of its midband value at f_{L_C}. For the network of Fig. 11.16 the ac equivalent network for the output section with $V_i = 0V$ appears in Fig. 11.20. The resulting value for R_o in Eq. (11.30) is then simply

$$R_o = R_C \| r_o \quad (11.31)$$

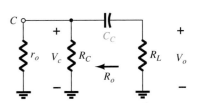

Figure 11.20 Localized ac equivalent for C_C with $V_i = 0V$.

f

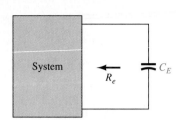

Figure 11.21 Determining the effect of C_E on the low-frequency response.

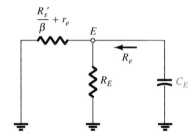

Figure 11.22 Localized ac equivalent of C_E.

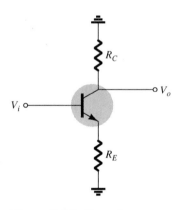

Figure 11.23 Network employed to describe the affect of C_E on the amplifier gain.

C_E

To determine f_{L_E} the network "seen" by C_E must be determined as shown in Fig. 11.21. Once the level of R_e is established, the cutoff frequency due to C_E can be determined using the following equation:

$$f_{L_E} = \frac{1}{2\pi R_e C_E}$$

✓ (11.32)

For the network of Fig. 11.16 the ac equivalent as "seen" by C_E appears in Fig. 11.22. The value of R_e is therefore determined by

$$R_e = R_E \| \left(\frac{R_s'}{\beta} + r_e \right)$$

(11.33)

where $R_s' = R_s \| R_1 \| R_2$.

The effect of C_E on the gain is best described in a quantitative manner by recalling that the gain for the configuration of Fig. 11.23 is given by

$$A_v = \frac{-R_C}{r_e + R_E}$$

The maximum gain is obviously available where R_E is zero ohms. At low frequencies, with the bypass capacitor C_E in its "open-circuit" equivalent state, all of R_E appears in the gain equation above, resulting in the minimum gain. As the frequency increases the reactance of the capacitor C_E will decrease, reducing the parallel impedance of R_E and C_E until the resistor R_E is effectively "shorted out" by C_E. The result is a maximum or midband gain determined by $A_v = -R_C/r_e$. At f_{L_E} the gain will be 3 dB below the midband value determined with R_E "shorted out."

Before continuing, keep in mind that C_s, C_C, and C_E will affect only the low-frequency response. At the midband frequency level the short-circuit equivalents for the capacitors can be inserted. Although each will affect the gain $A_v = V_o/V_i$ in a similar frequency range, the highest low-frequency cutoff determined by C_s, C_C, or C_E will have the greatest impact since it will be the last encountered before the midband level. If the frequencies are relatively far apart, the highest cutoff frequency will essentially determine the lower cutoff frequency for the entire system. If there are two or more "high" cutoff frequencies, the effect will be to raise the lower cutoff frequency and reduce the resulting bandwidth of the system. In other words, there is an interaction between capacitive elements that can affect the resulting low cutoff frequency. However, if the cutoff frequencies established by each capacitor are sufficiently separated, the effect of one on the other can be ignored with a high degree of accuracy—a fact that will be demonstrated by the printouts to appear in the following example:

EXAMPLE 11.9

(a) Determine the lower cutoff frequency for the network of Fig. 11.16 using the following parameters:

$C_s = 10 \ \mu\text{F}, \qquad C_E = 20 \ \mu\text{F}, \qquad C_C = 1 \ \mu\text{F}$

$R_s = 1 \ \text{k}\Omega, \quad R_1 = 40 \ \text{k}\Omega, \quad R_2 = 10 \ \text{k}\Omega, \quad R_E = 2 \ \text{k}\Omega, \quad R_C = 4 \ \text{k}\Omega,$
$R_L = 2.2 \ \text{k}\Omega$

$\beta = 100, \qquad r_o = \infty \ \Omega, \qquad V_{CC} = 20 \ \text{V}$

(b) Sketch the frequency response using a Bode plot.

Solution

(a) Determining r_e for dc conditions:

$$\beta R_E = (100)(2 \text{ k}\Omega) = 200 \text{ k}\Omega \gg 10R_2 = 100 \text{ k}\Omega$$

The result is:

$$V_B \cong \frac{R_2 V_{CC}}{R_2 + R_1} = \frac{10 \text{ k}\Omega(20 \text{ V})}{10 \text{ k}\Omega + 40 \text{ k}\Omega} = \frac{200 \text{ V}}{50} = 4 \text{ V}$$

with

$$I_E = \frac{V_E}{R_E} = \frac{4 \text{ V} - 0.7 \text{ V}}{2 \text{ k}\Omega} = \frac{3.3 \text{ V}}{2 \text{ k}\Omega} = 1.65 \text{ mA}$$

so that

$$r_e = \frac{26 \text{ mV}}{1.65 \text{ mA}} \cong \mathbf{15.76 \ \Omega}$$

and

$$\beta r_e = 100(15.76 \ \Omega) = 1576\Omega = \mathbf{1.576 \ k\Omega}$$

Midband Gain:

$$A_v = \frac{V_o}{V_i} = \frac{-R_C \| R_L}{r_e} = -\frac{(4 \text{ k}\Omega)\|(2.2 \text{ k}\Omega)}{15.76 \ \Omega} \cong -90$$

The input impedance

$$Z_i = R_i = R_1 \| R_2 \| \beta r_e$$
$$= 40 \text{ k}\Omega \| 10 \text{ k}\Omega \| 1.576 \text{ k}\Omega$$
$$\cong 1.32 \text{ k}\Omega$$

and from Fig. 11.24,

$$V_i = \frac{R_i V_s}{R_i + R_s}$$

or

$$\frac{V_i}{V_s} = \frac{R_i}{R_i + R_s} = \frac{1.32 \text{ k}\Omega}{1.32 \text{ k}\Omega + 1 \text{ k}\Omega} = 0.569$$

so that

$$A_{v_s} = \frac{V_o}{V_s} = \frac{V_o}{V_i} \frac{V_i}{V_s} = (-90)(0.569)$$
$$= \mathbf{-51.21}$$

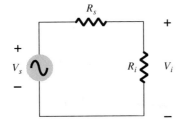

Figure 11.24 Determining the affect of R_s on the gain A_{v_s}.

C_s:

$$R_i = R_1 \| R_2 \| \beta r_e = 40 \text{ k}\Omega \| 10 \text{ k}\Omega \| 1.576 \text{ k}\Omega \cong 1.32 \text{ k}\Omega$$

$$f_{L_s} = \frac{1}{2\pi(R_s + R_i)C_s} = \frac{1}{(6.28)(1 \text{ k}\Omega + 1.32 \text{ k}\Omega)(10 \ \mu\text{F})}$$

$$f_{L_s} \cong \mathbf{6.86 \ Hz}$$

As verification of the result just calculated the network will be analyzed using PSpice and the nodes defined in Fig. 11.25. The input file of Fig. 11.26 reveals that the response is due solely to C_s with C_C and C_E set at very high levels of 1 Farad to ensure that they can be approximated by short-circuit equivalents. The level of V_s was set at 1 mV to provide a level for V_o easily compared to the gain of the system.

The PROBE response of Fig. 11.27 clearly reveals that the cutoff frequency determined by C_s is very close to 7 Hz. The cutoff level was determined from $(0.707)(52.21 \text{ mV}) = 36.21 \text{ mV}$. Note that the PROBE response used a log scale for frequency and a linear scale for the output voltage $V_o = V(7, 0)$.

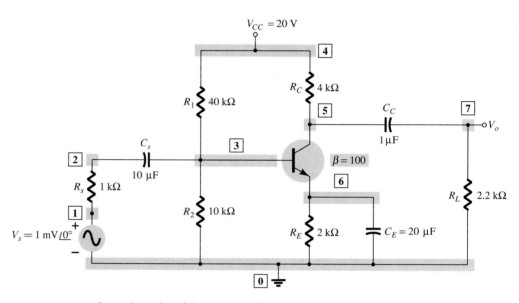

Figure 11.25 Defining the nodes of the network of Fig. 11.16 for a PSpice analysis.

```
**** 05/03/90 ******* Evaluation PSpice (January 1989) ******* 16:34:30 ****

 Frequency response of BJT circuit - Fig. 11.25 (Effect of Cs only)

****       CIRCUIT DESCRIPTION

************************************************************************

VCC 4 0 20V
RB1 4 3 40K
RB2 3 0 10K
RC 4 5 4K
RE 6 0 2K
RS 2 1 1K
RL 7 0 2.2K
*CE and CC made very large(so they have no effect)
CS 2 3 10UF
CE 6 0 1F
CC 5 7 1F
Q1 5 3 6 QN
.MODEL QN NPN(BF=100 IS=5E-15)
VS 1 0 AC 1M
.AC LIN 100 1HZ 100HZ
.PROBE
.OPTIONS NOPAGE
.END
```

Figure 11.26 Determining the effect of C_S on the low-frequency response of a BJT amplifer.

Chapter 11   BJT and JFET Frequency Response

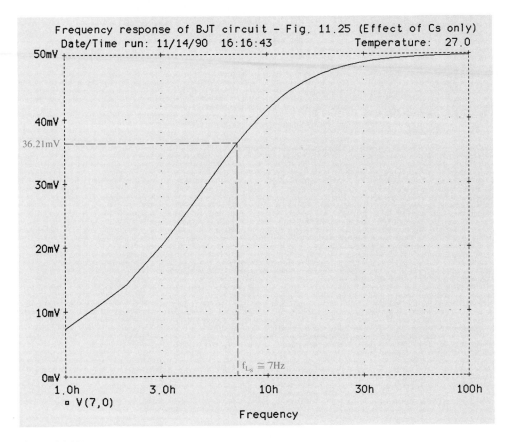

Figure 11.27

C_C:

$$f_{L_C} = \frac{1}{2\pi(R_C + R_L)C_C}$$

$$= \frac{1}{(6.28)(4 \text{ k}\Omega + 2.2 \text{ k}\Omega)(1 \ \mu\text{F})}$$

$$\cong \textbf{25.68 Hz}$$

To investigate the effects of C_C the input file of Fig. 11.26 is modified such that CC = 1 μF with CS = 1 F and CE = 1 F. The result is the output file of Fig. 11.28 substantiating the result obtained above.

C_E:

$$R'_s = R_s\|R_1\|R_2 = 1 \text{ k}\Omega\|40 \text{ k}\Omega\|10 \text{ k}\Omega \cong 0.889 \text{ k}\Omega$$

$$R_e = R_E \left\|\left(\frac{R'_s}{\beta} + r_e\right) = 2 \text{ k}\Omega\right\|\left(\frac{0.889 \text{ k}\Omega}{100} + 15.76 \ \Omega\right)$$

$$= 2 \text{ k}\Omega\|(8.89 \ \Omega + 15.76 \ \Omega) = 2 \text{ k}\Omega\|24.65 \ \Omega \cong 24.35 \ \Omega$$

$$f_{L_E} = \frac{1}{2\pi R_e C_E} = \frac{1}{(6.28)(24.35 \ \Omega)(20 \ \mu\text{F})} = \frac{10^6}{3058.36} \cong \textbf{327 Hz}$$

For C_E the input file of Fig. 11.26 is modified such that CE = 20 μF with CS = 1 F and CC = 1 F. The PROBE response of Fig. 11.29 clearly substantiates the theoretical result. The fact that f_{L_E} is significantly higher than f_{L_s} or f_{L_C} suggests that it will be the predominant factor in determining the low-frequency response for the complete

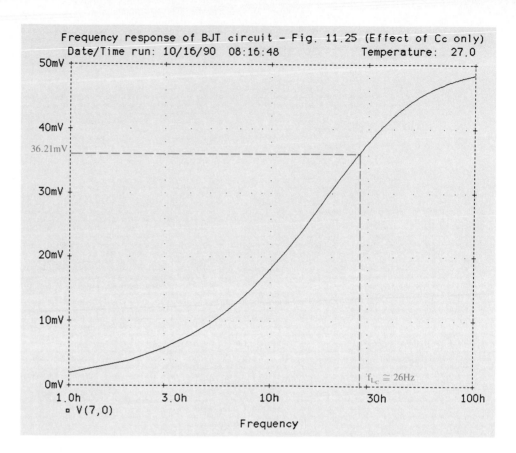

Figure 11.28 Effect of C_C on the low-frequency response of the BJT amplifier of Fig. 11.25.

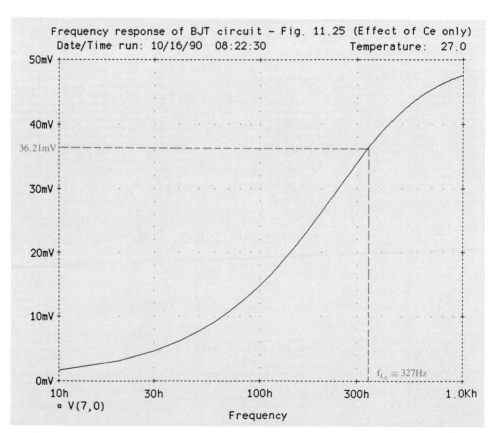

Figure 11.29 Effect of C_E on the low-frequency response of the BJT amplifier of Fig. 11.25.

Chapter 11   BJT and JFET Frequency Response

system. To test the accuracy of our hypothesis, the complete network and the plot of Fig. 11.30 was obtained. Note the very strong similarity with the plot of Fig. 11.29 with the only visible difference being the higher gain at lower frequencies on Fig. 11.29.

(b) It was mentioned earlier that dB plots are usually normalized by dividing the voltage gain A_v by the magnitude of the midband gain. For Fig. 11.16 the magnitude of the midband gain is 51.21 and naturally the ratio $|A_v/A_{v_{mid}}|$ will be 1 in the midband region. The result is a 0-dB asymptote in the midband region as shown in Fig. 11.31. Defining f_{L_E} as our lower cutoff frequency f_1, an asymptote at

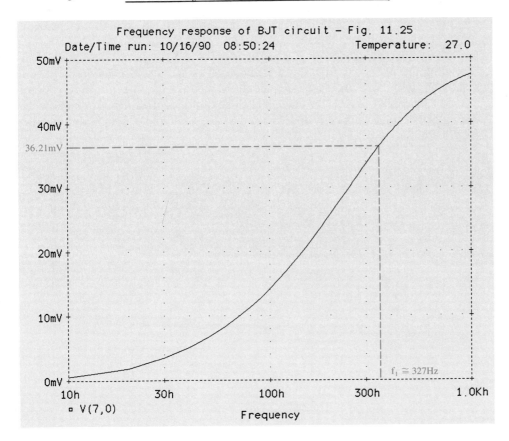

Figure 11.30 Net effect of C_S, C_C, and C_E on the low-frequency response of the BJT amplifer of Fig. 11.25.

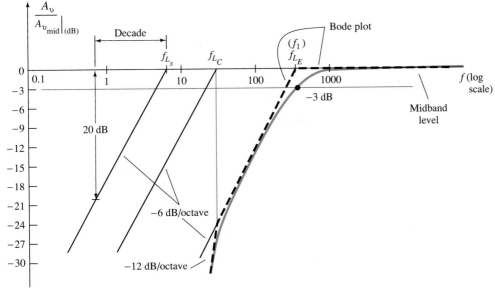

Figure 11.31 Low-frequency plot for the network of Example 11.9.

-6 dB/octave can be drawn as shown in Fig. 11.31 to form the Bode plot and our envelope for the actual response. At f_1 the actual curve is -3 dB down from the midband level as defined by the $0.707A_{V_{\text{mid}}}$ level, permitting a sketch of the actual frequency response curve as shown in Fig. 11.31. A -6-dB/octave asymptote was drawn at each frequency defined in the analysis above to demonstrate clearly that it is f_{L_E} for this network that will determine the -3-dB point. It is not until about -24 dB that f_{L_C} begins to affect the shape of the envelope. The magnitude plot shows that the slope of the resultant asymptote is the sum of the asymptotes having the same sloping direction in the same frequency interval. Note in Fig. 11.31 that the slope has dropped to -12 dB/octave for frequencies less than f_{L_C} and could drop to -18 dB/octave if the three defined cutoff frequencies of Fig. 11.31 were closer together.

Using PROBE a plot of $20\log_{10}|A_v/A_{v_{\text{mid}}}| = A_v/A_{v_{\text{mid}}}|_{\text{dB}}$ can be obtained by recalling that if $V_s = 1$ mV, the magnitude of $|A_v/A_{v_{\text{mid}}}|$ is the same as $|V_o/A_{v_{\text{mid}}}|$ since V_o will have the same numerical value as A_v. The resulting plot of Fig. 11.32 clearly reveals the change in slope of the asymptote at f_{L_C} and how the actual curve follows the envelope created by the Bode plot. In addition, note the 3-dB drop at f_1.

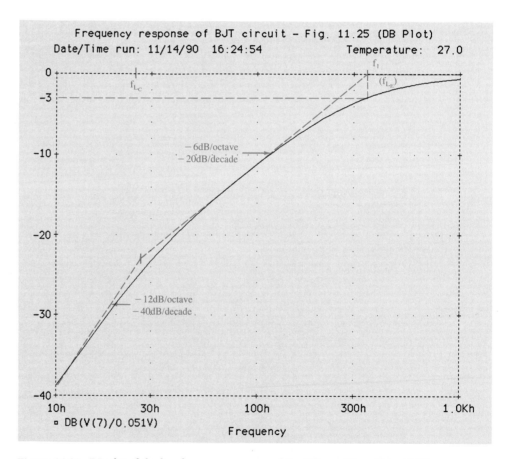

Figure 11.32 DB plot of the low-frequency response of the BJT amplifier of Fig. 11.25.

Keep in mind as we proceed to the next section that the analysis of this section is not limited to the network of Fig. 11.16. For any transistor configuration it is simply necessary to isolate each R-C combination formed by a capacitive element and determine the break frequencies. The resulting frequencies will then determine whether

Chapter 11   BJT and JFET Frequency Response

there is a strong interaction between capacitive elements in determining the overall response and which element will have the greatest impact on establishing the lower cutoff frequency. In fact, the analysis of the next section will parallel this section as we determine the low cutoff frequencies for the FET amplifier.

11.7 LOW-FREQUENCY RESPONSE— FET AMPLIFIER

The analysis of the FET amplifier in the low-frequency region will be quite similar to that of the BJT amplifier of Section 11.6. There are again three capacitors of primary concern as appearing in the network of Fig. 11.33: C_G, C_C, and C_S. Although Fig. 11.33 will be used to establish the fundamental equations, the procedure and conclusions can be applied to most FET configurations.

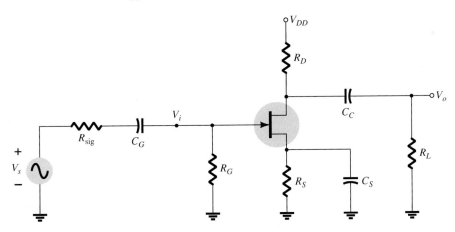

Figure 11.33 Capacitive elements that affect the low-frequency response of a JFET amplifier.

C_G

For the coupling capacitor between the source and the active device the ac equivalent network will appear as shown in Fig. 11.34. The cutoff frequency determined by C_G will then be

$$f_{L_G} = \frac{1}{2\pi(R_{\text{sig}} + R_i)C_G} \qquad (11.34)$$

which is an exact match of Eq. (11.26). For the network of Fig. 11.33,

$$R_i = R_G \qquad (11.35)$$

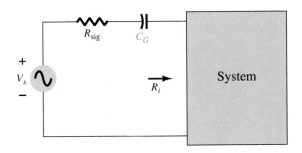

Figure 11.34 Determining the effect of C_G on the low-frequency response.

Typically, $R_G \gg R_{sig}$ and the lower cutoff frequency will be determined primarily by R_G and C_G. The fact that R_G is so large permits a relatively low level of C_G while maintaining a low cutoff frequency level for f_{L_G}.

C_C

For the coupling capacitor between the active device and the load the network of Fig. 11.35 will result, which is also an exact match of Fig. 11.19. The resulting cutoff frequency is

$$f_{L_C} = \frac{1}{2\pi(R_o + R_L)C_C}$$

(11.36)

For the network of Fig. 11.33,

$$R_o = R_D \| r_d$$

(11.37)

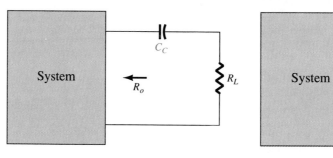

Figure 11.35 Determining the effect of C_C on the low-frequency response.

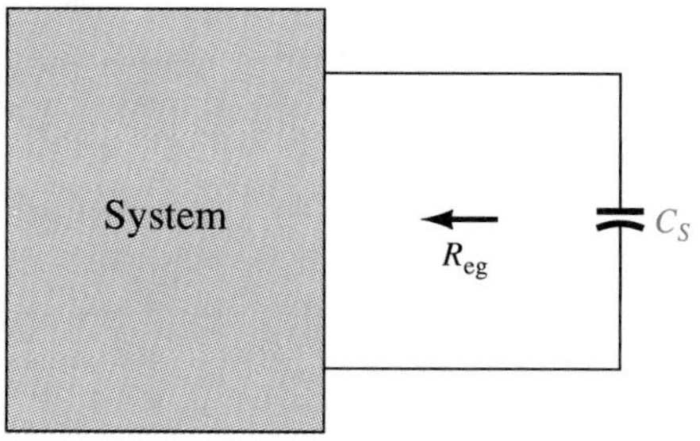

Figure 11.36 Determining the effect of C_S on the low-frequency response.

C_S

For the source capacitor C_S the resistance level of importance is defined by Fig. 11.36. The cutoff frequency will be defined by

$$f_{L_S} = \frac{1}{2\pi R_{eq} C_S}$$

(11.38)

For Fig. 11.33 the resulting value of R_{eq}:

$$R_{eq} = \frac{R_S}{1 + R_S(1 + g_m r_d)/(r_d + R_D \| R_L)}$$

(11.39)

which for $r_d \cong \infty \, \Omega$ becomes

$$R_{eq} = R_S \Big\| \frac{1}{g_m}$$

(11.40)

EXAMPLE 11.10

(a) Determine the lower cutoff frequency for the network of Fig. 11.33 using the following parameters:

$$C_G = 0.01 \, \mu F, \quad C_C = 0.5 \, \mu F, \quad C_S = 2 \, \mu F$$

$$R_{sig} = 10 \, k\Omega, \quad R_G = 1 \, M\Omega, \quad R_D = 4.7 \, k\Omega, \quad R_S = 1 \, k\Omega, \quad R_L = 2.2 \, k\Omega$$

$$I_{DSS} = 8 \, mA, \quad V_P = -4 \, V, \quad r_d = \infty \, \Omega$$

(b) Sketch the frequency response using a Bode plot.

Solution

(a) DC Analysis: Plotting the transfer curve of $I_D = I_{DSS}(1 - V_{GS}/V_P)^2$ and superimposing the curve defined by $V_{GS} = -I_D R_S$ will result in an intersection at $V_{GS_Q} = -2$ V and $I_{D_Q} = 2$ mA. In addition,

$$g_{mo} = \frac{2I_{DSS}}{|V_P|} = \frac{2(8\text{ mA})}{4\text{ V}} = 4\text{ mS}$$

$$g_m = g_{mo}\left(1 - \frac{V_{GS_Q}}{V_P}\right) = 4\text{ mS}\left(1 - \frac{-2\text{ V}}{-4\text{ V}}\right) = 2\text{ mS}$$

C_G:

Eq. (11.34): $f_{L_G} = \dfrac{1}{2\pi(10\text{ k}\Omega + 1\text{ M}\Omega)(0.01\ \mu\text{F})} \cong \mathbf{15.8\ Hz}$

C_C:

Eq. (11.36): $f_{L_C} = \dfrac{1}{2\pi(4.7\text{ k}\Omega + 2.2\text{ k}\Omega)(0.5\ \mu\text{F})} \cong \mathbf{46.13\ Hz}$

C_S:

Eq. (11.38): $R_{eq} = R_S\|\dfrac{1}{gm} = 1\text{ k}\Omega\|\dfrac{1}{2\text{ mS}} = 1\text{ k}\Omega\|0.5\text{ k}\Omega = 333.33\ \Omega$

$$f_{L_S} = \frac{1}{2\pi(333.33\ \Omega)(2\ \mu\text{F})} = \mathbf{238.73\ Hz}$$

Since f_{L_S} is the largest of the three cutoff frequencies, it defines the low cutoff frequency for the network of Fig. 11.33.

(b) The midband gain of the system is determined by

$$A_{v_{mid}} = \frac{V_o}{V_i} = -g_m(R_D\|R_L) = -(2\text{ mS})(4.7\text{ k}\Omega\|2.2\text{ k}\Omega)$$

$$= -(2\text{ mS})(1.499\text{ k}\Omega)$$

$$\cong \mathbf{-3}$$

Using the midband gain to normalize the response for the network of Fig. 11.33 will result in the frequency plot of Fig. 11.37, which is verified by the PROBE response of Fig. 11.38.

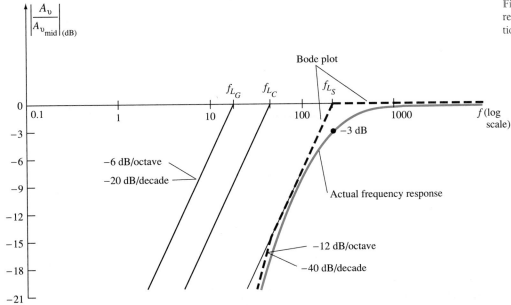

Figure 11.37 Low-frequency response for the JFET configuration of Example 11.10.

501

```
Low frequency response of JFET amplifier - Fig. 11.33
VDD 4 0 20V
RG 3 0 1MEG
RSIG 2 1 10K
RS 6 0 1K
RD 4 5 4.7K
RL 7 0 2.2K
C1 2 3 0.01UF
C2 5 7 0.5UF
CS 6 0 2UF
J1 5 3 6 JN
.MODEL JN NJF(VTO=-4V BETA=500E-6)
CW1 3 0 5PF
CW2 7 0 6PF
VSIG 1 0 AC 1MV
.AC DEC 10 10HZ 10KHZ
.OP
.PROBE
.OPTIONS NOPAGE
.END
```

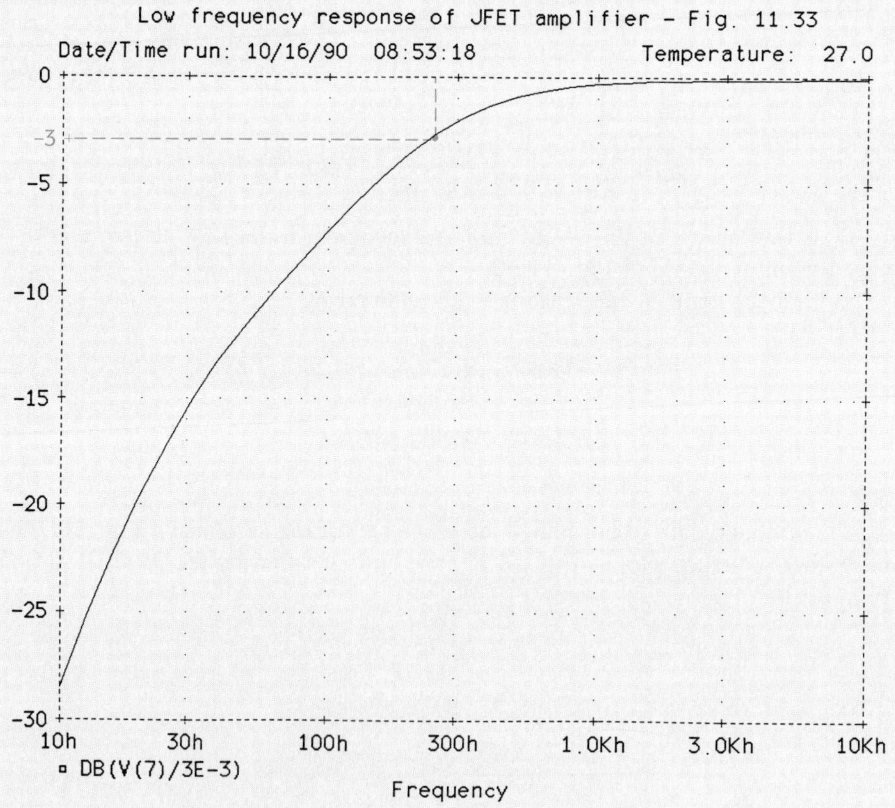

Low frequency response of JFET amplifier - Fig. 11.33

Date/Time run: 10/16/90 08:53:18   Temperature: 27.0

□ DB(V(7)/3E-3)

Frequency

Figure 11.38 PSpice analysis of the JFET amplifier of Example 11.10.

11.8 MILLER EFFECT CAPACITANCE

In the high-frequency region the capacitive elements of importance are the inter-electrode (between terminals) capacitances internal to the active device and the wiring capacitance between leads of the network. The large capacitors of the network that controlled the low-frequency response have all been replaced by their short-circuit equivalent due to their very low reactance levels.

Chapter 11 BJT and JFET Frequency Response

For *inverting* amplifiers (phase shift of 180° between input and output resulting in a negative value for A_v) the input and output capacitance is increased by a capacitance level sensitive to the interelectrode capacitance between the input and output terminals of the device and the gain of the amplifier. In Fig. 11.39 this "feedback" capacitance is defined by C_f.

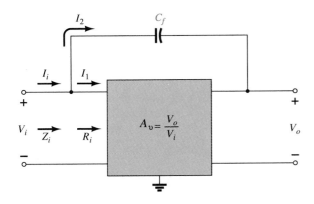

Figure 11.39 Network employed in the derivation of an equation for the Miller input capacitance.

Applying Kirchhoff's current law gives

$$I_i = I_1 + I_2$$

Using Ohm's law yields

$$I_i = \frac{V_i}{Z_i}, \qquad I_1 = \frac{V_i}{R_i}$$

and

$$I_2 = \frac{V_i - V_o}{X_{C_f}} = \frac{V_i - A_v V_i}{X_{C_f}} = \frac{(1 - A_v)V_i}{X_{C_f}}$$

Substituting, we obtain

$$\frac{V_i}{Z_i} = \frac{V_i}{R_i} + \frac{(1 - A_v)V_i}{X_{C_f}}$$

and

$$\frac{1}{Z_i} = \frac{1}{R_i} + \frac{1}{X_{C_f}/(1 - A_v)}$$

but

$$\frac{X_{C_f}}{1 - A_v} = \frac{1}{\underbrace{\omega(1 - A_v)C_f}_{C_M}} = X_{C_M}$$

and

$$\frac{1}{Z_i} = \frac{1}{R_i} + \frac{1}{X_{C_M}}$$

establishing the equivalent network of Fig. 11.40. The result is an equivalent input impedance to the amplifier of Fig. 11.39 that includes the same R_i that we have dealt with in previous chapters, with the addition of a feedback capacitor magnified by the

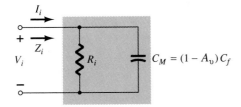

Figure 11.40 Demonstrating the impact of the Miller effect capacitance.

$$C_M = (1 - A_v)C_f$$

gain of the amplifier. Any interelectrode capacitance at the input terminals to the amplifier will simply be added in parallel with the elements of Fig. 11.40.

In general, therefore, the Miller effect input capacitance is defined by

$$C_{M_i} = (1 - A_v)C_f \qquad (11.41)$$

This shows us that:

For any inverting amplifier the input capacitance will be increased by a Miller effect capacitance sensitive to the gain of the amplifier and the inter-electrode capacitance connected between the input and output terminals of the active device.

The dilemma of an equation such as Eq. (11.41) is that at high frequencies the gain A_v will be a function of the level of C_{M_i}. However, since the maximum gain is the midband value, using the midband value will result in the highest level of C_{M_i} and at worst-case scenario. In general, therefore, the midband value is typically employed for A_v in Eq. (11.41).

The reason for the constraint that the amplifier be of the inverting variety is now more apparent when one examines Eq. (11.41). A positive value for A_v would result in a negative capacitance (for $A_v > 1$).

The Miller effect will also increase the level of output capacitance, which must also be considered when the high-frequency cutoff is determined. In Fig. 11.41 the parameters of importance to determine the output Miller effect are in place. Applying Kirchhoff's current law will result in

$$I_o = I_1 + I_2$$

with
$$I_1 = \frac{V_o}{R_o} \qquad \text{and} \qquad I_2 = \frac{V_o - V_i}{X_{C_f}}$$

The resistance R_o is usually sufficiently large to permit ignoring the first term of the equation compared to the second term and assuming that

$$I_o \cong \frac{V_o - V_i}{X_{C_f}}$$

Substituting $V_i = V_o/A_v$ from $A_v = V_o/V_i$ will result in

$$I_o = \frac{V_o - V_o/A_v}{X_{C_f}} = \frac{V_o(1 - 1/A_v)}{X_{C_f}}$$

and
$$\frac{I_o}{V_o} = \frac{1 - 1/A_v}{X_{C_f}}$$

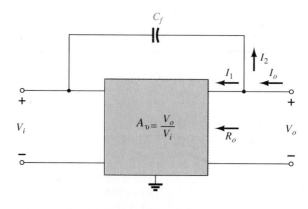

Figure 11.41 Network employed in the derivation of an equation for the Miller output capacitance.

or
$$\frac{V_o}{I_o} = \frac{X_{C_f}}{1 - 1/A_v} = \frac{1}{\omega C_f(1 - 1/A_v)} = \frac{1}{\omega C_{M_o}}$$

resulting in the following equation for the Miller output capacitance:

$$C_{M_o} = \left(1 - \frac{1}{A_v}\right)C_f \qquad\qquad (11.42a)$$

For the usual situation where $A_v \gg 1$ Eq. (11.42a) reduces to

$$C_{M_o} \cong C_f \Big|_{|A_v| \gg 1} \qquad\qquad (11.42b)$$

Examples in the use of Eq. (11.42) will appear in the next two sections as we investigate the high-frequency response of a BJT and FET amplifier.

11.9 HIGH-FREQUENCY RESPONSE— BJT AMPLIFIER

At the high-frequency end there are two factors that will define the -3-dB point: the network capacitance (parasitic and introduced) and the frequency dependence of $h_{fe}(\beta)$.

Network Parameters

In the high-frequency region the RC network of concern has the configuration appearing in Fig. 11.42. At increasing frequencies the reactance X_C will decrease in magnitude, resulting in a shorting effect across the output and a decrease in gain. The derivation leading to the corner frequency for this RC configuration follows along similar lines to that encountered for the low-frequency region. The most significant difference is in the general form of A_v appearing below:

$$A_v = \frac{1}{1 + j(f/f_2)} \qquad\qquad (11.43)$$

which results in a magnitude plot such as shown in Fig. 11.43 that drops off at 6 dB/octave with increasing frequency. Note that f_2 is in the denominator of the frequency ratio rather than the numerator as occurred for f_1 in Eq. (11.21).

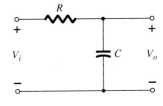

Figure 11.42 R-C combination that will define a high cut-off frequency.

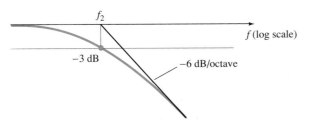

Figure 11.43 Asympotic plot as defined by Eq. (11.43).

In Fig. 11.44 the various parasitic capacitances (C_{be}, C_{bc}, C_{ce}) of the transistor have been included with the wiring capacitances (C_{W_i}, C_{W_o}) introduced during construction. The high-frequency equivalent model for the network of Fig. 11.44 appears in Fig. 11.45. Note the absence of the capacitors C_s, C_C, and C_E, which are all

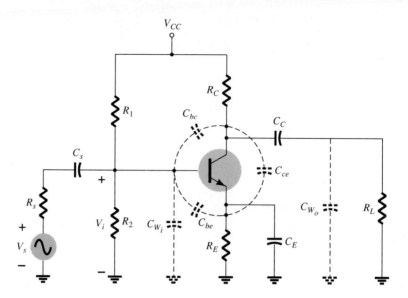

Figure 11.44 Network of Fig. 11.16 with the capacitors that affect the high-frequency response.

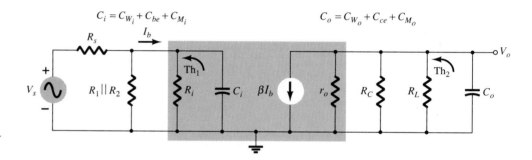

Figure 11.45 High-frequency ac equivalent model for the network of Fig. 11.44.

assumed to be in the short-circuit state at these frequencies. The capacitance C_i includes the input wiring capacitance C_{W_i}, the transition capacitance C_{be}, and the Miller capacitance C_{M_i}. The capacitance C_o includes the output wiring capacitance C_{W_o}, the parasitic capacitance C_{ce}, and the output Miller capacitance C_{M_o}. In general, the capacitance C_{be} is the largest of the parasitic capacitances with C_{ce} the smallest. In fact, most specification sheets simply provide the levels of C_{be} and C_{bc} and do not include C_{ce} unless it will affect the response of a particular type of transistor in a specific area of application.

Determining the Thévenin equivalent circuit for the input and output networks of Fig. 11.45 will result in the configurations of Fig. 11.46. For the input network the -3-dB frequency is defined by

$$f_{H_i} = \frac{1}{2\pi R_{Th_1} C_i} \tag{11.44}$$

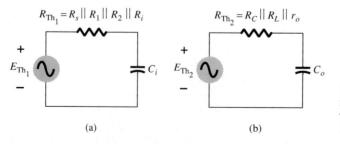

Figure 11.46 Thévenin circuits for the input and output networks of the network of Fig. 11.45.

with

$$R_{Th_1} = R_s\|R_1\|R_2\|R_i$$
(11.45)

and

$$C_i = C_{W_i} + C_{be} + C_{M_i} = C_{W_i} + C_{be} + (1 - A_v)C_{bc}$$
(11.46)

At very high frequencies the effect of C_i is to reduce the total impedance of the parallel combination of R_1, R_2, R_i, and C_i in Fig. 11.45. The result in a reduced level of voltage across C_i and a reduction in I_b establishing a reduced level of gain for the complete system.

For the output network

$$f_{H_o} = \frac{1}{2\pi R_{Th_2} C_o}$$
(11.47)

with

$$R_{Th_2} = R_C\|R_L\|r_o$$
(11.48)

and

$$C_o = C_{W_o} + C_{ce} + C_{M_o}$$
(11.49)

At very high frequencies the capacitive reactance of C_o will decrease and consequently reduce the total impedance of the output parallel branches of Fig. 11.45. The net result is that V_o will also decline toward zero as the reactance X_C becomes smaller. The frequencies f_{H_i} and f_{H_o} will each define a -6-dB/octave asymptote such as depicted in Fig. 11.43. If the parasitic capacitors were the only elements to determine the high cutoff frequency, the lowest frequency would be the determining factor. However, the decrease in h_{fe} (or β) with frequency must also be considered as to whether its break frequency is lower than f_{H_i} or f_{H_o}.

h_{fe} (or β) Variation

The variation of h_{fe} (or β) with frequency will approach, with some degree of accuracy, the following relationship:

$$h_{fe} = \frac{h_{fe_{mid}}}{1 + j(f/f_\beta)}$$
(11.50)

The use of h_{fe} rather than β in some of this descriptive material is due primarily to the fact that manufacturers typically use the hybrid parameters when covering this issue in their specification sheets, and so on.

The only undefined quantity, f_β, is determined by a set of parameters employed in the *hybrid* π or *Giacoletto* model frequently applied to best represent the transistor in the high-frequency region. It appears in Fig. 11.47. The various parameters warrant a moment of explanation. The resistance $r_{bb'}$ includes the base contact, base bulk, and base spreading resistance. The first is due to the actual connection to the base. The second includes the resistance from the external terminal to the active region of the

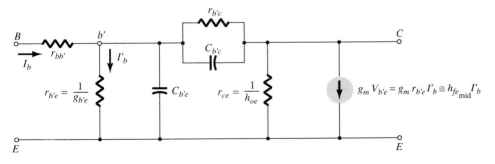

Figure 11.47 Giacoletto (or hybrid π) high-frequency transistor small-signal ac equivalent circuit.

transistors, while the last is the actual resistance within the active base region. The resistances $r_{b'e}$, r_{ce}, and $r_{b'c}$ are the resistances between the indicated terminals when the device is in the active region. The same is true for the capacitance $C_{b'c}$ and $C_{b'e}$, although the former is a transition capacitance while the latter is a diffusion capacitance. A more detailed explanation of the frequency dependence of each can be found in a number of readily available texts.

In terms of these parameters,

$$f_\beta(\text{sometimes appearing as } f_{h_{fe}}) = \frac{g_{b'e}}{2\pi(C_{b'e} + C_{b'c})}$$ (11.51)

or since the hybrid parameter h_{fe} is related to $g_{b'e}$ through $g_m = h_{fe_{\text{mid}}} g_{b'e}$,

$$f_\beta = \frac{1}{h_{fe_{\text{mid}}}} \frac{g_m}{2\pi(C_{b'e} + C_{b'c})}$$ (11.52)

Taking it a step further,

$$g_m = h_{fe_{\text{mid}}} g_{b'e} = h_{fe_{\text{mid}}} \frac{1}{r_{b'e}} \cong \frac{h_{fe_{\text{mid}}}}{h_{ie}} = \frac{\beta_{\text{mid}}}{\beta_{\text{mid}} r_e} = \frac{1}{r_e}$$

and using the approximations

$$C_{b'e} \cong C_{be} \quad \text{and} \quad C_{b'c} \cong C_{bc}$$

will result in the following form for Eq. (11.50):

$$f_\beta \cong \frac{1}{2\pi\beta_{\text{mid}} r_e(C_{be} + C_{bc})}$$ (11.53)

Equation (11.53) clearly reveals that since r_e is a function of the network design:

f_β *is a function of the bias conditions.*

The basic format of Eq. (11.50) is exactly the same as Eq. (11.43) if we extract the multiplying factor $h_{fe_{\text{mid}}}$, revealing that h_{fe} will drop off from its midband value with a 6-dB/octave slope as shown in Fig. 11.48. The same figure has a plot of h_{fb} (or α) versus frequency. Note the small change in h_{fb} for the chosen frequency range, revealing that the common-base configuration displays improved high-frequency characteristics over the common-emitter configuration. Recall also the absence of the Miller effect capacitance due to the noninverting characteristics of the common-base configuration. For this very reason, common-base high-frequency parameters, rather than common-emitter parameters, are often specified for a transistor—especially those designed specifically to operate in the high-frequency regions.

The following equation permits a direct conversion for determining f_β if f_α and α are specified.

$$f_\beta = f_\alpha(1 - \alpha)$$ (11.54)

A quantity called the *gain–bandwidth product* is defined for the transistor by the condition

$$\left| \frac{h_{fe_{\text{mid}}}}{1 + j(f/f_\beta)} \right| = 1$$

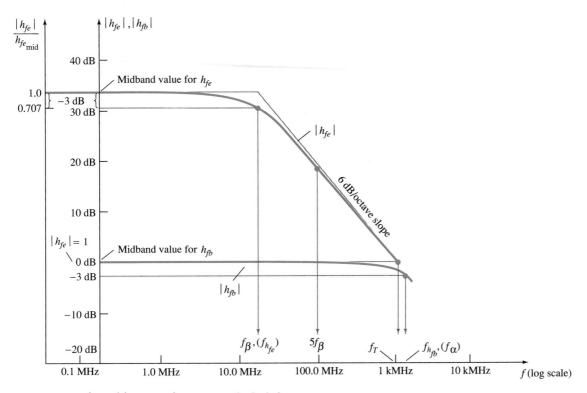

Figure 11.48 h_{fe} and h_{fb} versus frequency in the high-frequency region.

so that
$$|h_{fe}|_{dB} = 20 \log_{10} \left| \frac{h_{fe_{mid}}}{1 + j(f/f_\beta)} \right| = 20 \log_{10} 1 = 0 \text{ dB}$$

The frequency at which $|h_{fe}|_{dB} = 0$ dB is clearly indicated by f_T in Fig. 11.48. The magnitude of h_{fe} at the defined condition point $(f_T \gg f_\beta)$ is given by

$$\frac{h_{fe_{mid}}}{\sqrt{1 + (f_T/f_\beta)^2}} \cong \frac{h_{fe_{mid}}}{f_T/f_\beta} = 1$$

so that
$$\overset{(\cong \text{ BW})}{f_T \cong h_{fe_{mid}} \cdot f_\beta} \qquad \text{(gain–bandwidth product)} \qquad (11.55)$$

or
$$f_T \cong \beta_{mid} f_\beta \qquad (11.56)$$

with
$$f_\beta = \frac{f_T}{\beta_{mid}} \qquad (11.57)$$

Substituting Eq. 11.53 for f_β in Eq. (11.55) gives

$$f_T \cong \beta_{mid} \frac{1}{2\pi \beta_{mid} r_e (C_{be} + C_{bc})}$$

and
$$f_T \cong \frac{1}{2\pi r_e (C_{be} + C_{bc})} \qquad (11.58)$$

EXAMPLE 11.11   For the network of Fig. 11.44 with the same parameters as in Example 11.9, that is,

$$R_s = 1 \text{ k}\Omega, \ R_1 = 40 \text{ k}\Omega, \ R_2 = 10 \text{ k}\Omega, \ R_E = 2 \text{ k}\Omega, \ R_C = 4 \text{ k}\Omega, \ R_L = 2.2 \text{ k}\Omega$$

$$C_s = 10 \ \mu\text{F}, \ C_C = 1 \ \mu\text{F}, \ C_E = 20 \ \mu\text{F}$$

$$\beta = 100, \ r_o = \infty \ \Omega, \ V_{CC} = 20 \text{ V}$$

with the addition of

$$C_{be} = 36 \text{ pF}, \ C_{bc} = 4 \text{ pF}, \ C_{ce} = 1 \text{ pF}, \ C_{W_i} = 6 \text{ pF}, \ C_{W_o} = 8 \text{ pF}$$

(a) Determine f_{H_i} and f_{H_o}.
(b) Find f_β and f_T.
(c) Sketch the frequency response for the low- and high-frequency regions using the results of Example 11.9 and the results of parts (a) and (b).
(d) Obtain a PROBE response for the full frequency spectrum and compare with the results of part (c).

Solution

(a) From Example 11.9:

$$R_i = \beta r_e = 1.576 \text{ k}\Omega, \qquad A_{v_{mid}}(\text{amplifier}) = -90$$

and

$$R_{Th_1} = R_s \| R_1 \| R_2 \| R_i = 1 \text{ k}\Omega \| 40 \text{ k}\Omega \| 10 \text{ k}\Omega \| 1.576 \text{ k}\Omega$$

$$\cong 0.568 \text{ k}\Omega$$

with

$$C_i = C_{W_i} + C_{be} + (1 - A_v)C_{be}$$

$$= 6 \text{ pF} + 36 \text{ pF} + [1 - (-90)]4 \text{ pF}$$

$$= 406 \text{ pF}$$

$$f_{H_i} = \frac{1}{2\pi R_{Th_1} C_i} = \frac{1}{2\pi (0.568 \text{ k}\Omega)(406 \text{ pF})}$$

$$= \textbf{690.15 kHz}$$

$$R_{Th_2} = R_C \| R_L = 4 \text{ k}\Omega \| 2.2 \text{ k}\Omega = 1.419 \text{ k}\Omega$$

$$C_o = C_{W_o} + C_{ce} + C_{M_o} = 8 \text{ pF} + 1 \text{ pF} + \left(1 - \frac{1}{-90}\right)4 \text{ pF}$$

$$= 13.04 \text{ pF}$$

$$f_{H_o} = \frac{1}{2\pi R_{Th_2} C_o} = \frac{1}{2\pi (1.419 \text{ k}\Omega)(13.04 \text{ pF})}$$

$$= \textbf{8.6 MHz}$$

(b) Applying Eq. (11.53) gives

$$f_\beta = \frac{1}{2\pi \beta_{mid} r_e (C_{be} + C_{bc})}$$

$$= \frac{1}{2\pi (100)(15.76 \ \Omega)(36 \text{ pF} + 4 \text{ pF})} = \frac{1}{2\pi (100)(15.76 \ \Omega)(40 \text{ pF})}$$

$$= \textbf{2.52 MHz}$$

$$f_T = \beta_{mid} f_\beta = (100)(2.52 \text{ MHz})$$

$$= \textbf{252 MHz}$$

(c) See Fig. 11.49. Both f_β and f_{H_o} will lower the upper cutoff frequency below the level determined by f_{H_i}. f_β is closer to f_{H_i} and therefore will have a greater impact than f_{H_o}. In any event, the bandwidth will be less than that defined solely by f_{H_i}. In fact, for the parameters of this network the upper cutoff frequency will be relatively close to 600 kHz.

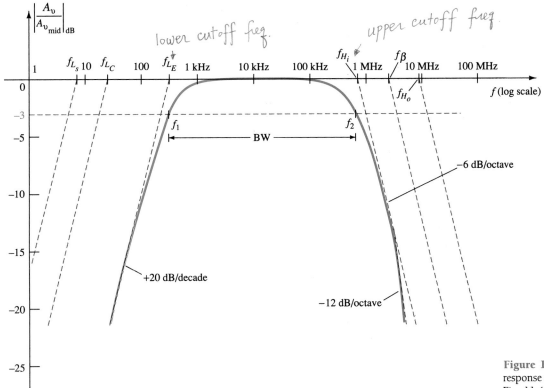

Figure 11.49 Full frequency response for the network of Fig. 11.44.

In general, therefore, the lowest of the upper-cutoff frequencies defines a maximum possible bandwidth for a system.

(d) The input file to obtain the PROBE response using PSpice appears in Fig. 11.50. The parasitic capacitance levels are not included in the model statement because

```
Full frequency response of BJT circuit - Fig. 11.44 (Q cap. & wiring cap.)
VCC 4 0 20V
RB1 4 3 40K
RB2 3 0 10K
RC 4 5 4K
RE 6 0 2K
RS 2 1 1K
RL 7 0 2.2K
CS 2 3 10UF
CE 6 0 20UF
CC 5 7 1UF
CBE 3 6 36PF
CBC 5 3 4PF
CCE 5 6 1PF
Q1 5 3 6 QN
.MODEL QN NPN(BF=100 IS=5E-15)
CW1 3 0 6PF
CW2 7 0 8PF
VS 1 0 AC 1MV
.AC DEC 10 10HZ 100MEGHZ
.PROBE
.OPTIONS NOPAGE
.END
```

Figure 11.50 PSpice analysis of the full frequency response of the network of Fig. 11.44.

the model statement calls for the zero-bias capacitance levels. The levels appearing in the example are at the bias conditions established by the network and are therefore included as network parameters. The DEC command in the .AC analysis statement specifies that the frequency be swept logarithmically from 10 Hz to 100 MHz in decade intervals to provide a sufficient number of data points for a good logarithmic plot.

The output response of Fig. 11.51 does not include the effects of f_β but clearly substantiates the analysis performed in parts (a) through (c) for the high-frequency region. The low cutoff frequency is near the 327 Hz defined by f_{L_E}, and the high cutoff frequency is in the neighborhood of 600 kHz. In other words, even though f_{H_o} is more than a decade higher than f_{H_i}, it will have an impact on the -3-dB cutoff frequency.

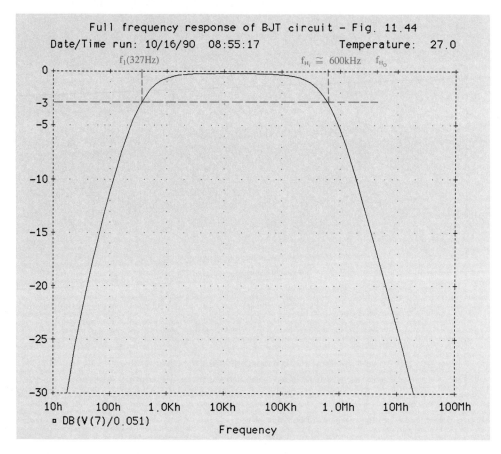

Figure 11.51 Full frequency response for the network of Fig. 11.44.

11.10   HIGH-FREQUENCY RESPONSE— FET AMPLIFIER

The analysis of the high-frequency response of the FET amplifier will proceed in a very similar manner to that encountered for the BJT amplifier. As shown in Fig. 11.52, there are interelectrode and wiring capacitances that will determine the high-frequency characteristics of the amplifier. The capacitors C_{gs} and C_{gd} typically vary from 1 to 10 pF while the capacitance C_{ds} is usually quite a bit smaller, ranging from 0.1 to 1 pF.

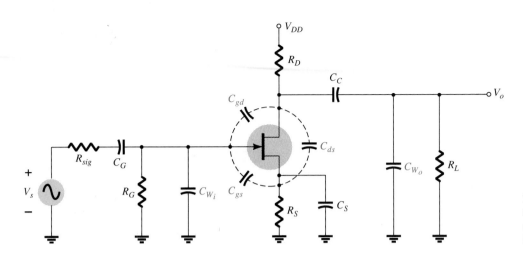

Figure 11.52 Capacitive elements that affect the high frequency response of a JFET amplifier.

Since the network of Fig. 11.52 is an inverting amplifier a Miller effect capacitance will appear in the high-frequency ac equivalent network appearing in Fig. 11.53. At high frequencies C_i will approach a short-circuit equivalent and V_{gs} will drop in value and reduce the overall gain. At frequencies where C_o approaches its short-circuit equivalent the parallel output voltage V_o will drop in magnitude.

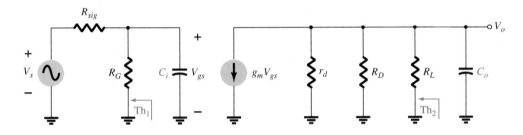

Figure 11.53 High-frequency ac equivalent circuit for Fig. 11.52.

The cutoff frequencies defined by the input and output circuits can be obtained by first finding the Thévenin equivalent circuits for each section as shown in Fig. 11.54. For the input circuit,

$$f_{H_i} = \frac{1}{2\pi R_{\text{Th}_1} C_i} \tag{11.59}$$

and

$$R_{\text{Th}_1} = R_{\text{sig}} \| R_G \tag{11.60}$$

with

$$C_i = C_{W_i} + C_{gs} + C_{M_i} \tag{11.61}$$

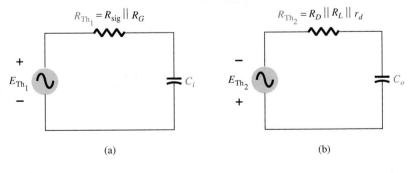

(a)

(b)

Figure 11.54 The Thévenin equivalent circuits for the (a) input circuit and (b) output circuit.

and

$$C_{M_i} = (1 - A_v)C_{gd} \qquad (11.62)$$

and for the output circuit:

$$f_{H_o} = \frac{1}{2\pi R_{\mathrm{Th}_2}C_o} \qquad (11.63)$$

with

$$R_{\mathrm{Th}_2} = R_D\|R_L\|r_d \qquad (11.64)$$

and

$$C_o = C_{W_o} + C_{ds} + C_{M_o}$$

$$\qquad (11.65)$$

and

$$C_{M_o} = \left(1 - \frac{1}{A_v}\right)C_{gd}$$

EXAMPLE 11.12

(a) Determine the high cutoff frequencies for the network of Fig. 11.52 using the same parameters as Example 11.10:

$$C_G = 0.01 \ \mu\mathrm{F}, \qquad C_C = 0.5 \ \mu\mathrm{F}, \qquad C_S = 2 \ \mu\mathrm{F}$$

$$R_{\mathrm{sig}} = 10 \ \mathrm{k\Omega}, \quad R_G = 1 \ \mathrm{M\Omega}, \quad R_D = 4.7 \ \mathrm{k\Omega}, \quad R_S = 1 \ \mathrm{k\Omega}, \quad R_L = 2.2 \ \mathrm{k\Omega}$$

$$I_{DSS} = 8 \ \mathrm{mA}, \qquad V_P = -4 \ \mathrm{V}, \qquad r_d = \infty \ \Omega$$

with the addition of

$$C_{gd} = 2 \ \mathrm{pF}, \quad C_{gs} = 4 \ \mathrm{pF}, \quad C_{ds} = 0.5 \ \mathrm{pF}, \quad C_{W_i} = 5 \ \mathrm{pF}, \quad C_{W_o} = 6 \ \mathrm{pF}$$

(b) Review a PROBE response for the full frequency range and note whether it supports the conclusions of Example 11.10 and the calculations above.

Solution

(a) $R_{\mathrm{Th}_1} = R_{\mathrm{sig}}\|R_G = 10 \ \mathrm{k\Omega}\|1 \ \mathrm{M\Omega} = 9.9 \ \mathrm{k\Omega}$
From Example 11.9,

$$A_v = -3$$

$$C_i = C_{W_i} + C_{gs} + (1 - A_v)C_{gd}$$

$$= 5 \ \mathrm{pF} + 4 \ \mathrm{pF} + (1 + 3)2 \ \mathrm{pF}$$

$$= 9 \ \mathrm{pF} + 8 \ \mathrm{pF}$$

$$= 17 \ \mathrm{pF}$$

$$f_{H_i} = \frac{1}{2\pi R_{\mathrm{Th}_1}C_i}$$

$$= \frac{1}{2\pi(9.9 \ \mathrm{k\Omega})(17 \ \mathrm{pF})} = \mathbf{945.67 \ kHz}$$

$$R_{\mathrm{Th}_2} = R_D\|R_L$$

$$= 4.7 \ \mathrm{k\Omega}\|2.2 \ \mathrm{k\Omega}$$

$$\cong 1.5 \ \mathrm{k\Omega}$$

$$C_o = C_{W_o} + C_{ds} + C_{M_o} = 6 \ \mathrm{pF} + 0.5 \ \mathrm{pF} + \left(1 - \frac{1}{-3}\right)2 \ \mathrm{pF} = 9.17 \ \mathrm{pF}$$

$$f_{H_o} = \frac{1}{2\pi(1.5 \text{ k}\Omega)(9.17 \text{ pF})} = \textbf{11.57 MHz}$$

The results above clearly indicate that the input capacitance with its Miller effect capacitance will determine the upper cutoff frequency. This is typically the case due to the smaller value of C_{ds} and the resistance levels encountered in the output circuit.

(b) The PROBE analysis of Figs. 11.55 and 11.56 clearly supports the results of Example 11.10 and the calculations above.

```
Full frequency response of JFET amplifier - Fig. 11.52
VDD 4 0 20V
RG 3 0 1MEG
RSIG 2 1 10K
RS 6 0 1K
RD 4 5 4.7K
RL 7 0 2.2K
C1 2 3 0.01UF
C2 5 7 0.5UF
CS 6 0 2UF
CGD 3 5 2PF
CGS 3 6 4PF
CDS 5 6 0.5PF
J1 5 3 6 JN
.MODEL JN NJF(VTO=-4V BETA=500E-6)
CW1 3 0 5PF
CW2 7 0 6PF
VSIG 1 0 AC 1MV
.AC DEC 10 10HZ 10MEGHZ
.OP
.PROBE
.OPTIONS NOPAGE
.END
```

Figure 11.55 PSpice full frequency analysis of the JFET amplifier of Fig. 11.52.

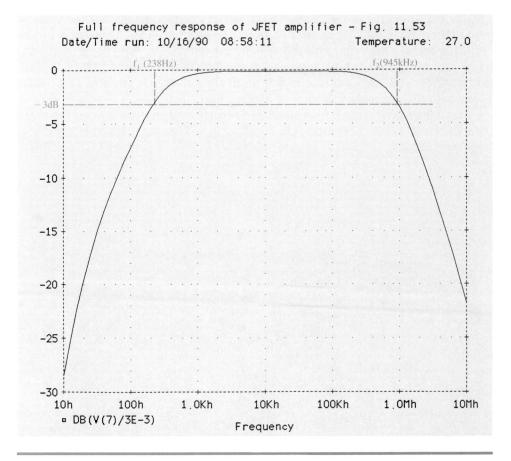

Figure 11.56 Frequency response for the network of Fig. 11.52.

Even though the analysis of the past few sections has been limited to two configurations the exposure to the general procedure for determining the cutoff frequencies should support the analysis of any other transistor configuration. Keep in mind that the Miller capacitance is limited to inverting amplifiers and that f_α is significantly greater than f_β if the common-base configuration is encountered. There is a great deal more literature on the analysis of single-stage amplifiers that goes beyond the coverage of this chapter. However, the content of this chapter should provide a firm foundation for any future analysis of frequency effects.

11.11 MULTISTAGE FREQUENCY EFFECTS

For a second transistor stage connected directly to the output of a first stage there will be a significant change in the overall frequency response. In the high-frequency region the output capacitance C_o must now include the wiring capacitance (C_{W_1}), parasitic capacitance (C_{be}), and Miller capacitance (C_{M_i}), of the following stage. Further, there will be additional low-frequency cutoff levels due to the second stage that will further reduce the overall gain of the system in this region. For each additional stage the upper cutoff frequency will be determined primarily by that stage having the lowest cutoff frequency. The low-frequency cutoff is primarily determined by that stage having the highest low-frequency cutoff frequency. Obviously, therefore, one poorly designed stage can offset an otherwise well-designed cascaded system.

The effect of increasing the number of *identical* stages can be clearly demonstrated by considering the situations indicated in Fig. 11.57. In each case the upper and lower cutoff frequencies of each of the cascaded stages are identical. For a single stage the cutoff frequencies are f_1 and f_2 as indicated. For two identical stages in cascade the drop-off rate in the high- and low-frequency regions has increased to -12 dB/octave or -40 dB/decade. At f_1 and f_2, therefore, the decibel drop is now -6 dB rather than the defined band frequency gain level of -3 dB. The -3-dB point has shifted to f_1' and f_2' as indicated with a resulting drop in the bandwidth. A -18-dB/octave or -60-dB/decade slope will result for a three-stage system of identical stages with the indicated reduction in bandwidth (f_1'' and f_2'').

Assuming identical stages, an equation for each band frequency as a function of the number of stages (n) can be determined in the following manner: For the low-frequency region,

$$A_{v_{\text{low, (overall)}}} = A_{v_{1_{\text{low}}}} A_{v_{2_{\text{low}}}} A_{v_{3_{\text{low}}}} \cdots A_{v_{n_{\text{low}}}}$$

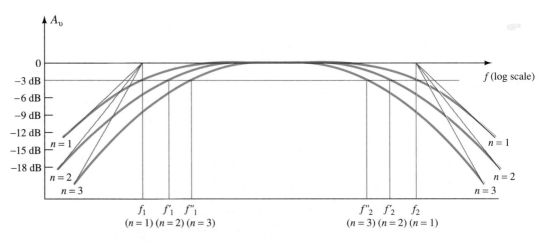

Figure 11.57 Effect of an increased number of stages on the cut-off frequencies and the bandwidth.

but since each stage is identical, $A_{v_{1_{low}}} = A_{v_{2_{low}}} =$ etc. and

$$A_{v_{low, (overall)}} = (A_{v_{1_{low}}})^n$$

or

$$\frac{A_{v_{low}}}{A_{v_{mid}}} (overall) = \left(\frac{A_{v_{1_{low}}}}{A_{v_{mid}}}\right)^n = \frac{1}{(1 - jf_1/f)^n}$$

Setting the magnitude of this result equal to $1/\sqrt{2}(-3$ dB level) results in

$$\frac{1}{[\sqrt{1 + (f_1/f_1')^2}]^n} = \frac{1}{\sqrt{2}}$$

or

$$\left\{\left[1 + \left(\frac{f_1}{f_1'}\right)^2\right]^{1/2}\right\}^n = \left\{\left[1 + \left(\frac{f_1}{f_1'}\right)^2\right]^n\right\}^{1/2} = (2)^{1/2}$$

so that

$$\left[1 + \left(\frac{f_1}{f_1'}\right)^2\right]^n = 2$$

and

$$1 + \left(\frac{f_1}{f_1'}\right)^2 = 2^{1/n}$$

with the result that

$$\boxed{f_1' = \frac{f_1}{\sqrt{2^{1/n} - 1}}}$$
(11.66)

In a similar manner, it can be shown that for the high-frequency region,

$$\boxed{f_2' = (\sqrt{2^{1/n} - 1})f_2}$$
(11.67)

Note the presence of the same factor $\sqrt{2^{1/n} - 1}$ in each equation. The magnitude of this factor for various values of n is listed below.

n	$\sqrt{2^{1/n} - 1}$
1	1
2	0.64
3	0.51
4	0.43
5	0.39

For $n = 2$, consider that the upper cutoff frequency $f_2' = 0.64f_2$ or 64% of the value obtained for a single stage, while $f_1' = (1/0.64)f_1 = 1.56f_1$. For $n = 3$, $f_2' = 0.51f_2$ or approximately $\frac{1}{2}$ the value of a single stage with $f_1' = (1/0.51)f_1 = 1.96f_1$ or approximately *twice* the single-stage value.

For the *RC*-coupled transistor amplifier, if $f_2 = f_\beta$, or if they are close enough in magnitude for both to affect the upper 3-dB frequency, the number of stages must be increased by a factor of 2 when determining f_2' due to the increased number of factors $1/(1 + jf/f_x)$.

A decrease in bandwidth is not always associated with an increase in the number of stages if the midband gain can remain fixed independent of the number of stages. For instance, if a single-stage amplifier produces a gain of 100 with a bandwidth of 10,000 Hz, the resulting gain–bandwidth product is $10^2 \times 10^4 = 10^6$. For a two-stage system the same gain can be obtained by having two stages with a gain of 10 since ($10 \times 10 = 100$). The bandwidth of each stage would then increase by a factor of 10 to 100,000 due to the lower gain requirement and fixed gain–bandwidth product of 10^6. Of course, the design must be such as to permit the increased bandwidth and establish the lower gain level.

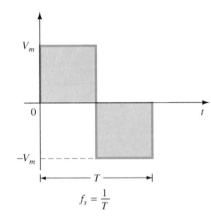

Figure 11.58 Square wave.

A sense for the frequency response of an amplifier can be determined experimentally by applying a square-wave signal to the amplifier and noting the output response. The shape of the output waveform will reveal whether the high or low frequencies are being properly amplified. The use of *square-wave testing* is significantly less time consuming than applying a series of sinusoidal signals at different frequencies and magnitudes to test the frequency response of the amplifier.

The reason for choosing a square-wave signal for the testing process is best described by examining the *Fourier series* expansion of a square wave composed of a series of sinusoidal components of different magnitudes and frequencies. The summation of the terms of the series will result in the original waveform. In other words, even though a waveform may not be sinusoidal, it can be reproduced by a series of sinusoidal terms of different frequencies and magnitudes.

The Fourier series expansion for the square wave of Fig. 11.58 is

$$v = \frac{4}{\pi}V_m\left(\sin 2\pi f_s t + \frac{1}{3}\sin 2\pi(3f_s)t + \frac{1}{5}\sin 2\pi(5f_s)t + \frac{1}{7}\sin 2\pi(7f_s)t \right.$$
$$\left. + \frac{1}{9}\sin 2\pi(9f_s)t + \cdots + \frac{1}{n}\sin 2\pi(nf_s)t\right) \quad (11.68)$$

The first term of the series is called the *fundamental* term and in this case has the same frequency, f_s as the square wave. The next term has a frequency equal to three times the fundamental and is referred to as the *third harmonic*. Its magnitude is one-third the magnitude of the fundamental term. The frequencies of the succeeding terms are odd multiples of the fundamental term and the magnitude decreases with each higher harmonic. Figure 11.59 demonstrates how the summation of terms of a Fourier series can result in a nonsinusoidal waveform. The generation of the square wave of Fig. 11.58 would require an infinite number of terms. However, the summation of just the fundamental term and the third harmonic in Fig. 11.59a clearly results in a waveform that is beginning to take on the appearance of a square wave. Including the fifth and seventh harmonics as in Fig. 11.59b takes us a step closer to the waveform of Fig. 11.58.

Since the ninth harmonic has a magnitude greater than 10% of the fundamental term [$\frac{1}{9}(100\%) = 11.1\%$], the fundamental term through the ninth harmonic are the major contributors to the Fourier series expansion of the square-wave function. It is therefore reasonable to assume that if the application of a square wave of a particular frequency results in a nice clean square wave at the output, then the frequency applied

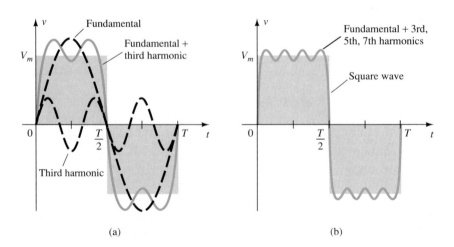

Figure 11.59 Harmonic content of a square wave.

through its ninth harmonic are being amplified without visual distortion by the amplifier. For instance, if an audio amplifier with a bandwidth of 20 kHz (audio range is from 20 Hz to 20 kHz) is to be tested, the frequency of the applied signal should be at least 20 kHz/9 = 2.22 kHz.

If the response of an amplifier to an applied square wave is an undistorted replica of the input, the frequency response (or BW) of the amplifier is obviously sufficient for the applied frequency. If the response is as shown in Fig. 11.60a and b, the low frequencies are not being amplified properly and the low cutoff frequency has to be investigated. If the waveform has the appearance of Fig. 11.60c and, the high-frequency components are not receiving sufficient amplification, and the high cutoff frequency (or BW) has to be reviewed.

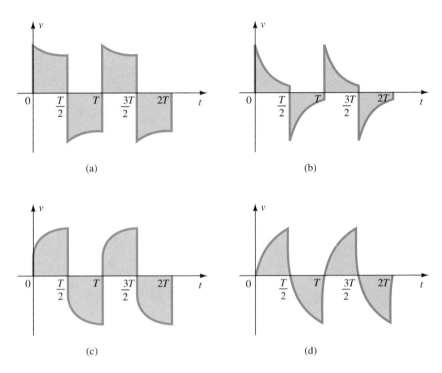

Figure 11.60 (a) Poor low-frequency response; (b) very poor low-frequency response; (c) poor high-frequency response; (d) very poor high-frequency response.

The actual high cutoff frequency (or BW) can be determined from the output waveform by carefully measuring the rise time defined between 10 and 90% of the peak value, as shown in Fig. 11.61. Substituting into the following equation will provide the upper cutoff frequency and since $BW = f_{H_i} - f_{L_o} \cong f_{H_i}$, the equation also provides an indication of the BW of the amplifier.

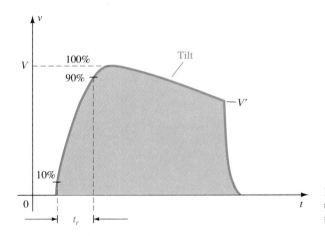

Figure 11.61 Defining the rise time and tilt of a square wave response.

$$\boxed{\text{BW} \cong f_{H_i} = \frac{0.35}{t_r}} \tag{11.69}$$

The low cutoff frequency can be determined from the output response by carefully measuring the tilt of Fig. 11.61 and substituting into one of the following equations:

$$\boxed{\% \text{ tilt} = P\% = \frac{V - V'}{V} \times 100\%} \tag{11.70}$$

$$\boxed{\text{tilt} = P = \frac{V - V'}{V}} \quad \text{(decimal form)} \tag{11.71}$$

The low cutoff frequency is then determined from

$$\boxed{f_{L_o} = \frac{P}{\pi} f_s} \tag{11.72}$$

EXAMPLE 11.13

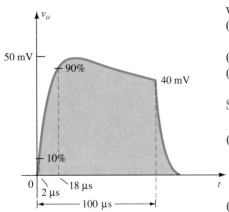

Figure 11.62 Example 11.13

The application of a 1-mV, 5-kHz square wave to an amplifier resulted in the output waveform of Fig. 11.62.
(a) Write the Fourier series expansion for the square wave through the ninth harmonic.
(b) Determine the bandwidth of the amplifier.
(c) Calculate the low cutoff frequency.

Solution

(a) $v_i = \dfrac{4 \text{ mV}}{\pi}\left(\sin 2\pi(5 \times 10^3)t + \dfrac{1}{3}\sin 2\pi(15 \times 10^3)t + \dfrac{1}{5}\sin 2\pi(25 \times 10^3)t\right.$

$\left. + \dfrac{1}{7}\sin 2\pi(35 \times 10^3)t + \dfrac{1}{9}\sin 2\pi(45 \times 10^3)t\right.$

(b) $t_r = 18 \ \mu s - 2 \ \mu s = 16 \ \mu s$

$$\text{BW} = \frac{0.35}{t_r} = \frac{0.35}{16 \ \mu s} = \textbf{21,875 Hz} \cong 4.4 \, f_s$$

(c) $P = \dfrac{V - V'}{V} = \dfrac{50 \text{ mV} - 40 \text{ mV}}{50 \text{ mV}} = 0.25$

$$f_{L_o} = \frac{P}{\pi} f_s = \left(\frac{0.25}{\pi}\right)(5 \text{ kHz}) = \textbf{397.89 Hz}$$

11.13 COMPUTER ANALYSIS

The computer analysis of this chapter was integrated into the chapter for emphasis and a clear demonstration of the power of the PSpice software package. The complete frequency response of a single or multistage system can be determined in a relatively short period of time to verify theoretical calculations or provide an immediate indication of the low and high cutoff frequencies of the system. The exercises in the chapter will provide an opportunity to apply the PSpice software package to a variety of networks.

§ 11.2 Logarithms

1. (a) Determine the common logarithm of the following numbers: 10^3, 50, and 0.707.
   (b) Determine the natural logarithm of the same numbers appearing in part (a).
   (c) Compare the solutions of parts (a) and (b).

2. (a) Determine the common logarithm of the number 2.2×10^3.
   (b) Determine the natural logarithm of the number of part (a) using Eq. (11.4).
   (c) Determine the natural logarithm of the number of part (a) using natural logarithms and compare with the solution of part (b).

3. Determine:
   (a) $20 \log_{10} \frac{40}{8}$ using Eq. (11.6) and compare with $20 \log_{10} 5$.
   (b) $10 \log_{10} \frac{1}{20}$ using Eq. (11.7) and compare with $10 \log_{10} 0.05$.
   (c) $\log_{10}(40)(0.125)$ using Eq. (11.8) and compare with $\log_{10} 5$.

4. Calculate the power gain in decibels for each of the following cases.
   (a) $P_o = 100$ W, $P_i = 5$ W.
   (b) $P_o = 100$ mW, $P_i = 5$ mW.
   (c) $P_o = 100$ μW, $P_i = 20$ μW.

5. Determine G_{dBm} for an output power level of 25 W.

6. Two voltage measurements made across the same resistance are $V_1 = 25$ V and $V_2 = 100$ V. Calculate the power gain in decibels of the second reading over the first reading.

7. Input and output voltage measurements of $V_i = 10$ mV and $V_o = 25$ V are made. What is the voltage gain in decibels?

* 8. (a) The total decibel gain of a three-stage system is 120 dB. Determine the decibel gain of each stage if the second stage has twice the decibel gain of the first and the third has 2.7 times the decibel gain of the first.
   (b) Determine the voltage gain of each stage.

* 9. If the applied ac power to a system is 5 μW at 100 mV and the output power is 48 W, determine:
   (a) The power gain in decibels.
   (b) The voltage gain in decibels if the output impedance is 40 kΩ.
   (c) The input impedance.
   (d) The output voltage.

§ 11.4 General Frequency Considerations

10. Given the characteristics of Fig. 11.63, sketch:
    (a) The normalized gain.
    (b) The normalized dB gain and determine the bandwidth and cutoff frequencies.

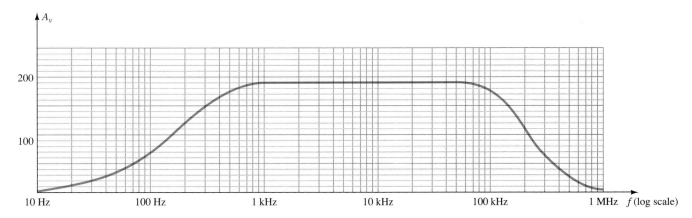

Figure 11.63 Problem 10

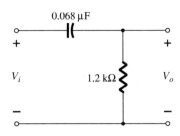

Figure 11.64 Problems 11, 12, 32

11. For the network of Fig. 11.64:
(a) Determine the mathematical expression for the magnitude of the ratio V_o/V_i.
(b) Using the results of part (a), determine V_o/V_i at 100 Hz, 1 kHz, 2 kHz, 5 kHz, and 10 kHz, and plot the resulting curve for the frequency range of 100 Hz to 10 kHz. Use a log scale.
(c) Determine the break frequency.
(d) Sketch the asymptotes and locate the −3-dB point.
(e) Sketch the frequency response for V_o/V_i and compare to the results of part (b).

12. For the network of Fig. 11.64:
(a) Determine the mathematical expression for the angle by which V_o leads V_i.
(b) Determine the phase angle at $f = 100$ Hz, 1 kHz, 2 kHz, 5 kHz, and 10 kHz and plot the resulting curve for the frequency range of 100 Hz to 10 kHz.
(c) Determine the break frequency.
(d) Sketch the frequency response of θ for the same frequency spectrum of part (b) and compare results.

13. (a) What frequency is 1 octave above 5 kHz?
(b) What frequency is 1 decade below 10 kHz?
(c) What frequency is 2 octaves below 20 kHz?
(d) What frequency is 2 decades above 1 kHz?

§ **11.6 Low-Frequency Response—BJT Amplifier**

14. Repeat the analysis of Example 11.9 with $r_o = 40$ kΩ. What is the effect on $A_{v_{mid}}, f_{L_S}, f_{L_C}, f_{L_E}$, and the resulting cutoff frequency.

15. For the network of Fig. 11.65:
(a) Determine r_e.
(b) Find $A_{v_{mid}} = V_o/V_i$.
(c) Calculate Z_i.
(d) Find $A_{v_{s\,mid}} = V_o/V_s$.
(e) Determine f_{L_S}, f_{L_C}, and f_{L_E}.
(f) Determine the low cutoff frequency.
(g) Sketch the asymptotes of the Bode plot defined by the cutoff frequencies of part (e).
(h) Sketch the low-frequency response for the amplifier using the results of part (5).

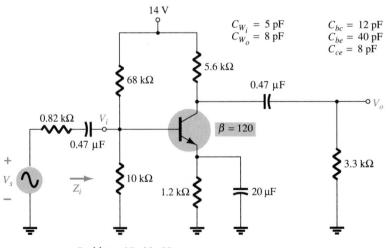

Figure 11.65 Problems 15, 22, 33

Chapter 11 **BJT and JFET Frequency Response**

*   **16.** Repeat Problem 15 for the emitter-stabilized network of Fig. 11.66.

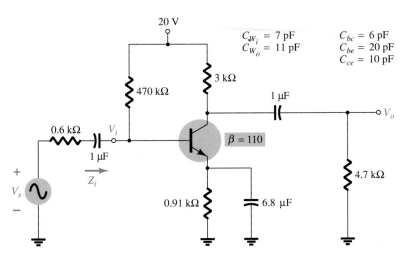

Figure 11.66 Problems 16, 23

*   **17.** Repeat Problem 15 for the emitter-follower network of Fig. 11.67.

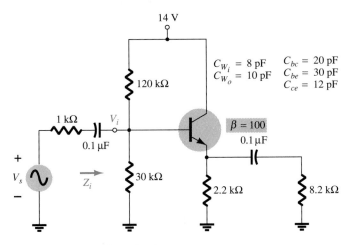

Figure 11.67 Problems 17, 24

*   **18.** Repeat Problem 15 for the common-base configuration of Fig. 11.68. Keep in mind that the common-base configuration is a noninverting network when you consider the Miller effect.

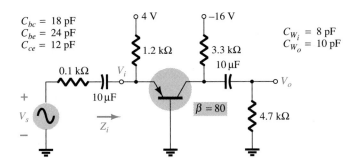

Figure 11.68 Problems 18, 25, 34

19. For the network of Fig. 11.69:

(a) Determine V_{GS_Q} and I_{D_Q}.

(b) Find g_{mo} and g_m.

(c) Calculate the midband gain of $A_v = V_o/V_i$.

(d) Determine Z_i.

(e) Calculate $A_{v_s} = V_o/V_s$.

(f) Determine f_{L_G}, f_{L_C}, and f_{L_S}.

(g) Determine the low cutoff frequency.

(h) Sketch the asymptotes of the Bode plot defined by part (f).

(i) Sketch the low-frequency response for the amplifier using the results of part (f).

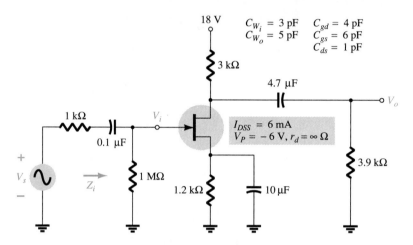

Figure 11.69 Problems 19, 20, 26, 35

* **20.** Repeat the analysis of Problem 19 with $r_d = 100$ kΩ. Does it have any impact of any consequence on the results? If so, which elements?

* **21.** Repeat the analysis of Problem 19 for the network of Fig. 11.70. What effect did the voltage-divider configuration have on the input impedance and the gain A_{v_s} compared to the biasing arrangement of Fig. 11.69?

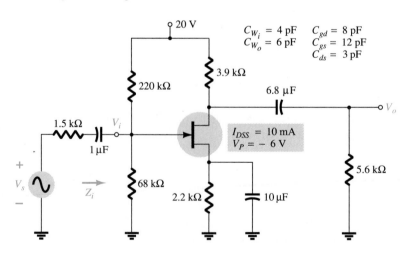

Figure 11.70 Problems 21, 27

§ 11.9 High-Frequency Response—BJT Amplifier

22. For the network of Fig. 11.65:

(a) Determine f_{H_i} and f_{H_o}.

(b) Assuming that $C_{b'e} = C_{be}$ and $C_{b'c} = C_{bc}$, find f_β and f_T.

(c) Sketch the frequency response for the high-frequency region using a Bode plot and determine the cutoff frequency.

* **23.** Repeat the analysis of Problem 22 for the network of Fig. 11.66.

* **24.** Repeat the analysis of Problem 22 for the network of Fig. 11.67.

* **25.** Repeat the analysis of Problem 22 for the network of Fig. 11.68.

§ 11.10 High-Frequency Response—FET Amplifier

26. For the network of Fig. 11.69:
(a) Determine g_{mo} and g_m.
(b) Find A_v and A_{v_s} in the mid-frequency range.
(c) Determine f_{H_i} and f_{H_o}.
(d) Sketch the frequency response for the high-frequency region using a Bode plot and determine the cutoff frequency.

* **27.** Repeat the analysis of Problem 26 for the network of Fig. 11.70.

§ 11.11 Multistage Frequency Effects

28. Calculate the overall voltage gain of four identical stages of an amplifier, each having a gain
of 20.

29. Calculate the overall upper 3-dB frequency for a four-stage amplifier having an individual stage
value of $f_2 = 2.5$ MHz.

30. A four-stage amplifier has a lower 3-dB frequency for an individual stage of $f_1 = 40$ Hz. What
is the value of f_1 for this full amplifier?

§ 11.12 Square-Wave Testing

* **31.** The application of a 10-mV, 100 kHz square wave to an amplifier resulted in the output
waveform of Fig. 11.71.
(a) Write the Fourier series expansion for the square wave through the ninth harmonic.
(b) Determine the bandwidth of the amplifier to the accuracy available by the waveform of
Fig. 11.71.
(c) Calculate the low cutoff frequency.

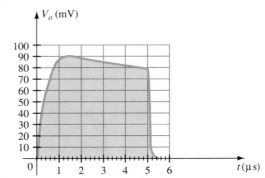

Figure 11.71 Problem 31

§ 11.13 Computer Analysis

32. (a) Write the input file for a PSpice analysis of the frequency response of V_o/V_i for the high-
pass filter of Fig. 11.64.
(b) Perform the analysis of part (a) and compare against the theoretical value of the cutoff
frequency.

33. (a) Write the input file for a PSpice analysis of the frequency response of V_o/V_s for the BJT
amplifier of Fig. 11.65.
(b) Perform the analysis of part (a) and compare against the theoretical solution.

34. Repeat Problem 33 for the network of Fig. 11.68.

35. Repeat Problem 33 for the JFET configuration of Fig. 11.69.

---

*Please Note: Asterisks indicate more difficult problems.

12 Compound Configurations

12.1 INTRODUCTION

In the present chapter we introduce a number of circuit connections which, although not standard common-emitter, common-collector, or common-base, are still quite important, being widely used in either discrete or in integrated circuits. The cascade connection provides stages in series, while the cascode connection places one transistor on top of another. Both these connection forms are found in practical circuits. The Darlington connection and the feedback pair connection provides multiple transistors connected for operation as a single transistor for improved performance, usually with much larger current gain.

The CMOS connection, using both p-type enhancement and n-type enhancement MOSFET transistors in a very low power operating circuit, is introduced in this chapter. Much of the newest digital circuitry uses CMOS circuits to either permit portable operation at very low battery power or to allow very high packing density in integrated circuits with lowest power dissipation in the small space used by an IC chip.

Both discrete circuits and integrated circuits use the current source connection. The current mirror connection provides constant current to various other circuits and is especially important in linear integrated circuits.

The differential amplifier is the basic part of operational amplifier circuits (to be covered fully in Chapter 14). The basic differential circuit connection and its operation is introduced in this chapter. Although placed at the end of the chapter it is nevertheless a most important circuit connection. A bipolar-JFET circuit used in ICs is the BiFET connection, while the bipolar-MOSFET connection is called a BiMOS connection. Both of these are used in linear integrated circuits.

12.2 CASCADE CONNECTION

A popular connection of amplifier stages is the cascade connection. Basically, a cascade connection is a series connection with the output of one stage then applied as input to the second stage. Figure 12.1 shows a cascade connection of two FET amplifier stages. The cascade connection provides a multiplication of the gain of each stage for a larger overall gain.

The gain of the overall cascade amplifier is the product of stage gains A_{v_1} and A_{v_2},

$$A_v = A_{v_1}A_{v_2} = (-g_{m_1}R_{D_1})(-g_{m_2}R_{D_2}) \tag{12.1}$$

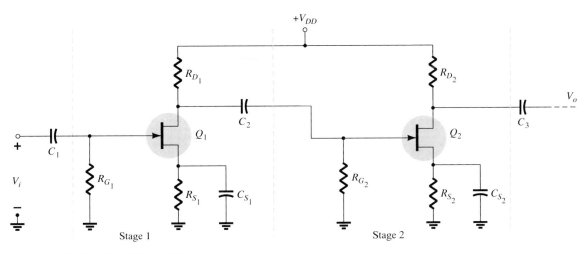

Figure 12.1 Cascaded FET amplifier.

The input impedance of the cascade amplifier is that of stage 1,

$$Z_i = R_{G_1} \tag{12.2}$$

while the output impedance is that of stage 2,

$$Z_o = R_{D_2} \tag{12.3}$$

The main function of cascading stages is the larger overall gain achieved. Since dc bias and ac calculations for a cascade amplifier follow those derived for the individual stages, an example will demonstrate the various calculations to determine dc bias and ac operation.

Calculate the dc bias, voltage gain, input impedance, output impedance, and the resulting output voltage for the cascade amplifier shown in Fig. 12.2. Calculate the load voltage if a 10-kΩ load is connected across the output.

EXAMPLE 12.1

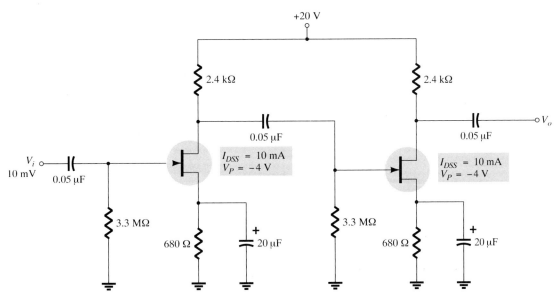

Figure 12.2 Cascade amplifier circuit for Example 12.1.

Solution

Both amplifier stages have the same dc bias. Using dc bias techniques from Chapter 6 results in

$$V_{GS_Q} = -1.9 \text{ V}, \qquad I_{D_Q} = 2.8 \text{ mA}$$

Both transistors have

$$g_{m0} = \frac{2I_{DSS}}{|V_P|} = \frac{2(10 \text{ mA})}{|-4 \text{ V}|} = 5 \text{ mS}$$

and at the dc bias point,

$$g_m = g_{m0}\left(1 - \frac{V_{GS_Q}}{V_P}\right) = (5 \text{ mS})\left(1 - \frac{-1.9 \text{ V}}{-4 \text{ V}}\right) = \textbf{2.6 mS}$$

The voltage gain of each stage is then

$$A_{v_1} = A_{v_2} = -g_m R_D = -(2.6 \text{ mS})(2.4 \text{ k}\Omega) = \textbf{-6.2}$$

The cascade amplifier voltage gain is then

$$\text{Eq. (12.1):} \quad A_v = A_{v_1} A_{v_2} = (-6.2)(-6.2) = 38.4$$

The output voltage is then

$$V_o = A_v V_i = (38.4)(10 \text{ mV}) = 384 \text{ mV}$$

The cascade amplifier input impedance is

$$Z_i = R_G = \textbf{3.3 M}\Omega$$

The cascade amplifier output impedance (assuming that $r_d = \infty$) is

$$Z_o = R_D = \textbf{2.4 k}\Omega$$

The output voltage across a 10-kΩ load would then be

$$V_L = \frac{R_L}{Z_o + R_L} V_o = \frac{10 \text{ k}\Omega}{2.4 \text{ k}\Omega + 10 \text{ k}\Omega}(384 \text{ mV}) = \textbf{310 mV}$$

BJT Cascade Amplifier

An *RC*-coupled cascade amplifier built using BJTs is shown in Fig. 12.3. As before, the advantage of cascading stages is the large overall voltage gain. Dc bias is obtained

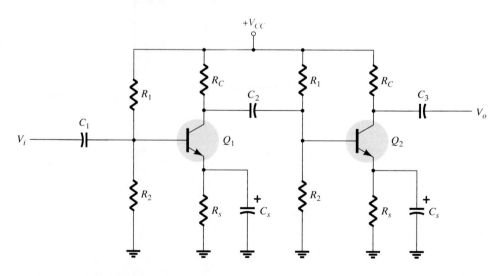

Figure 12.3 Cascaded BJT amplifier (*RC* coupled).

Chapter 12   Compound Configurations

using the procedures of Chapter 4. The voltage gain of each stage is

$$A_v = \frac{-R_C \| R_L}{r_e} \qquad (12.4)$$

The amplifier input impedance is that of stage 1,

$$Z_i = R_1 \| R_2 \| \beta r_e \qquad (12.5)$$

and the output impedance of the amplifier is that of stage 2,

$$Z_o = R_C \| r_o \qquad (12.6)$$

The next example demonstrates the analysis of a cascade BJT amplifier showing the large voltage gain achieved.

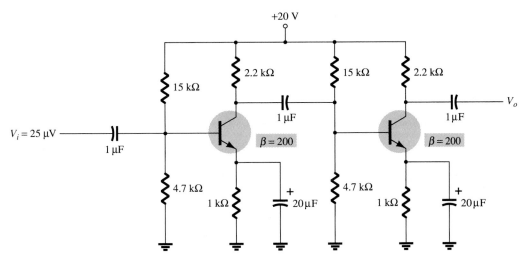

EXAMPLE 12.2

Calculate the voltage gain, output voltage, input impedance, and output impedance for the cascade BJT amplifier of Fig. 12.4. Calculate the output voltage resulting if a 10-kΩ load is connected to the output.

Figure 12.4   *RC-coupled BJT amplifier for Example 12.2.*

Solution

Dc bias analysis results in

$$V_B = 4.8 \text{ V}, \qquad V_E = 4.1 \text{ V}, \qquad V_C = 11 \text{ V}, \qquad I_C = 4.1 \text{ mA}$$

At the bias point,

$$r_e = \frac{26}{I_c} = \frac{26}{4.1} = 6.3 \ \Omega$$

The voltage gain of stage 1 is then

$$A_{v_1} = -\frac{R_C \| (R_1 \| R_2 \| \beta r_e)}{r_e}$$

$$= -\frac{(2.2 \text{ k}\Omega) \| [15 \text{ k}\Omega \| 4.7 \text{ k}\Omega \| (200)(6.3 \ \Omega)]}{6.3 \ \Omega}$$

$$= -\frac{654.6 \ \Omega}{6.3 \ \Omega} = -104$$

while the voltage gain of stage 2 is

$$A_{v_2} = -\frac{R_C}{r_e} = -\frac{2.2 \text{ k}\Omega}{6.3 \ \Omega} = -349$$

for an overall voltage gain of

$$A_v = A_{v_1}A_{v_2} = (-104)(-349) = \mathbf{36{,}296}$$

The output voltage is then

$$V_o = A_v V_i = (36{,}296)(25 \ \mu\text{V}) = \mathbf{0.9 \ V}$$

The amplifier input impedance is

$$Z_i = R_1 \| R_2 \| \beta r_e = 4.7 \text{ k}\Omega \| 15 \text{ k}\Omega \| (200)(6.3 \ \Omega)$$

$$= \mathbf{932 \ \Omega}$$

while the amplifier output impedance is

$$Z_o = R_C = \mathbf{2.2 \ k\Omega}$$

If a 10-kΩ load is connected to the amplifier output, the resulting voltage across the load is

$$V_L = \frac{R_L}{Z_o + R_L}V_o = \frac{10 \text{ k}\Omega}{2.2 \text{ k}\Omega + 10 \text{ k}\Omega}(0.9 \ \text{V}) = \mathbf{0.7 \ V}$$

A combination of FET and BJT stages can also be used to provide high voltage gain and high input impedance, as demonstrated by the next example.

EXAMPLE 12.3

For the cascade amplifier of Fig. 12.5, calculate dc bias and then input impedance, output impedance, voltage gain, and the resulting output voltage.

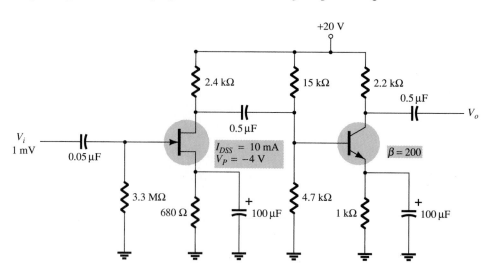

Figure 12.5 Cascaded JFET-BJT amplifier for Example 12.3.

Solution

Since R_i (stage 2) = 15 k$\Omega\|$4.7 k$\Omega\|$200(6.4 Ω) = 943 Ω, the gain of stage 1 (when loaded by stage 2) is

$$A_{v_1} = -g_m[R_D \| R_i(\text{stage 2})]$$

$$= -2.6 \text{ mS}(2.4 \text{ k}\Omega \| 943\Omega) = -1.8$$

530   Chapter 12 Compound Configurations

from Example 12.2, the voltage gain of stage 2 is $A_{v_2} = -349$. The overall voltage gain is then

$$A_v = A_{v_1} A_{v_2} = (-1.8)(-349) = \mathbf{628}$$

The output voltage is then

$$V_o = A_v V_1 = (628)(1 \text{ mV}) = \mathbf{0.63 \text{ V}}$$

The input impedance of the amplifier is that of stage 1,

$$Z_i = \mathbf{3.3 \text{ M}\Omega}$$

while the output impedance is that of stage 2,

$$Z_o = R_D = \mathbf{2.2 \text{ k}\Omega}$$

12.3 CASCODE CONNECTION

A cascode connection has one transistor on top of (in series with) another. Figure 12.6 shows a cascode configuration with a common-emitter (CE) stage feeding a common-base (CB) stage. This arrangement is designed to provide a high input impedance with low voltage gain to ensure that the input Miller capacitance (see Chapter 11) is at a minimum with the CB stage providing good high-frequency operation. A practical BJT version of a cascode amplifier is provided in Fig. 12.7.

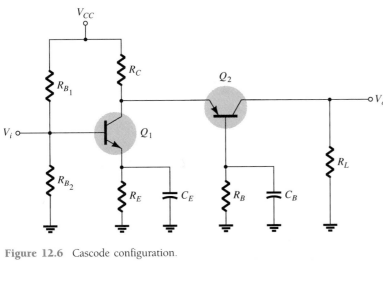

Figure 12.6 Cascode configuration.

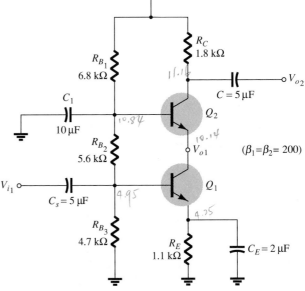

Figure 12.7 Practical cascode circuit for Example 12.4.

EXAMPLE 12.4

Calculate the voltage gain for the cascode amplifier of Fig. 12.7.

Solution

Dc bias analysis using procedures of Chapter 4 result in

$$V_{B_1} = 4.9 \text{ V}, \qquad V_{B_2} = 10.8 \text{ V}, \qquad I_{C_1} \approx I_{C_2} = 3.8 \text{ mA}$$

The dynamic resistance of each transistor is then

$$r_e = \frac{26}{I_E} = \frac{26}{3.8} = 6.8 \ \Omega$$

The voltage gain of stage 1 (common-emitter) is approximately

$$A_{v_1} = -\frac{R_C}{r_e} = -\frac{r_e}{r_e} = -1$$

The voltage gain of stage 2 (common-base) is

$$A_{v_2} = \frac{R_C}{r_e} = \frac{1.8 \text{ k}\Omega}{6.8 \ \Omega} = 265$$

resulting in an overall cascode amplifier gain of

$$A_v = A_{v_1} A_{v_2} = (-1)(265) = \boldsymbol{-265}$$

As expected, the CE stage with a gain of -1 provides the higher input impedance of a CE stage (over that of a CB stage). With a voltage gain of only -1, the Miller input capacitance is kept quite small. A large voltage gain is then provided by the CB stage, resulting in a large overall gain ($A_v = -265$).

12.4 DARLINGTON CONNECTION

A very popular connection of two bipolar junction transistors for operation as one ''superbeta'' transistor is the Darlington connection shown in Fig. 12.8. The main feature of the Darlington connection is that the composite transistor acts as a single unit with a current gain that is the product of the current gains of the individual transistors. If the connection is made using two separate transistors having current gains of β_1 and β_2, the Darlington connection provides a current gain of

$$\beta_D = \beta_1 \beta_2 \tag{12.7}$$

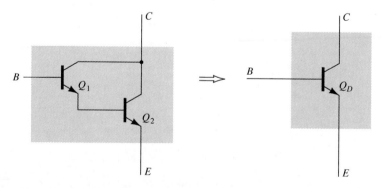

Figure 12.8 Makeup of Darlington transistor.

If the two transistors are matched so that $\beta_1 = \beta_2 = \beta$. The Darlington connection provides a current gain of

$$\beta_D = \beta^2 \qquad (12.8)$$

A Darlington transistor connection provides a transistor having a very large current gain, typically a few thousand.

EXAMPLE 12.5

What current gain is provided by a Darlington connection of two identical transistors each having a current gain of $\beta = 200$?

Solution

Eq. (12.8): $\beta_D = \beta^2 = (200)^2 = \mathbf{40,000}$

Packaged Darlington Transistor

Since the Darlington connection is popular, one can obtain a single package containing two BJTs internally connected as a Darlington transistor. Fig. 12.9 provides some specification sheet data on a typical Darlington pair. The current gain listed is that of the overall Darlington-connected transistor, the device providing externally only three terminals (base, emitter, and collector). One may consider the unit a single Darlington transistor having very high current gain when compared to other typical single transistors.

Type 2N999
N-P-N Darlington-Connected
   Silicon Transistor Package

Parameter	Test Conditions	Min.	Max.
V_{BE}	$I_C = 100$ mA		1.8 V
h_{FE}	$I_C = 10$ mA	4000	
(β_D)	$I_C = 100$ mA	7000	70,000

Figure 12.9 Specification information on Darlington transistor package (2N999).

DC Bias of Darlington Circuit

A basic Darlington circuit is shown in Fig. 12.10. A Darlington transistor having very high current gain, β_D, is used. The base current may be calculated from

$$I_B = \frac{V_{CC} - V_{BE}}{R_B + \beta_D R_E} \qquad (12.9)$$

While this equation is the same as for a regular transistor, the value of β_D is much greater, and the value of V_{BE} is larger, as indicated by the data in the spec sheet of Fig. 12.9. The emitter current is then

$$I_E = (\beta_D + 1)I_B \approx \beta_D I_B \qquad (12.10)$$

Dc voltages are

$$V_E = I_E R_E \qquad (12.11)$$

$$V_B = V_E + V_{BE} \qquad (12.12)$$

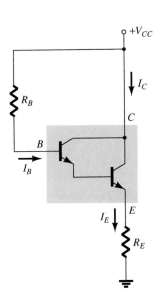

Figure 12.10 Basic Darlington bias circuit.

Calculate the dc bias voltages and currents in the circuit of Fig. 12.11.

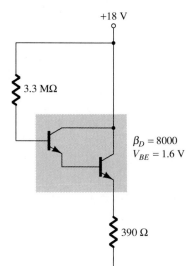

Figure 12.11 Circuit for Example 12.6.

Solution

The base current is

$$\text{Eq. (12.9):} \quad I_B = \frac{18 \text{ V} - 1.6 \text{ V}}{3.3 \text{ M}\Omega + 8000(390 \text{ }\Omega)} \approx \mathbf{2.56 \text{ }\mu A}$$

The emitter current is then

$$\text{Eq. (12.10):} \quad I_E \approx 8000(2.56 \text{ }\mu A) = \mathbf{20.48 \text{ mA}} \approx I_C$$

The emitter dc voltage is

$$\text{Eq. (12.11):} \quad V_E = 20.48 \text{ mA}(390 \text{ }\Omega) \approx \mathbf{8 \text{ V}}$$

and the base voltage is

$$\text{Eq. (12.12):} \quad V_B = 8 \text{ V} + 1.6 \text{ V} = \mathbf{9.6 \text{ V}}$$

The collector voltage is the supply value of

$$V_C = \mathbf{18 \text{ V}}$$

AC Equivalent Circuit

A Darlington emitter-follower circuit is shown in Fig. 12.12. The ac input signal is applied to the base of the Darlington transistor through capacitor C_1, with the ac output, V_o, obtained from the emitter through capacitor C_2. An ac equivalent circuit is drawn in Fig. 12.13. The Darlington transistor is replaced by an ac equivalent circuit comprised of an input resistance, r_i, and an output current source, $\beta_D I_b$.

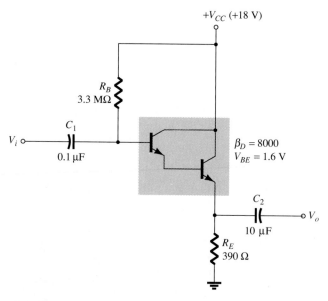

Figure 12.12 Darlington emitter-follower circuit.

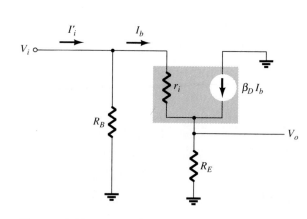

Figure 12.13 Ac equivalent circuit of Darlington emitter follower.

Chapter 12 Compound Configurations

AC INPUT IMPEDANCE

The ac base current through r_i is

$$I_b = \frac{V_i - V_o}{r_i} \qquad (12.13)$$

Since

$$V_o = (I_b + \beta_D I_b)R_E \qquad (12.14)$$

We can use Eq. (12.13) in Eq. (12.14) to obtain

$$I_b r_i = V_i - V_o = V_i - I_b(1 + \beta_D)R_E$$

Solving for V_i,

$$V_i = I_b[r_i + (1 + \beta_D)R_E] \approx I_b(r_i + \beta_D R_E)$$

The ac input impedance looking into the transistor base is then

$$\frac{V_i}{I_b} = r_i + \beta_D R_E$$

and that looking into the circuit is

$$Z_i = R_B \| (r_i + \beta_D R_E) \qquad (12.15)$$

Calculate the input impedance of the circuit of Fig. 12.12 if $r_i = 5$ kΩ.

EXAMPLE 12.7

Solution

Eq. (12.15): $Z_i = 3.3\ \text{M}\Omega \| [5\ \text{k}\Omega + (8000)(390\ \Omega)] = \mathbf{1.6\ M\Omega}$

AC CURRENT GAIN

The ac output current through R_E is (see Fig. 12.13)

$$I_o = I_b + \beta_D I_b = (\beta_D + 1)I_b \approx \beta_D I_b$$

The transistor current gain is then

$$\frac{I_o}{I_b} = \beta_D$$

The ac current gain of the circuit is

$$A_i = \frac{I_o}{I_i} = \frac{I_o}{I_b}\frac{I_b}{I_i}$$

We can use the current-divider rule to express I_b/I_i:

$$I_b = \frac{R_B}{(r_i + \beta_D R_E) + R_B}I_i \approx \frac{R_B}{R_B + \beta_D R_E}I_i$$

so that the ac circuit current gain is

$$A_i = \beta_D \frac{R_B}{R_B + \beta_D R_E} = \frac{\beta_D R_B}{R_B + \beta_D R_E} \qquad (12.16)$$

EXAMPLE 12.8

Calculate the ac current gain of the circuit in Fig. 12.12.

Solution

Eq. (12.16): $A_i = \dfrac{\beta_D R_B}{R_B + \beta_D R_E} = \dfrac{(8000)(3.3 \text{ M}\Omega)}{3.3 \text{ M}\Omega + (8000)(390 \text{ }\Omega)} = \textbf{4112}$

AC OUTPUT IMPEDANCE

The ac output impedance can be determined for the ac circuit shown in Fig. 12.14a. The output impedance seen by load R_L is determined by applying a voltage V_o and measuring the current I_o (with input V_s set to zero). Figure 12.14b shows this situation. Solving for I_o yields

$$I_o = \frac{V_o}{R_E} + \frac{V_o}{r_i} - \beta_D I_b = \frac{V_o}{R_E} + \frac{V_o}{r_i} - \beta_D\left(\frac{V_o}{r_i}\right)$$

$$= \left(\frac{1}{R_E} + \frac{1}{r_i} + \frac{\beta_D}{r_i}\right)V_o$$

Solving for Z_o gives

$$Z_o = \frac{V_o}{I_o} = \frac{1}{1/R_E + 1/r_i + \beta_D/r_i}$$

$$= R_E\|r_i\|\frac{r_i}{\beta_D} \approx \frac{r_i}{\beta_D} \tag{12.17}$$

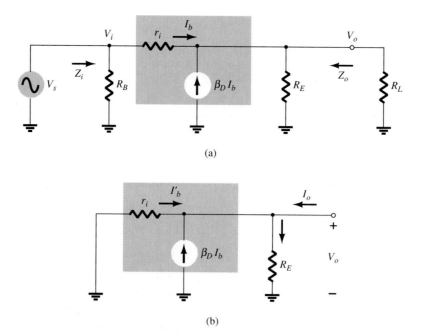

(a)

(b)

Figure 12.14 Ac equivalent circuit to determine z_o.

Calculate the output impedance of the circuit in Fig. 12.12.

EXAMPLE 12.9

Solution

$$\text{Eq. (12.17):} \quad Z_o = 390 \ \Omega \| 5 \ k\Omega \| \frac{5 \ k\Omega}{8000} \approx \frac{5 \ k\Omega}{8000} = \mathbf{0.625 \ \Omega}$$

AC VOLTAGE GAIN

The ac voltage gain for the circuit of Fig. 12.12 can be determined using the ac equivalent circuit of Fig. 12.15. Since

$$V_o = (I_b + \beta_D I_b)R_E = I_b(R_E + \beta_D R_E)$$

and
$$V_i = I_b r_i + (I_b + \beta_D I_b)R_E$$

from which we obtain

$$V_i = I_b(r_i + R_E + \beta_D R_E)$$

so that
$$V_o = \frac{V_i}{r_i + R_E + \beta_D R_E}(R_E + \beta_D R_E)$$

$$A_v = \frac{V_o}{V_i} = \frac{R_E + \beta_D R_E}{r_i + (R_E + \beta_D R_E)} \approx 1 \tag{12.18}$$

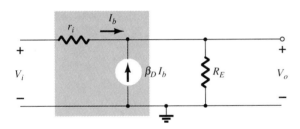

Figure 12.15 Ac equivalent circuit to determine A_v.

Calculate the ac voltage gain A_v for the circuit of Fig. 12.12.

EXAMPLE 12.10

Solution

$$A_v = \frac{390 \ \Omega + (8000)(390 \ \Omega)}{5 \ k\Omega + [390 \ \Omega + (8000)(390 \ \Omega)]} = \mathbf{0.998}$$

12.5 FEEDBACK PAIR

The feedback pair connection (see Fig. 12.16) is a two-transistor circuit that operates like the Darlington circuit. Notice that the feedback pair uses a *pnp* transistor driving an *npn* transistor, the two devices acting effectively much like one *pnp* transistor. As with a Darlington connection the feedback pair provides very high current gain (the product of the transistor current gains). A typical application (see Chapter 16) uses a Darlington connection and a feedback pair connection to provide complementary transistor operation. A practical circuit using a feedback pair is provided in Fig. 12.17. Some consideration of the dc bias and ac operation will provide better understanding of how the connection works.

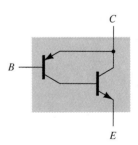

Figure 12.16 Feedback pair connection.

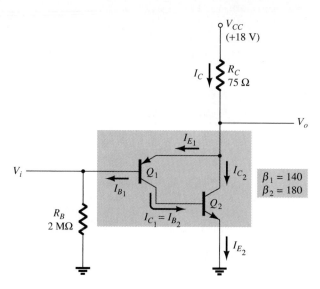

Figure 12.17 Operation of feedback pair.

DC Bias

The dc bias calculations that follow use practical simplifications wherever possible to provide simpler results. From the Q_1 base-emitter loop, one obtains

$$V_{CC} - I_C R_C - V_{EB_1} - I_{B_1} R_B = 0$$

$$V_{CC} - \beta_1 \beta_2 I_{B_1} R_C - V_{EB_1} - I_{B_1} R_B = 0$$

The base current is then

$$I_{B_1} = \frac{V_{CC} - V_{EB_1}}{R_B + \beta_1 \beta_2 R_C} \tag{12.19}$$

The collector current of Q_1 is

$$I_{C_1} = \beta_1 I_{B_1} = I_{B_2}$$

which is also the base Q_2 current. The transistor Q_2 collector current is

$$I_{C_2} = \beta_2 I_{B_2} \approx I_{E_2}$$

so that the current through R_C is

$$I_C = I_{E_1} + I_{C_2} \approx I_{C_1} + I_{C_2} \approx I_{C_2} \tag{12.20}$$

EXAMPLE 12.11

Calculate the dc bias currents and voltages for the circuit of Fig. 12.17 to provide V_o at one-half the supply voltage ($I_C R_C = 9$ V).

Solution

$$I_{B_1} = \frac{18 \text{ V} - 0.7 \text{ V}}{2 \text{ M}\Omega + (140)(180)(75 \text{ }\Omega)} = \frac{17.3 \text{ V}}{3.89 \times 10^6} = 4.45 \text{ }\mu\text{A}$$

The base Q_2 current is then

$$I_{B_2} = I_{C_1} = \beta_1 I_{B_1} = 140(4.45 \text{ }\mu\text{A}) = 0.623 \text{ mA}$$

resulting in a Q_2 collector current of

$$I_{C_2} = \beta_2 I_{B_2} = 180(0.623 \text{ mA}) = 112.1 \text{ mA}$$

and the current through R_C is then

Eq. (12.20): $I_C = I_{E_1} + I_{C_2} = 0.623$ mA $+ 112.1$ mA $\approx I_{C_2} = 112.1$ mA

The dc voltage at the output is thus

$$V_o(dc) = V_{CC} - I_C R_C = 18 \text{ V} - 112.1 \text{ mA}(75 \text{ }\Omega) = 9.6 \text{ V}$$

and $$V_i(dc) = V_o(dc) - V_{BE} = 9.6 \text{ V} - 0.7 \text{ V} = 8.9 \text{ V}$$

AC Operation

The ac equivalent circuit for that of Fig. 12.17 is drawn in Fig. 12.18. The circuit is first drawn in Fig. 12.18a to show clearly each transistor and the base and collector resistor placement. The ac equivalent circuit is then redrawn in Fig. 12.18b to permit analysis.

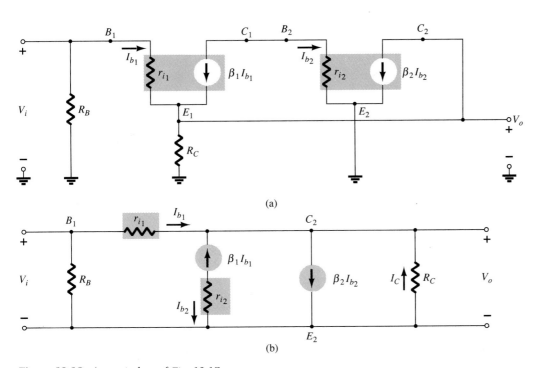

(a)

(b)

Figure 12.18 Ac equivalent of Fig. 12.17.

AC INPUT IMPEDANCE, Z_I

The ac input impedance seen looking into the base of transistor Q_1 is determined (refer to Fig. 12.18b) as follows:

$$I_{b_1} = \frac{V_i - V_o}{r_{i_1}}$$

where $$V_o = -I_C R_C \approx (-\beta_1 I_{b_1} + \beta_2 I_{b_2})R_C \approx (\beta_2 I_{b_2})R_C$$

so that $$I_{b_1} r_{i_1} = V_i - V_o \approx V_i - \beta_2 I_{b_2} R_C$$

$$I_{b_1} r_{i_1} + \beta_2(\beta_1 I_{b_1})R_C = V_i \qquad (\text{since } I_{b_2} = I_{C_1} = \beta_1 I_{b_1})$$

$$\frac{V_i}{I_{b_1}} = r_{i_1} + \beta_1 \beta_2 R_C$$

Including the base-bias resistance,

$$Z_i \approx R_B \| (r_{i_1} + \beta_1 \beta_2 R_C) \qquad (12.21)$$

AC CURRENT GAIN, A_I

The ac current gain can be determined as follows:

$$I_o = \beta_2 I_{b_2} - \beta_1 I_{b_1} - I_{b_1}$$
$$= \beta_2(\beta_1 I_{b_1}) - (1 + \beta_1)I_{b_1} \approx \beta_1 \beta_2 I_{b_1}$$

$$\frac{I_o}{I_{b_1}} = \beta_1 \beta_2$$

Including R_B, the current gain is

$$A_i = \frac{I_o}{I_i} = \frac{I_o}{I_{b_1}} \frac{I_{b_1}}{I_i} = \beta_1 \beta_2 \frac{R_B}{R_B + Z_i}$$

AC OUTPUT IMPEDANCE, Z_O

Z_o can be obtained by applying a voltage, V_o with V_i set to 0. The resulting analysis provides that

$$Z_o = \frac{V_o}{I_o} = R_C \| r_{i_1} \| \frac{r_{i_1}}{\beta_1} \| \frac{r_{i_1}}{\beta_1 \beta_2} \approx \frac{r_{i_1}}{\beta_1 \beta_2}$$

which results in a low output impedance.

AC VOLTAGE GAIN, A_V

The output voltage V_o is

$$V_o = -I_C R_C \approx \beta_1 \beta_2 I_{b_1} R_C$$

Since

$$I_{b_1} = \frac{V_i - V_o}{r_{i_1}}$$

$$V_o = V_i - I_{b_1} r_{i_1} = V_i - \frac{V_o}{\beta_1 \beta_2 R_C} r_{i_1}$$

$$A_v = \frac{V_o}{V_i} = \frac{1}{1 + r_{i_1}/(\beta_1 \beta_2 R_C)} = \frac{\beta_1 \beta_2 R_C}{\beta_1 \beta_2 R_C + r_{i_1}} \qquad (12.22)$$

EXAMPLE 12.12 Calculate the ac circuit values of Z_i, Z_o, A_i, and A_v for the circuit of Fig. 12.17. Assume that $r_{i_1} = 3$ kΩ.

Solution

$$Z_i \approx R_B \| (r_{i_1} + \beta_1 \beta_2 R_C) = 2 \text{ M}\Omega \| [3 \text{ k}\Omega + (140)(180)(75 \text{ }\Omega)]$$
$$\approx \mathbf{974 \text{ k}\Omega}$$

$$A_i = \beta_1 \beta_2 \frac{R_B}{R_B + Z_i} = (140)(180)\left(\frac{2 \text{ M}\Omega}{2 \text{ M}\Omega + 974 \text{ k}\Omega}\right)$$
$$= \mathbf{3.7 \times 10^6}$$

$$Z_o \approx \frac{r_{i_1}}{\beta_1 \beta_2} = \frac{3 \times 10^3}{(140)(180)} = \mathbf{0.12 \ \Omega}$$

and
$$A_v = \frac{\beta_1 \beta_2 R_C}{\beta_1 \beta_2 R_C + r_{i_1}} = \frac{(140)(180)(75 \ \Omega)}{(140)(180)(75 \ \Omega) + 3000 \ \Omega}$$

$$= \mathbf{0.9984 \approx 1}$$

Example 12.12 shows that the feedback pair connection provides operation with voltage gain very near 1 (just as with a Darlington emitter follower), a very high current gain, a very low output impedance, and a high input impedance.

12.6 CMOS CIRCUIT

A form of circuit popular in digital circuitry uses both n-channel and p-channel enhancement MOSFET transistors (see Fig. 12.19). This complementary MOSFET or CMOS circuit uses these opposite (or complementary)-type transistors. The input, V_i, is applied to both gates with the output taken from the connected drains. Before going into the operation of the CMOS circuit let's review the operation of the enhancement MOSFET transistors.

nMOS On/Off Operation

The drain characteristic of an n-channel enhancement MOSFET or nMOS transistor is shown in Fig. 12.20a. With 0 V applied to the gate–source, there is no drain current. Not until V_{GS} is raised past the device threshold level, V_T, does any current result. With an input of, say, +5 V, the nMOS device is fully on with current I_D present. In summary:

An input of 0 V leaves the n*MOS off, while an input of +5 V turns the* n*MOS on.*

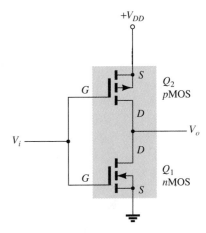

Figure 12.19 CMOS inverter circuit.

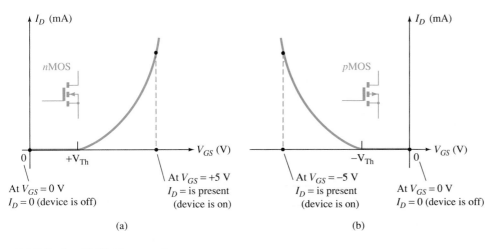

(a)   (b)

Figure 12.20 Enhancement MOSFET characteristic showing off and on conditions: (a) nMOS; (b) pMOS.

pMOS On/Off Operation

The drain characteristic for a p-channel MOSFET or pMOS transistor is shown in Fig. 12.20b. When 0 V is applied, the device is off (no drain current present), while for an input of -5 V (greater than the threshold voltage), the device is on with drain current present. In summary:

$V_{GS} = 0$ *V leaves p*MOS off; $V_{GS} = -5$ *V turns p*MOS on.*

Operation of CMOS Circuit

Consider next how the actual CMOS circuit of Fig. 12.21 operates for input of 0 V or input of +5 V.

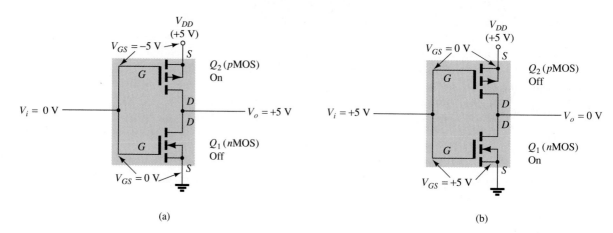

Figure 12.21 Operation of CMOS circuit: (a) output +5 V; (b) output 0 V.

0-V INPUT

When 0 V is applied as input to the CMOS circuit it provides 0 V to both nMOS and pMOS gates. Figure 12.21a shows that

$$\text{For } n\text{MOS } (Q_1): \quad V_{GS} = V_i - 0 \text{ V} = 0 \text{ V} - 0 \text{ V} = 0 \text{ V}$$

$$\text{For } p\text{MOS } (Q_2): \quad V_{GS} = V_i - (+5 \text{ V}) = 0 \text{ V} - 5 \text{ V} = -5 \text{ V}$$

Input of 0 V to an nMOS transistor Q_1 leaves that device off. The same 0-V input, however, results in the gate–source voltage of pMOS transistor Q_2 being -5 V (gate at 0 V is 5 V less than source at +5 V), resulting in that device turning on. The output, V_o, is then +5 V.

+5-V INPUT

When $V_i = +5$ V it provides +5 V to both gates. Figure 12.21b shows that

$$\text{For } n\text{MOS } (Q_1): \quad V_{GS} = V_i - 0 \text{ V} = +5 \text{ V} - 0 \text{ V} = +5 \text{ V}$$

$$\text{For } p\text{MOS } (Q_2): \quad V_{GS} = V_i - (+5 \text{ V}) = +5 \text{ V} - 5 \text{ V} = 0 \text{ V}$$

This input results in transistor Q_1 being turned on and transistor Q_2 remaining off, the output then near 0 V, through conducting transistor Q_2. The CMOS connection of Fig. 12.19 provides operation as a logic inverter with V_o the opposite of V_i, as shown in Table 12.1.

TABLE 12.1 Operation of CMOS Circuit

V_i(V)	Q_1	Q_2	V_o(V)
0	Off	On	+5
+5	On	Off	0

12.7 CURRENT SOURCE CIRCUITS

The concept of a power supply provides a start in our consideration of current source circuits. A practical voltage source (see Fig. 12.22a) is a voltage supply in series with a resistance. An ideal voltage source has $R = 0$, while a practical source includes some small resistance. A practical current source (see Fig. 12.22b) is a current supply in parallel with a resistance. An ideal current source has $R = \infty$, while a practical current source includes some very large resistance.

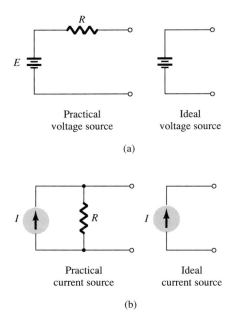

Practical voltage source   Ideal voltage source

(a)

Practical current source   Ideal current source

(b)

Figure 12.22 Voltage and current sources.

An ideal current source provides a constant current regardless of the load connected to it. There are many uses in electronics for a circuit providing a constant current at a very high impedance. Constant-current circuits can be built using FET devices, bipolar devices, and a combination of these components. There are circuits used in discrete form and others more suitable for operation in integrated circuits. We consider some forms of both types in this section and in Section 12.8.

JFET Current Source

A simple JFET current source is that of Fig. 12.23. With V_{GS} set to 0 V, the drain current is fixed at

$$I_D = I_{DSS} = 10 \text{ mA}$$

The device therefore operates like a current source of value 10 mA. While the actual JFET does have an output resistance, the ideal current source would be a 10-mA supply, as shown in Fig. 12.23.

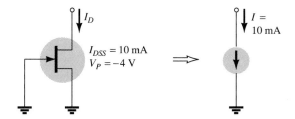

Figure 12.23 JFET constant-current source.

EXAMPLE 12.13

Determine the load current I_D and output voltage V_o for the circuit of Fig. 12.24 for:
(a) $R_D = 1.2 \text{ k}\Omega$
(b) $R_D = 3.3 \text{ k}\Omega$.

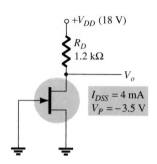

Figure 12.24 JFET current source for Example 12.13.

Solution

Since $V_{GS} = 0$ V, $I_D = I_{DSS} = $ **4 mA.**
(a) $V_o = V_{DD} - I_D R_D = 18 \text{ V} - (4 \text{ mA})(1.2 \text{ k}\Omega) = $ **13.2 V**
(b) $V_o = V_{DD} - I_D R_D = 18 \text{ V} - (4 \text{ mA})(3.3 \text{ k}\Omega) = $ **4.8 V**
Notice that the output voltage changes with R_D, but the current through R_D remains 4 mA, since the JFET operates as a constant-current source.

Bipolar Transistor Constant-Current Source

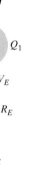

Figure 12.25 Discrete constant-current source.

Bipolar transistors can be connected in a circuit which acts as a constant-current source in a number of ways. Figure 12.25 shows a circuit using a few resistors and an *npn* transistor for operation as a constant-current circuit. The current through I_E can be determined as follows. Assuming that the base input impedance is much larger than R_1 or R_2,

$$V_B = \frac{R_1}{R_1 + R_2}(-V_{EE})$$

and

$$V_E = V_B - 0.7 \text{ V}$$

with

$$I_E = \frac{V_E - (-V_{EE})}{R_E} \approx I_C \qquad (12.23)$$

where I_C is the constant current provided by the circuit of Fig. 12.25.

EXAMPLE 12.14

Calculate the constant current I_C in the circuit of Fig. 12.26.

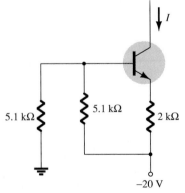

Figure 12.26 Constant-current source for Example 12.14.

Solution

$$V_B = \frac{R_1}{R_1 + R_2}(-V_{EE}) = \frac{5.1 \text{ k}\Omega}{5.1 \text{ k}\Omega + 5.1 \text{ k}\Omega}(-20 \text{ V}) = -10 \text{ V}$$

$$V_E = V_B - 0.7 \text{ V} = -10 \text{ V} - 0.7 \text{ V} = -10.7 \text{ V}$$

$$I = I_E = \frac{V_E - (-V_{EE})}{R_E} = \frac{-10.7 \text{ V} - (-20 \text{ V})}{2 \text{ k}\Omega}$$

$$= \frac{9.3 \text{ V}}{2 \text{ k}\Omega} = \textbf{4.65 mA}$$

Transistor/Zener Constant-Current Source

Replacing resistor R_2 with a Zener diode, as shown in Fig. 12.27, provides an improved constant-current source over that of Fig. 12.25. The Zener diode results in a constant current calculated using the base-emitter KVL equation. The value of I can be calculated using

$$I \approx I_E = \frac{V_Z - V_{BE}}{R_E} \tag{12.24}$$

A major point to consider is that the constant current depends on the Zener diode voltage, which remains quite constant and the emitter resistor R_E. The voltage supply V_{EE} has no effect on the value of I.

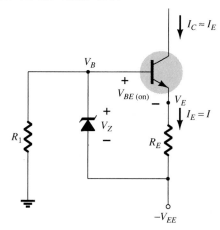

Figure 12.27 Constant-current circuit using Zener diode.

Calculate the constant current I in the circuit of Fig. 12.28.

EXAMPLE 12.15

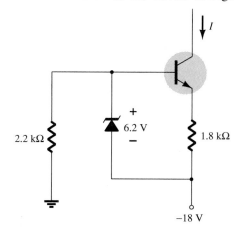

Figure 12.28 Constant-current circuit for Example 12.15.

Solution

Eq. (12.24): $I = \dfrac{V_Z - V_{BE}}{R_E} = \dfrac{6.2\ \text{V} - 0.7\ \text{V}}{1.8\ \text{k}\Omega} = 3.06\ \text{mA} \approx \mathbf{3\ mA}$

12.8 CURRENT MIRROR CIRCUITS

A current mirror circuit (see Fig. 12.29) provides a constant current and is used primarily in integrated circuits. The constant current is obtained from an output current which is the reflection or mirror of a constant current developed on one side of

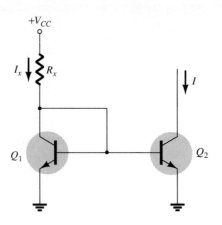

Figure 12.29 Current mirror circuit.

the circuit. The circuit is particularly suited to IC manufacture since the circuit requires that the transistors used have identical base-emitter voltage drops and identical values of beta—results best achieved when transistors are formed at the same time in IC manufacture. In Fig. 12.29 the current I_X, set by transistor Q_1 and resistor R_X is mirrored in the current I through transistor Q_2.

The currents I_X and I can be obtained using the circuit currents listed in Fig. 12.30. We assume that the emitter current (I_E) for both transistors are the same (Q_1 and Q_2 being fabricated near each other on the same chip). The two transistor base currents are then approximately

$$I_B = \frac{I_E}{\beta + 1} \approx \frac{I_E}{\beta}$$

The collector current of each transistor is then

$$I_C \approx I_E$$

Finally, the current through resistor R_X, I_X is

$$I_X = I_E + \frac{2I_E}{\beta} = \frac{\beta I_E}{\beta} + \frac{2I_E}{\beta} = \frac{\beta + 2}{\beta} I_E \approx I_E$$

In summary, the constant current provided at the collector of Q_2 mirrors that of Q_1. Since

$$I_X = \frac{V_{CC} - V_{BE}}{R_X} \tag{12.25}$$

the current I_X set by V_{CC} and R_X is mirrored in the current into the collector of Q_2.

Transistor Q_1 is referred to as a diode-connected transistor because the base and collector are shorted together.

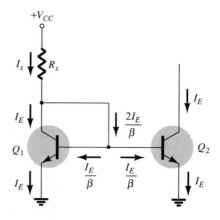

Figure 12.30 Circuit currents for current-mirror circuit.

Calculate the mirrored current, I, in the circuit of Fig. 12.31.

EXAMPLE 12.16

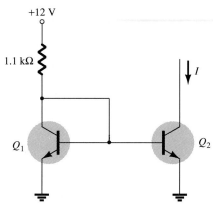

Figure 12.31   Current mirror circuit for Example 12.16.

Solution

Eq. (12.25): $I = I_X = \dfrac{V_{CC} - V_{BE}}{R_X} = \dfrac{12 \text{ V} - 0.7 \text{ V}}{1.1 \text{ k}\Omega} = \mathbf{10.27 \text{ mA}}$

Calculate the current, I, through each of the transistors Q_2 and Q_3 in the circuit of Fig. 12.32.

EXAMPLE 12.17

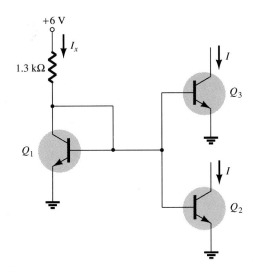

Figure 12.32   Current mirror circuit for Example 12.17.

Solution

The current I_X is

$$I_X = I_E + \frac{3I_E}{\beta} = \frac{\beta + 3}{\beta} I_E \approx I_E$$

Therefore,

$$I \approx I_X = \frac{V_{CC} - V_{BE}}{R_X} = \frac{6 \text{ V} - 0.7 \text{ V}}{1.3 \text{ k}\Omega} = \mathbf{4.08 \text{ mA}}$$

Figure 12.33 shows another form of current mirror to provide higher output impedance than that of Fig. 12.29. The current through R_X is

$$I_X = \frac{V_{CC} - 2V_{BE}}{R_X} \approx I_E + \frac{I_E}{\beta} = \frac{\beta + 1}{\beta} I_E \approx I_E$$

Assuming that Q_1 and Q_2 are well matched, the output current, I, is held constant at

$$I \approx I_E = I_X$$

Again we see that the output current I is a mirrored value of the current set by the fixed current through R_X.

Figure 12.34 shows still another form of current mirror. The JFET provides a constant-current set at the value of I_{DSS}. This current is mirrored, resulting in a current through Q_2 of the same value:

$$I = I_{DSS}$$

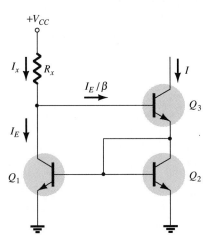

Figure 12.33 Current mirror circuit with higher output impedance.

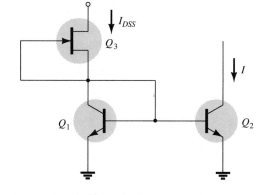

Figure 12.34 Current mirror connection.

12.9 DIFFERENTIAL AMPLIFIER CIRCUIT

The differential amplifier circuit is an extremely popular connection used in IC units. This connection can be described by considering the basic differential amplifier shown in Fig. 12.35. Notice that the circuit has two separate inputs, two separate outputs, and that the emitters are connected together. While most differential amplifier circuits use two separate voltage supplies, the circuit can also operate using a single supply.

A number of input signal combinations are possible.

If an input signal is applied to either input, with the other input connected to ground, the operation is referred to as *single-ended*.

If two opposite polarity input signals are applied, the operation is referred to as *double-ended*.

If the same input is applied to both inputs, the operation is called *common-mode*.

In single-ended operation a single input signal is applied. However, due to the common-emitter connection the input signal operates both transistors, resulting in output from *both* collectors.

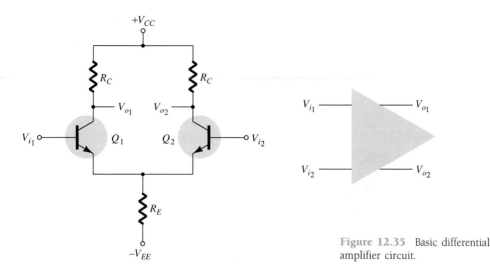

Figure 12.35 Basic differential amplifier circuit.

In double-ended operation two input signals are applied, the difference of the inputs resulting in outputs from both collectors due to the difference of the signals applied to both inputs.

In common-mode operation the common input signal results in opposite signals at each collector, these signals canceling so that the resulting output signal is zero. As a practical matter the opposite signals do not completely cancel and a small signal results.

The main feature of the differential amplifier is the very large gain when opposite signals are applied to the inputs as compared to the very small gain resulting from common inputs. The ratio of this difference gain to the common gain is called *common-mode rejection*. These concepts are discussed fully in Chapter 14. For the present, the operation of the differential amplifier circuit will be fully covered.

DC Bias

Let's first consider the dc bias operation of the circuit of Fig. 12.35. With ac inputs obtained from voltage sources, the dc voltage at each input is essentially connected to 0 V, as shown in Fig. 12.36. With each base voltage at 0 V, the common-emitter dc bias voltage is

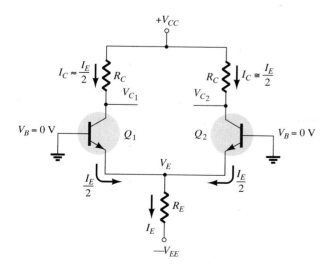

Figure 12.36 Dc bias of difference amplifier circuit.

$$V_E = 0 \text{ V} - V_{BE} = -0.7 \text{ V}$$

The emitter dc bias current is then

$$I_E = \frac{V_E - (-V_{EE})}{R_E} \approx \frac{V_{EE} - 0.7 \text{ V}}{R_E} \tag{12.26}$$

Assuming that the transistors are well matched (as would occur in an IC unit),

$$I_{C_1} = I_{C_2} = \frac{I_E}{2} \tag{12.27}$$

resulting in a collector voltage of

$$V_{C_1} = V_{C_2} = V_{CC} - I_C R_C = V_{CC} - \frac{I_E}{2} R_C \tag{12.28}$$

EXAMPLE 12.18

Calculate the dc voltages and currents in the circuit of Fig. 12.37.

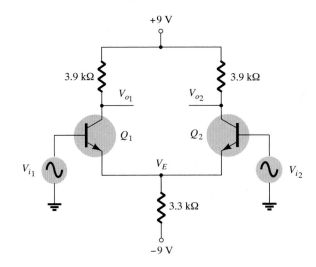

Figure 12.37 Difference amplifier circuit for Example 12.18.

Solution

$$\text{Eq. (12.26):} \quad I_E = \frac{V_{EE} - 0.7 \text{ V}}{R_E} = \frac{9 \text{ V} - 0.7 \text{ V}}{3.3 \text{ k}\Omega} \approx \textbf{2.5 mA}$$

The collector current is then

$$\text{Eq. (12.27):} \quad I_C = \frac{I_E}{2} = \frac{2.5 \text{ mA}}{2} = \textbf{1.25 mA}$$

resulting in a collector voltage of

$$\text{Eq. (12.28):} \quad V_C = V_{CC} - I_C R_C = 9 \text{ V} - (1.25 \text{ mA})(3.9 \text{ k}\Omega) \approx \textbf{4.1 V}$$

The common-emitter voltage is thus -0.7 V, while the collector bias voltage is near 4.1 V for both outputs.

AC Operation of Circuit

An ac connection of a differential amplifier is shown in Fig. 12.38. Separate input signals are applied as V_{i_1} and V_{i_2}, with separate outputs resulting as V_{o_1} and V_{o_2}. To carry out ac analysis the circuit is redrawn in Fig. 12.39. Each transistor is replaced by its ac equivalent.

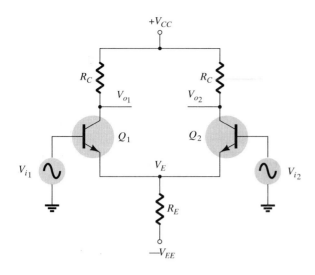

Figure 12.38 Ac connection of differential amplifier.

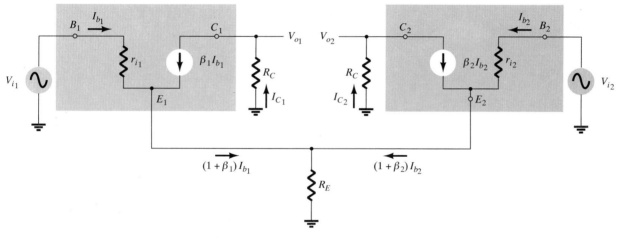

Figure 12.39 Ac equivalent of difference amplifier circuit.

SINGLE-ENDED AC VOLTAGE GAIN

To calculate the single-ended ac voltage gain, V_o/V_i, apply signal to one input with the other connected to ground, as shown in Fig. 12.40. The ac equivalent of this connection is drawn in Fig. 12.41. The ac base current can be calculated using the

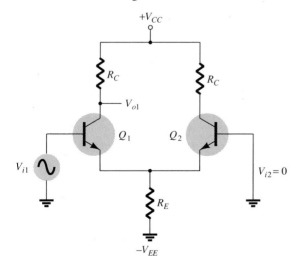

Figure 12.40 Connection to calculate $A_{v1} = V_{o1}/V_{i1}$.

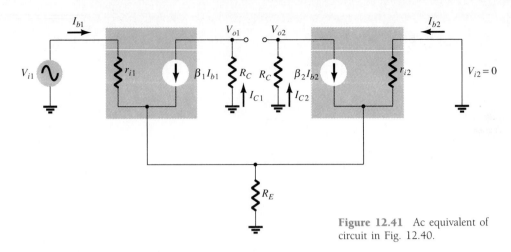

Figure 12.41 Ac equivalent of circuit in Fig. 12.40.

base 1 input KVL (Kirchhoff voltage loop) equation. If one assumes that the two transistors are well matched, then

$$I_{b_1} = I_{b_2} = I_b$$

$$r_{i_1} = r_{i_2} = r_i$$

With R_E very large (ideally infinite), the circuit for obtaining the KVL equation simplifies to that of Fig. 12.42, from which we can write

$$V_{i_1} - I_b r_i - I_b r_i = 0$$

so that

$$I_b = \frac{V_i}{2r_i}$$

If we also assume that

$$\beta_1 = \beta_2 = \beta$$

then

$$I_C = \beta I_b = \beta \frac{V_{i_1}}{2r_i}$$

and the output voltage magnitude at either collector is

$$V_o = I_C R_C = \beta \frac{V_{i_1}}{2r_1} R_C = \frac{\beta R_C}{2\beta r_e} V_i$$

for which the single-ended voltage gain magnitude at either collector is

$$\boxed{A_v = \frac{V_o}{V_{i_1}} = \frac{R_C}{2r_e}} \tag{12.29}$$

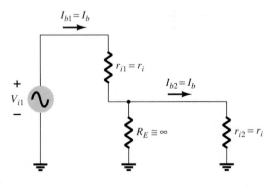

Figure 12.42 Partial circuit to calculate I_b.

Chapter 12 Compound Configurations

Calculate the single-ended output voltage, V_{o_1}, for the circuit of Fig. 12.43.

EXAMPLE 12.19

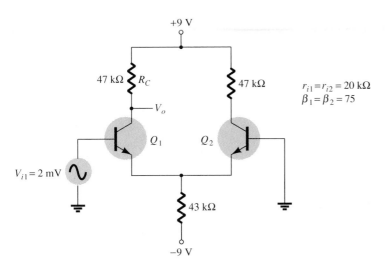

Figure 12.43 Circuit for Examples 12.19 and 12.20.

Solution

The dc bias calculations provide

$$I_E = \frac{V_{EE} - 0.7\text{ V}}{R_E} = \frac{9\text{ V} - 0.7\text{ V}}{43\text{ k}\Omega} = 193\ \mu\text{A}$$

The collector dc current is then

$$I_C = \frac{I_E}{2} = 96.5\ \mu\text{A}$$

so that $\qquad V_C = V_{CC} - I_C R_C = 9\text{ V} - (96.5\ \mu\text{A})(47\text{ k}\Omega) = 4.5\text{ V}$

The value of r_e is

$$r_e = \frac{26}{0.0965} \cong 269\ \Omega$$

The ac voltage gain magnitude can be calculated using Eq. (12.29):

$$A_v = \frac{R_C}{2r_e} = \frac{(47\text{ k}\Omega)}{2(269\ \Omega)} = 87.4$$

providing an output ac voltage of magnitude

$$V_o = A_v V_i = (87.4)(2\text{ mV}) = 174.8\text{ mV} = \mathbf{0.175\ V}$$

DOUBLE-ENDED AC VOLTAGE GAIN

A similar analysis could also be used to show that for the condition of signals applied to both inputs, the differential voltage gain magnitude would be

$$\boxed{A_d = \frac{V_o}{V_d} = \frac{\beta R_C}{2r_i}} \qquad\qquad Y_i = \beta Y_e \qquad (12.30)$$

where $V_d = V_{i_1} - V_{i_2}$.

Common-Mode Operation of Circuit

While a differential amplifier provides large amplification of the difference signal applied to both inputs, it should also provide as small an amplification of the signal common to both inputs. An ac connection showing common input to both transistors is shown in Fig. 12.44. The ac equivalent circuit is then drawn in Fig. 12.45, from which we can write

$$I_b = \frac{V_i - 2(\beta + 1)I_b R_E}{r_i}$$

which can be rewritten as

$$I_b = \frac{V_i}{r_i + 2(\beta + 1)R_E}$$

The output voltage magnitude is then

$$V_o = I_C R_C = \beta I_b R_C = \frac{\beta V_i R_C}{r_i + 2(\beta + 1)R_E}$$

providing a voltage gain magnitude of

$$\boxed{A_c = \frac{V_o}{V_i} = \frac{\beta R_C}{r_i + 2(\beta + 1)R_E}} \qquad (12.31)$$

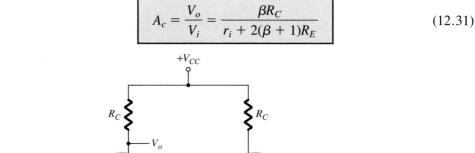

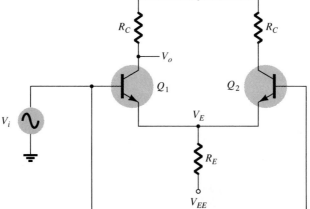

Figure 12.44 Common-mode connection.

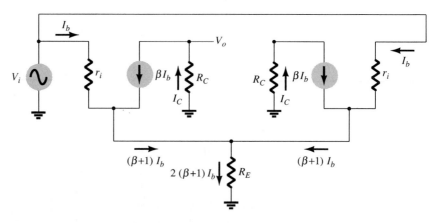

Figure 12.45 Ac circuit in common-mode connection.

Chapter 12   Compound Configurations

Calculate the common-mode gain for the amplifier circuit of Fig. 12.43.

EXAMPLE 12.20

Solution

Eq. (12.33): $A_c = \dfrac{V_o}{V_i} = \dfrac{\beta R_C}{r_i + 2(\beta + 1)R_E} = \dfrac{75(47 \text{ k}\Omega)}{20 \text{ k}\Omega + 2(76)(43 \text{ k}\Omega)} = \mathbf{0.54}$

Use of Constant-Current Source

A good differential amplifier has a very large difference gain, A_d, which is much larger than the common-mode gain, A_c. The common-mode rejection ability of the circuit can be considerably improved by making the common-mode gain as small as possible (ideally, 0). From Eq. (12.24) we see that the larger R_E, the smaller A_c. One popular method for increasing the ac value of R_E is using a constant-current source circuit. Figure 12.46 shows a differential amplifier with constant-current source to provide a large value of resistance from common emitter to ac ground. The major improvement of this circuit over that in Fig. 12.35 is the much larger ac impedance for R_E obtained using the constant-current source. Figure 12.47 shows the ac equivalent circuit for the circuit of Fig. 12.46. A practical constant-current source is shown as a high impedance, in parallel with the constant current.

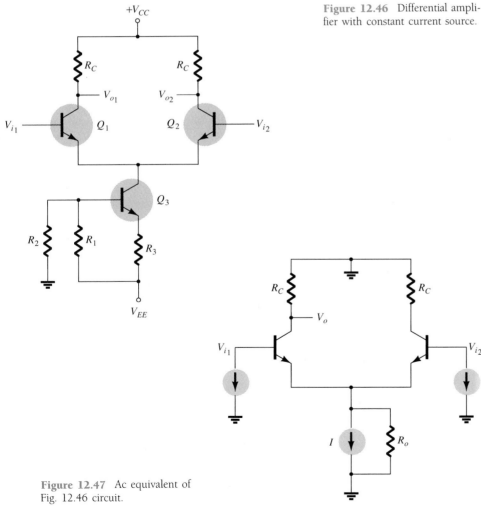

Figure 12.46 Differential amplifier with constant current source.

Figure 12.47 Ac equivalent of Fig. 12.46 circuit.

EXAMPLE 12.21

Calculate the common-mode gain for the differential amplifier of Fig. 12.48.

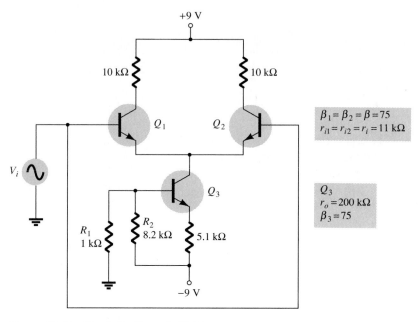

Figure 12.48 Circuit for Example 12.21.

Solution

Using $R_E = r_o = 200$ kΩ gives

$$A_c = \frac{\beta R_C}{r_i + 2(\beta + 1)R_E} = \frac{75(10 \text{ k}\Omega)}{11 \text{ k}\Omega + 2(76)(200 \text{ k}\Omega)} = \textbf{24.7} \times \textbf{10}^{-3}$$

12.10 BIFET, BIMOS, AND CMOS DIFFERENTIAL AMPLIFIER CIRCUITS

While the preceding section provided an introduction to the differential amplifier using bipolar devices, units commercially available also use JFET and MOSFET transistors to build these types of circuits. An IC unit containing a differential amplifier built using both bipolar (Bi) and junction field-effect (FET) transistors is referred to as a *BiFET circuit*. An IC unit made using both bipolar (Bi) and MOSFET (MOS) transistors is called a *BiMOS circuit*. Finally, a circuit built using opposite type MOSFET transistors is a *CMOS circuit*.

The circuits used below to show the various multidevice circuits are mostly symbolic, the actual circuits used in ICs being much more complex. Figure 12.49 shows a BiFET circuit with JFET transistors at the inputs and bipolar transistors to provide the current source (using a current mirror circuit). The current mirror ensures that each JFET is operated at the same bias current. For ac operation the JFET provides a high input impedance (much higher than provided using only bipolar transistors).

Figure 12.50 shows a circuit using MOSFET input transistors and bipolar transistors for the current sources, the BiMOS unit providing even higher input impedance than the BiFET, due to the use of MOSFET transistors.

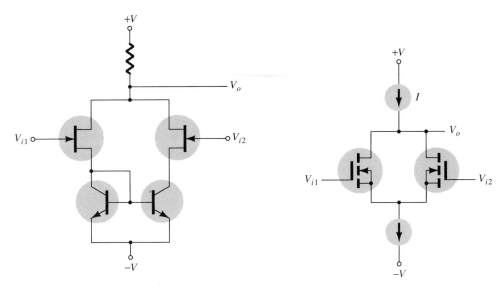

Figure 12.49 BiFET difference amplifier circuit.

Figure 12.50 BiMOS difference amplifier circuit.

Finally, a differential amplifier circuit can be built using complementary MOSFET transistors as shown in Fig. 12.51. The *p*MOS transistors provide the opposite inputs, while the *n*MOS transistors operate as constant-current source. A single output is taken from the common point between *n*MOS and *p*MOS transistors on one side of the circuit. This type of CMOS differential amplifier is particularly well suited for battery operation due to the low power dissipation of a CMOS circuit.

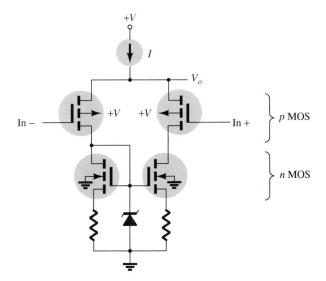

Figure 12.51 CMOS differential amplifier.

12.11 COMPUTER ANALYSIS

Computer analysis of various compound circuits can be easily obtained using PSpice. While BASIC or other programming languages could be used to obtain dc bias and ac operating features, each compound circuit would require writing a lengthy program to analyze only that specific circuit. Using PSpice, one must still describe the individual circuit—but a few minutes may be all that is necessary to produce a listing of the circuit and the results desired.

A number of PSpice lines can be used to specify the details of the desired ac analysis.

To specify the input ac signal:

$$VI \quad N1 \quad N2 \quad AC \quad VOLTAGE$$

e.g.,

$$VI \quad 1 \quad 2 \quad AC \quad 10MV \qquad V_i = 10 \text{ mV (ac)}$$

To specify input signal frequency:

$$.AC \quad LIN \quad NS \quad FS \quad FE$$

e.g.,

$$.AC \quad LIN \quad 1 \quad 10KH \quad 10KH \qquad (f_i = 10 \text{ kHz})$$

To specify the ac output: Having requested ac analysis one can then include a print line listing the ac circuit voltages or currents desired. The form of the print line is

$$.PRINT \ AC \ VOLTAGE\_LIST$$

e.g.,

$$.PRINT \ AC \ V(1) \quad V(6) \quad I(RD) \quad V(3,4)$$

MODEL LINES

1. For a BJT device the model line includes device beta.

$$.MODEL \quad DEV\_NAME \quad NPN \quad (BF = \underline{\quad\quad})$$

   e.g.,

$$.MODEL \quad TRAN1 \quad NPN \quad (BF = 200)$$

2. For a JFET device the model line includes V_P and I_{DSS}.

$$.MODEL \quad DEV\_NAME \quad NJF \quad VTO = \underline{\quad\quad} \quad BETA = \underline{\quad\quad}$$

   e.g.,

$$.MODEL \quad FET3 \quad NJF \quad VTO = -4 \quad BETA = 0.625E\text{-}3$$

   n-channel JFET: VTO $= V_P = -4$ V, BETA $= I_{DSS}/V_P^2$, so that $I_{DSS} = 10$ mA

3. For an enhancement MOSFET the model line includes V_T.

$$.MODEL \quad DEV\_NAME \quad PMOS \text{ or } NMOS \quad (VTO = \underline{\quad\quad})$$

   e.g.,

$$.MODEL \quad MOSA \quad PMOS \quad (VTO = -2V) \qquad p\text{MOS with } V_T = -2 \text{ V}$$

Program 12.1. Cascade JFET Amplifier

A PSpice listing to provide analysis of the JFET cascade amplifier of Fig. 12.2 is provided in Fig. 12.52; see Fig. 12.53 for the circuit showing all node points used. Looking at the PSpice listing, the supply voltage, resistor elements, and capacitor elements are described first. A load of $R_L = 1$ MΩ is added to complete the path from the output of capacitor C_3 to ground. The two JFETs are seen to be the same model having specified values of

$$VTO = V_P = -4 \text{ V, and } I_{DSS} = 10 \text{ mA (from BETA = 0.625 E-6)}$$

```
**** 01/25/90 ******* Evaluation PSpice (January 1989) ******* 21:15:59 ****

Cascade JFET Amplifier

***      CIRCUIT DESCRIPTION

*************************************************************************

VDD 8 0 20V
RG1 2 0 3.3MEG
RD1 3 8 2.4K
RS1 4 0 680
CS1 4 0 100UF
C1 1 2 0.05UF
C2 3 5 0.05UF
RG2 5 0 3.3MEG
RD2 6 8 2.4K
RS2 7 0 680
CS2 7 0 100UF
C3 6 9 0.05UF
RL 9 0 1MEG
J1 3 2 4 NFET
J2 6 5 7 NFET
.MODEL NFET NJF VTO=-4V BETA=0.625E-3
VI 1 0 AC 10MV
.AC LIN 1 10KH 10KH
.OP
.PRINT AC V(1) V(3) V(6) V(9)
.OPTIONS NOPAGE
.END


****      Junction FET MODEL PARAMETERS
             NJF
     VTO     -4
    BETA   625.000000E-06


****      SMALL SIGNAL BIAS SOLUTION          TEMPERATURE =   27.000 DEG C
  NODE    VOLTAGE       NODE    VOLTAGE       NODE    VOLTAGE       NODE    VOLTAGE
 (    1)    0.0000    (    2) 50.28E-06    (    3)   13.3270    (    4)    1.8908
 (    5) 50.28E-06    (    6)   13.3270    (    7)    1.8908    (    8)   20.0000
 (    9)    0.0000

        VOLTAGE SOURCE CURRENTS
        NAME        CURRENT
        VDD        -5.561E-03
        TOTAL POWER DISSIPATION    1.11E-01   WATTS

****      OPERATING POINT INFORMATION     *** JFETS
  NAME        J1           J2
  MODEL       NFET         NFET
  ID          2.78E-03     2.78E-03
  VGS        -1.89E+00    -1.89E+00
  VDS         1.14E+01     1.14E+01
  GM          2.64E-03     2.64E-03

****      AC ANALYSIS                          TEMPERATURE =   27.000 DEG C
   FREQ        V(1)         V(3)         V(6)         V(9)
  1.000E+04   1.000E-02    6.323E-02    3.992E-01    3.992E-01
```

Figure 12.52 PSpice output for circuit of Fig. 12.53.

Figure 12.53 Circuit for PSpice Program 1.

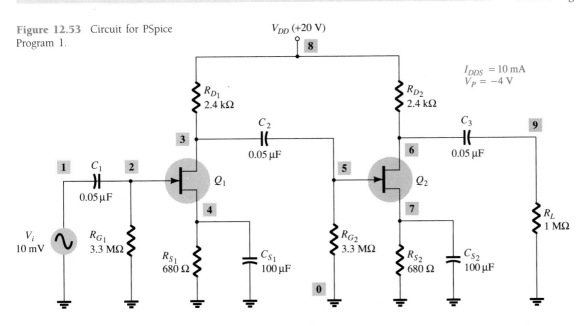

559

The input ac signal is $V_i = 10$ mV at $f = 10$ kHz. The .OP line calls for output of operating point information—dc bias values and transistor operating parameters. The output also provides listing of ac voltages at input and output of each stage. A summary of the results obtained is provided next.

Dc bias results (for each transistor):

$$V_G \approx 0 \text{ V}, \qquad V_D \approx 13.3 \text{ V}, \qquad V_S \approx 1.9 \text{ V}$$

JFET parameters (for each transistor):

$$I_{DQ} = 2.78 \text{ mA}, \qquad V_{GSQ} = -1.89 \text{ V}, \qquad g_m = 2.64 \text{ mS} \quad (g_m = 2.6 \text{ mS in Example 12.1})$$

Ac results:

$$A_{v_1} = \frac{V(3)}{V(1)} = \frac{6.323 \times 10^{-2}}{1 \times 10^{-2}} = 6.3 \qquad (-6.2 \text{ in Example 12.1})$$

$$A_{v_2} = \frac{V(6)}{V(3)} = \frac{3.992 \times 10^{-2}}{6.323 \times 10^{-2}} = 6.3 \qquad (-6.2 \text{ in Example 12.1})$$

$$V_o = V(9) = 3.992 \times 10^{-1} = 399 \text{ mV} \qquad (V_o = 384 \text{ mV in Example 12.1})$$

The ac voltage gain and output ac voltage obtained in Example 12.1 and calculated using PSpice compare quite well. Recall that PSpice uses a more sophisticated model than that used in Example 12.1, and that all steps in PSpice are carried out using more decimal places, causing the results to be somewhat different.

Program 12.2. Cascade BJT Amplifier

The cascade BJT amplifier of Example 12.2 is analyzed by the PSpice listing of Fig. 12.54 (the circuit is shown in Fig. 12.55). The BJT model provides for identical transistors

$$\text{.MODEL BJT NPN (BF = 200 IS = 7E - 15)}$$

```
****  01/25/90  *******  Evaluation PSpice (January 1989)  *******  21:25:11  ****

 Cascaded BJT Amplifier

****        CIRCUIT DESCRIPTION

********************************************************************************

VCC 8 0 20V
R1 8 2 15K
R2 2 0 4.7K
RC1 8 3 2.2K
RE1 4 0 1K
R3 8 5 15K
R4 5 0 4.7K
RC2 8 6 2.2K
RE2 7 0 1K
RL 9 0 1MEG
Q1 3 2 4 BJT
Q2 6 5 7 BJT
C1 1 2 10UF
C2 3 5 10UF
C3 6 9 10UF
CS1 4 0 500UF
CS2 7 0 500UF
VI 1 0  AC 25UV
.MODEL BJT NPN(BF=200 IS=7E-15)
.AC LIN 1 1KH 1KH
.PRINT AC V(1) V(3) V(6) V(9)
.OPTIONS NOPAGE
.END
```

Figure 12.54 PSpice output for circuit of Fig. 12.55.

```
****       BJT MODEL PARAMETERS
              NPN
      IS       7.000000E-15
      BF     200
      NF       1
      BR       1
      NR       1


****       SMALL SIGNAL BIAS SOLUTION        TEMPERATURE =    27.000 DEG C
   NODE   VOLTAGE      NODE   VOLTAGE      NODE   VOLTAGE      NODE   VOLTAGE
   (   1)   0.0000   (   2)    4.7004   (   3)   11.2430   (   4)    4.0003
   (   5)   4.7004   (   6)   11.2430   (   7)    4.0003   (   8)   20.0000
   (   9)   0.0000


      VOLTAGE SOURCE CURRENTS
      NAME            CURRENT
      VCC            -1.000E-02
      TOTAL POWER DISSIPATION    2.00E-01  WATTS


****       AC ANALYSIS                       TEMPERATURE =    27.000 DEG C
    FREQ        V(1)         V(3)          V(6)          V(9)
   1.000E+03   2.500E-05    2.558E-03    8.625E-01    8.625E-01
```

Figure 12.54 Continued.

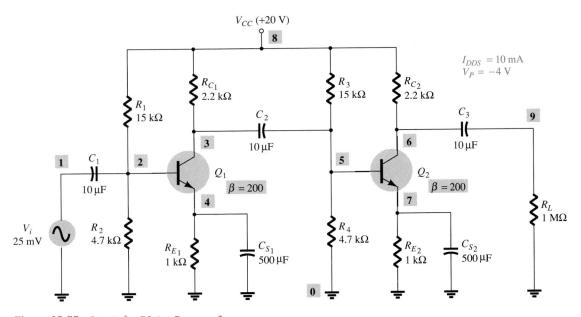

Figure 12.55 Circuit for PSpice Program 2.

where $\beta = 200$ and $I_S = 7 \times 10^{-7}$ causes $V_{BE} = 0.7$ V in the PSpice model. The ac input signal is

$$V_i = 25 \ \mu\text{V, at a frequency of 1 kHz [.AC LIN 1 1KH 1KH]}$$

A summary of the results obtained is provided next.

Dc bias (each transistor):

$$V_{B_Q} = 4.7 \text{ V,} \qquad V_{E_Q} = 4.0 \text{ V,} \qquad V_{C_Q} = 11.2 \text{ V}$$

BJT parameters (each transistor):

$$I_B = 19.9 \ \mu A$$

$$I_C = 3.98 \text{ mA} \qquad (\beta = I_C/I_B = 3.98 \text{ mA}/19.9 \ \mu A = 200)$$

$$V_{BE} = 0.7 \text{ V}$$

Ac results:

$$A_{v_1} = \frac{V_{o_1}}{V_{i_1}} = \frac{V(3)}{V(1)} = \frac{2.558 \times 10^{-3}}{2.5 \times 10^{-5}}$$

$$= 102.3 \qquad (-104 \text{ in Example 12.2})$$

$$A_{v_2} = \frac{V_{o_2}}{V_{i_2}} = \frac{V(6)}{V(3)} = \frac{8.625 \times 10^{-4}}{2.558 \times 10^{-5}}$$

$$= 337.2 \qquad (-349 \text{ in Example 12.2})$$

Another comparison of the results obtained by the two methods that can be made involves r_e. From the PSpice listing

$$\text{RPI} = 1.3 \times 10^3 = 1.3 \text{ k}\Omega$$

This is the input impedance looking into the BJT base. Since

$$\text{RPI} = r_i = \beta r_e$$

we can write

$$r_e = \frac{r_i}{\beta} = \frac{1.3 \times 10^3}{200} = 6.5 \ \Omega \qquad (6.3 \ \Omega \text{ in Example 12.2})$$

This difference accounts for the greatest difference in the output ac voltages computed by the example and the PSpice program. From the PSpice listing $V_o = V(9) = 0.86$ V, while Example 12.2 provides $V_o = 0.9$ V, the two values in good agreement (less than 5% difference).

Program 12.3. Darlington Circuit

The Darlington circuit of Fig. 12.12 is analyzed by the PSpice program of Fig. 12.56 (see also Fig. 12.57). Two identical BJT devices are connected as a Darlington device. A value of BF = 89.4 was used so that

$$\beta^2 = (89.4)^2 = 7992 \approx 8000$$

Dc bias:

$$V_{B_1} = V(2) = 9.65 \text{ V}$$

$$V_{E_2} = V(4) = 8.06 \text{ V}$$

providing V_{BE} (Darlington) = 1.59 V

Transistor parameters:

$$I_{B_1} = 2.53 \ \mu A, \quad I_{C_1} = 0.23 \text{ mA} \qquad (\beta_1 = 0.23 \text{ mA}/2.53 \ \mu A = 90.9)$$

$$I_{B_2} = 229 \ \mu A, \quad I_{C_2} = 20.4 \text{ mA} \qquad (\beta_2 = 20.4 \text{ mA}/229 \ \mu A = 89.1)$$

for a Darlington beta of

$$\beta_D = \beta_1 \beta_2 = (90.9)(89.1) \approx 8100$$

```
**** 01/25/90 ******* Evaluation PSpice (January 1989) ******* 21:29:28 ****

  Darlington Amplifier

***      CIRCUIT DESCRIPTION

*************************************************************************

VCC 6 0 18V
RB 6 2 3.3MEG
C1 1 2 0.5UF
RE 4 0 390
C2 4 5 0.5UF
RL 5 0 1MEG
Q1 6 2 3 BJT
Q2 6 3 4 BJT
.MODEL BJT NPN(BF=89.4)
VI 1 0 AC 100MV
.AC LIN 1 10KH 10KH
.PRINT AC V(1) V(4) V(5)
.OPTIONS NOPAGE
.END

****      BJT MODEL PARAMETERS
            NPN
       IS    100.000000E-18
       BF    89.4
       NF     1
       BR     1
       NR     1

****     SMALL SIGNAL BIAS SOLUTION      TEMPERATURE =   27.000 DEG C
   NODE    VOLTAGE     NODE   VOLTAGE     NODE   VOLTAGE     NODE    VOLTAGE
 (    1)    0.0000   (    2)   9.6513   (    3)   8.9155   (    4)    8.0632
 (    5)    0.0000   (    6)  18.0000

     VOLTAGE SOURCE CURRENTS
     NAME           CURRENT
     VCC          -2.068E-02
     TOTAL POWER DISSIPATION   3.72E-01   WATTS

****     AC ANALYSIS                      TEMPERATURE =   27.000 DEG C
   FREQ        V(1)        V(4)        V(5)
  1.000E+04   1.000E-01   9.936E-02   9.936E-02
```

Figure 12.56 PSpice output for circuit of Fig. 12.57.

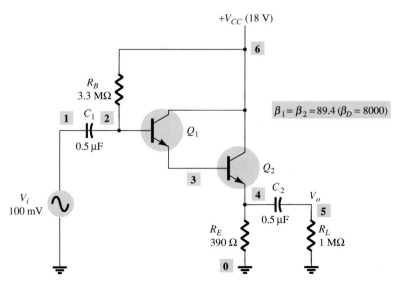

Figure 12.57 Circuit for PSpice Program 3.

It's difficult to force the PSpice transistor model to exactly match the ideal transistor model used in Fig. 12.12. Notice that PSpice results provide

$$V_{BE_1} = 0.736 \text{ V}, \qquad V_{BE_2} = 0.852 \text{ V}$$

while the model used in Fig. 12.12 specifies $V_{BE}(D) = 1.6$ V (about the same as 0.736 V + 0.852 V).

Ac operation: For an input of $V_i = 100$ mV, the output in the PSpice listing is

$$V_o = \text{V(5)} = 9.936E{-}2 = 99.36 \text{ mV}$$

providing an amplifier gain of

$$A_V = \frac{V_o}{V_i} = \frac{\text{V(5)}}{\text{V(1)}} = \frac{9.936 \times 10^{-2}}{1 \times 10^{-1}} = 0.9936$$

while the results in Example 12.10 provide $A_v = 0.998$, quite close.

Program 12.4. CMOS Inverter Circuit

A CMOS inverter circuit is analyzed in the listing provided in Fig. 12.58 (see also Fig. 12.59). A *p*-channel enhancement MOSFET, M1, and *n*-channel enhancement MOSFET, M2 are operated as a CMOS inverter circuit. With input varied from a dc value of 0 V to a dc value of +5 V, the calculated output voltage is listed by the PSpice program. This input voltage variation is provided by the line

$$\text{.DC VI 0 5 5}$$

which varies VI from 0 to 5 V, up to a final value of 5 V. The listing provides output data

$$\text{VI} = 0 \text{ V} \qquad \text{V(2)} = 5 \text{ V}$$
$$\text{VI} = 5 \text{ V} \qquad \text{V(2)} \approx 0 \text{ V}$$

to show that the circuit operates as a logic inverter, providing the opposite output voltage.

```
**** 01/25/90 ******* Evaluation PSpice (January 1989) ******* 21:36:09 ****

   CMOS Inverter Circuit

****       CIRCUIT DESCRIPTION

*********************************************************************************
VDD 5 0 5V
M1 5 1 2 5 PM
M2 2 1 0 0 NM
.MODEL PM PMOS (VTO=-2V)
.MODEL NM NMOS (VTO=2V)
VI 1 0 5V
.DC VI 0 5 5
.PRINT DC V(2)
.OPTIONS NOPAGE
.END


****       MOSFET MODEL PARAMETERS
              PMOS              NMOS
   LEVEL       1                 1
     VTO      -2                 2
      KP      20.000000E-06   20.000000E-06

****       DC TRANSFER CURVES                    TEMPERATURE =    27.000 DEG C
   VI           V(2)
   0.000E+00    5.000E+00
   5.000E+00    8.350E-08
```

Figure 12.58 PSpice output for circuit of Fig. 12.59.

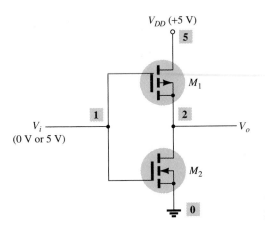

V_{DD} (+5 V)

Figure 12.59 Circuit for PSpice Program 4.

§ 12.2 Cascade Connection

1. For the JFET cascade amplifier in Fig. 12.60, calculate the dc bias conditions for the two identical stages, using JFETs with $I_{DSS} = 8$ mA and $V_P = -4.5$ V.

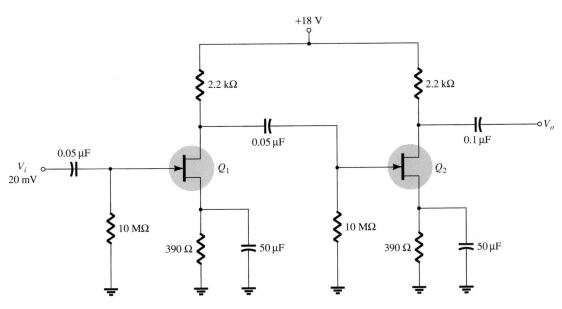

Figure 12.60 Problems 1–5, 30–31

2. For the JFET cascade amplifier of Fig. 12.60, using identical JFETs with $I_{DSS} = 8$ mA and $V_P = -4.5$ V, calculate the voltage gain of each stage, the overall gain of the amplifier, and the output voltage, V_o.

3. If both JFETs in the cascade amplifier of Fig. 12.60 are changed to those having specifications $I_{DSS} = 12$ mA and $V_P = -3$ V, calculate the resulting dc bias of each stage.

4. If both JFETs in the cascade amplifier of Fig. 12.60 are changed to those having the specifications $I_{DSS} = 12$ mA, $V_P = -3$ V, and $y_{os} = 25$ μS, calculate the resulting voltage gain for each stage, the overall voltage gain, and the output voltage, V_o.

5. For the cascade amplifier of Fig. 12.60 using JFETs with specifications $I_{DSS} = 12$ mA, $V_P = -3$ V, and $y_{os} = 25$ μS, calculate the circuit input impedance (Z_i) and output impedance (Z_o).

Problems

6. For the BJT cascade amplifier of Fig. 12.61, calculate the dc bias voltages and collector current for each stage.

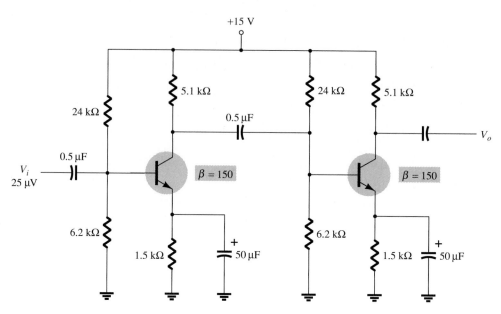

Figure 12.61 Problems 6–8, 32

7. Calculate the voltage gain of each stage and the overall ac voltage gain for the BJT cascade amplifier circuit of Fig. 12.61.

8. For the circuit of Fig. 12.61, calculate the input impedance (Z_i) and output impedance (Z_o).

9. For the cascade amplifier of Fig. 12.62, calculate the dc bias voltages and collector current of each stage.

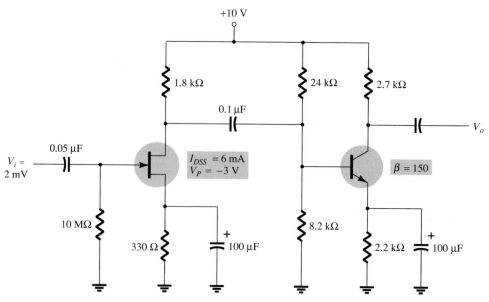

Figure 12.62 Problems 9–11

10. For the amplifier circuit of Fig. 12.62, calculate the voltage gain of each stage and the overall amplifier voltage gain.

Chapter 12 Compound Configurations

11. Calculate the input impedance (Z_i) and output impedance (Z_o) for the amplifier circuit of Fig. 12.62.

§ **12.3 Cascode Connection**

12. In the cascode amplifier circuit of Fig. 12.63, calculate the dc bias voltages V_{B_1}, V_{B_2}, and V_{C_2}.

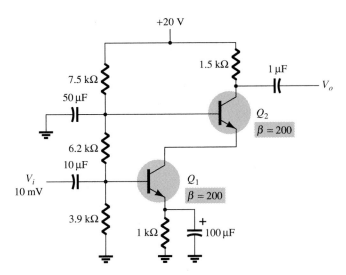

Figure 12.63 Problems 12–14

* **13.** For the cascode amplifier circuit of Fig. 12.63, calculate the voltage gain, A_v, and ouptut voltage, V_o.

14. Calculate the ac voltage across a 10-kΩ load connected at the output of the circuit in Fig. 12.63.

§ **12.4 Darlington Connection**

15. For the circuit of Fig. 12.64, calculate the dc bias voltage, V_{E_2}, and emitter current, I_{E_2}.

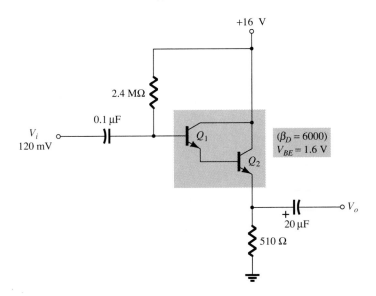

Figure 12.64 Problems 15–16, 33

* **16.** For the circuit of Fig. 12.64, calculate the amplifier voltage gain.

17. For the feedback pair circuit of Fig. 12.65, calculate the dc bias values of V_{B_1}, V_{C_2}, and I_C.

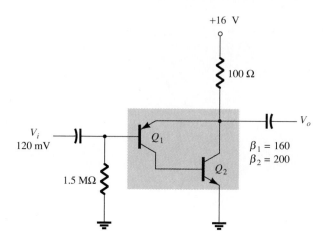

Figure 12.65 Problems 17–18

* **18.** Calculate the output ac voltage for the circuit of Fig. 12.65.

§ 12.6 CMOS Circuit

19. Determine which transistors are off and which are on on the circuit of Fig. 12.66, for an input of:
(a) $V_1 = 0$ V, $V_2 = 0$ V.
(b) $V_1 = +5$ V, $V_2 = +5$ V.
(c) $V_1 = 0$ V, $V_2 = +5$ V.

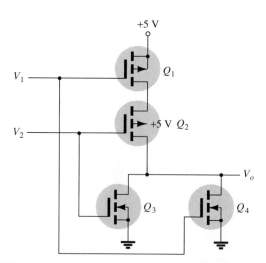

Figure 12.66 Problems 19–20, 34

20. For the circuit of Fig. 12.66, complete the voltage table below.

V_1	V_2	V_o
0 V	0 V	
0 V	+5 V	
+5 V	0 V	
+5 V	+5 V	

§ 12.7 **Current Source Circuits**

21. Calculate the current through the 2-kΩ load in the circuit of Fig. 12.67.

22. For the circuit of Fig. 12.68, calculate the current I.

* **23.** Calculate the current I in the circuit of Fig. 12.69.

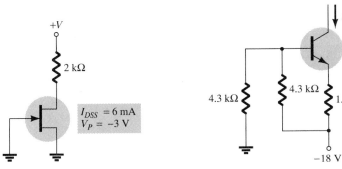

Figure 12.67 Problem 21

Figure 12.68 Problem 22

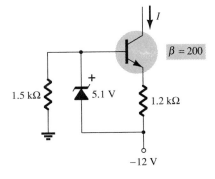

Figure 12.69 Problem 23

§ 12.8 **Current Mirror Circuits**

24. Calculate the mirrored current I in the circuit of Fig. 12.70.

* **25.** Calculate collector currents for Q_1 and Q_2 in Fig. 12.71.

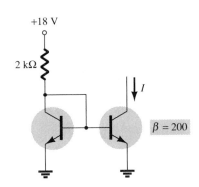

Figure 12.70 Problem 24

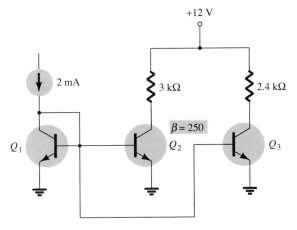

Figure 12.71 Problem 25

§ 12.9 **Differential Amplifier Circuit**

26. Calculate dc bias values of I_C and V_C for the matched transistors of Fig. 12.72.

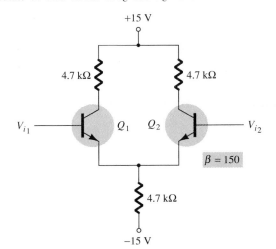

Figure 12.72 Problem 26

27. Calculate the dc bias values of I_C and V_C for the matched transistors of Fig. 12.73.

* 28. Calculate V_o in the circuit in Fig. 12.74.

* 29. Calculate V_o in the circuit of Fig. 12.75.

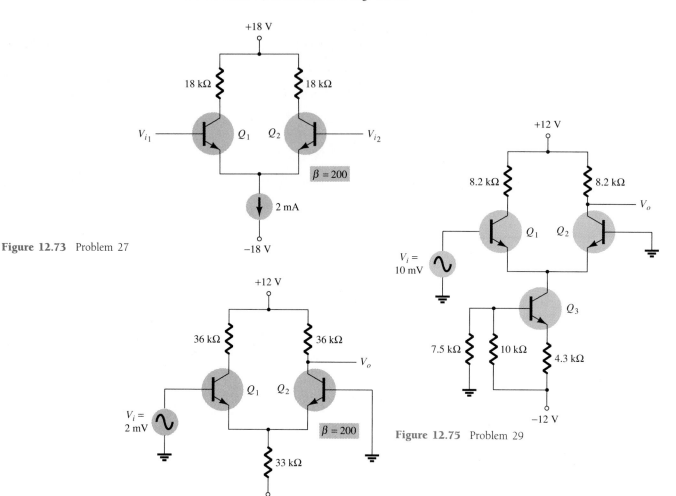

Figure 12.73 Problem 27

Figure 12.74 Problem 28

Figure 12.75 Problem 29

§ 12.10 Computer Analysis

* 30. Write a PSpice program to calculate dc bias voltage of the cascade JFET amplifier of Fig. 12.60, using $I_{DSS} = 12$ mA and $V_P = -3$ V.

* 31. Write a PSpice program to calcualte the output voltage, V_o for the cascade JFET circuit of Fig. 12.60, using $I_{DSS} = 12$ mA, $V_P = -3$ V, and $y_{os} = 25$ μS.

* 32. Write a PSpice program to calculate the output ac voltage of each stage of the BJT cascade amplifier of Fig. 12.61.

* 33. Write a PSpice program to calculate the transistor operating point information and the output ac voltage for the Darlington amplifier circuit of Fig. 12.64.

* 34. Write a PSpice program to list the dc voltages for the following input sets for the CMOS circuit of Fig. 12.66.
   (a) $V_1 = 0$ V, $V_2 = 0$ V.
   (b) $V_1 = 0$ V and $V_2 = +5$ V.
   (c) $V_1 = +5$ V, $V_2 = +5$ V.

---

*Please Note: Asterisks indicate more difficult problems.

Discrete and IC Manufacturing Techniques

13.1 INTRODUCTION

The techniques applied to the manufacture of semiconductor devices is continually being reviewed, modified, and upgraded. In recent years the primary emphasis has been in increasing the *yield rate* (number of good elements in a batch), expanding the automation level (less hands-on requirements), and increasing the density levels. The sequence of steps in the manufacturing of *discrete* units (single elements) or *integrated-circuits* (ICs) (high-density chips with thousands of elements) has not changed that dramatically. However, the manner in which each step is performed has experienced a tremendous change in the last decade.

This book is designed simply to develop an overall picture of the production cycle for discrete and IC units by introducing some of the important production phases and applied terminology. A detailed discussion of just one step of the cycle would require a book in itself.

13.2 SEMICONDUCTOR MATERIALS, Si AND Ge

The first step in the manufacture of any semiconductor device is to obtain semiconductor materials, such as germanium or silicon, of the desired purity level. Impurity levels of *less* than *one* part in *one billion* (1 in 1,000,000,000) are required for most semiconductor fabrications today.

The raw materials are first subjected to a series of chemical reactions and a *zone-refining process* to form a *polycrystalline crystal* of the desired purity level. The atoms of a polycrystalline crystal are haphazardly arranged, while in the *single crystal* desired the atoms are arranged in a symmetrical, uniform, geometrical lattice structure.

The zone-refining apparatus of Fig. 13.1 consists of a graphite or quartz boat for minimum contamination, a quartz container, and a set of RF (radio-frequency) induction coils. Either the coils or boat must be movable along the length of the quartz container. The same result will be obtained in either case, although the moving coils approach is introduced here since it appears to be the more popular method. The interior of the quartz container is filled with either an inert (little or no chemical reaction) gas, or vacuum, to reduce further the chance of contamination. In the zone refining process, a bar of germanium is placed in the boat with the coils at one end of the bar as shown in Fig. 13.1. The radio-frequency signal is then applied to the coil,

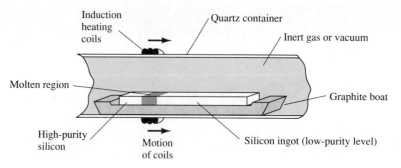

Figure 13.1 Zone refining process.

which will induce a flow of charge (eddy currents) in the germanium ingot. The magnitude of these currents is increased until sufficient heat is developed to melt that region of the semiconductor material. The impurities in the ingot will enter a more liquid state than the surrounding semiconductor material. If the induction coils of Fig. 13.1 are now slowly moved to the right to induce melting in the neighboring region, the "more fluidic" impurities will "follow" the molten region. The net result is that a large percentage of the impurities will appear at the right end of the ingot when the induction coils have reached this end. This end piece of impurities can then be cut off and the entire process repeated until the desired purity level is reached.

The next step in the fabrication sequence is the formation of a single crystal of a germanium or silicon. This is most commonly accomplished using the *Czochralski* technique. The apparatus employed in the Czochralski technique is shown in Fig. 13.2a. The polycrystalline material is first transformed to the molten state by the RF

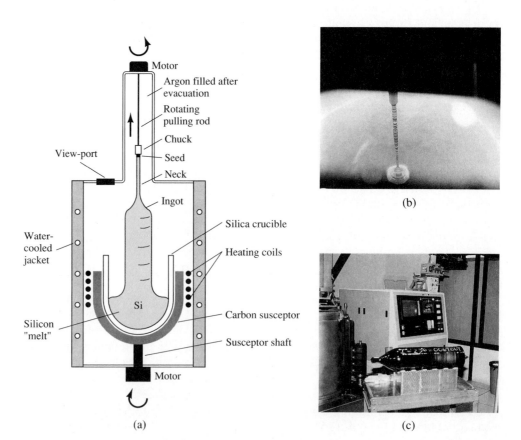

Figure 13.2 (a) Czochralski Oven; (b) Ingot "neck"; (c) Ingot cooling in front of a Czochralski Oven. (Courtesy of Texas Instruments, Inc.)

Chapter 13 Discrete and IC Manufacturing Techniques

induction coils. A single-crystal ''seed'' of the desired impurity level is then immersed in the molten silicon and gradually withdrawn while the shaft holding the seed is slowly turning. As the ''seed'' is withdrawn, a single-crystal silicon lattice structure will grow on the ''seed'' as demonstrated by the ''neck'' of the ingot in Fig. 13.2b. The resulting single-crystal ingots are typically 6 to 36 in. in length and 1 to 5 in. in diameter. Ingots having a length of 48 in. and a diameter of 3 in. have been grown. The weight of such a structure is about 28.5 lb. An ingot and a Czochralski oven appear in Fig. 13.2(c).

13.3 DISCRETE DIODES

Semiconductor diodes are normally one of the following types: *grown junction, alloy, diffusion* or *epitaxial growth*. A brief description of each process is provided in the next few paragraphs.

Grown Junction

Diodes of this type are formed during the Czochralski *crystal pulling* process. Impurities of *p*- and *n*-type can be alternately added to the molten semiconductor material in the crucible, resulting in a *p-n* junction, as indicated in Fig. 13.3 when the crystal is pulled. After slicing, the large-area device can then be cut into a large number (sometimes thousands) of smaller-area semiconductor diodes. The area of grown-junction diodes is sufficiently large to handle high currents (and therefore have high power ratings). The large area, however, will introduce undesired junction capacitive effects.

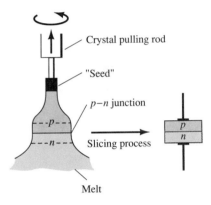

Figure 13.3 Grown junction diode.

Alloy

The alloy process will result in a junction-type semiconductor diode that will also have a high current rating and large PIV rating. The junction capacitance is also large, however, due to the large junction area.

The *p-n* junction is formed by first placing a *p*-type impurity on an *n*-type substrate and heating the two until liquefaction occurs where the two materials meet (Fig. 13.4). An alloy will result that, when cooled, will produce a *p-n* junction at the boundary of the alloy and substrate. The roles played by the *n*- and *p*-type materials can be interchanged.

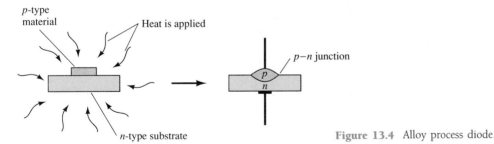

Figure 13.4 Alloy process diode.

Diffusion

The diffusion process of forming semiconductor junction diodes can employ either solid or gaseous diffusion. This process requires more time than the alloy process but it is relatively inexpensive and can be very accurately controlled. Diffusion is a process by which a heavy concentration of particles will ''diffuse'' into a surrounding region of lesser concentration. The primary difference between the diffusion and alloy process is the fact that liquefaction is not reached in the diffusion process. Heat is applied in the diffusion process only to increase the activity of the elements involved.

The process of solid diffusion begins with the "painting" of an acceptor impurity on an *n*-type substrate and heating the two until the impurity diffuses into the substrate to form the *p*-type layer (Fig. 13.5a).

In the process of gaseous diffusion, an *n*-type material is submerged in a gaseous atmosphere of acceptor impurities and then heated (Fig. 13.5b). The impurity diffuses into the substrate to form the *p*-type layer of the semiconductor diode. The roles of the *p*- and the *n*-type materials can also be interchanged in each case. The diffusion process is the most frequently used today in the manufacture of semiconductor diodes.

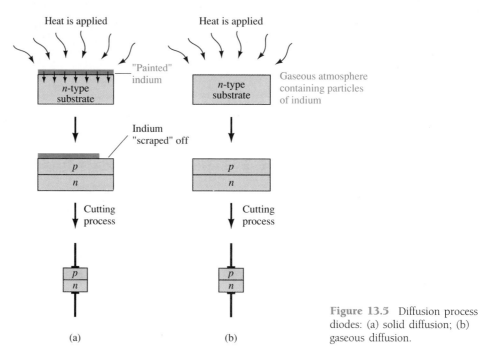

Figure 13.5 Diffusion process diodes: (a) solid diffusion; (b) gaseous diffusion.

Epitaxial Growth

The term *epitaxial* has its derivation from the Greek terms *epi* meaning "upon" and *taxis* meaning "arrangement." A base wafer of n^+ material is connected to a metallic conductor as shown in Fig. 13.6. The n^+ indicates a very high doping level for a reduced resistance characteristic. Its purpose is to act as a semiconductor extension of the conductor and not the *n*-type material of the *p-n* junction. The *n*-type layer is to be deposited on this layer shown in Fig. 13.6 using a diffusion process. This technique of using an n^+ base gives the manufacturer definite design advantages. The *p*-type silicon is then applied by using a diffusion technique and the anode metallic connector added as indicated in Fig. 13.6.

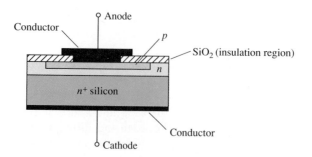

Figure 13.6 Epitaxially grown semiconductor diode.

13.4 TRANSISTOR FABRICATION

The majority of the methods used to fabricate transistors are simply extensions of the methods used to manufacture semiconductor diodes. The methods most frequently employed today include *alloy junction, grown junction, and diffusion*. The following discussion of each method will be brief, but the fundamental steps included in each will be presented.

Alloy Junction

The alloy junction technique is also an extension of the alloy method of manufacturing semiconductor diodes. For a transistor, however, two dots of the same impurity are deposited on each side of a semiconductor wafer having the opposite impurity as shown in Fig. 13.7. The entire structure is then heated until melting occurs and each dot is alloyed to the base wafer resulting in the *p-n* junctions indicated in Fig. 13.7 as described for semiconductor diodes.

The collector dot and resulting junction are larger, to withstand the heavy current and power dissipation at the collector-base junction. This method is not employed as much as the diffusion technique to be described shortly, but it is still used extensively in the manufacture of high-power diodes.

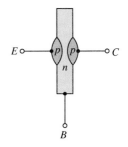

Figure 13.7 Alloy junction transistor.

Grown Junction

The Czochralski technique is used to form the two *p-n* junctions of a grown-junction transistor. The process, as depicted in Fig. 13.8, requires that the impurity control and withdrawal rate be such as to ensure the proper base width and doping levels of the *n*- and *p*-type materials. Transistors of this type are, in general, limited to less than $\frac{1}{4}$-W rating.

Diffusion

The most frequently employed method of manufacturing transistors today is the diffusion technique. The basic process was introduced in the discussion of semi-conductor diode fabrication. The diffusion technique is employed in the production of *mesa* and *planar* transistors, each of which can be of the *diffused* or *epitaxial* type.

In the *pnp*, diffusion-type mesa transistor the first process is an *n*-type diffusion into a *p*-type wafer, as shown in Fig. 13.9, to form the base region. Next, the *p*-type emitter is diffused or alloyed to the *n*-type base as shown in the figure. Etching is done to reduce the capacitance of the collector junction. The term "mesa" is derived from its similarities with the geographical formation. As mentioned earlier in the discussion of diode fabrication, the diffusion technique permits very tight control of the doping levels and thicknesses of the various regions.

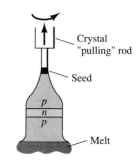

Figure 13.8 Grown junction transistor.

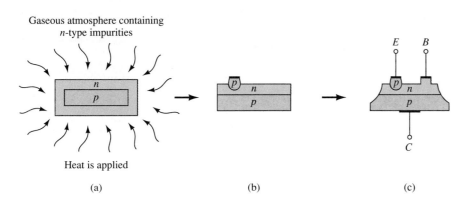

Figure 13.9 Mesa transistor: (a) diffusion process; (b) alloy process; (c) etching process.

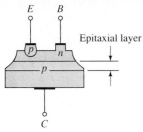

Figure 13.10 Epitaxial mesa transistor.

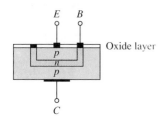

Figure 13.11 Planar transistor.

The major difference between the epitaxial mesa transistor and the mesa transistor is the addition of an epitaxial layer on the original collector substrate. The term epitaxial is derived from the Greek words *epi*—upon, and *taxi*—arrange, which describe the process involved in forming this additional layer. The original *p*-type substrate (collector of Fig. 13.10) is placed in a closed container having a vapor of the same impurity. Through proper temperature control, the atoms of the vapor will *fall upon* and *arrange* themselves on the original *p*-type substrate resulting in the epitaxial layer indicated in Fig. 13.10. Once this layer is established, the process continues, as described above for the mesa transistor, to form the base and emitter regions. The original *p*-type substrate will have a higher doping level and correspondingly less resistance than the epitaxial layer. The result is a low-resistance connection to the collector lead that will reduce the dissipation losses of the transistor.

The planar and epitaxial planar transistors are fabricated using two diffusion processes to form the base and emitter regions. The planar transistor, as shown in Fig. 13.11, has a flat surface, which accounts for the term *planar*. An oxide layer is added as shown in Fig. 13.11 to eliminate exposed junctions, which will reduce substantially the surface leakage loss (leakage currents on the surface rather than through the junction).

13.5 INTEGRATED CIRCUITS

During the past decade the *integrated circuit* (IC) has through expanded usage and the various media of advertising, become a product whose basic function and purpose are now understood by the layperson. The most noticeable characteristic of an IC is its size. It is typically thousands of times smaller than a semiconductor structure built in the usual manner with discrete components. For example, the integrated circuit shown in Fig. 13.12 has 275,000 transistors in addition to a multitude of other elements although it is only 280 × 250 mils or about $\frac{9}{32}''$ by $\frac{1}{4}''$. The MC68030 is a microprocessor unit that is the heart of a microcomputer manufactured by companies such as Apple, Hewlett Packard, Motorola and others.

Integrated circuits are seldom, if ever, repaired; that is, if a single component within an IC should fail, the entire structure (complete circuit) is replaced—a more economical approach. There are three types of ICs commercially available on a large scale today. They include the monolithic, thin (or thick) film, and hybrid integrated circuits. Each will be introduced in this chapter.

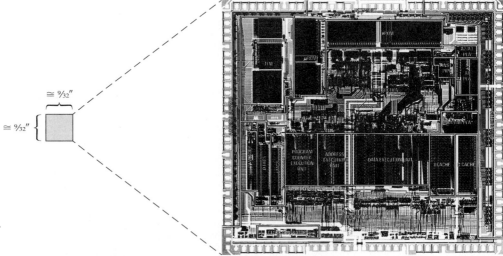

Figure 13.12 The MC68030 microprocessor and its actual external dimensions. (Courtesy Motorola, Inc.).

Recent Developments

Although the sequence of steps leading to the manufacture of an integrated circuit has not changed substantially during the past decade, the manner in which each step is performed has changed dramatically. In the early days the IC manufacturer designed, built, and maintained the equipment employed in the production cycle. Today, however, new industries have emerged that have assumed the responsibility of introducing the latest technological advances into the processing equipment. The result is that the manufacturer can concentrate on design, quality control, improved performance and reliability characteristics, and further miniaturization. The equipment available from the peripheral companies carries a high price tag (unit costs in excess of 1 million are not unusual), and 24-hour operation is almost a necessity to ensure sound economic policy. In an effort to ensure continuous operation, the larger IC manufacturers have their own service staff rather than having to rely on immediate response from the equipment manufacturer.

Automation continues to be important in the production cycle. A great deal of microprocessor control introduced in the form of "cassette addressment" has significantly reduced the possibility of error owing to incorrect transfer of information to the processing unit. It also has a sensitivity to the process being performed that is unavailable through the human response curve. A complete set of instructions is placed on a magnetic tape cassette and identified for use with the preparation of a particular IC wafer. The increased level of automation also reduces the amount of "handling" and contact with the wafer, thereby reducing the number of sources of contaminants and increasing the yield factor.

One of the continuing areas of concerns is the yield level. The average number of "good" die resulting from a wafer is improving but still remains at the 60 to 80% level. However, one must realize that as "feature" sizes decreases and density increases the yield level may not change significantly but the number of components produced in the same wafer area is increasing at a dramatic rate. In other words, if we utilized the improved production procedures of today on ICs manufactured 5 years ago the yield level would probably exceed 95%.

Developments of the last decade have resulted in a general acceptance by the industry that *IC density will just about double every two years*. At one time dimensions were provided in mils and square mils. Today, the micron or micrometer (1/millionth of a meter; μm) is the standard measure, with 1 mil $= \frac{1}{1000} = 25.4 \ \mu$m.

The increased density and improved yield levels are due to more sophisticated machinery in the production cycle, improved methods to detect and correct flaws, higher levels of cleanliness, increased purity levels of processing materials, improved manufacturing materials, and an increased number of processing steps.

Whereas class 100 rooms were common 5 years ago, class 10 is the current industry standard. A class 10 room is tenfold cleaner than a typical hospital environment. The class number indicates the number of particles 1 μm or larger per cubic foot. The cost of establishing such an environment can only be described as staggering. A continual laminar flow of filtered air is established between the floor and ceiling to maintain the high level of cleanliness. The white robe, boots, and hat ("bunny suits") that appear in some of the photographs in this chapter are required in production areas. Control is so tight that women working in many of these areas cannot wear makeup, to eliminate any possible introduction of foreign particles into the environment.

The water employed in the rinsing and cleaning operations is filtered at 0.2 μm and has an 18-MΩ resistivity level (recall the discussion of resistivity in Chapter 1). It is so free of organic contaminants that it will not support a culture growth. In addition, the purity of the processing materials, such as the chemicals, coatings, and other materials that "touch" the wafer, has improved to match the increased density levels.

The line widths of current manufacturing techniques extend from 0.8 to 2 μm for RF low power devices up to 10 to 15 μm for higher-power devices. The current state of the art is down to $\frac{1}{2}$-μm widths, but production is typically at the 1.5-μm level. It is expected that the production process will be down to 0.6 μm in the next 1 to 2 years.

Silicon has been the mainstay for the industry from the birth of the industry to today's production cycle. As density levels continue to increase and line widths decrease there may be a need to turn to materials such as GaAs (gallium arsenide) with its range of improved performance characteristics.

Due to large investments, it is an absolute necessity that the product processing be tightly controlled through a strong management system. The computer is now playing a very important role in providing the data required for such continuous surveillance of the production cycle. A number of improvements in the manufacturing process are described below as each production step is described.

13.6 MONOLITHIC INTEGRATED CIRCUIT

The term *monolithic* is derived from a combination of the Greek words *monos,* meaning single, and *lithos* meaning stone, which in combination result in the literal translation, single-stone, or more appropriately, single-solid structure. As this descriptive term implies, the monolithic IC is constructed within a *single* wafer of semiconductor material. Wafers as thin as 1/1000 inch ($\cong$ one-fifth of the thickness of this page) can be obtained using a slicing process as shown in Fig. 13.13. The greater portion of the wafer will simply act as a supporting structure for the very thin resulting IC. An overall view of the stages involved in the fabrication of monolithic ICs is provided in Fig. 13.14. The actual number of steps leading to a finished product is many times that appearing in Fig. 13.14. The figure does, however, point out the major production phases of forming a monolithic IC.

As indicated in the preceding section, the processing equipment and not the steps have changed significantly in recent years. As indicated in the figure, it is first necessary to design a circuit that will meet the specifications. The circuit must then be laid out in order to ensure optimum use of available space and a minimum of difficulty in performing the diffusion processes to follow. The appearance of the mask and its function in the sequence of stages indicated will be introduced in Section 13.8. For the moment, let it suffice to say that a mask has the appearance of a negative through

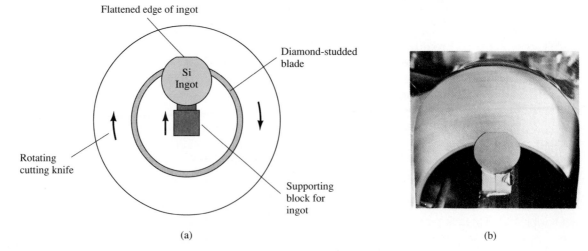

Figure 13.13 Slicing the single-crystal ingot into wafers. (Courtesy Texas Instruments, Inc.).

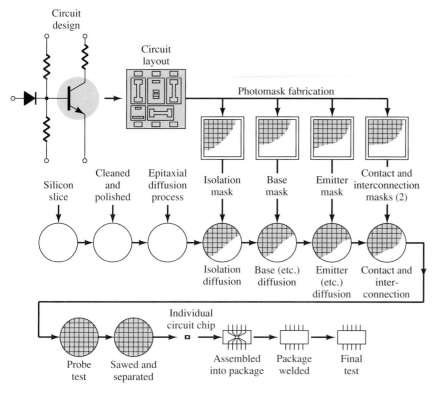

Figure 13.14 Monolithic integrated-circuit fabrication. (Courtesy Robert Hibberd.)

which impurities may be diffused (through the light areas) into the silicon slice. The actual diffusion process for each phase is similar to that applied in the fabrication of diffused transistors. The last mask of the series will control the placement of the interconnecting conducting pattern between the various elements. The wafer then goes through various testing procedures, is sawed and separated into individual chips, packaged and assembled as indicated. A processed silicon wafer appears in Fig. 13.15. The original wafer can be anywhere from 3 to 8 in. in diameter. The size of

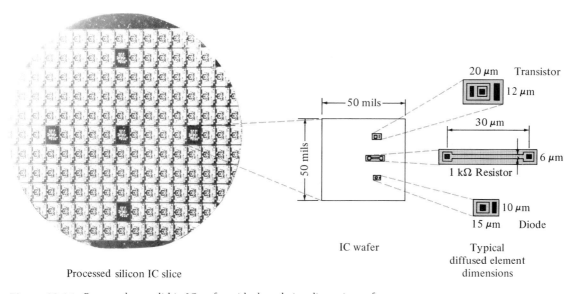

Figure 13.15 Processed monolithic IC wafer with the relative dimensions of the various elements. (Courtesy Robert Hibberd.)

each chip will, of course, determine the number of individual circuits resulting from a single wafer. The dimensions of each chip of the wafer in Fig. 13.15 are 50 × 50 mils. To point out the microminiature size of these chips, consider that 20 of them can be lined up along a 1-in. length. The average relative size of the elements of a monolithic IC appear in Fig. 13.15. Note the large area required for the 1-kΩ resistor as compared to the other elements indicated. In the next section we examine the basic construction of each of these elements.

A recent article indicated, by percentage, the relative costs of the various stages in the production of monolithic ICs as compared to discrete transistors. The resulting graphs appear in Fig. 13.16. The processing phase includes all stages leading up to the individual chips of Fig. 13.15. Note the difference in cost for the various phases of production as determined by the size and density of the chip.

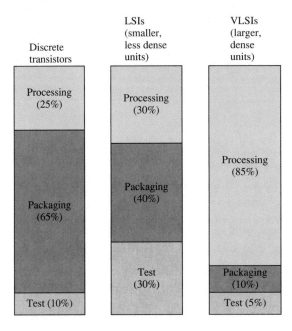

Figure 13.16 Cost breakdown for the manufacturing of discrete transistors and large-scale integrated circuits (LSI$_s$) and very large-scale integrated circuits (VLSI$_s$).

13.7 MONOLITHIC CIRCUIT ELEMENTS

The surface appearance of the transistor, diode, and resistor appear in Fig. 13.15. We shall now examine the basic construction of each in more detail.

Resistor

You will recall that the resistance of a material is determined by the resistivity, length, area, and temperature of the material. For the integrated circuit, each necessary element is present in the sheet of semiconductor material appearing in Fig. 13.17. As indicated in the figure, the semiconductor material can be either p- or n-type, although the p-type is most frequently employed.

The resistance of any bulk material is determined by

$$R = \rho \frac{l}{A}$$

For $l = w$, resulting in a square sheet,

$$R = \frac{\rho l}{yw} = \frac{\rho l}{yl}$$

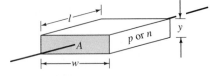

Figure 13.17 Parameters determining the resistance of a sheet of semiconductor material.

Chapter 13 Discrete and IC Manufacturing Techniques

and

$$R_S = \frac{\rho}{y} \qquad \text{ohms} \qquad (13.1)$$

where ρ is in ohm-centimeters and y is in centimeters. R_S is called the *sheet resistance* and has the units ohms per square. The equation clearly reveals that the sheet resistance is independent of the size of the square.

In general, where $l \neq w$,

$$R = R_S \frac{l}{w} \qquad \text{ohms} \qquad (13.2)$$

For the resistor appearing in Fig. 13.15, $w = 6$ μm, $l = 30$ μm, and $R_S = 200$ Ω/square:

$$R = R_S \frac{l}{w} = 200 \times \frac{30}{6} = 1 \text{ k}\Omega$$

A cross-sectional view of a monolithic resistor appears in Fig. 13.18 along with the surface appearance of two monolithic resistors. In Fig. 13.18a the sheet resistive material (p) is indicated with its aluminum terminal connections. The n-isolation region performs exactly that function indicated by its name; that is, it isolates the monolithic resistive elements from the other elements of the chip. Note in Fig. 13.18b the method employed to obtain a maximum l in a limited area. The resistors of Fig. 13.18 are called base-diffusion resistors since the p-material is diffused into the p-type substrate during the base-diffusion process indicated in Fig. 13.14.

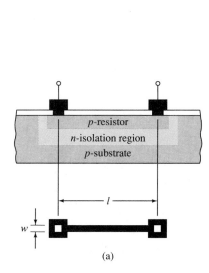

(a)

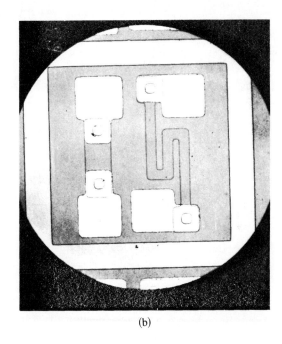

(b)

Figure 13.18 Monolithic resistors: (a) cross section and determining dimensions; (b) surface view of two monolithic resistances in a single die. (Courtesy Motorola, Inc.)

Capacitor

Monolithic capacitive elements are formed by making use of the transition capacitance of a reverse-biased *p-n* junction. At increasing reverse-bias potentials, there is an increasing distance at the junction between the *p*- and *n*-type impurities. The region between these oppositely doped layers is called the depletion region due to the absence of "free" carriers. The necessary elements of a capacitive element are therefore present—the depletion region has insulating characteristics that separate the two oppositely charged layers. The transition capacitance is related to the width (W) of the depletion region, the area (A) of the junction, and the permittivity (ϵ) of the material within the depletion region by

$$C_T = \epsilon \frac{A}{W} \tag{13.3}$$

The cross section and surface appearance of a monolithic capacitive element appear in Fig. 13.19. The reverse-biased junction of interest is J_2. The undesirable parasitic capacitance at junction J_1 is minimized through careful design. Due to the fact that aluminum is p-type impurity in silicon, a heavily doped n^+ region is diffused into the n-type region as shown to avoid the possibility of establishing an undesired p-n junction at the boundary between the aluminum contact and the n-type impurity region.

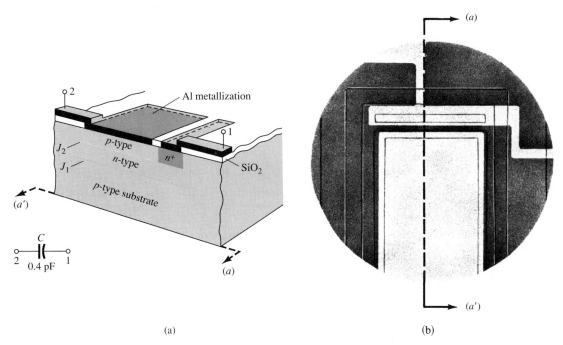

Figure 13.19 Monolithic capacitor: (a) cross section; (b) photograph. (Courtesy Motorola, Inc.)

Inductor

Whenever possible, inductors are avoided in the design of integrated circuits. An effective technique for obtaining nominal values of inductances has so far not been devised for monolithic integrated circuits. In many instances, the need for inductive elements can be eliminated through the use of a technique known as *RC* synthesis. Thin (or thick) film or hybrid integrated circuits have an option open to them that cannot be employed in monolithic integrated circuits: the addition of discrete inductive elements to the surface of the structure. Even with this option, however, they are seldom employed due to their relatively bulky nature.

Transistors

The cross section of a monolithic transistor appears in Fig. 13.20. Note again the presence of the n^+ region in the n-type epitaxial collector region. The vast majority of

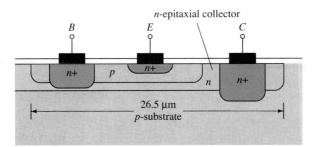

Figure 13.20 Cross section of a monolithic *npn* transistor.

monolithic IC transistors are *npn* rather than *pnp* for reasons to be found in more advanced texts on the subject. Keep in mind when examining Fig. 13.20 that the *p*-substrate is only a supporting and isolating structure forming no part of the active device itself. The base, emitter, and collector regions are formed during the corresponding diffusion processes of Fig. 13.14. The order of diffusion is *C*, *E*, and *B* as defined by the depth and area of diffusion.

Diodes

The diodes of a monolithic integrated circuit are formed by first diffusing the required regions of a transistor and then masking the diode rather than transistor terminal connections. There is, however, more than one way of hooking up a transistor to perform a basic diode action. The two most common methods applied to monolithic integrated circuits appear in Fig. 13.21. The structure of a *BC-E* diode appears in Fig. 13.22.

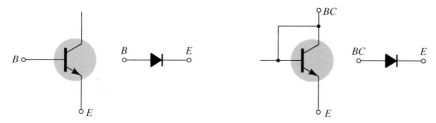

Figure 13.21 Transistor structure and two possible connections employed in the formation of monolithic diodes.

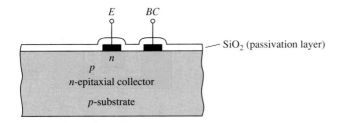

Figure 13.22 Cross-sectional view of a *BC-E* monolithic diode.

13.8 MASKS

The selective diffusion required in the formation of the various active and passive elements of an integrated circuit is accomplished through the use of masks such as that shown in Fig. 13.23. We shall find in Section 13.9 that the light areas are the only areas through which donor and acceptor impurities can pass. The dark areas will block the diffusion of impurities somewhat as a shade will prevent sunlight from changing the pigment of the skin. The next section will demonstrate the use of these masks in the formation of a computer logic circuit.

Figure 13.23 Mask. (Courtesy Motorola, Inc.)

The sequence of steps leading to the final mask is controlled by the micron width of the smallest features on the wafer. For the range 0.5 to 2 μm, electron beam lithography must be employed, whereas for the range 3 to 5 μm, less sophisticated (and less expensive) methods can be employed.

At one time the making of a mask first required a large-scale stabilene drawing of all layers. The artwork was then transferred to a clear Mylar coated with red plastic called *Rubylith*. Very precise cuts were made in the red material and sections peeled off to reveal the regions through which the diffusion of impurities could take place. The resulting pattern was then photographed and reduced by 500× (500 times the desired size for production) in a series of steps until the desired master (reticle) was obtained.

Today, the same stabilene drawing is taken to a CAD (computer-aided-design) area called Calma (the manufacturer of the equipment employed) and is mounted on a digitizing board as shown in Fig. 13.24. The operator performs a function called

Figure 13.24 Calma digitizing station. (Courtesy Calma, Inc.)

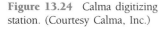

digitizing through which the geometrical features from the stabilene drawing are transferred to the computer memory of the Calma system. The same procedure is repeated for each layer of the integrated circuit. Any section of the design can then be recalled and placed on the CRT of an editing station to modify the pattern, check continuity, and so on, as shown in Fig. 13.25. One distinct advantage of the Calma system is that repetitive cells can be added from memory at the editing stage and not appear as part of the stabilene drawing. In addition, a library of cells could be established that could be introduced as required. The net result is that a skilled operator could design and develop a major portion of the complete reticle with a minimum of effort at the stabilene stage.

Figure 13.25 Circuit design/modification at a Calma (interactive computer graphics system) editing/tablet station. (Courtesy Calma, Inc.)

(a)

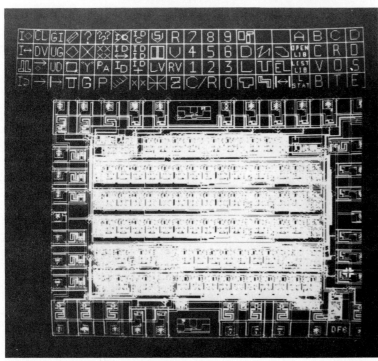

(b)

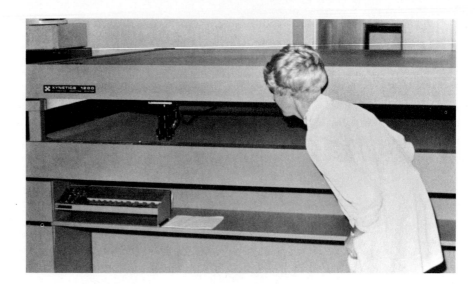

Figure 13.26 Xynetics 1200 Plotter (on line with the Calma system). (Courtesy Xynetics, Inc.)

At this juncture there are two methods that can be applied to produce the final mask. One method employs the Xynetics plotter shown in Fig. 13.26 to generate a precise line drawing from the descriptive data provided by the Calma system. It is an amazing piece of equipment to watch in operation, with its multicolor option and very quick precise movements. It can enlarge a region to $10,000\times$ if required. A photo-reduction process can then be applied to reduce the pattern to $100\times$ and then $10\times$. The $10\times$ pattern can be obtained in one step utilizing a *pattern generator*. The magnetic tape on which the pattern data are stored is loaded into the console, and a $10\times$ pattern is obtained through computer control of the exposure shapes and placement. For a complex LSI, 200,000 to 300,000 flashes may be required over a period of 20 to 36 hours. A programmed step-and-repeat machine will then generate the required format (multiple images on the same mask) from the $10\times$ pattern for the product group.

Although the procedure described above is still employed today, most production cycles are changing over to a "direct write" approach using an electron beam (E-beam) system such as appearing in Fig. 13.27. It saves a number of intermediate steps

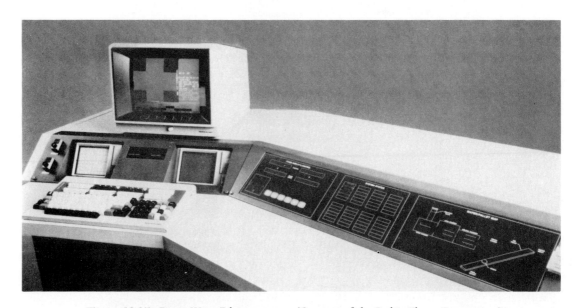

Figure 13.27 Direct-Write E-beam system. (Courtesy of the Perkin-Elmer Corporation.)

Chapter 13 Discrete and IC Manufacturing Techniques

by "cutting" the mask pattern directly from the Calma data using an electron beam as the "exposure tool." The reduced number of steps and the direct exposure of the mark reduce the number of flaws and omissions that may appear in the final product. For VLSI units, the time involved from initial design to mask availability may extend from a few days to 1 to 2 months.

13.9 MONOLITHIC INTEGRATED CIRCUIT—THE NAND GATE

This section is devoted to the sequence of production stages of a monolithic NAND-gate circuit for computer logic design. A detailed examination of each process would require many more pages than it is possible to include in this text. The description, however, should be sufficiently complete to establish a foundation for any future contact with this rapidly changing area. The circuit to be prepared appears in Fig. 13.28a. The criteria of space allocation, placement of pin connection, and so on require that the elements be situated in the relative positions indicated in Fig. 13.28b. The regions to be isolated from one another appear within the solid heavy lines. A set of masks for the various diffusion processes must then be made up for the circuit as it appears in Fig. 13.28b.

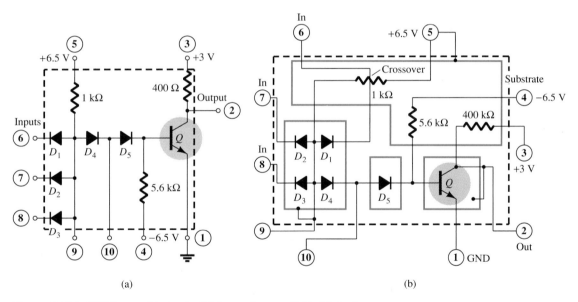

Figure 13.28 NAND gate: (a) circuit; (b) layout for monolithic fabrication.

Now we shall proceed slowly through the first diffusion process to demonstrate the natural sequence of steps that must be followed through each diffusion process indicated in Fig. 13.14.

p-Type Silicon Wafer Preparation

After being sliced from the grown ingot, a p-type silicon wafer is lapped and polished (Fig. 13.29a) and checked (Fig. 13.29) to produce the structure of Fig. 13.29. A chemical etch process is also applied to further smooth the surface and remove a layer of the wafer that may have been damaged during the lapping and polishing sequence.

(a)

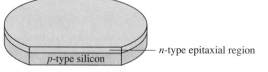

(b)

(c)

p-type
silicon wafer

Figure 13.29 (a) Lapping and polishing stage of wafer preparation; (b) checking wafer with computer; (c) *p*-type silicon wafer. (Courtesy of Texas Instruments, Inc.)

n-Type Epitaxial Region

An *n*-type epitaxial region is then grown on to the *p*-type substrate as shown in Fig. 13.30. It is deposited in a way that will result in a single-crystal structure having the same crystal structure and orientation as the substrate but with a different conductivity level. It is in this thin epitaxial layer that the active and passive elements will be diffused. The *p*-type area is essentially a supporting structure that adds some thickness to the structure to increase its strength and permit easier handling.

p-type silicon — *n*-type epitaxial region

Figure 13.30 *p*-Type silicon wafer after the *n*-type epitaxial diffusion process.

The apparatus most frequently employed in the deposition process is the radiantly heated barrel reactor of Fig. 13.31. The *susceptor* (silicon coated graphite) is a hexagonal cross-section structure upon which several wafers are placed on each face. The gases with the desired impurities are injected into the top of the chamber and exhausted from the bottom. The wafers are heated by water-cooled quartz lamps. By holding the wafers in a near vertical position (only 2.5° from the vertical) there is less chance of wafer contamination.

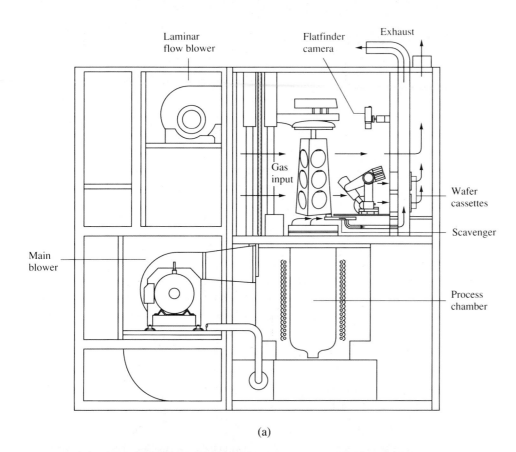

(a)

Figure 13.31 Radiantly heated barrel reactor (a) Schematic; (b) Non-contaminating wafer placement; (c) External control. (Courtesy Applied Materials, Inc.)

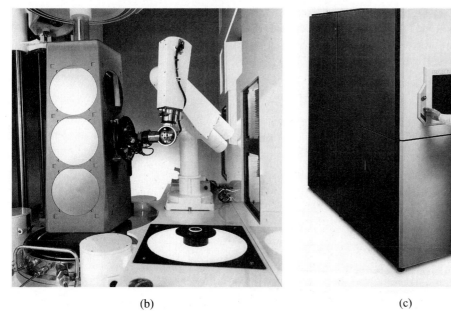

(b)   (c)

Silicon Oxidation (SiO$_2$)

The resulting wafer is then subjected to an oxidation process resulting in a surface layer of SiO$_2$ (silicon dioxide) as shown in Fig. 13.32. This surface layer will prevent any impurities from entering the *n*-type epitaxial layer. However, selective etching of this layer will permit the diffusion of the proper impurity into designated areas of the *n*-type epitaxial region of the silicon wafer.

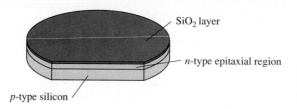

Figure 13.32 Wafer of Fig. 13.30 following the deposit of the SiO$_2$ layer.

The apparatus employed in the oxidation process is similar to that used to establish the epitaxial layer in that the wafers are placed in a boat (now made of quartz) and inserted into a quartz tube. Typically, about 200 wafers are introduced at the same time. In this case, however, a resistance heater is wound around the tube to raise the temperature to about 1100°C. Oxygen is introduced in the wet or dry form until the desired SiO$_2$ layer is established. Recent developments include raising the atmospheric pressure in the vessel to permit a measurable reduction in the processing temperature. For every 1 atm (atmosphere) increase in pressure, there is a 30°C reduction in required temperature. At 10 atm, the temperature can be reduced 300°C. At lower processing temperatures there is also an improved quality of oxide, a reduction in introduced stresses, and reduction or elimination of a number of device design limitations.

The time involved for the oxidation process can extend from a few hours to as long as 24 hours depending on the thickness of the oxide and the quality desired.

Photolithographic Process

The selective etching of the SiO$_2$ layer is accomplished through the use of a photolithographic process. The wafer is first coated with a thin layer of photosensitive material, commonly called *photoresist,* by the system appearing in Fig. 13.33. The application of the photoresist is entirely computer controlled. A deck of wafers is deposited into the receiving cassette appearing in the left-hand region of Fig. 13.33. The equipment automatically applies a high-pressure scrub, a dehydration process, the resist coating, and a soft bake. Similar equipment then develops and hard bakes the wafers.

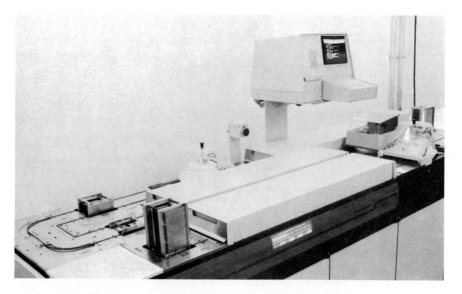

Figure 13.33 Microprocessor-controlled photoresist module. (Courtesy Motorola, Inc.)

The next step is to use one of the masks developed earlier to determine those areas of the SiO$_2$ layer to be removed in preparation for the isolation diffusion process. The mask can either be placed directly on the photoresist as shown in Fig. 13.34 or separated from the mask using the equipment of Fig. 13.35. When placed directly on the wafer an ultraviolet light is applied that will expose those regions of the photosensitive material not covered by the masking pattern (Fig. 13.34).

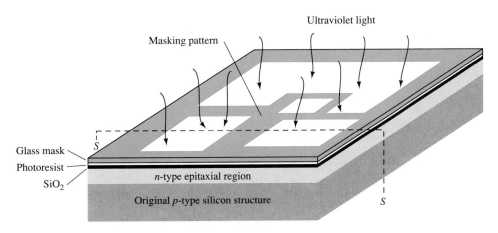

Figure 13.34 Photolithographic process: the application of ultraviolet light after the mask is properly set; the structure is only one of the 200, 400, or even 500 individual NAND gate circuits being formed on the wafer of Figs. 13.29 through 13.32.

Figure 13.35 600 HT Micralign Projection Printing System. (Courtesy Perkin-Elmer Corporation.)

The method employed in Fig. 13.35 is called *projection printing*. It employs optics to expose the various regions. The primary advantage of this approach is that the mask cannot introduce contaminants to the wafer surface. It is the most frequently used approach of recent years.

The resulting wafer is then subjected to a chemical solution that will remove the unexposed photosensitive material. A cross section of a chip (S-S of Fig. 13.34) will then appear as indicated in Fig. 13.36. A second solution will then etch away the SiO$_2$ layer from any region not covered by the photoresist material (Fig. 13.37). The final step before the diffusion process is the removal of the remaining photosensitive material. The structure will then appear as shown in Fig. 13.38.

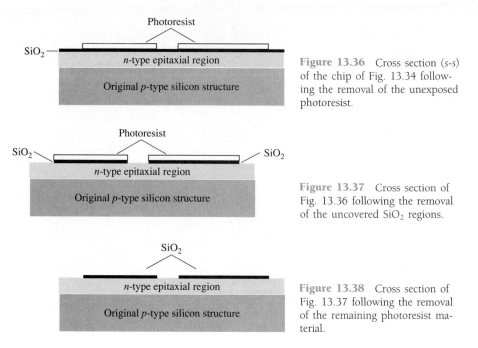

Figure 13.36 Cross section (s-s) of the chip of Fig. 13.34 following the removal of the unexposed photoresist.

Figure 13.37 Cross section of Fig. 13.36 following the removal of the uncovered SiO₂ regions.

Figure 13.38 Cross section of Fig. 13.37 following the removal of the remaining photoresist material.

Isolation Diffusion

The structure of Fig. 13.38 is then subjected to a p-type diffusion process resulting in the islands of n-type regions indicated in Fig. 13.39. The diffusion process ensures a heavily doped p-type region (indicated by p^+) between the n-type islands. The p^+ regions will result in improved *isolation* properties between the active and passive components to be formed in the n-type islands.

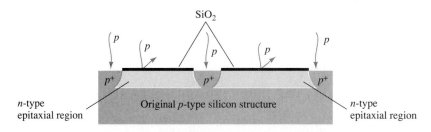

Figure 13.39 Cross section of Fig. 13.38 following the isolation diffusion process.

The apparatus employed includes a quartz boat and tube (to minimize the possibility of contaminating the processing environment) that is heated by a high-resistance wire wrapped around the tube. The diffusion operation normally occurs at a temperature neighboring on 1200°C. The system as appearing in Fig. 13.40 is totally microprocessor controlled. Three or four people can operate 16 furnaces; and the entire operation, from pulling the boats in and out of the furnaces to monitoring the temperature and dopant level, is computer controlled.

An alternative to the high-temperature diffusion process is *ion implantation*. A beam of dopant ions (about the size of a pencil) is directed toward a wafer at a very high velocity by an ion-accelerator. The ions will penetrate the medium to a level that can be controlled to within $\frac{1}{2}$ μm as compared to $2\frac{1}{2}$ μm using other methods. In addition to improved control, the processing temperature is lower, and a broader range of electrical parameters is now available. At present the primary use of this method is in establishing bases. In time, with the proper modifications, emitter diffusions will also be a possibility.

Chapter 13 Discrete and IC Manufacturing Techniques

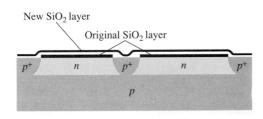

Figure 13.40 Microprocessor-controlled diffusion operation (with 5-in. wafers). (Courtesy Motorola, Inc.)

In preparation for the next masking and diffusion process, the entire surface of the wafer is coated with an SiO_2 layer as indicated in Fig. 13.41.

Figure 13.41 In preparation for the next diffusion process, the entire wafer is coated with an SiO_2 layer.

Base and Emitter Diffusion Processes

The isolation diffusion process is followed by the base and emitter diffusion cycles. The sequence of steps in either case is the same as that encountered in the description of the isolation diffusion process. Although "base" and "emitter" refer specifically to the transistor structure, necessary parts (layers) of each element (resistor, capacitor, and diodes) will be formed during each diffusion process. The surface appearance of the NAND gate after the isolation base and emitter diffusion processes appears in

Fig. 13.42. The mask employed in each process is also provided above each photograph. The cross section of the transistor will appear as shown in Fig. 13.43 after the base and emitter diffusion cycles.

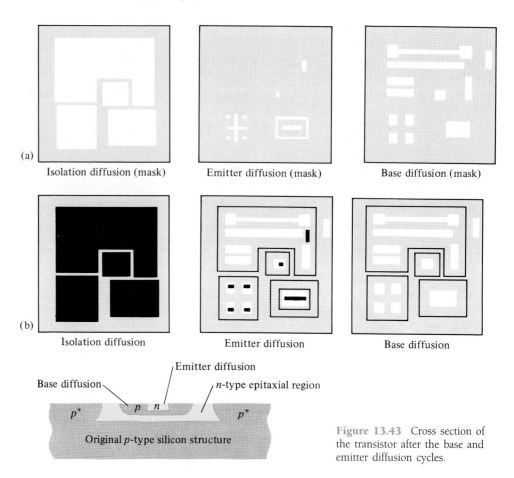

(a)

Isolation diffusion (mask)   Emitter diffusion (mask)   Base diffusion (mask)

Figure 13.42 Masks (a) employed in each diffusion process (b). The surface appearance of the monolithic NAND gate after the isolation, base, and emitter diffusion processes. (Courtesy Motorola Monitor.)

(b)

Isolation diffusion   Emitter diffusion   Base diffusion

Base diffusion   Emitter diffusion n-type epitaxial region

p^+ p n p^+

Original p-type silicon structure

Figure 13.43 Cross section of the transistor after the base and emitter diffusion cycles.

Preohmic Etch

In preparation for a good ohmic contact, n^+ regions (see Section 13.7) are diffused into the structure as indicated by the light areas of Fig. 13.44. Note the correspondence between the light areas and the mask pattern.

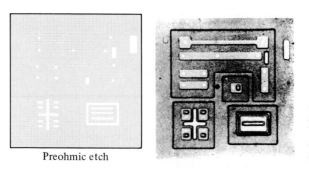

Preohmic etch

Figure 13.44 Surface appearance of the chip of Fig. 13.42 after the preohmic etch cycle. The mask employed is also included. (Courtesy Motorola Monitor.)

Metallization

A final masking pattern exposes those regions of each element to which a metallic contact must be made. The entire wafer is then coated with a thin layer of aluminum, gold, molybdenum, or tantalum (all high-conductivity, low-boiling-point metals) that after being properly etched will result in the desired interconnecting conduction pattern. A photograph of the completed metallization process appears in Fig. 13.45.

The two methods most commonly applied to establish the uniform layer of conducting material are *evaporation* and *sputtering*. In the former, the metal is either melted through the use of heating coils or bombarded with an electron gun (E-gun) to result in the evaporation of the source metal. The metallization material is thereby sprayed over the wafers, which are held by clips in a drum or hemisphere structure such as is shown in Fig. 13.46.

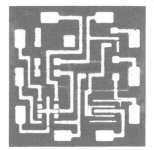

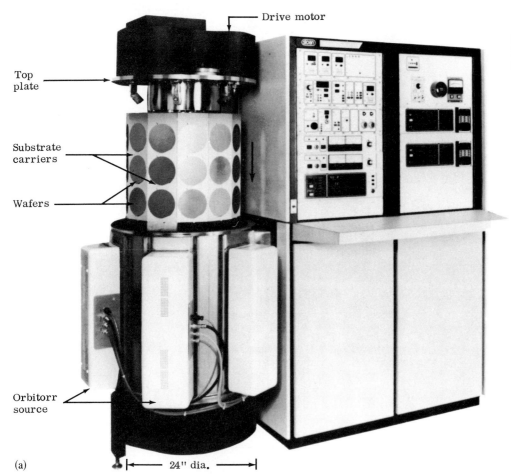

Metalization

Figure 13.45 Completed metallization process. (Courtesy Motorola Monitor.)

Figure 13.46 Deposition of the interconnect metal through metal evaporation. (Courtesy Motorola, Inc.)

An automated sputtering system that employs robotic units such as that shown in Fig. 13.47 places the source metal (at a very high negative potential) opposite, but not touching, an anode plate at a positive potential. An inert gas, such as argon, introduced between the plates will release positive ions that will bombard the negative plate and sputter some of the surface metal from the source. The ''free'' metal will then be deposited on the wafers on the surface of the anode.

Figure 13.47 Automated sputter coater. (Courtesy Perkin-Elmer Corporation.)

The sputtering technique is often preferred over the evaporation method because the coverage is less line of sight. There is therefore a more uniform layer of deposition over abrupt junctions. The complete structure with each element indicated appears in Fig. 13.48. Try to relate the interconnecting metallic pattern to the original circuit of Fig. 13.28a.

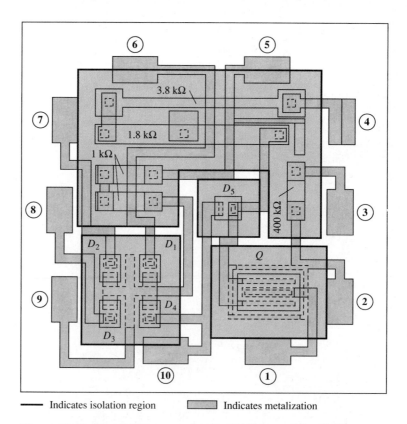

—— Indicates isolation region   ▨ Indicates metalization

Figure 13.48 Monolithic structure for the NAND gate of Fig. 13.28.

Passivation

An SiO_2 layer deposited on the surface of the entire structure will be an effective protection layer for water vapor and some contaminants. However, certain metal ions can migrate through the SiO_2 layer and disturb the device characteristics. In an effort to improve the passivation process, a layer (2000 to 5000 Å) of glass (plasma silicon nitride) is applied to further reduce the degradation problem.

Testing

Before sawing the wafer into the individual dies, an electrical test of each die is performed by the inspection system appearing in Fig. 13.49. The system automatically loads/unloads the wafers using carousels to cut down further on the degree of "handling." This process, as with most in the production cycle, is also computer controlled. There is a probe card for each IC that will permit not only rejection but categorization of the type of faults (open, short, gain, etc). The bad die are identified by a red dot automatically deposited by the inspection system.

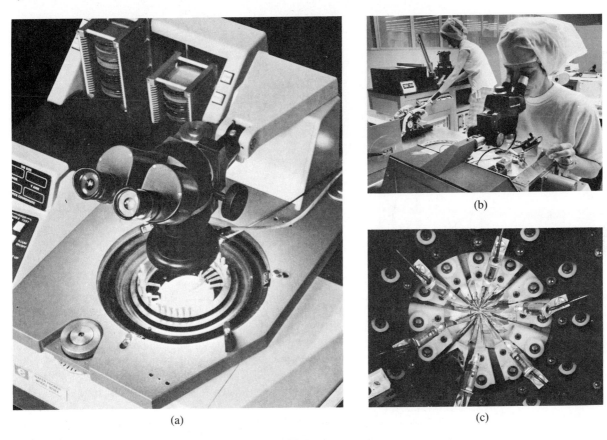

(a)

(b)

(c)

Figure 13.49 Electrical testing of the individual dies. [(a) Multi-probe test station, Courtesy Electroglas, Inc.; (b) Manual inspection, Courtesy Texas Instruments, Inc.; (c) Multi-probe contacts touch bond-pack on chips, Courtesy Autonetics, North American Rockwell Corporation.]

Packaging

Once the metallization and testing processes are complete, the wafer must be separated into its individual chips. This is accomplished through the sawing process. Each individual chip may then be packaged in one of the forms shown in Fig. 13.50.

Figure 13.50   Monolithic packaging techniques. (Courtesy Motorola Inc.)

13.10   THIN- AND THICK-FILM INTEGRATED CIRCUITS

The general characteristics, properties, and appearance of thin- and thick-film integrated circuits are similar, although they both differ in many respects from the monolithic integrated circuit. They are not formed within a semiconductor wafer but *on* the surface of an insulating substrate such as glass or an appropriate ceramic material. In addition, *only* passive elements (resistors, capacitors) are formed through thin or thick film techniques on the insulating surface. The active elements (transistors, diodes) are added as *discrete* elements to the surface of the structure after the passive elements have been formed. The discrete active devices are frequently produced using the monolithic process.

The primary difference between the thin- and thick-film techniques is the process employed for forming the passive components and the metallic conduction pattern. The thin-film circuit employs an evaporation or cathode-sputtering technique; the thick film employs silk-screen techniques. Priorities do not permit a detailed description of these processes here.

In general, the passive components of film circuits can be formed with a broader range of values and reduced tolerances as compared to the monolithic IC. The use of discrete elements also increases the flexibility of design of film circuits although, obviously, the resulting circuit will be that much larger. The cost of film circuits with a larger number of elements is also, in general, considerably higher than that of monolithic integrated circuits.

13.11 HYBRID INTEGRATED CIRCUITS

The term *hybrid integrated circuit* is applied to the wide variety of multichip integrated circuits and also those formed by a combination of the film and monolithic IC techniques. The multichip integrated circuit employs either the monolithic or film technique to form the various components, or set of individual circuits, which are then interconnected on an insulating substrate and packaged in the same container. Integrated circuits of this type appear in Fig. 13.51. In a more sophisticated type of hybrid integrated circuit the active devices are first formed within a semiconductor wafer which is subsequently covered with an insulating layer such as SiO_2. Film techniques are then employed to form the passive elements on the SiO_2 surface. Connections are made from the film to the monolithic structure through "windows" cut in the SiO_2 layer.

Figure 13.51 Hybrid integrated circuits.
(Courtesy Texas Instruments, Inc.)

14 Operational Amplifiers

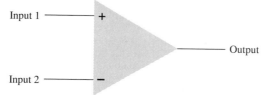

14.1 INTRODUCTION

An operational amplifier or op-amp is a very high gain differential amplifier with high input impedance and low output impedance. Typical uses of the operational amplifier are to provide voltage amplitude changes (amplitude and polarity), oscillators, filter circuits, and many types of instrumentation circuits. An op-amp contains a number of differential amplifier stages to achieve a very high voltage gain.

Figure 14.1 shows a basic op-amp with two inputs and one output as would result using a differential amplifier input stage. Recall from Chapter 12 that each input results in either the same or an opposite polarity (or phase) output, depending on whether the signal is applied to the plus $(+)$ or the minus $(-)$ input.

Input 1 ——— +

Output

Input 2 ——— −

Figure 14.1 Basic op-amp.

Single-Ended Input

Single-ended input operation results when the input signal is connected to one input with the other input connected to ground. Figure 14.2 shows the signals connected for

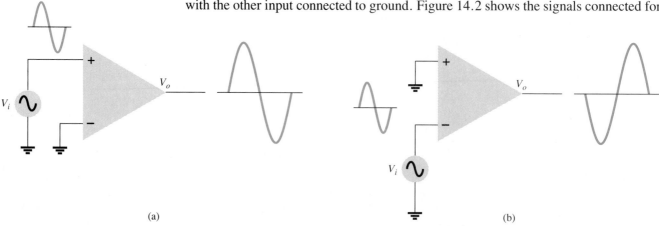

(a)

(b)

Figure 14.2 Single-ended operation.

this operation. In Fig. 14.2a the input is applied to the plus input (with minus input at ground), which results in an output having the same polarity as the applied input signal. Figure 14.2b shows an input signal applied to the minus input, the output then being opposite in phase to the applied signal.

Double-Ended (Differential) Input

In addition to using only one input, it is possible to apply signals at each input—this being a double-ended operation. Figure 14.3a shows an input, V_d, applied between the two input terminals (recall that neither input is at ground), with the resulting amplified output in phase with that applied between the plus and minus inputs. Figure 14.3b shows the same action resulting when two separate signals are applied to the inputs, the difference signal being $V_{i_1} - V_{i_2}$.

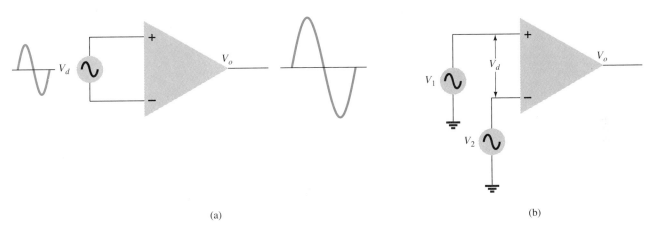

(a)   (b)

Figure 14.3 Double-ended (differential) operation.

Double-Ended Output

While the operation discussed so far had a single output, the op-amp can also be operated with opposite outputs, as shown in Fig. 14.4. An input applied to either input will result in outputs from both output terminals, these outputs always being opposite in polarity. Figure 14.5 shows a single-ended input with a double-ended output. As shown, the signal applied to the plus input results in two amplified outputs of opposite polarity. Figure 14.6 shows the same operation with a single output

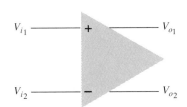

Figure 14.4 Double-ended output.

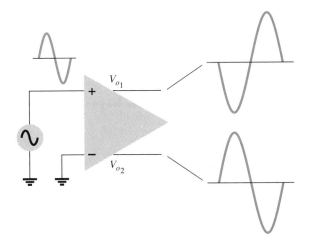

Figure 14.5 Double-ended output with single-ended input.

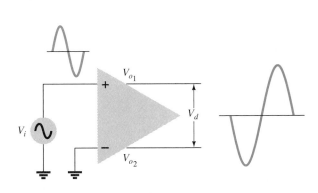

Figure 14.6 Double-ended output.

measured between output terminals (not with respect to ground). This difference output signal is $V_{o_1} - V_{o_2}$. The difference output is also referred to as a *floating signal* since neither output terminal is the ground (reference) terminal. Notice that the difference output is twice as large as either V_{o_1} or V_{o_2} since they are of opposite polarity and subtracting them results in twice their amplitude [i.e., $10 \text{ V} - (-10 \text{ V}) = 20 \text{ V}$]. Figure 14.7 shows differential input, differential output operation. The input is applied between the two input terminals and the output taken from between the two output terminals. This is fully differential operation.

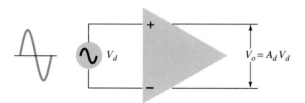

Figure 14.7 Differential-input, differential-output operation.

Common-Mode Operation

When the same input signals are applied to both inputs, common-mode operation results, as shown in Fig. 14.8. Ideally, the two inputs are equally amplified and since they result in opposite polarity signals at the output, these signals cancel, resulting in 0 V output. Practically, a small output signal will result.

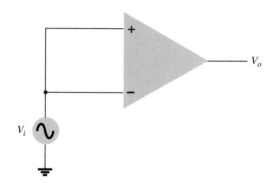

Figure 14.8 Common-mode operation.

Common-Mode Rejection

A significant feature of a differential connection is that the signals which are opposite at the inputs are highly amplified, while those which are common to the two inputs are only slightly amplified—the overall operation being to amplify the difference signal while rejecting the common signal at the two inputs. Since noise (any unwanted input signal) is generally common to both inputs, the differential connection tends to provide attenuation of this unwanted input while providing an amplified output of the difference signal applied to the inputs. This operating feature, referred to as common-mode rejection, is discussed more fully in the next section.

14.2 DIFFERENTIAL AND COMMON-MODE OPERATION

One of the more important features of a differential circuit connection, as provided in an op-amp, is the circuit's ability to greatly amplify signals that are opposite at the two inputs, while only slightly amplifying signals that are common to both inputs. An op-amp provides an output component that is due to the amplification of the difference of the signals applied to the plus and minus inputs and a component due to the signals common to both inputs. Since amplification of the opposite input signals is much greater than that of the common input signals, the circuit provides a common-mode rejection as described by a numerical value called the common-mode rejection ratio (CMRR).

Differential Inputs

When separate inputs are applied to the op-amp, the resulting difference signal is the difference between the two inputs.

$$V_d = V_{i_1} - V_{i_2} \qquad\qquad (14.1)$$

Common Inputs

When both input signals are the same, a common signal element due to the two inputs can be defined as the average of the sum of the two signals,

$$V_c = \tfrac{1}{2}(V_{i_1} + V_{i_2}) \qquad\qquad (14.2)$$

Output Voltage

Since any signals applied to an op-amp in general have both in-phase and out-of-phase components, the resulting output can be expressed as

$$V_o = A_d V_d + A_c V_c \qquad\qquad (14.3)$$

where V_d = difference voltage given by Eq. (14.1)
V_c = common voltage given by Eq. (14.2)
A_d = differential gain of the amplifier
A_c = common-mode gain of the amplifier

Opposite Polarity Inputs

If opposite polarity inputs applied to an op-amp are ideally opposite signals, $V_{i_1} = -V_{i_2} = V_s$, the resulting difference voltage is

$$\text{Eq. (14.1):} \quad V_d = V_{i_1} - V_{i_2} = V_s - (-V_s) = 2V_s$$

while the resulting common voltage is

$$\text{Eq. (14.2):} \quad V_c = \tfrac{1}{2}(V_{i_1} + V_{i_2}) = \tfrac{1}{2}[V_s + (-V_s)] = 0$$

so that the resulting output voltage is

$$\text{Eq. (14.3):} \quad V_o = A_d V_d + A_c V_c = A_d(2V_s) + 0 = 2A_d V_s$$

This shows that when the inputs are an ideal opposite signal (no common element), the output is the differential gain times twice the input signal applied to one of the inputs.

Same Polarity Inputs

If the same polarity inputs are applied to an op-amp, $V_{i_1} = V_{i_2} = V_s$, the resulting difference voltage is

$$\text{Eq. (14.1):} \quad V_d = V_{i_1} - V_{i_2} = V_s - V_s = 0$$

while the resulting common voltage is

$$\text{Eq. (14.2):} \quad V_c = \tfrac{1}{2}(V_{i_1} + V_{i_2}) = \tfrac{1}{2}(V_s + V_s) = V_s$$

so that the resulting output voltage is

$$\text{Eq. (14.3):} \quad V_o = A_d V_d + A_c V_c = A_d(0) + A_c V_s = A_c V_s$$

This shows that when the inputs are ideal in-phase signals (no difference signal), the output is the common-mode gain times the input signal, V_s, which shows that only common-mode operation occurs.

Common-Mode Rejection

The solutions above provide the relationships that can be used to measure A_d and A_c in op-amp circuits.

1. *To measure A_d:* Set $V_{i_1} = -V_{i_2} = V_s = 0.5$ V, so that

$$\text{Eq. (14.1):} \quad V_d = (V_{i_1} - V_{i_2}) = (0.5 \text{ V} - (-0.5 \text{ V}) = 1 \text{ V}$$

and $\quad$ $$\text{Eq. (14.2):} \quad V_c = \tfrac{1}{2}(V_{i_1} + V_{i_2}) = \tfrac{1}{2}[0.5 \text{ V} + (-0.5 \text{ V})] = 0 \text{ V}$$

Under these conditions the output voltage is

$$\text{Eq. (14.3):} \quad V_o = A_d V_d + A_c V_c = A_d(1 \text{ V}) + A_c(0) = A_d$$

Thus, setting the input voltages $V_{i_1} = -V_{i_2} = 0.5$ V results in an output voltage numerically equal to the value of A_d.

2. *To measure A_c:* Set $V_{i_1} = V_{i_2} = V_s = 1$ V, so that

$$\text{Eq. (14.1):} \quad V_d = (V_{i_1} - V_{i_2}) = (1 \text{ V} - 1 \text{ V}) = 0 \text{ V}$$

and $\quad$ $$\text{Eq. (14.2):} \quad V_c = \tfrac{1}{2}(V_{i_1} + V_{i_2}) = \tfrac{1}{2}(1 \text{ V} + 1 \text{ V}) = 1 \text{ V}$$

Under these conditions the output voltage is

$$\text{Eq. (14.3):} \quad V_o = A_d V_d + A_c V_c = A_d(0 \text{ V}) + A_c(1 \text{ V}) = A_c$$

Thus, setting the input voltages $V_{i_1} = V_{i_2} = 1$ V results in an output voltage numerically equal to the value of A_c.

Common-Mode Rejection Ratio

Having obtained A_d and A_c (as in the measurement procedure discussed above), we can now calculate a value for the common-mode rejection ratio (CMRR) which is defined by the following equation:

$$\text{CMRR} = \frac{A_d}{A_c} \tag{14.4}$$

The value of CMRR can also be expressed in logarithmic terms as

$$\text{CMRR (log)} = 20 \log_{10} \frac{A_d}{A_c} \quad (dB) \tag{14.5}$$

EXAMPLE 14.1

Calculate the CMRR for the circuit measurements shown in Fig. 14.9.

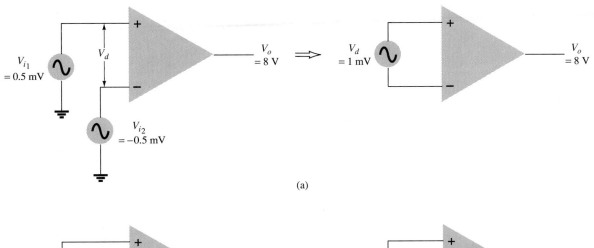

(a)

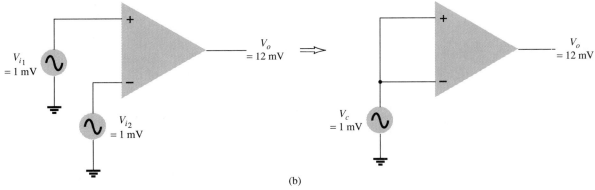

(b)

Figure 14.9 Differential and common-mode operation: (a) differential-mode; (b) common-mode.

Solution

From the measurement shown in Fig. 14.9a, using the procedure in step 1 above, we obtain

$$A_d = \frac{V_o}{V_d} = \frac{8\text{ V}}{1\text{ mV}} = 8000$$

From the measurement shown in Fig. 14.9b, using the procedure in step 2 above, gives us

$$A_c = \frac{V_o}{V_c} = \frac{12\text{ mV}}{1\text{ mV}} = 12$$

Using Eq. (14.4), the value of CMRR is

$$\text{CMRR} = \frac{A_d}{A_c} = \frac{8000}{12} = \textbf{666.7}$$

which can also be expressed as

$$\text{CMRR} = 20\ \log_{10}\frac{A_d}{A_c} = 20\ \log_{10} 666.7 = \textbf{56.48 dB}$$

It should be clear that the desired operation will have A_d very large with A_c very small. That is, the signal components of opposite polarity will appear greatly amplified at the output, whereas the signal components that are in phase will mostly cancel out so that the common-mode gain, A_c, is very small. Ideally, the value of the CMRR is infinite. Practically, the larger the value of CMRR, the better the circuit operation. We can express the output voltage in terms of the value of CMRR as follows.

Eq. (14.3): $\quad V_o = A_d V_d + A_c V_c = A_d V_d\left(1 + \dfrac{A_c V_c}{A_d V_d}\right) = A_d V_d\left(1 + \dfrac{A_c V_c}{A_d V_d}\right)$

Using Eq. (14.4), we can write the above as

$$V_o = A_d V_d\left(1 + \frac{1}{\text{CMRR}}\frac{V_c}{V_d}\right) \tag{14.6}$$

Even when both V_d and V_c components of signal are present, Eq. (14.6) shows that for large values of CMRR the output voltage will be due mostly to the difference signal, with the common-mode component greatly reduced or rejected. Some practical examples should help clarify this idea.

EXAMPLE 14.2

Determine the output voltage of an op-amp for input voltages of $V_{i_1} = 150\ \mu V$, $V_{i_2} = 140\ \mu V$. The amplifier has a differential gain of $A_d = 4000$ and the value of CMRR is:

(a) 100.
(b) 10^5.

Solution

Eq. (14.1): $\quad V_d = V_{i_1} - V_{i_2} = (150 - 140)\ \mu V = 10\ \mu V$

Eq. (14.2): $\quad V_c = \dfrac{1}{2}(V_{i_1} + V_{i_2}) = \dfrac{150\ \mu V + 140\mu V}{2} = 145\ \mu V$

(a) Eq. (14.6): $\quad V_o = A_d V_d\left(1 + \dfrac{1}{\text{CMRR}}\dfrac{V_c}{V_d}\right) = (4000)(10\ \mu V)$

$$\left(1 + \frac{1}{100}\frac{145\ \mu V}{10\ \mu V}\right)$$

$$= 40\ \text{mV}(1.145) = \mathbf{45.8\ mV}$$

(b) $V_o = (4000)(10\ \mu V)\left(1 + \dfrac{1}{10^5}\dfrac{145\ \mu V}{10\ \mu V}\right) = 40\ \text{mV}(1.000145) = \mathbf{40.006\ mV}$

Example 14.2 shows that the larger the value of CMRR, the closer the output voltage is to the difference input times the difference gain with the common-mode signal being rejected.

14.3   OP-AMP BASICS

An operational amplifier is a very high gain amplifier having very high input impedance (typically a few megohms) and low output impedance (less than 100 Ω). The basic circuit is made using a difference amplifier having two inputs (plus and minus) and at least one output. Figure 14.10 shows a basic op-amp unit. As discussed earlier,

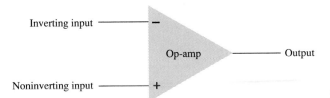

Figure 14.10 Basic op-amp.

the plus (+) input produces an output that is in phase with the signal applied, while an input to the minus (−) input results in an opposite polarity output. The ac equivalent circuit of the op-amp is shown in Fig. 14.11a. As shown, the input signal applied between input terminals sees an input impedance, R_i, typically very high. The output voltage is shown to be the amplifier gain times the input signal taken through an output impedance, R_o, which is typically very low. An ideal op-amp circuit, as shown in Fig. 14.11b, would have infinite input impedance, zero output impedance, and an infinite voltage gain.

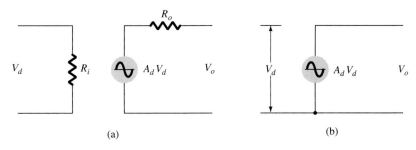

Figure 14.11 Ac equivalent of op-amp circuit: (a) practical; (b) ideal.

The basic circuit connection using an op-amp is shown in Fig. 14.12. As shown, the circuit provides operation as a constant-gain multiplier or scale changer. An input signal, V_1, is applied through resistor R_1 to the minus input. The output is then connected back to the same minus input through resistor R_f. The plus input is connected to ground. Since the signal V_1 is essentially applied to the minus input, the resulting output is opposite in phase to the input signal. Figure 14.13a shows the op-amp replaced by its ac equivalent circuit. If we use the ideal op-amp equivalent circuit, replacing R_i by an infinite resistance and R_o by zero resistance, the ac equivalent circuit is that shown in Fig. 14.13b. The circuit is then redrawn, as shown in Fig. 14.13c, from which circuit analysis is carried out.

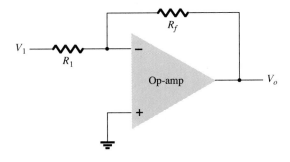

Figure 14.12 Basic op-amp connection.

14.3 Op-Amp Basics

607

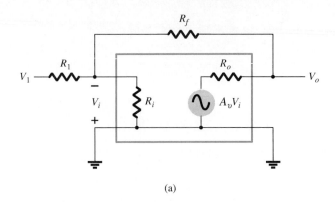

(a)

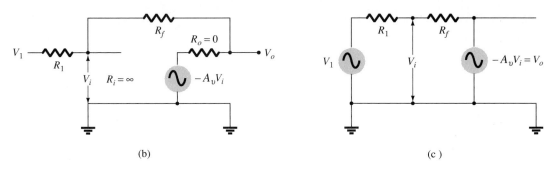

(b)   (c)

Figure 14.13 Operation of op-amp as constant-gain multiplier: (a) op-amp ac equivalent circuit; (b) ideal op-amp equivalent circuit; (c) redrawn equivalent circuit.

Using superposition we can solve for the voltage V_1 in terms of the components due to each of the sources. For source V_1 only ($-A_vV_i$ set to zero),

$$V_{i_1} = \frac{R_f}{R_1 + R_f} V_1$$

For source $-A_vV_i$ only (V_1 set to zero),

$$V_{i_2} = \frac{R_1}{R_1 + R_f} (-A_vV_i)$$

The total voltage V_i is then

$$V_i = V_{i_1} + V_{i_2} = \frac{R_f}{R_1 + R_f} V_1 + \frac{R_1}{R_1 + R_f} (-A_vV_i)$$

which can be solved for V_i as

$$V_i = \frac{R_f}{R_f + (1 + A_v)R_1} V_1 \qquad (14.7)$$

If $A_v \gg 1$ and $A_vR_1 \gg R_f$, as is usually true, then

$$V_i = \frac{R_f}{A_vR_1} V_1$$

Solving for V_o/V_i, we get

$$\frac{V_o}{V_i} = \frac{-A_vV_i}{V_i} = \frac{-A_v}{V_i} \frac{R_fV_1}{A_vR_1} = -\frac{R_f}{R_1} \frac{V_1}{V_i}$$

so that

$$\frac{V_o}{V_1} = -\frac{R_f}{R_1}$$ (14.8)

The result, in Eq. (14.8), shows that the ratio of overall output to input voltage is dependent only on the values of resistors R_1 and R_f—provided that A_v is very large.

Unity Gain

If $R_f = R_1$, the gain is

$$\text{voltage gain} = -\frac{R_f}{R_1} = -1$$

so that the circuit provides a unity voltage gain with 180° phase inversion. If R_f is exactly R_1, the voltage gain is exactly 1.

Constant Magnitude Gain

If R_f is some multiple of R_1, the overall amplifier gain is a constant. For example, if $R_f = 10R_1$, then

$$\text{voltage gain} = -\frac{R_f}{R_1} = -10$$

and the circuit provides a voltage gain of exactly 10 along with an 180° phase inversion from the input signal. If we select precise resistor values for R_f and R_1, we can obtain a wide range of gains, the gain being as accurate as the resistors used and is only slightly affected by temperature and other circuit factors.

Virtual Ground

The output voltage is limited by the supply voltage of, typically, a few volts. As stated before, voltage gains are very high. If, for example, $V_o = -10$ V and $A_v = 20,000$, the input voltage would then be

$$V_i = \frac{-V_o}{A_v} = \frac{10\ V}{20,000} = 0.5\ mV$$

If the circuit has an overall gain (V_o/V_1) of, say, 1, the value of V_1 would then be 10 V. Compared to all other input and output voltages, the value of V_i, is then small and may be considered 0 V.

Note that although $V_i \approx 0$ V, it is not exactly 0 V. (The output voltage is a few volts, due to the very small input V_i times a very large gain A_v.) The fact that $V_i \approx 0$ V leads to the concept that at the amplifier input there exists a virtual short circuit or virtual ground.

The concept of a virtual short implies that although the voltage is nearly 0 V, there is no current through the amplifier input to ground. Figure 14.14 depicts the virtual ground concept. The heavy line is used to indicate that we may consider that a

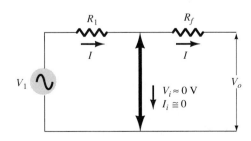

Figure 14.14 Virtual ground in an op-amp.

short exists with $V_i \approx 0$ V, but that this is a virtual short so that no current goes through the short to ground. Current goes only through resistors R_1 and R_f as shown.

Using the virtual ground concept, we can write equations for the current I as follows:

$$I = \frac{V_1}{R_1} = -\frac{V_o}{R_f}$$

which can be solved for V_o/V_1:

$$\frac{V_o}{V_1} = -\frac{R_f}{R_1}$$

The virtual ground concept, which depends on A_v being very large, allowed a simple solution to determine the overall voltage gain. It should be understood that although the circuit of Fig. 14.14 is not physically correct, it does allow an easy means for determining the overall voltage gain.

14.4 PRACTICAL OP-AMP CIRCUITS

The op-amp can be connected in a large number of circuits to provide various operating characteristics. In this section we cover a few of the most common of these circuit connections.

Inverting Amplifier

The most widely used constant-gain amplifier circuit is the inverting amplifier, shown in Fig. 14.15. The output is obtained by multiplying the input by a fixed or constant gain, set by the input resistor (R_1) and feedback resistor (R_f)—this output also being inverted from the input. Using Eq. (14.8) we can write

$$V_o = -\frac{R_f}{R_1} V_1$$

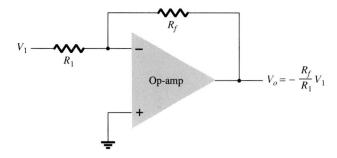

Figure 14.15 Inverting constant-gain multiplier.

EXAMPLE 14.3

If the circuit of Fig. 14.15 has $R_1 = 100$ kΩ and $R_f = 500$ kΩ, what output voltage results for an input of $V_1 = 2$ V?

Solution

$$\text{Eq. (14.8):} \quad V_o = -\frac{R_f}{R_1} V_1 = -\frac{500 \text{ k}\Omega}{100 \text{ k}\Omega} (2 \text{ V}) = \mathbf{-10 \text{ V}}$$

Noninverting Amplifier

The connection of Fig. 14.16a shows an op-amp circuit that works as a noninverting amplifier or constant-gain multiplier. It should be noted that the inverting amplifier connection is more widely used because it has better frequency stability (discussed later). To determine the voltage gain of the circuit, we can use the equivalent representation shown in Fig. 14.16b. Note that the voltage across R_1 is V_1 since $V_i \approx 0$ V. This must be equal to the output voltage, through a voltage divider of R_1 and R_f, so that

$$V_1 = \frac{R_1}{R_1 + R_f} V_o$$

which results in

$$\boxed{\frac{V_o}{V_1} = \frac{R_1 + R_f}{R_1} = 1 + \frac{R_f}{R_1}}$$

(14.9)

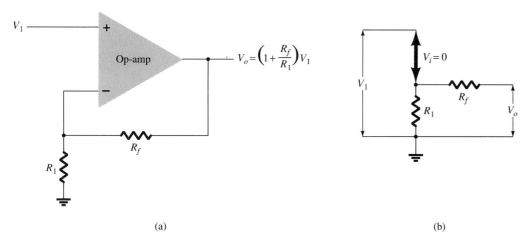

$$V_o = \left(1 + \frac{R_f}{R_1}\right)V_1$$

$V_i = 0$

(a) (b)

Figure 14.16 Noninverting constant-gain multiplier.

Calculate the output voltage of a noninverting amplifier (as in Fig. 14.16) for values of $V_1 = 2$ V, $R_f = 500$ kΩ, and $R_1 = 100$ kΩ.

EXAMPLE 14.4

Solution

Eq. (14.9): $V_o = \left(1 + \frac{R_f}{R_1}\right)V_1 = \left(1 + \frac{500 \text{ k}\Omega}{100 \text{ k}\Omega}\right)(2 \text{ V}) = 6(2 \text{ V}) = \mathbf{+12 \text{ V}}$

Unity Follower

The unity-follower circuit, shown in Fig. 14.17a, provides a gain of unity (1) with no polarity or phase reversal. From the equivalent circuit (see Fig. 14.17b) it is clear that

$$\boxed{V_o = V_1}$$

(14.10)

and that the output is the same polarity and magnitude as the input. The circuit operates like an emitter- or source-follower circuit except that the gain is exactly unity.

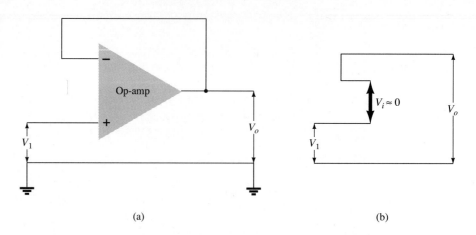

(a)   (b)

Figure 14.17 (a) Unity follower; (b) virtual-ground equivalent circuit.

Summing Amplifier

Probably the most used of the op-amp circuits is the summing amplifier circuit shown in Fig. 14.18a. The circuit shows a three-input summing amplifier circuit which provides a means of algebraically summing (adding) three voltages, each multiplied by a constant-gain factor. Using the equivalent representation shown in Fig. 14.18b, the output voltage can be expressed in terms of the inputs as

$$V_o = -\left(\frac{R_f}{R_1} V_1 + \frac{R_f}{R_2} V_2 + \frac{R_f}{R_3} V_3\right) \qquad (14.11)$$

In other words, each input adds a voltage to the output multiplied by its separate constant-gain multiplier. If more inputs are used, they each add an additional component to the output.

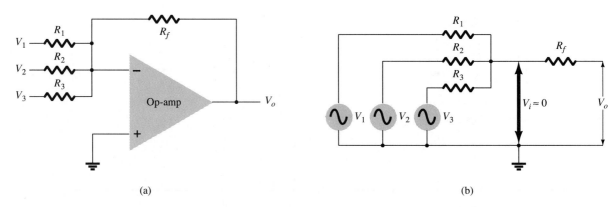

(a)   (b)

Figure 14.18 (a) Summing amplifer; (b) virtual-ground equivalent circuit.

EXAMPLE 14.5

Calculate the output voltage of an op-amp summing amplifier for the following sets of voltages and resistors. (Use $R_f = 1$ MΩ in all cases.)
(a) $V_1 = +1$ V, $V_2 = +2$ V, $V_3 = +3$ V, $R_1 = 500$ kΩ, $R_2 = 1$ MΩ, $R_3 = 1$ MΩ.
(b) $V_1 = -2$ V, $V_2 = +3$ V, $V_3 = +1$ V, $R_1 = 200$ kΩ, $R_2 = 500$ kΩ, $R_3 = 1$ MΩ.

Chapter 14 Operational Amplifiers

Solution

Using Eq. (14.11):

(a) $V_o = -\left[\dfrac{1000\ \text{k}\Omega}{500\ \text{k}\Omega}(+1\ \text{V}) + \dfrac{1000\ \text{k}\Omega}{1000\ \text{k}\Omega}(+2\ \text{V}) + \dfrac{1000\ \text{k}\Omega}{1000\ \text{k}\Omega}(+3\ \text{V})\right]$

$= -[2(1\ \text{V}) + 1(2\ \text{V}) + 1(3\ \text{V})] = \mathbf{-7\ V}$

(b) $V_o = -\left[\dfrac{1000\ \text{k}\Omega}{200\ \text{k}\Omega}(-2\ \text{V}) + \dfrac{1000\ \text{k}\Omega}{500\ \text{k}\Omega}(+3\ \text{V}) + \dfrac{1000\ \text{k}\Omega}{1000\ \text{k}\Omega}(+1\ \text{V})\right]$

$= -[5(-2\ \text{V}) + 2(3\ \text{V}) + 1(1\ \text{V})] = \mathbf{+3\ V}$

Integrator

So far the input and feedback components have been resistors. If the feedback component used is a capacitor, as shown in Fig. 14.19a, the resulting connection is called an *integrator*. The virtual-ground equivalent circuit (Fig. 14.19b) shows that an expression for the voltage between input and output can be derived in terms of the current I, from input to output. Recall that virtual ground means that we can consider the voltage at the junction of R and X_C to be ground (since $V_i \approx 0$ V) but that no current goes into ground at that point. The capacitive impedance can be expressed as

$$X_C = \frac{1}{j\omega C} = \frac{1}{sC}$$

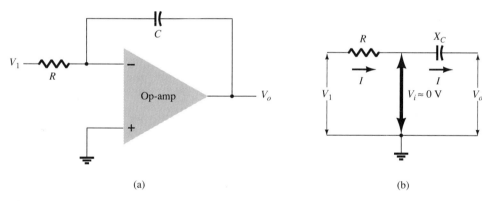

(a)　　　　　　　　(b)

Figure 14.19　Integrator.

where $s = j\omega$ is in the Laplace notation.[*] Solving for V_o/V_i yields

$$I = \frac{V_1}{R} = -\frac{V_o}{X_C} = \frac{-V_o}{1/sC} = -sCV_o$$

$$\frac{V_o}{V_1} = \frac{-1}{sCR} \tag{14.12}$$

The expression above can be rewritten in the time domain as

$$v_o(t) = -\frac{1}{RC}\int v_1(t)\,dt \tag{14.13}$$

---

[*]Laplace notation allows expressing differential or integral operations which are part of calculus in algebraic form using the operator **s**. Readers unfamiliar with calculus should ignore the steps leading to Eq. (14.13) and follow the physical meaning used thereafter.

Equation (14.13) shows that the output is the integral of the input, with an inversion and scale multiplier of $1/RC$. The ability to integrate a given signal provides the analog computer with the ability to solve differential equations and therefore provides the ability to electrically solve analogs of physical system operation.

The integration operation is one of summation, summing the area under a waveform or curve over a period of time. If a fixed voltage is applied as input to an integrator circuit, Eq. (14.13) shows that the output voltage grows over a period of time, providing a ramp voltage. Equation (14.13) can thus be understood to show that the output voltage ramp (for a fixed input voltage) is opposite in polarity to the input voltage and is multiplied by the factor $1/RC$. While the circuit of Fig. 14.19 can operate on many varied types of input signals, the following examples will use only a fixed input voltage, resulting in a ramp output voltage.

As an example, consider an input voltage, $V_1 = 1$ V, to the integrator circuit of Fig. 14.20a. The scale factor of $1/RC$ is

$$-\frac{1}{RC} = \frac{1}{(1 \text{ M}\Omega)(1 \text{ } \mu\text{F})} = -1$$

so that the output is a negative ramp voltage as shown in Fig. 14.20b. If the scale factor is changed by making $R = 100$ kΩ, for example, then

$$-\frac{1}{RC} = \frac{1}{(100 \text{ k}\Omega)(1 \text{ } \mu\text{F})} = -10$$

and the output is then a steeper ramp voltage, as shown in Fig. 14.20c.

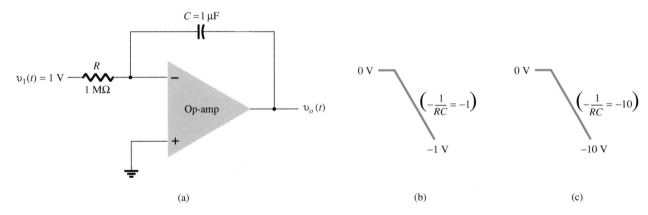

(a)   (b)   (c)

Figure 14.20 Operation of integrator with step input.

More than one input may be applied to an integrator, as shown in Fig. 14.21, with the resulting operation given by

$$v_o(t) = -\left[\frac{1}{R_1C}\int v_1(t) \, dt + \frac{1}{R_2C}\int v_2(t) \, dt + \frac{1}{R_3C}\int v_3(t) \, dt\right] \quad (14.14)$$

An example of a summing integrator as used in an analog computer is given in Fig. 14.21. The actual circuit is shown with input resistors and feedback capacitor, whereas the analog-computer representation indicates only the scale factor for each input.

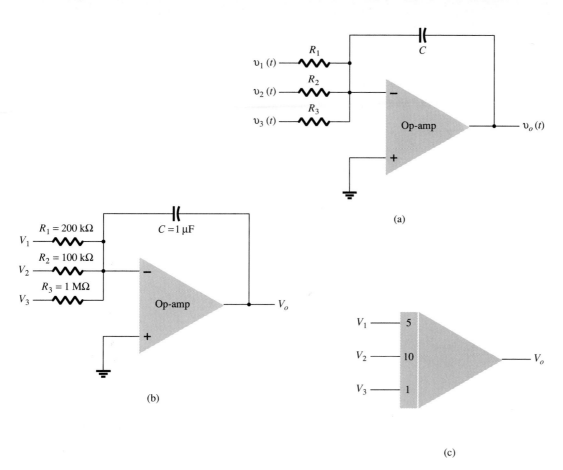

Figure 14.21 (a) Summing-integrator circuit; (b) component values; (c) analog-computer, integrator-circuit representation.

Differentiator

A differentiator circuit is shown in Fig. 14.22. While not as useful as the circuit forms covered above, the differentiator does provide a useful operation, the resulting relation for the circuit being

$$v_o(t) = -RC\,\frac{dv_1(t)}{dt} \tag{14.15}$$

where the scale factor is $-RC$.

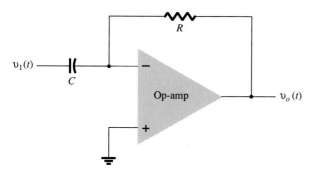

Figure 14.22 Differentiator circuit.

14.5   OP-AMP SPECIFICATIONS—DC OFFSET PARAMETERS

Before going into various practical applications using op-amps we should become familiar with some of the parameters used to define the operation of the unit. These specifications include both dc and transient or frequency operating features, as covered next.

Offset Currents and Voltages

While the op-amp output should be 0 V when the input is 0 V, in actual operation there is some offset voltage at the output. For example, if one connected 0 V to both op-amp inputs and then measured 26 mV(dc) at the output, this would represent 26 mV of unwanted voltage generated by the circuit and not by the input signal. Since the user may connect the amplifier circuit for various gain and polarity operations, however, the manufacturer specifies an input offset voltage for the op-amp. The output offset voltage is then determined by the input offset voltage and the gain of the amplifier, as connected by the user.

The output offset voltage can be shown to be affected by two separate circuit conditions. These are: (1) an input offset voltage, V_{IO}, and (2) an offset current due to the difference in currents resulting at the plus (+) and minus (−) inputs.

INPUT OFFSET VOLTAGE, V_{IO}

The manufacturer's specification sheet provides a value of V_{IO} for the op-amp. To determine the effect of this input voltage on the output, consider the connection shown in Fig. 14.23. Using $V_o = AV_i$, we can write

$$V_o = AV_i = A\left(V_{IO} - V_O \frac{R_1}{R_1 + R_f}\right)$$

Solving for V_o, we get

$$V_o = V_{IO} \frac{A}{1 + A[R_1/(R_1 + R_f)]} \approx V_{IO} \frac{A}{A[R_1/(R_1 + R_f)]}$$

from which we can write

$$\boxed{V_o(\text{offset}) = V_{IO} \frac{R_1 + R_f}{R_1}} \tag{14.16}$$

Equation (14.16) shows how the output offset voltage results from a specified input offset voltage for a typical amplifier connection of the op-amp.

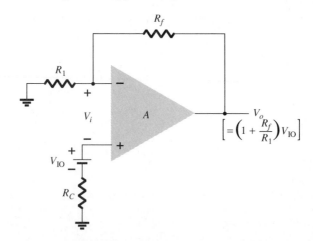

Figure 14.23   Operation showing effect of input offset voltage, V_{IO}.

Calculate the output offset voltage of the circuit in Fig. 14.24. The op-amp spec lists $V_{IO} = 1.2$ mV.

EXAMPLE 14.6

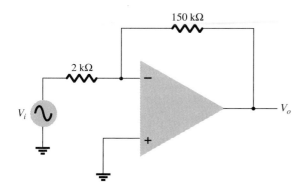

Figure 14.24 Op-amp connection for Examples 14.6 and 14.7.

Solution

Eq. (14.16): $V_o(\text{offset}) = V_{IO}\dfrac{R_1 + R_f}{R_1} = (1.2 \text{ mV})\left(\dfrac{2 \text{ k}\Omega + 150 \text{ k}\Omega}{2 \text{ k}\Omega}\right) = \mathbf{91.2 \text{ mV}}$

OUTPUT OFFSET VOLTAGE DUE TO INPUT OFFSET CURRENT, I_{IO}

An output offset voltage will also result due to any difference in dc bias currents at both inputs. Since the two input transistors are never exactly matched, each will operate at a slightly different current. For a typical op-amp connection, such as that shown in Fig. 14.25, an output offset voltage can be determined as follows. Replacing the bias currents through the input resistors by the voltage drop that each develops, as shown in Fig. 14.26, we can determine the expression for the resulting output voltage. Using superposition, the output voltage due to input bias current I_{IB}^+, denoted by V_o^+, is

$$V_o^+ = I_{IB}^+ R_C\left(1 + \frac{R_f}{R_1}\right)$$

while the output voltage due to only I_{IB}^-, denoted by V_o^-, is

$$V_o^- = I_{IB}^- R_1\left(-\frac{R_f}{R_1}\right)$$

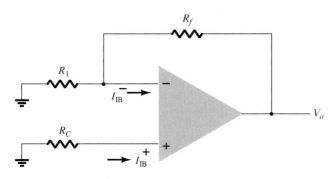

Figure 14.25 Op-amp connection showing input bias currents.

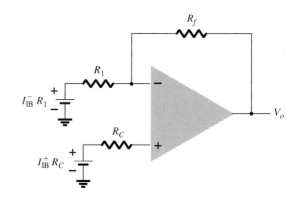

Figure 14.26 Redrawn circuit of Fig. 14.25.

for a total output offset voltage of

$$V_o(\text{offset due to } I_{IB}^+ \text{ and } I_{IB}^-) = I_{IB}^+ R_C\left(1 + \frac{R_f}{R_1}\right) - I_{IB}^- R_1\frac{R_f}{R_1} \qquad (14.17)$$

Since the main consideration is the difference between the input bias currents rather than each value, we define the offset current I_{IO} by

$$I_{IO} = I_{IB}^+ - I_{IB}^-$$

Since the compensating resistance R_C is usually approximately equal to the value of R_1, using $R_C = R_1$ in Eq. (14.17) we can write

$$V_o(\text{offset}) = I_{IB}^+(R_1 + R_f) - I_{IB}^- R_f$$
$$= I_{IB}^+ R_f - I_{IB}^- R_f = R_f(I_{IB}^+ - I_{IB}^-)$$

resulting in

$$\boxed{V_o(\text{offset due to } I_{IO}) = I_{IO}R_f} \qquad (14.18)$$

EXAMPLE 14.7

Calculate the offset voltage for the circuit of Fig. 14.24 for op-amp specification listing $I_{IO} = 100$ nA.

Solution

$$\text{Eq. (14.18):} \quad V_o = I_{IO}R_f = (100 \text{ nA})(150 \text{ k}\Omega) = \textbf{15 mV}$$

TOTAL OFFSET DUE TO V_{IO} AND I_{IO}

Since the op-amp output may have an output offset voltage due to both factors covered above, the total output offset voltage can be expressed as

$$|V_o(\text{offset})| = |V_o(\text{offset due to } V_{IO})| + |V_o(\text{offset due to } I_{IO})| \qquad (14.19)$$

The absolute magnitude is used to accommodate the fact that the offset polarity may be either positive or negative.

EXAMPLE 14.8

Calculate the total offset voltage for the circuit of Fig. 14.27 for an op-amp with specified values of input offset voltage, $V_{IO} = 4$ mV and input offset current $I_{IO} = 150$ nA.

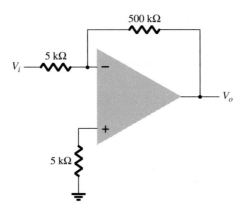

Figure 14.27 Op-amp circuit for Example 14.8.

Solution

The offset due to V_{IO} is

Eq. (14.16): $\quad V_o(\text{offset due to } V_{IO}) = V_{IO}\dfrac{R_1 + R_f}{R_1} = (4 \text{ mV})\left(\dfrac{5 \text{ k}\Omega + 500 \text{ k}\Omega}{5 \text{ k}\Omega}\right)$

$$= 404 \text{ mV}$$

Eq. (14.18): $\quad V_o(\text{offset due to } I_{IO}) = I_{IO}R_f = (150 \text{ nA})(500 \text{ k}\Omega) = 75 \text{ mV}$

resulting in a total offset

Eq. (14.19): $\quad V_o(\text{total offset}) = V_o(\text{offset due to } V_{IO}) + V_o(\text{offset due to } I_{IO})$

$$= 404 \text{ mV} + 75 \text{ mV} = \textbf{479 mV}$$

INPUT BIAS CURRENT, I_{IB}

A parameter related to I_{IO} and the separate input bias currents I_{IB}^{+} and I_{IB}^{-} is the average bias current defined as

$$I_{IB} = \frac{I_{IB}^{+} + I_{IB}^{-}}{2} \tag{14.20}$$

One could determine the separate input bias currents using the specified values I_{IO} and I_{IB}. It can be shown that for $I_{IB}^{+} > I_{IB}^{-}$

$$I_{IB}^{+} = I_{IB} + \frac{I_{IO}}{2}$$

$$\tag{14.21}$$

$$I_{IB}^{-} = I_{IB} - \frac{I_{IO}}{2}$$

Calculate the input bias currents at each input of an op-amp having specified values of $I_{IO} = 5$ nA and $I_{IB} = 30$ nA.

EXAMPLE 14.9

Solution

Using Eq. (14.21):

$$I_{IB}^{+} = I_{IB} + \frac{I_{IO}}{2} = 30 \text{ nA} + \frac{5 \text{ nA}}{2} = \textbf{32.5 nA}$$

$$I_{IB}^{-} = I_{IB} - \frac{I_{IO}}{2} = 30 \text{ nA} - \frac{5 \text{ nA}}{2} = \textbf{27.5 nA}$$

14.6 OP-AMP SPECIFICATIONS— FREQUENCY PARAMETERS

An op-amp is designed to be a high-gain, wide-bandwidth amplifier. This operation tends to be unstable (oscillate) due to positive feedback (see Chapter 18). To ensure stable operation, op-amps are built with internal compensation circuitry, which also causes the very high open-loop gain to diminish with increasing frequency. This gain reduction is referred to as *roll-off*. In most op-amps roll-off occurs at a rate of 20 dB per decade (−20 dB/decade) or 6 dB per octave (−6 dB/octave). (Refer to Chapter 11 for introductory coverage of dB and frequency response.)

Note that while op-amp specifications list an open-loop voltage gain (A_{VD}), the user typically connects the op-amp using feedback resistors to reduce the circuit voltage gain to a much smaller value (closed-loop voltage gain, A_{CL}). A number of circuit improvements result from this gain reduction. First, the amplifier voltage gain is a more stable, precise value set by the external resistors; second, the input impedance of the circuit is increased over that of the op-amp alone; third, the circuit output impedance is reduced from that of the op-amp alone; and finally, the frequency response of the circuit is increased over that of the op-amp alone.

Gain–Bandwidth

Because of the internal compensation circuitry included in an op-amp, the voltage gain drops off as frequency increases. Op-amp specifications provide a description of the gain versus bandwidth. Figure 14.28 provides a plot of gain versus frequency for a typical op-amp. At low frequency down to dc operation the gain is that value listed by the manufacturer's specification A_{VD} (voltage differential gain) and is typically a very large value. As the frequency of the input signal increases the open-loop gain drops off until it finally reaches the value of 1 (unity). The frequency at this gain value is specified by the manufacturer as the unity-gain bandwidth, B_1. While this value is a frequency (see Fig. 14.28) at which the gain becomes 1, it can be considered a bandwidth, since the frequency band from 0 Hz to the unity-gain frequency is also a bandwidth. One could therefore refer to the point at which the gain reduces to 1 as the unity-gain frequency (f_1) or unity-gain bandwidth (B_1).

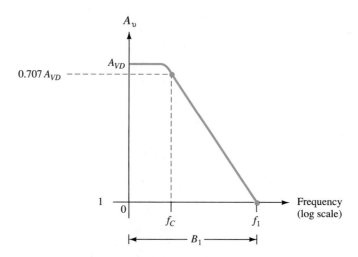

Figure 14.28 Gain versus frequency plot.

Another frequency of interest is that shown in Fig. 14.28, at which the gain drop by 3 dB (or to 0.707 the dc gain, A_{VD}), this being the cutoff frequency of the op-amp, f_C. In fact, the unity-gain frequency and cutoff frequency are related by

$$f_1 = A_{VD}f_C \tag{14.22}$$

Equation (14.22) shows that the unity-gain frequency may also be called the gain–bandwidth product of the op-amp.

Determine the cutoff frequency of an op-amp having specified values $B_1 = 1$ MHz and $A_{VD} = 200$ V/mV.

EXAMPLE 14.10

Solution

Since $f_1 = B_1 = 1$ MHz, we can use Eq. (14.22) to calculate

$$f_C = \frac{f_1}{A_{VD}} = \frac{1 \text{ MHz}}{200 \text{ V/mV}} = \frac{1 \times 10^6}{200 \times 10^3} = \mathbf{5 \text{ Hz}}$$

Slew Rate, SR

Another parameter reflecting the op-amp's ability to handling varying signals is slew rate, defined as

slew rate = maximum rate at which amplifier output can change in volts per microsecond (V/μs)

$$\boxed{SR = \frac{\Delta V_o}{\Delta t} \quad V/\mu s}, \text{ with } t \text{ in } \mu s. \tag{14.23}$$

The slew rate provides a parameter specifying the maximum rate of change of the output voltage when driven by a large step-input signal.[*] If one tried to drive the output at a rate of voltage change greater than the slew rate, the output would not be able to change fast enough and would not vary over the full range expected, resulting in signal clipping or distortion. In any case the output would not be an amplified duplicate of the input signal if the op-amp slew rate is exceeded.

For an op-amp having a slew rate of SR = 2 V/μs, what is the maximum closed-loop voltage gain that can be used when the input signal varies by 0.5 V in 10 μs?

EXAMPLE 14.11

Solution

Since $V_o = A_{CL}V_i$, we can use

$$\frac{\Delta V_o}{\Delta t} = A_{CL}\frac{\Delta V_i}{\Delta t}$$

from which we get

$$A_{CL} = \frac{\Delta V_o/\Delta t}{\Delta V_i/\Delta t} = \frac{SR}{\Delta V_i/\Delta t} = \frac{2 \text{ V/}\mu s}{0.5 \text{ V/10 } \mu s} = \mathbf{40}$$

Any closed-loop voltage gain of magnitude greater than 40 would drive the output at a rate greater than the slew rate allows, so the maximum closed-loop gain is $\Delta V_i/\Delta t$ 40.

---

[*]The closed-loop gain is that obtained with the output connected back to the input in some way.

Maximum Signal Frequency

The maximum frequency that an op-amp may operate at depends on both the bandwidth (B) and slew rate (SR) parameters of the op-amp. For a sinusoidal signal of general form

$$v_o = K \sin(2\pi ft)$$

the maximum voltage rate of change can be shown to be

$$\text{signal maximum rate of change} = 2\pi fK \qquad \text{V/s}$$

To prevent distortion at the ouput the rate of change must also be less than the slew rate that is,

$$2\pi fK \leq SR$$

$$\omega K \leq SR$$

so that

$$\boxed{f \leq \frac{SR}{2\pi K} \qquad \text{Hz}\\[2mm] \omega \leq \frac{SR}{K} \qquad \text{rad/s}}$$

(14.24)

Additionally, the maximum frequency, f, in Eq. (14.24), is also limited by the unity-gain bandwidth.

EXAMPLE 14.12

For the signal and circuit of Fig. 14.29 determine the maximum frequency that may be used. (Op-amp slew rate is SR = 0.5 V/μs.)

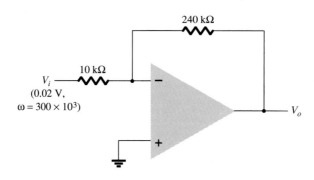

Figure 14.29 Op-amp circuit for Example 14.12.

Solution

For a gain of magnitude

$$A_{CL} = \left| \frac{R_f}{R_1} \right| = \frac{240 \text{ k}\Omega}{10 \text{ k}\Omega} = 24$$

the output voltage provides

$$K = A_{CL}V_i = 24(0.02 \text{ V}) = 0.48 \text{ V}$$

$$\text{Eq. (14.24):} \quad \omega_S \leq \frac{SR}{K} = \frac{0.5 \text{ V/}\mu s}{0.48 \text{ V}} = \mathbf{1.1 \times 10^6 \text{ rad/s}}$$

Since the signal's frequency, $\omega = 300 \times 10^3$ rad/s, is less than the maximum value determined above, no output distortion will result.

Chapter 14   Operational Amplifiers

14.7 OP-AMP UNIT SPECIFICATIONS

In this section we discuss how the manufacturer's specifications are read for a typical op-amp unit. A popular bipolar op-amp IC is the 741 described by the information provided in Fig. 14.30. The op-amp is available in a number of packages, an 8-pin DIP and a 10-pin flatpack being among the more usual forms.

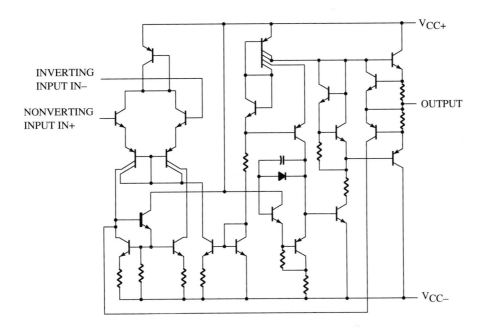

absolute maximum ratings over operating free-air temperature range (unless otherwise noted)

		uA741M	uA741C	UNIT
Supply voltage V_{CC+} (see Note 1)		22	18	V
Supply voltage V_{CC-} (see Note 1)		− 22	− 18	V
Differential input voltage (see Note 2)		± 30	± 30	V
Input voltage any input (see Notes 1 and 3)		± 15	± 15	V
Voltage between either offset null terminal (N1/N2) and V_{CC-}		± 0.5	± 0.5	V
Duration of output short-circuit (see Note 4)		unlimited	unlimited	
Continuous total power dissipation at (or below) 25°C free-air temperature (see Note 5)		500	500	mW
Operating free-air temperature range		− 55 to 125	0 to 70	°C
Storage temperature range		− 65 to 150	− 65 to 150	°C
Lead temperature 1,6 mm (1/16 inch) from case for 60 seconds	FH, FK, J, JG, or U package	300	300	°C
Lead temperature 1,6 mm (1/16 inch) from case for 10 seconds	D, N, or P package		260	°C

NOTES: 1. All voltage values, unless otherwise noted, are with respect to the midpoint between V_{CC+} and V_{CC-}.

2. Differential voltages are at the noninverting input terminal with respect to the inverting input terminal.

3. The magnitude of the input voltage must never exceed the magnitude of the supply voltage or 15 volts, whichever is less.

4. The output may be shorted to ground or either power supply. For the uA741M only, the unlimited duration of the short-circuit applies at (or below) 125°C case temperature ot 75°C free-air temperature.

5. For operation above 25°C free-air temperature, refer to Dissipation Derating Curves, Section 2. In the J and JG packages, uA741M chips are alloy mounted; uA741C chips are glass mounted.

Figure 14.30 741 op-amp specifications.

PARAMETER		TEST CONDITIONS †		uA741M			uA741C			UNIT
				MIN	TYP	MAX	MIN	TYP	MAX	
V_{IO}	Input offset voltage	$V_O = 0$	25°C		1	5		1	6	mV
			Full range			6			7.5	
$\Delta V_{IO(adj)}$	Offset voltage adjust range	$V_O = 0$	25°C		± 15			± 15		mV
I_{IO}	Input offset current	$V_O = 0$	25°C		20	200		20	200	nA
			Full range			500			300	
I_{IB}	Input bias current	$V_O = 0$	25°C		80	500		80	500	nA
			Full range			1500			800	
V_{ICR}	Common-mode input voltage range		25°C	± 12	± 13		± 12	± 13		V
			Full range	± 12			± 12			
V_{OM}	Maximum peak output voltage swing	$R_L = 10$ kΩ	25°C	± 12	± 14		± 12	± 14		V
		$R_L \geq 10$ kΩ	Full range	± 12			± 12			
		$R_L = 2$ kΩ	25°C	± 10	± 13		± 10	± 13		
		$R_L \geq 2$ kΩ	Full range	± 10			± 10			
A_{VD}	Large-signal differential voltage amplification	$R_L \geq 2$ kΩ	25°C	50	200		20	200		V/mV
		$V_O = \pm 10$ V	Full range	25			15			
r_i	Input resistance		25°C	0.3	2		0.3	2		MΩ
r_o	Output resistance	$VO = 0$ See note 6	25°C		75			75		Ω
C_i	Input capacitance		25°C		1.4			1.4		pF
CMRR	Common-mode rejection ratio	$V_{IC} = V_{ICR}$ min	25°C	70	90		70	90		dB
			Full range	70			70			
k_{SVS}	Supply voltage sensitivity $\Delta V_{IO}/\Delta V_{CC}$)	$V_{CC} = \pm 9$ V to ± 15 V	25°C		30	150		30	150	μV/V
			Full range			150			150	
I_{OS}	Short-circuit output current		25°C	± 25	± 40		± 25	± 40		mA
I_{CC}	Supply current	No load, $V_O = 0$	25°C		1.7	2.8		1.7	2.8	mA
			Full range			3.3			3.3	
P_D	Total power dissipation	No load, $V_O = 0$	25°C		50	85		50	85	mW
			Full range			100			100	

operating characteristics, $V_{CC+} = 15$ V, $V_{CC-} = -15$ V, $T_A = 25$°C

PARAMETER		TEST CONDITIONS		uA741M			uA741C			UNIT
				MIN	TYP	MAX	MIN	TYP	MAX	
t_r	Rise time	$V_I = 20$ mV, $R_L = 2$ kΩ,			0.3			0.3		μs
	Overshoot factor	$C_L = 100$ pF, See Figure 1			5%			5%		
SR	Slew rate at unity gain	$V_I = 10$ V, $R_L = 2$ kΩ, $C_L = 100$ pF, See Figure 1			0.5			0.5		V/μs

Figure 14.30 Continued.

Absolute Maximum Ratings

The absolute maximum ratings provide information on what largest voltage supplies may be used, how large the input signal swing may be, and at how much power the device is capable of operating. Depending on the particular version of 741 used, the largest supply voltage is a dual supply of ±18 V or ±22 V. In addition, the IC can internally dissipate from 310 to 570 mW, depending on the IC package used. Table 14.1 summarizes some typical values to use in examples and problems.

TABLE 14.1 Absolute Maximum Ratings	
Supply voltage	±22 V
Internal power dissipation	500 mW
Differential input voltage	±30 V
Input voltage	±15 V

EXAMPLE 14.13   Determine the current draw from a dual power supply of ±12 V if the IC dissipates 500 mW.

Solution

If we assume that each supply provides half the total power to the IC, then

$$P = VI$$

$$250 \text{ mW} = 12 \text{ V}(I)$$

so that each supply must provide a current of

$$I = \frac{250 \text{ mW}}{12 \text{ V}} = \textbf{20.83 mA}$$

Electrical Characteristics

Electrical characteristics include many of the parameters covered earlier in this chapter. The manufacturer provides some combination of typical, minimum, or maximum values for various parameters as deemed most useful to the user. A summary is provided in Table 14.2.

TABLE 14.2 μA741 Electrical Characteristics: $V_{CC} = \pm 15$ V, $T_A = 25°C$

Characteristic	MIN	TYP	MAX	Unit
V_{IO} Input offset voltage		1	6	mV
I_{IO} Input offset current		20	200	nA
I_{IB} Input Bias Current		80	500	nA
V_{ICR} Common-mode input voltage range	± 12	± 13		V
V_{OM} Maximum peak output voltage swing	± 12	± 14		V
A_{VD} Large-signal differential voltage amplification	20	200		V/mV
r_i Input resistance	0.3	2		MΩ
r_O Output resistance		75		Ω
C_i Input capacitance		1.4		pF
CMRR Common-mode rejection ratio	70	90		dB
I_{CC} Supply current		1.7	2.8	mA
P_D Total power dissipation		50	85	mW

V_{IO} **Input offset voltage:** The input offset voltage is seen to be typically 1 mV, but can go as high as 6 mV. The output offset voltage is then computed based on the circuit used. If the worst condition possible is of interest, the maximum value should be used. Typical values are those more commonly expected when using the op-amp.

I_{IO} **Input offset current:** The input offset current is listed to be typically 20 nA, while the largest value expected is 200 nA.

I_{IB} **Input bias current:** The input bias current is typically 80 nA and may be as large as 500 nA.

V_{ICR} **Common-mode input voltage range:** This parameter lists the range that the input voltage may vary over (using a supply of ± 15 V), about ± 12 to ± 13 V. Inputs larger in amplitude than this value will probably result in output distortion and should be avoided.

V_{OM} **Maximum peak output voltage swing:** This parameter lists the largest value the output may vary (using a ± 15-V supply). Depending on the circuit closed-loop gain, the input signal should be limited to keep the output from varying by an amount no larger than ± 12 V in the worst case or by ± 14 V, typically.

A_{VD} **Large-signal differential voltage amplification:** This is the open-loop voltage gain of the op-amp. While a minimum value of 20 V/mV, or 20,000 V/V is listed, the manufacturer also lists a typical value of 200 V/mV or 200,000 V/V.

r_i **Input resistance:** The input resistance of the op-amp when measured under open-loop, is typically 2 MΩ, but could be as little as 0.3 MΩ or 300 kΩ. In a closed-loop circuit this input impedance can be much larger, as discussed previously.

r_o **Output resistance:** The op-amp output resistance is listed as typically 75 Ω. No minimum or maximum value is given by the manufacturer for this op-amp. Again, in a closed-loop circuit the output impedance can be lower, depending on the circuit gain.

C_i **Input capacitance:** For high-frequency considerations it is helpful to know that the input to the op-amp has typically 1.4 pF of capacitance, a generally small value compared even to stray wiring.

CMRR **Common-mode rejection ratio:** The op-amp parameter is seen to be typically 90 dB but could go as low as 70 dB. Since 90 dB is equivalent to 31622.78, the op-amp amplifies noise (common inputs) by over 30,000 times less than difference inputs.

I_{CC} **Supply current:** The op-amp draws a total of 2.8 mA, typically from the dual voltage supply, but the current drawn could be as little as 1.7 mA. This parameter helps the user determine the size of the voltage supply to use. It also can be used to calculate the power dissipated by the IC ($P_D = 2V_{CC}I_{CC}$).

P_D **Total power dissipation:** The total power dissipated by the op-amp is typically 50 mW, but could go as high as 85 mW. Referring to the previous parameter, the op-amp will dissipate about 50 mW when drawing about 1.7 mA using a dual 15-V supply. At smaller supply voltages the current drawn will be less and the total power dissipated will also be less.

EXAMPLE 14.14

Using the specifications listed in Table 14.2, calculate the typical output offset voltage for the circuit connection of Fig. 14.31.

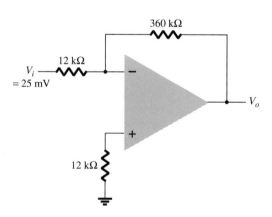

Figure 14.31 Op-amp circuit for Example 14.14.

Solution

The output offset due to V_{10} is calculated to be

Eq. (14.16): $\quad V_o(\text{offset}) = V_{IO}\dfrac{R_1 + R_f}{R_1} = (1 \text{ mV})\left(\dfrac{12 \text{ k}\Omega + 360 \text{ k}\Omega}{12 \text{ k}\Omega}\right) = 31 \text{ mV}$

The output voltage due to I_{IO} is calculated to be

Eq. (14.18): $V_o(\text{offset}) = I_{IO}R_f = 20 \text{ nA}(360 \text{ k}\Omega) = 7.2 \text{ mV}$

Assuming that these two offsets are the same polarity at the output, the total output offset voltage is then

$$V_o(\text{offset}) = 31 \text{ mV} + 7.2 \text{ mV} = \mathbf{38.2 \text{ mV}}$$

For the typical characteristics of the 741 op-amp ($r_o = 25\Omega$, $A = 200$ k) calculate the following values for the circuit of Fig. 14.31.

EXAMPLE 14.15

(a) A_{CL}.
(b) Z_i.
(c) Z_o.

Solution

(a) Eq. (14.8): $A_{CL} = \dfrac{V_o}{V_i} = -\dfrac{R_f}{R_1} = -\dfrac{360 \text{ k}\Omega}{12 \text{ k}\Omega} = \mathbf{-30} \cong \dfrac{1}{\beta}$

(b) $Z_i = R_1 = \mathbf{12 \text{ k}\Omega}$

(c) $Z_o = \dfrac{r_o}{(1 + \beta A)} = \dfrac{75\Omega}{1 + \left(\dfrac{1}{30}\right)(200 \text{ k})} = \mathbf{0.011\Omega}$

Operating Characteristics

Another group of values used to describe the operation of the op-amp over varying signals are provided in Table 14.3.

TABLE 14.3 Operating Characteristics: $V_{CC} = \pm 15$ V, $T_A = 25°C$

Parameter	MIN	TYP	MAX	Unit
SR Slew rate at unity gain		0.5		V/μs
B_1 Unity gain bandwidth		1		MHz
t_r Rise time		0.3		μs

Calculate the cutoff frequency of an op-amp having characteristics given in Tables 14.2 and 14.3.

EXAMPLE 14.16

Solution

$$\text{Eq. (14.22):} \quad f_C = \frac{f_1}{A_{VD}} = \frac{B_1}{A_{VD}} = \frac{1 \text{ MHz}}{20,000} = \mathbf{50 \text{ Hz}}$$

Calculate the maximum frequency of the input signal for the circuit in Fig. 14.31, with an input of $V_i = 25$ mV.

EXAMPLE 14.17

Solution

For a closed-loop gain of $A_{CL} = -30$ and an input of $V_i = 25$ mV, the output gain factor is calculated to be

$$K = A_{CL}V_i = 30(25 \text{ mV}) = 750 \text{ mV} = 0.750 \text{ V}$$

Using Eq. (14.24), the maximum signal frequency, f_{max}, is

$$f_{max} = \frac{SR}{2\pi K} = \frac{0.5 \text{ V}/\mu s}{2\pi(0.750 \text{ V})} = \textbf{106 kHz}$$

Op-Amp Performance

The manufacturer provides a number of graphical descriptions to describe the performance of the op-amp. Figure 14.32 includes some typical performance curves comparing various characteristics as a function of supply voltage. The open-loop voltage gain is seen to get larger with a larger supply voltage value. While the previous tabular information provided information at a particular supply voltage, the performance curve shows how the voltage gain is affected by using a range of supply voltage values.

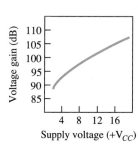

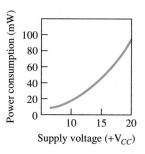

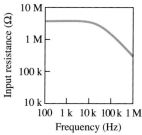

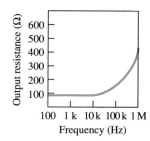

Figure 14.32 Performance curves.

EXAMPLE 14.18

Using Fig. 14.32, determine the open-loop voltage gain for a supply voltage of $V_{CC} = \pm 12$ V.

Solution

From the curve in Fig. 14.32, $A_{VD} \approx 104$ dB. This is a linear voltage gain of

$$A_{VD}(\text{dB}) = 20 \log_{10} A_{VD}$$

$$104 \text{ dB} = 20 \log A_{VD}$$

$$A_{VD} = \text{antilog} \frac{104}{20} = 158.5 \times 10^3$$

Another performance curve shows how power consumption varies as a function of supply voltage. As shown, the power consumption increases with larger values of supply voltage. For example, while the power dissipation is about 50 mW at $V_{CC} = \pm 15$ V, it drops to about 5 mW with $V_{CC} = \pm 5$ V. Two other curves show how the input and output resistances are affected by frequency, the input impedance dropping and the output resistance increasing at higher frequency.

Chapter 14   Operational Amplifiers

14.8 COMPUTER ANALYSIS

An op-amp can be modeled using PSpice so that the various connections can then be made for inverting op-amps, noninverting op-amps, summing op-amps, and unity followers. Let's start by describing an op-amp model that can be used in PSpice.

PSpice Op-amp Model

An op-amp can be described in a PSpice listing to represent a device having an input impedance, R_i, and output impedance, R_o, and a voltage gain, A_V. Figure 14.33 shows the circuit of this basic connection, with terminals identified as

Terminal 2:   the noninverting input

Terminal 1:   the inverting input

Terminal 3:   the op-amp output

Terminal 4:   an internal terminal

Terminal 0:   the ground reference

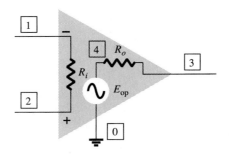

```
RI    2  1  1E12
RO    4  3  1M
EOP   4  0  2  1  1E10
```

Figure 14.33   PSpice ideal op-amp model.

For an ideal op-amp: The input impedance is listed as

$$\text{RI } 2 \; 1 \; 1\text{E}12 \qquad (R_i = 10^{12})$$

which is essentially an infinite impedance, or an open circuit. The output impedance, R_o, is listed as

$$\text{RO } 4 \; 3 \; 1\text{M} \qquad (R_o = 0.001 \; \Omega)$$

which is essentially zero ohms impedance. The voltage gain of the amplifier is given by

$$\text{EOP } 4 \; 0 \; 2 \; 1 \; 1\text{E}10 \qquad (A_v = 10^{10})$$

which is essentially an infinite voltage gain. The last line provides a voltage source at the output (from terminals 4 to 0) which is 1E10 times as large as the voltage at the input (across terminals 2 to 1).

For a practical op-amp, such as the 741, the PSpice lines would be as follows:

$$\text{RI } 2 \; 1 \; 2\text{E}6$$

$$\text{RO } 4 \; 3 \; 75$$

$$\text{EOP } 4 \; 0 \; 2 \; 1 \; 200\text{E}3$$

The 741 is thereby described as having

$$R_i = 2 \; \text{M}\Omega$$

$$R_o = 75 \; \Omega$$

and $$A_V = 200,000 \; \text{V/V} = 200 \; \text{V/mV}$$

Program 14.1: Inverting Op-Amp

An inverting op-amp of the type described in Example 14.3 and shown in Fig. 14.15 is considered first. An inverting op-amp circuit showing terminals with numbering consistent with the ideal op-amp described in Fig. 14.33 is provided in Fig. 14.34. The PSpice description is provided in the Program 14.1 listing (Fig. 14.35). For circuit values of V1 = 2 V, RF = 500 kΩ, and R1 = 100 kΩ, the output should be the 2 V amplified by a gain of

$$A_V = -\frac{R_f}{R_1} = -\frac{500 \text{ k}\Omega}{100 \text{ k}\Omega} = -5$$

so that $V_o = -5(2 \text{ V}) = -10 \text{ V}$. The component RP is used to provide a connection from terminal 2 to ground, allowing the op-amp terminal numbers to stay the same as in Fig. 14.33.

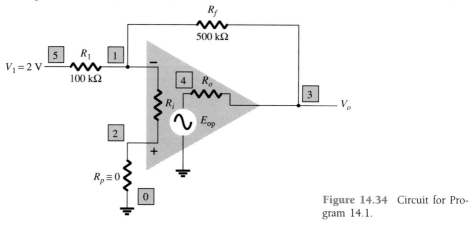

Figure 14.34 Circuit for Program 14.1.

Figure 14.35 PSpice output for Program 1.

```
**** 01/27/90 ******* Evaluation PSpice (January 1989) ******* 15:51:31 ****

Inverting op-amp circuit

****        CIRCUIT DESCRIPTION

*******************************************************************************

V1 5 0  2V
R1 5 1 100K
RF 3 1 500K
RP 2 0 1M
*OPAMP Description
RI 2 1 1E12
RO 4 3 1M
EOP 4 0 2 1 1E10
*Output
.DC V1 2 2 2
.PRINT DC V(5) V(3)
.TF V(3) V1
.OPTIONS NOPAGE
.END


****        DC TRANSFER CURVES                  TEMPERATURE =    27.000 DEG C
    V1            V(5)          V(3)
     2.000E+00    2.000E+00   -1.000E+01


****        SMALL SIGNAL BIAS SOLUTION          TEMPERATURE =    27.000 DEG C
  NODE    VOLTAGE      NODE    VOLTAGE     NODE    VOLTAGE     NODE    VOLTAGE
(    1) 1.000E-09  (     2) 1.000E-24  (     3)  -10.0000  (     4)  -10.0000
(    5)    2.0000


      VOLTAGE SOURCE CURRENTS
      NAME            CURRENT
      V1             -2.000E-05


****        SMALL-SIGNAL CHARACTERISTICS
      V(3)/V1 = -5.000E+00
      INPUT RESISTANCE AT V1 =   1.000E+05
      OUTPUT RESISTANCE AT V(3) =   6.000E-13
```

An additional feature provided in the output analysis is the transfer function of the circuit provided by the PSpice line

.TF V(3) V1

which provides the transfer function for the output at V(3) due to the input voltage V1. The results show a voltage gain of −5, an input resistance of 100 mΩ (that of R_1), and an output resistance of nearly 0 (due to R_o).

The op-amp used in Program 14.1 used the same terminal identification as that described in Fig. 14.33. In Program 14.1, the values of RI, RO, and EOP represent those of a nearly ideal op-amp.

Program 14.2: Noninverting Op-Amp

A noninverting circuit such as that of Example 14.4 is shown in Fig. 14.36. and described in Fig. 14.37. For the noninverting connection (using the same ideal op-

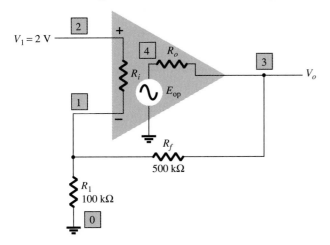

Figure 14.36 Circuit for Program 14.2.

Figure 14.37 PSpice output for Program 14.2.

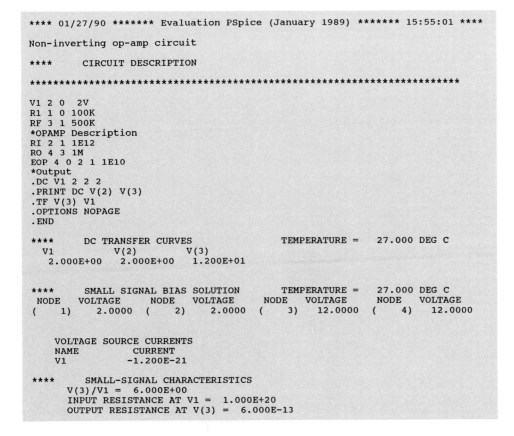

```
**** 01/27/90 ******* Evaluation PSpice (January 1989) ******* 15:55:01 ****

Non-inverting op-amp circuit

****      CIRCUIT DESCRIPTION

*********************************************************************

V1 2 0  2V
R1 1 0 100K
RF 3 1 500K
*OPAMP Description
RI 2 1 1E12
RO 4 3 1M
EOP 4 0 2 1 1E10
*Output
.DC V1 2 2 2
.PRINT DC V(2) V(3)
.TF V(3) V1
.OPTIONS NOPAGE
.END

****      DC TRANSFER CURVES              TEMPERATURE =   27.000 DEG C
  V1          V(2)          V(3)
  2.000E+00    2.000E+00    1.200E+01

****      SMALL SIGNAL BIAS SOLUTION      TEMPERATURE =   27.000 DEG C
  NODE    VOLTAGE       NODE    VOLTAGE      NODE    VOLTAGE      NODE    VOLTAGE
 (   1)    2.0000   (    2)    2.0000   (    3)   12.0000   (    4)   12.0000


     VOLTAGE SOURCE CURRENTS
     NAME            CURRENT
     V1              -1.200E-21

****      SMALL-SIGNAL CHARACTERISTICS
     V(3)/V1 =   6.000E+00
     INPUT RESISTANCE AT V1 =   1.000E+20
     OUTPUT RESISTANCE AT V(3) =   6.000E-13
```

amp as described in Fig. 14.33) the voltage gain is

$$A_V = 1 + \frac{R_f}{R_1} = 1 + \frac{500 \text{ k}\Omega}{100 \text{ k}\Omega} = +6$$

with the output voltage then

$$V_o = (+6)(2 \text{ V}) = 12 \text{ V}$$

Program 14.3: Summing Op-Amp Circuit

A summing op-amp circuit such as that in Example 14.5 is shown in Fig. 14.38. Figure 14.39 provides the PSpice output. The output is V(3) = 3 V, as expected, since the ideal op-amp model is used.

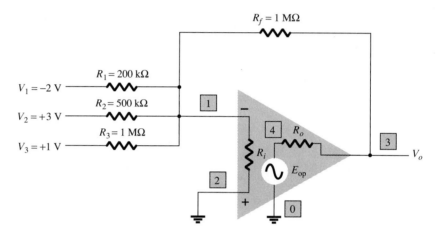

Figure 14.38 Circuit for Program 14.3.

```
**** 01/27/90 ****** Evaluation PSpice (January 1989) ****** 15:59:30 ****

Summing op-amp circuit

****      CIRCUIT DESCRIPTION

*********************************************************************

V1 5 0 -2V
V2 6 0 +3V
V3 7 0 +1V
R1 5 1 200K
R2 6 1 500K
R3 7 1 1MEG
RF 1 3 1MEG
RP 2 0 1M
*OPAMP Description
RI 2 1 1E12
RO 4 3 1M
EOP 4 0 2 1 1E10
*Output
.DC V1 -2 -2 1
.PRINT DC V(5) V(6) V(7) V(3)
.OPTIONS NOPAGE
.END

****      DC TRANSFER CURVES              TEMPERATURE =    27.000 DEG C
  V1          V(5)          V(6)        V(7)          V(3)
-2.000E+00   -2.000E+00   3.000E+00   1.000E+00   3.000E+00
```

Figure 14.39 PSpice output for Program 14.3.

Program 14.4: Unity-Follower Op-Amp Circuit

Figure 14.40 provides the circuit of an op-amp connected as a unity follower. For an input of V1 = 2 V, the output (see listing in Fig. 14.41) is seen to be V(3) = 2 V, the same as the input. A transfer function calculation provides that the circuit voltage gain is

$$\frac{V(3)}{V1} = 1$$

the input impedance seen by source V_1 is 1E20, essentially infinite, and the output impedance at terminal 3 is 1E-13, essentially 0.

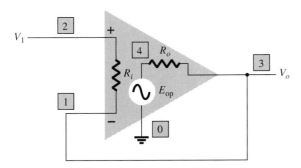

Figure 14.40 Circuit for Program 14.4.

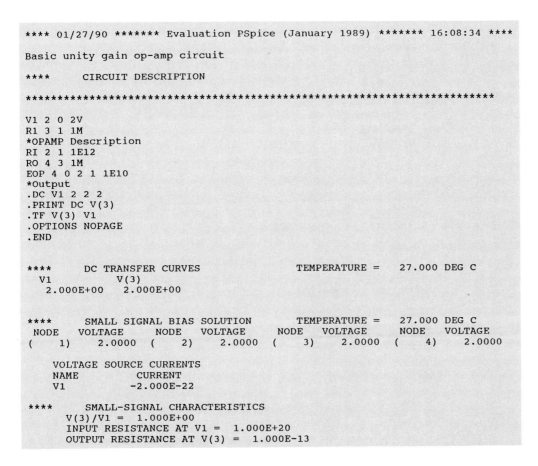

```
**** 01/27/90 ******* Evaluation PSpice (January 1989) ******* 16:08:34 ****

Basic unity gain op-amp circuit

****        CIRCUIT DESCRIPTION

*******************************************************************

V1 2 0 2V
R1 3 1 1M
*OPAMP Description
RI 2 1 1E12
RO 4 3 1M
EOP 4 0 2 1 1E10
*Output
.DC V1 2 2 2
.PRINT DC V(3)
.TF V(3) V1
.OPTIONS NOPAGE
.END


****        DC TRANSFER CURVES              TEMPERATURE =    27.000 DEG C
   V1           V(3)
    2.000E+00    2.000E+00



****        SMALL SIGNAL BIAS SOLUTION      TEMPERATURE =    27.000 DEG C
  NODE    VOLTAGE        NODE    VOLTAGE      NODE    VOLTAGE      NODE    VOLTAGE
 (    1)     2.0000  (     2)     2.0000  (     3)     2.0000  (     4)     2.0000

      VOLTAGE SOURCE CURRENTS
      NAME            CURRENT
      V1             -2.000E-22

****        SMALL-SIGNAL CHARACTERISTICS
          V(3)/V1 =   1.000E+00
          INPUT RESISTANCE AT V1 =   1.000E+20
          OUTPUT RESISTANCE AT V(3) =   1.000E-13
```

Figure 14.41 PSpice output for Program 14.4.

PROBLEMS

§ 14.2 Differential and Common-Mode Operation

1. Calculate the CMRR (in dB) for the circuit measurements of $V_d = 1$ mV, $V_o = 120$ mV, and $V_C = 1$ mV, $V_o = 20$ μV.

2. Determine the output voltage of an op-amp for input voltages of $V_{i_1} = 200$ μV and $V_{i_2} = 140$ μV. The amplifier has a differential gain of $A_d = 6000$ and the value of CMRR is:
(a) 200.
(b) 10^5.

§ 14.4 Practical Op-Amp Units

3. What is the output voltage in the circuit of Fig. 14.42?

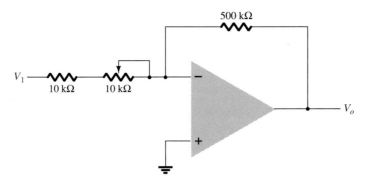

Figure 14.42 Problems 3, 25

4. What is the range of the voltage-gain adjustment in the circuit of Fig. 14.43?

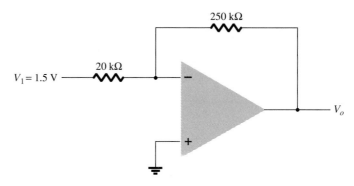

Figure 14.43 Problem 4

5. What input voltage results in an output of 2 V in the circuit of Fig. 14.44?

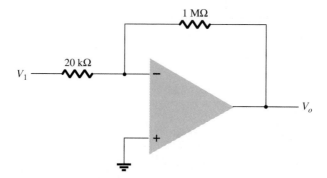

Figure 14.44 Problem 5

6. What is the range of the output voltage in the circuit of Fig. 14.45 if the input can vary from 0.1 to 0.5 V?

7. What output voltage results in the circuit of Fig. 14.46 for an input of $V_1 = -0.3$ V?

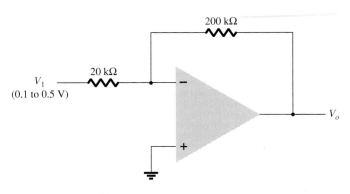

Figure 14.45 Problem 6

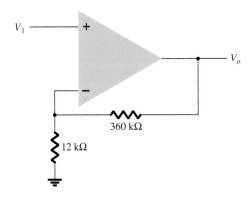

Figure 14.46 Problems 7–8, 26

8. What input must be applied to the input of Fig. 14.46 to result in an output of 2.4 V?

9. What range of output voltage is developed in the circuit of Fig. 14.47?

10. Calculate the output voltage developed by the circuit of Fig. 14.48 for $R_f = 330$ kΩ.

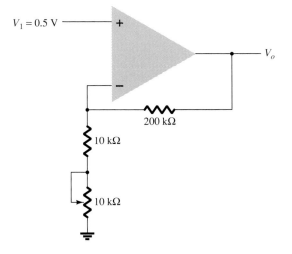

Figure 14.47 Problem 9

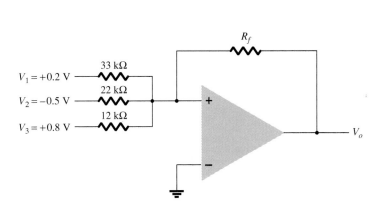

Figure 14.48 Problems 10–11, 27

11. Calculate the output voltage of the circuit in Fig. 14.48 for $R_f = 68$ kΩ.

12. Sketch the output waveform resulting in Fig. 14.49.

13. What output voltage results in the circuit of Fig. 14.50 for $V_1 = +0.5$ V?

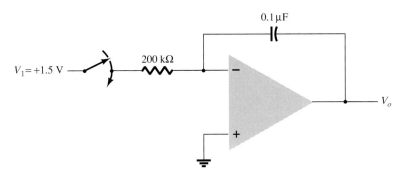

Figure 14.49 Problem 12

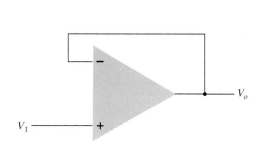

Figure 14.50 Problem 13

14. Calculate the output voltage for the circuit of Fig. 14.51.

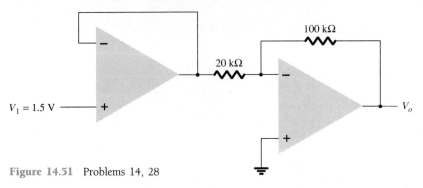

Figure 14.51 Problems 14, 28

15. Calculate the output voltages V_2 and V_3 in the circuit of Fig. 14.52.

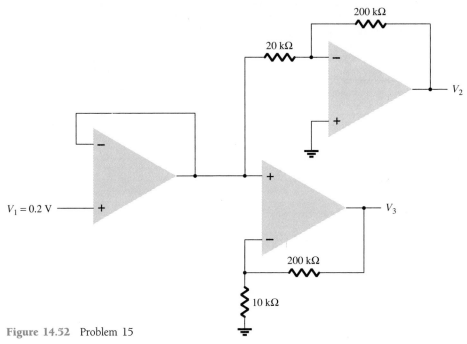

Figure 14.52 Problem 15

16. Calculate the output voltage, V_o, in the circuit of Fig. 14.53.

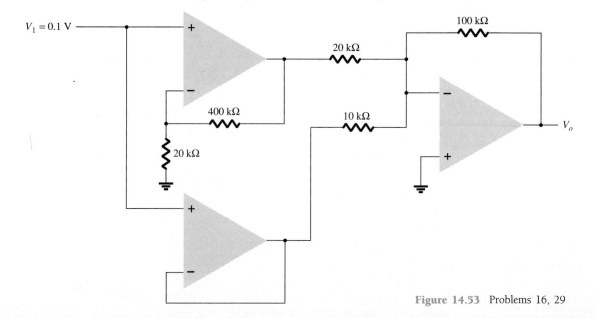

Figure 14.53 Problems 16, 29

17. Calculate V_o in the circuit of Fig. 14.54.

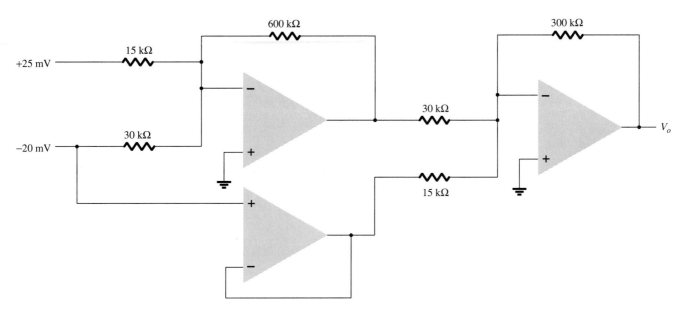

Figure 14.54 Problem 17

§ 14.5 Op-Amp Specifications—DC Offset Parameters

* **18.** Calculate the total offset voltage for the circuit of Fig. 14.55 for an op-amp with specified values of input offset voltage, $V_{IO} = 6$ mV and input offset current $I_{IO} = 120$ nA.

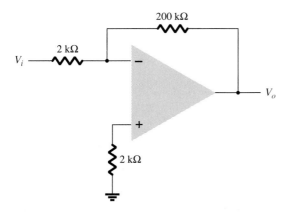

Figure 14.55 Problems 18, 22, 23, 24

* **19.** Calculate the input bias current at each input of an op-amp having specified values of $I_{IO} = 4$ nA and $I_{IB} = 20$ nA.

§ 14.6 Op-Amp Specifications—Frequency Parameters

20. Determine the cutoff frequency of an op-amp having specified values $B_1 = 800$ kHz and $A_{v_d} = 150$ V/mV.

* **21.** For an op-amp having a slew rate of SR $= 2.4$ V/μs, what is the maximum closed-loop voltage gain that can be used when the input signal varies by 0.3 V in 10 μs?

* **22.** For an input of $V_1 = 50$ mV in the circuit of Fig. 14.55, determine the maximum frequency that may be used. The op-amp slew rate SR $= 0.4$ V/μs.)

23. Using the specifications listed in Table 14.2, calculate the typical offset voltage for the circuit connection of Fig. 14.55.

* **24.** For the typical characteristics of the 741 op-amp, calculate the following values for the circuit of Fig. 14.55.
  (a) A_{CL}.
  (b) Z_i.
  (c) Z_o.

§ 14.8 Computer Analysis

* **25.** Write a PSpice program to determine the output voltage in the circuit of Fig. 14.42.

* **26.** Write a PSpice program to calculate the output voltage in the circuit of Fig. 14.46 for the input $V_i = 0.5$ V.

* **27.** Write a PSpice program to calculate the output voltage of the circuit in Fig. 14.48 for $R_f = 68$ kΩ.

* **28.** Write a PSpice program to calculate the output voltage in the circuit of Fig. 14.51.

* **29.** Write a PSpice program to calculate the output voltage in the circuit of Fig. 14.53.

---

*Please Note: Asterisks indicate more difficult problems.

Op-Amp Applications

CHAPTER

15

15.1 CONSTANT-GAIN MULTIPLIER

One of the most common op-amp circuits is the inverting constant-gain multiplier, which provides a precise gain or amplification. Figure 15.1 shows a standard circuit connection with the resulting gain being given by

$$A = -\frac{R_f}{R_1}$$

(15.1)

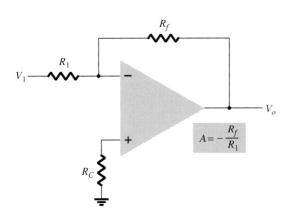

Figure 15.1 Fixed-gain amplifier.

Determine the output voltage for the circuit of Fig. 15.2 with a sinusoidal input of 2.5 mV.

EXAMPLE 15.1

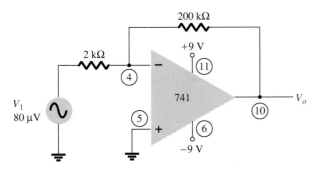

Figure 15.2 Circuit for Example 15.1.

Solution

The circuit of Fig. 15.2 uses a 741 op-amp to provide a constant or fixed gain, calculated from Eq. (15.1) to be

$$A = -\frac{R_f}{R_1} = -\frac{200 \text{ k}\Omega}{2 \text{ k}\Omega} = -100$$

The output voltage is then

$$V_o = AV_i = -100(2.5 \text{ mV}) = -250 \text{ mV} = \mathbf{-0.25 \text{ V}}$$

A noninverting constant-gain multiplier is provided by the circuit of Fig. 15.3, with the gain given by

$$\boxed{A = 1 + \frac{R_f}{R_1}} \tag{15.2}$$

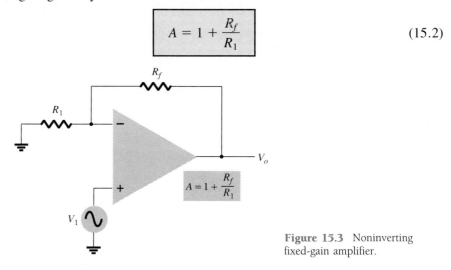

Figure 15.3 Noninverting fixed-gain amplifier.

EXAMPLE 15.2

Calculate the output voltage from the circuit of Fig. 15.4 for an input of 120 μV.

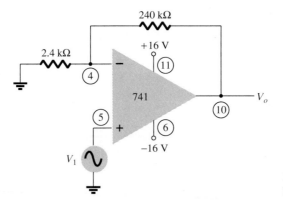

Figure 15.4 Circuit for Example 15.2.

Solution

The gain of the op-amp circuit is calculated using Eq. (15.2) to be

$$A = 1 + \frac{R_f}{R_1} = 1 + \frac{240 \text{ k}\Omega}{2.4 \text{ k}\Omega} = 1 + 100 = 101$$

The output voltage is then

$$V_o = AV_i = 101(120 \ \mu\text{V}) = \mathbf{12.12 \text{ mV}}$$

Multiple-Stage Gains

When a number of stages are connected in series the overall gain is the product of the individual stage gains. Figure 15.5 shows a connection of three stages. The first stage is connected to provide noninverting gain as given by Eq. (15.2). The next two stages provide an inverting gain given by Eq. (15.1). The overall circuit gain is then noninverting and calculated by

$$A = A_1A_2A_3$$

where $A_1 = 1 + R_f/R_1$, $A_2 = -R_f/R_2$, and $A_3 = -R_f/R_3$.

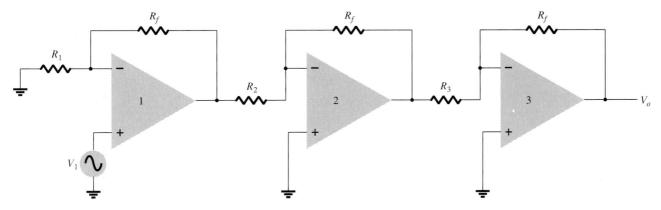

Figure 15.5 Constant-gain connection with multiple stages.

Calculate the output voltage using the circuit of Fig. 15.5 for resistor components of value: $R_f = 470$ kΩ, $R_1 = 4.3$ kΩ, $R_2 = 33$ kΩ, and $R_3 = 33$ kΩ, for an input of 80 μV.

EXAMPLE 15.3

Solution

The amplifier gain is calculated to be

$$A = A_1A_2A_3 = \left(1 + \frac{R_f}{R_1}\right)\left(-\frac{R_f}{R_2}\right)\left(-\frac{R_f}{R_3}\right)$$

$$= \left(1 + \frac{470 \text{ k}\Omega}{4.3 \text{ k}\Omega}\right)\left(-\frac{470 \text{ k}\Omega}{33 \text{ k}\Omega}\right)\left(-\frac{470 \text{ k}\Omega}{33 \text{ k}\Omega}\right)$$

$$= (110.3)(-14.2)(-14.2) = 22.2 \times 10^3$$

so that $\quad V_o = AV_i = 22.2 \times 10^3(80 \text{ μV}) = \mathbf{1.78 \ V}$

Show the connection of an LM124 quad op-amp as a three-stage amplifier with gains of $+10$, -18, and -27. Use a 270-kΩ feedback resistor for all three circuits. What output voltage will result for an input of 150 μV?

EXAMPLE 15.4

Solution

For the gain of $+10$:

$$A_1 = 1 + \frac{R_f}{R_1} = +10$$

$$\frac{R_f}{R_1} = 10 - 1 = 9$$

$$R_1 = \frac{R_f}{9} = \frac{270 \text{ k}\Omega}{9} = 30 \text{ k}\Omega$$

For the gain of -18:

$$A_2 = -\frac{R_f}{R_2} = -18$$

$$R_2 = \frac{R_f}{18} = \frac{270 \text{ k}\Omega}{18} = 15 \text{ k}\Omega$$

For the gain of -27:

$$A_3 = -\frac{R_f}{R_3} = -27$$

$$R_3 = -\frac{R_f}{27} = \frac{270 \text{ k}\Omega}{27} = 10 \text{ k}\Omega$$

The circuit showing the pin connections and all components used is in Fig. 15.6. For an input of $V_1 = 150 \ \mu\text{V}$, the output voltage will be

$$V_o = A_1 A_2 A_3 V_1 = (10)(-18)(-27)(150 \ \mu\text{V}) = 4860(150 \ \mu\text{V})$$

$$= \mathbf{0.729 \ V}$$

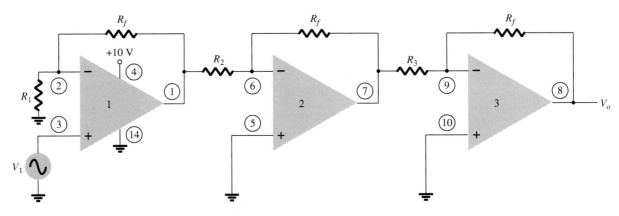

Figure 15.6 Circuit for Example 15.4 (using LM124).

A number of op-amp stages could also be used to provide separate gains, as demonstrated in the next example.

EXAMPLE 15.5

Show the connection of three op-amp stages using an LM348 IC to provide outputs that are 10, 20, and 50 times larger than the input. Use a feedback resistor of $R_f = 500 \text{ k}\Omega$ in all stages.

Solution

The resistor component for each stage is calculated to be

$$R_1 = -\frac{R_f}{A_1} = -\frac{500 \text{ k}\Omega}{-10} = 50 \text{ k}\Omega$$

$$R_2 = -\frac{R_f}{A_2} = -\frac{500 \text{ k}\Omega}{-20} = 25 \text{ k}\Omega$$

$$R_3 = -\frac{R_f}{A_3} = -\frac{500 \text{ k}\Omega}{-50} = 10 \text{ k}\Omega$$

The resulting circuit is drawn in Fig. 15.7.

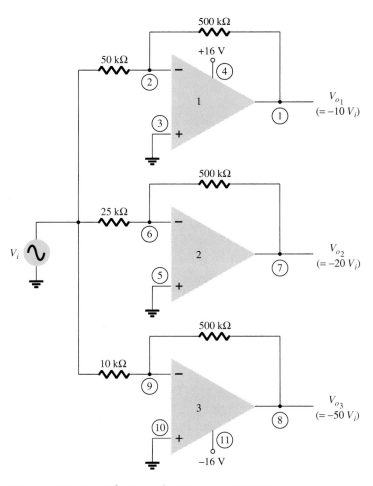

Figure 15.7 Circuit for Example 15.5 (using LM348).

15.2 VOLTAGE SUMMING

Another popular use of an op-amp is as a summing amplifier. Figure 15.8 shows the connection with the output being the sum of the three inputs, each multiplied by a different gain. The output voltage is

$$V_o = -\left(\frac{R_f}{R_1} V_1 + \frac{R_f}{R_2} V_2 + \frac{R_f}{R_3} V_3\right) \tag{15.3}$$

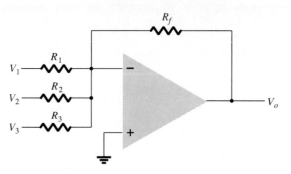

Figure 15.8 Summing amplifier.

EXAMPLE 15.6

Calculate the output voltage for the circuit of Fig. 15.9. The inputs are $V_1 =$ 50 mV sin(1000t) and $V_2 = $ 10 mV sin(3000t).

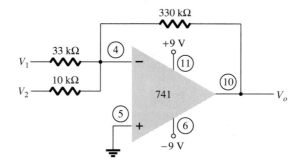

Figure 15.9 Circuit for Example 15.6.

Solution

The output voltage is

$$V_o = \left(\frac{330\ k\Omega}{33\ k\Omega}\ V_1 + \frac{330\ k\Omega}{10\ k\Omega}\ V_2 \right) = -(10V_1 + 33V_2)$$

$$= -[10(50\ mV)\ \sin(1000t) + 33(10\ mV)\ \sin(3000t)]$$

$$= -[0.5\ \sin(1000t) + 0.33\ \sin(3000t)]$$

Voltage Subtraction

Two signals can be subtracted, one from the other, in a number of ways. Figure 15.10 shows two op-amp stages used to provide subtraction of input signals. The resulting output is given by

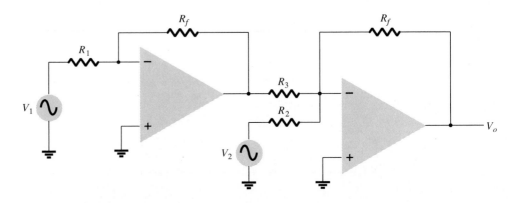

Figure 15.10 Circuit to subtract two signals.

Chapter 15 Op-Amp Applications

$$V_o = -\left[\frac{R_f}{R_3}\left(-\frac{R_f}{R_1}V_1\right) + \frac{R_f}{R_2}V_2\right]$$

$$V_o = -\left(\frac{R_f}{R_2}V_2 - \frac{R_f}{R_3}\frac{R_f}{R_1}V_1\right) \tag{15.4}$$

Determine the output for the circuit of Fig. 15.10 with components $R_f = 1$ MΩ, $R_1 = 100$ kΩ, $R_2 = 50$ kΩ, and $R_3 = 500$ kΩ.

EXAMPLE 15.7

Solution

The output voltage is calculated to be

$$V_o = \left(\frac{1\ \text{M}\Omega}{50\ \text{k}\Omega}V_2 - \frac{1\ \text{M}\Omega}{500\ \text{k}\Omega}\frac{1\ \text{M}\Omega}{100\ \text{k}\Omega}V_1\right) = 20V_2 - 20V_1 = 20(V_2 - V_1)$$

The output is seen to be the difference of V_2 and V_1 multiplied by a gain factor of 20.

Another connection to provide subtraction of two signals is shown in Fig. 15.11. This connection uses only one op-amp stage to provide subtracting two input signals. Using superposition the output can be shown to be

$$V_o = \frac{R_3}{R_1 + R_3}\frac{R_2 + R_4}{R_2}V_1 - \frac{R_4}{R_2}V_2 \tag{15.5}$$

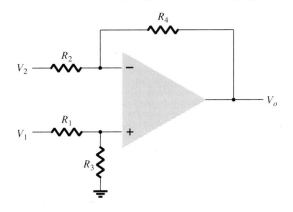

Figure 15.11 Subtraction circuit.

Determine the output voltage for the circuit of Fig. 15.12.

EXAMPLE 15.8

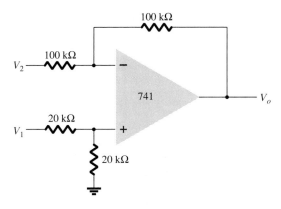

Figure 15.12 Circuit for Example 15.8.

15.2 Voltage Summing 645

Solution

The resulting output voltage can be expressed as

$$V_o = \left(\frac{20 \text{ k}\Omega}{20 \text{ k}\Omega + 20 \text{ k}\Omega}\right)\left(\frac{100 \text{ k}\Omega + 100 \text{ k}\Omega}{100 \text{ k}\Omega}\right)V_1 - \frac{100 \text{ k}\Omega}{100 \text{ k}\Omega}V_2$$

$$= V_1 - V_2$$

The resulting output voltage is seen to be the difference of the two input voltages.

15.3   VOLTAGE BUFFER

A voltage buffer circuit provides a means of isolating an input signal from a load by using a stage having unity voltage gain, with no phase or polarity inversion, and acting as an ideal circuit with very high input impedance and low output impedance. Figure 15.13 shows an op-amp connected to provide this buffer amplifier operation. The output voltage is determined by

$$\boxed{V_o = V_1} \tag{15.6}$$

Figure 15.14 shows how an input signal can be provided to two separate outputs. The advantage of this connection is that the load connected across one output has no (or little) effect on the other output. In effect, the outputs are buffered or isolated from each other.

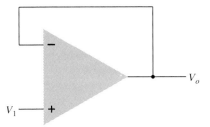

Figure 15.13   Unity-gain (buffer) amplifier.

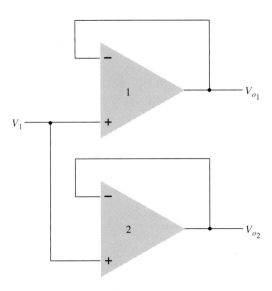

Figure 15.14   Use of buffer amplifier to provide output signals.

Show the connection of a 741 as a unity-gain circuit.

EXAMPLE 15.9

Solution

The connection is shown in Fig. 15.15.

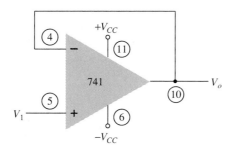

Figure 15.15 Connection for Example 15.9.

15.4 CONTROLLED SOURCES

Operational amplifiers can be used to form various types of controlled sources. An input voltage can be used to control an output voltage or current, or an input current can be used to control an output voltage or current. These types of connections are suitable for use in various instrumentation circuits. A form of each type of controlled source is provided next.

Voltage-Controlled Voltage Source

An ideal form of a voltage source whose output V_o is controlled by an input voltage V_1 is shown in Fig. 15.16. The output voltage is seen to be dependent on the input voltage (times a scale factor k). This type of circuit can be built using an op-amp as shown in Fig. 15.17. Two versions of the circuit are shown, one using the inverting input, the other the noninverting input. For the connection of Fig. 15.17a the output voltage is

$$V_o = -\frac{R_f}{R_1} V_1 = kV_1 \qquad (15.7)$$

Figure 15.16 Ideal voltage-controlled voltage source.

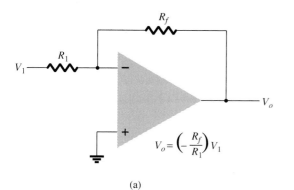

(a)

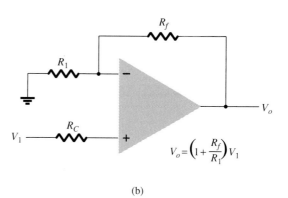

(b)

Figure 15.17 Practical voltage-controlled voltage source circuits.

while that of Fig. 15.17b results in

$$V_o = \left(1 + \frac{R_f}{R_1}\right)V_1 = kV_1 \qquad (15.8)$$

Voltage-Controlled Current Source

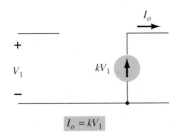

$I_o = kV_1$

Figure 15.18 Ideal voltage-controlled current source.

An ideal form of circuit providing an output current controlled by an input voltage is that of Fig. 15.18. The output current is dependent on the input voltage. A practical circuit can be built, as in Fig. 15.19, with the output current through load resistor R_L controlled by the input voltage V_1. The current through load resistor R_L can be seen to be

$$I_o = \frac{V_1}{R_1} = kV_1 \qquad (15.9)$$

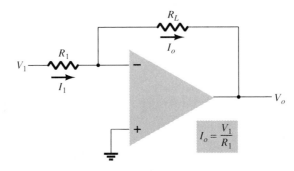

$I_o = \frac{V_1}{R_1}$

Figure 15.19 Practical voltage-controlled current source.

Current-Controlled Voltage Source

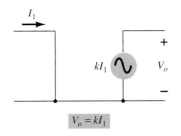

$V_o = kI_1$

Figure 15.20 Ideal current-controlled voltage source.

An ideal form of a voltage source controlled by an input current is shown in Fig. 15.20. The output voltage is dependent on the input current. A practical form of the circuit is built using an op-amp as shown in Fig. 15.21. The output voltage is seen to be

$$V_o = -I_1 R_L = kI_1 \qquad (15.10)$$

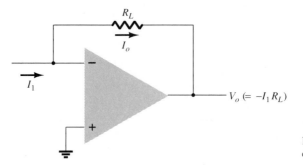

$V_o (= -I_1 R_L)$

Figure 15.21 Practical form of current-controlled voltage source.

Current-Controlled Current Source

An ideal form of a circuit providing an output current dependent on an input current is shown in Fig. 15.22. In this type of circuit an output current is provided dependent on the input current. A practical form of the circuit is shown in Fig. 15.23. The input current I_1 can be shown to result in the output current I_o so that

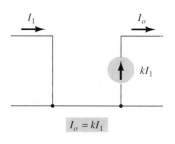

Figure 15.22 Ideal current-controlled current source.

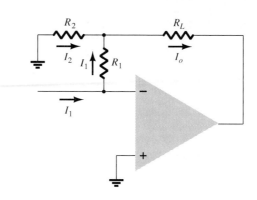

Figure 15.23 Practical form of current-controlled current source.

$$I_o = I_1 + I_2 = I_1 + \frac{I_1 R_1}{R_2} = \left(1 + \frac{R_1}{R_2}\right)I_1 = kI_1 \qquad (15.11)$$

(a) For the circuit of Fig. 15.24a, calculate I_L.
(b) For the circuit of Fig. 15.24b, calculate V_o.

EXAMPLE 15.10

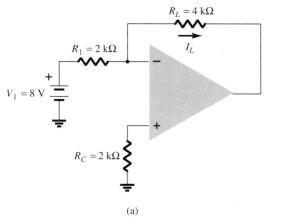

(a)

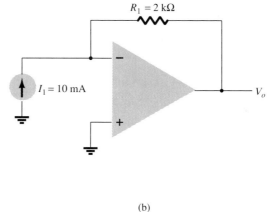

(b)

Figure 15.24 Circuits for Example 15.10.

Solution

(a) For the circuit of Fig. 15.24a,

$$I_L = \frac{V_1}{R_1} = \frac{8 \text{ V}}{2 \text{ k}\Omega} = \mathbf{4 \text{ mA}}$$

(b) For the circuit of Fig. 15.24b,

$$V_o = -I_1 R_1 = -(10 \text{ mA})(2 \text{ k}\Omega) = \mathbf{-20 \text{ V}}$$

15.5 INSTRUMENTATION CIRCUITS

A popular area of op-amp application is in instrumentation circuits such as dc or ac voltmeters. A few typical circuits will demonstrate how op-amps can be used.

DC Millivoltmeter

Figure 15.25 shows a 741 op-amp used as the basic amplifier in a dc millivoltmeter. The amplifier provides a meter with high input impedance and scale factors dependent only on resistor value and accuracy. Notice that the meter reading represents millivolts of signal at the circuit input. An analysis of the op-amp circuit provides the circuit transfer function

$$\left|\frac{I_o}{V_1}\right| = \frac{R_f}{R_1}\left(\frac{1}{R_S}\right) = \left(\frac{100 \text{ k}\Omega}{100 \text{ k}\Omega}\right)\left(\frac{1}{10 \text{ }\Omega}\right) = \frac{1 \text{ mA}}{10 \text{ mV}}$$

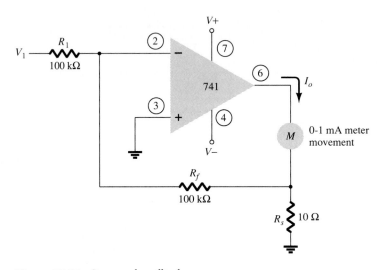

Figure 15.25 Op-amp dc millivoltmeter.

Thus an input of 10 mV will result in a current through the meter of 1 mA. If the input is 5 mV, the current through the meter will be 0.5 mA, which is half-scale deflection. Changing R_f to 200 kΩ, for example, would result in a circuit scale factor of

$$\left|\frac{I_o}{V_1}\right| = \left(\frac{200 \text{ k}\Omega}{100 \text{ k}\Omega}\right)\left(\frac{1}{10 \text{ }\Omega}\right) = \frac{1 \text{ mA}}{5 \text{ mV}}$$

showing that the meter now reads 5 mV, full scale. It should be kept in mind that building such a millivoltmeter requires purchasing an op-amp, a few resistors, diodes, capacitors and a meter movement.

AC Millivoltmeter

Another example of an instrumentation circuit is the ac millivoltmeter shown in Fig. 15.26. The circuit transfer function is

$$\left|\frac{I_o}{V_1}\right| = \frac{R_f}{R_1}\left(\frac{1}{R_S}\right) = \left(\frac{100 \text{ k}\Omega}{100 \text{ k}\Omega}\right)\left(\frac{1}{10 \text{ }\Omega}\right) = \frac{1 \text{ mA}}{10 \text{ mV}}$$

which appears the same as the dc millivoltmeter, except that in this case the signal handled is an ac signal. The meter indication provides a full-scale deflection, for an ac input voltage of 10 mV, while an ac input of 5 mV will result in half-scale deflection, with the meter reading interpreted in millivolt units.

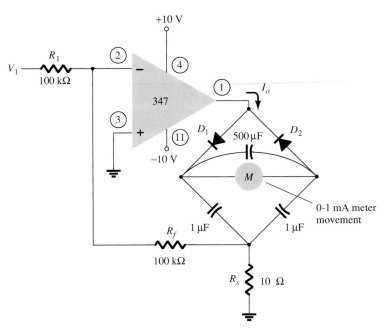

Figure 15.26 Ac millivoltmeter using op-amp.

Display Driver

Figure 15.27 shows op-amp circuits that can be used to drive a lamp display or LED display. When the noninverting input to the circuit in Fig. 15.27a goes above the inverting input, the output at terminal 1 goes to the positive saturation level (near +5 V in this example), and the lamp is driven on when transistor Q_1 conducts. As shown in the circuit, the output of the op-amp provides 30 mA of current to the base of transistor Q_1, which then drives 600 mA through a suitably selected transistor (with $\beta > 20$) capable of handling that amount of current. Figure 15.27b shows an op-amp circuit that can supply 20 mA to drive an LED display when the non-inverting input goes positive compared to the inverting input.

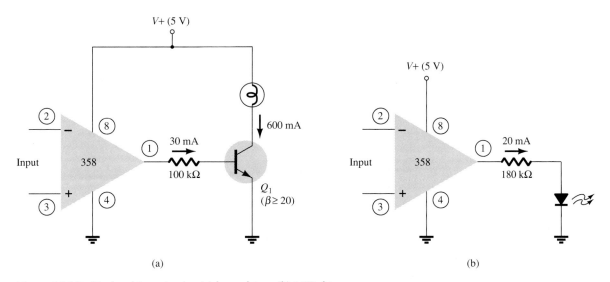

Figure 15.27 Display driver circuits: (a) lamp driver; (b) LED driver.

Instrumentation Amplifier

A circuit providing an output based on the difference between two inputs (times a scale factor) is shown in Fig. 15.28. A potentiometer is provided to permit adjusting the scale factor of the circuit. While three op-amps are used, a single-quad op-amp IC is all that is necessary (other than the resistor components). The output voltage can be shown to be

$$\frac{V_o}{V_1 - V_2} = 1 + \frac{2R}{R_P}$$

so that the output can be obtained from

$$V_o = \left(1 + \frac{2R}{R_P}\right)(V_1 - V_2) = k(V_1 - V_2) \qquad (15.12)$$

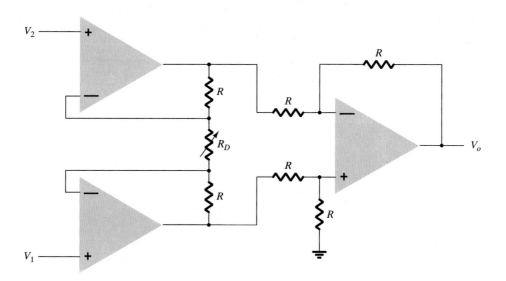

Figure 15.28 Instrumentation amplifier.

EXAMPLE 15.11 Calculate the output voltage expression for the circuit of Fig. 15.29.

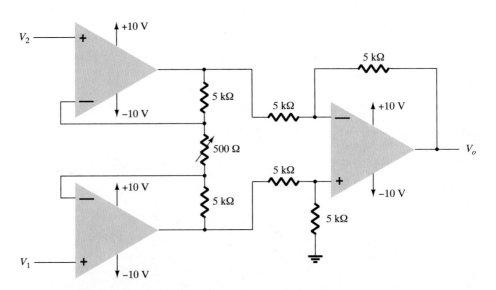

Figure 15.29 Circuit for Example 15.11.

Solution

The output voltage can then be expressed using Eq. (15.3) as

$$V_o = \left(1 + \frac{2R}{R_P}\right)(V_1 - V_2) = \left[1 + \frac{2(5000)}{500}\right](V_1 - V_2)$$

$$= 21(V_1 - V_2)$$

15.6 ACTIVE FILTERS

A popular application uses op-amps to build active filter circuits. A filter circuit can be constructed using passive components: resistors and capacitors. An active filter additionally uses an amplifier to provide voltage amplification and signal isolation or buffering.

A filter that provides a constant output from dc up to a cutoff frequency f_{OH} and then passes no signal above that frequency is called an ideal low-pass filter. The ideal response of a low-pass filter is shown in Fig. 15.30a. A filter that provides or passes signals above a cutoff frequency f_{OL} is a high-pass filter, as idealized in Fig. 15.30b. When the filter circuit passes signals that are above one ideal cutoff frequency and below a second cutoff frequency, it is called a bandpass filter as idealized in Fig. 15.30c.

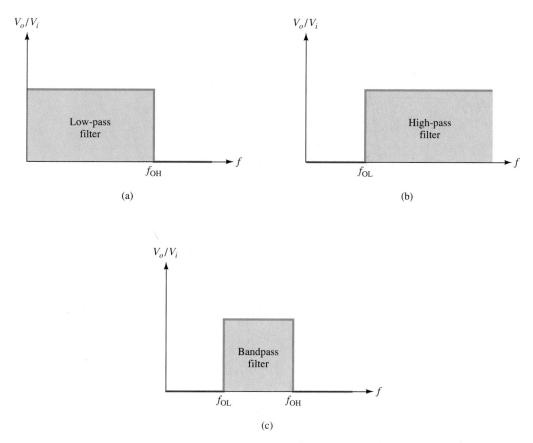

Figure 15.30 Ideal filter response: (a) low-pass; (b) high-pass; (c) bandpass.

Low-Pass Filter

A first-order, low-pass filter using a single resistor and capacitor as in Fig. 15.31a has a practical slope of 20 dB per decade, as shown in Fig. 15.31b (rather than the ideal response of Fig. 15.30a). The voltage gain below the cutoff frequency is constant at

$$A_v = 1 + \frac{R_f}{R_1} \tag{15.13}$$

at a cutoff frequency of

$$f_{OH} = \frac{1}{2\pi R_1 C_1} \tag{15.14}$$

Connecting two sections of filter as in Fig. 15.32 results in a second-order low-pass filter with cutoff at 40 dB/decade—closer to the ideal characteristic of Fig. 15.30a.

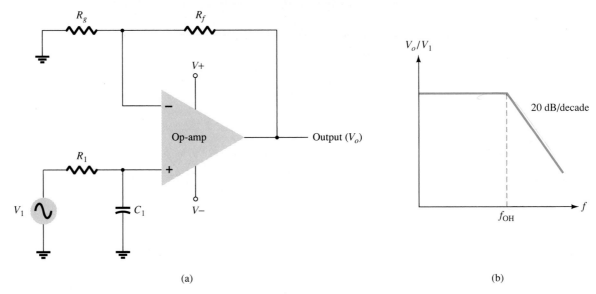

(a)   (b)

Figure 15.31 First-order low-pass active filter.

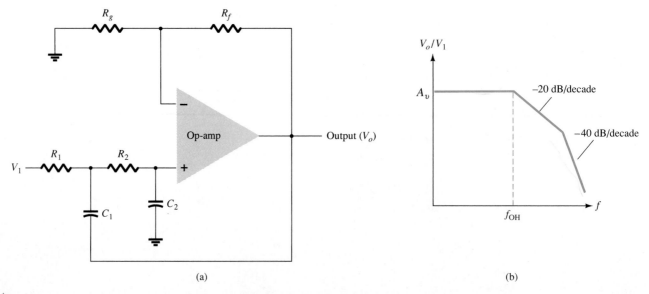

(a)   (b)

Figure 15.32 Second-order low-pass active filter.

The circuit voltage gain and cutoff frequency are the same for the second-order circuit as for the first-order filter circuit except that the filter response drops at a faster rate for a second-order filter circuit.

EXAMPLE 15.12

Calculate the cutoff frequency of a first-order low-pass filter for $R_1 = 1.2$ kΩ and $C_1 = 0.02$ μF.

Solution

$$f_{OH} = \frac{1}{2\pi R_1 C_1} = \frac{1}{2\pi(1.2 \times 10^3)(0.02 \times 10^{-6})} = \mathbf{6.63 \ kHz}$$

High-Pass Active Filter

First- and second-order high-pass active filters can be built as shown in Fig. 15.33. The amplifier gain is calculated using Eq. (15.13), with cutoff frequency

$$\boxed{f_{OL} = \frac{1}{2\pi R_1 C_1}} \tag{15.15}$$

(With a second-order filter $R_1 = R_2$, and $C_1 = C_2$ results in the same cutoff frequency.)

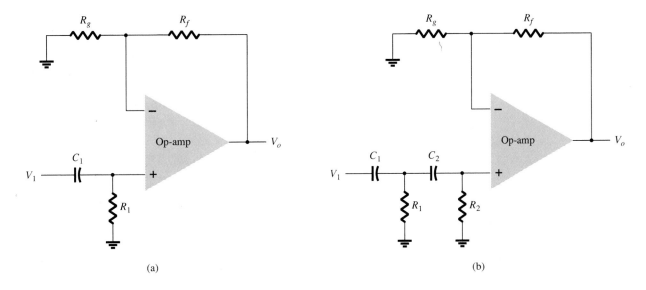

(a)

(b)

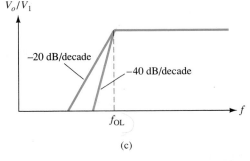

(c)

Figure 15.33 High-pass filter: (a) first order; (b) second order; (c) response plot.

EXAMPLE 15.13

Calculate the cutoff frequency of a second-order high-pass filter as in Fig. 15.33b for $R_1 = R_2 = 2.1$ kΩ, $C_1 = C_2 = 0.05$ μF, and $R_{o1} = 10$ kΩ, $R_{of} = 50$ kΩ.

Solution

$$\text{Eq. (15.13):} \quad A_v = 1 + \frac{R_{of}}{R_{o1}} = 1 + \frac{50 \text{ k}\Omega}{10 \text{ k}\Omega} = 6$$

The cutoff frequency is then

$$\text{Eq. (15.15):} \quad f_{OL} = \frac{1}{2\pi R_1 C_1} = \frac{1}{2\pi (2.1 \times 10^3)(0.05 \times 10^{-6})} \approx \mathbf{1.5 \text{ kHz}}$$

Bandpass Filter

Figure 15.34 shows a bandpass filter using two stages, the first a high-pass filter and the second a low-pass filter, the combined operation being the desired bandpass response.

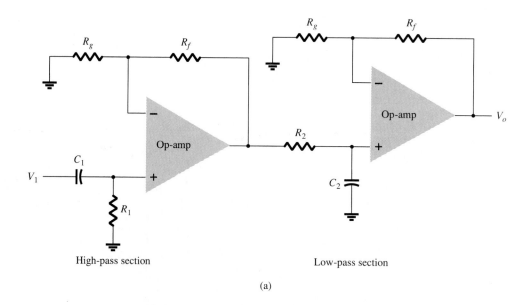

High-pass section   Low-pass section

(a)

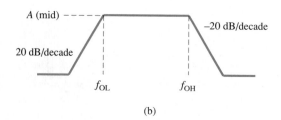

(b)

Figure 15.34 Bandpass active filter.

Calculate the cutoff frequencies of the bandpass filter circuit of Fig. 15.34 with $R_1 = R_2 = 10$ kΩ, $C_1 = 0.1$ μF, and $C_2 = 0.002$ μF.

EXAMPLE 15.14

Solution

$$f_{OL} = \frac{1}{2\pi R_1 C_1} = \frac{1}{2\pi(10 \times 10^3)(0.1 \times 10^{-6})} = \mathbf{159.15\ Hz}$$

$$f_{OH} = \frac{1}{2\pi R_2 C_2} = \frac{1}{2\pi(10 \times 10^3)(0.002 \times 10^{-6})} = \mathbf{7.96\ kHz}$$

15.7 COMPUTER ANALYSIS

Practical op-amp circuits can be analyzed using PSpice. While the material in this section will use the ideal or practical op-amp description defined in Chapter 14, a commercial version of PSpice is available with detailed models of various practical IC op-amp devices. While Chapter 14 provided examples using mainly an ideal op-amp, the problems presented in this section will use practical op-amps, based on the specifications provided in the devices specification sheets.

Program 15.1: Summing Op-Amp

A summing op-amp using a μA709 IC is shown in Fig. 15.35. Figure 15.36 provides a description of the circuit, a description of the op-amp, and specifies output of the resulting voltage, V_o. The output voltage is calculated to be

$$V_o = -\left[\frac{1\ M\Omega}{2\ M\Omega}(+2\ V) + \frac{1\ M\Omega}{5\ M\Omega}(-3\ V) + \frac{1\ M\Omega}{1\ M\Omega}(+1\ V)\right]$$

$$= -(1\ V - 0.6\ V + 1V) = -1.4\ V$$

as calculated in Fig. 15.36, V(3) = -1.4 V.

The 741 op-amp has specified values of $R_i = 2$ MΩ, $R_o = 75$ Ω, and voltage gain of $A_{v_d} = 200$ V/mV. A resistor RP of about 0 Ω is used to connect the op-amp input at pin 2 to ground to maintain pin 2 for the positive input.

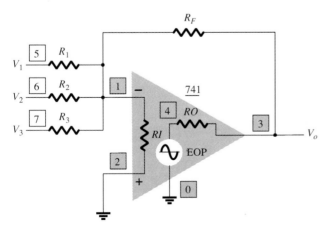

Figure 15.35 Summing amplifier using 741 op-amp.

```
**** 01/27/90 ******* Evaluation PSpice (January 1989) ******* 16:35:47 ****

Summing op-amp circuit

****       CIRCUIT DESCRIPTION

*********************************************************************

V1 5 0 2V
V2 6 0 -3V
V3 7 0 +1V
R1 5 1 2MEG
R2 6 1 5MEG
R3 7 1 1MEG
RF 1 3 1MEG
RP 2 0 1M
*OPAMP Description = 741 IC
RI 2 1 2MEG
RO 4 3 75
EOP 4 0 2 1 200E3
*Output
.DC V1 2 2 1
.PRINT DC V(5) V(6) V(7) V(3)
.OPTIONS NOPAGE
.END


****       DC TRANSFER CURVES                    TEMPERATURE =   27.000 DEG C
  V1          V(5)          V(6)         V(7)        V(3)
  2.000E+00   2.000E+00   -3.000E+00   1.000E+00   -1.400E+00
```

Figure 15.36 PSpice output for circuit of Fig. 15.35.

Program 15.2: DC Millivoltmeter

A dc millivoltmeter built using a 741 op-amp is provided in Fig. 15.37. A list of the given circuit, the op-amp itself, and the desired current through the meter movement is provided in Fig. 15.38. In particular, note that a dc milliameter movement connected from the op-amp output to the junction of resistors R_F and R_S is provided in the PSpice listing by

$$\text{VSENSE } 6\ 5\ 0$$

which is a dead voltage source (VSENSE = 0 V) serving as the current meter. The output prints the current through the meter

$$I(\text{VSENSE}) = -0.2 \text{ A}$$

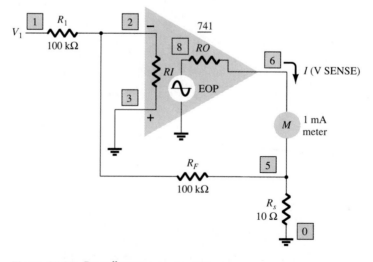

Figure 15.37 Dc milliammeter using 709 op-amp.

Chapter 15 Op-Amp Applications

```
**** 01/27/90 ******* Evaluation PSpice (January 1989) ******* 16:37:47 ****

Op-amp dc Millivoltmeter

****       CIRCUIT DESCRIPTION

**********************************************************************

V1 0 1 2V
R1 1 2 100K
RF 2 5 100K
RS 5 0 10
RP 3 0 1M
*741 Op-amp
RI 2 3 2MEG
RO 8 6 75
EOP 8 0 3 2 200E3
*DC 1 ma meter movement
VSENSE 6 5 0
.DC V1 2 2 1
.PRINT DC I(VSENSE)
.OPTIONS NOPAGE
.END


****       DC TRANSFER CURVES              TEMPERATURE =    27.000 DEG C
   V1            I(VSENSE)
   2.000E+00     2.000E-01
```

Figure 15.38 PSpice output for circuit of Fig. 15.37.

the negative sign indicating the current direction from the negative to the positive of the meter (from pin 5 to pin 6).* The value of resistors R_1, R_F, and R_S could be changed to obtain different full-scale settings of the millivoltmeter as described in Section 15.5.

Program 15.3: Low-Pass Active Filter

Figures 15.39 and 15.40 describe a low-pass active filter. The resulting plot of the output voltage as a function of the input signal frequency is also provided in Fig. 15.39 and is obtained using PSpice's plot output. In particular, notice that the ac frequency variation is made logarithmically spaced by the line

$$.AC\ DEC\ 5\ 1\ 1E5$$

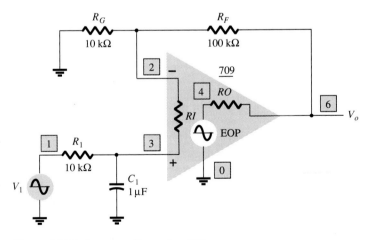

Figure 15.39 Low-pass filter using 709 op-amp.

---

*Recall that in PSpice, positive current is defined as that from the positive node to the negative node *through the source*.

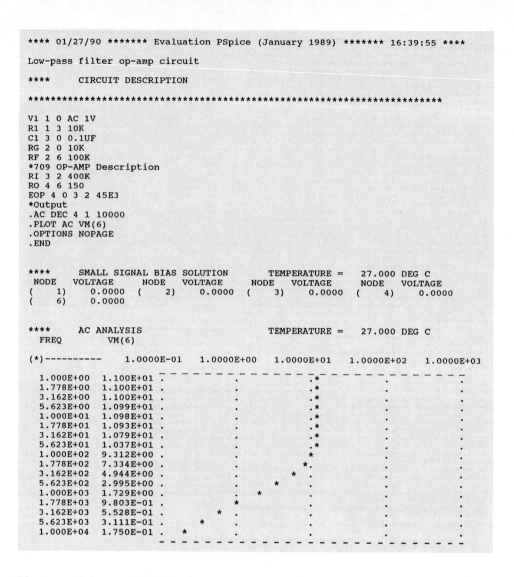

```
**** 01/27/90 ******* Evaluation PSpice (January 1989) ******* 16:39:55 ****

Low-pass filter op-amp circuit

****      CIRCUIT DESCRIPTION

********************************************************************

V1 1 0 AC 1V
R1 1 3 10K
C1 3 0 0.1UF
RG 2 0 10K
RF 2 6 100K
*709 OP-AMP Description
RI 3 2 400K
RO 4 6 150
EOP 4 0 3 2 45E3
*Output
.AC DEC 4 1 10000
.PLOT AC VM(6)
.OPTIONS NOPAGE
.END


****      SMALL SIGNAL BIAS SOLUTION          TEMPERATURE =   27.000 DEG C
    NODE    VOLTAGE     NODE    VOLTAGE     NODE    VOLTAGE     NODE    VOLTAGE
(     1)     0.0000  (    2)     0.0000  (    3)     0.0000  (    4)     0.0000
(     6)     0.0000


****      AC ANALYSIS                          TEMPERATURE =   27.000 DEG C
    FREQ         VM(6)

(*)----------   1.0000E-01   1.0000E+00   1.0000E+01   1.0000E+02   1.0000E+03

    1.000E+00   1.100E+01 .                      .*                   .            .
    1.778E+00   1.100E+01 .                      .*                   .            .
    3.162E+00   1.100E+01 .                      .*                   .            .
    5.623E+00   1.099E+01 .                      .*                   .            .
    1.000E+01   1.098E+01 .                      .*                   .            .
    1.778E+01   1.093E+01 .                      .*                   .            .
    3.162E+01   1.079E+01 .                      .*                   .            .
    5.623E+01   1.037E+01 .                      .*                   .            .
    1.000E+02   9.312E+00 .                      *                    .            .
    1.778E+02   7.334E+00 .                    *.                     .            .
    3.162E+02   4.944E+00 .               *    .                      .            .
    5.623E+02   2.995E+00 .          *         .                      .            .
    1.000E+03   1.729E+00 .       *  .                                .            .
    1.778E+03   9.803E-01 .    *     .                                .            .
    3.162E+03   5.528E-01 .  *       .                                .            .
    5.623E+03   3.111E-01 .*         .                                .            .
    1.000E+04   1.750E-01 *          .                                .            .
```

Figure 15.40 PSpice output for circuit of Fig. 15.39.

The general form of this line is

$$.AC \ DEC \ ND \ FSTART \ FSTOP$$

where   ND = number of frequency points per decade
    FSTART = starting frequency
    FSTOP = last (highest) frequency

For Fig. 15.40, a plot of the output amplitude is provided over a range from 1 Hz to 100,000 Hz (10^5 Hz). Notice that the output starts at a voltage of 11 V (amplifier voltage gain is $1 + R_f/R_1 = 1 + 100 \ \text{k}\Omega/10 \ \text{k}\Omega = 11$), the gain remaining about the same at low frequencies, dropping off at higher frequencies. Since the low-pass filter's critical frequency causes the voltage to drop to 0.707 of that at low frequencies, the plot shows this frequency to be about 160 Hz.

$$\text{Eq. (15.14):} \quad f_{\text{OH}} = \frac{1}{2\pi R_1 C_1} = \frac{1}{2\pi (10 \ \text{k}\Omega)(0.1 \ \mu\text{F})} = 159 \ \text{Hz}$$

Program 15.4: High-pass Active Filter

Figure 15.41 shows a high-pass active filter circuit. The resulting plot of the output voltage as a function of the input signal frequency is provided in Fig. 15.42. The

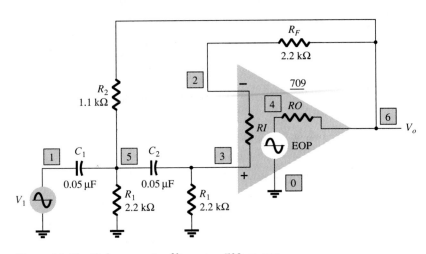

Figure 15.41 High-pass active filter using 709 op-amp.

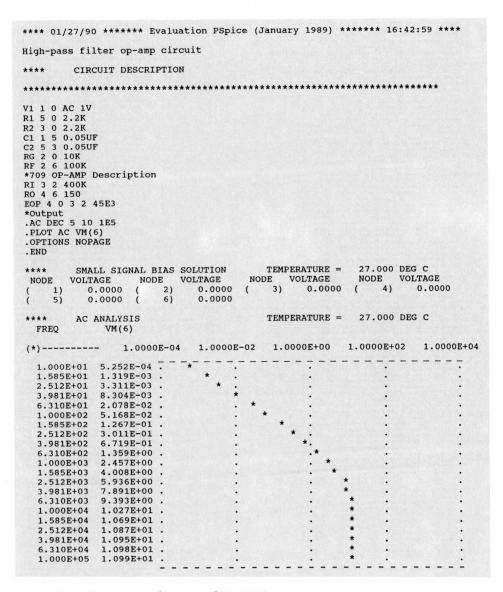

```
**** 01/27/90 ******* Evaluation PSpice (January 1989) ******* 16:42:59 ****

High-pass filter op-amp circuit

****      CIRCUIT DESCRIPTION

***********************************************************************

V1 1 0 AC 1V
R1 5 0 2.2K
R2 3 0 2.2K
C1 1 5 0.05UF
C2 5 3 0.05UF
RG 2 0 10K
RF 2 6 100K
*709 OP-AMP Description
RI 3 2 400K
RO 4 6 150
EOP 4 0 3 2 45E3
*Output
.AC DEC 5 10 1E5
.PLOT AC VM(6)
.OPTIONS NOPAGE
.END

****      SMALL SIGNAL BIAS SOLUTION       TEMPERATURE =    27.000 DEG C
   NODE   VOLTAGE       NODE   VOLTAGE      NODE   VOLTAGE      NODE   VOLTAGE
(    1)    0.0000  (     2)    0.0000  (    3)    0.0000  (    4)    0.0000
(    5)    0.0000  (     6)    0.0000

****      AC ANALYSIS                       TEMPERATURE =    27.000 DEG C
   FREQ        VM(6)

(*)----------  1.0000E-04   1.0000E-02   1.0000E+00   1.0000E+02   1.0000E+04

   1.000E+01  5.252E-04 .- - *     .           .            .           .
   1.585E+01  1.319E-03 .      *   .           .            .           .
   2.512E+01  3.311E-03 .        * .           .            .           .
   3.981E+01  8.304E-03 .          *           .            .           .
   6.310E+01  2.078E-02 .          .  *        .            .           .
   1.000E+02  5.168E-02 .          .     *     .            .           .
   1.585E+02  1.267E-01 .          .        *  .            .           .
   2.512E+02  3.011E-01 .          .           * .          .           .
   3.981E+02  6.719E-01 .          .           .*.          .           .
   6.310E+02  1.359E+00 .          .           .  *         .           .
   1.000E+03  2.457E+00 .          .           .     *      .           .
   1.585E+03  4.008E+00 .          .           .        *   .           .
   2.512E+03  5.936E+00 .          .           .          * .           .
   3.981E+03  7.891E+00 .          .           .          *             .
   6.310E+03  9.393E+00 .          .           .         *              .
   1.000E+04  1.027E+01 .          .           .         *              .
   1.585E+04  1.069E+01 .          .           .         *              .
   2.512E+04  1.087E+01 .          .           .         *              .
   3.981E+04  1.095E+01 .          .           .         *              .
   6.310E+04  1.098E+01 .          .           .         *              .
   1.000E+05  1.099E+01 .          .           .         *              .
                       - - - - - - - - - - - - - - - - - - - - - - - - -
```

Figure 15.42 PSpice output for circuit of Fig. 15.41.

high-pass cutoff frequency is calculated to be

$$f_{OL} = \frac{1}{2\pi RC} = \frac{1}{2\pi(2.2 \times 10^3)(0.5 \times 10^{-6})} = 1447 \text{ Hz}$$

The plot in Fig. 15.42 shows the amplitude rising to a pass value of 11 V (amplifier gain is $1 + R_f/R_g = 1 + 100 \text{ k}\Omega/10 \text{ k}\Omega = 11$).

Program 15.5:   Bandpass Active Filter

Figure 15.43 shows a bandpass active filter circuit. The resulting plot of the output voltage as a function of the input signal frequency is provided in Fig. 15.44. As shown, the output signal amplitude first increases to the pass level, holds a while, then drops off. Two separate stages are used, the first a high-pass and the second a low-pass, each stage having gain of 2 for a passband amplitude of 4 V.

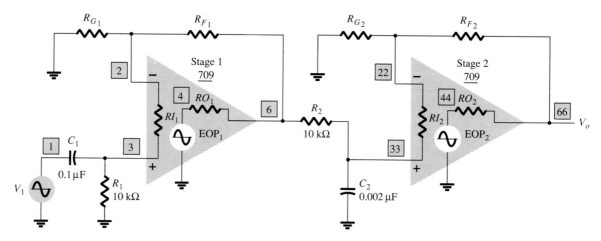

Figure 15.43 Bandpass filter using 709 op-amps.

```
**** 01/27/90 ******* Evaluation PSpice (January 1989) ******* 16:45:43 ****

Band-pass filter op-amp circuit

****        CIRCUIT DESCRIPTION

*********************************************************************

V1 1 0 AC 1V
*1st stage
R1 3 0 10K
C1 1 3 0.1UF
RG1 2 0 10K
RF1 2 6 10K
*709 OP-AMP Description
RI1 3 2 400K
RO1 4 6 150
EOP1 4 0 3 2 45E3
*2nd stage
R2 6 33 10K
C2 33 0 0.002UF
RG2 22 0 10K
RF2 22 66 10K
*709 OP-AMP Description
RI2 33 22 400K
RO2 44 66 150
EOP2 44 0 33 22 45E3
.AC DEC 5 1 1E7
.PLOT AC VM(66)
.OPTIONS NOPAGE
.END
```

Figure 15.44 SPICE output for circuit of Fig. 15.43.

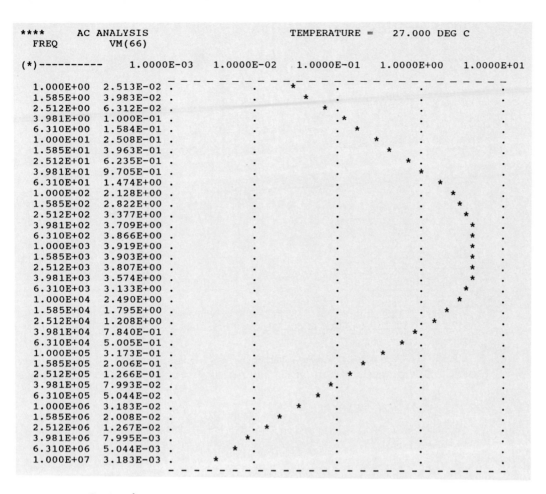

```
****     AC ANALYSIS                    TEMPERATURE =   27.000 DEG C
   FREQ         VM(66)

(*)----------    1.0000E-03   1.0000E-02   1.0000E-01   1.0000E+00   1.0000E+01

 1.000E+00   2.513E-02 .                        *       .                .            .
 1.585E+00   3.983E-02 .                      *   .                .            .
 2.512E+00   6.312E-02 .                    *   .                .            .
 3.981E+00   1.000E-01 .                        .*              .                .            .
 6.310E+00   1.584E-01 .                        .  *            .                .            .
 1.000E+01   2.508E-01 .                        .       *       .                .            .
 1.585E+01   3.963E-01 .                        .          *    .                .            .
 2.512E+01   6.235E-01 .                        .             * .                .            .
 3.981E+01   9.705E-01 .                        .               *                .            .
 6.310E+01   1.474E+00 .                        .               .  *             .            .
 1.000E+02   2.128E+00 .                        .               .     *          .            .
 1.585E+02   2.822E+00 .                        .               .       *        .            .
 2.512E+02   3.377E+00 .                        .               .         *      .            .
 3.981E+02   3.709E+00 .                        .               .          *     .            .
 6.310E+02   3.866E+00 .                        .               .           *    .            .
 1.000E+03   3.919E+00 .                        .               .           *    .            .
 1.585E+03   3.903E+00 .                        .               .           *    .            .
 2.512E+03   3.807E+00 .                        .               .          *     .            .
 3.981E+03   3.574E+00 .                        .               .         *      .            .
 6.310E+03   3.133E+00 .                        .               .        *       .            .
 1.000E+04   2.490E+00 .                        .               .     *          .            .
 1.585E+04   1.795E+00 .                        .               .   *            .            .
 2.512E+04   1.208E+00 .                        .               . *              .            .
 3.981E+04   7.840E-01 .                        .             *..                .            .
 6.310E+04   5.005E-01 .                        .          *    .                .            .
 1.000E+05   3.173E-01 .                        .       *       .                .            .
 1.585E+05   2.006E-01 .                        .     *         .                .            .
 2.512E+05   1.266E-01 .                        .   *           .                .            .
 3.981E+05   7.993E-02 .                        . *             .                .            .
 6.310E+05   5.044E-02 .                      *   .             .                .            .
 1.000E+06   3.183E-02 .                    *     .             .                .            .
 1.585E+06   2.008E-02 .                  *       .             .                .            .
 2.512E+06   1.267E-02 .               . *        .             .                .            .
 3.981E+06   7.995E-03 .             *.           .             .                .            .
 6.310E+06   5.044E-03 .           *              .             .                .            .
 1.000E+07   3.183E-03 .         *                .             .                .            .
```

Figure 15.44 Continued.

---

§ **15.1 Constant-Gain Multiplier**

PROBLEMS

1. Calculate the output voltage for the circuit of Fig. 15.45 for an input of $V_i = 3.5$ mV rms.

2. Calculate the output voltage of the circuit of Fig. 15.46 for input of 150 mV rms.

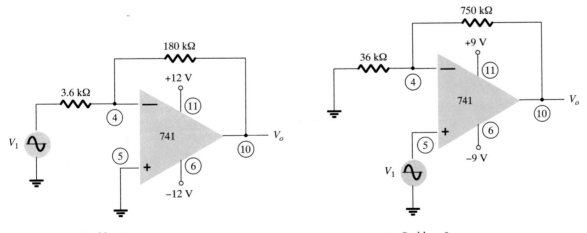

Figure 15.45 Problem 1

Figure 15.46 Problem 2

* **3.** Calculate the output voltage in the circuit of Fig. 15.47.

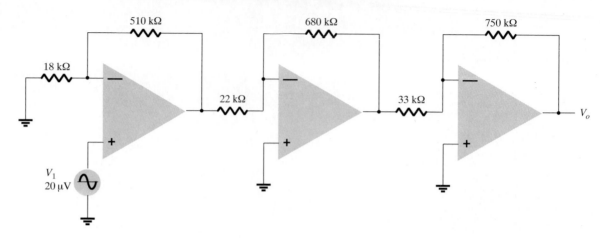

Figure 15.47 Problem 3

* **4.** Show the connection of a LM124 quad op-amp as a three-stage amplifier with gains of +15, −22, and −30. Use a 420-kΩ feedback resistor for all stages. What output voltage results for an input of $V_1 = 80$ μV?

5. Show the connection of two op-amp stages using an LM358 IC to provide outputs that are 15, and −30, times larger than the input. Use a feedback resistor, $R_f = 150$ kΩ in all stages.

§ **15.2 Voltage Summing**

6. Calculate the output voltage for the circuit of Fig. 15.48 with inputs of $V_1 = 40$ mV rms and $V_2 = 20$ mV rms.

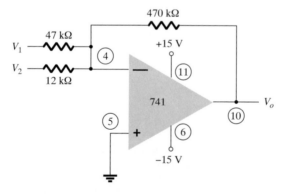

Figure 15.48 Problem 6

7. Determine the output voltage for the circuit of Fig. 15.49.

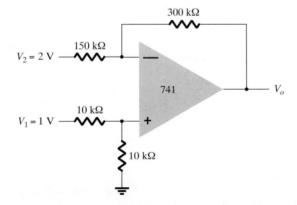

Figure 15.49 Problem 7

8. Determine the output voltage for the circuit of Fig. 15.50.

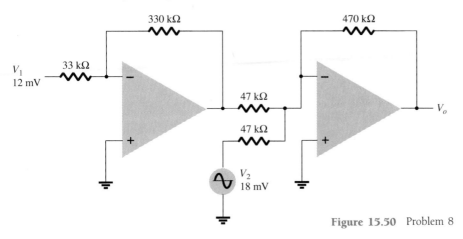

Figure 15.50 Problem 8

§ 15.3 Voltage Buffer

9. Show the connection (including pin information) of an LM124 IC stage connected as a unity-gain amplifier.

10. Show the connection (including pin information) of two LM358 stages connected as unity-gain amplifiers to provide the same output.

§ 15.4 Controlled Sources

11. For the circuit of Fig. 15.51, calculate I_L.

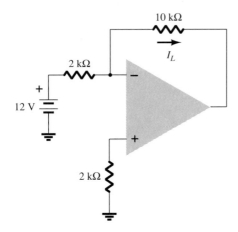

Figure 15.51 Problem 11

12. Calculate V_o for the circuit of Fig. 15.52.

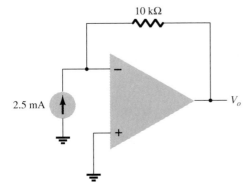

Figure 15.52 Problem 12

13. Calculate the output current I_o, in the circuit of Fig. 15.53.

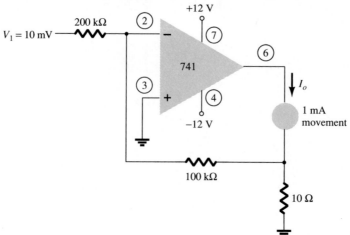

Figure 15.53 Problem 13

* **14.** Calculate V_o in the circuit of Fig. 15.54.

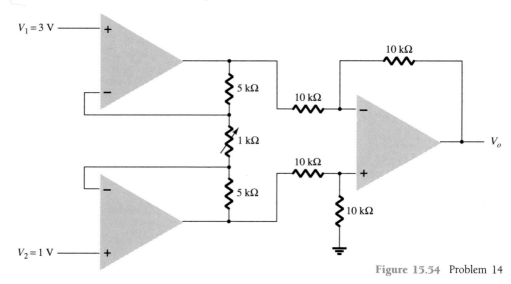

Figure 15.54 Problem 14

§ 15.6 Active Filters

15. Calculate the cutoff frequency of a first-order low-pass filter in the circuit of Fig. 15.55.

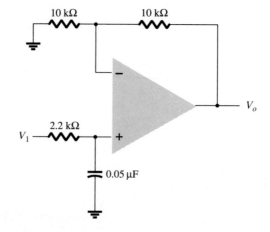

Figure 15.55 Problem 15

16. Calculate the cutoff frequency of the high-pass filter circuit in Fig. 15.56.

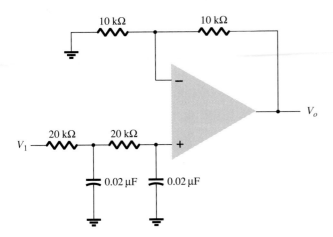

Figure 15.56 Problem 16

17. Calculate the lower and upper cutoff frequencies of the bandpass filter circuit in Fig. 15.57.

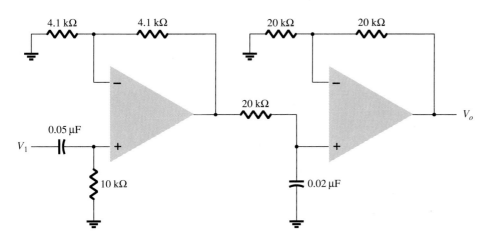

Figure 15.57 Problem 17

§ **15.7 Computer Analysis**

* **18.** Write a PSpice program to calculate V_o in the circuit of Fig. 15.58.

* **19.** Write a PSpice program to calculate I(VSENSE) in the circuit of Fig. 15.59.

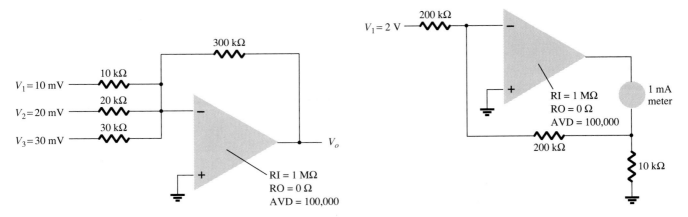

Figure 15.58 Problem 18

Figure 15.59 Problem 19

* **20.** Write a PSpice program to plot the response of the low-pass filter circuit in Fig. 15.60.

* **21.** Write a PSpice program to plot the response of the high-pass filter circuit in Fig. 15.61.

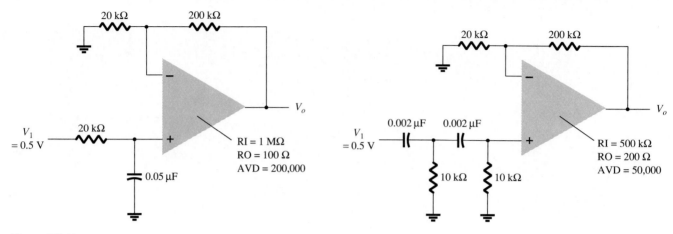

Figure 15.60 Problem 20

Figure 15.61 Problem 21

* **22.** Write a PSpice program to plot the response of the bandpass filter circuit in Fig. 15.62.

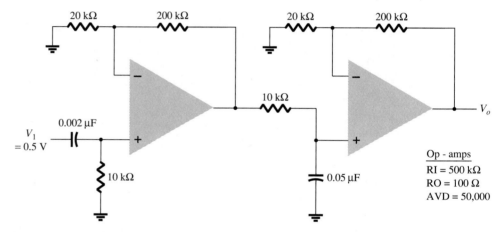

Figure 15.62 Problem 22

*Please Note: Asterisks indicate more difficult problems.

Power Amplifiers

P_L

16.1 INTRODUCTION—DEFINITIONS AND AMPLIFIER TYPES

An amplifier receives a signal from some pickup transducer or other input source and provides a larger version of the signal to some output device or to another amplifier stage. An input transducer signal is generally small (a few millivolts from a cassette or CD input, or a few microvolts from an antenna) and needs to be amplified sufficiently to operate an output device (speaker or other power-handling device). In small-signal amplifiers the main factors are usually amplification linearity and magnitude of gain. Since signal voltage and current are small in a small-signal amplifier, the amount of power-handling capacity and power efficiency are of little concern. A voltage amplifier provides voltage amplification primarily to increase the voltage of the input signal. Large-signal or power amplifiers, on the other hand, primarily provide sufficient power to an output load to drive a speaker or other power device, typically a few watts to tens of watts. In the present chapter we concentrate on those amplifier circuits used to handle large-voltage signals at moderate to high current levels. The main features of a large-signal amplifier are the circuit's power efficiency, the maximum amount of power that the circuit is capable of handling, and the impedance matching to the output device.

One method used to categorize amplifiers is by class. Basically, amplifier classes represent the amount the output signal varies over one cycle of operation for a full cycle of input signal. A brief description of amplifier classes is provided next.

Class A: The output signal varies for a full 360° of the cycle. Figure 16.1a

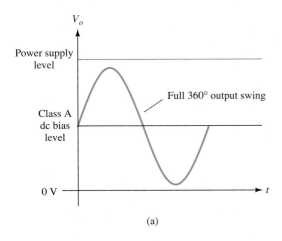

(a)

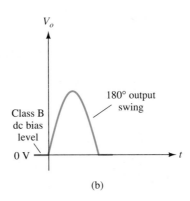

(b)

Figure 16.1 Amplifier operating classes.

shows that this requires the Q-point is biased at a level so that at least half the signal swing of the output may vary up and down without going to a high-enough voltage to be limited by the supply voltage level, or too low to approach the lower supply level, or 0-V in this description.

Class B: A class B circuit provides an output signal varying over one-half the input signal cycle, or for 180° of signal, as shown in Fig. 16.1b. The dc bias point for class B is therefore at 0 V, with the output then varying from this bias point for a half-cycle. Obviously, the output is not a faithful reproduction of the input if only one half-cycle is present. Two class B operations—one to provide output on the positive-output half-cycle and another to provide operation on the negative-output half-cycle are necessary. The combined half-cycles then provide an output for a full 360° of operation. This type of connection is referred to as push-pull operation, which is discussed later in this chapter. Note that class B operation, by itself, creates a very distorted output signal since reproduction of the input takes place for only 180° of the output signal swing.

Class AB: An amplifier may be biased at a dc level above the zero base current level of class B and above one-half the supply voltage level of class A; this bias condition is class AB. Class AB operation still requires a push-pull connection to achieve a full output cycle, but the dc bias level is usually closer to the zero base current level for better power efficiency, as described shortly. For class AB operation the output signal swing occurs between 180 and 360° and is neither class A nor class B operation.

Class C: The output of a class C amplifier is biased for operation at less than 180° of the cycle and will operate only with a tuned (resonant) circuit which provides a full cycle of operation for the tuned or resonant frequency. This operating class is therefore used in special areas of tuned circuits, such as radio or communications.

Class D: This operating class is a form of amplifier operation using pulse (digital) signals which are on for a short interval and off for a longer interval. Using digital techniques makes it possible to obtain a signal that varies over the full cycle (using sample-and-hold circuitry) to recreate the output from many pieces of input signal. The major advantage of class D operation is that the amplifier is on (using power) only for short intervals and the overall efficiency can practically be very high, as described next.

Amplifier Efficiency

The power efficiency of an amplifier, defined as the ratio of power output to power input, improves (gets higher) going from class A to class D. In general terms we see that a class A amplifier, with dc bias at one-half the supply voltage level, uses a good amount of power to maintain bias, even with no input signal applied. This results in very poor efficiency, especially with small input signals, when very little ac power is delivered to the load. In fact, the maximum efficiency of a class A circuit, occurring for the largest output voltage and current swing, is only 25% with a direct or series-fed load connection, and 50% with a transformer connection to the load. Class B operation, with no dc bias power for no input signal, can be shown to provide a maximum efficiency that reaches 78.5%. Class D operation can achieve power efficiency over 90% and provides the most efficient operation of all the operating classes. Since class AB falls between class A and class B in bias, it also falls between their efficiency ratings—between 25% (or 50%) and 78.5%. Table 16.1 summarizes the operation of the various amplifier classes. This table provides a relative comparison of the output cycle operation and power efficiency for the various class types. In class B operation a push-pull connection is obtained using either a transformer coupling or by using complementary (or quasi-complementary) operation with *npn* and *pnp* transistors to provide operation on opposite polarity cycles. While transformer

TABLE 16.1 Comparison of Amplifier Classes

			Class		
	A	*AB*	*B*	*C**	*D*
Operating cycle	360°	180° to 360°	180°	Less than 180°	Pulse operation
Power efficiency	25% to 50%	Between 25% (50%) and 78.5%	78.5%		Typically over 90%

*Class C is usually not used for delivering large amounts of power, thus the efficiency is not given here.

operation can provide opposite cycle signals, the transformer itself is quite large in many applications. A transformerless circuit using complementary transistors provides the same operation in a much smaller package. Circuits and examples are provided later in this chapter.

16.2 SERIES-FED CLASS A AMPLIFIER

The simple fixed-bias circuit connection shown in Fig. 16.2 can be used to discuss the main features of a class A series-fed amplifier. The only difference between this circuit and the small-signal version considered previously is that the signals handled by the large-signal circuit are in the range of volts, and the transistor used is a power transistor that is capable of operating in the range of a few to tens of watts. As will be shown in this section, this circuit is not the best to use as a large-signal amplifier because of its poor power efficiency. The beta of a power transistor is generally less than 100, the overall amplifier circuit using power transistors being capable of handling large power or current while not providing much voltage gain.

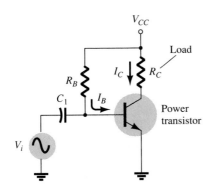

Figure 16.2 Series-fed class A large-signal amplifier.

DC Bias Operation

The dc bias set by V_{CC} and R_B fixes the dc base-bias current at

$$I_B = \frac{V_{CC} - 0.7\text{ V}}{R_B} \tag{16.1}$$

with the collector current then being

$$I_C = \beta I_B \tag{16.2}$$

with the collector-emitter voltage then

$$V_{CE} = V_{CC} - I_C R_C \tag{16.3}$$

To appreciate the importance of the dc bias on the operation of the power amplifier, consider the collector characteristic shown in Fig. 16.3. An ac load line is drawn using the values of V_{CC} and R_C. The intersection of the dc bias value of I_B with the dc load line then determines the operating point (Q-point) for the circuit. The quiescent-point values are those calculated using Eqs. (16.1) through (16.3). If the dc bias collector current is set at one-half the possible signal swing (between 0 and V_{CC}/R_C) the largest collector current swing will be possible. Additionally, if the quiescent collector–emitter voltage is set at one-half the supply voltage, the largest voltage swing will be possible. With the Q-point set at this optimum bias point, the power considerations for the circuit of Fig. 16.2 are determined as described below.

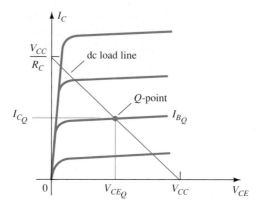

Figure 16.3 Transistor characteristic showing load line and Q-point.

AC Operation

When an input ac signal is applied to the amplifier of Fig. 16.2, the output will vary from its dc bias operating voltage and current. A small input signal, as shown in Fig. 16.4, will cause the base current to vary above and below the dc bias point, which will then cause the collector current (output) to vary from the dc bias point set as well

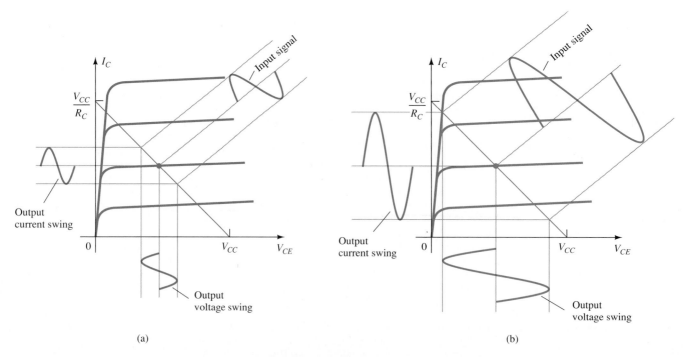

(a)

(b)

Figure 16.4 Amplifier input and output signal variation.

Chapter 16 Power Amplifiers

as the collector–emitter voltage to vary around its dc bias value. As the input signal is made larger, the output will vary further around the established dc bias point until either the current or the voltage reaches a limiting condition. For the current this limiting condition is either zero current at the low end or V_{CC}/R_C at the high end of its swing. For the collector–emitter voltage the limit is either 0 V or the supply voltage, V_{CC}.

Power Considerations

The power into an amplifier is provided by the supply. With no input signal the dc current drawn is the collector bias current, I_{CQ}. The power then drawn from the supply is

$$P_i(\text{dc}) = V_{CC}I_{C_Q}$$

(16.4)

Even with an ac signal applied, the average current drawn from the supply remains the same, so that Eq. (16.4) represents the input power supplied to the class A series-fed amplifier.

OUTPUT POWER

The output voltage and current varying around the bias point provide ac power to the load. This ac power is delivered to the load, R_C, in the circuit of Fig. 16.2. The ac signal, V_i, causes the base current to vary around the dc bias current and the collector current around its quiescent level, I_{CQ}. As shown in Fig. 16.4, the ac input signal results in ac current and ac voltage signals. The larger the input signal, the larger the output swing, up to the maximum set by the circuit. The ac power delivered to the load (R_C) can be expressed in a number of ways.

Using rms signals: The ac power delivered to the load (R_C) may be expressed using

$$P_o(\text{ac}) = V_{CE}(\text{rms})I_C(\text{rms})$$

(16.5a)

$$P_o(\text{ac}) = I_C^2(\text{rms})R_C$$

(16.5b)

$$P_o(\text{ac}) = \frac{V_C^2(\text{rms})}{R_C}$$

(16.5c)

Using peak signals: The ac power delivered to the load may be expressed using

$$P_o(\text{ac}) = \frac{V_{CE}(\text{p})I_C(\text{p})}{2}$$

(16.6a)

$$P_o(\text{ac}) = \frac{I_C^2(\text{p})}{2}R_C$$

(16.6b)

$$P_o(\text{ac}) = \frac{V_{CE}^2(\text{p})}{2R_C}$$

(16.6c)

Using peak-to-peak signals: The ac power delivered to the load may be expressed using

$$P_o(\text{ac}) = \frac{V_{CE}(\text{p-p})I_C(\text{p-p})}{8}$$

(16.7a)

$$P_o(\text{ac}) = \frac{I_C^2(\text{p-p})}{8}R_C \qquad (16.7b)$$

$$P_o(\text{ac}) = \frac{V_{CE}^2(\text{p-p})}{8R_C} \qquad (16.7c)$$

Efficiency

The efficiency of an amplifier represents the amount of ac power delivered (transferred) from the dc source. The efficiency of the amplifier is calculated using

$$\% \ \eta = \frac{P_o(\text{ac})}{P_i(\text{dc})} \times 100\% \qquad (16.8)$$

MAXIMUM EFFICIENCY

For the class A series-fed amplifier the maximum efficiency can be determined using the maximum voltage and current swings. For the voltage swing it is

$$\text{maximum } V_{CE}(\text{p-p}) = V_{CC}$$

For the current swing it is

$$\text{maximum } I_C(\text{p-p}) = \frac{V_{CC}}{R_C}$$

Using the maximum voltage swing in Eq. (16.7a) yields

$$\text{maximum } P_o(\text{ac}) = \frac{V_{CC}(V_{CC}/R_C)}{8}$$

$$= \frac{V_{CC}^2}{(8R_C)}$$

The maximum power input can be calculated using the dc bias current set to half the maximum value

$$\text{maximum } P_i(\text{dc}) = V_{CC}(\text{maximum } I_C) = V_{CC}\frac{V_{CC}/R_C}{2}$$

$$= \frac{V_{CC}^2}{2R_C}$$

We can then use Eq. (16.8) to calculate the maximum efficiency:

$$\text{maximum } \% \ \eta = \frac{\text{maximum } P_o(\text{ac})}{\text{maximum } P_i(\text{dc})} \times 100\%$$

$$= \frac{V_{CC}^2/8R_C}{V_{CC}^2/2R_C} \times 100\%$$

$$= 25\%$$

The maximum efficiency of a class A series-fed amplifier is thus seen to be 25%. Since this maximum efficiency will occur only for ideal conditions of both voltage swing and current swing, most series-fed circuits will provide efficiencies of much less than 25%.

Calculate the input power, output power, and efficiency of the amplifier circuit in Fig. 16.5 for an input voltage that results in a base current of 10 mA peak.

EXAMPLE 16.1

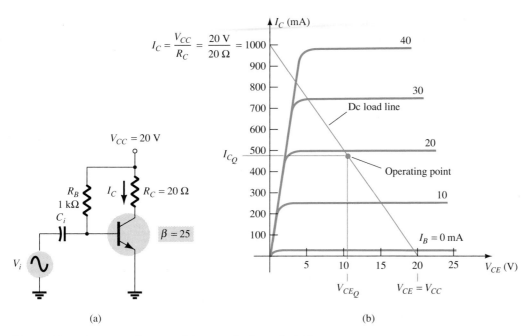

Figure 16.5 Operation of a series-fed circuit for Example 16.1.

Solution

Using Eqs. (16.1) through (16.3), the Q-point can be determined to be

$$I_B = \frac{V_{CC} - 0.7\ \text{V}}{R_B} = \frac{20\ \text{V} - 0.7\ \text{V}}{1\ \text{k}\Omega} = 19.3\ \text{mA}$$

$$I_C = \beta I_B = 25(19.3\ \text{mA}) = 482.5\ \text{mA} \cong 0.48\ \text{A}$$

$$V_{CE} = V_{CC} - I_C R_C = 20\ \text{V} - (0.48)(20) = 10.4\ \text{V}$$

This bias point is marked on the transistor collector characteristic of Fig. 16.5b. The ac variation of the output signal can be obtained graphically using the dc load line drawn on Fig. 16.5b by connecting $V_{CE} = V_{CC} = 20$ V with $I_C = V_{CC}/R_C = 1000$ mA = 1 A, as shown. When the input ac base current increases from its dc bias level, the collector current rises by

$$I_C(p) = \beta I_B(p) = 25(10\ \text{mA peak}) = 250\ \text{mA peak}$$

Using Eq. (16.6b) yields

$$P_o(\text{ac}) = \frac{I_C^2(p)}{2}R_C = \frac{(250 \times 10^{-3})^2}{2}(20) = \mathbf{0.625\ W}$$

Using Eq. (16.4) results in

$$P_i(\text{dc}) = V_{CC}I_{C_Q} = (20\ \text{V})(0.48\ \text{A}) = \mathbf{9.6\ W}$$

The amplifier's power efficiency can then be calculated using Eq. (16.8):

$$\%\ \eta = \frac{P_o(\text{ac})}{P_i(\text{dc})} \times 100\% = \frac{0.625\ \text{W}}{9.6\ \text{W}} \times 100\% = \mathbf{6.5\%}$$

16.3 TRANSFORMER-COUPLED CLASS A AMPLIFIER

A form of class A amplifier having maximum efficiency of 50% uses a transformer to couple the output signal to the load as shown in Fig. 16.6. This is a simple circuit form to use in presenting a few basic concepts. More practical circuit versions are covered later. Since the circuit uses a transformer to step voltage or current, a review of voltage and current step-up and step-down is presented next.

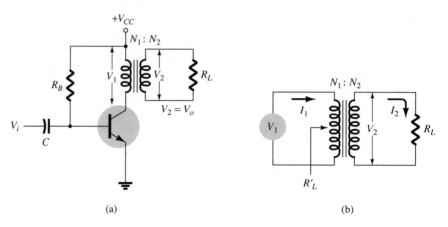

(a)   (b)

Figure 16.6 Transformer-coupled audio power amplifier.

Transformer Action

A transformer can increase or decrease voltage or current levels according to the turns ratio, as explained below. In addition, the impedance connected to one side of a transformer can be made to appear either larger or smaller (step up or step down) at the other side of the transformer, depending on the square of the transformer winding turns ratio. The following discussion assumes ideal (100%) power transfer from primary to secondary; that is, no power losses are considered.

VOLTAGE TRANSFORMATION

As shown in Fig. 16.7a, the transformer can step up or step down a voltage applied to one side directly as the ratio of the turns (or number of windings) on each side. The voltage transformation is given by

$$\frac{V_2}{V_1} = \frac{N_2}{N_1} \tag{16.9}$$

Equation (16.9) shows that if the number of turns of wire on the secondary side is larger than on the primary, the voltage at the secondary side is larger than the voltage at the primary side.

CURRENT TRANSFORMATION

The current in the secondary winding is inversely proportional to the number of turns in the windings. The current transformation is given by

$$\frac{I_2}{I_1} = \frac{N_1}{N_2} \tag{16.10}$$

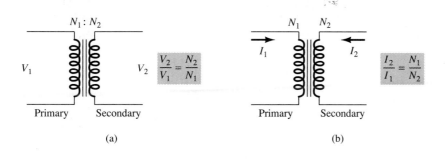

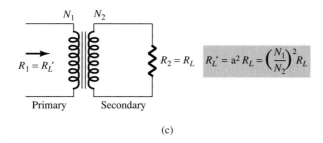

Figure 16.7 Transformer operation: (a) voltage transformation; (b) current transformation; (c) impedance transformation.

This relationship is shown in Fig. 16.7b. If the number of turns of wire on the secondary is greater than that on the primary, the secondary current will be less than the current in the primary.

IMPEDANCE TRANSFORMATION

Since the voltage and current can be changed by a transformer, an impedance "seen" from either side (primary or secondary) can also be changed. As shown in Fig. 16.7c, an impedance R_L is connected across the transformer secondary. This impedance is changed by the transformer when viewed at the primary side (R'_L). This can be shown as follows:

$$\frac{R_L}{R'_L} = \frac{R_2}{R_1} = \frac{V_2/I_2}{V_1/I_1} = \frac{V_2}{I_2}\frac{I_1}{V_1} = \frac{V_2}{V_1}\frac{I_1}{I_2} = \frac{N_2}{N_1}\frac{N_2}{N_1} = \left(\frac{N_2}{N_1}\right)^2$$

If we define $a = N_1/N_2$, where a is the turns ratio of the transformer, the above equation becomes

$$\frac{R'_L}{R_L} = \frac{R_1}{R_2} = \left(\frac{N_1}{N_2}\right)^2 = a^2 \qquad (16.11)$$

We can express the load resistance reflected to the primary side as:

$$R_1 = a^2 R_2 \qquad \text{or} \qquad R'_L = a^2 R_L \qquad (16.12)$$

where R'_L is the reflected impedance. As shown in Eq. (16.12), the reflected impedance is related directly to the square of the turns ratio. If the number of turns of the secondary is smaller than that of the primary, the impedance seen looking into the primary is larger than that of the secondary by the square of the turns ratio.

Calculate the effective resistance seen looking into the primary of a 15:1 transformer connected to an 8-Ω load.

EXAMPLE 16.2

Solution

Eq. (16.12): $R'_L = a^2 R_L = (15)^2(8 \ \Omega) = 1800 \ \Omega = \textbf{1.8 k}\boldsymbol{\Omega}$

EXAMPLE 16.3

What transformer turns ratio is required to match a 16-Ω speaker load so that the effective load resistance seen at the primary is 10 kΩ?

Solution

$$\text{Eq. (16.12):} \quad \left(\frac{N_1}{N_2}\right)^2 = \frac{R'_L}{R_L} = \frac{10 \text{ k}\Omega}{16 \text{ }\Omega} = 625$$

$$\frac{N_1}{N_2} = \sqrt{625} = \mathbf{25:1}$$

OPERATION OF AMPLIFIER STAGE

DC LOAD LINE

The transformer (dc) winding resistance determines the dc load line for the circuit of Fig. 16.6. Typically, this dc resistance is small (ideally 0 Ω) and, as shown in Fig. 16.8, a 0-Ω dc load line is a straight vertical line. A practical transformer winding resistance would be a few ohms, but only the ideal case will be considered in this discussion. There is no dc voltage drop across the 0-Ω dc load resistance and the load line is drawn straight vertically from the voltage point, $V_{CE_Q} = V_{CC}$.

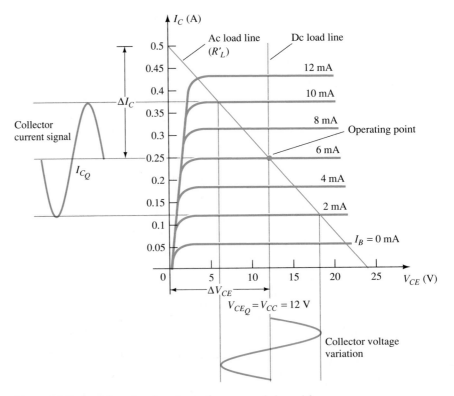

Figure 16.8 Load lines for class A transformer-coupled amplifier.

QUIESCENT OPERATING POINT

The operating point in the characteristic curve of Fig. 16.8 can be obtained graphically at the point of intersection of the dc load line and the base current set by the circuit. The collector quiescent current can then be obtained from the operating point. In class A operation keep in mind that the dc bias point sets the conditions for the maximum undistorted signal swing for both collector current and collector–emitter voltage. If the input signal produces a voltage swing less than the maximum possible, the efficiency of the circuit at that time will be less than 25%. The dc bias point is therefore important in setting the operation of a class A series-fed amplifier.

AC LOAD LINE

To carry out ac analysis it is necessary to calculate the ac load resistance "seen" looking into the primary side of the transformer, then draw the ac load line on the collector characteristic. The reflected load resistance (R'_L) is calculated using Eq. (16.12) using the value of the load connected across the secondary (R_L) and the turns ratio of the transformer. The graphical analysis technique then proceeds as follows. Draw the ac load line so that it passes through the operating point and has a slope equal to $-1/R'_L$ (the reflected load resistance), the load line slope being the negative reciprocal of the ac load resistance. Notice that the ac load line shows that the output signal swing can exceed the value of V_{CC}. In fact, the voltage developed across the transformer primary can be quite large. It is therefore necessary after obtaining the ac load line to check that the possible voltage swing does not exceed transistor maximum ratings.

SIGNAL SWING AND OUTPUT AC POWER

Figure 16.9 shows the voltage and current signal swings resulting in the circuit of Fig. 16.6. From the signal variations shown in Figs. 16.9, the values of the peak-to-peak signal swings are

$$V_{CE}(\text{p-p}) = V_{CE_{max}} - V_{CE_{min}}$$
$$I_C(\text{p-p}) = I_{C_{max}} - I_{C_{min}}$$

The ac power developed across the transformer primary can then be calculated using

$$P_o(\text{ac}) = \frac{(V_{CE_{max}} - V_{CE_{min}})(I_{C_{max}} - I_{C_{min}})}{8} \qquad (16.13)$$

The ac power calculated is that developed across the primary of the transformer. Assuming an ideal transformer (a highly efficient transformer has an efficiency of well over 90%), the power delivered by the secondary to the load is approximately that calculated using Eq. (16.13). The output ac power can also be determined using the voltage delivered to the load.

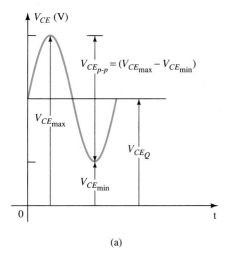

(a)

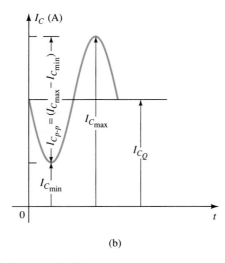

(b)

Figure 16.9 Graphical operation of transformer-coupled class A amplifier.

For the ideal transformer the voltage delivered to the load can be calculated using Eq. (16.9):

$$V_L = V_2 = \frac{N_2}{N_1}V_1$$

The power across the load can then be expressed as

$$P_L = \frac{V_L^2(\text{rms})}{R_L}$$

and equals the power calculated using Eq. (16.5c).

Using Eq. (16.10) to calculate the load current yields

$$I_L = I_2 = \frac{N_1}{N_2}I_C$$

with the output ac power then calculated using

$$P_L = I_L^2(\text{rms})R_L$$

EXAMPLE 16.4

Calculate the ac power delivered to the 8-Ω speaker for the circuit of Fig. 16.10. The circuit component values result in a dc base current of 6 mA, and the input signal (V_i) results in a peak base current swing of 4 mA.

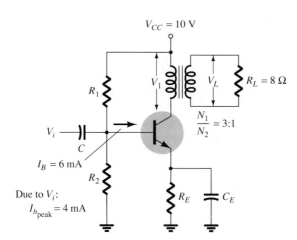

$V_{CC} = 10$ V

$R_L = 8\ \Omega$

$\frac{N_1}{N_2} = 3:1$

$I_B = 6$ mA

Due to V_i:
$I_{b_{\text{peak}}} = 4$ mA

Figure 16.10 Transformer-coupled class A amplifier for Example 16.4.

Solution

The dc load line is drawn vertically (see Fig. 16.11) from the voltage point:

$$V_{CE_Q} = V_{CC} = 10\ \text{V}$$

For $I_B = 6$ mA the operating point on Fig. 16.11 is

$$V_{CE_Q} = 10\ \text{V} \qquad \text{and} \qquad I_{C_Q} = 140\ \text{mA}$$

The effective ac resistance seen at the primary is

$$R_L' = \left(\frac{N_2}{N_1}\right)^2 = (3)^2(8) = 72\ \Omega$$

The ac load line can then be drawn of slope $-1/72$ going through the indicated operating point. To help draw the load line, consider the following procedure. For a current swing of

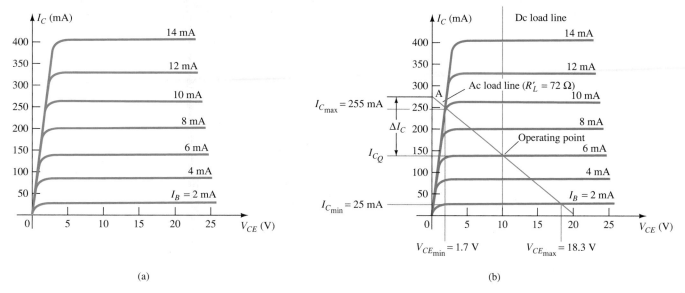

Figure 16.11 Transformer-coupled class A transistor characteristic for Examples 16.4 and 16.5: (a) device characteristic; (b) dc and ac load lines.

$$I_C = \frac{V_{CE}}{R'_L} = \frac{10 \text{ V}}{72 \text{ }\Omega} = 139 \text{ mA}$$

mark a point (A):

$$I_{CE_Q} + I_C = 140 \text{ mA} + 139 \text{ mA} = 279 \text{ mA along the } y\text{-axis}$$

Connect point A through the Q-point to obtain the ac load line. For the given base current swing of 4 mA peak, the maximum and minimum collector current and collector-emitter voltage obtained from Fig. 16.11 are

$$V_{CE_{min}} = 1.7 \text{ V} \qquad I_{C_{min}} = 25 \text{ mA}$$
$$V_{CE_{max}} = 18.3 \text{ V} \qquad I_{C_{max}} = 255 \text{ mA}$$

The ac power delivered to the load can then be calculated using Eq. (16.13):

$$P_o(\text{ac}) = \frac{(V_{CE_{max}} - V_{CE_{min}})(I_{C_{max}} - I_{C_{min}})}{8}$$

$$= \frac{(18.3 \text{ V} - 1.7 \text{ V})(255 \text{ mA} - 25 \text{ mA})}{8} = \mathbf{0.477 \text{ W}}$$

Efficiency

So far we have considered calculating the ac power delivered to the load. We next consider the input power from the battery, power losses in the amplifier, and the overall power efficiency of the transformer-coupled class A amplifier.

The input (dc) power obtained from the supply is calculated from the supply dc voltage and the average power drawn from the supply:

$$P_i(\text{dc}) = V_{CC}I_{C_Q} \qquad (16.14)$$

For the transformer-coupled amplifier the power dissipated by the transformer is small (due to the small dc resistance of a coil) and will be ignored in the present calculations. Thus the only power loss considered here is that dissipated by the power transistor and calculated using

$$P_Q = P_i(\text{dc}) - P_o(\text{ac}) \qquad (16.15)$$

where P_Q is the power dissipated as heat. While the equation is simple, it is nevertheless significant when operating a class A amplifier. The amount of power dissipated by the transistor is the difference between that drawn from the dc supply (set by the bias point) and the amount delivered to the ac load. When the input signal is very small, with very little ac power delivered to the load, the maximum power is dissipated by the transistor. When the input signal is larger and power delivered to the load is larger, less power is dissipated by the transistor. In other words, the transistor of a class A amplifier has to work hardest (dissipate the most power) when the load is disconnected from the amplifier, and the transistor dissipates least power when the load is drawing maximum power from the circuit.

EXAMPLE 16.5

For the circuit of Fig. 16.10 and results of Example 16.4, calculate the dc input power, power dissipated by the transistor, and efficiency of the circuit for the input signal of Example 16.4.

Solution

$$\text{Eq. (16.14):} \quad P_i(\text{dc}) = V_{CC}I_{C_Q} = (10 \text{ V})(140 \text{ mA}) = \textbf{1.4 W}$$

$$\text{Eq. (16.15):} \quad P_Q = P_i(\text{dc}) - P_o(\text{ac}) = 1.4 \text{ W} - 0.477 \text{ W} = \textbf{0.92 W}$$

The efficiency of the amplifier is then

$$\% \ \eta = \frac{P_o(\text{ac})}{P_i(\text{dc})} \times 100\% = \frac{0.477 \text{ W}}{1.4 \text{ W}} \times 100\% = \textbf{34.3\%}$$

MAXIMUM THEORETICAL EFFICIENCY

For a class A transformer-coupled amplifier the maximum theoretical efficiency goes up to 50%. Based on the signals obtained using the amplifier, the efficiency can be expressed as

$$\% \ \eta = 50 \left(\frac{V_{CE_{\max}} - V_{CE_{\min}}}{V_{CE_{\max}} + V_{CE_{\min}}} \right)^2 \quad \% \qquad (16.16)$$

The larger the value of $V_{CE_{\max}}$ and the smaller the value of $V_{CE_{\min}}$, the closer the efficiency approaches the theoretical limit of 50%.

EXAMPLE 16.6

Calculate the efficiency of a transformer-coupled class A amplifier for a supply of 12 V and outputs of:
(a) $V(\text{p}) = 12$ V.
(b) $V(\text{p}) = 6$ V.
(c) $V(\text{p}) = 2$ V.

Solution

Since $V_{CE_Q} = V_{CC} = 12$ V, the maximum and minimum of the voltage swing are
(a) $V_{CE_{\max}} = V_{CE_Q} + V(\text{p}) = 12 \text{ V} + 12 \text{ V} = 24 \text{ V}$

$\quad V_{CE_{\min}} = V_{CE_Q} - V(\text{p}) = 12 \text{ V} - 12 \text{ V} = 0 \text{ V}$
resulting in

$$\% \; \eta = 50\left(\frac{24 \text{ V} - 0 \text{ V}}{24 \text{ V} + 0 \text{ V}}\right)^2 \qquad \% = \mathbf{50\%}$$

(b) $V_{CE_{max}} = V_{CE_Q} + V(\text{p}) = 12 \text{ V} + 6 \text{ V} = 18 \text{ V}$

$\quad V_{CE_{min}} = V_{CE_Q} - V(\text{p}) = 12 \text{ V} - 6 \text{ V} = 6 \text{ V}$

resulting in

$$\% \; \eta = 50\left(\frac{18 \text{ V} - 6 \text{ V}}{18 \text{ V} + 6 \text{ V}}\right)^2 \qquad \% = \mathbf{12.5\%}$$

(c) $V_{CE_{max}} = V_{CE_Q} + V(\text{p}) = 12 \text{ V} + 2 \text{ V} = 14 \text{ V}$

$\quad V_{CE_{min}} = V_{CE_Q} - V(\text{p}) = 12 \text{ V} - 2 \text{ V} = 10 \text{ V}$

resulting in

$$\% \; \eta = 50\left(\frac{14 \text{ V} - 10 \text{ V}}{14 \text{ V} + 10 \text{ V}}\right)^2 \qquad \% = \mathbf{1.39\%}$$

Notice how dramatically the amplifier efficiency drops from a maximum of 50% for $V(\text{p}) = V_{CC}$ to slightly over 1% for $V(\text{p}) = 2$ V.

16.4 CLASS B AMPLIFIER OPERATION

Class B operation is provided when the dc bias leaves the transistor biased just off, the transistor turning on when the ac signal is applied. This is essentially no bias and the transistor conducts current for only one-half of the signal cycle. To obtain output for the full cycle of signal, it is necessary to use two transistors and have each conduct on opposite half-cycles, the combined operation providing a full cycle of output signal. Since one part of the circuit pushes the signal high during one half-cycle and the other part pulls the signal low during the other half-cycle, the circuit is referred to as a *push-pull circuit*. Figure 16.12 shows a diagram for push-pull operation. An ac input signal is applied to the push-pull circuit, with each half operating on alternate half-cycles, the load then receiving a signal for the full ac cycle. The power transistors used in the push-pull circuit are capable of delivering the desired power to the load, and the class B operation of these transistors provides greater efficiency than was possible using a single transistor in class A operation.

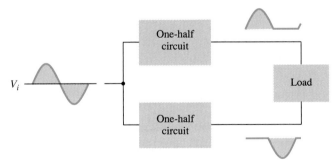

Figure 16.12 Block representation of push-pull operation.

Input (DC) Power

The power supplied to the load by an amplifier is drawn from the power supply (or power supplies; see Fig. 16.13) which provides the input or dc power. The amount of this input power can be calculated using

$$P_i(\text{dc}) = V_{CC}I_{\text{dc}} \qquad (16.17)$$

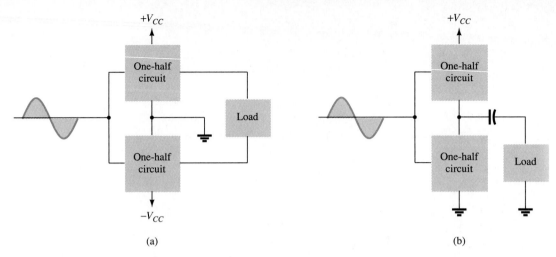

Figure 16.13 Connection of push-pull amplifier to load: (a) using two voltage supplies; (b) using one voltage supply.

where I_{dc} is the average or dc current drawn from the power supplies. In class B operation the current drawn from a single power supply has the form of a full-wave rectified signal, while that drawn from two power supplies has the form of a half-wave rectified signal from each supply. In either case the value of the average current drawn can be expressed as

$$I_{dc} = \frac{2}{\pi}I(p)$$

(16.18)

where $I(p)$ is the peak value of the output current waveform. Using Eq. (16.18) in the power input equation (16.17) results in

$$P_i(dc) = V_{CC}\left(\frac{2}{\pi}I(p)\right)$$

(16.19)

Output (AC) Power

The power delivered to the load (usually referred to as a resistance, R_L) can be calculated using any one of a number of equations. If one is using an rms meter to measure the voltage across the load, the output power can be calculated as

$$P_o(ac) = \frac{V_L^2(rms)}{R_L}$$

(16.20)

If one is using an oscilloscope, the peak, or peak-to-peak, output voltage measured can be used:

$$P_o(ac) = \frac{V_L^2(p\text{-}p)}{8R_L} = \frac{V_L^2(p)}{2R_L}$$

(16.21)

The larger the rms or peak output voltage, the larger the power delivered to the load.

Efficiency

The efficiency of the class B amplifier can be calculated using the basic equation:

$$\% \ \eta = \frac{P_o(ac)}{P_i(dc)} \times 100\%$$

Using Eqs. (16.19) and (16.21) in the efficiency equation above results in

$$\% \ \eta = \frac{P_o(\text{ac})}{P_i(\text{dc})} \times 100\% = \frac{V_L^2(\text{p})/2R_L}{V_{CC}[(2/\pi)I(\text{p})]} \times 100\% = \frac{\pi}{4} \frac{V_L(\text{p})}{V_{CC}} \times 100\% \qquad (16.22)$$

(using $I(\text{p}) = V_L(\text{p})/R_L$). Equation (16.22) shows that the larger the peak voltage, the higher the circuit efficiency, up to a maximum value when $V_L(\text{p}) = V_{CC}$, this maximum efficiency then being

$$\text{maximum efficiency} = \frac{\pi}{4} \times 100\% = 78.5\%$$

Power Dissipated by Output Transistors

The power dissipated (as heat) by the output power transistors is the difference between the input power delivered by the supplies and the output power delivered to the load.

$$P_{2Q} = P_i(\text{dc}) - P_o(\text{ac}) \qquad (16.23)$$

where P_{2Q} is the power dissipated by the two output power transistors. The dissipated power handled by each transistor is then

$$P_Q = \frac{P_{2Q}}{2} \qquad (16.24)$$

For a class B amplifier providing a 20-V peak signal to a 16-Ω load (speaker) and a power supply of $V_{CC} = 30$ V, determine the input power, output power, and circuit efficiency.

EXAMPLE 16.7

Solution

A 20-V peak signal across a 16-Ω load provides a peak load current of

$$I_L(\text{p}) = \frac{V_L(\text{p})}{R_L} = \frac{20 \text{ V}}{16 \ \Omega} = 1.25 \text{ A}$$

The dc value of the current drawn from the power supply is then

$$I_{\text{dc}} = \frac{2}{\pi} I_L(\text{p}) = \frac{2}{\pi}(1.25 \text{ A}) = 0.796 \text{ A}$$

and the input power delivered by the supply voltage is

$$P_i(\text{dc}) = V_{CC}I_{\text{dc}} = (30 \text{ V})(0.796 \text{ A}) = \mathbf{23.9 \ W}$$

The output power delivered to the load is

$$P_o(\text{ac}) = \frac{V_L^2(\text{p})}{2R_L} = \frac{(20 \text{ V})^2}{2(16 \ \Omega)} = \mathbf{12.5 \ W}$$

for a resulting efficiency of

$$\% \ \eta = \frac{P_o(\text{ac})}{P_i(\text{dc})} \times 100\% = \frac{12.5 \text{ W}}{23.9 \text{ W}} \times 100\% = \mathbf{52.3\%}$$

Maximum Power Considerations

For class B operation the maximum output power is delivered to the load when $V_L(p) = V_{CC}$:

$$\boxed{\text{maximum } P_o(\text{ac}) = \frac{V_{CC}^2}{2R_L}} \tag{16.25}$$

The corresponding peak ac current $I(p)$ is then

$$I(p) = \frac{V_{CC}}{R_L}$$

so that the maximum value of average current from the power supply is

$$\text{maximum } I_{dc} = \frac{2}{\pi}I(p) = \frac{2\,V_{CC}}{\pi\,R_L}$$

Using this current to calculate the maximum value of input power results in

$$\boxed{\text{maximum } P_i(\text{dc}) = V_{CC}(\text{maximum } I_{dc}) = V_{CC}\left(\frac{2\,V_{CC}}{\pi\,R_L}\right) = \frac{2V_{CC}^2}{\pi R_L}} \tag{16.26}$$

The maximum circuit efficiency for class B operation is then

$$\text{maximum } \%\ \eta = \frac{P_o(\text{ac})}{P_i(\text{dc})} \times 100\% = \frac{V_{CC}^2/2R_L}{V_{CC}[(2/\pi)(V_{CC}/R_L)]} \times 100\%$$

$$= \frac{\pi}{4} \times 100\% = \mathbf{78.54\%} \tag{16.27}$$

When the input signal results in less than the maximum output signal swing, the circuit efficiency is less than 78.5%.

For class B operation the maximum power dissipated by the output transistors does not occur at the maximum power input or output condition. The maximum power dissipated by the two output transistors occurs when the output voltage across the load is

$$V_L(p) = 0.636V_{CC} \qquad \left(= \frac{2}{\pi}V_{CC}\right)$$

for a maximum transistor power dissipation of

$$\boxed{\text{maximum } P_{2Q} = \frac{2}{\pi^2}\frac{V_{CC}^2}{R_L}} \tag{16.28}$$

EXAMPLE 16.8

For a class B amplifier using a supply of $V_{CC} = 30$ V and driving a load of 16 Ω, determine the maximum input power, output power, and transistor dissipation.

Solution

The maximum output power is

$$\text{maximum } P_o(\text{ac}) = \frac{V_{CC}^2}{2R_L} = \frac{(30\text{ V})^2}{2(16\text{ }\Omega)} = \mathbf{28.125\ W}$$

The maximum input power drawn from the voltage supply is

$$\text{maximum } P_i(\text{dc}) = \frac{2V_{CC}^2}{\pi R_L} = \frac{2(30 \text{ V})^2}{\pi(16 \ \Omega)} = \textbf{35.81 W}$$

The circuit efficiency is then

$$\text{maximum } \% \ \eta = \frac{P_o(\text{ac})}{P_i(\text{dc})} \times 100\% = \frac{28.125 \text{ W}}{35.81 \text{ W}} \times 100\% = 78.54\%$$

as expected. The maximum power dissipated by each transistor is

$$\text{maximum } P_Q = \frac{\text{maximum } P_{2Q}}{2} = 0.5\left(\frac{2}{\pi^2} \frac{V_{CC}^2}{R_L}\right) = 0.5\left[\frac{2}{\pi^2} \frac{(30 \text{ V})^2}{16 \ \Omega}\right] = \textbf{5.7 W}$$

Under maximum conditions a pair of transistors, each handling 5.7 W at most, can deliver 28.125 W to a 16-Ω load, while drawing 35.81 W from the supply.

The maximum efficiency of a class B amplifier can also be expressed as follows:

$$P_o(\text{ac}) = \frac{V_L^2(\text{p})}{2R_L}$$

$$P_i(\text{dc}) = V_{CC}I_{\text{dc}} = V_{CC}\left[\frac{2}{\pi} \frac{V_L(\text{p})}{R_L}\right]$$

so that
$$\% \ \eta = \frac{P_o(\text{ac})}{P_i(\text{dc})} \times 100\% = \frac{V_L^2(\text{p})/2R_L}{V_{CC}[(2/\pi)(V_L(\text{p})/R_L)]} \times 100\%$$

$$= 78.54\frac{V_L(\text{p})}{V_{CC}}\% \tag{16.29}$$

Calculate the efficiency of a class B amplifier for a supply voltage of $V_{CC} = 24$ V with peak output voltages of:
(a) $V_L(\text{p}) = 22$ V.
(b) $V_L(\text{p}) = 6$ V.

EXAMPLE 16.9

Solution

Using Eq. (16.29) gives
(a) $\% \ \eta = 78.54 \dfrac{V_L(\text{p})}{V_{CC}}\% = 78.54\left(\dfrac{22 \text{ V}}{24 \text{ V}}\right) = \textbf{72\%}$

(b) $\% \ \eta = 78.54\left(\dfrac{6 \text{ V}}{24 \text{ V}}\right)\% = \textbf{19.6\%}$

Notice that a voltage near the maximum [22 V in part (a)] results in an efficiency near the maximum, while a small voltage swing [6 V in part (b)] still provides an efficiency near 20%. Similar power supply and signal swings would have resulted in much poorer efficiency in a class A amplifier.

16.5 CLASS B AMPLIFIER CIRCUITS

A number of circuit arrangements for obtaining class B operation are possible. We will consider the advantages and disadvantages of a number of the more popular circuits in this section. The input signals to the amplifier could be a single signal, the circuit then providing two different output stages, each operating for one-half the cycle. If the input is in the form of two opposite polarity signals, two similar stages could be used, each operating on the alternate cycle because of the input signal. One means of obtaining polarity or phase inversion is using a transformer, the transformer-coupled amplifier having been very popular for a long time. Opposite polarity inputs can easily be obtained using an op-amp having two opposite outputs, or using a few op-amp stages to obtain two opposite polarity signals. An opposite polarity operation can also be achieved using a single input and complementary transistors (*npn* and *pnp*, or ηMOS and *p*MOS).

Figure 16.14 shows different ways to obtain phase-inverted signals from a single input signal. Figure 16.14a shows a center-tapped transformer to provide opposite

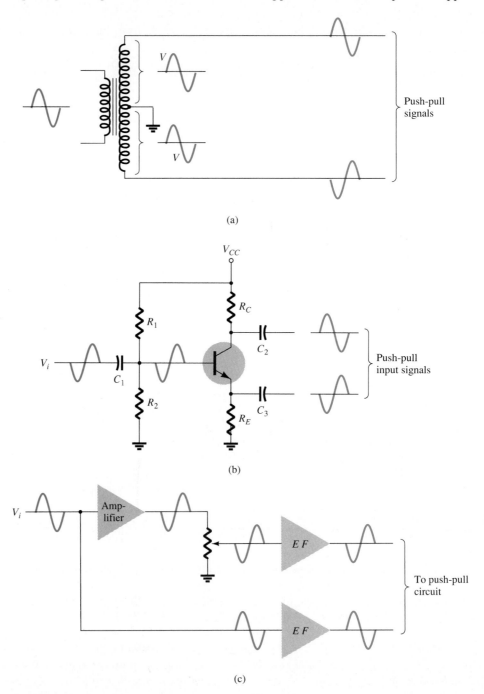

(a)

(b)

(c)

Figure 16.14 Phase-splitter circuits.

Chapter 16 **Power Amplifiers**

phase signals. If the transformer is exactly center-tapped, the two signals are exactly opposite in phase and of the same magnitude. The circuit of Fig. 16.14b uses a BJT stage with in-phase output from the emitter and opposite phase output from the collector. If the gain is made nearly 1 for each output, the same magnitude results. Probably most common would be using op-amp stages, one to provide an inverting gain of unity, the other a noninverting gain of unity, to provide two outputs of the same magnitude but of opposite phase.

Transformer-Coupled Push–Pull Circuits

The circuit of Fig. 16.15 uses a center-tapped input transformer to produce opposite polarity signals to the two transistor inputs and an output transformer to drive the load in a push-pull mode of operation described next.

During the first half-cycle of operation, transistor Q_1 is driven into conduction, whereas transistor Q_2 is driven off. The current i_1 through the transformer results in the first half-cycle of signal to the load. During the second half-cycle of the input signal, Q_2 conducts, whereas Q_1 stays off, the current i_2 through the transformer resulting in the second half-cycle to the load. The overall signal developed across the load then varies over the full cycle of signal operation.

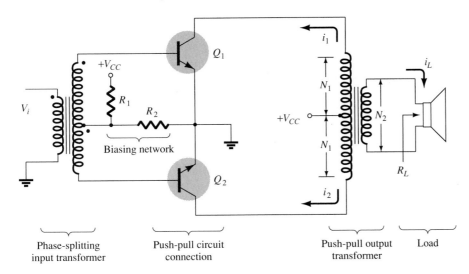

Figure 16.15 Push-pull circuit.

Complementary-Symmetry Circuits

Using complementary transistors (*npn* and *pnp*) it is possible to obtain a full cycle output across a load using half-cycles of operation from each transistor, as shown in Fig. 16.16a. While a single input signal is applied to the base of both transistors, the transistors, being of opposite type, will conduct on opposite half-cycles of the input. The *npn* transistor will be biased into conduction by the positive half-cycle of signal, with a resulting half-cycle of signal across the load as shown in Fig. 16.16b. During the negative half-cycle of signal the *pnp* transistor is biased into conduction when the input goes negative, as shown in Fig. 16.16c.

During a complete cycle of the input a complete cycle of output signal is developed across the load. One disadvantage of the circuit is the need for two separate voltage supplies. Another, less obvious disadvantage with the complementary circuit is shown in the resulting crossover distortion in the output signal (see Fig. 16.16d). *Crossover distortion* refers to the fact that during the signal crossover from positive to negative (or vice versa) there is some nonlinearity in the output signal. This results from the fact that the circuit does not provide exact switching of one transistor off and the other on at the zero-voltage condition. Both transistors may be partially off so that

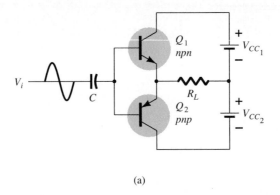

(a)

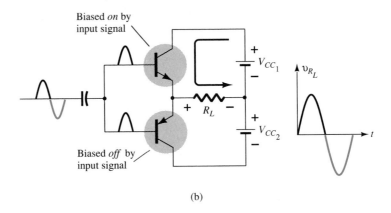

(b)

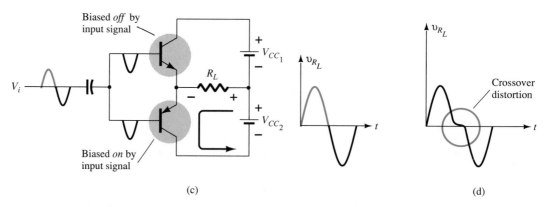

(c) (d)

Figure 16.16 Complementary-symmetry push-pull circuit.

the output voltage does not follow the input around the zero voltage condition. Biasing the transistors in class AB improves this operation by biasing both transistors to be on for more than half a cycle.

A more practical version of a push-pull circuit using complementary transistors is shown in Fig. 16.17. Note that the load is driven as the output of an emitter-follower so that the load resistance of the load is matched by the low output resistance of the driving source. The circuit uses complementary Darlington-connected transistors to provide higher output current and lower output resistance.

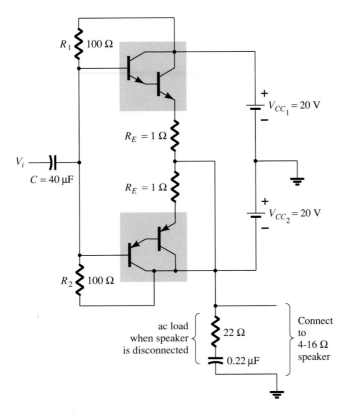

Figure 16.17 Complementary-symmetry push-pull circuit using Darlington transistors.

Quasi-complementary Push–Pull Amplifier

In practical power amplifier circuits it is preferable to use *npn* transistors for both high-current-output devices. Since the push-pull connection requires complementary devices, a *pnp* high-power transistor must be used. A practical means of obtaining complementary operation while using the same, matched *npn* transistors for the output is provided by a quasi-complementary circuit, as shown in Fig. 16.18. The push-

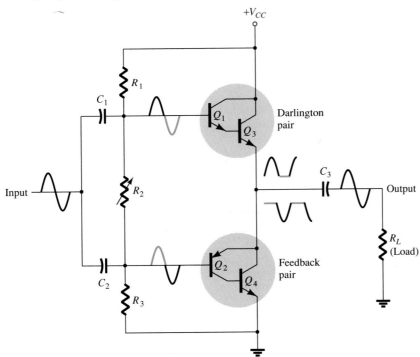

Figure 16.18 Quasi-comple-mentary push-pull transformerless power amplifier.

pull operation is achieved by using complementary transistors (Q_1 and Q_2) before the matched *npn* output transistors (Q_3 and Q_4). Notice that transistors Q_1 and Q_3 form a Darlington connection that provides output from a low-impedance emitter-follower. The connection of transistors Q_2 and Q_4 forms a feedback pair, which similarly provides a low-impedance drive to the load. Resistor R_2 can be adjusted to minimize crossover distortion by adjusting the dc bias condition. The single input signal applied to the push-pull stage then results in a full cycle output to the load. The quasi-complementary push-pull amplifier is presently the most popular form of power amplifier.

EXAMPLE 16.10

For the circuit of Fig. 16.19 calculate the input power, output power, and power handled by each output transistor and the circuit efficiency for an input of 12 V rms.

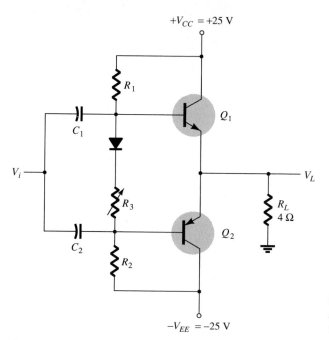

Figure 16.19 Class B power amplifier for Examples 16.10 thru 16.12.

Solution

The peak input voltage is

$$V_i(\text{p}) = \sqrt{2}\, V_i\,(\text{rms}) = \sqrt{2}\,(12\text{ V}) = 16.69\text{ V} \approx 17\text{ V}$$

Since the resulting voltage across the load is ideally the same as the input signal (the amplifier has, ideally, a voltage gain of unity),

$$V_L(\text{p}) = 17\text{ V}$$

and the output power developed across the load is

$$P_o(\text{ac}) = \frac{V_L^2(\text{p})}{2R_L} = \frac{(17\text{ V})^2}{2(4\ \Omega)} = \mathbf{36.125\text{ W}}$$

The peak load current is

$$I_L(\text{p}) = \frac{V_L(\text{p})}{R_L} = \frac{17\text{ V}}{4\ \Omega} = 4.25\text{ A}$$

from which the dc current from the supplies is calculated to be

$$I_{\text{dc}} = \frac{2}{\pi} I_L(\text{p}) = \frac{2(4.25\text{ A})}{\pi} = 2.71\text{ A}$$

so that the power supplied to the circuit is

$$P_i(\text{dc}) = V_{CC}I_{\text{dc}} = (25 \text{ V})(2.71 \text{ A}) = \textbf{67.75 W}$$

The power dissipated by each output transistor is

$$P_Q = \frac{P_{2Q}}{2} = \frac{P_i - P_o}{2} = \frac{67.75 \text{ W} - 36.125 \text{ W}}{2} = \textbf{15.8 W}$$

The circuit efficiency (for the input of 12 V, rms) is then

$$\% \; \eta = \frac{P_o}{P_i} \times 100\% = \frac{36.125 \text{ W}}{67.75 \text{ W}} \times 100\% = \textbf{53.3\%}$$

For the circuit of Fig. 16.19, calculate the maximum input power, maximum output power, input voltage for maximum power operation, and the power dissipated by the output transistors at this voltage.

EXAMPLE 16.11

Solution

The maximum input power is

$$\text{maximum } P_i(\text{dc}) = \frac{2 \; V_{CC}^2}{\pi \; R_L} = \frac{2 \; (25 \text{ V})^2}{\pi \quad 4 \; \Omega} = \textbf{99.47 W}$$

The maximum output power is

$$\text{maximum } P_o(\text{ac}) = \frac{V_{CC}^2}{2R_L} = \frac{(25 \text{ V})^2}{2(4 \; \Omega)} = \textbf{78.125 W}$$

[Note that the maximum efficiency is achieved:]

$$\% \; \eta = \frac{P_o}{P_i} \times 100\% = \frac{78.125 \text{ W}}{99.47 \text{ W}} \; 100\% = 78.54\%$$

To achieve maximum power operation the output voltage must be

$$V_L(\text{p}) = V_{CC} = 25 \text{ V}$$

and the power dissipated by the output transistors is then

$$P_{2Q} = P_i - P_o = 99.47 \text{ W} - 78.125 \text{ W} = \textbf{21.3 W}$$

For the circuit of Fig. 16.19, determine the maximum power dissipated by the output transistors and the input voltage at which this occurs.

EXAMPLE 16.12

Solution

The maximum power dissipated by both output transistors is

$$\text{maximum } P_{2Q} = \frac{2}{\pi^2} \frac{V_{CC}^2}{R_L} = \frac{2}{\pi^2} \frac{(25 \text{ V})^2}{4 \; \Omega} = \textbf{31.66 W}$$

This maximum dissipation occurs at

$$V_L = 0.636V_L(\text{p}) = 0.636(25 \text{ V}) = \textbf{15.9 V}$$

(Notice that at $V_L = 15.9$ V the circuit required the output transistors to dissipate 31.66 W, while at $V_L = 25$ V they only had to dissipate 21.3 W.)

16.6 AMPLIFIER DISTORTION

A pure sinusoidal signal has a single frequency at which the voltage varies positive and negative by equal amounts. Any signal varying over less than the full 360° cycle is considered to have distortion. An ideal amplifier is capable of amplifying a pure sinusoidal signal to provide a larger version, the resulting waveform being a pure single-frequency sinusoidal signal. When distortion occurs the output will not be an exact duplicate (except for magnitude) of the input signal.

Distortion can occur because the device characteristic is not linear, in which case nonlinear or amplitude distortion occurs. This can occur with all classes of amplifier operation. Distortion can also occur because the circuit elements and devices respond to the input signal differently at various frequencies, this being frequency distortion.

One technique for describing distorted but period waveforms uses Fourier analysis, a method that describes any periodic waveform in terms of its fundamental frequency component and frequency components at integer multiples—these components are called *harmonic components* or *harmonics*. For example, a signal that is originally 1000 Hz could result, after distortion, in a frequency component at 1000 Hz (1 kHz) and harmonic components at 2 kHz (2 × 1 kHz), 3 kHz (3 × 1 kHz), 4 kHz (4 × 1 kHz), and so on. The original frequency of 1 kHz is called the *fundamental frequency;* those at integer multiples are the harmonics. The 2-kHz component is therefore called a *second harmonic*, that at 3 kHz is the *third harmonic*, and so on. The fundamental frequency is not considered a harmonic. Fourier analysis does not allow for fractional harmonic frequencies—only integer multiples of the fundamental.

Harmonic Distortion

A signal is considered to have harmonic distortion when there are harmonic frequency components (not just the fundamental component). If the fundamental frequency has an amplitude, A_1, and the ηth frequency component has an amplitude, A_η, a harmonic distortion can be defined as

$$\% \ \eta\text{th harmonic distortion} = \% \ D_\eta = \frac{|A_\eta|}{|A_1|} \times 100\% \qquad (16.30)$$

The fundamental component is typically larger than any harmonic component.

EXAMPLE 16.13

Calculate the harmonic distortion components for an output signal having fundamental amplitude of 2.5 V, second harmonic amplitude of 0.25 V, third harmonic amplitude of 0.1 V, and fourth harmonic amplitude of 0.05 V.

Solution

Using Eq. (16.30) yields

$$\% \ D_2 = \frac{|A_2|}{|A_1|} \times 100\% = \frac{0.25 \text{ V}}{2.5 \text{ V}} \times 100\% = \mathbf{10\%}$$

$$\% \ D_3 = \frac{|A_3|}{|A_1|} \times 100\% = \frac{0.1 \text{ V}}{2.5 \text{ V}} \times 100\% = \mathbf{4\%}$$

$$\% \ D_4 = \frac{|A_4|}{|A_1|} \times 100\% = \frac{0.05 \text{ V}}{2.5 \text{ V}} \times 100\% = \mathbf{2\%}$$

TOTAL HARMONIC DISTORTION

When an output signal has a number of individual harmonic distortion components, the signal can be seen to have a total harmonic distortion based on the individual elements as combined by the relationship of the following equation:

$$\% \text{ THD} = \sqrt{D_2^2 + D_3^2 + D_4^2 + \cdots} \times 100\% \qquad (16.31)$$

where THD is total harmonic distortion.

Calculate the total harmonic distortion for the amplitude components given in Example 16.13.

EXAMPLE 16.14

Solution

Using the computed values of $D_2 = 0.10$, $D_3 = 0.04$, and $D_4 = 0.02$ in Eq. (16.31),

$$\begin{aligned}
\% \text{ THD} &= \sqrt{D_2^2 + D_3^2 + D_4^2} \times 100\% \\
&= \sqrt{(0.10)^2 + (0.04)^2 + (0.02)^2} \times 100\% = 0.1095 \times 100\% \\
&= \mathbf{10.95\%}
\end{aligned}$$

An instrument such as a spectrum analyzer would allow measurement of the harmonics present in the signal by providing a display of the fundamental component of a signal and a number of its harmonics on a display screen. Similarly, a wave analyzer instrument allows more precise measurement of the harmonic components of a distorted signal by filtering out each of these components and providing a reading of these components. In any case the technique of considering any distorted signal as containing a fundamental component and harmonic components is practical and useful. For a signal occurring in class AB or class B, the distortion may be mainly even harmonics, of which the second harmonic component is the largest. Thus although the distorted signal theoretically contains all harmonic components from the second harmonic up, the most important in terms of the amount of distortion in the classes presented above is the second harmonic.

SECOND HARMONIC DISTORTION

Figure 16.20 shows a waveform to use for obtaining second-harmonic distortion. A collector current waveform is shown with the quiescent, minimum, and maximum signal levels, and the time at which they occur marked on the waveform. The signal

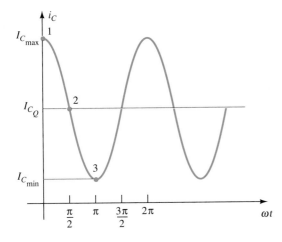

Figure 16.20 Waveform for obtaining second-harmonic distortion.

shown indicates that some distortion is present. An equation that approximately describes the distorted signal waveform is

$$i_C \approx I_{cq} + I_0 + I_1 \cos \omega t + I_2 \cos \omega t \qquad (16.32)$$

The current waveform contains the original quiescent current I_{CQ}, which occurs with zero input signal; an additional dc current I_0, due to the nonzero average of the distorted signal; the fundamental component of the distorted ac signal, I_1; and a second harmonic component I_2, at twice the fundamental frequency. Although other harmonics are also present, only the second is considered here. Equating the resulting current from Eq. (16.32) at a few points in the cycle to that shown on the current waveform provides the following three relations:

At point 1 ($\omega t = 0$):

$$i_C = I_{C_{max}} = I_{C_Q} + I_0 + I_1 \cos 0 + I_2 \cos 0$$

$$I_{C_{max}} = I_{C_Q} + I_0 + I_1 + I_2$$

At point 2 ($\omega t = \pi/2$):

$$i_C = I_{C_Q} = I_{C_Q} + I_0 + I_1 \cos \frac{\pi}{2} + I_2 \cos \frac{2\pi}{2}$$

$$I_{C_Q} = I_{C_Q} + I_0 - I_2$$

At point 3 ($\omega t = \pi$):

$$i_C = I_{C_{min}} = I_{C_Q} + I_0 + I_1 \cos \pi + I_2 \cos 2\pi$$

$$I_{C_{min}} = I_{C_Q} + I_0 - I_1 + I_2$$

Solving the preceding three equations simultaneously gives the following results:

$$I_0 = I_2 = \frac{I_{C_{max}} + I_{C_{min}} - 2I_{C_Q}}{4}, \qquad I_1 = \frac{I_{C_{max}} - I_{C_{min}}}{2}$$

Referring to Eq. (16.30), the definition of second harmonic distortion may be expressed as

$$D_2 = \left| \frac{I_2}{I_1} \right| \times 100\%$$

Inserting the values of I_1 and I_2 determined above gives

$$D_2 = \left| \frac{\frac{1}{2}(I_{C_{max}} + I_{C_{min}}) - I_{C_Q}}{I_{C_{max}} - I_{C_{min}}} \right| \times 100\% \qquad (16.33)$$

In a similar manner, the second harmonic distortion can be expressed in terms of measured collector-emitter voltages:

$$D_2 = \left| \frac{\frac{1}{2}(V_{CE_{max}} + V_{CE_{min}}) - V_{CE_Q}}{V_{CE_{max}} - V_{CE_{min}}} \right| \times 100\% \qquad (16.34)$$

EXAMPLE 16.15

An output waveform displayed on an oscilloscope provides the following measurements:

(a) $V_{CE_{min}} = 1$ V, $V_{CE_{max}} = 22$ V, $V_{CE_Q} = 12$ V.

(b) $V_{CE_{min}} = 4$ V, $V_{CE_{max}} = 20$ V, $V_{CE_Q} = 12$ V.

Solution

Using Eq. (16.34), we get

(a) $D_2 = \left| \dfrac{\frac{1}{2}(22 \text{ V} + 1 \text{ V}) - 12 \text{ V}}{22 \text{ V} - 1 \text{ V}} \right| \times 100\% = \mathbf{2.38\%}$

(b) $D_2 = \left| \dfrac{\frac{1}{2}(20 \text{ V} + 4 \text{ V}) - 12 \text{ V}}{20 \text{ V} - 4 \text{ V}} \right| \times 100\% = \mathbf{0\%}$ (no distortion)

Power of Signal Having Distortion

When distortion does occur, the output power calculated for the undistorted signal is no longer correct. When distortion is present, the output power delivered to the load resistor R_C due to the fundamental component of the distorted signal is

$$P_1 = \frac{I_1^2 R_C}{2} \tag{16.35}$$

The total power due to all the harmonic components of the distorted signal can then be calculated using

$$P = (I_1^2 + I_2^2 + I_3^2 + \cdots) \frac{R_C}{2} \tag{16.36}$$

The total power can also be expressed in terms of the total harmonic distortion,

$$P = (1 + D_2^2 + D_3^2 + \cdots) I_1^2 \frac{R_C}{2} = (1 + \text{THD}^2)P_1 \tag{16.37}$$

EXAMPLE 16.16

For harmonic distortion reading of $D_2 = 0.1$, $D_3 = 0.02$, and $D_4 = 0.01$, with $I_1 = 4$ A and $R_C = 8 \ \Omega$, calculate the total harmonic distortion, fundamental power component, and total power.

Solution

The total harmonic distortion is

$$\text{THD} = \sqrt{D_2^2 + D_3^2 + D_4^2} = \sqrt{(0.1)^2 + (0.02)^2 + (0.01)^2} \approx \mathbf{0.1}$$

The fundamental power, using Eq. (16.35), is

$$P_1 = \frac{I_1^2 R_C}{2} = \frac{(4 \text{ A})^2 (8 \ \Omega)}{2} = \mathbf{64 \ W}$$

The total power calculated using Eq. (16.37) is then

$$P = (1 + \text{THD}^2)P_1 = [1 + (0.1)^2]64 = (1.01)64 = \mathbf{64.64 \ W}$$

(Note that the total power is due mainly to the fundamental component even with 10% second harmonic distortion.)

Graphical Description of Harmonic Components of Distorted Signal

A distorted waveform such as that which occurs in class B operation can be represented using Fourier analysis as a fundamental with harmonic components. Figure 16.21a shows a positive half-cycle such as the type that would result in one side of a

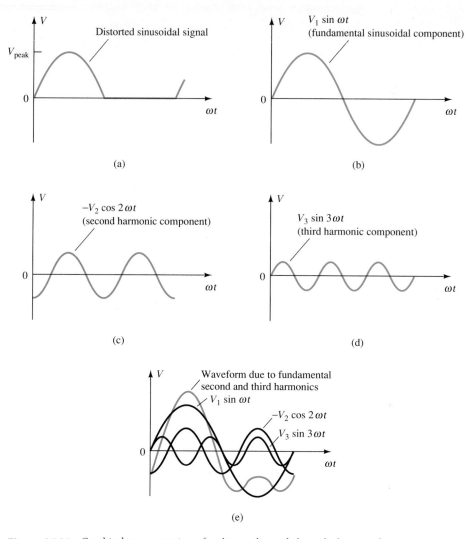

Figure 16.21 Graphical representation of a distorted signal through the use of harmonic components.

class B amplifier. Using Fourier analysis techniques the fundamental component of the distorted signal can be obtained, as shown in Fig. 16.21b. Similarly, the second and third harmonic components can be obtained and are shown in Fig. 16.21c and d, respectively. Using the Fourier technique, the distorted waveform can be made by adding the fundamental and harmonic components, as shown in Fig. 16.21e. In general, any periodic distorted waveform can be represented by adding a fundamental component and all harmonic components, each of varying amplitude and at various phase angles.

16.7 POWER TRANSISTOR HEAT SINKING

While integrated circuits are used for small-signal and low-power applications, most high-power applications still require individual power transistors. Improvements in production techniques have provided higher power ratings in small-sized packaging cases, have increased the maximum transistor breakdown voltage, and have provided faster-switching power transistors.

The maximum power handled by a particular device and the temperature of the transistor junctions are related since the power dissipated by the device causes an increase in temperature at the junction of the device. Obviously, a 100-W transistor will provide more power capability than a 10-W transistor. On the other hand, proper heat-sinking techniques will allow operation of a device at about one-half its maximum power rating.

We should note that of the two types of bipolar transistors—germanium and silicon—silicon transistors provide greater maximum temperature ratings. Typically, the maximum junction temperature of these types of power transistors is

Silicon: 150–200°C

Germanium: 100–110°C

For many applications the average power dissipated may be approximated by

$$P_D = V_{CE}I_C \tag{16.38}$$

This power dissipation, however, is allowed only up to a maximum temperature. Above this temperature the device power dissipation capacity must be reduced (or derated) so that at higher case temperatures the power-handling capacity is reduced, down to 0 W at the device maximum case temperature.

The greater the power handled by the transistor, the higher the case temperature. Actually, the limiting factor in power handling by a particular transistor is the temperature of the device's collector junction. Power transistors are mounted in large metal cases to provide a large area from which the heat generated by the device may radiate (be transferred). Even so, operating a transistor directly into air (mounting it on a plastic board, for example) severely limits the device power rating. If, instead (as is usual practice), the device is mounted on some form of heat sink, its power-handling capacity can approach the rated maximum value more closely. A few heat sinks are shown in Fig. 16.22. When the heat sink is used, the heat produced by the transistor dissipating power has a larger area from which to radiate (transfer) the heat into the air, thereby holding the case temperature to a much lower value than would result without the heat sink. Even with an infinite heat sink (which, of course, is not available), for which the case temperature is held at the ambient (air) temperature, the junction will be heated above the case temperature and a maximum power rating must be considered.

Since even a good heat sink cannot hold the transistor case temperature at ambient (which, by the way, could be more than 25°C if the transistor circuit is in a confined area where other devices are also radiating a good amount of heat), it is necessary to derate the amount of maximum power allowed for a particular transistor as a function of increased case temperature.

Figure 16.23 shows a typical power derating curve for a silicon transistor. The curve shows that the manufacturer will specify an upper temperature point (not neces-

Figure 16.22 Typical power heat sinks.

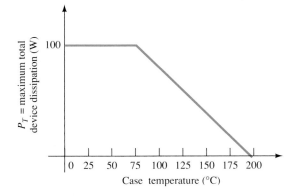

Figure 16.23 Typical power derating curve for silicon transistors.

sarily 25°C), after which a linear derating takes place. For silicon the maximum power that should be handled by the device does not reduce to 0 W until the case temperature is 200°C.

It is not necessary to provide a derating curve since the same information could be given simply as a listed derating factor on the device specification sheet. Stated mathematically, we have

$$P_D(\text{temp}_1) = P_D(\text{temp}_0) - (\text{Temp}_1 - \text{Temp}_0)(\text{derating factor}) \qquad (16.39)$$

where the value of Temp_0 is the temperature at which derating should begin, the value of Temp_1 is the particular temperature of interest (above the value Temp_0), $P_D(\text{temp}_0)$ and $P_D(\text{temp}_1)$ are the maximum power dissipations at the temperatures specified, and the derating factor is the value given by the manufacturer in units of watts (or milliwatts) per degree of temperature.

EXAMPLE 16.17

Determine what maximum dissipation will be allowed for an 80-W silicon transistor (rated at 25°C) if derating is required above 25°C by a derating factor of 0.5 W/°C at a case temperature of 125°C.

Solution

$$P_D(125°C) = P_D(25°C) - (125°C - 25°C)(0.5 \text{ W/°C})$$

$$= 80 \text{ W} - 100°C(0.5 \text{ W/°C}) = \mathbf{30 \text{ W}}$$

It is interesting to note what power rating results using a power transistor without a heat sink. For example, a silicon transistor rated at 100 W at (or below) 100°C is rated only 4 W at (or below) 25°C, the free-air temperature. Thus, operated without a heat sink, the device can handle a maximum of only 4 W at the room temperature of 25°C. Using a heat sink large enough to hold the case temperature to 100°C at 100 W allows operating at the maximum power rating.

Thermal Analogy of Power Transistor

Selection of a suitable heat sink requires a considerable amount of detail which is not appropriate to our present basic considerations of the power transistor. However, more detail about the thermal characteristics of the transistor and its relation to the power dissipation of the transistor may help provide a clearer understanding of power as limited by temperature. The following discussion should prove useful.

A picture of how the junction temperature (T_J), case temperature (T_C), and ambient (air) temperature (T_A) are related by the device heat-handling capacity—a temperature coefficient usually called thermal resistance—is presented in the thermal-electric analogy shown in Fig. 16.24.

In providing a thermal-electrical analogy, the term *thermal resistance* is used to describe heat effects by an electrical term. The terms in Fig. 16.24 are defined as follows:

$$\Theta_{JA} = \text{total thermal resistance (junction to ambient)}$$

$$\Theta_{JC} = \text{transistor thermal resistance (junction to case)}$$

$$\Theta_{CS} = \text{insulator thermal resistance (case to heat sink)}$$

$$\Theta_{SA} = \text{heat-sink thermal resistance (heat sink to ambient)}$$

Using the electrical analogy for thermal resistances, we can write

$$\boxed{\Theta_{JA} = \Theta_{JC} + \Theta_{CS} + \Theta_{SA}} \qquad (16.40)$$

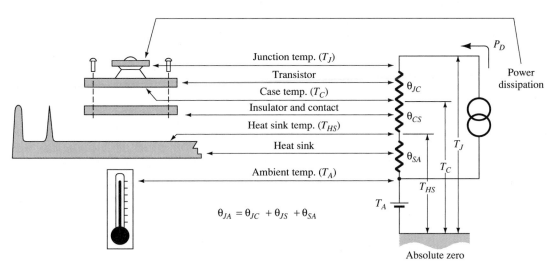

Figure 16.24 Thermal-to-electrical analogy.

The analogy can also be used in applying Kirchhoff's law to obtain

$$T_J = P_D\Theta_{JA} + T_A \qquad (16.41)$$

The last relation shows that the junction temperature "floats" on the ambient temperature and that the higher the ambient temperature, the lower the allowed value of device power dissipation.

The thermal factor Θ provides information about how much temperature drop (or rise) results for a given amount of power dissipation. For example, the value of Θ_{JC} is usually about 0.5°C/W. This means that for a power dissipation of 50 W, the difference in temperature between case temperature (as measured by a thermocouple) and the inside junction temperature is only

$$T_J - T_C = \Theta_{JC}P_D = (0.5°C/W)(50 \text{ W}) = 25°C$$

Thus, if the heat sink can hold the case at, say, 50°C, the junction is then only at 75°C. This is a relatively small temperature difference, especially at lower power-dissipation levels.

The value of thermal resistance from junction to free air (using no heat sink) is, typically,

$$\Theta_{JA} = 40°C/W \qquad \text{(into free air)}$$

For this thermal resistance only 1 W of power dissipation results in a junction temperature 40°C greater than the ambient.

A heat sink can now be seen to provide a low thermal resistance between case and air—much less than the 40°C/W value of the transistor case alone. Using a heat sink having

$$\Theta_{SA} = 2°C/W$$

and with an insulating thermal resistance (from case to heat sink) of

$$\Theta_{CS} = 0.8°C/W$$

and finally, for the transistor,

$$\Theta_{CJ} = 0.5°C/W$$

we can obtain

$$\Theta_{JA} = \Theta_{SA} + \Theta_{CS} + \Theta_{CJ}$$
$$= 2.0°C/W + 0.8°C/W + 0.5°C/W = 3.3°C/W$$

So with a heat sink, the thermal resistance between air and the junction is only 3.3°C/W, compared to 40°C/W for the transistor operating directly into free air. Using the value of Θ_{JA} above for a transistor operated at, say, 2 W, we calculate

$$T_J - T_A = \Theta_{JA}P_D = (3.3°C/W)(2\ W) = 6.6°C$$

In other words, the use of a heat sink in this example provides only a 6.6°C increase in junction temperature as compared to an 80°C rise without a heat sink.

EXAMPLE 16.18

A silicon power transistor is operated with a heat sink ($\Theta_{SA} = 1.5°C/W$). The transistor, rated at 150 W (25°C), has $\Theta_{JC} = 0.5°C/W$ and the mounting insulation has $\Theta_{CS} = 0.6°C/W$. What maximum power can be dissipated if the ambient temperature is 40°C and $T_{J_{max}} = 200°C$?

Solution

$$P_D = \frac{T_J - T_A}{\Theta_{JC} + \Theta_{CS} + \Theta_{SA}} = \frac{200°C - 40°C}{0.5°C/W + 0.6°C/W + 1.5°C/W} \approx \mathbf{61.5\ W}$$

16.8   CLASS C AND CLASS D AMPLIFIERS

Although class A, class AB, and class B amplifiers are most used as power amplifiers, class D amplifiers are popular because of their very high efficiency. Class C amplifiers, while not used as audio amplifiers, do find use in tuned circuits as used in communications.

Class C Amplifier

A class C amplifier, as that shown in Fig. 16.25, is biased to operate for less than 180° of the input signal cycle. The tuned circuit in the output, however, will provide a full cycle of output signal for the fundamental or resonant frequency of the tuned circuit (*L* and *C* tank circuit) of the output. This type of operation is therefore limited to use at one fixed frequency, as occurs in a communications circuit, for example. Operation of a class C circuit is not intended primarily for large-signal or power amplifiers.

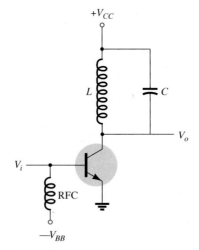

Figure 16.25 Class C amplifier circuit.

Class D Amplifier

A class D amplifier is designed to operate with digital or pulse-type signals. An efficiency of over 90% is achieved using this type of circuit, making it quite desirable in power amplifiers. It is necessary, however, to convert any input signal into a pulse-type waveform before using it to drive a large power load and to convert the signal back to a sinusoidal-type signal to recover the original signal. Figure 16.26 shows how a sinusoidal signal may be converted into a pulse-type signal using some form of sawtooth or chopping waveform to be applied with the input into a comparator-type op-amp circuit so that a representative pulse-type signal is produced. While the letter D is used to describe the next type of bias operation after class C, the D could also be considered to stand for "Digital," since that is the nature of the signals provided to the class D amplifier.

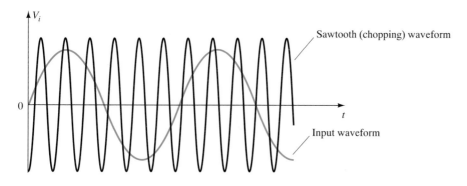

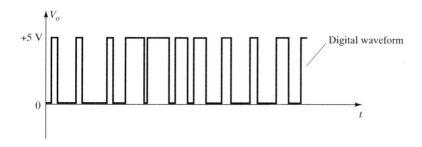

Figure 16.26 Chopping of sinusoidal waveform to produce digital waveform.

Figure 16.27 shows a block diagram of the unit needed to amplify the class D signal and then convert back to the sinusoidal-type signal using a low-pass filter. Since the amplifier's transistor devices used to provide the output are basically either off or on, they provide current only when they are turned on, with little power loss

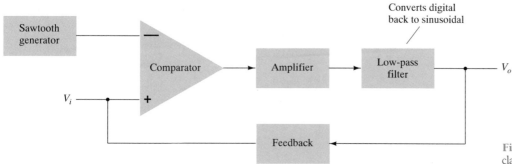

Figure 16.27 Block diagram of class D amplifier.

due to their low on-voltage. Since most of the power applied to the amplifier is transferred to the load, the efficiency of the circuit is typically very high. Power MOSFET devices have been quite popular as the driver devices for the class D amplifier.

16.9 COMPUTER ANALYSIS

Program 16.1: Series-Fed Class A Amplifier

The dc bias and ac operation of a series-fed, class A amplifier is provided by the circuit in Fig. 16.28 and the program and output results in Fig. 16.29. Dc data provide that

$$P_i(\text{dc}) = V_{CC}I_C = (22 \text{ V})(105.5 \text{ mA}) = 2.3 \text{ W}$$

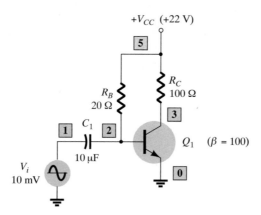

Figure 16.28 Class-A, series-fed amplifier.

```
**** 01/24/90 ******* Evaluation PSpice (January 1989) ******* 21:27:45 ****

 Series-fed Class-A Amplifier

****      CIRCUIT DESCRIPTION

******************************************************************

VCC 5 0 22V
RB 5 2 20K
RC 5 3 100
C1 1 2 10UF
VI 1 0 AC 10MV
Q1 3 2 0 QN
.MODEL QN NPN(BF=100)
.AC LIN 1 2KH 2KH
.PRINT AC VM(1) VM(3,5) IM(RC)
.DC VCC 22 22 1
.PRINT DC I(RC)
.OPTIONS NOPAGE
.END

****      DC TRANSFER CURVES               TEMPERATURE =   27.000 DEG C
  VCC          I(RC)
   2.200E+01   1.055E-01


****      SMALL SIGNAL BIAS SOLUTION       TEMPERATURE =   27.000 DEG C
  NODE    VOLTAGE        NODE    VOLTAGE        NODE    VOLTAGE        NODE    VOLTAGE
(    1)   0.0000  (    2)     .8947  (    3)   11.4470  (    5)   22.0000

TOTAL POWER DISSIPATION   2.34E+00  WATTS
****      AC ANALYSIS                      TEMPERATURE =   27.000 DEG C
   FREQ        VM(1)        VM(3,5)       IM(RC)
   2.000E+03   1.000E-02    3.880E+00    3.880E-02
```

Figure 16.29 PSpice output for circuit of Fig. 16.28.

Ac data provide

$$P_o(\text{ac}) = V_o I_o = (3.88 \text{ V})(3.88 \times 10^{-2}) = 150.5 \text{ MW}$$

The efficiency of the amplifier for the given input signal is then

$$\% \ \eta = \frac{P_o}{P_i} \times 100\% = \frac{150.5 \text{ mW}}{2.3 \text{ W}} \times 100\% = 6.5\%$$

A larger input signal would increase the ac power delivered to the load and increase the efficiency (the maximum being 25%).

Program 16.2: Quasi-complementary Push–Pull Amplifier

Figure 16.30 shows a quasi-complementary amplifier. The circuit is described in the PSpice listing of Fig. 16.31, along with the output results of dc bias and ac voltages.

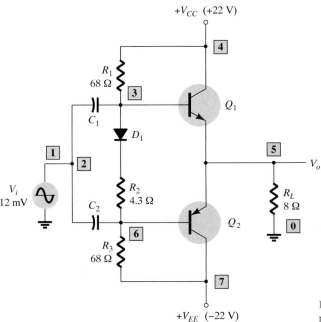

Figure 16.30 Quasi-complementary class-B power amplifier.

```
**** 01/24/90 ******* Evaluation PSpice (January 1989) ******* 21:31:44 ****

Quasi-complementary Class-B power amplifier

****        CIRCUIT DESCRIPTION

*******************************************************************

VCC 4 0 22V
VEE 0 7 22V
R1 4 3 68
D1 3 2 DA
C1 1 2 50UF
R2 2 6 4.3
R3 6 7 68
RL 5 0 8
Q1 4 3 5 QN
Q2 5 6 7 QP
.MODEL DA D
.MODEL QN NPN(BF=50)
.MODEL QP PNP(BF=50)
VI 1 0 AC 12V
.AC LIN 1 1KH 1KH
.PRINT AC VM(1) VM(RL)
.OPTIONS NOPAGE
.END
```

Figure 16.31 PSpice output for circuit of Fig. 16.30.

```
****        SMALL SIGNAL BIAS SOLUTION
    NODE     VOLTAGE      NODE    VOLTAGE      NODE    VOLTAGE      NODE    VOLTAGE
(    1)      0.0000  (     2)     3.7677  (     3)     4.5648  (     4)    22.0000
(    5)      3.6219  (     6)     2.7237  (     7)   -22.0000

        VOLTAGE SOURCE CURRENTS
        NAME              CURRENT
        VCC             -9.371E-01
        VEE             -4.844E-01
TOTAL POWER DISSIPATION      3.13E+01   WATTS

****        AC ANALYSIS
    FREQ          VM(1)          VM(RL)
    1.000E+03     1.200E+01      1.189E+01
```

Figure 16.31 Continued.

The particular resistor values result in class AB operation, the dc voltage at the load resistor being about 3.6 V. Input power from the supplies is seen to be about 31 W. The ac power delivered to the 8-Ω load is

$$P_o(\text{ac}) = \frac{V_o^2}{R_L} = \frac{(11.89)^2}{8} = 17.7 \text{ W}$$

so that the circuit is presently operating at an efficiency of

$$\% \ \eta = \frac{P_o(\text{ac})}{P_i(\text{dc})} \times 100\% = \frac{17.7 \text{ W}}{31 \text{ W}} \times 100\% = 57.1\%$$

Varying the values of resistors R_1 to R_3 can alter the dc bias and thereby the dc power supplied to the circuit. The ac input signal can be varied to change the amount of the ac power delivered to the load. At best, the efficiency should be 78.5%.

PROBLEMS

§ 16.2 Series-Fed Class A Amplifier

1. Calculate the input and output power for the circuit of Fig. 16.32. The input signal results in a base current of 5 mA rms.

2. Calculate the input power dissipated by the circuit of Fig. 16.32 if R_B is changed to 1.5 kΩ.

3. What maximum output power can be delivered by the circuit of Fig. 16.32 if R_B is changed to 1.5 kΩ?

4. If the circuit of Fig. 16.32 is biased at its center voltage and center collector operating point, what is the input power for a maximum output power of 1.5 W?

§ 16.3 Transformer-Coupled Class A Amplifier

5. A class A transformer-coupled amplifier uses a 25:1 transformer to drive a 4-Ω load. Calculate the effective ac load (seen by the transistor connected to the larger turns side of the transformer).

6. What turns ratio transformer is needed to couple to an 8-Ω load so that it appears as an 8-kΩ effective load?

7. Calculate the transformer turns ratio required to connect four parallel 16-Ω speakers so that they appear as an 8-kΩ effective load.

* 8. A transformer-coupled class A amplifier drives a 16-Ω speaker through a 3.87:1 transformer. Using a power supply of $V_{CC} = 36$ V, the circuit delivers 2 W to the load. Calculate:
(a) $P(\text{ac})$ across transformer primary.
(b) $V_L(\text{ac})$.
(c) $V(\text{ac})$ at transformer primary.
(d) The rms values of load and primary current.

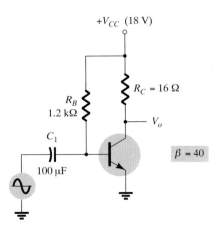

Figure 16.32 Problems 1–4

Chapter 16 Power Amplifiers

9. Calculate the efficiency of the circuit of Problem 8 if the bias current is $I_{CQ} = 150$ mA.

10. Draw the circuit diagram of a class A transformer-coupled amplifier using an *npn* transistor.

§ 16.4 Class B Amplifier Operation

11. Draw the circuit diagram of a class B *npn* push-pull power amplifier using transformer-coupled input.

12. For a class B amplifier providing a 22-V peak signal to an 8-Ω load and a power supply of $V_{CC} = 25$ V, determine:
   (a) Input power.
   (b) Output power.
   (c) Circuit efficiency.

13. For a class B amplifier with $V_{CC} = 25$ V driving an 8-Ω load, determine:
   (a) Maximum input power.
   (b) Maximum output power.
   (c) Maximum circuit efficiency.

* **14.** Calculate the efficiency of a class B amplifier for a supply voltage of $V_{CC} = 22$ V driving a 4Ω load with peak output voltages of:
   (a) $V_L(\text{p}) = 20$ V.
   (b) $V_L(\text{p}) = 4$ V.

§ 16.5 Class B Amplifier Circuits

15. Sketch the circuit diagram of a quasi-complementary amplifier, showing voltage waveforms in the circuit.

16. For the class B power amplifier of Fig. 16.33, calculate:
   (a) Maximum $P_o(\text{ac})$.
   (b) Maximum $P_i(\text{dc})$.
   (c) Maximum $\%\eta$.
   (d) Maximum power dissipated by both transistors.

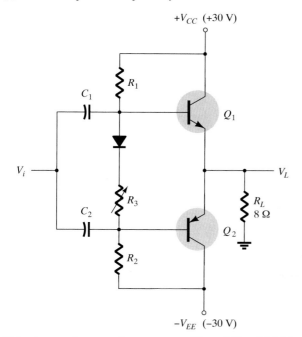

$+V_{CC}$ (+30 V)

$-V_{EE}$ (−30 V)   **Figure 16.33** Problems 16, 17

* **17.** If the input voltage to the power amplifier of Fig. 16.33 is 8-V rms, calculate:
   (a) $P_i(\text{dc})$.
   (b) $P_o(\text{ac})$.
   (c) $\%\eta$.
   (d) Power dissipated by both power output transistors.

* **18.** For the power amplifier of Fig. 16.34, calculate:
  (a) $P_o(\text{ac})$.
  (b) $P_i(\text{dc})$.
  (c) $\%\eta$.
  (d) Power dissipated by both output transistors.

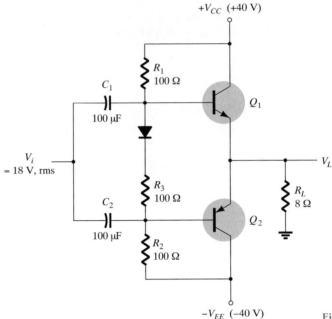

Figure 16.34 Problem 18

§ **16.6 Amplifier Distortion**

19. Calculate the harmonic distortion components for an output signal having fundamental amplitude of 2.1 V, second harmonic amplitude of 0.3 V, third harmonic component of 0.1 V, and fourth harmonic component of 0.05 V.

20. Calculate the total harmonic distortion for the amplitude components of Problem 19.

21. Calculate the second harmonic distortion for an output waveform having measured values of $V_{CE_{min}} = 2.4$ V, $V_{CE_Q} = 10$ V, and $V_{CE_{max}} = 20$ V.

22. For distortion readings of $D_2 = 0.15$, $D_3 = 0.01$, and $D_4 = 0.05$, with $I_1 = 3.3$ A and $R_C = 4\ \Omega$, calculate the total harmonic distortion fundamental power component and total power.

§ **16.7 Power Transistor Heat Sinking**

23. Determine the maximum dissipation allowed for a 100-W silicon transistor (rated at 25°C) for a derating factor of 0.6 W/°C at a case temperature of 150°C.

* **24.** A 160-W silicon power transistor operated with a heat sink ($\Theta_{SA} = 1.5$°C/W) has $\Theta_{JC} = 0.5$°C/W and a mounting insulation of $\Theta_{CS} = 0.8$°C/W. What maximum power can be handled by the transistor at an ambient temperature of 80°C? (The junction temperature should not exceed 200°C.)

25. What maximum power can a silicon transistor ($T_{J_{max}} = 200$°C) dissipate into free air at an ambient temperature of 80°C?

§ **16.9 Computer Analysis**

* **26.** Write a PSpice program to calculate V_o for $V_i = 9.1$ mV in the circuit of Fig. 16.32.

* **27.** Write a PSpice program to calculate the voltage across R_L in the circuit of Fig. 16.33.

Please Note: Asterisks indicate more difficult problems.

Linear-Digital ICs

17

17.1 INTRODUCTION

While there are many ICs containing only digital circuits, and many that contain only linear circuits, there are a number of units that contain both linear and digital circuits. Among the linear/digital ICs are comparator circuits, digital/analog converters, interface circuits, timer circuits, voltage-controlled oscillator (VCO) circuits, and phase-locked loops (PLLs).

The comparator circuit is one to which a linear input voltage is compared to another reference voltage, the output being a digital condition representing whether the input voltage exceeded the reference voltage.

Circuits that convert digital signals into an analog or linear voltage, and those that convert a linear voltage into a digital value, are popular in aerospace equipment, automotive equipment, and compact disk (CD) players, among many others.

Interface circuits are used to enable connecting signals of different digital voltage levels, from different types of output devices, or from different impedances so that both the driver stage and the receiver stage operate properly.

Timer ICs provide linear and digital circuits to use in various timing operations, as in a car alarm, a home timer to turn lights on or off, and a circuit in electromechanical equipment to provide proper timing to match the intended unit operation. The 555 timer has long been a popular IC unit. A voltage-controlled oscillator provides an output clock signal whose frequency can be varied or adjusted by an input voltage. One popular application of a VCO is in a phase-locked loop unit as used in various communication transmitters and receivers.

17.2 COMPARATOR UNIT OPERATION

A comparator circuit accepts input of linear voltages and provides a digital output that indicates when one input is less than or greater than the second. A basic comparator circuit can be represented as in Fig. 17.1a. The output is a digital signal that stays at a high voltage level when the noninverting($+$) input is greater than the voltage at the inverting($-$) input, and switches to a lower voltage level when the noninverting input voltage goes below the inverting input voltage.

Figure 17.1b shows a typical connection with one input (the inverting input in this example) connected to a reference voltage, the other connected to the input signal voltage. As long as V_{in} is less than the reference voltage level of $+2$ V, the output remains at a low voltage level (near -10 V). When the input rises just above $+2$ V,

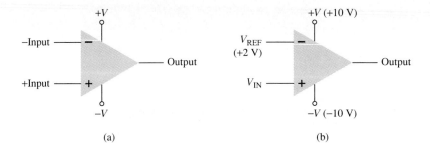

Figure 17.1 Comparator unit: (a) basic unit; (b) typical application.

(a)   (b)

the output quickly switches to a high-voltage level (near $+10$ V). Thus the high output indicates that the input signal is greater than $+2$ V.

Since the internal circuit used to build a comparator contains essentially an op-amp circuit with very high voltage gain, we can examine the operation of a comparator using a 741 op-amp, as shown in Fig. 17.2. With reference input (at pin 2) set to 0 V, a sinusoidal signal applied to the noninverting input (pin 3) will cause the output to switch between its two output states, as shown in Fig. 17.2b. The input V_i going even a fraction of a millivolt above the 0-V reference level will be amplified by the very high voltage gain (typically over 100,000) so that the output rises to its positive output saturation level and remains there while the input stays above $V_{ref} = 0$ V. When the input drops just below the 0-V reference level, the output is driven to its lower saturation level and stays there while the input remains below $V_{ref} = 0$ V. Figure 17.2b clearly shows that the input signal is linear while the output is digital.

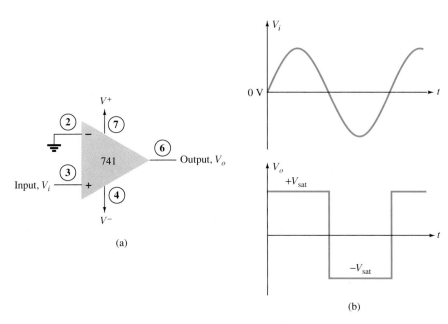

Figure 17.2 Operation of 741 op-amp as comparator.

In general use, the reference level need not be 0 V but can be any desired positive or negative voltage. Also, the reference voltage may be connected to either plus or minus input and the input signal then applied to the other input.

Use of Op-Amp as Comparator

Figure 17.3a shows a circuit operating with a positive reference voltage connected to the minus input and the output connected to an indicator LED. The reference voltage level is set at

$$V_{ref} = \frac{10 \text{ k}\Omega}{10 \text{ k}\Omega + 10 \text{ k}\Omega}\,(+12 \text{ V}) = +6 \text{ V}$$

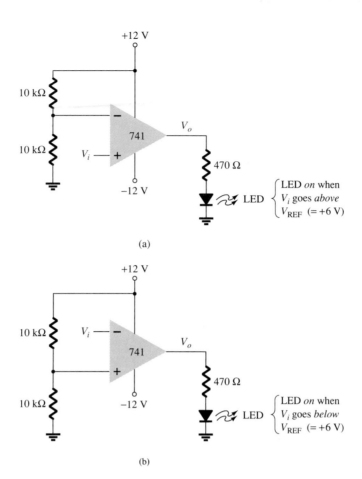

Figure 17.3 A 741 op-amp used as a comparator.

(a)

(b)

Since the reference voltage is connected to the inverting input, the output will switch to its positive saturation level when the input, V_i, goes more positive than the +6-V reference voltage level. The output, V_o, then drives the LED on, as an indication that the input is more positive than the reference level.

As an alternative connection, the reference voltage could be connected to the noninverting input as shown in Fig. 17.3b. With this connection the input signal going below the reference level would cause the output to drive the LED on. The LED can thus be made to go on when the input signal goes above or below the reference level, depending on which input is connected as signal input and which as reference input.

Using Comparator IC Units

While op-amps can be used as comparator circuits, separate IC comparator units are more suitable. Some of the improvements built into a comparator IC are faster switching between the two output levels, built-in noise immunity to prevent the output from oscillating when the input passes by the reference level, and outputs capable of directly driving a variety of loads. A few popular IC comparators are covered next, describing their pin connections and how they may be used.

311 COMPARATOR

The 311 voltage comparator shown in Fig. 17.4 contains a comparator circuit that can operate as well from dual power supplies of ±15 V as from a single +5-V supply (as used in digital logic circuits). The output can provide a voltage at one of two distinct levels or can be used to drive a lamp or a relay. Notice that the output is taken

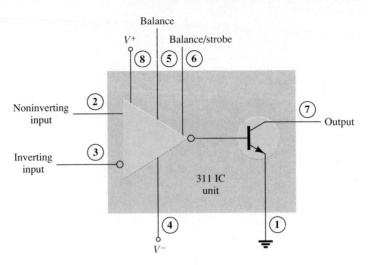

Figure 17.4 A 311 comparator (eight-pin DIP unit).

from a bipolar transistor to allow driving a variety of loads. The unit also has balance and strobe inputs, the strobe input allowing gating of the output. A few examples will show how this comparator unit can be used in some common applications.

A zero-crossing detector that senses (detects) the input voltage crossing through 0 V is shown using the 311 IC in Fig. 17.5. The inverting input is connected to ground (the reference voltage). The input signal going positive drives the output transistor on, with the output then going low (-10 V in this case). The input signal going negative (below 0 V) will drive the output transistor off, the output then going high (to $+10$ V). The output is thus an indication of whether the input is above or below 0 V. When the input is any positive voltage, the output is low, while any negative voltage will result in the output going to a high voltage level.

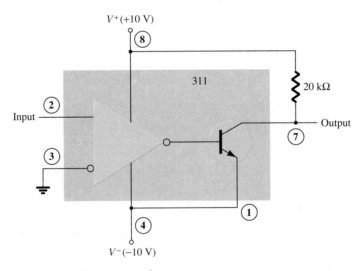

Figure 17.5 Zero-crossing detector using a 311 IC.

Figure 17.6 shows how a 311 comparator can be used with strobing. In this example the output will go high when the input goes above the reference level—but only if the TTL strobe input is off (or 0 V). If the TTL strobe input goes high, it drives the 311 strobe input at pin 6 low, causing the output to remain in the off state (with output high) regardless of the input signal. In effect, the output remains high

Chapter 17   **Linear-Digital ICs**

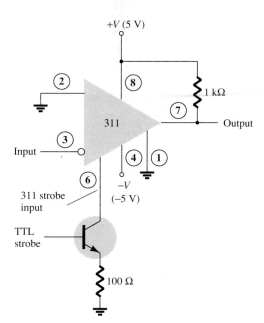

Figure 17.6 Operation of a 311 comparator with strobe input.

unless strobed. If strobed, the output then acts normally, switching from high to low, depending on the input signal level. In operation, the comparator output will respond to the input signal only during the time the strobe signal allows such operation.

Figure 17.7 shows the comparator output driving a relay. When the input goes below 0 V, driving the output low, the relay is activated, closing the normally open (N.O.) contacts at that time. These contact can then be connected to operate a large variety of devices. For example, a buzzer or bell wired to the contacts can be driven on whenever the input voltage drops below 0 V. As long as the voltage is present at the input terminal, the buzzer will remain off.

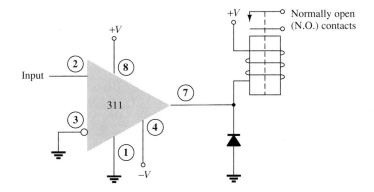

Figure 17.7 Operation of a 311 comparator with relay output.

339 COMPARATOR

The 339 IC is a quad comparator containing four independent voltage comparator circuits, connected to external pins as shown in Fig. 17.8. Each comparator has inverting and noninverting inputs and a single output. The supply voltage applied to a pair of pins powers all four comparators. Even if one wishes to use one comparator, all four will be drawing power.

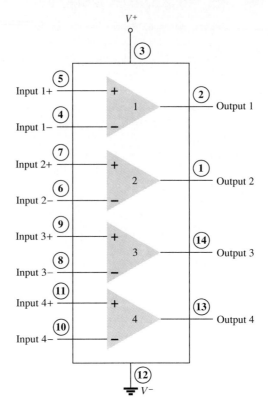

Figure 17.8 Quad comparator IC (339).

To see how these comparator circuits can be used, Fig. 17.9 shows one of the 339 comparator circuits connected as a zero-crossing detector. Whenever the input signal goes above 0 V, the output switches to V^+. The input switches to V^- only when the input goes below 0 V.

A reference level other than 0 V can also be used and either input terminal could be used as the reference, the other terminal then being connected to the input signal. The operation of one of the comparator circuits is described next.

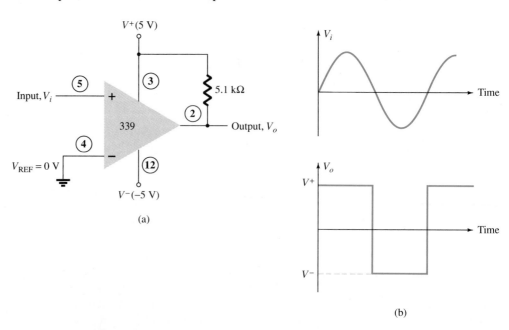

Figure 17.9 Operation of one 339 comparator circuit as a zero-crossing detector.

The differential input voltage (difference voltage across input terminals) going positive drives the output transistor off (open circuit), while a negative differential input voltage drives the output transistor on—the output then at the supply low level.

If the negative input is set at a reference level V_{ref}, the positive input goes above V_{ref} results in a positive differential input with output driven to the open-circuit state. When the noninverting input goes below V_{ref}, resulting in a negative differential input, the output will be driven to V^-.

If the positive input is set at the reference level, the inverting input going below V_{ref} results in the output open circuit, while the inverting input going above V_{ref} results in the output at V^-. This operation is summarized in Fig. 17.10.

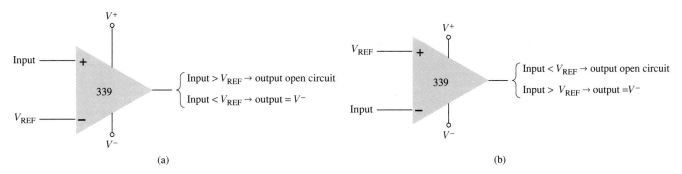

Figure 17.10 Operation of a 339 comparator circuit with reference input at (a) minus input; (b) plus input.

Since the output of one of these comparator circuits is from an open-circuit collector, applications in which the outputs from more than one circuit can be wire-ORed are possible. Figure 17.11 shows two comparator circuits connected with common output and also with common input. Comparator 1 has a +5-V reference voltage

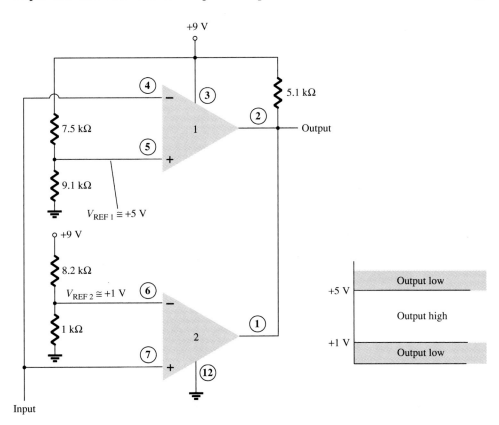

Figure 17.11 Operation of two 339 comparator circuits as a window detector.

input connected to the noninverting input. The output will be driven low by comparator 1 when the input signal goes above +5 V. Comparator 2 has a reference voltage of +1 V connected to the inverting input. The output of comparator 2 will be driven low when the input signal goes below +1 V. In total, the output will go low whenever the input is below +1 V or above +5 V, as shown in Fig. 17.11, the overall operation being that of a voltage window detector. The high output indicates that the input is within a voltage window of +1 to +5 V (these values being set by the reference voltage levels used).

17.3 DIGITAL–ANALOG CONVERTERS

Many voltages and currents in electronics vary continuously over some range of values. In digital circuitry the signals are at either one of two levels, representing the binary values of 1 or zero. An analog–digital converter (ADC) obtains a digital value representing an input analog voltage, while a digital–analog converter (DAC) changes a digital value back into an analog voltage.

Digital-to-Analog Conversion

LADDER NETWORK CONVERSION

Digital-to-analog conversion can be achieved using a number of different methods. One popular scheme uses a network of resistors, called a *ladder network*. A ladder network accepts inputs of binary values at, typically, 0 V or V_{ref}, and provides an output voltage proportional to the binary input value. Figure 17.12a shows a ladder network with four input voltages, representing 4 bits of digital data and a dc voltage output. The output voltage is proportional to the digital input value as given by the relation

$$V_o = \frac{D_0 \times 2^0 + D_1 \times 2^1 + D_2 \times 2^2 + D_3 \times 2^3}{2^4} V_{ref} \qquad (17.1)$$

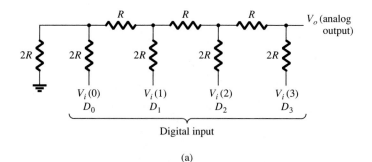

(a)

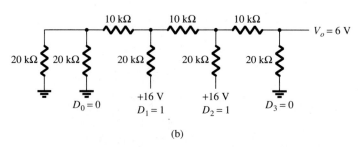

(b)

Figure 17.12 Four-stage ladder network used as D/A converter: (a) basic circuit; (b) circuit example with 0110 input.

In the example shown in Fig. 17.12b, the output voltage resulting should be

$$V_o = \frac{0 \times 1 + 1 \times 2 + 1 \times 4 + 0 \times 8}{16} \,(16 \text{ V}) = 6 \text{ V}$$

Therefore, 0110_2, digital, converts to 6 V, analog.

The function of the ladder network is to convert the 16 possible binary values from 0000 to 1111 into one of 16 voltage levels in steps of $V_{ref}/16$. Using more sections of ladder allows having more binary inputs and greater quantization for each step. For example, a 10-stage ladder network could extend the number of voltage steps or the voltage resolution to $V_{ref}/2^{10}$ or $V_{ref}/1024$. A reference voltage of $V_{ref} = 10$ V would then provide output voltage steps of 10 V/1024 or approximately 10 mV. More ladder stages provide greater voltage resolution. In general, the voltage resolution for n ladder stages is

$$\boxed{\frac{V_{ref}}{2^n}} \qquad (17.2)$$

Figure 17.13 shows a block diagram of a typical DAC using a ladder network. The ladder network, referred in the diagram as an *R-2R ladder*, is sandwiched between the reference current supply and current switches connected to each binary input, the resulting output current proportional to the input binary value. The binary input turns on selected legs of the ladder, the output current being a weighted summing of the reference current. Connecting the output current through a resistor will produce an analog voltage, if desired.

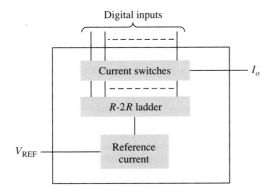

Figure 17.13 D/A converter IC using *R-2R* ladder network.

Analog-to-Digital Conversion

DUAL-SLOPE CONVERSION

A popular method for converting an analog voltage into a digital value is the dual-slope method. Figure 17.14a shows a block diagram of the basic dual-slope converter. The analog voltage to be converted is applied through an electronic switch to an integrator or ramp-generator circuit (essentially a constant current charging a capacitor to produce a linear ramp voltage). The digital output is obtained from a counter operated during both positive and negative slope intervals of the integrator.

The method of conversion proceeds as follows. For a fixed time interval (usually the full count range of the counter), the analog voltage connected to the integrator raises the voltage at the comparator input to some positive level. Figure 17.14b shows that at the end of the fixed time interval the voltage from the integrator is greater for the larger input voltage. At the end of the fixed count interval, the count is set to zero and the electronic switch connects the integrator to a reference or fixed input voltage.

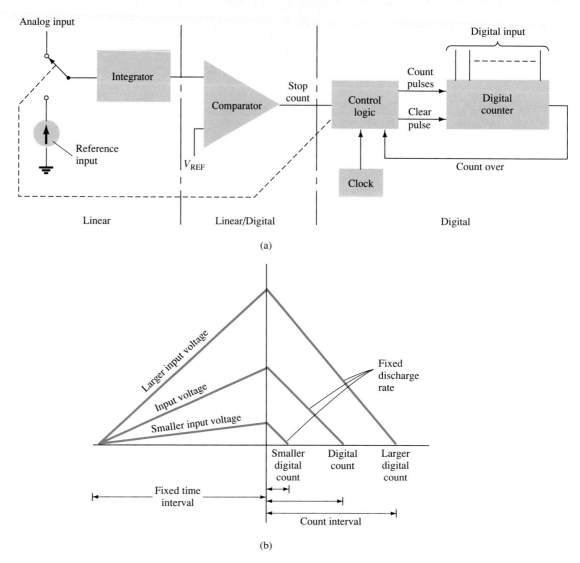

Figure 17.14 A/D conversion using dual-slope method: (a) logic diagram; (b) waveform.

The integrator output (or capacitor input) then decreases at a fixed rate. The counter advances during this time, while the integrator's output decreases at a fixed rate until it drops below the comparator reference voltage, at which time the control logic receives a signal (the comparator output) to stop the count. The digital value stored in the counter is then the digital output of the converter.

Using the same clock and integrator to perform the conversion during positive and negative slope intervals tends to compensate for clock frequency drift and integrator accuracy limitations. Setting the reference input value and clock rate can scale the counter output as desired. The counter can be a binary, BCD, or other form of digital counter, if desired.

LADDER-NETWORK CONVERSION

Another popular method of analog-to-digital conversion uses a ladder network along with counter and comparator circuits (see Fig. 17.15). A digital counter advances from a zero count while a ladder network driven by the counter outputs a staircase voltage, as shown in Fig. 17.15b, which increases one voltage increment for each count step. A comparator circuit, receiving both staircase voltage and analog

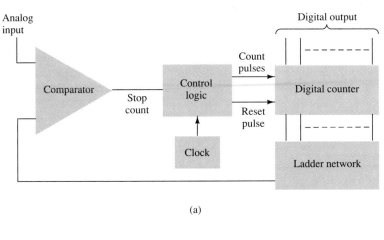

(a)

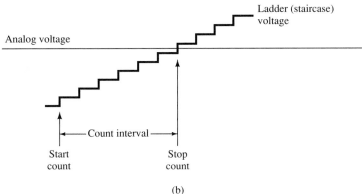

(b)

Figure 17.15 A/D conversion using ladder network: (a) logic diagram; (b) waveform.

input voltage, provides a signal to stop the count when the staircase voltage rises above the input voltage. The counter value at that time is the digital output.

The amount of voltage change stepped by the staircase signal depends on the number of count bits used. A 12-stage counter operating a 12-stage ladder network using a reference voltage of 10 V would step each count by a voltage of

$$\frac{V_{ref}}{2^{12}} = \frac{10 \text{ V}}{4096} = 2.4 \text{ mV}$$

This would result in a conversion resolution of 2.4 mV. The clock rate of the counter would affect the time required to carry out a conversion. A clock rate of 1 MHz operating a 12-stage counter would need a maximum conversion time of

$$4096 \times 1\mu s = 4096 \ \mu s \approx 4.1 \text{ ms}$$

The minimum number of conversions that could be carried out each second would then be

$$\text{number of conversions} = 1/4.1 \text{ ms} \approx 244 \text{ conversions/second}$$

Since on the average, with some conversions requiring little count time and others near maximum count time, a conversion time of 4.1 ms/2 = 2.05 ms would be needed, and the average number of conversions would be 2 × 244 = 488 conversions/second. A slower clock rate would result in fewer conversions per second. A converter using fewer count stages (and less conversion resolution) would carry out more conversions per second. The conversion accuracy depends on the accuracy of the comparator.

17.3 **Digital–Analog Converters**

719

17.4 TIMER IC UNIT OPERATION

Another popular analog-digital integrated circuit is the versatile 555 timer. The IC is made of a combination of linear comparators and digital flip-flops as described in Fig. 17.16. The entire circuit is usually housed in an eight-pin package as specified in Fig. 17.16. A series connection of three resistors sets the reference voltage levels to the two comparators at $2V_{CC}/3$ and $V_{CC}/3$, the output of these comparators setting or resetting the flip-flop unit. The output of the flip-flop circuit is then brought out through an output amplifier stage. The flip-flop circuit also operates a transistor inside the IC, the transistor collector usually being driven low to discharge a timing capacitor.

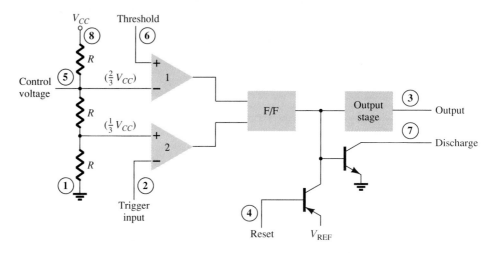

Figure 17.16 Details of 555 timer IC.

Astable Operation

One popular application of the 555 timer IC is as an astable multivibrator or clock circuit. The following analysis of the operation of the 555 as an astable circuit includes details of the different parts of the unit and how the various inputs and outputs are utilized. Figure 17.17 shows an astable circuit built using an external resistor and capacitor to set the timing interval of the output signal.

Figure 17.17 Astable multivibrator using 555 IC.

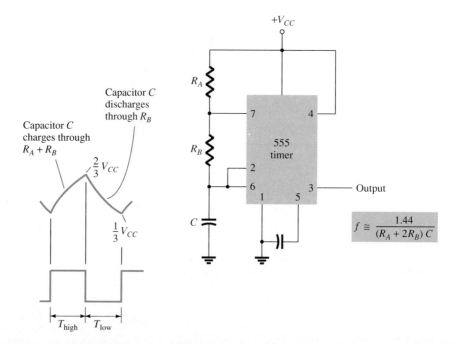

$$f \cong \frac{1.44}{(R_A + 2R_B)\,C}$$

Capacitor C charges toward V_{CC} through external resistors R_A and R_B. Referring to Fig. 17.17, the capacitor voltage rises until it goes above $2V_{CC}/3$. This voltage is the threshold voltage at pin 6, which drives comparator 1 to trigger the flip-flop so that the output at pin 3 goes low. In addition, the discharge transistor is driven on, causing the output at pin 7 to discharge the capacitor through resistor R_B. The capacitor voltage then decreases until it drops below the trigger level ($V_{CC}/3$). The flip-flop is triggered so that the output goes back high and the discharge transistor is turned off, so that the capacitor can again charge through resistors R_A and R_B toward V_{CC}.

Figure 17.18a shows the capacitor and output waveforms resulting from the astable circuit. Calculation of the time intervals during which the output is high and low can be made using the relations

$$\boxed{T_{\text{high}} \approx 0.7(R_A + R_B)C} \qquad (17.3)$$

$$\boxed{T_{\text{low}} \approx 0.7R_BC} \qquad (17.4)$$

The total period is

$$T = \text{period} = T_{\text{high}} + T_{\text{low}} \qquad (17.5)$$

The frequency of the astable circuit is then calculated using*

$$\boxed{f = \frac{1}{T} \approx \frac{1.44}{(R_A + 2R_B)C}} \qquad (17.6)$$

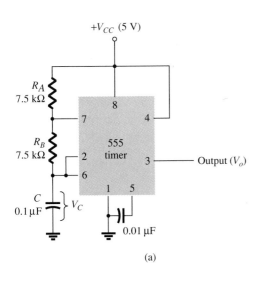

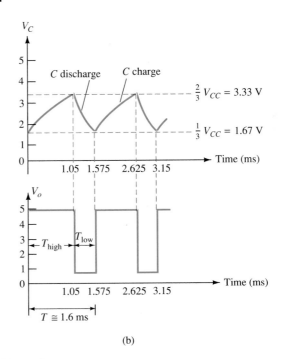

Figure 17.18 Astable multivibrator for Example 17.1: (a) circuit; (b) waveforms.

---

*The period can be directly calculated from

$$T = 0.693(R_A + 2R_B)C \approx 0.7(R_A + 2R_B)C$$

and the frequency from

$$f \approx \frac{1.44}{(R_A + 2R_B)C}$$

EXAMPLE 17.1

Determine the frequency and draw the output waveform for the circuit of Fig. 17.18a.

Solution

Using Eqs. (17.3) through (17.6) yields

$$T_{\text{high}} = 0.7(R_A + R_B)C = 0.7(7.5 \times 10^3 + 7.5 \times 10^3)(0.1 \times 10^{-6})$$

$$= 1.05 \text{ ms}$$

$$T_{\text{low}} = 0.7R_BC = 0.7(7.5 \times 10^3)(0.1 \times 10^{-6}) = 0.525 \text{ ms}$$

$$T = T_{\text{high}} + T_{\text{low}} = 1.05 \text{ ms} + 0.525 \text{ ms} = 1.575 \text{ ms}$$

$$f = \frac{1}{T} = \frac{1}{1.575 \times 10^{-3}} \approx \mathbf{635 \text{ Hz}}$$

The waveforms are drawn in Fig. 17.18b.

Monostable Operation

The 555 timer can also be used as a one-shot or monostable multivibrator circuit, as shown in Fig. 17.19. When the trigger input signal goes negative, it triggers the one-shot with output at pin 3 then going high for a time period

$$T_{\text{high}} = 1.1R_AC \qquad (17.7)$$

Referring back to Fig. 17.16, the negative edge of the trigger input causes comparator 2 to trigger the flip-flop with the output at pin 3 going high. Capacitor C charges toward V_{CC} through resistor R_A. During the charge interval the output remains high. When the voltage across the capacitor reaches the threshold level of $2V_{CC}/3$, comparator 1 triggers the flip-flop with output going low. The discharge transistor also goes low, causing the capacitor to remain at near 0 V until triggered again.

Figure 17.19b shows the input trigger signal and the resulting output waveform for the 555 timer operated as a one-shot. Time periods for this circuit can range from microseconds to many seconds, making this IC useful for a range of applications.

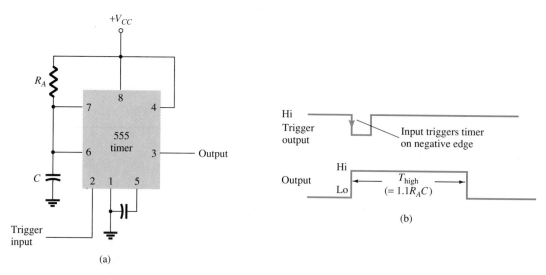

Figure 17.19 Operation of 555 timer as one-shot: (a) circuit; (b) waveforms.

Determine the period of the output waveform for the circuit of Fig. 17.20 when
triggered by a negative pulse.

EXAMPLE 17.2

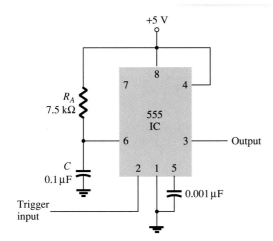

Figure 17.20 Monostable circuit for Example 17.2.

Solution

Using Eq. (17.7), we obtain

$$T_{\text{high}} = 1.1R_AC = 1.1(7.5 \times 10^3)(0.1 \times 10^{-6}) = \textbf{0.825 ms}$$

17.5 VOLTAGE-CONTROLLED OSCILLATOR

A voltage-controlled oscillator (VCO) is a circuit that provides a varying output
signal (typically of square-wave or triangular-wave form) whose frequency can be
adjusted over a range controlled by a dc voltage. An example of a VCO is the 566 IC
unit, which contains circuitry to generate both square-wave and triangular-wave signals
whose frequency is set by an external resistor and capacitor and then varied by an
applied dc voltage. Figure 17.21a shows that the 566 contains current sources to
charge and discharge an external capacitor C_1 at a rate set by external resistor R_1 and
the modulating dc input voltage. A Schmitt trigger circuit is used to switch the current
sources between charging and discharging the capacitor, and the triangular voltage
developed across the capacitor and square wave from the Schmitt trigger are provided
as outputs through buffer amplifiers.

Figure 17.21b shows the pin connection of the 566 unit and a summary of formula
and value limitations. The oscillator can be programmed over a 10-to-1 frequency
range by proper selection of an external resistor and capacitor, and then modulated
over a 10-to-1 frequency range by a control voltage, V_C.

A free-running or center-operating frequency, f_O, can be calculated from

$$f_O = \frac{2}{R_1C_1}\left(\frac{V^+ - V_C}{V^+}\right) \tag{17.8}$$

with the following practical circuit value restrictions:

1. R_1 should be within the range $2 \text{ k}\Omega \leq R_1 \leq 20 \text{ k}\Omega$.
2. V_C should be within range $\frac{3}{4}V^+ \leq V_C \leq V^+$.

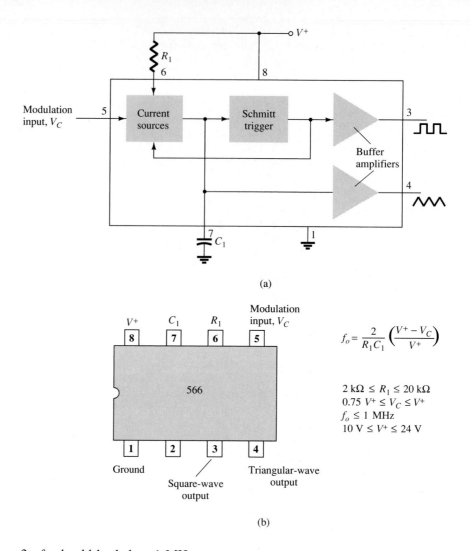

(a)

$$f_o = \frac{2}{R_1 C_1}\left(\frac{V^+ - V_C}{V^+}\right)$$

$2\ \text{k}\Omega \le R_1 \le 20\ \text{k}\Omega$
$0.75\ V^+ \le V_C \le V^+$
$f_o \le 1\ \text{MHz}$
$10\ \text{V} \le V^+ \le 24\ \text{V}$

Ground

Square-wave
output

Triangular-wave
output

Figure 17.21 A 566 function generator: (a) block diagram; (b) pin configuration and summary of operating data.

(b)

3. f_O should be below 1 MHz.

4. V^+ should range between 10 and 24 V.

Figure 17.22 shows an example in which the 566 function generator is used to provide both square-wave and triangular-wave signals at a fixed frequency set by R_1, C_1, and V_C. A resistor divider R_2 and R_3 sets the dc modulating voltage at a fixed value

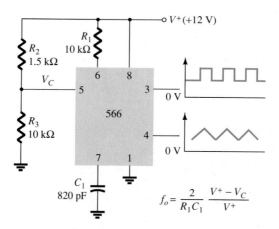

$$f_o = \frac{2}{R_1 C_1}\,\frac{V^+ - V_C}{V^+}$$

Figure 17.22 Connection of 566 VCO unit.

$$V_C = \frac{R_3}{R_2 + R_3} V^+ = \frac{10 \text{ k}\Omega}{1.5 \text{ k}\Omega + 10 \text{ k}\Omega}(12 \text{ V}) = 10.4 \text{ V}$$

(which falls properly in the voltage range $0.75V^+ = 9$ V and $V^+ = 12$ V). Using Eq. (17.8) yields

$$f_O = \frac{2}{(10 \times 10^3)(820 \times 10^{-12})}\left(\frac{12 - 10.4}{12}\right) \approx 32.5 \text{ kHz}$$

The circuit of Fig. 17.23 shows how the output square-wave frequency can be adjusted using the input voltage, V_C, to vary the signal frequency. Potentiometer R_3 allows varying V_C from about 9 V to near 12 V, over the full 10-to-1 frequency range. With the potentiometer wiper set at the top, the control voltage is

$$V_C = \frac{R_3 + R_4}{R_2 + R_3 + R_4}(V^+) = \frac{5 \text{ k}\Omega + 18 \text{ k}\Omega}{510 \text{ }\Omega + 5 \text{ k}\Omega + 18 \text{ k}\Omega}(+12 \text{ V}) = 11.74 \text{ V}$$

resulting in a lower output frequency of

$$f_O = \frac{2}{(10 \times 10^3)(220 \times 10^{-12})}\left(\frac{12 - 11.74}{12}\right) \approx 19.7 \text{ kHz}$$

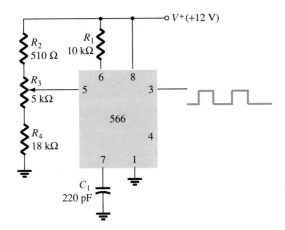

Figure 17.23 Connection of 566 as a VCO unit.

With the wiper arm of R_3 set at the bottom, the control voltage is

$$V_C = \frac{R_4}{R_2 + R_3 + R_4}(V^+) = \frac{18 \text{ k}\Omega}{510 \text{ }\Omega + 5 \text{ k}\Omega + 18 \text{ k}\Omega}(+12 \text{ V}) = 9.19 \text{ V}$$

resulting in an upper frequency of

$$f_O = \frac{2}{(10 \times 10^3)(220 \times 10^{-12})}\left(\frac{12 - 9.19}{12}\right) \approx 212.9 \text{ kHz}$$

The frequency of the output square wave can then be varied using potentiometer R_3 over a frequency range of at least 10 to 1.

Rather than varying a potentiometer setting to change the value of V_C, an input modulating voltage, V_{in}, can be applied as shown in Fig. 17.24. The voltage divider sets V_C at about 10.4 V. An input ac voltage of about 1.4 V peak can drive V_C around the bias point between voltages of 9 and 11.8 V, causing the output frequency to vary over about a 10-to-1 range. The input signal V_{in} thus frequency-modulates the output voltage around the center frequency set by the bias value of $V_C = 10.4$ V ($f_O = 121.2$ kHz).

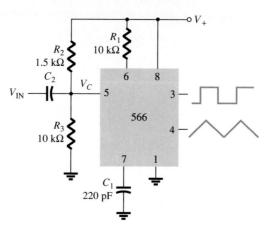

Figure 17.24 Operation of VCO with frequency-modulating input.

17.6 PHASE-LOCKED LOOP

A phase-locked loop (PLL) is an electronic circuit that consists of a phase detector, a low-pass filter, and a voltage-controlled oscillator connected as shown in Fig. 17.25. Common applications of a PLL include: (1) frequency synthesizers that provide multiples of a reference signal frequency [e.g., the carrier frequency for the multiple channels of a citizens' band (CB) unit or marine-radio-band unit can be generated using a single-crystal-controlled frequency and its multiples generated using a PLL]; (2) FM demodulation networks for FM operation with excellent linearity between the input signal frequency and the PLL output voltage; (3) demodulation of the two data transmission or carrier frequencies in digital-data transmission used in frequency-shift keying (FSK) operation; and (4) a wide variety of areas including modems, telemetry receivers and transmitters, tone decoders, AM detectors, and tracking filters.

An input signal, V_i, and that from a VCO, V_o, are compared by a phase comparator (refer to Fig. 17.25) providing an output voltage, V_e, that represents the phase difference between the two signals. This voltage is then fed to a low-pass filter that

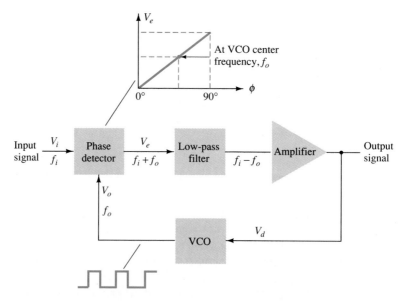

Figure 17.25 Block diagram of basic phase-locked loop (PLL).

provides an output voltage (amplified if necessary) that can be taken as the output voltage from the PLL and is used internally as the voltage to modulate the VCO's frequency. The closed-loop operation of the circuit is to maintain the VCO frequency locked to that of the input signal frequency.

Basic PLL Operation

The basic operation of a PLL circuit can be explained using the circuit of Fig. 17.25 as reference. We will first consider the operation of the various circuits in the phase-locked loop when the loop is operating in lock (the input signal frequency and the VCO frequency are the same). When the input signal frequency is the same as that from the VCO to the comparator, the voltage, V_d, taken as output is the value needed to hold the VCO in lock with the input signal. The VCO then provides output of a fixed-amplitude square-wave signal at the frequency of the input. Best operation is obtained if the VCO center frequency, f_O, is set with the dc bias voltage midway in its linear operating range. The amplifier allows this adjustment in dc voltage from that obtained as output of the filter circuit. When the loop is in lock, the two signals to the comparator are of the same frequency, although not necessarily in phase. A fixed phase difference between the two signals to the comparator results in a fixed dc voltage to the VCO. Changes in the input signal frequency then result in change in the dc voltage to the VCO. Within a capture-and-lock frequency range, the dc voltage will drive the VCO frequency to match that of the input.

While the loop is trying to achieve lock, the output of the phase comparator contains frequency components at the sum and difference of the signals compared. A low-pass filter passes only the lower-frequency component of the signal so that the loop can obtain lock between input and VCO signals.

Owing to the limited operating range of the VCO and the feedback connection of the PLL circuit, there are two important frequency bands specified for a PLL. The capture range of a PLL is the frequency range centered about the VCO free-running frequency, f_O, over which the loop can acquire lock with the input signal. Once the PLL has achieved capture, it can maintain lock with the input signal over a somewhat wider frequency range called the *lock range*.

Applications

The PLL can be used in a wide variety of applications, including (1) frequency demodulation, (2) frequency synthesis, and (3) FSK decoders. Examples of each of these follow.

FREQUENCY DEMODULATION

FM demodulation or detection can be directly achieved using the PLL circuit. If the PLL center frequency is selected or designed at the FM carrier frequency, the filtered or output voltage of the circuit of Fig. 17.25 is the desired demodulated voltage, varying in value proportional to the variation of the signal frequency. The PLL circuit thus operates as a complete intermediate-frequency (IF) strip, limiter, and demodulator as used in FM receivers.

One popular PLL unit is the 565, shown in Fig. 17.26a. The 565 contains a phase detector, amplifier, and voltage-controlled oscillator, which are only partially connected internally. An external resistor and capacitor, R_1 and C_1, are used to set the free-running or center frequency of the VCO. Another external capacitor, C_2, is used to set the low-pass filter passband, and the VCO output must be connected back as input to the phase detector to close the PLL loop. The 565 typically uses two power supplies, V^+ and V^-.

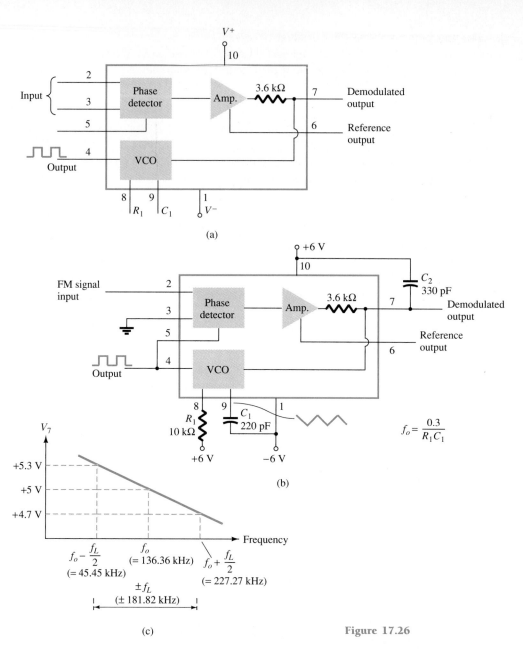

(a)

(b)

$$f_o = \frac{0.3}{R_1 C_1}$$

(c)

Figure 17.26

Figure 17.26b shows the PLL connected to work as an FM demodulator. Resistor R_1 and capacitor C_1 set the free-running frequency, f_O,

$$f_O = \frac{0.3}{R_1 C_1} \tag{17.9}$$

$$= \frac{0.3}{(10 \times 10^3)(220 \times 10^{-12})} = 136.36 \text{ kHz}$$

with limitation $2 \text{ k}\Omega \le R_1 \le 20 \text{ k}\Omega$. The lock range is

$$f_L = \pm\frac{8f_O}{V}$$

$$= \pm\frac{8(136.36 \times 10^3)}{6} = \pm 181.8 \text{ kHz}$$

for supply voltages $V = \pm 6$ V. The capture range is

$$f_C = \pm \frac{1}{2\pi} \sqrt{\frac{2\pi f_L}{R_2 C_2}} \tag{17.10}$$

$$= \pm \frac{1}{2\pi} \sqrt{\frac{2\pi(181.8 \times 10^3)}{(3.6 \times 10^3)(330 \times 10^{-12})}} = 156.1 \text{ kHz}$$

The signal at pin 4 is a 136.36-kHz square wave. An input within the lock range of 181.8 kHz will result in the output at pin 7 varying around its dc voltage level set with input signal at f_O. Figure 17.26c shows the output at pin 7 as a function of the input signal frequency. The dc voltage at pin 7 is linearly related to the input signal frequency within the frequency range $f_L = 181.8$ kHz around the center frequency 136.36 kHz. The output voltage is the demodulated signal that varies with frequency within the operating range specified.

FREQUENCY SYNTHESIS

A frequency synthesizer can be built around a PLL as shown in Fig. 17.27. A frequency divider is inserted between the VCO output and the phase comparator so that the loop signal to the comparator is at frequency f_O while the VCO output is Nf_O. This output is a multiple of the input frequency as long as the loop is in lock. The input signal can be stabilized at f_1 with the resulting VCO output at Nf_1 if the loop is

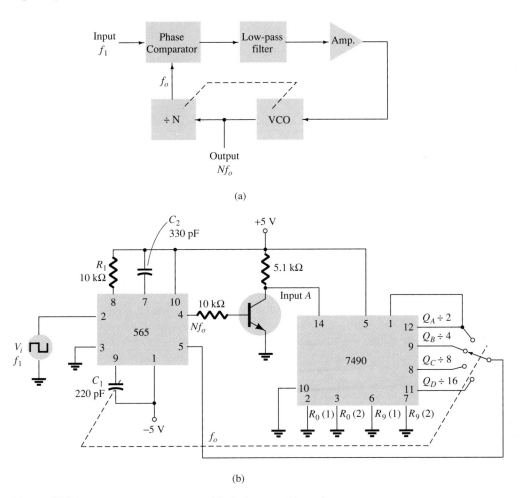

Figure 17.27 Frequency synthesizer: (a) block diagram; (b) implementation using 565 PLL unit.

set up to lock at the fundamental frequency (when $f_O = f_1$). Figure 17.27b shows an example using a 565 PLL as frequency multiplier and a 7490 as divider. The input V_i at frequency f_1 is compared to the input (frequency f_O) at pin 5. An output at Nf_O ($4f_O$ in the present example) is connected through an inverter circuit to provide an input at pin 14 of the 7490, which varies between 0 and +5 V. Using the output at pin 9, which is divided by 4 from that at the input to the 7490, the signal at pin 4 of the PLL is four times the input frequency as long as the loop remains in lock. Since the VCO can only vary over a limited range from its center frequency, it may be necessary to change the VCO frequency whenever the divider value is changed. As long as the PLL circuit is in lock, the VCO output frequency will be exactly N times the input frequency. It is only necessary to readjust f_O to be within the capture-and-lock range, the closed loop then resulting in the VCO output becoming exactly Nf_1 at lock.

FSK DECODERS

An FSK (frequency-shift keyed) signal decoder can be built as shown in Fig. 17.28. The decoder receives a signal at one of two distinct carrier frequencies, 1270 Hz or 1070 Hz, representing the RS-232C logic levels or mark (-5 V) or space ($+14$ V), respectively. As the signal appears at the input, the loop locks to the input frequency and tracks it between two possible frequencies with a corresponding dc shift at the output.

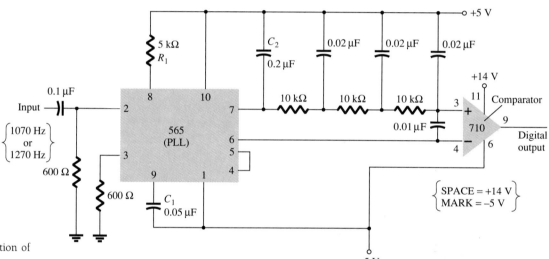

Figure 17.28 Connection of 565 as FSK decoder.

The *RC* ladder filter (three sections of $C = 0.02$ μF and $R = 10$ kΩ) is used to remove the sum frequency component. The free-running frequency is adjusted with R_1 so that the dc voltage level at the output (pin 7) is the same as that at pin 6. Then an input at frequency 1070 Hz will drive the decoder output voltage to a more positive voltage level, driving the digital output to the high level (space or $+14$ V). An input at 1270 Hz will correspondingly drive the 565 dc output less positive with the digital output, which then drops to the low level (mark or -5 V).

17.7   INTERFACING CIRCUITRY

Connecting different types of circuits, either in digital or analog circuits, may require some sort of interfacing circuit. An interface circuit may be used to drive a load or to obtain a signal as a receiver circuit. A driver circuit provides the output signal at voltage or current level suitable to operate a number of loads, or operate such devices as relays, displays, or power units. A receiver circuit essentially accepts an input signal, providing high input impedance to minimize loading of the input signal. Furthermore, the interface circuits may include strobing, which provides connecting the interface signals during specific times intervals established by the strobe.

Figure 17.29a shows a dual-line driver, each driver accepting input of TTL signals, providing output capable of driving TTL or MOS device circuits. This type of interface circuit comes in various forms, some as inverting and others as noninverting units. The circuit of Fig. 17.29b shows a dual-line receiver having both inverting and noninverting inputs so that either operating condition can be selected. As an example, connection of an input signal to the inverting input would result in an inverted output from the receiver unit. Connecting the input to the noninverting input would provide the same interfacing except that the output obtained would have the same polarity as the received signal. The driver-receiver unit of Fig. 17.29 provides an output when the strobe signal is present (high in this case).

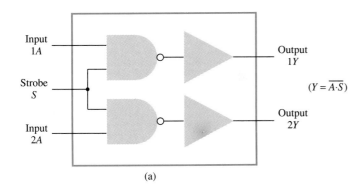

(a)

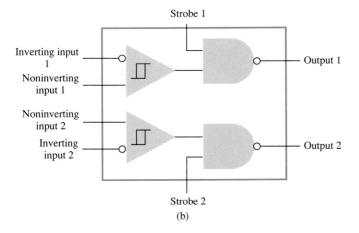

(b)

Figure 17.29 Interface units: (a) dual-line drivers (SN75150); (b) dual-line receivers (SN75152).

Another type of interface circuit is that used to connect various digital input and output units, signals with devices such as keyboards, video terminals, and printers. One of the EIA electronic industry standards is referred to as RS-232C. This standard states that a digital signal represents a mark (logic-1) and a space (logic-0). The definitions of mark and space vary with the type of circuit used (although a full reading of the standard will spell out the acceptable limits of mark and space signals).

RS-232C-to-TTL Converter

For TTL circuits +5 V is a mark and 0 V is a space. For RS-232C a mark could be −12 V and a space +12 V. Figure 17.30a provides a tabulation of some mark and space definitions. For a unit having outputs defined by RS-232C that is to operate into another unit operating with a TTL signal level, an interface circuit as shown in Fig. 30b could be used. A mark output from the driver (at −12 V) would be clipped by the diode so that the input to the inverter circuit is near 0 V, resulting in an output of +5 V (TTL mark). A space output at +12 V would drive the inverter output low for a 0-V output (a space).

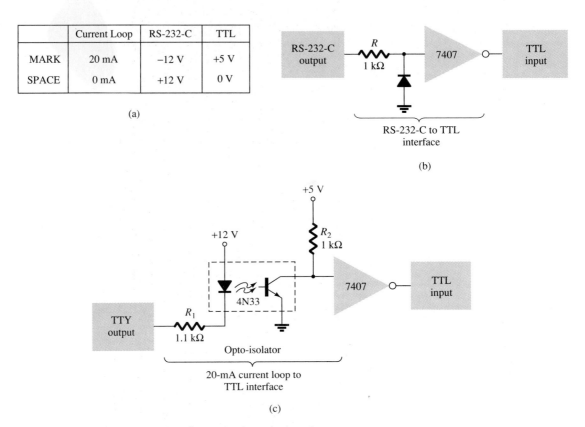

	Current Loop	RS-232-C	TTL
MARK	20 mA	−12 V	+5 V
SPACE	0 mA	+12 V	0 V

(a)

RS-232-C to TTL interface

(b)

Opto-isolator

20-mA current loop to TTL interface

(c)

Figure 17.30 Interfacing signal standards and converter circuits.

Another example of an interface circuit converts the signals from a TTY current loop into TTL levels as shown in Fig. 17.30c. An input mark results when 20 mA of current is drawn from the source through the output line of the teletype (TTY). This current then goes through the diode element of an opto-isolator, driving the output transistor on. The input to the inverter going low results in a +5-V signal from the 7407 inverter output, so that a mark from the teletype results in a mark to the TTL input. A space from the teletype current loop provides no current, with the opto-isolator transistor remaining off and the inverter output then 0 V, which is a TTL space signal.

Another means of interfacing digital signals is made using open-collector output or tri-state output. When a signal is output from a transistor collector (see Fig. 17.31) which is not connected to any other electronic component, the output is open-collector. This permits connecting a number of signals to the same wire or bus. Any transistor going on then provides a low output voltage, while all transistors remaining off provides a high output voltage.

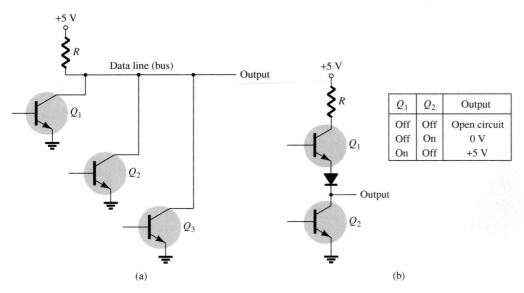

Figure 17.31   Connections to data lines: (a) open-collector output; (b) tri-state output.

Q_1	Q_2	Output
Off	Off	Open circuit
Off	On	0 V
On	Off	+5 V

17.8   COMPUTER ANALYSIS

A comparator is essentially a high-gain op-amp with some built-in hysteresis which provides a digital output with one or two saturation voltage levels, depending on the input applied. A good comparator can be obtained in PSpice using a practical op-amp subcircuit. The subcircuit contains all the details describing the components of the desired circuit, this subcircuit then used in any desired connection. The PSpice program supplied by MicroSim Corp. provides the detailed description of a 741 op-amp.

Program 17.1: Comparator Circuit Used to Drive an LED

A comparator circuit with output driving an LED is shown in Fig. 17.32. The PSpice listing and output of the comparator circuit are provided in Fig. 17.33.

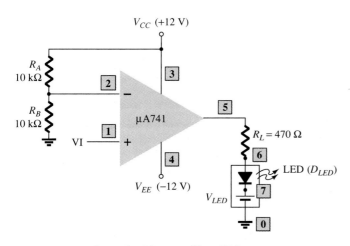

Figure 17.32   Circuit for PSpice problem 17.1.

```
 Comparator Driving LED
 ****      CIRCUIT DESCRIPTION
 **************************************************************
 * connections:   non-inverting input
 *                 |  inverting input
 *                 |   |  positive power supply
 *                 |   |   |  negative power supply
 *                 |   |   |   |  output
 *                 |   |   |   |   |
 .subckt uA741     1   2   3   4   5
   c1    11 12 8.661E-12
   c2     6  7 30.00E-12
   dc     5 53 dx
   de    54  5 dx
   dlp   90 91 dx
   dln   92 90 dx
   dp     4  3 dx
   egnd  99  0 poly(2) (3,0) (4,0) 0 .5 .5
   fb     7 99 poly(5) vb vc ve vlp vln 0 10.61E6 -10E6 10E6 10E6 -10E6
   ga     6  0 11 12 188.5E-6
   gcm    0  6 10 99 5.961E-9
   iee   10  4 dc 15.16E-6
   hlim  90  0 vlim 1K
   q1    11  2 13 qx
   q2    12  1 14 qx
   r2     6  9 100.0E3
   rc1    3 11 5.305E3
   rc2    3 12 5.305E3
   re1   13 10 1.836E3
   re2   14 10 1.836E3
   ree   10 99 13.19E6
   ro1    8  5 50
   ro2    7 99 100
   rp     3  4 18.16E3
   vb     9  0 dc 0
   vc     3 53 dc 1
   ve    54  4 dc 1
   vlim   7  8 dc 0
   vlp   91  0 dc 40
   vln    0 92 dc 40
 .model dx D(Is=800.0E-18 Rs=1)
 .model qx NPN(Is=800.0E-18 Bf=93.75)
 .ends
 XOPAMP 1 2 3 4 5 UA741
 VCC 3 0 +12V
 VEE 4 0 -12V
 VI 1 0 8V
 RA 3 2 10K
 RB 2 0 10K
 RL 5 6 470
 DLED 6 7 DL
 .MODEL DL D(IS=1E-14)
 VLED 7 0 1.5V
 .DC VI 4 8 1
 .PRINT DC I(DLED) V(5)
 .OPTIONS NOPAGE
 .END


 ****      DC TRANSFER CURVES                TEMPERATURE =    27.000 DEG C
   VI            I(DLED)      V(5)
    4.000E+00   -1.312E-11   -1.161E+01
    5.000E+00   -1.312E-11   -1.161E+01
    6.000E+00    1.962E-02    1.145E+01
    7.000E+00    1.996E-02    1.161E+01
    8.000E+00    1.997E-02    1.161E+01
```

Figure 17.33 PSpice output for circuit of Fig. 17.32.

Subcircuit: The subcircuit of the 741 op-amp is described in the lines from

.SUBCKT UA741 1 2 3 4 5 6

to

*End of library file

Since these lines were provided to us, we need only put our attention on how to use the op-amp as a comparator driving the LED.

A voltage divider provides a voltage to the minus input of 6 V, so that any input V_i below 6 V will result in the output at the minus saturation voltage (near -10 V), while inputs above $+6$ V result in the output going to the positive saturations level (near $+10$ V). The LED will therefore be driven on by any input above the reference level of $+6$ V. As output, a table of LED current for inputs from 4 to 8 V is provided to show that the LED current is near 0 for inputs up to $+6$ V and that about 16.5 mA lights the LED for inputs above $+6$ V.

Program 17.2: Comparator Operation

The operation of a comparator IC can be demonstrated using an LM111 unit (see Fig. 17.34). The description of the IC is obtained from the MicroSim library. Figure 17.35

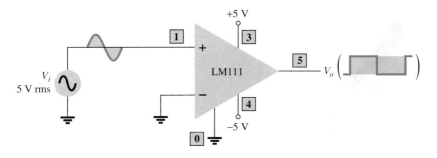

Figure 17.34 Circuit for PSpice problem 17.2.

```
**** 10/28/90 ******* Evaluation PSpice (January 1989) *******15:40:57 ****

 Comparator Operation Using LM111

****        CIRCUIT DESCRIPTION


*********************************************************************
* connections:    non-inverting input
*                 | inverting input
*                 | | positive power supply
*                 | | | negative power supply
*                 | | | | open collector output
*                 | | | | | output ground
*                 | | | | | |
.subckt LM111    1 2 3 4 5 6
*
    f1      9   3 v1 1
    iee     3   7 dc 100.0E-6
    vi1    21   1 dc .45
    vi2    22   2 dc .45
    q1      9  21   7 qin
    q2      8  22   7 qin
    q3      9   8   4 qmo
    q4      8   8   4 qmi
.model qin PNP(Is=800.0E-18 Bf=833.3)
.model qmi NPN(Is=800.0E-18 Bf=1002)
.model qmo NPN(Is=800.0E-18 Bf=1000 Cjc=1E-15 Tr=118.8E-9)
    e1     10   6   9   4 1
    v1     10  11 dc 0
    q5      5  11   6 qoc
.model qoc NPN(Is=800.0E-18 Bf=34.49E3 Cjc=1E-15 Tf=364.6E-12
Tr=79.34E-9)
    dp      4   3 dx
    rp      3   4 6.122E3
.model dx  D(Is=800.0E-18 Rs=1)
*
.ends
xcomp 1 0 3 4 5 0 lm111
VCC 3 0 +5V
VEE 4 0 -5V
RL 5 3 1K
VI 1 0 SIN(0 5)
.TRAN 1MS 20MS
.PLOT TRAN V(1) V(5)
.OPTIONS NOPAGE
.END
```

Figure 17.35 PSpice output for circuit of Fig. 17.34.

```
       LEGEND:
       *: V(1)
       +: V(5)
       TIME         V(1)
       (*)----------    -1.0000E+01   -5.0000E+00    0.0000E+00     5.0000E+00    1.0000E+01
       (+)----------     0.0000E+00    2.0000E+00    4.0000E+00     6.0000E+00    8.0000E+00

       0.000E+00    0.000E+00  .  _ _ _ _ _ +  . _ _ _ _ _ _  *  _ _ _ _ _ _ _ _ _ _ _ _  .
       1.000E-03    1.545E+00  .                    .          .       *    +      .         .
       2.000E-03    2.933E+00  .                    .          .            +*    .          .
       3.000E-03    4.044E+00  .                    .          .            +    *.          .
       4.000E-03    4.746E+00  .                    .          .            +     *.         .
       5.000E-03    4.999E+00  .                    .          .            +      *          .
       6.000E-03    4.746E+00  .                    .          .            +     *.         .
       7.000E-03    4.044E+00  .                    .          .            +    *.          .
       8.000E-03    2.933E+00  .                    .          .            +*    .          .
       9.000E-03    1.545E+00  .                    .          .       *    +      .         .
       1.000E-02    7.086E-05  .                    .          *          +       .          .
       1.100E-02   -1.543E+00  .+                   .       *  .                   .          .
       1.200E-02   -2.934E+00  .+                   .    *     .                   .          .
       1.300E-02   -4.039E+00  .+                 . *          .                   .          .
       1.400E-02   -4.748E+00  .+                .*            .                   .          .
       1.500E-02   -4.993E+00  .+                *             .                   .          .
       1.600E-02   -4.748E+00  .+                .*            .                   .          .
       1.700E-02   -4.039E+00  .+                 . *          .                   .          .
       1.800E-02   -2.935E+00  .+                   .    *     .                   .          .
       1.900E-02   -1.543E+00  .+                   .       *  .                   .          .
       2.000E-02    6.283E-10  .  +                 .          *                   .          .
                               _ _ _ _ _ _ _ _ _ _ _ _ _ _ _ _ _ _ _ _ _ _ _ _ _ _ _ - _ _
```

Figure 17.35 Continued.

provides the subcircuit description of the LM111 IC. The subcircuit is used to show how a comparator operates on a sinusoidal input signal. The input sinusoidal signal is a 5-V peak waveform

$$VI \ 1 \ 0 \ SIN(0 \ 5)$$

Since it is applied to the noninverting input, the output is in phase with the input. When the input goes above 0 V, the output goes to the positive saturation level, +5 V. When the input goes below 0 V, the output goes to the lower saturation level, 0 V. To show the input and output signals, Fig. 17.35 provides a plot of input, V(1), and output, V(5).

The output plot can be seen using the PROBE operation of PSpice as provided in the graphical display shown in Fig. 17.36.

Figure 17.36 Probe output for circuit of Fig. 17.34.

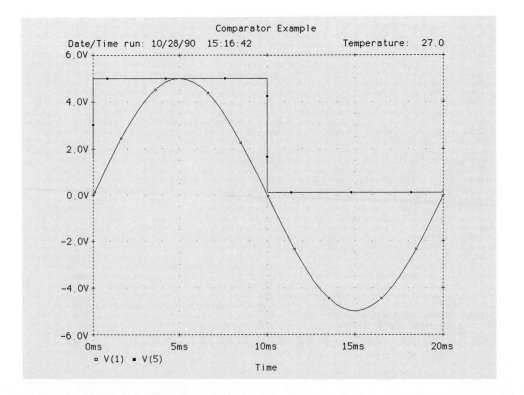

Program 17.3: Ladder Network

A ladder network as used in D/A conversion is shown in Fig. 17.37. This example of a four-stage ladder circuit converts four input bits (+10 V or 0 V) into an analog dc voltage. The listing in Fig. 17.38a shows that for all inputs of +10 V, the output is

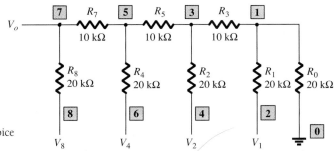

Figure 17.37 Circuit for PSpice problem 17.3.

Figure 17.38 PSpice output for circuit of Fig. 17.37.

```
**** 01/30/90 ******* Evaluation PSpice (January 1989) ******* 08:29:23 ****

 Ladder Network

****        CIRCUIT DESCRIPTION

***********************************************************************

R0 1 0 20K
R3 1 3 10K
R5 3 5 10K
R7 5 7 10K
R1 1 2 20K
R2 3 4 20K
R4 5 6 20K
R8 7 8 20K
V1 2 0 10V
V2 4 0 10V
V4 6 0 10V
V8 8 0 10V
.DC V8 10 10 1
.PRINT DC V(2) V(4) V(6) V(8) V(7)
.OPTIONS NOPAGE
.END


***      DC TRANSFER CURVES              TEMPERATURE =   27.000 DEG C
  V8          V(2)         V(4)       V(6)        V(8)        V(7)
   1.000E+01   1.000E+01   1.000E+01   1.000E+01   1.000E+01   9.375E+00
```

(a)

```
**** 01/30/90 ******* Evaluation PSpice (January 1989) ******* 08:29:23 ****

 Ladder Network

****        CIRCUIT DESCRIPTION

***********************************************************************

R0 1 0 20K
R3 1 3 10K
R5 3 5 10K
R7 5 7 10K
R1 1 2 20K
R2 3 4 20K
R4 5 6 20K
R8 7 8 20K
V1 2 0 10V
V2 4 0 0V
V4 6 0 0V
V8 8 0 10V
.DC V8 10 10 1
.PRINT DC V(2) V(4) V(6) V(8) V(7)
.OPTIONS NOPAGE
.END


***      DC TRANSFER CURVES              TEMPERATURE =   27.000 DEG C
  V8          V(2)         V(4)       V(6)        V(8)        V(7)
   1.000E+01   1.000E+01   0.000E+00   0.000E+00   1.000E+01   5.625E+00
```

(b)

9.375 V. For input 1　1　1　1 = 15, the output is

$$\frac{15}{16}(+10 \text{ V}) = 9.375 \text{ V}$$

Figure 17.38b shows that for inputs of 1　0　0　1 = 9, the output voltage is

$$\frac{9}{16}(+10 \text{ V}) = 5.625 \text{ V}$$

PROBLEMS

§ 17.2 Comparator Unit Operation

1. Draw the diagram of a 741 op-amp operated from ±15-V supplies with $V_i(-) = 0$ V and $V_i(+) = +5$ V. Include terminal pin connections.

2. Sketch the output waveform for the circuit of Fig. 17.39.

3. Draw a circuit diagram of a 311 op-amp showing an input of 10 V rms applied to the inverting input and the plus input to ground. Identify all pin numbers.

4. Draw the resulting output waveform for the circuit of Fig. 17.40.

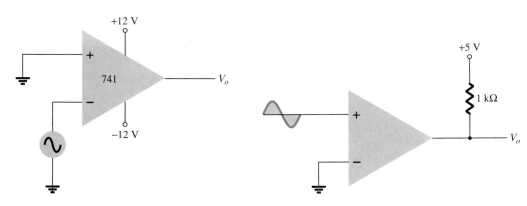

Figure 17.39　Problem 2　　　　　**Figure 17.40**　Problem 4

5. Draw the circuit diagram of a zero-crossing detector using a 339 comparator stage with ±12-V supplies.

6. Sketch the output waveform for the circuit of Fig. 17.41.

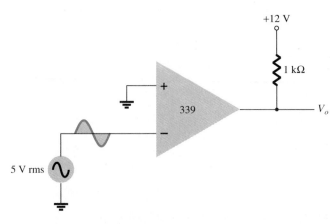

Figure 17.41　Problem 6

　　　　Chapter 17　**Linear-Digital ICs**

* 7. Describe the operation of the circuit in Fig. 17.42.

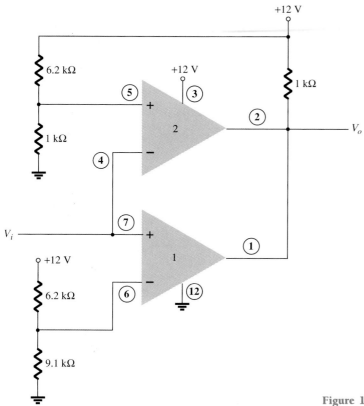

Figure 17.42 Problems 7, 27

§ 17.3 Digital–Analog Converters

8. Sketch a five-stage ladder network using 15-kΩ and 30-kΩ resistors.

9. For a reference voltage of 16 V, calculate the output voltage for an input of 11010 to the circuit of Problem 8.

10. What voltage resolution is possible using a 12-stage ladder network with a 10-V reference voltage?

11. For a dual-slope converter, describe what occurs during the fixed time interval and the count interval.

12. How many count steps occur using a 12-stage digital counter at the output of an A/D converter?

13. What is the maximum count interval using a 12-stage counter operated at a clock rate of 20 MHz?

§ 17.4 Timer IC Unit Operation

14. Sketch the circuit of a 555 timer connected as an astable multivibrator for operation at 350 kHz. Determine the value of capacitor, C, needed using $R_A = R_B = 7.5$ kΩ.

15. Draw the circuit of a one-shot using a 555 timer to provide one time period of 20 μs. If $R_A = 7.5$ kΩ, what value of C is needed?

16. Sketch the input and output waveforms for a one-shot using a 555 timer triggered by a 10-kHz clock for $R_A = 5.1$ kΩ and $C = 5$ nF.

§ 17.5 Voltage-Controlled Oscillator

17. Calculate the center frequency of a VCO using a 566 IC as in Fig. 17.22 for $R_1 = 4.7$ kΩ, $R_2 = 1.8$ kΩ, $R_3 = 11$ kΩ, and $C_1 = 0.001$ μF.

* **18.** What frequency range results in the circuit of Fig. 17.23 for $C_1 = 0.001 \ \mu F$?

19. Determine the capacitor needed in the circuit of Fig. 17.22 to obtain a 200-kHz output.

§ 17.6 Phase-Locked Loop

20. Calculate the VCO free-running frequency for the circuit of Fig. 17.26b with $R_1 = 4.7 \ k\Omega$ and $C_1 = 0.001 \ \mu F$.

21. What value capacitor, C_1, is required in the circuit of Fig. 17.26b to obtain a center frequency of 100 kHz?

22. What is the lock range of the PLL circuit in Fig. 17.26b for $R_1 = 4.7 \ k\Omega$ and $C_1 = 0.001 \ \mu F$?

§ 17.7 Interfacing Circuitry

23. Describe the signal conditions for current-loop and RS-232C interfaces.

24. What is a data bus?

25. What is the difference between open-collector and tri-state output?

§ 17.8 Computer Analysis

* **26.** Write a PSpice program to provide the output waveform for a circuit using an LM111 with $V_i = 5$ V, rms applied to minus ($-$) input and $+5$ V, rms applied to plus ($+$) input.

* **27.** Write a PSpice program to provide output listing for inputs from $V_i = 1$ V to 12 V (in 0.5 V steps) in the circuit of Fig. 17.42. Use IC with $R_i = 1 \ M\Omega$, $R_o = 10 \ k\Omega$, and $A = 100$ V/mV.

* **28.** Write a PSpice program to list the output voltage for a 2-stage ladder for all 4 input values (use 10 kΩ, 20 kΩ & 10 V as in Fig. 17.37).

---

*Please Note: Asterisks indicate more difficult problems.

Feedback and Oscillator Circuits

18

A_f

18.1 FEEDBACK CONCEPTS

Feedback has been mentioned previously. In particular, feedback was used in op-amp circuits as described in Chapters 14 and 15. Depending on the relative polarity of the signal being fed back into a circuit, one may have negative or positive feedback. Negative feedback results in decreased voltage gain, for which a number of circuit features are improved as summarized below. Positive feedback drives a circuit into oscillation as in various types of oscillator circuits.

A typical feedback connection is shown in Fig. 18.1. The input signal, V_s, is applied to a mixer network, where it is combined with a feedback signal, V_f. The difference of these signals, V_i, is then the input voltage to the amplifier. A portion of the amplifier output, V_o, is connected to the feedback network (β), which provides a reduced portion of the output as feedback signal to the input mixer network.

If the feedback signal is of opposite polarity to the input signal, as shown in Fig. 18.1, negative feedback results. While negative feedback results in reduced overall voltage gain, a number of improvements are obtained, among them being

1. Higher input impedance
2. Better stabilized voltage gain
3. Improved frequency response
4. Higher input impedance
5. Lower output impedance
6. Reduced noise
7. More linear operation

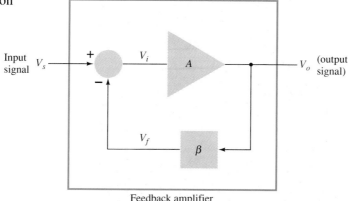

Feedback amplifier

Figure 18.1 Simple block diagram of feedback amplifier.

741

18.2 FEEDBACK CONNECTION TYPES

There are four basic ways of connecting the feedback signal. Both *voltage* and *current* can be fed back to the input either in *series* or *parallel*. Specifically, there can be

1. Voltage-series feedback (Fig. 18.2a)
2. Voltage-shunt feedback (Fig. 18.2b)
3. Current-series feedback (Fig. 18.2c)
4. Current-shunt feedback (Fig. 18.2d)

In the list above, *voltage* refers to connecting the output voltage as input to the feedback network; *current* refers to tapping off some output current through the feedback network. *Series* refers to connecting the feedback signal in series with the input signal voltage; *shunt* refers to connecting the feedback signal in shunt (parallel) with an input current source.

Series feedback connections tend to *increase* the input resistance while shunt feedback connections tend to *decrease* the input resistance. Voltage feedback tends to *decrease* the output impedance while current feedback tends to *increase* the output impedance. Typically, higher input and lower output impedances are desired for most

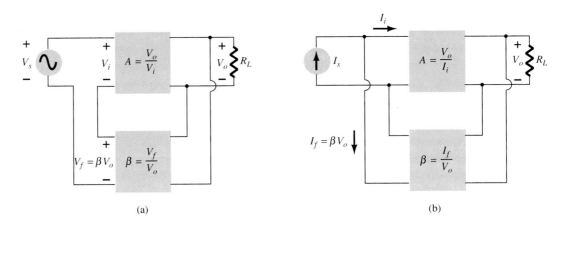

(a)　　　　　　　　　　　　　　　(b)

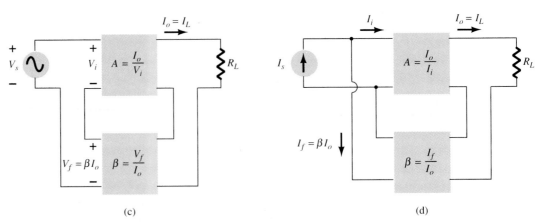

(c)　　　　　　　　　　　　　　　(d)

Figure 18.2 Feedback amplifier types: (a) voltage-series feedback, $A_f = V_o/V_s$; (b) voltage-shunt feedback, $A_f = V_o/I_s$; (c) current-series feedback, $A_f = I_o/V_s$; (d) current-shunt feedback, $A_f = I_o/I_s$.

cascade amplifiers. Both of these are provided using the voltage-series feedback connection. We shall therefore concentrate first on this amplifier connection.

Gain with Feedback

In this section we examine the gain of each of the feedback circuit connections of Fig. 18.2. The gain without feedback, A, is that of the amplifier stage. With feedback, β, the overall gain of the circuit is reduced by a factor $(1 + \beta A)$, as detailed below. A summary of the gain, feedback factor, and gain with feedback of Fig. 18.2 is provided for reference in Table 18.1.

TABLE 18.1 Summary of Gain, Feedback, and Gain with Feedback from Fig. 18.2

		Voltage-Series	Voltage-Shunt	Current-Series	Current-Shunt
Gain without feedback	A	$\dfrac{V_o}{V_i}$	$\dfrac{V_o}{I_i}$	$\dfrac{I_o}{V_i}$	$\dfrac{I_o}{I_i}$
Feedback	β	$\dfrac{V_f}{V_o}$	$\dfrac{I_f}{V_o}$	$\dfrac{V_f}{I_o}$	$\dfrac{I_f}{I_o}$
Gain with feedback	A_f	$\dfrac{V_o}{V_s}$	$\dfrac{V_o}{I_s}$	$\dfrac{I_o}{V_s}$	$\dfrac{I_o}{I_s}$

VOLTAGE SERIES

Figure 18.2a shows the voltage-series feedback connection with a part of the output voltage fed back in series with the input signal, resulting in an overall gain reduction. If there is no feedback ($V_f = 0$), the voltage gain of the amplifier stage is

$$A = \frac{V_o}{V_s} = \frac{V_o}{V_i} \tag{18.1}$$

If a feedback signal, V_f, is connected in series with the input, then

$$V_i = V_s - V_f \tag{18.2}$$

Since $\qquad V_o = AV_i = A(V_s - V_f) = AV_s - AV_f = AV_s - A(\beta V_o)$

then $\qquad (1 + \beta A)V_o = AV_s$

so that the overall voltage gain *with* feedback is

$$A_f = \frac{V_o}{V_s} = \frac{A}{1 + \beta A} \tag{18.3}$$

Equation (18.3) shows that the gain *with* feedback is the amplifier gain reduced by the factor $(1 + \beta A)$. This factor will be seen also to affect input and output impedance among other circuit features.

VOLTAGE SHUNT

The gain with feedback for the network of Fig. 18.2b is

$$A_f = \frac{V_o}{I_s} = \frac{AI_i}{I_i + I_f} = \frac{AI_i}{I_i + \beta V_o} = \frac{AI_i}{I_i + \beta AI_i}$$

$$= \frac{A}{1 + \beta A} \tag{18.4}$$

Input Impedance with Feedback

VOLTAGE-SERIES FEEDBACK

A more detailed voltage-series feedback connection is shown in Fig. 18.3. The input impedance can be determined as follows:

$$I_i = \frac{V_i}{Z_i} = \frac{V_s - V_f}{Z_i} = \frac{V_s - \beta V_o}{Z_i} = \frac{V_s - \beta A V_i}{Z_i}$$

$$I_i Z_i = V_s - \beta A V_i$$

$$V_s = I_i Z_i + \beta A V_i = I_i Z_i + \beta A I_i Z_i$$

$$Z_{if} = \frac{V_s}{I_i} = Z_i + (\beta A)Z_i = Z_i(1 + \beta A)$$

The input impedance with series feedback is seen to be the value of the input impedance without feedback multiplied by the factor $(1 + \beta A)$ and applies to both voltage-series (Fig. 18.2a) and current-series (Fig. 18.2c) configurations.

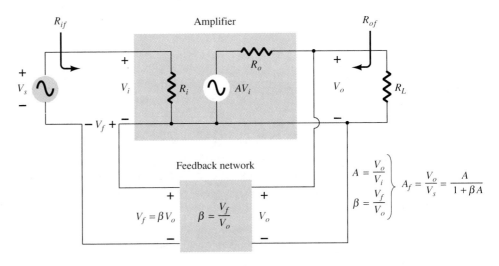

$$A = \frac{V_o}{V_i}$$
$$\beta = \frac{V_f}{V_o}$$
$$A_f = \frac{V_o}{V_s} = \frac{A}{1 + \beta A}$$

Figure 18.3 Voltage-series feedback connection.

VOLTAGE-SHUNT FEEDBACK

A more detailed voltage-shunt feedback connection is shown in Fig. 18.4. The input impedance can be determined to be

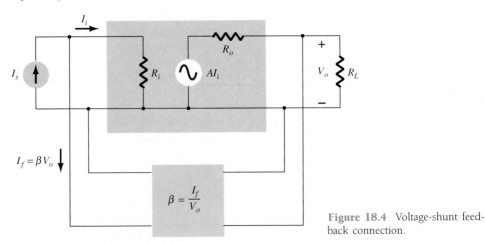

Figure 18.4 Voltage-shunt feedback connection.

Chapter 18 Feedback and Oscillator Circuits

$$Z_{if} = \frac{V_i}{I_s} = \frac{V_i}{I_i + I_f} = \frac{V_i}{I_i + \beta V_o}$$

$$= \frac{V_i/I_i}{I_i/I_i + \beta V_o/I_i}$$

$$= \frac{Z_i}{1 + \beta A} \tag{18.5}$$

This reduced input impedance applies to the voltage-series connection of Fig. 18.2a and the voltage-shunt connection of Fig. 18.2b.

Output Impedance with Feedback

The output impedance for the connections of Fig. 18.2 are dependent on whether voltage or current feedback is used. For voltage feedback the output impedance is decreased, while current feedback increases the output impedance.

VOLTAGE-SERIES FEEDBACK

The voltage-series feedback circuit of Fig. 18.3 provides sufficient circuit detail to determine the output impedance with feedback. The output impedance is determined by applying a voltage, V, resulting in a current, I, with V_s shorted out ($V_s = 0$). The voltage V is then

$$V = IZ_o + AV_i$$

For $V_s = 0$, $\qquad V_i = -V_f$

so that $\qquad V = IZ_o - AV_f = IZ_o - A(\beta V)$

Rewriting the equation as

$$V + \beta A V = IZ_o$$

allows solving for the output resistance with feedback:

$$Z_{of} = \frac{V}{I} = \frac{Z_o}{1 + \beta A} \tag{18.6}$$

Equation (18.6) shows that with voltage-series feedback the output impedance is reduced from that without feedback by the factor $(1 + \beta A)$.

CURRENT-SERIES FEEDBACK

The output impedance with current-series feedback can be determined by applying a signal V to the output with V_s shorted out, resulting in a current I, the ratio of V to I being the output impedance. Figure 18.5 shows a more detailed connection with current-series feedback. For the output part of a current-series feedback connection shown in Fig. 18.5, the resulting output impedance is determined as follows. With $V_s = 0$,

$$V_i = V_f$$

$$I = \frac{V}{Z_o} - AV_i = \frac{V}{Z_o} - AV_f = \frac{V}{Z_o} - A\beta I$$

$$Z_o(1 + \beta A)I = V$$

$$Z_{of} = \frac{V}{I} = Z_o(1 + \beta A) \tag{18.7}$$

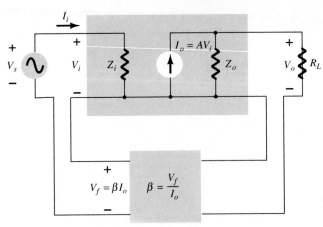

Figure 18.5 Current-series feedback connection.

A summary of the effect of feedback on input and output impedance is provided in Table 18.2.

TABLE 18.2 Effect of Feedback Connection on Input and Output Impedance

	Voltage-Series	Current-Series	Voltage-Shunt	Current-Shunt
Z_{if}	$Z_i(1 + \beta A)$	$Z_i(1 + \beta A)$	$\dfrac{Z_i}{1 + \beta A}$	$\dfrac{Z_i}{1 + \beta A}$
	(increased)	(increased)	(decreased)	(decreased)
Z_{of}	$\dfrac{Z_o}{1 + \beta A}$	$Z_o(1 + \beta A)$	$\dfrac{Z_o}{1 + \beta A}$	$Z_o(1 + \beta A)$
	(decreased)	(increased)	(decreased)	(increased)

EXAMPLE 18.1

Determine the voltage gain, input, and output impedance with feedback for voltage-series feedback having $A = -100$, $R_i = 10$ kΩ, $R_o = 20$ kΩ for feedback of (a) $\beta = -0.1$ and (b) $\beta = -0.5$.

Solution

Using Eqs. (18.3), (18.5), and (18.7), we obtain

(a) $A_f = \dfrac{A}{1 + \beta A} = \dfrac{-100}{1 + (-0.1)(-100)} = \dfrac{-100}{11} = \mathbf{-9.09}$

$Z_{if} = Z_i(1 + \beta A) = 10 \text{ k}\Omega(11) = \mathbf{110 \text{ k}\Omega}$

$Z_{of} = \dfrac{Z_o}{1 + \beta A} = \dfrac{20 \times 10^3}{11} = \mathbf{1.82 \text{ k}\Omega}$

(b) $A_f = \dfrac{A}{1 + \beta A} = \dfrac{-100}{1 + (0.5)(100)} = \dfrac{-100}{51} = \mathbf{-1.96}$

$Z_{if} = Z_i(1 + \beta A) = 10 \text{ k}\Omega(51) = \mathbf{510 \text{ k}\Omega}$

$Z_{of} = \dfrac{Z_o}{1 + \beta A} = \dfrac{20 \times 10^3}{51} = \mathbf{392.16 \text{ }\Omega}$

Example 18.1 demonstrates the trade-off of gain for improved input and output resistance. Reducing the gain by a factor of 11 (from 100 to 9.09) is complimented by a reduced output resistance and increased input resistance by the same factor of 11. Reducing the gain by a factor of 51 provides a gain of only 2 but with input resistance increased by the factor of 51 (to over 500 kΩ) and output resistance reduced from 20 kΩ to under 400 Ω. Feedback offers the designer the choice of trading away some of the available amplifier gain for other improved circuit features.

Reduction in Frequency Distortion

For a negative-feedback amplifier having $\beta A \gg 1$, the gain with feedback is $A_f \cong 1/\beta$. It follows from this that if the feedback network is purely resistive, the gain with feedback is not dependent on frequency even though the basic amplifier gain is frequency dependent. Practically, the frequency distortion arising because of varying amplifier gain with frequency is considerably reduced in a negative-voltage feedback amplifier circuit.

Reduction in Noise and Nonlinear Distortion

Signal feedback tends to hold down the amount of noise signal (such as power-supply hum) and nonlinear distortion. The factor $(1 + \beta A)$ reduces both input noise and resulting nonlinear distortion for considerable improvement. However, it should be noted that there is a reduction in overall gain (the price required for the improvement in circuit performance). If additional stages are used to bring the overall gain up to the level without feedback, it should be noted that the extra stage(s) might introduce as much noise back into the system as that reduced by the feedback amplifier. This problem can be somewhat alleviated by readjusting the gain of the feedback-amplifier circuit to obtain higher gain while also providing reduced noise signal.

Effect of Negative Feedback on Gain and Bandwidth

In Eq. (18.3) the overall gain with negative feedback is shown to be

$$A_f = \frac{A}{1 + \beta A} \cong \frac{A}{\beta A} = \frac{1}{\beta} \qquad \text{for } \beta A \gg 1$$

As long as $\beta A \gg 1$ the overall gain is approximately $1/\beta$. We should realize that for a practical amplifier (for single low- and high-frequency breakpoints) the open-loop gain drops off at high frequencies due to the active device and circuit capacitances. Gain may also drop off at low frequencies for capacitively coupled amplifier stages. Once the open-loop gain A drops low enough and the factor βA is no longer much larger than 1, the conclusion of Eq. (18.3) that $A_f \cong 1/\beta$ no longer holds true.

Figure 18.6 shows that the amplifier with negative feedback has more bandwidth (B_f) than the amplifier without feedback (B). The feedback amplifier has a higher upper 3-dB frequency and smaller lower 3-dB frequency.

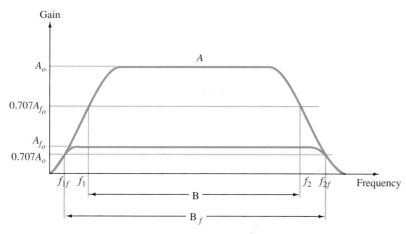

Figure 18.6 Effect of negative feedback on gain and bandwidth.

18.2 Feedback Connection Types

It is interesting to note that the use of feedback, while resulting in a lowering of voltage gain, has provided an increase in B and in the upper 3-dB frequency, particularly. In fact, the product of gain and frequency remains the same so that the gain–bandwidth product of the basic amplifier is the same value for the feedback amplifier. However, since the feedback amplifier has lower gain, the net operation was to *trade* gain for bandwidth (we use bandwidth for the upper 3-dB frequency since typically $f_2 \gg f_1$).

Gain Stability with Feedback

In addition to the β factor setting a precise gain value, we are also interested in how stable the feedback amplifier is compared to an amplifier without feedback. Differentiating Eq. (18.3) leads to

$$\left| \frac{dA_f}{A_f} \right| = \frac{1}{|1 + \beta A|} \left| \frac{dA}{A} \right| \tag{18.8}$$

$$\left| \frac{dA_f}{A_f} \right| \cong \left| \frac{1}{\beta A} \right| \left| \frac{dA}{A} \right| \qquad \text{for } \beta A \gg 1 \tag{18.9}$$

This shows that magnitude of the relative change in gain $\left| \dfrac{dA_f}{A_f} \right|$ is reduced by the factor $|\beta A|$ compared to that without feedback $\left(\left| \dfrac{dA}{A} \right| \right)$.

EXAMPLE 18.2

If an amplifier with gain of -1000 and feedback of $\beta = -0.1$ has a gain change of 20% due to temperature, calculate the change in gain of the feedback amplifier.

Solution

Using Eq. (18.9), we get

$$\left| \frac{dA_f}{A_f} \right| \cong \left| \frac{1}{\beta A} \right| \left| \frac{dA}{A} \right| = \left| \frac{1}{-0.1(-1000)} (20\%) \right| = \mathbf{0.2\%}$$

The improvement is 100 times. Thus, while the amplifier gain changes from $|A| = 1000$ by 20%, the gain with feedback changes from $|A_f| = 100$ by only 0.2%.

18.3 PRACTICAL FEEDBACK CIRCUITS

Examples of practical feedback circuits will provide a means of demonstrating the effect feedback has on the various connection types. This section provides only a basic introduction to this topic.

Voltage-Series Feedback

Figure 18.7 shows a FET amplifier stage with voltage-series feedback. A part of the output signal (V_o) is obtained using a feedback network of resistors R_1 and R_2. The feedback voltage V_f is connected in series with the source signal V_s, their difference being the input signal V_i.

Without feedback the amplifier gain is

$$A = \frac{V_o}{V_i} = -g_m R_L \tag{18.10}$$

where R_L is the parallel combination of resistors

Chapter 18 Feedback and Oscillator Circuits

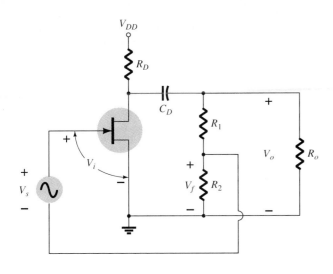

Figure 18.7 FET amplifier stage with voltage-series feedback.

$$R_L = R_D \| R_o \| (R_1 + R_2) \tag{18.11}$$

The feedback network provides a feedback factor of

$$\beta = \frac{V_f}{V_o} = \frac{-R_2}{R_1 + R_2} \tag{18.12}$$

Using the values of A and β above in Eq. (18.3), we find the gain with negative feedback to be

$$A_f = \frac{A}{1 + \beta A} = \frac{-g_m R_L}{1 + [R_2 R_L/(R_1 + R_2)]g_m} \tag{18.13}$$

If $\beta A \gg 1$, we have

$$A_f \cong \frac{1}{\beta} = -\frac{R_1 + R_2}{R_2} \tag{18.14}$$

Calculate the gain without and with feedback for the FET amplifier circuit of Fig. 18.7 and the following circuit values: $R_1 = 80$ kΩ, $R_2 = 20$ kΩ, $R_o = 10$ kΩ, $R_D = 10$ kΩ, and $g_m = 4000$ μS.

EXAMPLE 18.3

Solution

$$R_L \cong \frac{R_o R_D}{R_o + R_D} = \frac{10 \text{ k}\Omega \ (10 \text{ k}\Omega)}{10 \text{ k}\Omega + 10 \text{ k}\Omega} = 5 \text{ k}\Omega$$

(neglecting 100 kΩ resistance of R_1 and R_2 in series)

$$A = -g_m R_L = -(4000 \times 10^{-6})(5 \text{ k}\Omega) = \mathbf{-20}$$

The feedback factor is

$$\beta = \frac{-R_2}{R_1 + R_2} = \frac{-20}{80 + 20} = -0.2$$

The gain with feedback is

$$A_f = \frac{A}{1 + \beta A} = \frac{-20}{1 + (-0.2)(-20)} = \frac{-20}{5} = \mathbf{-4}$$

Figure 18.8 shows a voltage-series feedback connection using an op-amp. The gain of the op-amp, A, without feedback, is reduced by the feedback factor

$$\beta = \frac{-R_2}{R_1 + R_2} \qquad (18.15)$$

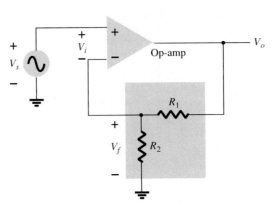

Figure 18.8 Voltage-series feedback in an op-amp connection.

EXAMPLE 18.4

Calculate the amplifier gain of the circuit of Fig. 18.8 for op-amp gain $A = 100,000$ and resistances $R_1 = 1.8$ kΩ and $R_2 = 200$ Ω.

Solution

$$\beta = \frac{R_2}{R_1 + R_2} = \frac{200}{200\ \Omega + 1.8\ \text{k}\Omega} = 0.1$$

$$A_f = \frac{A}{1 + \beta A} = \frac{100,000}{1 + (0.1)(100,000)}$$

$$= \frac{100,000}{10,001} = 9.999$$

Note that since $\beta A \gg 1$,

$$A_f \cong \frac{1}{\beta} = \frac{1}{0.1} = \mathbf{10}$$

The emitter-follower circuit of Fig. 18.9 provides voltage-series feedback. The signal voltage, V_s, is the input voltage, V_i. The output voltage, V_o, is also the feed-

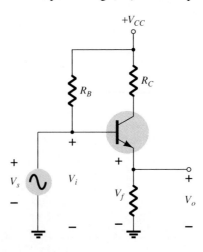

Figure 18.9 Voltage-series feedback circuit (emitter-follower).

back voltage in series with the input voltage. The amplifier, as shown in Fig. 18.9, provides the operation *with* feedback. The operation of the circuit without feedback provides $V_f = 0$, so that

$$A = \frac{V_o}{V_s} = \frac{h_{fe}I_bR_E}{V_s} = \frac{h_{fe}R_E(V_s/h_{ie})}{V_s} = \frac{h_{fe}R_E}{h_{ie}}$$

and

$$\beta = \frac{V_f}{V_o} = 1$$

The operation with feedback then provides that

$$A_f = \frac{V_o}{V_s} = \frac{A}{1 + \beta A} = \frac{h_{fe}R_E/h_{ie}}{1 + (1)(h_{fe}R_E/h_{ie})}$$

$$= \frac{h_{fe}R_E}{h_{ie} + h_{fe}R_E}$$

For $h_{fe}R_E \gg h_{ie}$,

$$A_f \cong 1$$

Current-Series Feedback

Another feedback technique is to sample the output current (I_o) and return a proportional voltage in series with the input. While stabilizing the amplifier gain, the current-series feedback connection increases input resistance.

Figure 18.10 shows a single transistor amplifier stage. Since the emitter of this stage has an unbypassed emitter, it effectively has current-series feedback. The current through resistor R_E results in a feedback voltage that opposes the source signal applied so that the output voltage V_o is reduced. To remove the current-series feedback the emitter resistor must be either removed or bypassed by a capacitor (as is usually done).

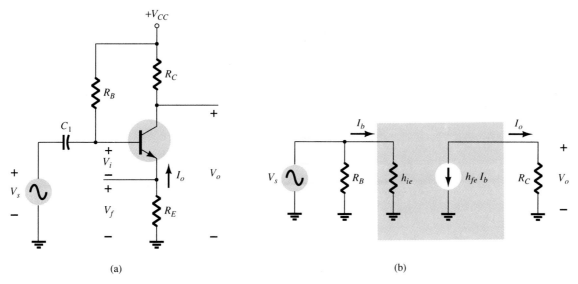

(a)   (b)

Figure 18.10 Transistor amplifier with unbypassed emitter resistor (R_E) for current-series feedback: (a) amplifier circuit; (b) ac equivalent circuit without feedback.

18.3   **Practical Feedback Circuits**

WITHOUT FEEDBACK

Referring to the basic format of Fig. 18.2a and summarized in Table 18.1, we have

$$A = \frac{I_o}{V_i} = \frac{-I_b h_{fe}}{I_b h_{ie} + R_E} = \frac{-h_{fe}}{h_{ie} + R_E} \qquad (18.16)$$

$$\beta = \frac{V_f}{I_o} = \frac{-I_o R_E}{I_o} = -R_E \qquad (18.17)$$

The input and output impedances are

$$Z_i = R_B \parallel (h_{ie} + R_E) \cong h_{ie} + R_E \qquad (18.18)$$

$$Z_o = R_C \qquad (18.19)$$

WITH FEEDBACK

$$A_f = \frac{I_o}{V_s} = \frac{A}{1 + \beta A} = \frac{-h_{fe}/h_{ie}}{1 + (-R_E)\left(\dfrac{-h_{fe}}{h_{ie} + R_E}\right)} \cong \frac{-h_{fe}}{h_{ie} + h_{fe}R_E} \qquad (18.20)$$

The input and output impedance is calculated as specified in Table 18.2.

$$Z_{if} = Z_i(1 + \beta A) \cong h_{ie}\left(1 + \frac{h_{fe}R_E}{h_{ie}}\right) = h_{ie} + h_{fe}R_E \qquad (18.21)$$

$$Z_{of} = Z_o(1 + \beta A) = R_C\left(1 + \frac{h_{fe}R_E}{h_{ie}}\right) \qquad (18.22)$$

The voltage gain (A) with feedback is

$$A = \frac{V_o}{V_s} = \frac{I_o R_C}{V_s} = \left(\frac{I_o}{V_s}\right)R_C = A_f R_C \cong \frac{-h_{fe}R_C}{h_{ie} + h_{fe}R_E} \qquad (18.23)$$

EXAMPLE 18.5

Calculate the voltage gain of the circuit of Fig. 18.11.

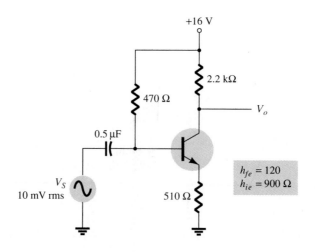

Figure 18.11 BJT amplifier with current-series feedback for Example 18.5.

Solution

Without feedback,

$$A = \frac{I_o}{V_i} = \frac{-h_{fe}}{h_{ie} + R_E} = \frac{-120}{900 + 510} = -0.085$$

$$\beta = \frac{V_f}{I_o} = -R_E = -510$$

The factor $(1 + \beta A)$ is then

$$1 + \beta A = 1 + (-0.085)(-510) = 43.35$$

The gain with feedback is then

$$A_f = \frac{I_o}{V_s} = \frac{A}{1 + \beta A} = \frac{-0.085}{43.35} = -1.96 \times 10^{-3}$$

and the voltage gain with feedback A_{vf} is

$$A_{vf} = \frac{V_o}{V_s} = A_f R_C = (-1.96 \times 10^{-3})(2.2 \times 10^3) = \mathbf{-4.3}$$

Without feedback ($R_E = 0$) the voltage gain is

$$A_v = \frac{-R_C}{r_e} = \frac{-2.2 \times 10^3}{7.5} = -293.3$$

Voltage-Shunt Feedback

The constant-gain op-amp circuit of Fig. 18.12a provides voltage-shunt feedback. Referring to Fig. 18.2b and Table 18.1 and the op-amp ideal characteristics $I_i = 0$, $V_i = 0$, and voltage gain of infinity; we have

$$A = \frac{V_o}{I_i} = \text{infinity} \tag{18.24}$$

$$\beta = \frac{I_f}{V_o} = \frac{-1}{R_o} \tag{18.25}$$

The gain with feedback is then

$$A_f = \frac{V_o}{I_s} = \frac{V_o}{I_i} = \frac{A}{1 + \beta A} = \frac{1}{\beta} = -R_o \tag{18.26}$$

This is a transfer resistance gain. The more usual gain is the voltage gain with feedback,

$$A_{vf} = \frac{V_o}{I_s} \frac{I_s}{V_1} = (-R_o)\frac{1}{R_1} = \frac{-R_o}{R_1} \tag{18.27}$$

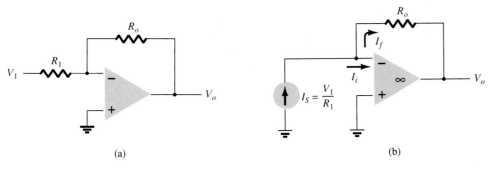

(a) (b)

Figure 18.12 Voltage-shunt negative feedback amplifier: (a) constant-gain circuit; (b) equivalent circuit.

The circuit of Fig. 18.13 is a voltage-shunt feedback amplifier using an FET with no feedback, $V_f = 0$.

$$A = \frac{V_o}{I_i} \cong -g_m R_D R_S \qquad (18.28)$$

The feedback is

$$\beta = \frac{I_f}{V_o} = \frac{-1}{R_F} \qquad (18.29)$$

With feedback, the gain of the circuit is

$$A_f = \frac{V_o}{I_s} = \frac{A}{1 + \beta A} = \frac{-g_m R_D R_S}{1 + (-1/R_F)(-g_m R_D R_S)}$$

$$= \frac{-g_m R_D R_S R_F}{R_F + g_m R_D R_S} \qquad (18.30)$$

The voltage gain of the circuit with feedback is then

$$A_{vf} = \frac{V_o}{I_s}\frac{I_s}{V_s} = \frac{-g_m R_D R_S R_F}{R_F + g_m R_D R_S}\left(\frac{1}{R_S}\right)$$

$$= \frac{-g_m R_D R_F}{R_F + g_m R_D R_S} = (-g_m R_D)\frac{R_F}{R_F + g_m R_D R_S} \qquad (18.31)$$

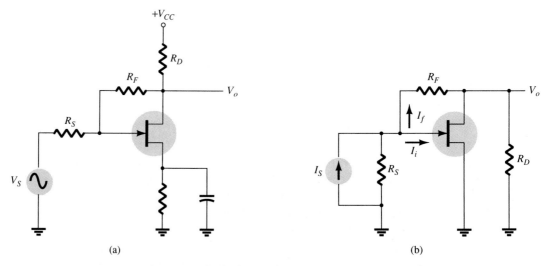

Figure 18.13 Voltage-shunt feedback amplifier using an FET: (a) circuit; (b) equivalent circuit.

EXAMPLE 18.6

Calculate the voltage gain with and without feedback for the circuit of Fig. 18.13a with values of $g_m = 5$ mS, $R_D = 5.1$ kΩ, $R_S = 1$ kΩ, and $R_F = 20$ kΩ.

Solution

Without feedback the voltage gain is

$$A_v = -g_m R_D = -(5 \times 10^{-3})(5.1 \times 10^3) = \mathbf{-25.5}$$

With feedback the gain is reduced to

$$A_{vf} = (-g_m R_D)\frac{R_F}{R_F + g_m R_D R_S}$$

$$= (-25.5)\frac{20 \times 10^3}{(20 \times 10^3) + (5 \times 10^{-3})(5.1 \times 10^3)(1 \times 10^3)}$$

$$= -25.5(0.44) = \mathbf{-11.2}$$

18.4 FEEDBACK AMPLIFIER—PHASE AND FREQUENCY CONSIDERATIONS

So far we have considered the operation of a feedback amplifier in which the feedback signal was *opposite* to the input signal—negative feedback. In any practical circuit this condition occurs only for some midfrequency range of operation. We know that an amplifier gain will change with frequency, dropping off at high frequencies from the midfrequency value. In addition, the phase shift of an amplifier will also change with frequency.

If, as the frequency increases, the phase shift changes then some of the feedback signal will *add* to the input signal. It is then possible for the amplifier to break into oscillations due to positive feedback. If the amplifier oscillates at some low or high frequency, it is no longer useful as an amplifier. Proper feedback-amplifier design requires that the circuit be stable at *all* frequencies, not merely those in the range of interest. Otherwise, a transient disturbance could cause a seemingly stable amplifier to suddenly start oscillating.

Nyquist Criterion

In judging the stability of a feedback amplifier, as a function of frequency, the βA product and the phase shift between input and output are the determining factors. One of the most popular techniques used to investigate stability is the Nyquist method. A Nyquist diagram is used to plot gain and phase shift as a function of frequency on a complex plane. The Nyquist plot, in effect, combines the two Bode plots of gain versus frequency and phase-shift versus frequency on a single plot. A Nyquist plot is used to quickly show whether an amplifier is stable for all frequencies and how stable the amplifier is relative to some gain or phase-shift criteria.

As a start, consider the *complex plane* shown in Fig. 18.14. A few points of various gain (βA) values are shown at a few different phase-shift angles. By using the

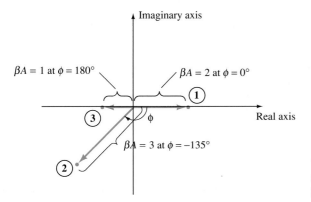

Figure 18.14 Complex plane showing typical gain-phase points.

positive real axis as reference (0°) a magnitude of $\beta A = 2$ is shown at a phase shift of 0° at point 1. Additionally, a magnitude of $\beta A = 3$ at a phase shift of $-135°$ is shown at point 2 and a magnitude/phase of $\beta A = 1$ at 180° is shown at point 3. Thus points on this plot can represent *both* gain magnitude of βA and phase shift. If the points representing gain and phase shift for an amplifier circuit are plotted at increasing frequency, then a Nyquist plot is obtained as shown by the plot in Fig. 18.15. At the origin the gain is 0 at a frequency of 0 (for *RC*-type coupling). At increasing frequency points f_1, f_2, and f_3 and the phase shift increased as did the magnitude of βA. At a representative frequency f_4 the value of A is the vector length from the origin to point f_4 and the phase shift is the angle ϕ. At a frequency f_5 the phase shift is 180°. At higher frequencies the gain is shown to decrease back to 0.

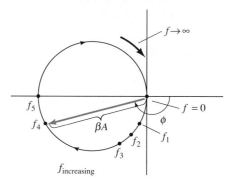

Figure 18.15 Nyquist plot.

The Nyquist criterion for stability can be stated as follows: *The amplifier is unstable if the Nyquist curve plotted encloses (encircles) the -1 point, and it is stable otherwise.*

An example of the Nyquist criterion is demonstrated by the curves in Fig. 18.16. The Nyquist plot in Fig. 18.16a is stable since it does not encircle the -1 point, whereas that shown in Fig. 18.16b is unstable since the curve does encircle the -1 point. Keep in mind that encircling the -1 point means that at a phase shift of 180° the loop gain (βA) is greater than 1; therefore, the feedback signal is in phase with the input and large enough to result in a larger input signal than that applied, with the result that oscillation occurs.

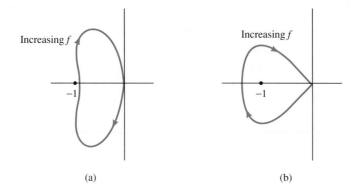

(a) (b)

Figure 18.16 Nyquist plots showing stability conditions; (a) stable; (b) unstable.

Gain and Phase Margins

From the Nyquist criterion we know that a feedback amplifier is stable if the loop gain (βA) is less than unity (0 dB) when its phase angle is 180°. We can additionally determine some margins of stability to indicate how close to instability the amplifier is. That is, if the gain (βA) is less than unity but, say 0.95 in value, this would not be as relatively stable as another amplifier having, say, $(\beta A) = 0.7$ (both measured at 180°). Of course, amplifiers with loop gains 0.95 and 0.7 are both stable, but one is closer to instability, if the loop gain increases, than the other. We can define the following terms:

Gain margin (GM) is defined as the negative of the value of $|\beta A|$ in decibels at the frequency at which the phase angle is 180°. Thus, 0 dB, equal to a value of $\beta A = 1$, is on the border of stability and any negative decibel value is stable. The GM may be evaluated in decibels from the curve of Fig. 18.17.

Phase margin (PM) is defined as the angle of 180° minus the magnitude of the angle at which the value $|\beta A|$ is unity (0 dB). The PM may also be evaluated directly from the curve of Fig. 18.17.

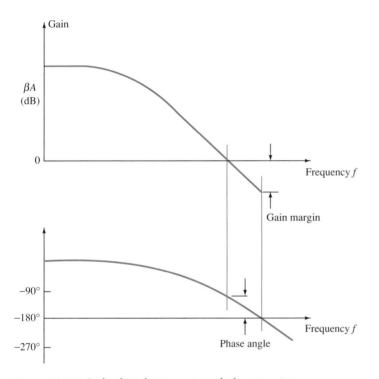

Figure 18.17 Bode plots showing gain and phase margins.

18.5 OSCILLATOR OPERATION

The use of positive feedback which results in a feedback amplifier having closed-loop gain $|A_f|$ greater than 1 and satisfies the phase conditions will result in operation as an oscillator circuit. An oscillator circuit then provides a varying output signal. If the output signal varies sinusoidally, the circuit is referred to as a *sinusoidal oscillator*. If the output voltage rises quickly to one voltage level and later drops quickly to another voltage level, the circuit is generally referred to as a *pulse* or *square-wave oscillator*.

To understand how a feedback circuit performs as an oscillator consider the feedback circuit of Fig. 18.18. When the switch at the amplifier input is open, no oscillation occurs. Consider that we have a *fictitious* voltage at the amplifier input (V_i). This results in an output voltage $V_o = AV_i$ after the amplifier stage and in a voltage $V_f = \beta(AV_i)$ after the feedback stage. Thus, we have a feedback voltage $V_f = \beta AV_i$, where βA is referred to as the *loop gain*. If the circuits of the base amplifier and feedback network provide βA of a correct magnitude and phase, V_f can be made equal to V_i.

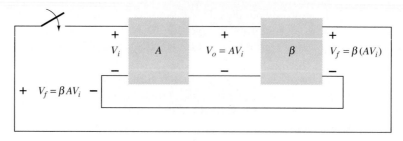

Figure 18.18 Feedback circuit used as an oscillator.

Then, when the switch is closed and fictitious voltage V_i is removed, the circuit will continue operating since the feedback voltage is sufficient to drive the amplifier and feedback circuits resulting in a proper input voltage to sustain the loop operation. The output waveform will still exist after the switch is closed if the condition

$$\beta A = 1 \tag{18.32}$$

is met. This is known as the *Barkhausen criterion* for oscillation.

In reality, no input signal is needed to start the oscillator going. Only the condition $\beta A = 1$ must be satisfied for self-sustained oscillations to result. In practice βA is made greater than 1, and the system is started oscillating by amplifying noise voltage which is always present. Saturation factors in the practical circuit provide an "average" value of βA of 1. The resulting waveforms are never exactly sinusoidal. However, the closer the value βA is to exactly 1 the more nearly sinusoidal is the waveform. Figure 18.19 shows how the noise signal results in a buildup of a steady-state oscillation condition.

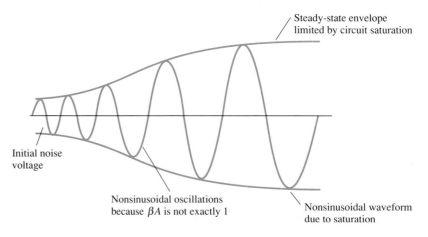

Figure 18.19 Buildup of steady-state oscillations.

Another way of seeing how the feedback circuit provides operation as an oscillator is obtained by noting the denominator in the basic feedback equation (18.3), $A_f = A/(1 + \beta A)$. When $\beta A = -1$ or magnitude 1 at a phase angle of 180°, the denominator becomes 0 and the gain with feedback, A_f, becomes infinite. Thus, an infinitesimal signal (noise voltage) can provide a measurable output voltage, and the circuit acts as an oscillator even without an input signal.

The remainder of this chapter is devoted to various oscillator circuits that use a variety of components. Practical considerations are included so that workable circuits in each of the various cases are discussed.

Chapter 18 Feedback and Oscillator Circuits

18.6 PHASE-SHIFT OSCILLATOR

An example of an oscillator circuit that follows the basic development of a feedback circuit is the *phase-shift oscillator*. An idealized version of this circuit is shown in Fig. 18.20. Recall that the requirements for oscillation are that the loop gain, βA, is greater than unity *and* that the phase shift around the feedback network is 180° (providing positive feedback). In the present idealization we are considering the feedback network to be driven by a perfect source (zero source impedance) and the output of the feedback network is connected into a perfect load (infinite load impedance). The idealized case will allow development of the theory behind the operation of the phase-shift oscillator. Practical circuit versions will then be considered.

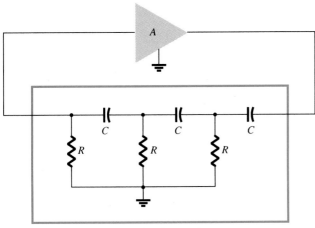

Feedback network

Figure 18.20 Idealized phase-shift oscillator.

Concentrating our attention on the phase-shift network we are interested in the attenuation of the network at the frequency at which the phase shift is exactly 180°. Using classical network analysis, we find that

$$f = \frac{1}{2\pi RC\sqrt{6}} \qquad (18.33)$$

$$\beta = \frac{1}{29} \qquad (18.34)$$

and the phase shift is 180°.

For the loop gain βA to be greater than unity the gain of the amplifier stage must be greater than $1/\beta$ or 29

$$A > 29 \qquad (18.35)$$

When considering the operation of the feedback network one might naively select the values of R and C to provide (at a specific frequency) 60°-phase shift per section for three sections, resulting in a 180° phase shift, as desired. This, however, is not the case, since each section of the RC in the feedback network loads down the previous one. The net result that the *total* phase shift be 180° is all that is important. The frequency given by Eq. (18.33) is that at which the *total* phase shift is 180°. If one measured the phase shift per RC section, each section would not provide the same phase shift (although the overall phase shift is 180°). If it were desired to obtain exactly a 60° phase shift for each of three stages, then emitter-follower stages would be needed for each RC section to prevent each from being loaded from the following circuit.

FET Phase-Shift Oscillator

A practical version of a phase-shift oscillator circuit is shown in Fig. 18.21a. The circuit is drawn to show clearly the amplifier and feedback network. The amplifier stage is self-biased with a capacitor bypassed source resistor R_S and a drain bias resistor R_D. The FET device parameters of interest are g_m and r_d. From FET amplifier theory the amplifier gain magnitude is calculated from

$$|A| = g_m R_L \tag{18.36}$$

where R_L in this case is the parallel resistance of R_D and r_d

$$R_L = \frac{R_D r_d}{R_D + r_d} \tag{18.37}$$

We shall assume as a very good approximation that the input impedance of the FET amplifier stage is infinite. This assumption is valid as long as the oscillator operating frequency is low enough so that FET capacitive impedances can be neglected. The output impedance of the amplifier stage given by R_L should also be small compared to the impedance seen looking into the feedback network so that no attenuation due to loading occurs. In practice, these considerations are not always negligible, and the amplifier stage gain is then selected somewhat larger than the needed factor of 29 to assure oscillator action.

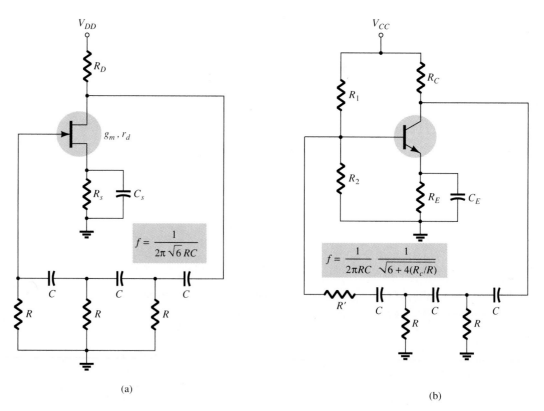

Figure 18.21 Practical phase-shift oscillator circuits: (a) FET version; (b) BJT version.

EXAMPLE 18.7

It is desired to design a phase-shift oscillator (as in Fig. 18.21a) using an FET having $g_m = 5000\ \mu S$, $r_d = 40\ k\Omega$, and feedback circuit value of $R = 10\ k\Omega$. Select the value of C for oscillator operation at 1 kHz and R_D for $A > 29$ to ensure oscillator action.

Chapter 18 Feedback and Oscillator Circuits

Solution

Equation (18.33) is used to solve for the capacitor value. Since $f = 1/2\pi RC\sqrt{6}$, we can solve for C:

$$C = \frac{1}{2\pi Rf\sqrt{6}} = \frac{1}{(6.28)(10 \times 10^3)(1 \times 10^3)(2.45)} = \mathbf{6.5\ nF}$$

Using Eq. (18.36), we solve for R_L to provide a gain of, say, $A = 40$ (this allows for some loading between R_L and the feedback network input impedance):

$$|A| = g_m R_L$$

$$R_L = \frac{|A|}{g_m} = \frac{40}{5000 \times 10^{-6}} = 8\ k\Omega$$

Using Eq. (18.37), we solve for $R_D = \mathbf{10\ k\Omega.}$

Transistor Phase-Shift Oscillator

If a transistor is used as the active element of the amplifier stage, the output of the feedback network is loaded appreciably by the relatively low input resistance (h_{ie}) of the transistor. Of course, an emitter-follower input stage followed by a common-emitter amplifier stage could be used. If a single transistor stage is desired, however, the use of voltage-shunt feedback (as shown in Fig. 18.21b) is more suitable. In this connection, the feedback signal is coupled through the feedback resistor R' in *series* with the amplifier stage input resistance (R_i).

Analysis of the ac circuit provides the following equation for the resulting oscillator frequency:

$$f = \frac{1}{2\pi RC} \frac{1}{\sqrt{6 + 4(R_C/R)}} \qquad (18.38)$$

For the loop gain to be greater than unity, the requirement on the current gain of the transistor is found to be

$$h_{fe} > 23 + 29\frac{R}{R_C} + 4\frac{R_C}{R} \qquad (18.39)$$

IC Phase-Shift Oscillator

As IC circuits have become more popular they have been adapted to operate in oscillator circuits. One need buy only an op-amp to obtain an amplifier circuit of stabilized gain setting and incorporate some means of signal feedback to produce an oscillator circuit. For example, a phase-shift oscillator is shown in Fig. 18.22. The

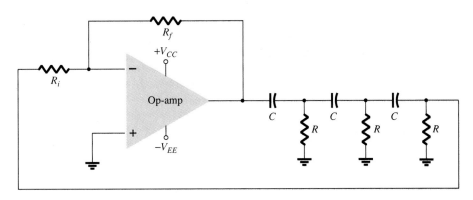

Figure 18.22 Phase-shift oscillator using op-amp.

output of the op-amp is fed to a three-stage RC network which provides the needed 180° of phase shift (at an attenuation factor of 1/29). If the op-amp provides gain (set by resistors R_i and R_f) of greater than 29, a loop gain greater than unity results and the circuit acts as an oscillator [oscillator frequency is given by Eq. (18.33)].

18.7 WIEN BRIDGE OSCILLATOR

A practical oscillator circuit uses an op-amp and RC bridge circuit, with the oscillator frequency set by the R and C components. Figure 18.23 shows a basic version of a Wien bridge oscillator circuit. Note the basic bridge connection. Resistors R_1, R_2 and capacitors C_1, C_2 form the frequency-adjustment elements, while resistors R_3 and R_4 form part of the feedback path. The op-amp output is connected as the bridge input at points a and c. The bridge circuit output at points b and d is the input to the op-amp.

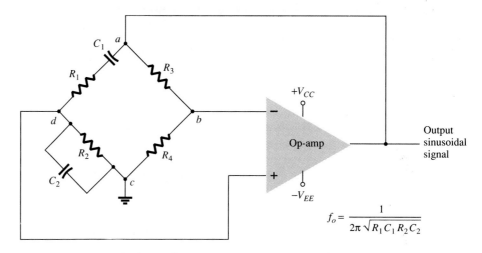

Figure 18.23 Wien bridge oscillator circuit using op-amp amplifier.

Neglecting loading effects of the op-amp input and output impedances, the analysis of the bridge circuit results in

$$\frac{R_3}{R_4} = \frac{R_1}{R_2} + \frac{C_2}{C_1} \tag{18.40}$$

and

$$\boxed{f_o = \frac{1}{2\pi\sqrt{R_1 C_1 R_2 C_2}}} \tag{18.41}$$

If, in particular, the values are $R_1 = R_2 = R$ and $C_1 = C_2 = C$, the resulting oscillator frequency is

$$f_o = \frac{1}{2\pi RC} \tag{18.42}$$

and

$$\frac{R_3}{R_4} = 2 \tag{18.43}$$

Thus a ratio of R_3 to R_4 greater than 2 will provide sufficient loop gain for the circuit to oscillate at the frequency calculated using Eq. (18.42).

Chapter 18   Feedback and Oscillator Circuits

Calculate the resonant frequency of the Wien bridge oscillator of Fig. 18.24.

EXAMPLE 18.8

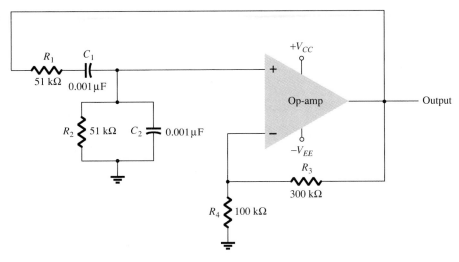

Figure 18.24 Wien bridge oscillator circuit for Example 18.8.

Solution

Using Eq. (18.42) yields

$$f_o = \frac{1}{2\pi RC} = \frac{1}{2\pi(51 \times 10^3)(0.001 \times 10^{-6})} = \mathbf{3120.7 \ Hz}$$

Design the RC elements of a Wien bridge oscillator as in Fig. 18.24 for operation at $f_o = 10$ kHz.

EXAMPLE 18.9

Solution

Using equal values of R and C we can select $R = 100$ kΩ and calculate the required value of C using Eq. (18.42):

$$C = \frac{1}{2\pi f_o R} = \frac{1}{6.28(10 \times 10^3)(100 \times 10^3)} = \frac{10^{-9}}{6.28} = \mathbf{159 \ pF}$$

We can use $R_3 = 300$ kΩ and $R_4 = 100$ kΩ to provide a ratio R_3/R_4 greater than 2 for oscillation to take place.

18.8 TUNED OSCILLATOR CIRCUIT

Tuned-Input, Tuned-Output Oscillator Circuits

A variety of circuits can be built using that shown in Fig. 18.25 by providing tuning in both the input and output sections of the circuit. Analysis of the circuit of Fig. 18.25 reveals that the following types of oscillators are obtained when the reactance elements are as designated:

A_f

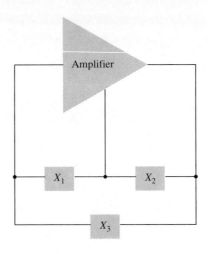

Figure 18.25 Basic configuration of resonant circuit oscillator.

	Reactance Element		
Oscillator Type	X_1	X_2	X_3
Colpitts oscillator	C	C	L
Hartley oscillator	L	L	C
Tuned input, tuned output	LC	LC	—

Colpitts Oscillators

FET COLPITTS OSCILLATOR

A practical version of an FET Colpitts oscillator is shown in Fig. 18.26. The circuit is basically the same form as shown in Fig. 18.25 with the addition of the components needed for dc bias of the FET amplifier. The oscillator frequency can be found to be

$$f_o = \frac{1}{2\pi\sqrt{LC_{eq}}}$$

(18.44)

where

$$C_{eq} = \frac{C_1 C_2}{C_1 + C_2}$$

(18.45)

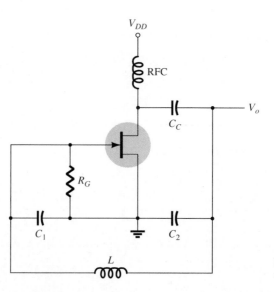

Figure 18.26 FET Colpitts oscillator.

TRANSISTOR COLPITTS OSCILLATOR

A transistor Colpitts oscillator circuit can be made as shown in Fig. 18.27. The circuit frequency of oscillation is given by Eq. (18.44).

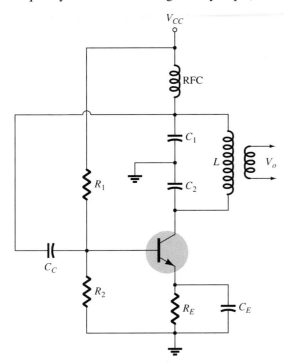

Figure 18.27 Transistor Colpitts oscillator.

IC COLPITTS OSCILLATOR

An op-amp Colpitts oscillator circuit is shown in Fig. 18.28. Again, the op-amp provides the basic amplification needed while the oscillator frequency is set by an *LC* feedback network of a Colpitt configuration. The oscillator frequency is given by Eq. (18.44).

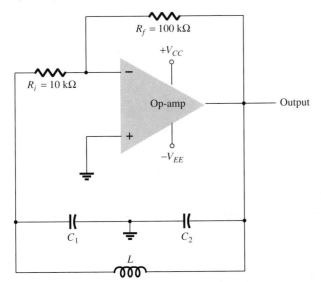

Figure 18.28 Op-amp Colpitts oscillator.

Hartley Oscillator

If the elements in the basic resonant circuit of Fig. 18.25 are X_1 and X_2 (inductors), and X_3 (capacitor), the circuit is a Hartley oscillator.

FET HARTLEY OSCILLATOR

An FET Hartley oscillator circuit is shown in Fig. 18.29. The circuit is drawn so that the feedback network conforms to the form shown in the basic resonant circuit (Fig. 18.25). Note, however, that inductors L_1 and L_2 have a mutual coupling, M, which must be taken into account in determining the equivalent inductance for the resonant tank circuit. The circuit frequency of oscillation is then given approximately by

$$f_o = \frac{1}{2\pi\sqrt{L_{eq}C}} \tag{18.46}$$

with
$$L_{eq} = L_1 + L_2 + 2\,M \tag{18.47}$$

TRANSISTOR HARTLEY OSCILLATOR

Figure 18.30 shows a transistor Hartley oscillator circuit. The circuit operates at a frequency given by Eq. (18.46).

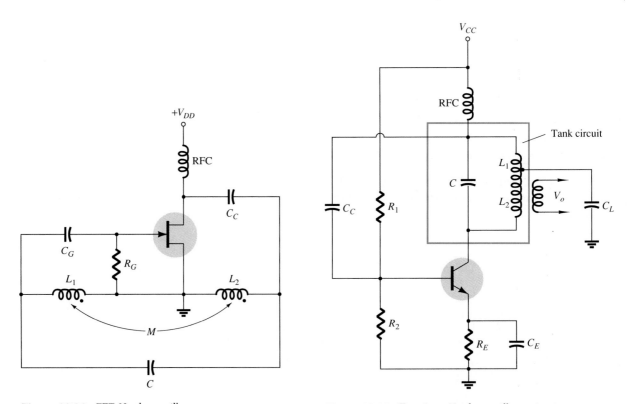

Figure 18.29 FET Hartley oscillator.

Figure 18.30 Transistor Hartley oscillator circuit.

18.9 CRYSTAL OSCILLATOR

A crystal oscillator is basically a tuned-circuit oscillator using a piezoelectric crystal as a resonant tank circuit. The crystal (usually quartz) has a greater stability in holding constant at whatever frequency the crystal is originally cut to operate. Crystal oscillators are used whenever great stability is required, for example, in communication transmitters and receivers.

Chapter 18 Feedback and Oscillator Circuits

Characteristics of a Quartz Crystal

A quartz crystal (one of a number of crystal types) exhibits the property that when mechanical stress is applied across the faces of the crystal, a difference of potential develops across opposite faces of the crystal. This property of a crystal is called the *piezoelectric effect*. Similarly, a voltage applied across one set of faces of the crystal causes mechanical distortion in the crystal shape.

When alternating voltage is applied to a crystal, mechanical vibrations are set up—these vibrations having a natural resonant frequency dependent on the crystal. Although the crystal has electromechanical resonance, we can represent the crystal action by an equivalent electrical resonant circuit as shown in Fig. 18.31. The inductor L and capacitor C represent electrical equivalents of crystal mass and compliance while resistance R is an electrical equivalent of the crystal structure's internal friction. The shunt capacitance C_M represents the capacitance due to mechanical mounting of the crystal. Because the crystal losses, represented by R, are small, the equivalent crystal Q (quality factor) is high—typically 20,000. Values of Q up to almost 10^6 can be achieved by using crystals.

The crystal as represented by the equivalent electrical circuit of Fig. 18.31 can have two resonant frequencies. One resonant condition occurs when the reactances of the series RLC leg are equal (and opposite). For this condition the *series-resonant* impedance is very low (equal to R). The other resonant condition occurs at a higher frequency when the reactance of the series-resonant leg equals the reactance of capacitor C_M. This is a parallel resonance or antiresonance condition of the crystal. At this frequency the crystal offers a very high impedance to the external circuit. The impedance versus frequency of the crystal is shown in Fig. 18.32. In order to use the crystal properly it must be connected in a circuit so that its low impedance in the series-resonant operating mode or high impedance in the antiresonant operating mode is selected.

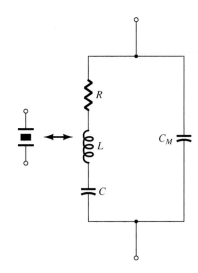

Figure 18.31 Electrical equivalent circuit of a crystal.

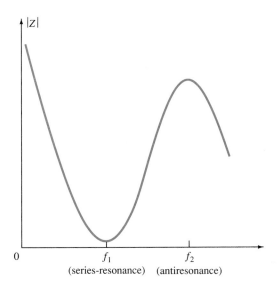

Figure 18.32 Crystal impedance versus frequency.

Series-Resonant Circuits

To excite a crystal for operation in the series-resonant mode it may be connected as a series element in a feedback path. At the series-resonant frequency of the crystal its impedance is smallest and the amount of (positive) feedback is largest. A typical transistor circuit is shown in Fig. 18.33. Resistors R_1, R_2, and R_E provide a voltage-divider stabilized dc bias circuit. Capacitor C_E provides ac bypass of the emitter

18.9   Crystal Oscillator   767

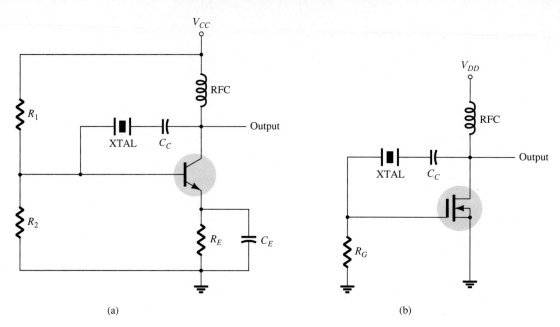

(a)   (b)

Figure 18.33 Crystal-controlled oscillator using crystal in series-feedback path:
(a) BJT circuit; (b) FET circuit.

resistor and the RFC coil provides for dc bias while decoupling any ac signal on the power lines from affecting the output signal. The voltage feedback from collector to base is a maximum when the crystal impedance is minimum (in series-resonant mode). The coupling capacitor C_C has negligible impedance at the circuit operating frequency but blocks any dc between collector and base.

The resulting circuit frequency of oscillation is set, then, by the series-resonant frequency of the crystal. Changes in supply voltage, transistor device parameters, and so on, have no effect on the circuit operating frequency which is held stabilized by the crystal. The circuit frequency stability is set by the crystal frequency stability, which is good.

Parallel-Resonant Circuits

Since the parallel-resonant impedance of a crystal is a maximum value, it is connected in shunt. At the parallel-resonant operating frequency a crystal appears as an inductive reactance of largest value. Figure 18.34 shows a crystal connected as the

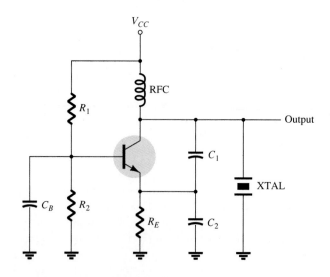

Figure 18.34 Crystal-controlled oscillator operating in parallel-resonant mode.

inductor element in a modified Colpitts circuit. The basic dc bias circuit should be evident. Maximum voltage is developed across the crystal at its parallel-resonant frequency. The voltage is coupled to the emitter by a capacitor voltage divider—capacitors C_1 and C_2.

A *Miller crystal-controlled oscillator* circuit is shown in Fig. 18.35. A tuned *LC* circuit in the drain section is adjusted near the crystal parallel-resonant frequency. The maximum gate–source signal occurs at the crystal antiresonant frequency controlling the circuit operating frequency.

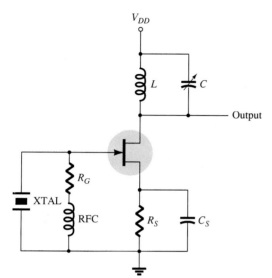

Figure 18.35 Miller crystal-controlled oscillator.

Crystal Oscillator

An op-amp can be used in a crystal oscillator as shown in Fig. 18.36. The crystal is connected in the series-resonant path and operates at the crystal series-resonant frequency. The present circuit has a high gain so that an output square-wave signal results as shown in the figure. A pair of Zener diodes is shown at the output to provide output amplitude at exactly the Zener voltage (V_Z).

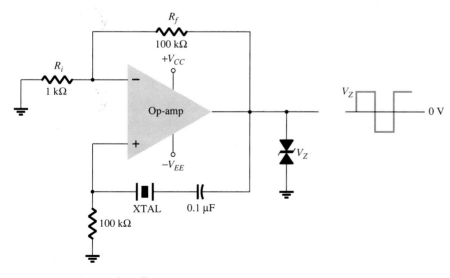

Figure 18.36 Crystal oscillator using op-amp.

18.10 UNIJUNCTION OSCILLATOR

A particular device, the unijunction transistor can be used in a single-stage oscillator circuit to provide a pulse signal suitable for digital-circuit applications. The unijunction transistor can be used in what is called a *relaxation oscillator* as shown by the basic circuit of Fig. 18.37. Resistor R_T and capacitor C_T are the timing components that set the circuit oscillating rate. The oscillating frequency may be calculated using Eq. (18.48) which includes the unijunction transistor *intrinsic stand-off ratio* η, as a factor (in addition to R_T and C_T) in the oscillator operating frequency.

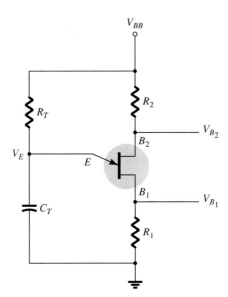

Figure 18.37 Basic unijunction oscillator circuit.

$$f_o \cong \frac{1}{R_T C_T \ln[1/(1 - \eta)]}$$ (18.48)

Typically, a unijunction transistor has a stand-off ratio from 0.4 to 0.6. Using a value of $\eta = 0.5$, we get

$$f_o \cong \frac{1}{R_T C_T \ln[1/(1 - 0.5)]} = \frac{1}{R_T C_T \ln 2} = \frac{1.44}{R_T C_T}$$

$$\cong \frac{1.5}{R_T C_T}$$ (18.49)

Capacitor C_T is charged through resistor R_T toward supply voltage V_{BB}. As long as the capacitor voltage V_E is below a stand-off voltage (V_P) set by the voltage across $B_1 - B_2$, and the transistor stand-off ratio η

$$V_P = \eta V_{B_1} V_{B_2} - V_D$$ (18.50)

the unijunction emitter lead appears as an open circuit. When the emitter voltage across capacitor C_T exceeds this value (V_P), the unijunction circuit fires, discharging the capacitor, after which a new charge cycle begins. When the unijunction fires, a voltage rise is developed across R_1 and a voltage drop is developed across R_2 as shown in Fig. 18.38. The signal at the emitter is a sawtooth voltage waveform, that at base 1 is a positive-going pulse, and that at base 2 is a negative-going pulse. A few circuit variations of the unijunction oscillator are provided in Fig. 18.39.

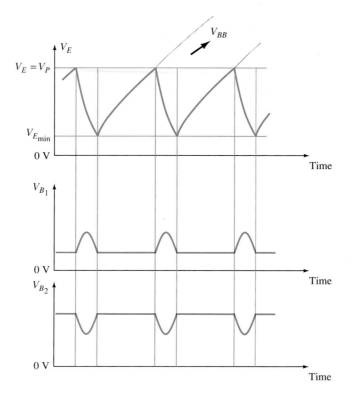

Figure 18.38 Unijunction oscillator waveforms.

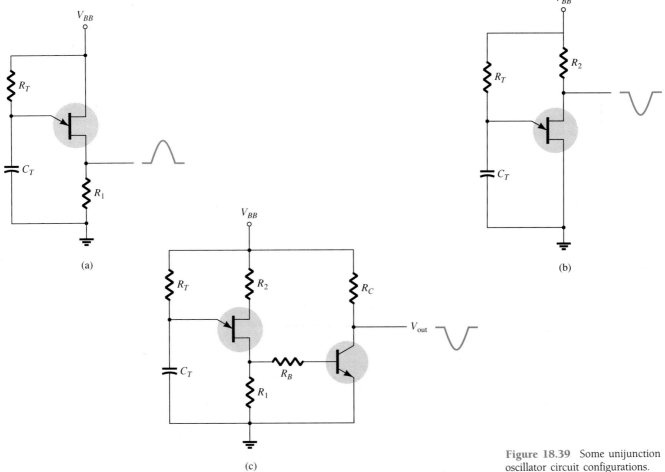

Figure 18.39 Some unijunction oscillator circuit configurations.

§ 18.2 Feedback Connection Types

1. Calculate the gain of a negative-feedback amplifier having $A = -2000$, $\beta = -1/10$.

2. If the gain of an amplifier changes from a value of -1000 by 10%, calculate the gain change if the amplifier is used in a feedback circuit having $\beta = -1/20$.

3. Calculate the gain, input, and output impedances of a voltage-series feedback amplifier having $A = -300$, $R_i = 1.5$ kΩ, $R_o = 50$ kΩ, and $\beta = -1/15$.

§ 18.3 Practical Feedback Circuits

* 4. Calculate the gain with and without feedback for an FET amplifier as in Fig. 18.5 for circuit values $R_1 = 800$ kΩ, $R_2 = 200$ Ω, $R_o = 40$ kΩ, $R_D = 8$ kΩ, and $g_m = 5000$ μS.

5. For a circuit as in Fig. 18.11 and the following circuit values, calculate the circuit gain and the input and output impedance with and without feedback: $R_B = 600$ kΩ, $R_E = 1.2$ kΩ, $R_C = 4.7$ kΩ, and $\beta = 75$. Use $V_{CC} = 16$ V.

§ 18.6 Phase-Shift Oscillator

6. An FET phase-shift oscillator having $g_m = 6000$ μS, $r_d = 36$ kΩ, and feedback resistor $R = 12$ kΩ is to operate at 2.5 kHz. Select C for specified oscillator operation.

7. Calculate the operating frequency of a BJT phase-shift oscillator as in Fig. 18.21b for $R = 6$ kΩ, $C = 1500$ pF, and $R_C = 18$ kΩ.

§ 18.7 Wien Bridge Oscillator

8. Calculate the frequency of a Wien bridge oscillator circuit (as in Fig. 18.23) when $R = 10$ kΩ and $C = 2400$ pF.

§ 18.8 Tuned Oscillator Circuit

9. For an FET Colpitts oscillator as in Fig. 18.26 and the following circuit values determine the circuit oscillation frequency: $C_1 = 750$ pF, $C_2 = 2500$ pF, $L = 40$ μH.

10. For the transistor Colpitts oscillator of Fig. 18.27 and the following circuit values, calculate the oscillation frequency: $L = 100$ μH, $L_{RFC} = 0.5$ mH, $C_1 = 0.005$ μF, $C_2 = 0.01$ μF, $C_C = 10$ μF.

11. Calculate the oscillator frequency for an FET Hartley oscillator as in Fig. 18.29 for the following circuit values: $C = 250$ pF, $L_1 = 1.5$ mH, $L_2 = 1.5$ mH, $M = 0.5$ mH.

12. Calculate the oscillation frequency for the transistor Hartley circuit of Fig. 18.30 and the following circuit values: $L_{RFC} = 0.5$ mH, $L_1 = 750$ μH, $L_2 = 750$ μH, $M = 150$ μH, $C = 150$ pF.

§ 18.9 Crystal Oscillator

13. Draw circuit diagrams of (a) a series-operated crystal oscillator and (b) a shunt-excited crystal oscillator.

§ 18.10 Unijunction Oscillator

14. Design a unijunction oscillator circuit for operation at (a) 1 kHz and (b) 150 kHz.

---

*Please Note: Asterisks indicate more difficult problems.

Chapter 18   Feedback and Oscillator Circuits

Power Supplies (Voltage Regulators)

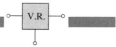

19.1 INTRODUCTION

The present chapter introduces the operation of power supply circuits built using filters, rectifiers, and then voltage regulators. (Refer to Chapter 2 for the initial description of diode rectifier circuits.) Starting with an ac voltage, a steady dc voltage is obtained by rectifying the ac voltage, then filtering to a dc level, and finally, regulating to obtain a desired fixed dc voltage. The regulation is usually obtained from an IC voltage regulator unit, which takes a dc voltage and provides a somewhat lower dc voltage which remains the same even if the input dc voltage varies or the output load connected to the dc voltage changes.

A block diagram containing the parts of a typical power supply and the voltage at various points in the unit is shown in Fig. 19.1. The ac voltage, typically 120 V rms, is connected to a transformer which steps that ac voltage down to the level for the desired dc output. A diode rectifier then provides a full-wave rectified voltage that is initially filtered by a simple capacitor filter to produce a dc voltage. This resulting dc voltage usually has some ripple or ac voltage variation. A regulator circuit can use this dc input to provide a dc voltage that not only has much less ripple voltage but also remains the same dc value even if the input dc voltage varies somewhat, or the load connected to the output dc voltage changes. This voltage regulation is usually obtained using one of a number of popular voltage regulator IC units.

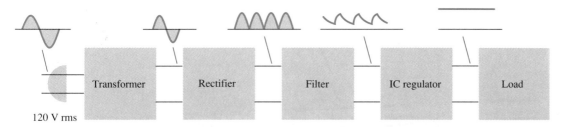

Figure 19.1 Block diagram showing parts of a power supply.

19.2 GENERAL FILTER CONSIDERATIONS

A rectifier circuit is necessary to convert a signal having zero average value into one that has a nonzero average. The output resulting from a rectifier is a pulsating dc voltage and not yet suitable as a battery replacement. Such a voltage could be used in, say, a battery charger, where the average dc voltage is large enough to provide a charging current for the battery. For dc supply voltages, as those used in a radio, stereo system, computer, and so on, the pulsating dc voltage from a rectifier is not good enough. A filter circuit is necessary to provide a more steady dc voltage.

Filter Voltage Regulation and Ripple Voltage

Before going into the details of a filter circuit it would be appropriate to consider the usual methods of rating filter circuits so that we can compare a circuit's effectiveness as a filter. Figure 19.2 shows a typical filter output voltage, which will be used to define some of the signal factors. The filtered output of Fig. 19.2 has a dc value and some ac variation (ripple). Although a battery has essentially a constant or dc output voltage, the dc voltage derived from an ac source signal by rectifying and filtering will have some ac variation (ripple). The smaller the ac variation with respect to the dc level, the better the filter circuit's operation.

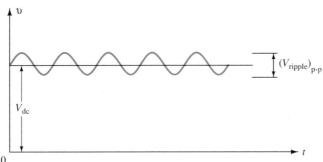

Figure 19.2 Filter voltage waveform showing dc and ripple voltages.

Consider measuring the output voltage of a filter circuit using a dc voltmeter and an ac (rms) voltmeter. The dc voltmeter will read only the average or dc level of the output voltage. The ac (rms) meter will read only the rms value of the ac component of the output voltage (assuming the ac signal is coupled through a capacitor to block out the dc level).

Definition: Ripple

$$r = \frac{\text{ripple voltage (rms)}}{\text{dc voltage}} = \frac{V_r(\text{rms})}{V_{dc}} \times 100\% \qquad (19.1)$$

EXAMPLE 19.1

Using a dc and ac voltmeter to measure the output signal from a filter circuit, we obtain readings of 25 V dc and 1.5 V rms. Calculate the ripple of the filter output voltage.

Solution

$$r = \frac{V_r(\text{rms})}{V_{dc}} \times 100\% = \frac{1.5 \text{ V}}{25 \text{ V}} \times 100\% = \mathbf{6\%}$$

VOLTAGE REGULATION

Another factor of importance in a power supply is the amount the dc output voltage changes over a range of circuit operation. The voltage provided at the output under no-load condition (no current drawn from the supply) is reduced when load current is drawn from the supply (under load). The amount the dc voltage changes between the no-load and load conditions is described by a factor called voltage regulation.

Definition: Voltage regulation

$$\text{Voltage regulation} = \frac{\text{no-load voltage} - \text{full-load voltage}}{\text{full-load voltage}}$$

$$\%\text{V.R.} = \frac{V_{NL} - V_{FL}}{V_{FL}} \times 100\% \tag{19.2}$$

A dc voltage supply provides 60 V when the output is unloaded. When connected to a load the output drops to 56 V. Calculate the value of voltage regulation.

EXAMPLE 19.2

Solution

Eq. (19.2): $\quad \%\text{V.R.} = \dfrac{V_{NL} - V_{FL}}{V_{FL}} \times 100\% = \dfrac{60\text{ V} - 56\text{ V}}{56\text{ V}} \times 100\% = \mathbf{7.1\%}$

If the value of full-load voltage is the same as the no-load voltage, the voltage regulation calculated is 0%, which is the best expected. This means that the supply is a perfect voltage source for which the output voltage is independent of the current drawn from the supply. The smaller the voltage regulation, the better the operation of the voltage supply circuit.

RIPPLE FACTOR OF RECTIFIED SIGNAL

Although the rectified voltage is not a filtered voltage, it nevertheless contains a dc component and a ripple component. We will see that the full-wave rectified signal has a larger dc component and less ripple than the half-wave rectified voltage.

For a half-wave rectified signal the output dc voltage is

$$V_{dc} = 0.318V_m \tag{19.3}$$

The rms value of the ac component of the output signal can be calculated (see Appendix B) to be

$$V_r(\text{rms}) = 0.385V_m \tag{19.4}$$

The percent ripple of a half-wave rectified signal can then be calculated as

$$r = \frac{V_r(\text{rms})}{V_{dc}} \times 100\% = \frac{0.385V_m}{0.318V_m} \times 100\% = 121\% \tag{19.5}$$

For a full-wave rectified voltage the dc value is

$$V_{dc} = 0.636V_m \tag{19.6}$$

The rms value of the ac component of the output signal can be calculated (see Appendix B) to be

$$V_r(\text{rms}) = 0.308V_m \tag{19.7}$$

The percent ripple of a full-wave rectified signal can then be calculated as

$$r = \frac{V_r(\text{rms})}{V_{dc}} \times 100\% = \frac{0.308V_m}{0.636V_m} \times 100\% = 48\% \tag{19.8}$$

In summary, a full-wave rectified signal has less ripple than a half-wave rectified signal and is thus better to apply to a filter.

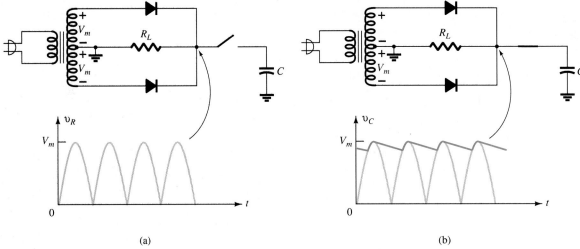
19.3 CAPACITOR FILTER

A most popular filter circuit is the capacitor-filter circuit shown in Fig. 19.3. A capacitor is connected at the rectifier output, and a dc voltage is obtained across the capacitor. Figure 19.4a shows the output voltage of a full-wave rectifier before the signal is filtered, while Fig. 19.4b shows the resulting waveform after the filter capacitor is connected at the rectifier output. Notice that the filtered waveform is essentially a dc voltage with some ripple (or ac variation).

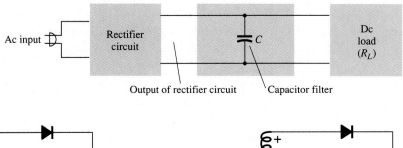

Figure 19.3 Simple capacitor filter.

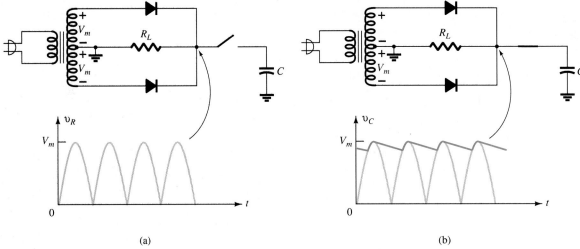

Figure 19.4 Capacitor filter operation: (a) full-wave rectifier voltage; (b) filtered output voltage.

Figure 19.5a shows a full-wave bridge rectifier and the output waveform obtained from the circuit when connected to a load (R_L). If no load were connected across the capacitor, the output waveform would ideally be a constant dc level equal in value to the peak voltage (V_m) from the rectifier circuit. However, the purpose of obtaining a

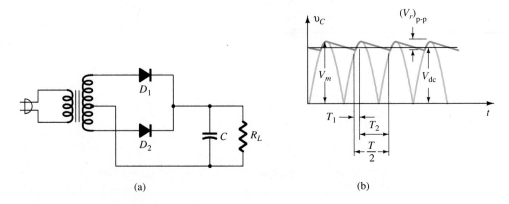

Figure 19.5 Capacitor filter: (a) capacitor filter circuit; (b) output voltage waveform.

Chapter 19   Power Supplies (Voltage Regulators)

dc voltage is to provide this voltage for use by various electronic circuits, which then constitute a load on the voltage supply. Since there will always be a load on the filter output, we must consider this practical case in our discussion.

Output Waveform Times

Figure 19.5b shows the waveform across a capacitor filter. Time T_1 is the time during which diodes of the full-wave rectifier conduct, charging the capacitor up to the peak rectifier voltage, V_m. Time T_2 is the time interval during which the rectifier voltage drops below the peak voltage, and the capacitor discharges through the load. Since the charge–discharge cycle occurs for each half-cycle for a full-wave rectifier, the period of the rectified waveform is $T/2$, one-half the input signal frequency. The filtered voltage, as shown in Fig. 19.6, shows the output waveform to have a dc level V_{dc} and a ripple voltage $V_r(rms)$ as the capacitor charges and discharges. Some details of these waveforms and the circuit elements are considered next.

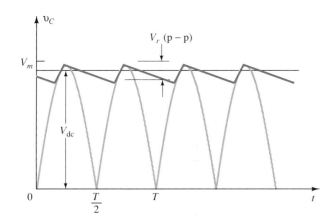

Figure 19.6 Approximate output voltage of capacitor filter circuit.

RIPPLE VOLTAGE, V_r(rms)

Appendix B provides the details for determining the value of the ripple voltage in terms of the other circuit parameters. The ripple voltage can be calculated from

$$V_r(\text{rms}) = \frac{I_{dc}}{4\sqrt{3}\,fC} = \frac{2.4I_{dc}}{C} = \frac{2.4V_{dc}}{R_L C} \tag{19.9}$$

where I_{dc} is in milliamperes, C is in microfarads, and R_L is in kilohms.

Calculate the ripple voltage of a full-wave rectifier with a 100-μF filter capacitor connected to a load drawing 50 mA.

EXAMPLE 19.3

Solution

$$\text{Eq. (19.9):} \quad V_r(\text{rms}) = \frac{2.4(50)}{100} = \textbf{1.2 V}$$

DC VOLTAGE, V_{dc}

From Appendix B we can express the dc value of the waveform across the filter capacitor as

$$V_{dc} = V_m - \frac{I_{dc}}{4fC} = V_m - \frac{4.17 I_{dc}}{C} \qquad (19.10)$$

where V_m is the peak rectifier voltage, I_{dc} is the load current in milliamperes, and C is the filter capacitor in microfarads.

EXAMPLE 19.4

If the peak rectified voltage for the filter circuit of Example 19.3 is 30 V, calculate the filter dc voltage.

Solution

$$\text{Eq. (19.10):} \quad V_{dc} = V_m - \frac{4.17 I_{dc}}{C} = 30 - \frac{4.17(50)}{100} = \mathbf{27.9 \ V}$$

Filter Capacitor Ripple

Using the definition of ripple [Eq. (19.1)], Eq. (19.9), and Eq. (19.10), with $V_{dc} \approx V_m$, we can obtain the expression for the output waveform ripple of a full-wave rectifier and filter-capacitor circuit.

$$r = \frac{V_r(\text{rms})}{V_{dc}} \times 100\% = \frac{2.4 I_{dc}}{C V_{dc}} \times 100\% = \frac{2.4}{R_L C} \times 100\% \qquad (19.11)$$

where I_{dc} is in milliamperes, C is in microfarads, V_{dc} is in volts, and R_L is in kilohms.

EXAMPLE 19.5

Calculate the ripple of a capacitor filter for a peak rectified voltage of 30 V, capacitor $C = 50 \ \mu\text{F}$, and a load current of 50 mA.

Solution

$$\text{Eq. (19.11):} \quad r = \frac{2.4 I_{dc}}{C V_{dc}} \times 100\% = \frac{2.4(50)}{100(27.9)} \times 100\% = \mathbf{4.3\%}$$

We could also calculate the ripple using the basic definition

$$r = \frac{V_r(\text{rms})}{V_{dc}} \times 100\% = \frac{1.2 \ \text{V}}{27.9 \ \text{V}} \times 100\% = \mathbf{4.3\%}$$

Diode Conduction Period and Peak Diode Current

From the previous discussion it should be clear that larger values of capacitance provide less ripple and higher average voltage, thereby providing better filter action. From this one might conclude that to improve the performance of a capacitor filter it is only necessary to increase the size of the filter capacitor. The capacitor, however, also affects the peak current drawn through the rectifying diodes, and as will be shown next, the larger the value of the capacitor, the larger the peak current drawn through the rectifying diodes.

Recall that the diodes conduct during period T_1 (see Fig. 19.5), during which time the diode must provide the necessary average current to charge the capacitor. The shorter this time interval, the larger the amount of the charging current. Figure 19.7 shows this relation for a half-wave rectified signal (it would be the same basic opera-

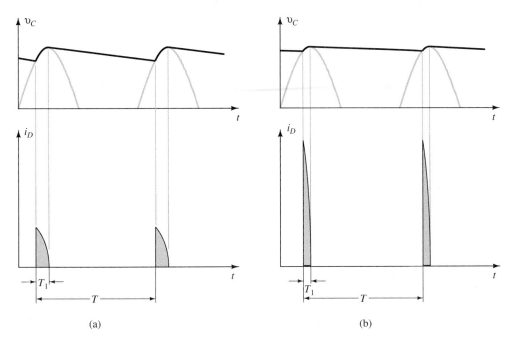

Figure 19.7 Output voltage and diode current waveforms: (a) small C; (b) large C.

tion for full-wave). Notice that for smaller values of capacitor, with T_1 larger, the peak diode current is less than for larger values of filter capacitor.

Since the average current drawn from the supply must equal the average diode current during the charging period, the following relation can be used (assuming constant diode current during charge time):

$$I_{dc} = \frac{T_1}{T} I_{peak}$$

from which we obtain

$$I_{peak} = \frac{T}{T_1} I_{dc} \qquad (19.12)$$

where T_1 = diode conduction time
   $T = 1/f$ ($f = 2 \times 60$ for full-wave)
   I_{dc} = average current drawn from filter
   I_{peak} = peak current through conducting diodes

19.4 RC FILTER

It is possible to further reduce the amount of ripple across a filter capacitor by using an additional *RC* filter section as shown in Fig. 19.8. The purpose of the added *RC* section is to pass most of the dc component while attenuating (reducing) as much of

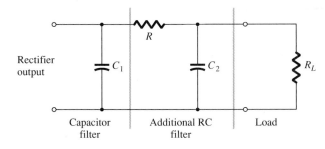

Figure 19.8 *RC* filter stage.

the ac component as possible. Figure 19.9 shows a full-wave rectifier with capacitor filter, followed by an *RC* filter section. The operation of the filter circuit can be analyzed using superposition for the dc and ac components of signal.

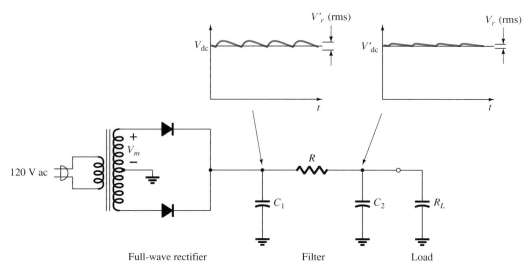

Figure 19.9 Full-wave rectifier and *RC* filter circuit.

DC Operation of *RC* Filter Section

Figure 19.10a shows the dc equivalent circuit to use in analyzing the *RC* filter circuit of Fig. 19.9. Since both capacitors are open-circuit for dc operation, the resulting output dc voltage is

$$V'_{dc} = \frac{R_L}{R + R_L} V_{dc} \qquad (19.13)$$

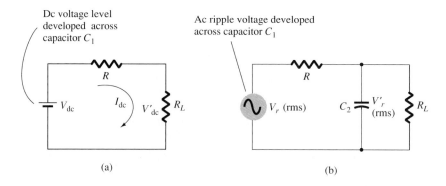

Figure 19.10 (a) Dc and (b) ac equivalent circuits of *RC* filter.

(a)

(b)

EXAMPLE 19.6

Calculate the dc voltage across a 1-kΩ load for an *RC* filter section ($R = 120\ \Omega$, $C = 10\ \mu$F). The dc voltage across the initial filter capacitor is $V_{dc} = 60$ V.

Solution

$$\text{Eq. (19.11):} \quad V'_{dc} = \frac{R_L}{R + R_L} V_{dc} = \frac{1000}{120 + 1000}(60\ \text{V}) = \textbf{53.6 V}$$

Chapter 19   Power Supplies (Voltage Regulators)

AC Operation of *RC* Filter Section

Figure 19.10b shows the ac equivalent circuit of the *RC* filter section. Due to the voltage-divider action of the capacitor ac impedance and the load resistor, the ac component of voltage resulting across the load is

$$V_r'(\text{rms}) \approx \frac{X_C}{R} V_r(\text{rms}) \qquad (19.14)$$

For a full-wave rectifier, with ac ripple at 120 Hz, the impedance of a capacitor can be calculated using

$$X_C = \frac{1.3}{C} \qquad (19.15)$$

where *C* is in microfarads and X_C is in kilohms.

Calculate the dc and ac components of the output signal across load R_L in the circuit of Fig. 19.11. Calculate the ripple of the output waveform.

EXAMPLE 19.7

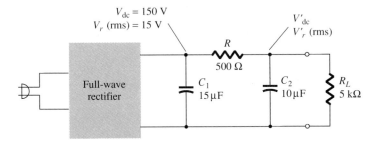

Figure 19.11 *RC* filter circuit for Example 19.7.

Solution

DC Calculation:

Eq. (19.11): $V_{dc}' = \dfrac{R_L}{R + R_L} V_{dc} = \dfrac{5 \text{ k}\Omega}{500 + 5 \text{ k}\Omega}(150 \text{ V}) = \mathbf{136.4 \ V}$

AC Calculation: The *RC* section capacitive impedance is

Eq. (19.13): $X_C = \dfrac{1.3}{C} = \dfrac{1.3}{10} = 0.13 \text{ k}\Omega = 130 \text{ }\Omega$

The ac component of the output voltage, calculated using Eq. (19.12), is

$$V_r'(\text{rms}) = \frac{X_C}{R} V_r(\text{rms}) = \frac{130}{500}(15 \text{ V}) = \mathbf{3.9 \ V}$$

The ripple of the output waveform is then

$$r = \frac{V_r'(\text{rms})}{V_{dc}'} \times 100\% = \frac{3.9 \text{ V}}{136.4 \text{ V}} \times 100\% = \mathbf{2.86\%}$$

19.5 DISCRETE TRANSISTOR VOLTAGE REGULATION

Two types of transistor voltage regulators are the series voltage regulator and the shunt voltage regulator. Each type of circuit can provide an output dc voltage that is regulated or maintained at a set value even if the input voltage varies or if the load connected to the output changes.

Series Voltage Regulation

The basic connection of a series regulator circuit is shown in the block diagram of Fig. 19.12. The series element controls the amount of the input voltage that gets to the output. The output voltage is sampled by a circuit that provides a feedback voltage to be compared to a reference voltage.

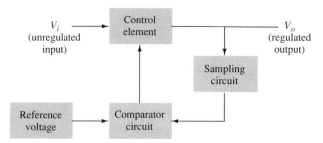

Figure 19.12 Series regulator block diagram.

1. If the output voltage increases, the comparator circuit provides a control signal to cause the series control element to decrease the amount of the output voltage—thereby maintaining the output voltage.

2. If the output voltage decreases, the comparator circuit provides a control signal to cause the series control element to increase the amount of the output voltage.

SERIES REGULATOR CIRCUIT

A simple series regulator circuit is shown in Fig. 19.13. Transistor Q_1 is the series control element and Zener diode D_Z provides the reference voltage. The regulating operation can be described as follows:

1. If the output voltage decreases, the increased base-emitter voltage causes transistor Q_1 to conduct more, thereby raising the output voltage—maintaining the output constant.

2. If the output voltage increases, the decreased base-emitter voltage causes transistor Q_1 to conduct less, thereby reducing the output voltage—maintaining the output constant.

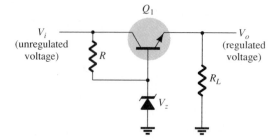

Figure 19.13 Series regulator circuit.

Calculate the output voltage and Zener current in the regulator circuit of Fig. 19.14 for $R_L = 1$ kΩ.

EXAMPLE 19.8

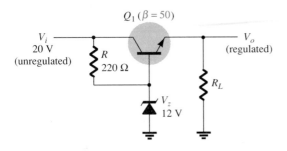

Figure 19.14 Circuit for Example 19.8.

Solution

$$V_o = V_Z - V_{BE} = 12 \text{ V} - 0.7 \text{ V} = \textbf{11.3 V}$$

$$V_{CE} = V_i - V_o = 20 \text{ V} - 11.3 \text{ V} = 8.7 \text{ V}$$

$$I_R = \frac{20 \text{ V} - 12 \text{ V}}{220 \ \Omega} = \frac{8 \text{ V}}{220 \ \Omega} = 36.4 \text{ mA}$$

For $R_L = 1$ kΩ,

$$I_L = \frac{V_o}{R_L} = \frac{11.3 \text{ V}}{1 \text{ k}\Omega} = 11.3 \text{ mA}$$

$$I_B = \frac{I_C}{\beta} = \frac{11.3 \text{ mA}}{50} = 226 \ \mu\text{A}$$

$$I_Z = I_R - I_B = 36.4 \text{ mA} - 226 \ \mu\text{A} \approx \textbf{36 mA}$$

IMPROVED SERIES REGULATOR

An improved series regulator circuit is that of Fig. 19.15. Resistors R_1 and R_2 act as a sampling circuit, Zener diode D_Z providing a reference voltage, and transistor Q_2 then controls the base current to transistor Q_1 to vary the current passed by transistor Q_1 to maintain the output voltage constant.

If the output voltage tries to increase, the increased voltage sampled by R_1 and R_2, increased voltage V_2, causes the base-emitter voltage of transistor Q_2 to go up (since

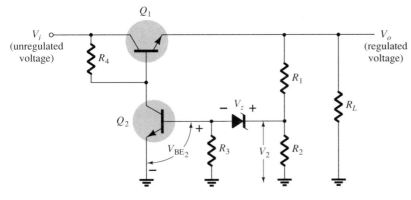

Figure 19.15 Improved series regulator circuit.

V_Z remains fixed). If Q_2 conducts more current, less goes to the base of transistor Q_1, which then passes less current to the load, reducing the output voltage—thereby maintaining the output voltage constant. The opposite takes place if the output voltage tries to decrease, causing less current to be supplied to the load, to keep the voltage from decreasing.

The voltage V_2 provided by sensing resistors R_1 and R_2 must equal the sum of the base-emitter voltage of Q_2 and the Zener diode, that is,

$$V_{BE_2} + V_Z = V_2 = \frac{R_2}{R_1 + R_2} V_o \tag{19.16}$$

Solving Eq. (19.16) for the regulated output voltage, V_o,

$$V_o = \frac{R_1 + R_2}{R_2}(V_Z + V_{BE_2}) \tag{19.17}$$

EXAMPLE 19.9

What regulated output voltage is provided by the circuit of Fig. 19.15 for circuit elements: $R_1 = 20$ kΩ, $R_2 = 30$ kΩ, and, $V_Z = 8.3$ V?

Solution

From Eq. (19.17) the regulated output voltage will be

$$V_o = \frac{30 \text{ k}\Omega}{30 \text{ k}\Omega + 20 \text{ k}\Omega}(8.3 \text{ V} + 0.7 \text{ V}) = \textbf{15 V}$$

OP-AMP SERIES REGULATOR

Another version of series regulator is that shown in Fig. 19.16. The op-amp compares the Zener diode reference voltage with the feedback voltage from sensing resistors R_1 and R_2. If the output voltage varies, the conduction of transistor Q_1 is controlled to maintain the output voltage constant. The output voltage will be maintained at a value of

$$V_o = \left(1 + \frac{R_1}{R_2}\right)V_Z \tag{19.18}$$

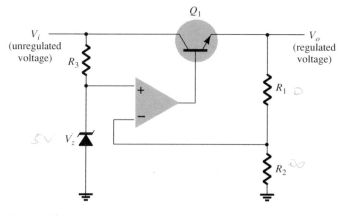

Figure 19.16 Op-amp series regulator circuit.

Calculate the regulated output voltage in the circuit of Fig. 19.17.

EXAMPLE 19.10

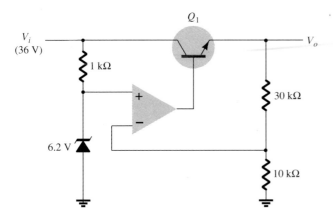

Figure 19.17 Circuit for Example 19.10.

Solution

$$\text{Eq. (19.18):} \quad V_o = \left(\frac{1 + 30 \text{ k}\Omega}{10 \text{ k}\Omega}\right)6.2 \text{ V} = \textbf{24.8 V}$$

CURRENT-LIMITING CIRCUIT

One form of short-circuit or overload protection is current limiting, as shown in Fig. 19.18. As load current I_L increases, the voltage drop across the short-circuit sensing resistor R_{SC} increases. When the voltage drop across R_{SC} becomes large enough, it will drive Q_2 on, diverting current from the base of transistor Q_1, thereby reducing the load current through transistor Q_1, preventing any additional current to load R_L. The action of components R_{SC} and Q_2 provides limiting of the maximum load current.

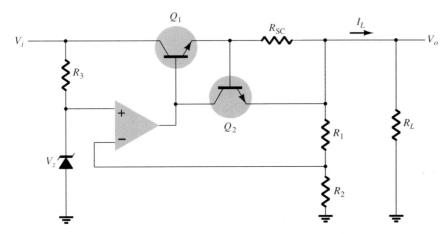

Figure 19.18 Current-limiting voltage regulator.

FOLDBACK LIMITING

Current limiting reduces the load voltage when the current becomes larger than the limiting value. The circuit of Fig. 19.19 provides foldback limiting, which reduces both the output voltage and output current protecting the load from overcurrent, as well as protecting the regulator.

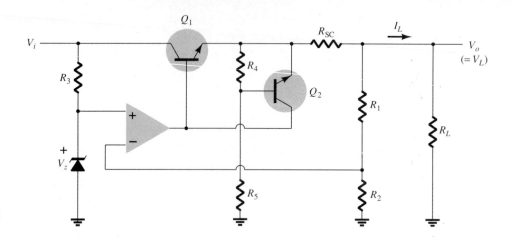

Figure 19.19 Foldback-limiting series regulator circuit.

Foldback limiting is provided by the additional voltage-divider network of R_4 and R_5 in the circuit of Fig. 19.19 (over that of Fig. 19.17). The divider circuit senses the voltage at the output (emitter) of Q_1. When I_L increases to its maximum value, the voltage across R_{SC} becomes large enough to drive Q_2 on, thereby providing current limiting. If the load resistance is made smaller, the voltage driving Q_2 on becomes less, so that I_L drops when V_L also drops in value—this action being foldback limiting. When the load resistance is returned to its rated value, the circuit resumes its voltage regulation action.

Shunt Voltage Regulation

A shunt voltage regulator provides regulation by shunting current away from the load to regulate the output voltage. Figure 19.20 shows the block diagram of such a voltage regulator. The input unregulated voltage provides current to the load. Some of the current is pulled away by the control element to maintain the regulated output voltage across the load. If the load voltage tries to change due to a change in the load, the sampling circuit provides a feedback signal to a comparator which then provides a control signal to vary the amount of the current shunted away from the load. As the output voltage tries to get larger, for example, the sampling circuit provides a feedback signal to the comparator circuit which then provides a control signal to draw increased shunt current, providing less load current, thereby keeping the regulated voltage from rising.

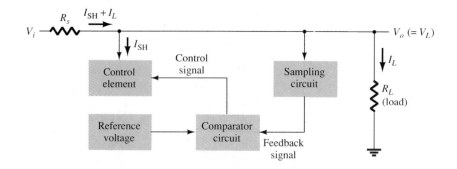

Figure 19.20 Block diagram of shunt voltage regulator.

BASIC TRANSISTOR SHUNT REGULATOR

A simple shunt regulator circuit is shown in Fig. 19.21. Resistor R_S drops the unregulated voltage by an amount that depends on the current supplied to the load, R_L. The voltage across the load is set by the Zener diode and transistor base-emitter

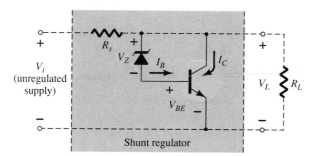

Figure 19.21 Transistor shunt voltage regulator.

voltage. If the load resistance decreases, a reduced drive current to the base of Q_1 results, shunting less collector current. The load current is thus larger, thereby maintaining the regulated voltage across the load. The output voltage to the load is

$$V_L = V_Z + V_{BE} \qquad (19.19)$$

Determine the regulated voltage and circuit currents for the shunt regulator of Fig. 19.22.

EXAMPLE 19.11

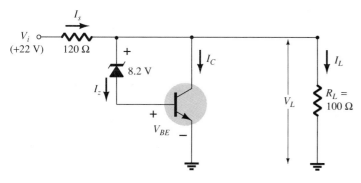

Figure 19.22 Circuit for Example 19.11.

Solution

The load voltage is

Eq. (19.19): $V_L = 8.2 \text{ V} + 0.7 \text{ V} = 8.9 \text{ V}$

For the given load,

$$I_L = \frac{V_L}{R_L} = \frac{8.9 \text{ V}}{100 \ \Omega} = 89 \text{ mA}$$

With the unregulated input voltage at 22 V, the current through R_S is

$$I_S = \frac{V_i - V_L}{R_S} = \frac{22 \text{ V} - 8.9 \text{ V}}{120} = 109 \text{ mA}$$

so that the collector current is

$$I_C = I_S - I_L = 109 \text{ mA} - 89 \text{ mA} = 20 \text{ mA}$$

(The current through the Zener and transistor base–emitter is smaller than I_C by the transistor beta.)

IMPROVED SHUNT REGULATOR

The circuit of Fig. 19.23 shows an improved shunt voltage regulator circuit. The zener diode provides a reference voltage so that the voltage across R_1 senses the output voltage. As the output voltage tries to change, the current shunted by transistor Q_1 is varied to maintain the output voltage constant. Transistor Q_2 provides a larger base current to transistor Q_1 than the circuit of Fig. 19.21, so that the regulator handles a larger load current. The output voltage is set by the zener voltage and that across the two transistor base–emitters,

$$V_o = V_L = V_Z + V_{BE_2} + V_{BE_1} \tag{19.20}$$

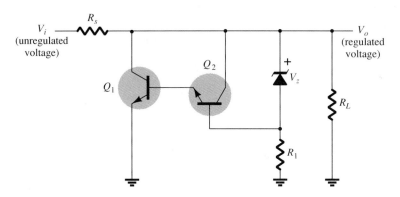

Figure 19.23 Improved shunt voltage regulator circuit.

SHUNT VOLTAGE REGULATOR USING OP-AMP

Figure 19.24 shows another version of a shunt voltage regulator using an op-amp as voltage comparator. The zener voltage is compared to the feedback voltage obtained from voltage divider R_1 and R_2 to provide the control drive current to shunt element Q_1. The current through resistor R_S is thus controlled to drop a voltage across R_S so that the output voltage is maintained.

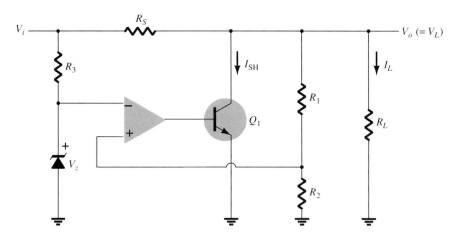

Figure 19.24 Shunt voltage regulator using op-amp.

Switching Regulation

A type of regulator circuit that is quite popular for its efficient transfer of power to the load is the switching regulator. Basically, a switching regulator passes voltage to the load in pulses which are then filtered to provide a smooth dc voltage. Figure 19.25 shows the basic components of such a voltage regulator. The added circuit complexity is well worth the improved operating efficiency obtained.

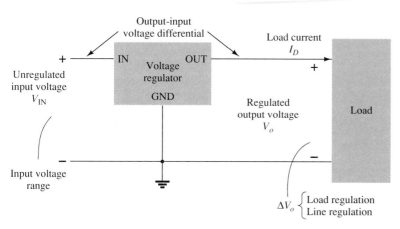

Figure 19.25 Block representation of three-terminal voltage regulator.

19.6 IC VOLTAGE REGULATORS

Voltage regulators comprise a class of widely used ICs. Regulator IC units contain the circuitry for reference source, comparator amplifier, control device, and overload protection all in a single IC. Although the internal construction of the IC is somewhat different from that described for discrete voltage regulator circuits, the external operation is much the same. IC units provide regulation of either a fixed positive voltage, a fixed negative voltage, or an adjustably set voltage.

A power supply can be built using a transformer connected to the ac supply line to step the ac voltage to a desired amplitude, then rectifying that ac voltage, filtering with a capacitor and RC filter, if desired, and finally regulating the dc voltage using an IC regulator. The regulators can be selected for operation with load currents from hundreds of milliamperes to tens of amperes, corresponding to power ratings from milliwatts to tens of watts.

Three-Terminal Voltage Regulators

Figure 19.25 shows the basic connection of a three-terminal voltage regulator IC to a load. The fixed voltage regulator has an unregulated dc input voltage, V_i, applied to one input terminal, a regulated output dc voltage, V_o, from a second terminal, with the third terminal connected to ground. For a selected regulator, IC device specifications list a voltage range over which the input voltage can vary to maintain a regulated output voltage over a range of load current. The specifications also list the amount of output voltage change resulting from a change in load current (load regulation) or in input voltage (line regulation).

Fixed Positive Voltage Regulators

The series 78 regulators provide fixed regulated voltages from 5 to 24 V. Figure 19.26 shows how one such IC, a 7812, is connected to provide voltage regulation with output from this unit of $+12$ V dc. An unregulated input voltage V_i is filtered by capacitor C_1 and connected to the IC's IN terminal. The IC's OUT terminal provides a regulated $+12$ V which is filtered by capacitor C_2 (mostly for any high-frequency

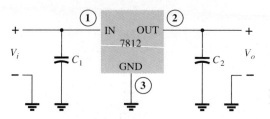

Figure 19.26 Connection of 7812 voltage regulator.

noise). The third IC terminal is connected to ground (GND). While the input voltage may vary over some permissible voltage range, and the output load may vary over some acceptable range, the output voltage remains constant within specified voltage variation limits. These limitations are spelled out in the manufacturer's specification sheets. A table of positive voltage regulator ICs is provided in Table 19.1.

TABLE 19.1 Positive Voltage Regulators in 7800 Series		
IC Part	Output Voltage (V)	Minimum V_i (V)
7805	+5	7.3
7806	+6	8.3
7808	+8	10.5
7810	+10	12.5
7812	+12	14.6
7815	+15	17.7
7818	+18	21.0
7824	+24	27.1

The connection of a 7812 in a complete voltage supply is shown in the connection of Fig. 19.27. The ac line voltage (120 V rms) is stepped down to 18 V rms across each half of the center-tapped transformer. A full wave-rectifier and capacitor filter then provides an unregulated dc voltage, shown as a dc voltage of about 22 V, with ac ripple of a few volts as input to the voltage regulator. The 7812 IC then provides an output that is a regulated +12 V dc.

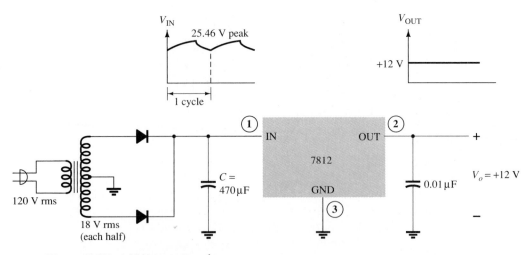

Figure 19.27 +12 V power supply.

POSITIVE VOLTAGE REGULATOR SPECIFICATIONS

The specifications sheet of voltage regulators is typified by that shown in Fig. 19.28 for the group of series 7800 positive voltage regulators. Some consideration of a few of the more important parameters should be considered.

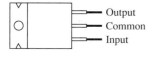

Nominal output voltage	Regulator
5 V	7805
6 V	7806
8 V	7808
10 V	7810
12 V	7812
15 V	7815
18 V	7818
24 V	7824

Absolute maximum ratings:

Input voltage 40 V
Continuous total dissipation 2 W
Operating free-air
   temperature range −65 to 150°C

μA 7812C electrical characteristics:

Parameter	Min.	Typ.	Max.	Units
Output voltage	11.5	12	12.5	V
Input regulation		3	120	mV
Ripple rejection	55	71		dB
Output regulation		4	120	mV
Output resistance		0.018		Ω
Dropout voltage		2.0		V
Short-circuit output current		350		mA
Peak output current		2.2		A

Figure 19.28 Specification sheet data for voltage regulator ICs.

Output voltage: The specifications for the 7812 shows that the output voltage is typically +12 V but could be as low as 11.5 V or as high as 12.5 V.

Output regulation: The output voltage regulation is seen to be typically 4 mV, to a maximum of 100 mV (at output currents from 0.25 to 0.75 A). This information specifies that the output voltage can typically vary only 4 mV from the rated 12 V dc.

Short-circuit output current: The amount of current is limited to typically 0.35 A if the output were to be short-circuited (presumably by accident or by another faulty component).

Peak output current: While the rated maximum current is 1.5 A for this series of IC, the typical peak output current that might be drawn by a load is 2.2 A. This shows that although the manufacturer rates the IC as capable of providing 1.5 A, one could draw somewhat more current (possibly for a short period of time).

Dropout voltage: The dropout voltage, typically 2 V, is the minimum amount of voltage across the input–output terminals that must be maintained if the IC is to operate as a regulator. If the input voltage drops too low or the output rises so that at least 2 V is not maintained across the IC input–output, the IC will no longer provide voltage regulation. One therefore maintains an input voltage large enough to assure that the dropout voltage is provided.

Fixed Negative Voltage Regulators

The series 7900 ICs provide negative voltage regulators, similar to those providing positive voltages. A list of negative voltage regulator ICs is provided in Table 19.2. As shown, IC regulators are available for a range of fixed negative voltages, the selected IC providing the rated output voltage as long as the input voltage is maintained greater than the minimum input value. For example, the 7912 provides an output of -12 V as long as the input to the regulator IC is more negative than -14.6 V.

TABLE 19.2 Negative Voltage Regulators in 7900 Series		
IC Part	*Output Voltage (V)*	*Minimum V_i (V)*
7905	-5	-7.3
7906	-6	-8.4
7908	-8	-10.5
7909	-9	-11.5
7912	-12	-14.6
7915	-15	-17.7
7918	-18	-20.8
7924	-24	-27.1

EXAMPLE 19.12

Draw a voltage supply using a full-wave bridge rectifier, capacitor filter, and IC regulator to provide an output of $+5$ V.

Solution

The resulting circuit is shown in Fig. 19.29.

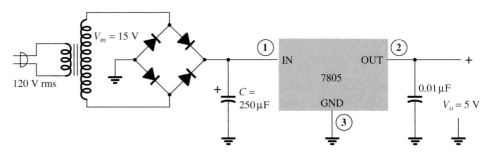

Figure 19.29 Positive 5-V power supply.

EXAMPLE 19.13

For a transformer output of 15 V and a filter capacitor of 250 μF, calculate the minimum input voltage when connected to a load drawing 400 mA.

Solution

The voltages across the filter capacitor are

$$V_r(\text{peak}) = \sqrt{3}\, V_r(\text{rms}) = \sqrt{3}\frac{2.4 I_{dc}}{C} = \sqrt{3}\frac{2.4(400)}{250} = 6.65 \text{ V}$$

$$V_{dc} = V_m - V_r(\text{peak}) = 15 \text{ V} - 6.65 \text{ V} = 8.35 \text{ V}$$

Since the input swings around this dc level, the minimum input voltage can drop to as low as

$$V_i(\text{low}) = V_{dc} - V_r(\text{peak}) = 15 \text{ V} - 6.65 \text{ V} = \textbf{8.35 V}$$

Since this voltage is greater than the minimum required for the IC regulator (from Table 19.1 $V_i = 7.3$ V), the IC can provide a regulated voltage to the given load.

Determine the maximum value of load current at which regulation is maintained for the circuit of Fig. 19.29.

EXAMPLE 19.14

Solution

To maintain $V_i(\text{min}) \geq 7.3$ V,

$$V_r(\text{peak}) \leq V_m - V_i(\text{min}) = 15 \text{ V} - 7.3 \text{ V} = 7.7 \text{ V}$$

so that

$$V_r(\text{rms}) = \frac{V_r(\text{peak})}{\sqrt{3}} = \frac{7.7 \text{ V}}{1.73} = 4.4 \text{ V}$$

The value of load current is then

$$I_{dc} = \frac{V_r(\text{rms})C}{2.4} = \frac{(4.4 \text{ V})(250)}{2.4} = \textbf{458 mA}$$

Any current above this value is too large for the circuit to maintain the regulator output at +5 V.

Adjustable Voltage Regulators

Voltage regulators are also available in circuit configurations that allow the user to set the output voltage to a desired regulated value. The LM317, for example, can be operated with the output voltage regulated at any setting over the range of voltage from 1.2 V to 37 V. Figure 19.30 shows how the regulated output voltage of an LM317 can be set.

Resistors R_1 and R_2 set the output to any desired voltage over the adjustment range (1.2 to 37 V). The output voltage desired can be calculated using

$$V_o = V_{\text{ref}}\left(1 + \frac{R_2}{R_1}\right) + I_{\text{adj}}R_2 \tag{19.21}$$

with typical IC values of

$$V_{\text{ref}} = 1.25 \text{ V} \qquad \text{and} \qquad I_{\text{adj}} = 100 \text{ }\mu\text{A}$$

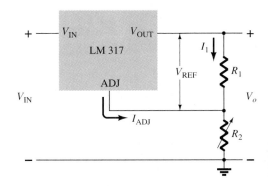

Figure 19.30 Connection of LM317 adjustable-voltage regulator.

EXAMPLE 19.15

Determine the regulated voltage in the circuit of Fig. 19.30 with $R_1 = 240$ Ω and $R_2 = 2.4$ kΩ.

Solution

Eq. (19.21): $V_o = 1.25$ V$\left(1 + \dfrac{2.4 \text{ k}\Omega}{240 \text{ }\Omega}\right) + (100 \text{ }\mu\text{A})(2.4 \text{ k}\Omega)$

$= 13.75$ V $+ 0.24$ V $= \mathbf{13.99}$ **V**

EXAMPLE 19.16

Determine the regulated output voltage of the circuit in Fig. 19.31.

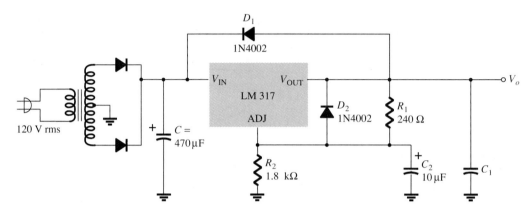

Figure 19.31   Positive adjustable-voltage regulator for Example 19.16.

Solution

The output voltage calculated using Eq. (19.21) is

$$V_o = 1.25 \text{ V}\left(1 + \frac{1.8 \text{ k}\Omega}{240 \text{ }\Omega}\right) + (100 \text{ }\mu\text{A})(1.8 \text{ k}\Omega) \approx \mathbf{10.8 \text{ V}}$$

A check of the filter capacitor voltage shows that an input–output difference of 2 V can be maintained up to at least 200 mA load current.

PROBLEMS

§ **19.2 General Filter Considerations**

1. What is the ripple factor of a sinusoidal signal having peak ripple of 2 V on an average of 50 V?

2. A filter circuit provides an output of 28 V unloaded and 25 V under full-load operation. Calculate the percent voltage regulation.

3. A half-wave rectifier develops 20 V dc. What is the value of the ripple voltage?

4. What is the rms ripple voltage of a full-wave rectifier with output voltage 8 V dc?

§ **19.3 Capacitor Filter**

5. A simple capacitor filter fed by a full-wave rectifier develops 14.5 V dc at 8.5% ripple factor. What is the output ripple voltage (rms)?

6. A full-wave rectified signal of 18 V peak is fed into a capacitor filter. What is the voltage regulation of the filter if the output is 17 V dc at full load?

7. A full-wave rectified voltage of 18 V peak is connected to a 400-μF filter capacitor. What are the ripple and dc voltages across the capacitor at a load of 100 mA?

8. A full-wave rectifier operating from the 60-Hz ac supply produces a 20-V peak rectified voltage. If a 200-μF capacitor is used, calculate the ripple at a load of 120 mA.

9. A full-wave rectifier (operating from a 60-Hz supply) drives a capacitor-filter circuit ($C = 100\ \mu$F) which develops 12 V dc when connected to a 2.5-kΩ load. Calculate the output voltage ripple.

10. Calculate the size of the filter capacitor needed to obtain a filtered voltage having 15% ripple at a load of 150 mA. The full-wave rectified voltage is 24 V dc and the supply is 60 Hz.

* 11. A 500-μF capacitor provides a load current of 200 mA at 8% ripple. Calculate the peak rectified voltage obtained from the 60-Hz supply and the dc voltage across the filter capacitor.

12. Calculate the size of the filter capacitor needed to obtain a filtered voltage with 7% ripple at a load of 200 mA. The full-wave rectified voltage is 30 V dc and supply is 60 Hz.

13. Calculate the percent ripple for the voltage developed across a 120-μF filter capacitor when providing a load current of 80 mA. The full-wave rectifier operating from the 60-Hz supply develops a peak rectified voltage of 25 V.

§ 19.4 RC Filter

14. An *RC* filter stage is added after a capacitor filter to reduce the percent of ripple to 2%. Calculate the ripple voltage at the output of the *RC* filter stage providing 80 V dc.

* 15. An *RC* filter stage ($R = 33\ \Omega$, $C = 120\ \mu$F) is used to filter a signal of 24 V dc with 2 V rms operating from a full-wave rectifier. Calculate the percent ripple at the output of the *RC* section for a 100-mA load. Also, calculate the ripple of the filtered signal applied to the *RC* stage.

* 16. A simple capacitor filter has an input of 40 V dc. If this voltage is fed through an *RC* filter section ($R = 50\ \Omega$, $C = 40\ \mu$F), what is the load current for a load resistance of 500 Ω?

17. Calculate the rms ripple voltage at the output of an *RC* filter section that feeds a 1-kΩ load when the filter input is 50 V dc with 2.5-V rms ripple from a full-wave rectifier and capacitor filter. The *RC* filter section components are $R = 100\ \Omega$ and $C = 100\ \mu$F.

18. If the no-load output voltage for Problem 17 is 50 V, calculate the percent voltage regulation with a 1-kΩ load.

§ 19.5 Discrete Transistor Voltage Regulation

* 19. Calculate the output voltage and Zener diode current in the regulator circuit of Fig. 19.32.

20. What regulated output voltage results in the circuit of Fig. 19.33?

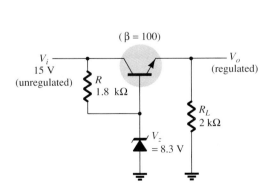

Figure 19.32 Problem 19

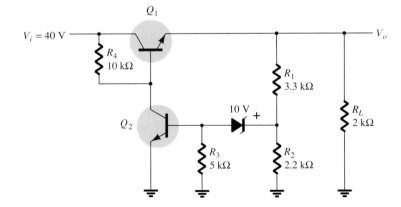

Figure 19.33 Problem 20

21. Calculate the regulated output voltage in the circuit of Fig. 19.34.

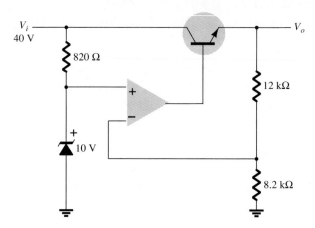

Figure 19.34 Problem 21

22. Determine the regulated voltage and circuit currents for the shunt regulator of Fig. 19.35.

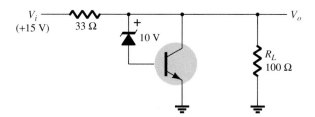

Figure 19.35 Problem 22

§ **19.6 IC Voltage Regulators**

23. Draw the circuit of a voltage supply comprised of a full-wave bridge rectifier, capacitor filter, and IC regulator to provide an output of +12 V.

* **24.** Calculate the minimum input voltage of the full-wave rectifier and filter capacitor network in Fig. 19.36 when connected to a load drawing 250 mA.

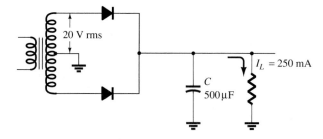

Figure 19.36 Problem 24

* **25.** Determine the maximum value of load current at which regulation is maintained for the circuit of Fig. 19.37.

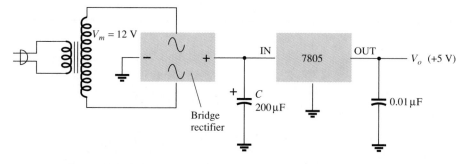

Figure 19.37 Problem 25

Chapter 19 Power Supplies (Voltage Regulators)

26. Determine the regulated voltage in the circuit of Fig. 19.30 with $R_1 = 240\ \Omega$ and $R_2 = 1.8\ k\Omega$.

27. Determine the regulated output voltage from the circuit of Fig. 19.38.

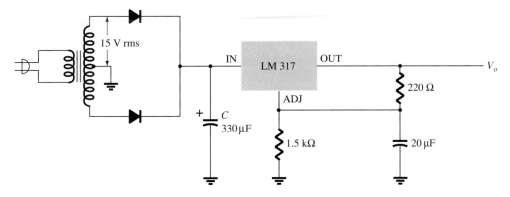

Figure 19.38 Problem 27

---

*Please Note: Asterisks indicate more difficult problems.

20
Other Two-Terminal Devices

20.1 INTRODUCTION

There are a number of two-terminal devices having a single *p-n* junction like the semiconductor or Zener diode but with different modes of operation, terminal characteristics, and areas of application. A number, including the Schottky, tunnel, varactor, photodiode, and solar cell will be introduced in this chapter. In addition, two-terminal devices of a different construction, such as the photoconductive cell, LCD (liquid-crystal display), and thermistor will be examined.

20.2 SCHOTTKY BARRIER (HOT-CARRIER) DIODES

In recent years there has been increasing interest in a two-terminal device referred to as a *Schottky-barrier, surface-barrier,* or *hot-carrier* diode. Its areas of application were first limited to the very high frequency range due to its quick response time (especially important at high frequencies) and a lower noise figure (a quantity of real importance in high-frequency applications). In recent years, however, it is appearing more and more in low-voltage/high-current power supplies and ac-to-dc converters. Other areas of application of the device include radar systems, Schottky TTL logic for computers, mixers and detectors in communication equipment, instrumentation, and analog-to-digital converters.

Its construction is quite different from the conventional *p-n* junction in that a metal-semiconductor junction is created such as shown in Fig. 20.1. The semicon-

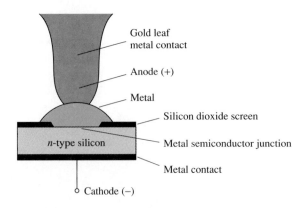

Figure 20.1 Passivated hot-carrier diode.

Labels: Gold leaf metal contact, Anode (+), Metal, Silicon dioxide screen, *n*-type silicon, Metal semiconductor junction, Metal contact, Cathode (−)

ductor is normally *n*-type silicon (although *p*-type silicon is sometimes used), while a host of different metals, such as molybdenum, platinum, chrome, or tungsten, are used. Different construction techniques will result in a different set of characteristics for the device, such as increased frequency range, lower forward bias, and so on. Priorities do not permit an examination of each technique here, but information will usually be provided by the manufacturer. In general, however, Schottky diode construction results in a more uniform junction region and a high level of ruggedness.

In both materials, the electron is the majority carrier. In the metal, the level of minority carriers (holes) is insignificant. When the materials are joined the electrons in the *n*-type silicon semiconductor material immediately flow into the adjoining metal, establishing a heavy flow of majority carriers. Since the injected carriers have a very high kinetic energy level compared to the electrons of the metal, they are commonly called "hot carriers." In the conventional *p-n* junction there was the injection of minority carriers into the adjoining region. Here the electrons are injected into a region of the same electron plurality. Schottky diodes are therefore unique, in that conduction is entirely by majority carriers. The heavy flow of electrons into the metal creates a region near the junction surface depleted of carriers in the silicon material—much like the depletion region in the *p-n* junction diode. The additional carriers in the metal establish a "negative wall" in the metal at the boundary between the two materials. The net result is a "surface barrier" between the two materials, preventing any further current. That is, any electrons (negatively charged) in the silicon material face a carrier-free region and a "negative wall" at the surface of the metal.

The application of a forward bias as shown in the first quadrant of Fig. 20.2 will reduce the strength of the negative barrier through the attraction of the applied positive potential for electrons from this region. The result is a return to the heavy flow of electrons across the boundary, the magnitude of which is controlled by the level of the applied bias potential. The barrier at the junction for a Schottky diode is less than that of the *p-n* junction device in both the forward- and reverse-bias regions. The result is therefore a higher current at the same applied bias in the forward- and reverse-bias regions. This is a desirable effect in the forward-bias region but highly undesirable in the reverse-bias region.

The exponential rise in current with forward bias is described by Eq. (1.4) but with η dependent on the construction technique (1.05 for the metal whisker type of

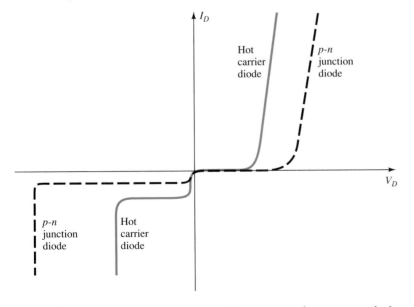

Figure 20.2 Comparison of characteristics of hot-carrier and *p-n* junction diodes.

construction, which is somewhat similar to the germanium diode). In the reverse-bias region the current I_s is due primarily to those electrons in the metal passing into the semiconductor material. One of the areas of continuing research on the Schottky diode centers on reducing the high leakage currents that result with temperatures over 100°C. Through design, improvement units are now becoming available that have a temperature range from -65 to $+150°C$. At room temperature, I_s is typically in the microampere range for low-power units and milliampere range for high-power devices, although it is typically larger than that encountered using conventional p-n junction devices with the same current limits. In addition, the PIV of Schottky diodes is usually significantly less than that of a comparable p-n junction unit. Typically, for a 50-A unit, the PIV of the Schottky diode would be about 50 V as compared to 150 V for the p-n junction variety. Recent advances, however, have resulted in Schottky diodes with PIVs greater than 100 V at this current level. It is obvious from the characteristics of Fig. 20.2 that the Schottky diode is closer to the ideal set of characteristics than the point contact and has levels of V_T less than the typical silicon semiconductor p-n junction. The level of V_T for the "hot-carrier" diode is controlled to a large measure by the metal employed. There exists a required trade-off between temperature range and level of V_T. An increase in one appears to correspond to a resulting increase in the other. In addition, the lower the range of allowable current levels, the lower the value of V_T. For some low-level units, the value of V_T can be assumed to be essentially zero on an approximate basis. For the middle and high range, however, a value of 0.2 V would appear to be a good representative value.

The maximum current rating of the device is presently limited to about 75 A, although 100-A units appear to be on the horizon. One of the primary areas of application of this diode is in *switching power supplies* that operate in the frequency range of 20 kHz or more. A typical unit at 25°C may be rated at 50 A at a forward voltage of 0.6 V with a recovery time of 10 ns for use in one of these supplies. A p-n junction device with the same current limit of 50 A may have a forward voltage drop of 1.1 V and a recovery time of 30 to 50 ns. The difference in forward voltage may not appear significant, but consider the power dissipation difference: $P_{\text{hot carrier}} = (0.6 \text{ V})(50 \text{ A}) = 30 \text{ W}$ compared to $P_{p\text{-}n} = (1.1 \text{ V})(50 \text{ A}) = 55 \text{ W}$, which is a measurable difference when efficiency criteria must be met. There will, of course, be a higher dissipation in the reverse-bias region for the Schottky diode due to the higher leakage current, but the total power loss in the forward- and reverse-bias regions is still significantly improved as compared to the p-n junction device.

Recall from our discussion of reverse recovery time for the semiconductor diode that the injected minority carriers accounted for the high level of t_{rr} (the reverse recovery time). The absence of minority carriers at any appreciable level in the Schottky diode results in a reverse recovery time of significantly lower levels, as indicated above. This is the primary reason Schottky diodes are so effective at frequencies approaching 20 GHz, where the device must switch states at a very high rate. For higher frequencies the point-contact diode, with its very small junction area, is still employed.

The equivalent circuit for the device (with typical values) and a commonly used symbol appear in Fig. 20.3. A number of manufacturers prefer to use the standard diode symbol for the device, since its function is essentially the same. The inductance L_P and capacitance C_P are package values, and r_B is the series resistance, which includes the contact and bulk resistance. The resistance r_d and capacitance C_J are values defined by equations introduced in earlier sections. For many applications, an excellent approximate equivalent circuit simply includes an ideal diode in parallel with the junction capacitance as shown in Fig. 20.4.

A number of hot-carrier rectifiers manufactured by Motorola Semiconductor Products, Inc., appear in Fig. 20.5 with their specifications and terminal identifica-

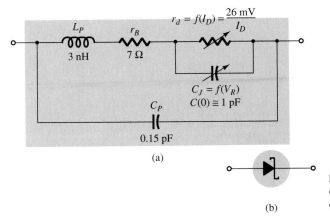

$$r_d = f(I_D) = \frac{26\ \text{mV}}{I_D}$$

L_P 3 nH r_B 7 Ω

$C_J = f(V_R)$
$C(0) \cong 1$ pF

C_P 0.15 pF

(a)

(b)

Figure 20.3 Schottky (hot-carrier) diode: (a) equivalent circuit; (b) symbol.

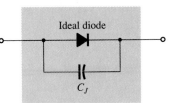

Ideal diode

C_J

Figure 20.4 Approximate equivalent circuit for the Schottky diode.

V_{RRM} (Volts)	I_O, Average rectified forward current (amperes)													
	0.5 A	1.0 A		3.0 A		3.0 A	5.0 A	15 A		25 A		40 A		
Case	51-02 (DO-7) Glass	59-04 Plastic		267 Plastic		60 Metal		257 (DO-4) Metal				257 (DO-5) Metal		430-2 (DO-21) Metal
20	MBR020	IN5817	MBR120P	IN5820	MBR320P	MBR320M	IN5823	IN5826	MBR1520	IN5829	MBR2520	IN5832	MBR4020	MBR4020PF
30	MBR030	IN5818	MBR130P	IN5821	MBR330P	MBR330M	IN5824	IN5827	MBR1530	IN5830	MBR2530	IN5833	MBR4030	MBR4030PF
35			MBR135P		MBR335P	MBR335M			MBR1535		MBR2535		MBR4035	MBR4035PF
40		IN5819	MBR140P	IN5822	MBR340P	MBR340M	IN5825	IN5828	MBR1540	IN5831	MBR2540	IN5834	MBR4040	
I_{FSM} (Amps)	5.0	100	50	250	200	500	500	500	500	800	800	800	800	800
T_C @ Rated I_O (°C)								85	80	85	80	75	70	50
T_J Max	125°C	125°C	125°C	125°C	125°C	125°C	125°C	125°C	125°C	125°C	125°C	125°C	125°C	125°C
Max V_F @ $I_{FM} = I_O$	0.50 V	*0.60 V	0.65 V	*0.525 V	0.60 V	0.45 V@5A	*0.38 V	*0.50 V	0.55 V	*0.48 V	0.55 V	*0.59 V	0.63 V	0.63 V

... Schottky barrier devices, ideal for use in low-voltage, high-frequency power supplies and as free-wheeling diodes. These units feature very low forward voltages and switching times estimated at less than 10 ns. They are offered in current ranges of 0.5 to 5.0 amperes and in voltages to 40.

V_{RRM} —respective peak reverse voltage
I_{FSM} —forward current, surge peak
I_{FM} —forward current, maximum

Figure 20.5 Motorola Schottky barrier devices. (Courtesy Motorola Semiconductor Products, Incorporated.)

tion. Note that the maximum forward voltage drop V_F does not exceed 0.65 V for any of the devices, while this was essentially V_T for a silicon diode.

Three sets of curves for the Hewlett-Packard 5082-2300 series of general-purpose Schottky barrier diodes are provided in Fig. 20.6. Note at $T = 100°C$ in Fig. 20.6a that V_F is only 0.1 V at a current of 0.01 mA. Note also that the reverse current has been limited to nanoamperes in Fig. 20.6b and the capacitance to 1 pF in Fig. 20.6c to ensure a high switching rate.

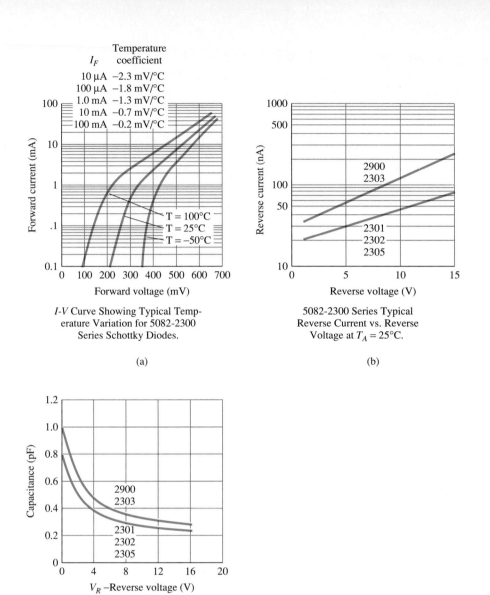

I-V Curve Showing Typical Temperature Variation for 5082-2300
Series Schottky Diodes.

(a)

5082-2300 Series Typical
Reverse Current vs. Reverse
Voltage at $T_A = 25°C$.

(b)

5082-2300 Series Typical
Capacitance vs. Reverse
Voltage at $T_A = 25°C$.

(c)

Figure 20.6 Characteristic curves for Hewlett-
Packard 5082-2300 series of general-purpose
Schottky barrier diodes. (Courtesy Hewlett-Packard
Corporation.)

20.3 VARACTOR (VARICAP) DIODES

Varactor [also called varicap, VVC (voltage-variable capacitance), or tuning] diodes
are semiconductor, voltage-dependent, variable capacitors. Their mode of operation
depends on the capacitance that exists at the *p-n* junction when the element is reverse-
biased. Under reverse-bias conditions, it was established that there is a region of
uncovered charge on either side of the junction that together make up the depletion
region and define the depletion width W_d. The transition capacitance (C_T) established
by the isolated uncovered charges is determined by

$$C_T = \epsilon \frac{A}{W_d}$$ (20.1)

where ϵ is the permittivity of the semiconductor materials, A the p-n junction area, and W_d the depletion width.

As the reverse-bias potential increases, the width of the depletion region increases, which in turn reduces the transition capacitance. The characteristics of a typical commercially available varicap diode appear in Fig. 20.7. Note the initial sharp decline in C_T with increase in reverse bias. The normal range of V_R for VVC diodes is limited to about 20 V. In terms of the applied reverse bias, the transition capacitance is given approximately by

$$C_T = \frac{K}{(V_T + V_R)^n} \qquad (20.2)$$

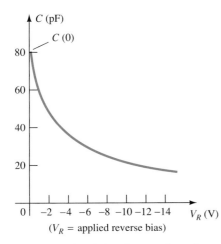

(VR = applied reverse bias)

Figure 20.7 Varicap characteristics: C (pF) versus V_R.

where K = constant determined by the semiconductor material and construction technique

V_T = knee potential as defined in Section 1.6

V_R = magnitude of the applied reverse-bias potential

$n = \frac{1}{2}$ for alloy junctions and $\frac{1}{3}$ for diffused junctions

In terms of the capacitance at the zero-bias condition $C(0)$, the capacitance as a function of V_R is given by

$$C_T(V_R) = \frac{C(0)}{(1 + |V_R/V_T|)^n} \qquad (20.3)$$

The symbols most commonly used for the varicap diode and a first approximation for its equivalent circuit in the reverse-bias region are shown in Fig. 20.8. Since we

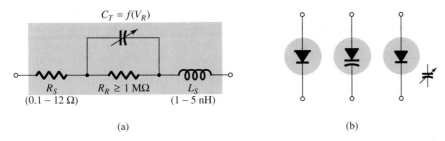

(a)

(b)

Figure 20.8 Varicap diode: (a) equivalent circuit in the reverse-bias region; (b) symbols.

are in the reverse-bias region, the resistance in the equivalent circuit is very large in magnitude—typically 1 MΩ or larger—while R_S, the geometric resistance of the diode, is, as indicated in Fig. 20.8, very small. The magnitude of C_T will vary from about 2 to 100 pF depending on the varicap considered. To ensure that R_R is as large (for minimum leakage current) as possible, silicon is normally used in varicap diodes. The fact that the device will be employed at very high frequencies requires that we include the inductance L_s even though it is measured in nanohenries. Recall that $X_L = 2\pi fL$ and a frequency of 10 GHz with $L_s = 1$ nH will result in an $X_{L_s} = 2\pi fL = (6.28)(10^{10}\text{ Hz})(10^{-9}\text{ F}) = 62.8\ \Omega$. There is obviously, therefore, a frequency limit associated with the use of each varicap diode.

Assuming the proper frequency range, a low value of R_S and X_{L_s} compared to the other series elements, then the equivalent circuit for the varicap of Fig. 20.8a can be replaced by the variable capacitor alone. The complete data sheet and its characteristic curves appear in Figs. 20.9 and 20.10, respectively. The C_3/C_{25} ratio in Fig. 20.9

BB 139

VHF/FM VARACTOR DIODE
DIFFUSED SILICON PLANAR

- C_3/C_{25} . . . 5.0-6.5
- **MATCHED SETS** (Note 2)

ABSOLUTE MAXIMUM RATINGS (Note 1)

Temperatures

Storage Temperature Range	−55°C to +150°C
Maximum Junction Operating Temperature	+150°C
Lead Temperature	+260°C

Maximum Voltage

WIV	Working Inverse Voltage	30 V

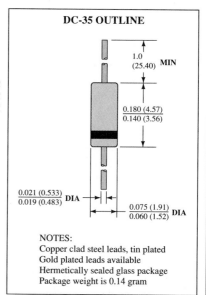

DC-35 OUTLINE

1.0 (25.40) **MIN**

$\frac{0.180\ (4.57)}{0.140\ (3.56)}$

$\frac{0.021\ (0.533)}{0.019\ (0.483)}$ **DIA**

$\frac{0.075\ (1.91)}{0.060\ (1.52)}$ **DIA**

NOTES:
Copper clad steel leads, tin plated
Gold plated leads available
Hermetically sealed glass package
Package weight is 0.14 gram

ELECTRICAL CHARACTERISTICS (25°C Ambient Temperature unless otherwise noted)

SYMBOL	CHARACTERISTIC	MIN	TYP	MAX	UNITS	TEST CONDITIONS
BV	Breakdown Voltage	30			V	$I_R = 100\ \mu A$
I_R	Reverse Current		10 0.1	50 0.5	nA μA	$V_R = 28$ V $V_R = 28$ V, $T_A = 60°C$
C	Capacitance	 4.3	29 5.1	 6.0	pF pF	$V_R = 3.0$ V, f = 1 MHz $V_R = 25$ V, f = 1 MHz
C_3/C_{25}	Capacitance Ratio	5.0	5.7	6.5		$V_R = 3$ V/25 V, f = 1 MHz
Q	Figure of Merit		150			$V_R = 3.0$ V, f = 100 MHz
R_S	Series Resistance		0.35		Ω	C = 10 pF, f = 600 MHz
L_S	Series Inductance		2.5		nH	1.5 mm from case
f_o	Series Resonant Frequency		1.4		GHz	$V_R = 25$ V

NOTES;
1. These ratings are limiting values above which the serviceability of the diodes may be impaired.
2. The capacitance diffrence between any two diodes in one set is less than 3% over the reverse voltage range of 0.5 V to 28 V

Figure 20.9 Electrical characteristics for a VHF/FM Fairchild varactor diode. (Courtesy Fairchild Camera and Instrument Corporation.)

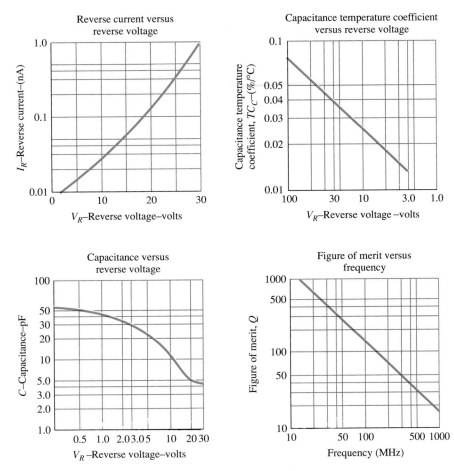

Figure 20.10 Characteristic curves for a VHF/FM Fairchild varactor diode. (Courtesy Fairchild Camera and Instrument Corporation.)

is the ratio of capacitance levels at reverse-bias potentials of 3 and 25 V. It provides a quick estimate of how much the capacitance will change with reverse-bias potential. The figure of merit is a quantity of consideration in the application of the device and is a measure of the ratio of energy stored by the capacitive device per cycle to the energy dissipated (or lost) per cycle. Since energy loss is seldom considered a positive attribute, the higher its relative value the better. The resonant frequency of the device is determined by $f_o = 1/2\pi\sqrt{LC}$ and affects the range of application of the device.

In Fig. 20.10, most quantities are self-explanatory. However, the *capacitance temperature coefficient* is defined by

$$TC_C = \frac{\Delta C}{C_0(T_1 - T_0)} \times 100\% \qquad \%/°C \qquad (20.4)$$

where ΔC is the change in capacitance due to the temperature change $T_1 - T_0$ and C_0 is the capacitance at T_0 for a particular reverse-bias potential. For example, Fig. 20.9 indicates that $C_0 = 29$ pF with $V_R = 3$ V and $T_0 = 25°C$. A change in capacitance ΔC could then be determined using Eq. (20.4) simply by substituting the new temperature T_1 and the TC_C as determined from the graph (= 0.013). At a new V_R the value of TC_C would change accordingly. Returning to Fig. 20.9, note that the maximum frequency appearing is 600 MHz. At this frequency,

20.3 Varactor (Varicap) Diodes **805**

$$X_L = 2\pi fL = (6.28)(600 \times 10^6 \text{ Hz})(2.5 \times 10^{-9} \text{ F}) = 9.42 \ \Omega$$

normally a quantity of sufficiently small magnitude to be ignored.

Some of the high-frequency (as defined by the small capacitance levels) areas of application include FM modulators, automatic-frequency-control devices, adjustable bandpass filters, and parametric amplifiers.

In Fig. 20.11 the varactor diode is employed in a tuning network. That is, the resonant frequency of the parallel L-C combination is determined by $f_p = 1/2\pi\sqrt{L_2 C_T'}$ (high-Q system) with the level of $C_T' = C_T + C_C$ determined by the applied reverse-bias potential V_{DD}. The coupling capacitor C_C is present to provide isolation between the shorting effect of L_2 and the applied bias. The selected frequencies of the tuned network are then passed on to the high input amplifier for further amplification.

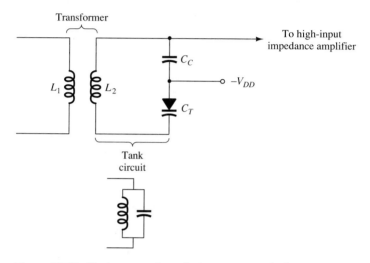

Figure 20.11 Tuning network employing a varactor diode.

20.4 POWER DIODES

There are a number of diodes designed specifically to handle the high-power and high-temperature demands of some applications. The most frequent use of power diodes occurs in the rectification process, in which ac signals (having zero average value) are converted to ones having an average or dc level. As noted in Chapter 2, when used in this capacity, diodes are normally referred to as *rectifiers*.

The majority of the power diodes are constructed using silicon because of its higher current, temperature, and PIV ratings. The higher current demands require that the junction area be larger, to ensure that there is a low forward diode resistance. If the forward resistance were too large, the I^2R losses would be excessive. The current capability of power diodes can be increased by placing two or more in parallel and the PIV rating can be increased by stacking the diodes in series.

Various types of power diodes and their current rating have been provided in Fig. 20.12a. The high temperatures resulting from the heavy current require, in many cases, that heat sinks be used to draw the heat away from the element. A few of the various types of heat sinks available are shown in Fig. 20.12b. If heat sinks are not employed, stud diodes are designed to be attached directly to the chassis, which in turn will act as the heat sink.

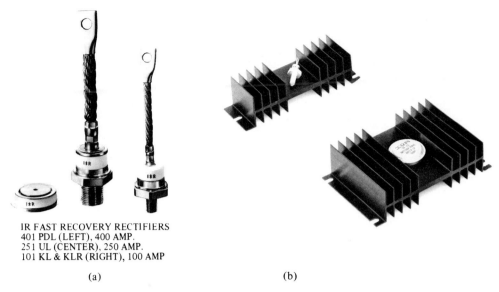

IR FAST RECOVERY RECTIFIERS
401 PDL (LEFT), 400 AMP.
251 UL (CENTER), 250 AMP.
101 KL & KLR (RIGHT), 100 AMP

(a)   (b)

Figure 20.12 Power diodes and heat sinks. (Courtesy International Rectifier Corporation.)

20.5 TUNNEL DIODES

The tunnel diode was first introduced by Leo Esaki in 1958. Its characteristics, shown in Fig. 20.13, are different from any diode discussed thus far in that it has a negative-resistance region. In this region, an increase in terminal voltage results in a reduction in diode current.

The tunnel diode is fabricated by doping the semiconductor materials that will form the *p-n* junction at a level one hundred to several thousand times that of a typical semiconductor diode. This will result in a greatly reduced depletion region, of the order of magnitude of 10^{-6} cm, or typically about $\frac{1}{100}$ the width of this region for a typical semiconductor diode. It is this thin depletion region that many carriers can "tunnel" through, rather than attempt to surmount, at low forward-bias potentials that accounts for the peak in the curve of Fig. 20.13. For comparison purposes, a

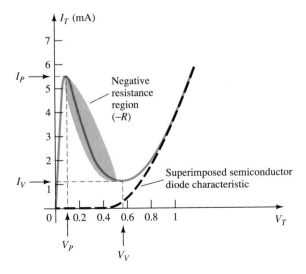

Figure 20.13 Tunnel diode characteristics.

typical semiconductor diode characteristic has been superimposed on the tunnel-diode characteristic of Fig. 20.13.

This reduced depletion region results in carriers "punching through" at velocities that far exceed those available with conventional diodes. The tunnel diode can therefore be used in high-speed applications such as in computers, where switching times in the order of nanoseconds or picoseconds are desirable.

You will recall from Section 1.14 that an increase in the doping level will drop the Zener potential. Note the effect of a very high doping level on this region in Fig. 20.13. The semiconductor materials most frequently used in the manufacture of tunnel diodes are germanium and gallium arsenide. The ratio I_p/I_v is very important for computer applications. For germanium it is typically 10:1, while for gallium arsenide it is closer to 20:1.

The peak current, I_p, of a tunnel diode can vary from a few microamperes to several hundred amperes. The peak voltage, however, is limited to about 600 mV. For this reason, a simple VOM with an internal dc battery potential of 1.5 V can severely damage a tunnel diode if applied improperly.

The tunnel diode equivalent circuit in the negative-resistance region is provided in Fig. 20.14, with the symbols most frequently employed for tunnel diodes. The values for each parameter are for the 1N2939 GE tunnel diode whose specifications appear in Table 20.1. The inductor L_S is due mainly to the terminal leads. The resistor R_S is due to the leads, ohmic contact at the lead–semiconductor junction, and the semiconductor materials themselves. The capacitance C is the junction diffusion capacitance and the R is the negative resistance of the region. The negative resistance finds application in oscillators to be described later.

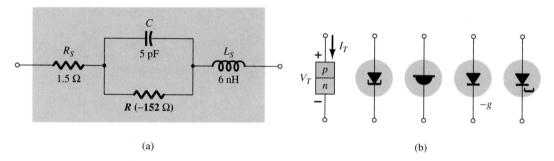

(a)   (b)

Figure 20.14 Tunnel diode: (a) equivalent circuit; (b) symbols.

TABLE 20.1 Specifications: Ge 1N2939

	Minimum	Typical	Maximum	
Absolute maximum ratings (25°C)				
Forward current (−55 to +100°C)		5 mA		
Reverse current (−55 to +100°C)		10 mA		
Electrical characteristics (25°C)($\frac{1}{8}$-in. leads)				
I_P	0.9	1.0	1.1	mA
I_V		0.1	0.14	mA
V_P	50	60	65	mV
V_V		350		mV
Reverse voltage ($I_R = 1.0$ mA)			30	mV
Forward peak point current voltage, V_{fp}	450	500	600	mV
I_p/I_v		10		
$-R$		−152		Ω
C		5	15	pF
L_S		6		nH
R_S		1.5	4.0	Ω

Chapter 20 Other Two-Terminal Devices

Note the lead length of $\frac{1}{8}$ in. included in the specifications. An increase in this length will cause L_S to increase. In fact, it was given for this device that L_S will vary 1 to 12 nH, depending on lead length. At high frequencies ($X_{L_S} = 2\pi f L_S$) this factor can take its toll.

The fact that $V_{fp} = 500$ mV (typ.) and $I_{forward}$ (max.) = 5 mA indicates that tunnel diodes are low-power devices [$P_D = (0.5$ V$)(5$ mA$) = 2.5$ mW], which is also excellent for computer applications. A rendering of the device appears in Fig. 20.15.

Although the use of tunnel diodes in present-day high-frequency systems has been dramatically stalled by manufacturing techniques that suggest alternatives to the tunnel diode, its simplicity, linearity, low power drain, and reliability ensure its continued life and application. The basic construction of an advance design tunnel diode appears in Fig. 20.16 with a photograph of the actual junction.

Figure 20.15 A Ge In2939 tunnel diode. (Courtesy Powerex, Inc.)

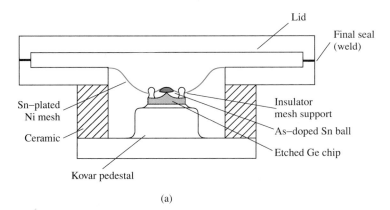

(a)

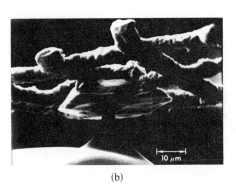

(b)

Figure 20.16 Tunnel diode: (a) construction; (b) photograph. (Courtesy *COM SAT Technical Review*, P. F. Varadi and T. D. Kirkendall.)

In Fig. 20.17 the chosen supply voltage and load resistance have defined a load line that intersects the tunnel diode characteristics at three points. Keep in mind that the load line is determined solely by the network and the characteristics by the device. The intersections at a and b are referred to as *stable* operating points, due to the positive resistance characteristic. That is, at either of these operating points, a slight disturbance in the network will not set the network into oscillations or result in a significant change in the location of the Q-point. For instance, if the defined operating point is at b, a slight increase in supply voltage E will move the operating point up the curve since the voltage across the diode will increase. Once the disturbance has

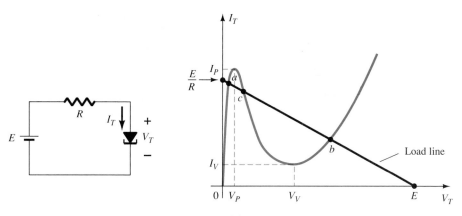

Figure 20.17 Tunnel diode and resulting load line.

passed, the voltage across the diode and the associated diode current will return to the levels defined by the Q-point at b. The operating point defined by c is an *unstable* one because a slight change in the voltage across or current through the diode will result in the Q-point moving to either a or b. For instance, the slightest increase in E will cause the voltage across the tunnel diode to increase above its level at c. In this region, however, an increase in V_T will cause a decrease in I_T and a further increase in V_T. This increased level in V_T will result in a continuing decrease in I_T, and so on. The result is an increase in V_T and a change in I_T until the stable operating point at b is established. A slight drop in supply voltage would result in a transition to stability at point a. In other words, point c can be defined as the operating point using the load-line technique, but once the system is energized it will eventually stabilize at location a or b.

The availability of a negative resistance region can be put to good use in the design of oscillators, switching networks, pulse generators, and amplifiers.

In Fig. 20.18a a *negative-resistance oscillator* was constructed using a tunnel diode. The choice of network elements is designed to establish a load line such as shown in Fig. 20.18b. Note that the only intersection with the characteristics is in the unstable negative-resistance region—a stable operating point is not defined. When the power is turned on, the terminal voltage of the supply will build up from 0 V to a final value of E volts. Initially, the current I_T will increase from 0 mA to I_P, resulting in a storage of energy in the inductor in the form of a magnetic field. However, once I_P is reached, the diode characteristics suggest that the current I_T must now decrease with increase in voltage across the diode. This is a contradiction to the fact that

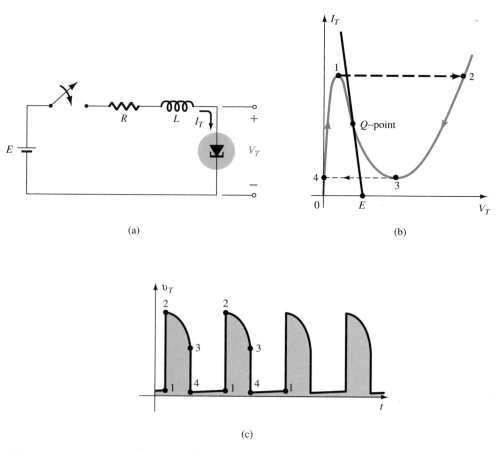

(a)

(b)

(c)

Figure 20.18 Negative-resistance oscillator.

Chapter 20 Other Two-Terminal Devices

$$E = I_T R + I_T(-R_T)$$

and

$$E = \underbrace{I_T}_{\text{less}} \underbrace{(R - R_T)}_{\text{less}}$$

If both elements of the equation above decrease, it would be impossible for the supply voltage to reach its set value. Therefore, for the current I_T to continuing rising, the point of operation must shift from point 1 to point 2. However, at point 2 the voltage V_T has jumped to a value greater than the applied voltage (point 2 is to the right of any point on the network load line). To satisfy Kirchhoff's voltage law, the polarity of the transient voltage across the coil must reverse and the current begin to decrease as shown from 2 to 3 on the characteristics. When V_T drops to V_V, the characteristics suggest that the current I_T will begin to increase again. This is unacceptable since V_T is still more than the applied voltage and the coil is discharging through the series circuit. The point of operation must shift to point 4 to permit a continuation of the decrease in I_T. However, once at point 4 the potential levels are such that the tunnel current can again increase from 0 mA to I_P as shown on the characteristics. The process will repeat itself again and again, never settling in on the operating point defined for the unstable region. The resulting voltage across the tunnel diode appears in Fig. 20.18c and will continue as long as the dc supply is energized. The result is an oscillatory output established by a fixed supply and a device with a negative-resistance characteristic. The waveform of Fig. 20.18c has extensive application in timing and computer logic circuitry.

A tunnel diode can also be used to generate a sinusoidal voltage using simply a dc supply and a few passive elements. In Fig. 20.19a the closing of the switch will result in a sinusoidal voltage that will decrease in amplitude with time. Depending on the elements employed, the time period can be almost instantaneous to one measurable in minutes using typical parameter values. This *damping* of the oscillatory output with time is due to the dissipative characteristics of the resistive elements. By placing a tunnel diode in series with the tank circuit as shown in Fig. 20.19c, the negative resistance of the tunnel diode will offset the resistive characteristics of the tank circuit, resulting in the *undamped* response appearing in the same figure. The design

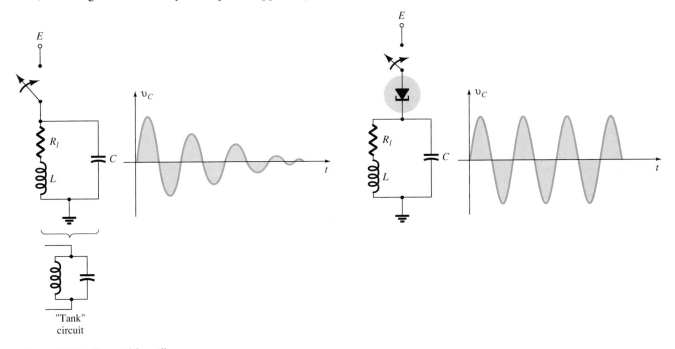

Figure 20.19 Sinusoidal oscillator.

must continue to result in a load line that will intersect the characteristics only in the negative-resistance region. In another light, the sinusoidal generator of Fig. 20.19 is simply an extension of the pulse oscillator of Fig. 20.18, with the addition of the capacitor to permit an exchange of energy between the inductor and the capacitor during the various phases of the cycle depicted in Fig. 20.18b.

20.6 PHOTODIODES

The interest in light-sensitive devices has been increasing at an almost exponential rate in recent years. The resulting field of *optoelectronics* will be receiving a great deal of research interest as efforts are made to improve efficiency levels. Through the advertising media, the layperson has become quite aware that light sources offer a unique source of energy. This energy, transmitted as discrete packages called *photons,* has a level directly related to the frequency of the traveling light wave as determined by the following equation:

$$W = hf \qquad \text{joules} \tag{20.5}$$

where h is called Planck's constant and is equal to 6.624×10^{-34} joule-second. It clearly states that since h is a constant, the energy associated with incident light waves is directly related to the frequency of the traveling wave.

The frequency is, in turn, directly related to the wavelength (distance between successive peaks) of the traveling wave by following equation:

$$\lambda = \frac{v}{f} \tag{20.6}$$

where λ = wavelength, meters
  v = velocity of light, 3×10^8 m/s
  f = frequency of the traveling wave, hertz

The wavelength is usually measured in angstrom units (Å) or micrometers (μm), where

$$1 \text{ Å} = 10^{-10} \text{ m} \qquad \text{and} \qquad 1 \text{ } \mu\text{m} = 10^{-6} \text{ m}$$

The wavelength is important because it will determine the material to be used in the optoelectronic device. The relative spectral response for Ge, Si, and selenium is provided in Fig. 20.20. The visible-light spectrum has also been included with an indication of the wavelength associated with the various colors.

The number of free electrons generated in each material is proportional to the *intensity* of the incident light. Light intensity is a measure of the amount of *luminous flux* falling in a particular surface area. Luminous flux is normally measured in *lumens* (lm) or watts. The two units are related by

$$1 \text{ lm} = 1.496 \times 10^{-10} \text{ W}$$

The light intensity is normally measured in lm/ft^2, footcandles (fc), or W/m^2, where

$$1 \text{ lm/ft}^2 = 1 \text{ fc} = 1.609 \times 10^{-9} \text{ W/m}^2$$

The photodiode is a semiconductor *p-n* junction device whose region of operation is limited to the reverse-bias region. The basic biasing arrangement, construction, and symbol for the device appear in Fig. 20.21.

Recall from Chapter 1 that the reverse saturation current is normally limited to a few microamperes. It is due solely to the thermally generated minority carriers in the

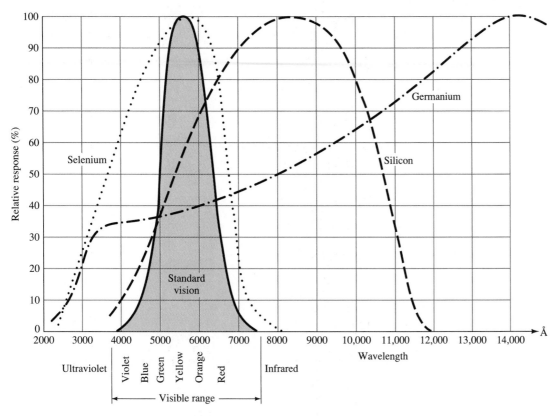

Figure 20.20 Relative spectral response for Si, Ge, and selenium as compared to the human eye.

n- and *p*-type materials. The application of light to the junction will result in a transfer of energy from the incident traveling light waves (in the form of photons) to the atomic structure, resulting in an increased number of minority carriers and an increased level of reverse current. This is clearly shown in Fig. 20.22 for different intensity levels. The *dark* current is that current that will exist with no applied illumination. Note that the current will only return to zero with a positive applied bias equal to V_T. In addition, Fig. 20.21 demonstrates the use of a lens to concentrate the light on the junction region. Commercially available photodiodes appear in Fig. 20.23.

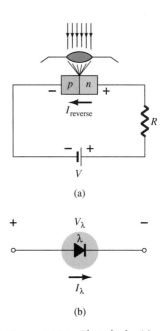

Figure 20.21 Photodiode: (a) basic-biasing arrangement and construction; (b) symbol.

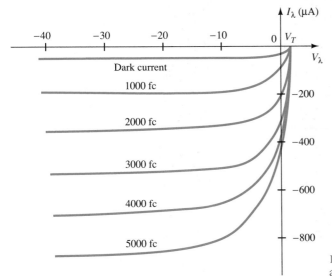

Figure 20.22 Photodiode characteristics.

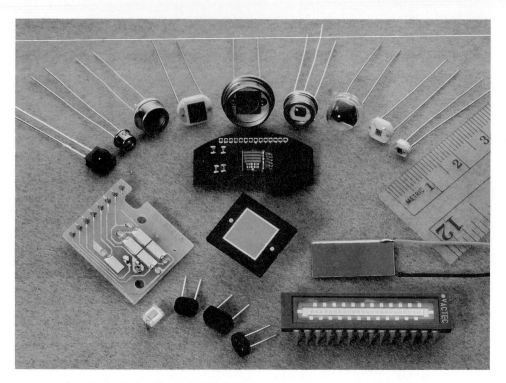

Figure 20.23 Photochiodes (Courtesy EG&G VACTEC Inc.)

The almost equal spacing between the curves for the same increment in luminous flux reveals that the reverse current and luminous flux are almost linearly related. In other words, an increase in light intensity will result in a similar increase in reverse current. A plot of the two to show this linear relationship appears in Fig. 20.24 for a fixed voltage V_λ of 20 V. On the relative basis we can assume that the reverse current is essentially zero in the absence of incident light. Since the rise and fall times (change-of-state parameters) are very small for this device (in the nanosecond range), the device can be used for high-speed counting or switching applications. Returning to Fig. 20.20, we note that Ge encompasses a wider spectrum of wavelengths than Si. This would make it suitable for incident light in the infrared region as provided by lasers and IR (infrared) light sources to be described shortly. Of course, Ge has a higher dark current than silicon, but it also has a higher level of reverse current. The level of current generated by the incident light on a photodiode is not such that it could be used as a direct control, but it can be amplified for this purpose.

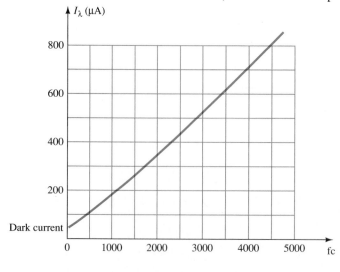

Figure 20.24 $I_\lambda (\mu A)$ versus f_c (at $V_A = 20$ V) for the photodiode of Fig. 20.22.

In Fig. 20.25 the photodiode is employed in an alarm system. The reverse current I_λ will continue to flow as long as the light beam is not broken. If interrupted, I_λ drops to the dark current level and sounds the alarm. In Fig. 20.26 a photodiode is used to count items on a conveyor belt. As each item passes the light beam is broken, I_λ drops to the dark current level and the counter is increased by one.

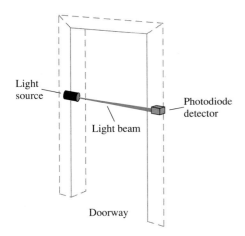

Figure 20.25 Use of a photodiode in an alarm system.

Figure 20.26 Using a photodiode in a counter operation.

20.7 PHOTOCONDUCTIVE CELLS

The photoconductive cell is a two-terminal semiconductor device whose terminal resistance will vary (linearly) with the intensity of the incident light. For obvious reasons, it is frequently called a photoresistive device. A typical photoconductive cell and the most widely used graphic symbol for the device appear in Fig. 20.27.

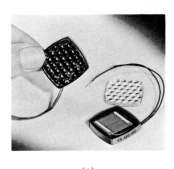

(a)

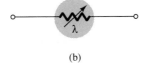

(b)

Figure 20.27 Photoconductive cell: (a) appearance; (b) symbol. [(a) Courtesy Internal Rectifier Corporation.]

The photoconductive materials most frequently used include cadmium sulfide (CdS) and cadmium selenide (CdSe). The peak spectral response of CdS occurs at approximately 5100 Å and for CdSe at 6150 Å, as shown in Fig. 20.20. The response time of CdS units is about 100 ms and 10 ms for CdSe cells. The photoconductive cell does not have a junction like the photodiode. A thin layer of the material connected between terminals is simply exposed to the incident light energy.

As the illumination on the device increases in intensity, the energy state of a larger number of electrons in the structure will also increase, because of the increased availability of the photon packages of energy. The result is an increasing number of relatively "free" electrons in the structure and a decrease in the terminal resistance. The sensitivity curve for a typical photoconductive device appears in Fig. 20.28. Note the linearity (when plotted using a log-log scale) of the resulting curve and the large change in resistance (100 kΩ → 100 Ω) for the indicated change in illumination.

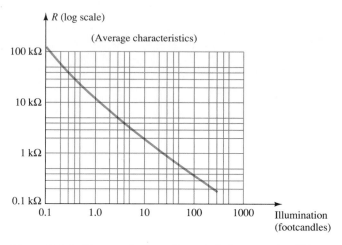

Figure 20.28 Photoconductive cell-terminal characteristics (GE type B425).

One rather simple, but interesting, application of the device appears in Fig. 20.29. The purpose of the system is to maintain V_o at a fixed level even though V_i may fluctuate from its rated value. As indicated in the figure, the photoconductive cell, bulb, and resistor all form part of this voltage-regulator system. If V_i should drop in magnitude for any number of reasons, the brightness of the bulb would also decrease. The decrease in illumination would result in an increase in the resistance (R_λ) of the photoconductive cell to maintain V_o at its rate level as determined by the voltage-divider rule; that is,

$$V_o = \frac{R_\lambda V_i}{R_\lambda + R_1} \qquad (20.7)$$

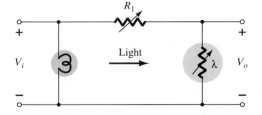

Figure 20.29 Voltage regulator employing a photoconductive cell.

In an effort to demonstrate the wealth of material available on each device from manufacturers, consider the CdS (cadmium sulfide) photoconductive cell described in Fig. 20.30. Note again the concern with temperature and response time.

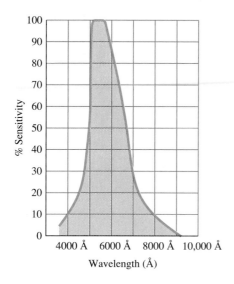

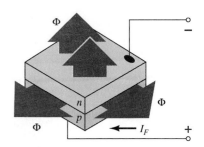

Figure 20.31 General structure of a semiconductor IR-emitting diode. (Courtesy RCA Solid State Division.)

Variation of Conductance With Temperature and Light					
Footcandles	.01	0.1	1.0	10	100
Temperature			% Conductance		
−25°C	103	104	104	102	106
0	98	102	102	100	103
25°C	100	100	100	100	100
50°C	98	102	103	104	99
75°C	90	106	108	109	104

Response Time Versus Light					
Footcandles	0.01	0.1	1.0	10	100
Rise (seconds)	0.5	0.095	0.022	0.005	0.002
Decay (seconds)	0.125	0.021	0.005	0.002	0.001

Figure 20.30 Characteristics of a Clairex CdS photoconductive cell. (Courtesy Clairex Electronics.)

20.8 IR EMITTERS

Infrared-emitting diodes are solid-state gallium arsenide devices that emit a beam of radiant flux when forward biased. The basic construction of the device is shown in Fig. 20.31. When the junction is forward biased, electrons from the *n*-region will recombine with excess holes of the *p*-material in a specially designed recombination region sandwiched between the *p*- and *n*-type materials. During this recombination process, energy is radiated away from the device in the form of photons. The generated photons will either be reabsorbed in the structure or leave the surface of the device as radiant energy, as shown in Fig. 20.31.

The radiant flux in mW versus the dc forward current for a typical device appears in Fig. 20.32. Note the almost linear relationship between the two. An interesting pattern for such devices is provided in Fig. 20.33. Note the very narrow pattern for

Case temperature $(T_C) = 27°C$

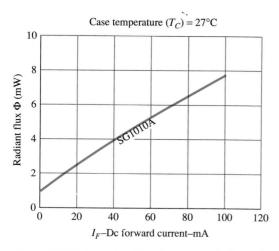

Figure 20.32 Typical radiant flux versus dc forward current for an IR-emitting diode. (Courtesy RCA Solid State Division.)

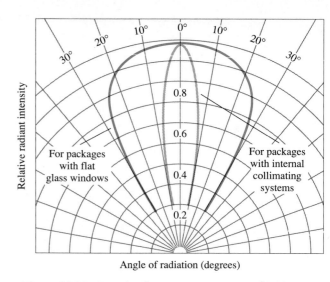

Figure 20.33 Typical radiant intensity patterns of RCA IR-emitting diodes. (Courtesy RCA Solid State Division.)

devices with an internal collimating system. One such device appears in Fig. 20.34, with its internal construction and graphic symbol. A few areas of application for such devices include card and paper-tape readers, shaft encoders, data-transmission systems, and intrusion alarms.

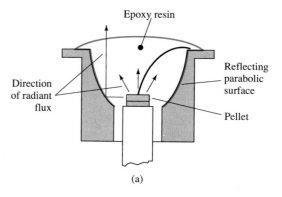

(a)

Figure 20.34 RCA IR-emitting diode: (a) construction; (b) photo; (c) symbol. (Courtesy RCA Solid State Division.)

(c)

Approx. 2X actual size

(b)

20.9 LIQUID-CRYSTAL DISPLAYS

The liquid-crystal display (LCD) has the distinct advantage of having a lower power requirement than the LED. It is typically in the order of microwatts for the display, as compared to the same order of milliwatts for LEDs. It does, however, require an external or internal light source, is limited to a temperature range of about 0° to 60°C, and lifetime is an area of concern because LCDs can chemically degrade. The types receiving the major interest today are the field-effect and dynamic-scattering units. Each will be covered in some detail in this section.

A liquid crystal is a material (normally organic for LCDs) that will flow like a liquid but whose molecular structure has some properties normally associated with solids. For the light-scattering units, the greatest interest is in the *nematic liquid crystal,* having the crystal structure shown in Fig. 20.35. The individual molecules have a rodlike appearance as shown in the figure. The indium oxide conducting surface is transparent and, under the condition shown in the figure, the incident light will simply pass through and the liquid-crystal structure will appear clear. If a voltage (for commercial units the threshold level is usually between 6 and 20 V) is applied across the conducting surfaces, as shown in Fig. 20.36, the molecular arrangement is disturbed, with the result that regions will be established with different indices of refraction. The incident light is, therefore, reflected in different directions at the interface between regions of different indices of refraction (referred to as *dynamic scattering*—first studied by RCA in 1968) with the result that the scattered light has a frosted-glass appearance. Note in Fig. 20.36, however, that the frosted look occurs

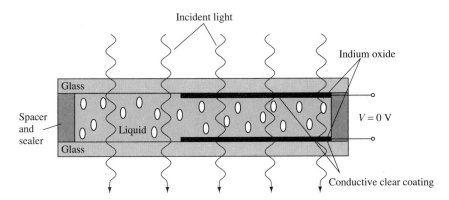

Figure 20.35 Nematic liquid crystal with no applied bias.

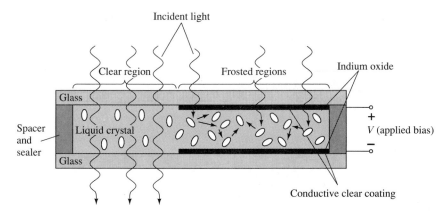

Figure 20.36 Nematic liquid crystal with applied bias.

20.9 Liquid-Crystal Displays   819

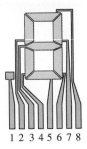

Figure 20.37 LCD eight-segment digit display.

1 2 3 4 5 6 7 8

only where the conducting surfaces are opposite each other and that the remaining areas remain translucent.

A digit on an LCD display may have the segment appearance shown in Fig. 20.37. The black area is actually a clear conducting surface connected to the terminals below for external control. Two similar masks are placed on opposite sides of a sealed thick layer of liquid-crystal material. If the number 2 were required, the terminals 8, 7, 3, 4, and 5 would be energized and only those regions would be frosted while the other areas would remain clear.

As indicated earlier, the LCD does not generate its own light but depends on an external or internal source. Under dark conditions it would be necessary for the unit to have its own internal light source either behind or to the side of the LCD. During the day, or in lighted areas, a reflector can be put behind the LCD to reflect the light back through the display for maximum intensity. For optimum operation, current watch manufacturers are using a combination of the transmissive (own light source) and reflective modes called *transflective*.

The *field-effect* or *twisted nematic* LCD has the same segment appearance and thin layer of encapsulated liquid crystal, but its mode of operation is very different. Similar to the dynamic-scattering LCD, the field effect can be operated in the reflective or transmissive mode with an internal source. The transmissive display appears in Fig. 20.38. The internal light source is on the right and the viewer is on the left. This figure is most noticeably different from Fig. 20.35 in that there is an addition of a *light polarizer*. Only the vertical component of the entering light on the right can pass through the vertical-light polarizer on the right. In the field-effect LCD, either the clear conducting surface to the right is chemically etched or an organic film is applied to orient the molecules in the liquid crystal in the vertical plane, parallel to the cell wall. Note the rods to the far right in the liquid crystal. The opposite conducting surface is also treated to ensure that the molecules are 90° out of phase in the direction shown (horizontal) but still parallel to the cell wall. In between the two walls of the liquid crystal there is a general drift from one polarization to the other, as shown in the figure. The left-hand light polarizer is also such that it permits the passage of only the vertically polarized incident light. If there is no applied voltage to the conducting surfaces, the vertically polarized light enters the liquid-crystal region and follows the 90° bending of the molecular structure. Its horizontal polarization at the left-hand vertical light polarizer does not allow it to pass through, and the viewer sees a uniformly dark pattern across the entire display. When a threshold voltage is applied (for commercial units from 2 to 8 V), the rodlike molecules align themselves with the field (perpendicular to the wall) and the light passes directly through without the 90° shift. The vertically incident light can then press directly through the second vertically polarized screen and a light area is seen by the viewer. Through proper excita-

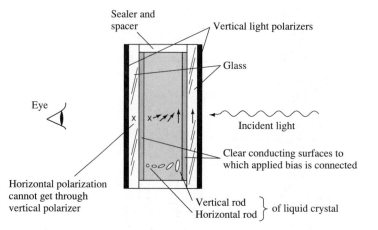

Figure 20.38 Transmissive field-effect LCD with no applied bias.

tion of the segments of each digit the pattern will appear as shown in Fig. 20.39. The reflective type field effect is shown in Fig. 20.40. In this case the horizontally polarized light at the far left encounters a horizontally polarized filter and passes through to the reflector, where it is reflected back into the liquid crystal, bent back to the other vertical polarization, and returned to the observer. If there is no applied voltage, there is a uniformly lit display. The application of a voltage results in a vertically incident light encountering a horizontally polarized filter at the left which will not be able to pass through and be reflected. A dark area results on the crystal, and the pattern as shown in Fig. 20.41 appears.

Figure 20.39 Reflective-type LCD. (Courtesy RCA Solid State Division.)

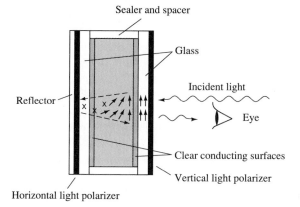

Sealer and spacer

Glass

Incident light

Reflector

Eye

Clear conducting surfaces

Vertical light polarizer

Horizontal light polarizer

Figure 20.40 Reflective field-effect LCD with no applied bias.

Figure 20.41 Transmissive-type LCD. (Courtesy RCA Solid State Division.)

Field-effect LCDs are normally used when a source of energy is a prime factor (e.g., in watches, portable instrumentation, etc.) since they absorb considerably less power than the light-scattering types—the microwatt range compared to the low-milliwatt range. The cost is typically higher for field-effect units, and their height is limited to about 2 in., while light-scattering units are available up to 8 in. in height.

A further consideration in displays is turn-on and turn-off time. LCDs are characteristically much slower then LEDs. LCDs typically have response times in the range 100 to 300 ms, while LEDs are available with response times below 100 ns. However, there are numerous applications, such as in a watch, where the difference between 100 ns and 100 ms ($\frac{1}{10}$ of a second) is of little consequence. For such applications the lower power demand of LCDs is a very attractive characteristic. The lifetime of LCD units is steadily increasing beyond the 10,000+ hours limit. Since the color generated by LCD units is dependent on the source of illumination, there is a greater range of color choice.

20.10 SOLAR CELLS

In recent years there has been increasing interest in the solar cell as an alternative source of energy. When we consider that the power density received from the sun at sea level is about 100 mW/cm^2 (1 kW/m^2), it is certainly an energy source that requires further research and development to maximize the conversion efficiency from solar to electrical energy.

The basic construction of a silicon *p-n* junction solar cell appears in Fig. 20.42. As shown in the top view, every effort is made to ensure that the surface area perpendicular to the sun is a maximum. Also, note that the metallic conductor connected to the *p*-type material and the thickness of the *p*-type material are such that they ensure that a maximum number of photons of light energy will reach the junction. A photon of light energy in this region may collide with a valence electron and impart to it sufficient energy to leave the parent atom. The result is a generation of free electrons and holes. This phenomenon will occur on each side of the junction. In the *p*-type

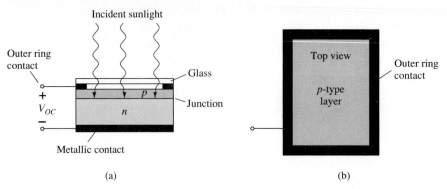

Figure 20.42 Solar cell: (a) cross section; (b) top view.

(a)

(b)

material the newly generated electrons are minority carriers and will move rather freely across the junction as explained for the basic *p-n* junction with no applied bias. A similar discussion is true for the holes generated in the *n*-type material. The result is an increase in the minority-carrier flow which is opposite in direction to the conventional forward current of a *p-n* junction. This increase in reverse current is shown in Fig. 20.43. Since $V = 0$ anywhere on the vertical axis and represents a short-circuit condition, the current at this intersection is called the *short-circuit current* and is represented by the notation I_{sc}. Under open-circuit conditions ($i_d = 0$) the *photovoltaic* voltage V_{oc} will result. This is a logarithmic function of the illumination, as shown in Fig. 20.44. V_{oc} is the terminal voltage of a battery under no-load (open-circuit) conditions. Note, however, in the same figure that the short-circuit current is a linear function of the illumination. That is, it will double for the same increase in illumination (f_{C_1} and $2f_{C_1}$ in Fig. 20.44) while the change in V_{oc} is less for this region. The major increase in V_{oc} occurs for lower-level increases in illumination. Eventually, a further increase in illumination will have very little effect on V_{oc}, although I_{sc} will increase, causing the power capabilities to increase.

Selenium and silicon are the most widely used materials for solar cells, although gallium arsenide, indium arsenide, and cadmium sulfide, among others, are also

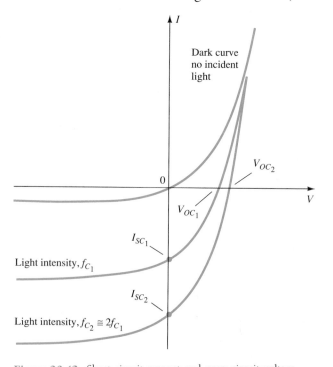

Figure 20.43 Short-circuit current and open-circuit voltage versus light intensity for a solar cell.

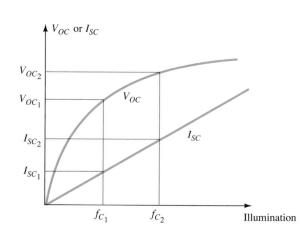

Figure 20.44 V_{oc} and I_{sc} versus illumination for a solar cell.

Chapter 20 Other Two-Terminal Devices

used. The wavelength of the incident light will affect the response of the *p-n* junction to the incident photons. Note in Fig. 20.45 how closely the selenium cell response curve matches that of the eye. This fact has widespread application in photographic equipment such as exposure meters and automatic exposure diaphragms. Silicon also overlaps the visible spectrum but has its peak at the 0.8 μm (8000 Å) wavelength, which is in the infrared region. In general, silicon has a higher conversion efficiency and greater stability and is less subject to fatigue. Both materials have excellent temperature characteristics. That is, they can withstand extreme high or low temperatures without a significant drop-off in efficiency. Typical solar cells, with their electrical characteristics, appear in Fig. 20.46.

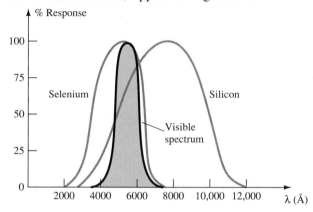

Figure 20.45 Spectral response of Se, Si, and the naked eye.

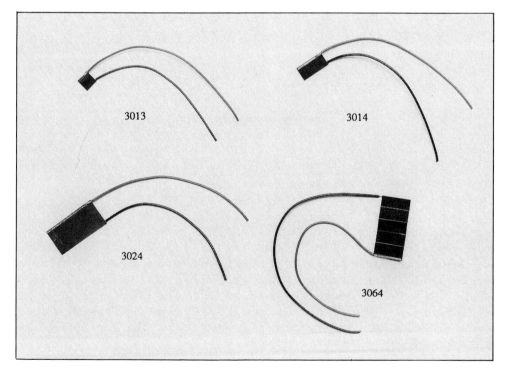

Figure 20.46 Typical solar cell and its electrical characteristics. (Courtesy EG&G VACTEC Inc.)

Electrical Characteristics

Part No.	Active Area	Test Voltage	Minimum current Test voltage
3013	0.032 in² (0.21 cm²)	0.4 V	4.2 mA
3014	0.065 in² (0.42 cm²)	0.4 V	8.4 mA
3024	0.29 in² (1.87 cm²)	0.4 V	38 mA
3064	0.325 in² (2.1 cm²)	2 V	8.4 mA

A very recent innovation in the use of solar cells appears in Fig. 20.47. The series arrangement of solar cells permits a voltage beyond that of a single element. The performance of a typical four-cell array appears in the same figure. At a current of approximately 2.6 mA the output voltage is about 1.6 V, resulting in an output power of 4.16 mW. The Schottky barrier diode is included to prevent battery current drain through the power converter. That is, the resistance of the Schottky diode is so high to charge flowing down through (+ to −) the power converter that it will appear as an open circuit to the rechargeable battery and not draw current from it.

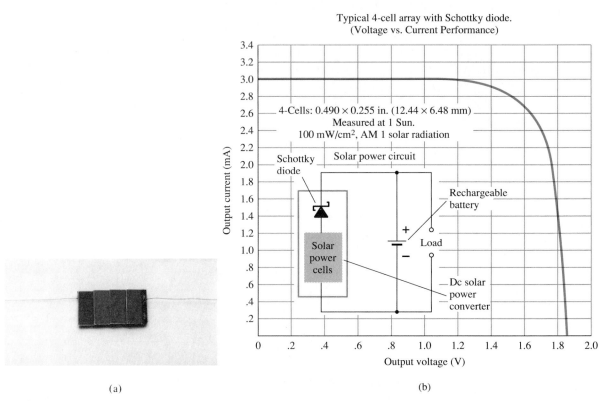

(a)   (b)

Figure 20.47 International Rectifier four-cell array: (a) appearance; (b) characteristics. (Courtesy International Rectifier Corporation.)

It might be of interest to note that the Lockheed Missiles and Space Company has been awarded a grant from the National Aeronautics and Space Administration to develop a massive solar-array wing for the space shuttle. The wing will measure 13.5 ft by 105 ft when extended and will contain 41 panels, each carrying 3060 silicon solar cells. The wing can generate a total of 12.5 kW of electrical power.

The efficiency of operation of a solar cell is determined by the electrical power output divided by the power provided by the light source. That is,

$$\eta = \frac{P_{o(\text{electrical})}}{P_{i(\text{light energy})}} \times 100\% = \frac{P_{\text{max(device)}}}{(\text{area in cm}^2)(100 \text{ mW/cm}^2)} \times 100\% \qquad (20.8)$$

Typical levels of efficiency range from 10 to 40%—a level that should improve measurably if the present interest continues. A typical set of output characteristics for silicon solar cells of 10% efficiency with an active area of 1 cm^2 appears in Fig. 20.48. Note the optimum power locus and the almost linear increase in output current with luminous flux for a fixed voltage.

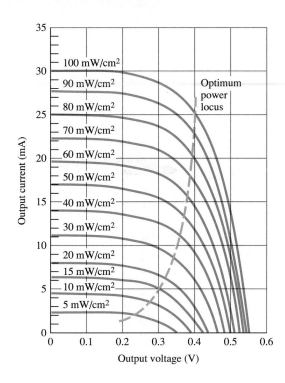

Figure 20.48 Typical output characteristics for silicon solar cells of 10% efficiency having an active area of 1 cm². Cell temperature is 30°C.

20.11 THERMISTORS

The thermistor is, as the name implies, a temperature-sensitive resistor; that is, its terminal resistance is related to its body temperature. It is not a junction device and is constructed of Ge, Si, or a mixture of oxides of cobalt, nickel, strontium, or manganese. The compound employed will determine whether the device has a positive or negative temperature coefficient.

The characteristics of a representative thermistor with a negative temperature coefficient are provided in Fig. 20.49, with the commonly used symbol for the device. Note, in particular, that at room temperature (20°C) the resistance of the thermistor is approximately 5000 Ω, while at 100°C (212°F) the resistance has decreased to 100 Ω. A temperature span of 80°C has therefore resulted in a 50:1 change in resist-

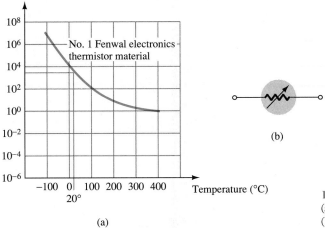

Figure 20.49 Thermistor:
(a) typical set of characteristics;
(b) symbol.

ance. It is typically 3 to 5% per degree change in temperature. There are, fundamentally, two ways to change the temperature of the device: internally and externally. A simple change in current through the device will result in an internal change in temperature. A small applied voltage will result in a current too small to raise the body temperature above that of the surroundings. In this region, as shown in Fig. 20.50, the thermistor will act like a resistor and have a positive temperature coefficient. However, as the current increases, the temperature will rise to the point where the negative temperature coefficient will appear as shown in Fig. 20.50. The fact that the rate of internal flow can have such an effect on the resistance of the device introduces a wide vista of applications in control, measuring techniques, and so on. An external change would require changing the temperature of the surrounding medium or immersing the device in a hot or cold solution.

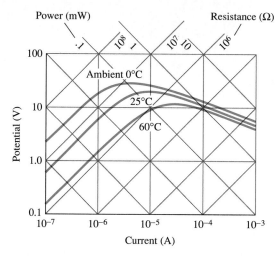

Figure 20.50 Steady-state voltage–current characteristics of Fenwal Electronics BK65VI Thermistor. (Courtesy Fenwal Electronics, Incorporated.)

A photograph of a number of commercially available thermistors is provided in Fig. 20.51. A simple temperature-indicating circuit appears in Fig. 20.52. Any increase in the temperature of the surrounding medium will result in a decrease in the resistance of the thermistor and an increase in the current I_T. An increase in I_T will

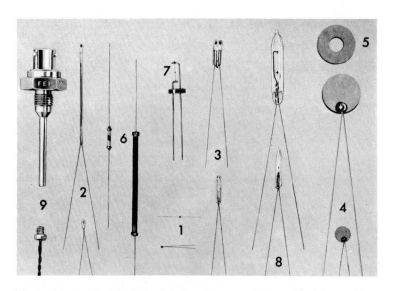

Figure 20.51 Various types of thermistors: (1) beads; (2) glass probes; (3) iso-curve interchangeable probes and beads; (4) disks; (5) washers; (6) rods; (7) specially mounted beads; (8) vacuum and gas-filled probes; (9) special probe assemblies. (Courtesy Fenwal Electronics, Incorporated.)

Chapter 20 Other Two-Terminal Devices

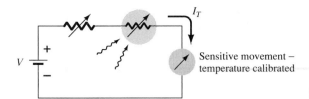

Figure 20.52 Temperature-indicating circuit.

produce an increased movement deflection, which when properly calibrated will accurately indicate the higher temperature. The variable resistance was added for calibration purposes.

§ 20.2 Schottky Barrier (Hot-Carrier) Diodes

1. (a) Describe your own words how the construction of the hot-carrier diode is significantly different from the conventional semiconductor diode.
   (b) In addition, describe its mode of operation.

2. (a) Consult Fig. 20.2. How would you compare the dynamic resistances of the diodes in the forward-bias regions?
   (b) How do the levels of I_s and V_Z compare?

3. Referring to Fig. 20.5, how does the maximum surge current I_{FSM} relate to the average rectified forward current? Is it typically greater than 20:1? Why is it possible to have such high levels of current? What noticeable difference is there in construction as the current rating increases?

4. Referring to Fig. 20.6a, at what temperature is the forward voltage drop 300 mV at a current of 1 mA? Which current levels have the highest levels of temperature coefficients? Assume a linear progression between temperature levels.

* 5. For the curve of Fig. 20.6b denoted 2900/2303, determine the percent change in I_R for a change in reverse voltage from 5 to 10 V. At what reverse voltage would you expect to reach a reverse current of 1 μA? Note the log scale for I_R.

* 6. Determine the percent change in capacitance between 0 and 2 V for the 2900/2303 curve of Fig. 20.6c. How does this compare to the change between 8 and 10 V?

§ 20.3 Varactor (Varicap) Diodes

7. (a) Determine the transition capacitance of a diffused junction varicap diode at a reverse potential of 4.2 V if $C(0) = 80$ pF and $V_T = 0.7$ V.
   (b) From the information of part (a), determine the constant K in Eq. (20.2).

8. (a) For a varicap diode having the characteristics of Fig. 20.7, determine the difference in capacitance between reverse-bias potentials of -3 and -12 V.
   (b) Determine the incremental rate of change ($\Delta C/\Delta V_r$) at $V = -8$ V. How does this value compare with the incremental change determined at -2 V?

* 9. (a) The resonant frequency of a series RLC network is determined by $f_0 = 1/(2\pi\sqrt{LC})$. Using the value of f_0 and L_S provided in Fig. 20.9, determine the value of C.
   (b) How does the value calculated in part (a) compare with that determined by the curve in Fig. 20.10 at $V_R = 25$ V?

10. Referring to Fig. 20.10, determine the ratio of capacitance at $V_R = 3$ V to $V_R = 25$ V and compare to the value of C_3/C_{25} given in Fig. 20.9 (maximum = 6.5).

11. Determine T_1 for a varactor diode if $C_0 = 22$ pF, $TC_C = 0.02\%/°C$, and $\Delta C = 0.11$ pF due to an increase in temperature above $T_0 = 25°C$.

12. What region of V_R would appear to have the greatest change in capacitance per change in reverse voltage for the BB139 varactor diode of Figs. 20.9 and 20.10? Be aware that the scales are nonlinear.

* 13. If $Q = X_L/R = 2\pi f L/R$, determine the figure of merit (Q) at 600 MHz using the fact that $R_S = 0.35$ Ω and $L_S = 2.5$ nH. Comment on the change in Q with frequency and the support or nonsupport of the curve in Fig. 20.10.

§ 20.4 Power Diodes

14. Consult a manufacturer's data book and compare the general characteristics of a high-power device (>10 A) to a low-power unit (<100 mA). Is there a significant change in the data and characteristics provided? Why?

§ 20.5 Tunnel Diodes

15. What are the essential differences between a semiconductor junction diode and a tunnel diode?

* **16.** Note in the equivalent circuit of Fig. 20.14 that the capacitor appears in parallel with the negative resistance. Determine the reactance of the capacitor at 1 MHz and 100 MHz if $C = 5$ pF, and determine the total impedance of the parallel combination (with R = −152 Ω) at each frequency. Is the magnitude of the inductive reactance anything to be overly concerned about at either of these frequencies if $L_S = 6$ nH?

* **17.** Why do you believe the maximum reverse current rating for the tunnel diode can be greater than the forward current rating? (*Hint:* Note the characteristics and consider the power rating.)

18. Determine the negative resistance for the tunnel diode of Fig. 20.13 between $V_T = 0.1$ V and $V_T = 0.3$ V.

19. Determine the stable operating points for the network of Fig. 20.17 if $E = 2$ V, $R = 0.39$ kΩ, and the tunnel diode of Fig. 20.13 is employed. Use typical values from Table 20.1.

* **20.** For $E = 0.5$ V and $R = 51$ Ω, sketch v_T for the network of Figure 20.18 and the tunnel diode of Fig. 20.13.

21. Determine the frequency of oscillation for the network of Fig. 20.19 if $L = 5$ mH, $R_l = 10$ Ω, and $C = 1$ μF.

§ 20.6 Photodiodes

22. Determine the energy associated with the photons of green light if the wavelength is 5000 Å. Give your answer in joules and electron volts.

23. (a) Referring to Fig. 20.20, what would appear to be the frequencies associated with the upper and lower limits of the visible spectrum?
(b) What is the wavelength in microns associated with the peak relative response of silicon?
(c) If we define the bandwidth of the spectral response of each material to occur at 70% of its peak level, what is the bandwidth of silicon?

24. Referring to Fig. 20.22, determine I_λ if $V_\lambda = 30$ V and the light intensity is 4×10^{-9} W/m^2.

25. (a) Which material of Fig. 20.20 would appear to provide the best response to yellow, red, green, and infrared (less than 11,000 Å) light sources?
(b) At a frequency of 0.5×10^{15} Hz, which color has the maximum spectral response?

* **26.** Determine the voltage drop across the resistor of Fig. 20.21 if the incident flux is 3000 fc, $V_\lambda = 25$ V, and $R = 100$ kΩ. Use the characteristics of Fig. 20.22.

§ 20.7 Photoconductive Cells

* **27.** What is the approximate rate of change of resistance with illumination for a photoconductive cell with the characteristics of Fig. 20.28 for the ranges (a) $0.1 \rightarrow 1$ kΩ, (b) $1 \rightarrow 10$ kΩ, and (c) $10 \rightarrow$ kΩ? (Note that this is a log scale.) Which region has the greatest rate of change in resistance with illumination?

28. What is the "dark current" of a photodiode?

29. If the illumination on the photoconductive diode in Fig. 20.29 is 10 fc, determine the magnitude of V_i to establish 6 V across the cell if R_1 is equal to 5 kΩ. Use the characteristics of Fig. 20.28.

* **30.** Using the data provided in Fig. 20.30, sketch a curve of percent conductance versus temperature for 0.01, 1.0, and 100 fc. Are there any noticeable effects?

* **31.** (a) Sketch a curve of rise time versus illumination using the data from Fig. 20.30.
(b) Repeat part (a) for the decay time.
(c) Discuss any noticeable effects of illumination in parts (a) and (b).

Chapter 20 Other Two-Terminal Devices

32. Which colors is the CdS unit of Fig. 20.30 most sensitive to?

§ 20.8 IR Emitters

33. (a) Determine the radiant flux at a dc forward current of 70 mA for the device of Fig. 20.32.
(b) Determine the radiant flux in lumens at a dc forward current of 45 mA.

* 34. (a) Through the use of Fig. 20.33, determine the relative radiant intensity at an angle of 25° for a package with a flat glass window.
(b) Plot a curve of relative radiant intensity versus degrees for the flat package.

* 35. If 60 mA of dc forward current is applied to an SG1010A IR emitter, what will be the incident radiant flux in lumens 5° off the center if the package has an internal collimating system? Refer to Figs. 20.32 and 20.33.

§ 20.9 Liquid-Crystal Displays

36. Referring to Fig. 20.37, which terminals must be energized to display number 7?

37. In your own words, describe the basic operation of an LCD.

38. Discuss the relative differences in mode of operation between an LED and an LCD display.

39. What are the relative advantages and disadvantages of an LCD display as compared to an LED display?

§ 20.10 Solar Cells

40. A 1 cm by 2 cm solar cell has a conversion efficiency of 9%. Determine the maximum power rating of the device.

* 41. If the power rating of a solar cell is determined on a very rough scale by the product $V_{oc}I_{sc}$, is the greatest rate of increase obtained at lower or higher levels of illumination? Explain your reasoning.

42. (a) Referring to Fig. 20.48, what power density is required to establish a current of 24 mA at an output voltage of 0.25 V?
(b) Why is 100 mW/cm^2 the maximum power density in Fig. 20.48?
(c) Determine the output current if the power is 40 mW/cm^2 and the output voltage is 0.3 V.

* 43. (a) Sketch a curve of output current versus power density at an output voltage of 0.15 V using the characteristics of Fig. 20.48.
(b) Sketch a curve of output voltage versus power density at a current of 19 mA.
(c) Is either of the curves from parts (a) and (b) linear within the limits of the maximum power limitation?

§ 20.11 Thermistors

* 44. For the thermistor of Fig. 20.49, determine the dynamic rate of change in specific resistance with temperature at $T = 20°C$. How does this compare to the value determined at $T = 300°C$? From the results, determine whether the greatest change in resistance per unit change in temperature occurs at lower or higher levels of temperature. Note the vertical log scale.

45. Using the information provided in Fig. 20.49, determine the total resistance of a 2-cm length of the material having a perpendicular surface area of 1 cm^2 at a temperature of 0°C. Note the vertical log scale.

46. (a) Referring to Fig. 20.50, determine the current at which a 25°C sample of the material changes from a positive to negative temperature coefficient. (Figure 20.50 is a log scale.)
(b) Determine the power and resistance levels of the device (Fig. 20.50) at the peak of the 0°C curve.
(c) At a temperature of 25°C, determine the power rating if the resistance level is 1 MΩ.

47. In Fig. 20.52, $V = 0.2$ V and $R_{variable} = 10$ Ω. If the current through the sensitive movement is 2 mA and the voltage drop across the movement is 0 V, what is the resistance of the thermistor?

---

*Please Note: Asterisks indicate more difficult problems.

21

pnpn and Other Devices

21.1 INTRODUCTION

In this chapter a number of important devices not discussed in detail in earlier chapters are introduced. The two-layer semiconductor diode has led to three-, four-, and even five-layer devices. A family of four-layer *pnpn* devices will first be considered: SCR (silicon-controlled rectifier), SCS (silicon-controlled switch), GTO (gate turn-off switch), LASCR (light-activated SCR), followed by an increasingly important device—the UJT (unijunction transistor). Those four-layer devices with a control mechanism are commonly referred to as *thyristors*, although the term is most frequently applied to the SCR (silicon-controlled rectifier). The chapter closes with an introduction to the phototransistor, opto-isolators, and the PUT (programmable unijunction transistor).

pnpn DEVICES

21.2 SILICON-CONTROLLED RECTIFIER

Within the family of *pnpn* devices the silicon-controlled rectifier (SCR) is unquestionably of the greatest interest today. It was first introduced in 1956 by Bell Telephone Laboratories. A few of the more common areas of application for SCRs include relay controls, time-delay circuits, regulated power suppliers, static switches, motor controls, choppers, inverters, cycloconverters, battery chargers, protective circuits, heater controls, and phase controls.

In recent years, SCRs have been designed to *control* powers as high as 10 MW with individual ratings as high as 2000 A at 1800 V. Its frequency range of application has also been extended to about 50 kHz, permitting some high-frequency applications such as induction heating and ultrasonic cleaning.

21.3 BASIC SILICON-CONTROLLED RECTIFIER OPERATION

As the terminology indicates, the SCR is a rectifier constructed of silicon material with a third terminal for control purposes. Silicon was chosen because of its high temperature and power capabilities. The basic operation of the SCR is different from the fundamental two-layer semiconductor diode in that a third terminal, called a *gate,* determines when the rectifier switches from the open-circuit to short-circuit state. It is not enough to simply forward-bias the anode-to-cathode region of the device. In the conduction region the dynamic resistance of the SCR is typically 0.01 to 0.1 Ω. The reverse resistance is typically 100 kΩ or more.

The graphic symbol for the SCR is shown in Fig. 21.1 with the corresponding connections to the four-layer semiconductor structure. As indicated in Fig. 21.1a, if forward conduction is to be established, the anode must be positive with respect to the cathode. This is not, however, a sufficient criterion for turning the device on. A pulse of sufficient magnitude must also be applied to the gate to establish a turn-on gate current, represented symbolically by I_{GT}.

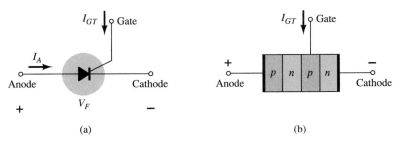

Figure 21.1 (a) SCR symbol; (b) basic construction.

A more detailed examination of the basic operation of an SCR is best effected by splitting the four-layer *pnpn* structure of Fig. 21.1b into two three-layer transistor structures as shown in Fig. 21.2a and then considering the resultant circuit of Fig. 21.2b.

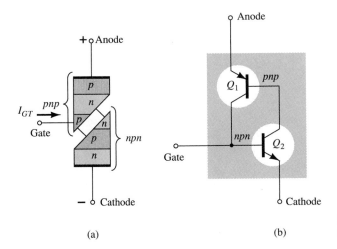

Figure 21.2 SCR two-transistor equivalent circuit.

Note that one transistor for Fig. 21.2 is an *npn* device while the other is a *pnp* transistor. For discussion purposes, the signal shown in Fig. 21.3a will be applied to the gate of the circuit of Fig. 21.2b. During the interval $0 \rightarrow t_1, V_{\text{gate}} = 0$ V, the circuit of Fig. 21.2b will appear as shown in Fig. 21.3b ($V_{\text{gate}} = 0$ V is equivalent to the gate terminal being grounded as shown in the figure). For $V_{BE_2} = V_{\text{gate}} = 0$ V, the base current $I_{B_2} = 0$ and I_{C_2} will be approximately I_{CO}. The base current of Q_1, $I_{B_1} = I_{C_2} = I_{CO}$, is too small to turn Q_1 on. Both transistors are therefore in the "off" state, resulting in a high impedance between the collector and emitter of each transistor and the open-circuit representation for the controlled rectifier as shown in Fig. 21.3c.

21.3 **Basic Silicon-Controlled Rectifier Operation**

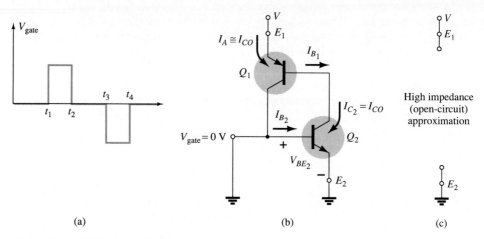

(a)　　　　　　　　　　　　(b)　　　　　　　　　　(c)

Figure 21.3 "Off" state of the SCR.

At $t = t_1$ a pulse of V_G volts will appear at the SCR gate. The circuit conditions established with this input are shown in Fig. 21.4a. The potential V_G was chosen sufficiently large to turn Q_2 on $(V_{BE_2} = V_G)$. The collector current of Q_2 will then rise to a value sufficiently large to turn Q_1 on $(I_{B_1} = I_{C_2})$. As Q_1 turns on, I_{C_1} will increase, resulting in a corresponding increase in I_{B_2}. The increase in base current for Q_2 will result in a further increase in I_{C_2}. The net result is a regenerative increase in the collector current of each transistor. The resulting anode-to-cathode resistance $[R_{\text{SCR}} = V/(I_A - \text{large})]$ is then very small, resulting in the short-circuit representation for the SCR as indicated in Fig. 21.4b. The regenerative action described above results in SCRs having typical turn-on times of 0.1 to 1 μs. However, high-power devices in the range 100 to 400 A may have 10- to 25-μs turn-on times.

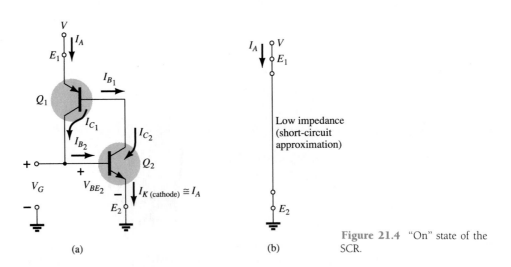

(a)　　　　　　　　　　　　(b)

Figure 21.4 "On" state of the SCR.

In addition to gate triggering, SCRs can also be turned on by significantly raising the temperature of the device or raising the anode-to-cathode voltage to the breakover value shown on the characteristics of Fig. 21.7.

The next question of concern is: How long is the turn-off time and how is turn-off accomplished? An SCR *cannot* be turned off by simply removing the gate signal, and only a special few can be turned off by applying a negative pulse to the gate terminal as shown in Fig. 21.3a at $t = t_3$.

The two general methods for turning off an SCR are categorized as the anode-current interruption and the forced-commutation technique.

The two possibilities for current interruption are shown in Fig. 21.5.

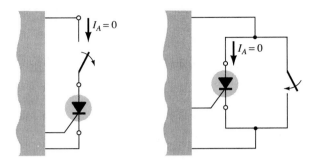

Figure 21.5 Anode current interruption.

In Fig. 21.5a, I_A is zero when the switch is opened (series interruption) while in Fig. 21.5b the same condition is established when the switch is closed (shunt interruption). Forced commutation is the "forcing" of current through the SCR in the direction opposite to forward conduction. There are a wide variety of circuits for performing this function, a number of which can be found in the manuals of major manufacturers in this area. One of the more basic types is shown in Fig. 21.6. As indicated in the figure, the turn-off circuit consists of an *npn* transistor, a dc battery V_B, and a pulse generator. During SCR conduction the transistor is in the "off" state; that is, $I_B = 0$ and the collector-to-emitter impedance is very high (for all practical purposes an open circuit). This high impedance will isolate the turn-off circuitry from affecting the operation of the SCR. For turn-off conditions, a positive pulse is applied to the base of the transistor, turning it heavily on, resulting in a very low impedance from collector to emitter (short-circuit representation). The battery potential will then appear directly across the SCR as shown in Fig. 21.6b, forcing current through it in the reverse direction for turn-off. Turn-off times of SCRs are typically 5 to 30 μs.

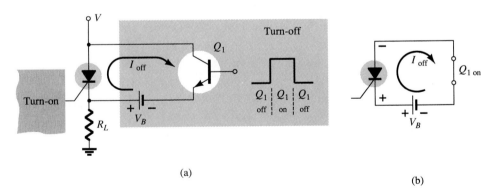

(a)

(b)

Figure 21.6 Forced-commutation technique.

21.4 SCR CHARACTERISTICS AND RATINGS

The characteristics of an SCR are provided in Fig. 21.7 for various values of gate current. The currents and voltages of usual interest are indicated on the characteristic. A brief description of each follows.

21.4 SCR Characteristics and Ratings   833

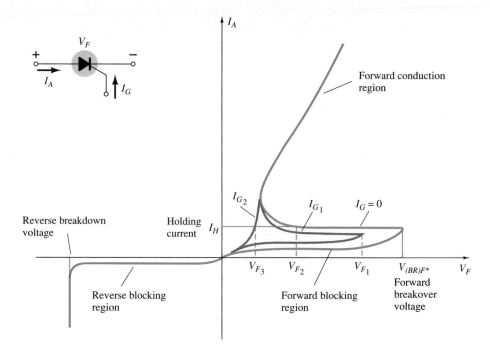

Figure 21.7 SCR characteristics.

1. *Forward breakover voltage* $V_{(BR)F*}$ is that voltage above which the SCR enters the conduction region. The asterisk (*) is a letter to be added that is dependent on the condition of the gate terminal as follows:

$$O = \text{open circuit from } G \text{ to } K$$

$$S = \text{short circuit from } G \text{ to } K$$

$$R = \text{resistor from } G \text{ to } K$$

$$V = \text{fixed bias (voltage) from } G \text{ to } K$$

2. *Holding current* (I_H) is that value of current below which the SCR switches from the conduction state to the forward blocking region under stated conditions.

3. *Forward and reverse blocking regions* are the regions corresponding to the open-circuit condition for the controlled rectifier which *block* the flow of charge (current) from anode to cathode.

4. *Reverse breakdown voltage* is equivalent to the Zener or avalanche region of the fundamental two-layer semiconductor diode.

It should be immediately obvious that the SCR characteristics of Fig. 21.7 are very similar to those of the basic two-layer semiconductor diode except for the horizontal offshoot before entering the conduction region. It is this horizontal jutting region that gives the gate control over the response of the SCR. For the characteristic having the solid line in Fig. 21.7 ($I_G = 0$), V_F must reach the largest required breakover voltage ($V_{(BR)F*}$) before the ''collapsing'' effect will result and the SCR can enter the conduction region corresponding to the *on* state. If the gate current is increased to I_{G_1}, as shown in the same figure by applying a bias voltage to the gate terminal, the value of V_F required for the conduction (V_{F_1}) is considerably less. Note also that I_H drops with increase in I_G. If increased to I_{G_2} the SCR will fire at very low values of voltage (V_{F_3}) and the characteristics begin to approach those of the basic p-n junction diode. Looking at the characteristics in a completely different sense, for a particular V_F voltage, say V_{F_2} (Fig. 21.7), if the gate current is increased from $I_G = 0$ to I_{G_1} or more, the SCR will fire.

The gate characteristics are provided in Fig. 21.8. The characteristics of Fig. 21.8b are an expanded version of the shaded region of Fig. 21.8a. In Fig. 21.8a the three gate ratings of greatest interest, P_{GFM}, I_{GFM}, and V_{GFM} are indicated. Each is included on the characteristics in the same manner employed for the transistor. Except for portions of the shaded region, any combination of gate current and voltage that falls within this region will fire any SCR in the series of components for which these characteristics are provided. Temperature will determine which sections of the shaded region must be avoided. At $-65°C$ the minimum current that will trigger the series of SCRs is 80 mA, while at $+150°C$ only 20 mA are required. The effect of temperature on the minimum gate voltage is usually not indicated on curves of this type since gate potentials of 3 V or more are usually obtained easily. As indicated on Fig. 21.8b, a minimum of 3 V is indicated for all units for the temperature range of interest.

Other parameters usually included on the specification sheet of an SCR are the turn-on time (t_{on}), turn-off time (t_{off}), junction temperature (T_J), and case temperature (T_C), all of which should by now be, to some extent, self-explanatory.

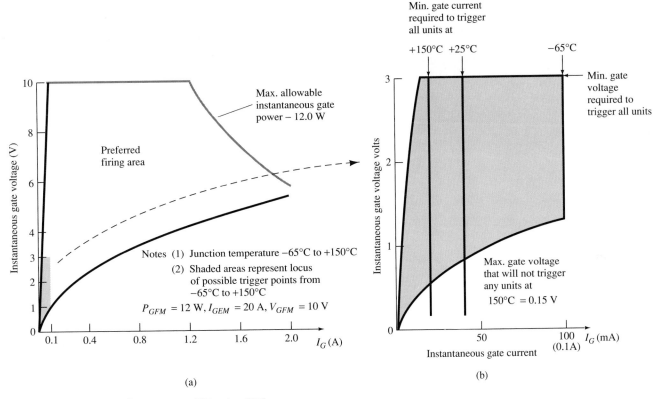

Figure 21.8 SCR gate characteristics (GE series C38).

21.5 SCR CONSTRUCTION AND TERMINAL IDENTIFICATION

The basic construction of the four-layer pellet of an SCR is shown in Fig. 21.9a. The complete construction of a thermal-fatigue-free, high-current SCR is shown in Fig. 21.9b. Note the position of the gate, cathode, and anode terminals. The pedestal acts as a heat sink by transferring the heat developed to the chassis on which the SCR is mounted. The case construction and terminal identification of SCRs will vary with the application. Other case-construction techniques and the terminal identification of each are indicated in Fig. 21.10.

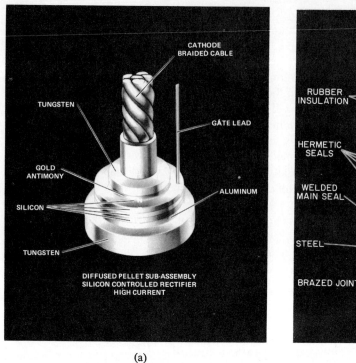

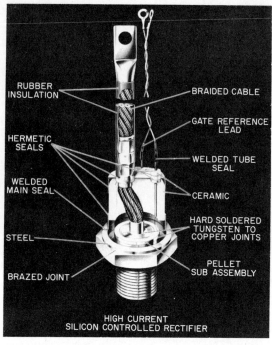

Figure 21.9 (a) Alloy-diffused SCR pellet; (b) thermal fatigue-free SCR construction. (Courtesy General Electric Company.)

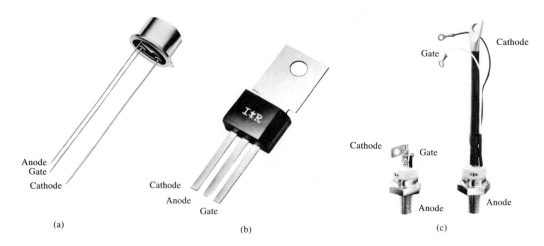

Figure 21.10 SCR case construction and terminal identification. [(a) Courtesy General Electric Company; (b) and (c) courtesy International Rectifier Corporation.]

21.6 SCR APPLICATIONS

A few of the possible applications for the SCR are listed in the introduction to the SCR (Section 21.2). In this section we consider five: a static switch, phase-control system, battery charger, temperature controller, and single-source emergency-lighting system.

A half-wave *series static switch* is shown in Fig. 21.11a. If the switch is closed as shown in Fig. 21.11b, a gate current will flow during the positive portion of the input signal, turning the SCR on. Resistor R_1 limits the magnitude of the gate current.

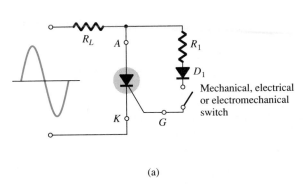

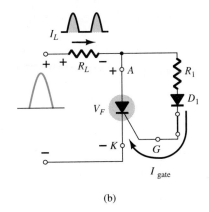

(a)   (b)

Figure 21.11 Half-wave series static switch.

When the SCR turns on, the anode-to cathode voltage (V_F) will drop to the conduction value, resulting in a greatly reduced gate current and very little loss in the gate circuitry. For the negative region of the input signal the SCR will turn off, since the anode is negative with respect to the cathode. The diode D_1 is included to prevent a reversal in gate current.

The waveforms for the resulting load current and voltage are shown in Fig. 21.11b. The result is a half-wave-rectified signal through the load. If less than 180° conduction is desired, the switch can be closed at any phase displacement during the positive portion of the input signal. The switch can be electronic, electromagnetic, or mechanical, depending on the application.

A circuit capable of establishing a conduction angle between 90 and 180° is shown in Fig. 21.12a. The circuit is similar to that of Fig. 21.11a except for the addition of a variable resistor and the elimination of the switch. The combination of the resistors R and R_1 will limit the gate current during the positive portion of the input signal. If R_1 is set to its maximum value, the gate current may never reach turn-on magnitude. As R_1 is decreased from the maximum the gate current will increase from the same input voltage. In this way, the required turn-on gate current can be established in any point between 0 and 90° as shown in Fig. 21.12b. If R_1 is low, the SCR will fire almost immediately, resulting in the same action as that obtained from the circuit of Fig. 21.11a (180° conduction). However, as indicated above, if R_1 is increased, a larger input voltage (positive) will be required to fire the SCR. As shown in Fig. 21.12b, the control cannot be extended past a 90° phase displacement since the input is its maximum at this point. If it fails to fire at this and lesser values of input voltage on the positive slope of the input, the same response must be expected from the negatively sloped portion of the signal waveform. The operation here is normally referred to in technical terms as *half-wave variable-resistance phase con-*

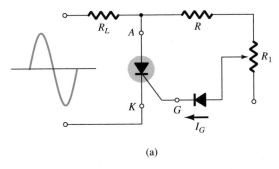

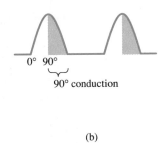

(a)   (b)

Figure 21.12 Half-wave variable-resistance phase control.

trol. It is an effective method of controlling the rms current and therefore power to the load.

A third popular application of the SCR is in a *battery-charging regulator*. The fundamental components of the circuit are shown in Fig. 21.13. You will note that the control circuit has been blocked off for discussion purposes.

As indicated in the figure, D_1 and D_2 establish a full-wave-rectified signal across SCR_1 and the 12-V battery to be charged. At low battery voltages SCR_2 is in the "off" state for reasons to be explained shortly. With SCR_2 open, the SCR_1 controlling circuit is exactly the same as the series static switch control discussed earlier in this section. When the full-wave-rectified input is sufficiently large to produce the required turn-on gate current (controlled by R_1), SCR_1 will turn on and charging of the battery will commence. At the start of charging, the low battery voltage will result in a low voltage V_R as determined by the simple voltage-divider circuit. Voltage V_R is in turn too small to cause 11.0-V Zener conduction. In the off state, the Zener is effectively an open circuit maintaining SCR_2 in the "off" state since the gate current is zero. The capacitor C_1 is included to prevent any voltage transients in the circuit from accidentally turning on SCR_2. Recall from your fundamental study of circuit analysis that the voltage cannot instantaneously change across a capacitor. In this way C_1 prevents transient effects from affecting the SCR.

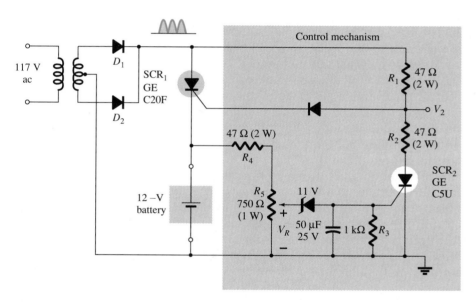

Figure 21.13 Battery-charging regulator.

As charging continues, the battery voltage rises to a point where V_R is sufficiently high to both turn on the 11.0-V Zener and fire SCR_2. Once SCR_2 has fired, the short-circuit representation for SCR_2 will result in a voltage-divider circuit determined by R_1 and R_2 that will maintain V_2 at a level too small to turn SCR_1 on. When this occurs, the battery is fully charged and the open-circuit state of SCR_1 will cut off the charging current. Thus the regulator recharges the battery whenever the voltage drops and prevents overcharging when fully charged.

The schematic diagram of a 100-W heater control using an SCR appears in Fig. 21.14. It is designed such that the 100-W heater will turn on and off as determined by thermostats. Mercury-in-glass thermostats are very sensitive to temperature change. In fact, they can sense changes as small as 0.1°C. It is limited in application, however, in that it can only handle very low levels of current—below 1 mA. In this application, the SCR serves as a current amplifier in a load-switching element. It is

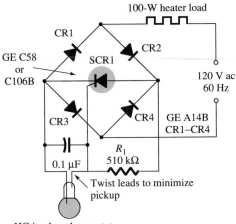

100-W heater load

CR1

GE C58 or C106B

SCR1

CR2

120 V ac
60 Hz

CR3

CR4

GE A14B
CR1–CR4

R_1
510 kΩ

0.1 μF

Twist leads to minimize pickup

HG in glass thermostat
(such as vap. air div. 206-44
series; princo #T141, or
equivalent)

Figure 21.14 Temperature controller. (Courtesy General Electric Semiconductor Products Division.)

not an amplifier in the sense that it magnifies the current level of the thermostat. Rather it is a device whose higher current level is controlled by the behavior of the thermostat.

It should be clear that the bridge network is connected to the ac supply through the 100-W heater. This will result in a full-wave-rectified voltage across the SCR. When the thermostat is open, the voltage across the capacitor will charge to a gate-firing potential through each pulse of the rectified signal. The charging time constant is determined by the RC product. This will trigger the SCR during each half-cycle of the input signal, permitting a flow of charge (current) to the heater. As the temperature rises, the conductive thermostat will short-circuit the capacitor, eliminating the possibility of the capacitor charging to the firing potential and triggering the SCR. The 510-kΩ resistor will then contribute to maintaining a very low current (less than 250 μA) through the thermostat.

The last application for the SCR to be described is shown in Fig. 21.15. It is a single-source emergency-lighting system that will maintain the charge on a 6-V battery to ensure its availability and also provide dc energy to a bulb if there is a power shortage. A full-wave-rectified signal will appear across the 6-V lamp due to diodes D_2 and D_1. The capacitor C_1 will charge to a voltage slightly less than a difference between the peak value of the full-wave-rectified signal and the dc voltage across R_2 established by the 6-V battery. In any event, the cathode of SCR$_1$ is higher than the

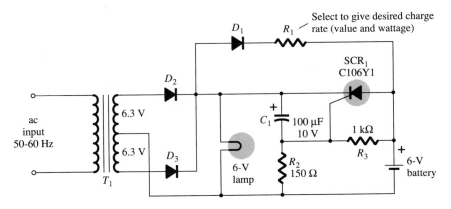

Figure 21.15 Single-source emergency lighting system. (Courtesy General Electric Semiconductor Products Division.)

anode and the gate-to-cathode voltage is negative, ensuring that the SCR is nonconducting. The battery is being charged through R_1 and D_1 at a rate determined by R_1. Charging will only take place when the anode of D_1 is more positive than its cathode. The dc level of the full-wave-rectified signal will ensure that the bulb is lit when the power is on. If the power should fail, the capacitor C_1 will discharge through D_1, R_1, and R_3 until the cathode of SCR_1 is less positive than the anode. At the same time the junction of R_2 and R_3 will become positive and establish sufficient gate-to-cathode voltage to trigger the SCR. Once fired, the 6-V battery would discharge through the SCR_1 and energize the lamp and maintain its illumination. Once power is restored the capacitor C_1 will recharge and reestablish the nonconducting state of SCR_1 as described above.

21.7 SILICON-CONTROLLED SWITCH

The silicon-controlled switch (SCS), like the silicon-controlled rectifier, is a four-layer *pnpn* device. All four semiconductor layers of the SCS are available due to the addition of an anode gate, as shown in Fig. 21.16a. The graphic symbol and transistor equivalent circuit are shown in the same figure. The characteristics of the device are essentially the same as those for the SCR. The effect of an anode gate current is very similar to that demonstrated by the gate current in Fig. 21.7. The higher the anode gate current, the lower the required anode-to-cathode voltage to turn the device on.

The anode gate connection can be used to turn the device either on or off. To turn on the device, a negative pulse must be applied to the anode gate terminal, while a positive pulse is required to turn off the device. The need for the type of pulse indicated above can be demonstrated using the circuit of Fig. 21.16c. A negative pulse at the anode gate will forward-bias the base-to-emitter junction of Q_1, turning it on. The resulting heavy collector current I_{C_1} will turn on Q_2, resulting in a regenerative action and the on state for the SCS device. A positive pulse at the anode gate will reverse-bias the base-to-emitter junction of Q_1, turning it off, resulting in the open-circuit "off" state of the device. In general, the triggering (turn-on) anode gate current is larger in magnitude than the required cathode gate current. For one representative SCS device, the triggering anode gate current is 1.5 mA while the required cathode gate current is 1 μA. The required turn-on gate current at either terminal is affected by many factors. A few include the operating temperature, anode-to-cathode voltage, load placement, and type of cathode, gate-to-cathode or anode gate-to-anode connection (short-circuit, open-circuit, bias, load, etc.). Tables, graphs, and curves are normally available for each device to provide the type of information indicated above.

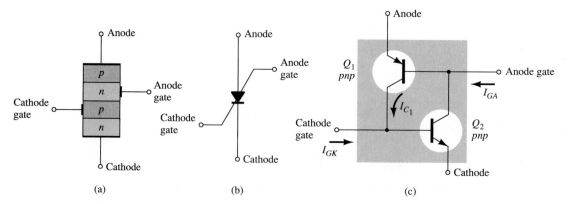

Figure 21.16 Silicon-controlled switch (SCS): (a) basic construction; (b) graphic symbol; (c) equivalent transistor circuit.

Three of the more fundamental types of turn-off circuits for the SCS are shown in Fig. 21.17. When a pulse is applied to the circuit of Fig. 21.17a, the transistor conducts heavily, resulting in a low-impedance ($\cong$ short-circuit) characteristic between collector and emitter. This low-impedance branch diverts anode current away from the SCS, dropping it below the holding value and consequently turning it off. Similarly, the positive pulse at the anode gate of Fig. 21.17b will turn the SCS off by the mechanism described earlier in this section. The circuit of Fig. 21.17c can be turned either off *or* on by a pulse of the proper magnitude at the cathode gate. The turn-off characteristic is possible only if the correct value of R_A is employed. It will control the amount of regenerative feedback, the magnitude of which is critical for this type of operation. Note the variety of positions in which the load resistor R_L can be placed. There are a number of other possibilities that can be found in any comprehensive semiconductor handbook or manual.

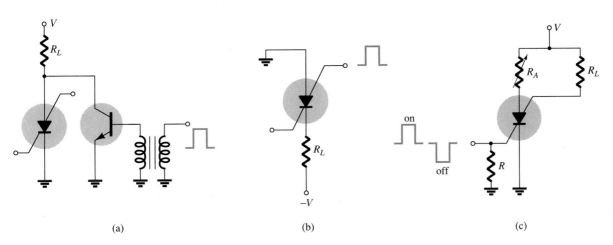

Figure 21.17 SCS turn-off techniques.

An advantage of the SCS over a corresponding SCR is the reduced turn-off time, typically within the range 1 to 10 μs for the SCS and 5 to 30 μs for the SCR. Some of the remaining advantages of the SCS over an SCR include increased control and triggering sensitivity and a more predictable firing situation. At present, however, the SCS is limited to low power, current, and voltage ratings. Typical maximum anode currents range from 100 to 300 mA with dissipation (power) ratings of 100 to 500 mW.

A few of the more common areas of application include a wide variety of computer circuits (counters, registers, and timing circuits), pulse generators, voltage sensors, and oscillators. One simple application for an SCS as a voltage-sensing device is shown in Fig. 21.18. It is an alarm system with n inputs from various stations. Any single input will turn that particular SCS on, resulting in an energized alarm relay and light in the anode gate circuit to indicate the location of the input (disturbance).

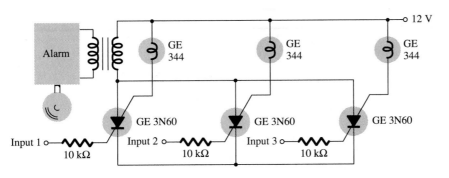

Figure 21.18 SCS alarm circuit.

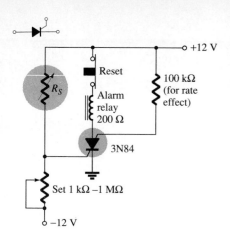

One additional application of the SCS is in the alarm circuit of Fig. 21.19. R_S represents a temperature-, light-, or radiation-sensitive resistor, that is, an element whose resistance will decrease with the application of any of the three energy sources listed above. The cathode gate potential is determined by the divider relationship established by R_S and the variable resistor. Note that the gate potential is at approximately 0 volts if R_S equals the value set by the variable resistor, since both resistors will have 12 V across them. However, if R_S decreases, the potential of the junction will increase until the SCS is forward biased, causing the SCS to turn on and energize the alarm relay.

The 100-kΩ resistor is included to reduce the possibility of accidental triggering of the device through a phenomena known as *rate effect*. It is caused by the stray capacitance levels between gates. A high-frequency transient can establish sufficient base current to turn the SCS on accidentally. The device is reset by pressing the reset button, which in turn opens the conduction path of the SCS and reduces the anode current to zero.

Sensitivity to resistors R_S that increase in resistance due to the application of any of the three energy sources described above can be accommodated by simply interchanging the location of R_S and the variable resistor. The terminal identification of an SCS is shown in Fig. 21.20 with a packaged SCS.

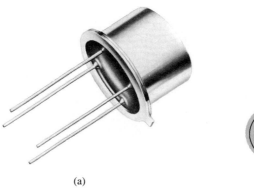

(a)   (b)

Figure 21.20 Silicon-controlled switch (SCS): (a) device; (b) terminal identification. (Courtesy General Electric Company.)

21.8 GATE TURN-OFF SWITCH

The gate turn-off switch (GTO) is the third *pnpn* device to be introduced in this chapter. Like the SCR, however, it has only three external terminals, as indicated in Fig. 21.21a. Its graphic symbol is also shown in Fig. 21.21b. Although the graphic symbol is different from either the SCR or the SCS, the transistor equivalent is exactly the same and the characteristics are similar.

The most obvious advantage of the GTO over the SCR or SCS is the fact that it can be turned on *or* off by applying the proper pulse to the cathode gate (without the

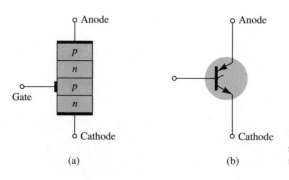

(a)   (b)

Figure 21.21 Gate turn-off switch (GTO): (a) basic construction; (b) symbol.

anode gate and associated circuitry required for the SCS). A consequence of this turn-off capability is an increase in the magnitude of the required gate current for triggering. For an SCR and GTO of similar maximum rms current ratings, the gate-triggering current of a particular SCR is 30 μA, while the triggering current of the GTO is 20 mA. The turn-off current of a GTO is slightly larger than the required triggering current. The maximum rms current and dissipation ratings of GTOs manufactured today are limited to about 3 A and 20 W, respectively.

A second very important characteristic of the GTO is improved switching characteristics. The turn-on time is similar to the SCR (typically 1 μs), but the turn-off time of about the *same* duration (1 μs) is much smaller than the typical turn-off time of an SCR (5 to 30 μs). The fact that the turn-off time is similar to the turn-on time rather than considerably larger permits the use of this device in high-speed applications.

A typical GTO and its terminal identification are shown in Fig. 21.22. The GTO gate input characteristics and turn-off circuits can be found in a comprehensive manual or specification sheet. The majority of the SCR turn-off circuits can also be used for GTOs.

Some of the areas of application for the GTO include counters, pulse generators, multivibrators, and voltage regulators. Figure 21.23 is an illustration of a simple sawtooth generator employing a GTO and a Zener diode.

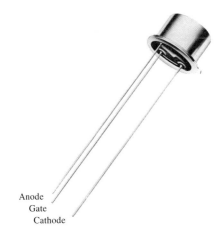

Figure 21.22 Typical GTO and its terminal identification. (Courtesy General Electric Company.)

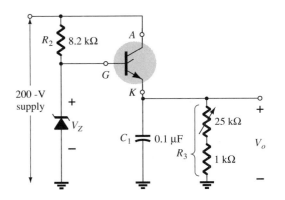
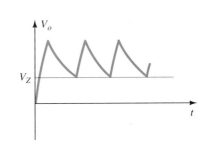

Figure 21.23 GTO sawtooth generator.

When the supply is energized, the GTO will turn on, resulting in the short-circuit equivalent from anode to cathode. The capacitor C_1 will then begin to charge toward the supply voltage as shown in Fig. 21.23. As the voltage across the capacitor C_1 charges above the Zener potential, a reversal in gate-to-cathode voltage will result, establishing a reversal in gate current. Eventually, the negative gate current will be large enough to turn the GTO off. Once the GTO turns off, resulting in the open-circuit representation, the capacitor C_1 will discharge through the resistor R_3. The discharge time will be determined by the circuit time constant $\tau = R_3 C_1$. The proper choice of R_3 and C_1 will result in the sawtooth waveform of Fig. 21.23. Once the output potential V_o drops below V_Z, the GTO will turn on and the process will repeat.

21.9 LIGHT-ACTIVATED SCR

The next in the series of *pnpn* devices is the light-activated SCR (LASCR). As indicated by the terminology, it is an SCR whose state is controlled by the light falling upon a silicon semiconductor layer of the device. The basic construction of an LASCR is shown in Fig. 21.24a. As indicated in Fig. 21.24a, a gate lead is also provided to permit triggering the device using typical SCR methods. Note also in the

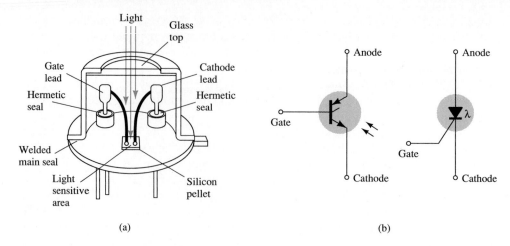

Figure 21.24 Light-activated SCR (LASCR): (a) basic construction; (b) symbols.

figure that the mounting surface for the silicon pellet is the anode connection for the device. The graphic symbols most commonly employed for the LASCR are provided in Fig. 21.24b. The terminal identification and a typical LASCR appear in Fig. 21.25a.

Some of the areas of application for the LASCR include optical light controls, relays, phase control, motor control, and a variety of computer applications. The

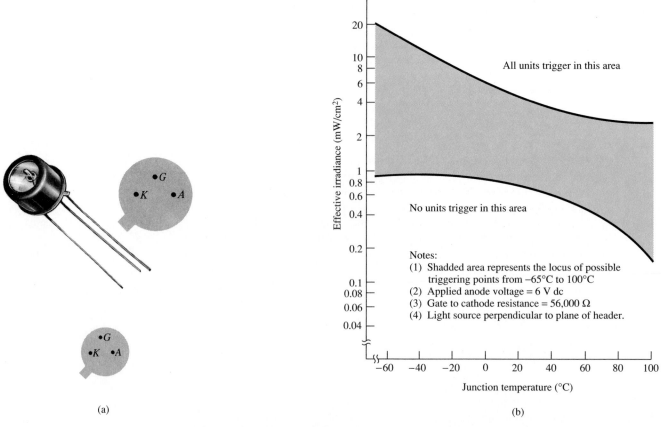

Figure 21.25 LASCR: (a) appearance and terminal identification; (b) light-triggering characteristics. (Courtesy General Electric Company.)

maximum current (rms) and power (gate) ratings for LASCRs commercially available today are about 3 A and 0.1 W. The characteristics (light triggering) of a representative LASCR are provided in Fig. 21.25b. Note in this figure that an increase in junction temperature results in a reduction in light energy required to activate the device.

One interesting application of an LASCR is in the AND and OR circuits of Fig. 21.26. Only when light falls on LASCR$_1$ *and* LASCR$_2$ will the short-circuit representation for each be applicable and the supply voltage appear across the load. For the OR circuit, light energy applied to LASCR$_1$ *or* LASCR$_2$ will result in the supply voltage appearing across the load.

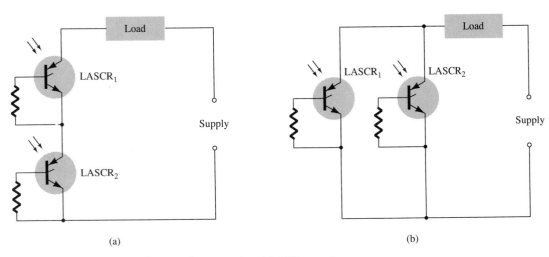

Figure 21.26 LASCR optoelectronic logic circuitry: (a) AND gate: input to LASCR$_1$ *and* LASCR$_2$ required for energization of the load; (b) OR gate: input to either LASCR$_1$ *or* LASCR$_2$ will energize the load.

The LASCR is most sensitive to light when the gate terminal is open. Its sensitivity can be reduced and controlled somewhat by the insertion of a gate resistor, as shown in Fig. 21.26.

A second application of the LASCR appears in Fig. 21.27. It is the semiconductor analog of an electromechanical relay. Note that it offers complete isolation between the input and switching element. The energizing current can be passed through a light-emitting diode or a lamp, as shown in the figure. The incident light will cause the LASCR to turn on and permit a flow of charge (current) through the load as established by the dc supply. The LASCR can be turned off using the reset switch S_1. This system offers the additional advantages over an electromechanical switch of long life, microsecond response, small size, and the elimination of contact bounce.

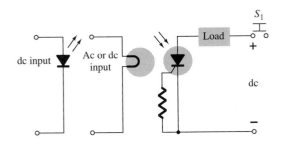

Figure 21.27 Latching relay. (Courtesy Powerex Inc.)

21.10 SHOCKLEY DIODE

The Shockley diode is a four-layer *pnpn* diode with only two external terminals, as shown in Fig. 21.28a with its graphic symbol. The characteristics (Fig. 21.28b) of the device are exactly the same as those encountered for the SCR with $I_G = 0$. As indicated by the characteristics, the device is in the off state (open-circuit representation) until the breakover voltage is reached, at which time avalanche conditions develop and the device turns on (short-circuit representation).

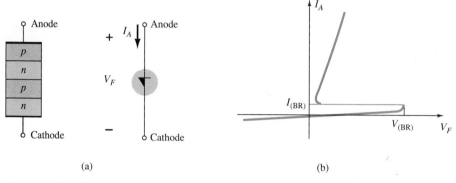

(a)   (b)

Figure 21.28 Shockley diode: (a) basic construction and symbol; (b) characteristics.

One common application of the Shockley diode is shown in Fig. 21.29, where it is employed as a trigger switch for an SCR. When the circuit is energized, the voltage across the capacitor will begin to change toward the supply voltage. Eventually, the voltage across the capacitor will be sufficiently high to first turn on the Shockley diode and then the SCR.

21.11 DIAC

The diac is basically a two-terminal parallel-inverse combination of semiconductor layers that permits triggering in either direction. The characteristics of the device, presented in Fig. 21.30a, clearly demonstrate that there is a breakover voltage in

Figure 21.29 Shockley diode application—trigger switch for an SCR.

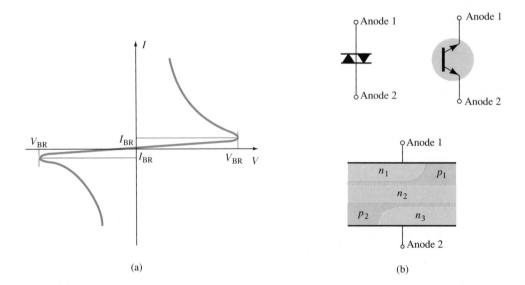

(a)   (b)

Figure 21.30 Diac: (a) characteristics; (b) symbols and basic construction. (Courtesy General Electric Company.)

Chapter 21 *pnpn* and Other Devices

either direction. This possibility of an *on* condition in either direction can be used to its fullest advantage in ac applications.

The basic arrangement of the semiconductor layers of the diac is shown in Fig. 21.30b, along with its graphic symbol. Note that neither terminal is referred to as the cathode. Instead, there is an anode 1 (or electrode 1) and an anode 2 (or electrode 2). When anode 1 is positive with respect to anode 2, the semiconductor layers of particular interest are $p_1 n_2 p_2$ and n_3. For anode 2 positive with respect to anode 1 the applicable layers are $p_2 n_2 p_1$ and n_1.

For the unit appearing in Fig. 21.30, the breakdown voltages are very close in magnitude but may vary from a minimum of 28 V to a maximum of 42 V. They are related by the following equation provided in the specification sheet:

$$V_{BR_1} = V_{BR_2} \pm 10\% \; V_{BR_2} \qquad (21.1)$$

The current levels (I_{BR_1} and I_{BR_2}) are also very close in magnitude for each device. For the unit of Fig. 21.30, both current levels are about 200 μA = 0.2 mA.

The use of the diac in a proximity detector appears in Fig. 21.31. Note the use of an SCR in series with the load and the programmable unijunction transistor (to be described in Section 21.13) connected directly to the sensing electrode.

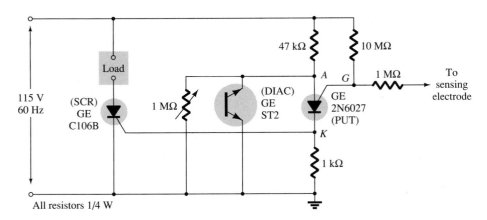

Figure 21.31 Proximity detector or touch switch. (Courtesy Powerex, Inc.)

As the human body approaches the sensing electrode, the capacitance between the electrode and ground will increase. The programmable UJT (PUT) is a device that will fire (enter the short-circuit state) when the anode voltage (V_A) is at least 0.7 V (for silicon) greater than the gate voltage (V_G). Before the programmable device turns on, the system is essentially as shown in Fig. 21.32. As the input voltage rises, the

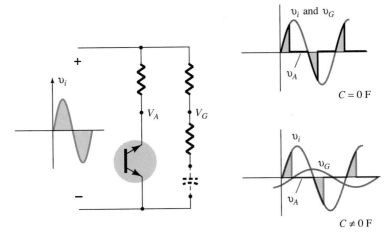

Figure 21.32 Effect of capacitive element on the behavior of the network of Fig. 21.31.

diac voltage V_G will follow as shown in the figure until the firing potential is reached. It will then turn on and the diac voltage will drop substantially, as shown. Note that the diac is in essentially an open-circuit state until it fires. Before the capacitive element is introduced, the voltage V_G will be the same as the input. As indicated in the figure, since both V_A and V_G follow the input, V_A can never be greater than V_G by 0.7 V and turn on the device. However, as the capacitive element is introduced, the voltage V_G will begin to lag the input voltage by an increasing angle, as indicated in the figure. There is therefore a point established where V_A can exceed V_G by 0.7 V and cause the programmable device to fire. A heavy current is established through the PUT at this point, raising the voltage V_K and turning on the SCR. A heavy SCR current will then exist through the load, reacting to the presence of the approaching person.

A second application of the diac appears in the next section (Fig. 21.34) as we consider an important power-control device: the triac.

21.12 TRIAC

The triac is fundamentally a diac with a gate terminal for controlling the turn-on conditions of the bilateral device in either direction. In other words, for either direction the gate current can control the action of the device in a manner very similar to that demonstrated for an SCR. The characteristics, however, of the triac in the first and third quadrants are somewhat different from those of the diac, as shown in Fig. 21.33c. Note the holding current in each direction not present in the characteristics of the diac.

The graphic symbol for the device and the distribution of the semiconductor layers are provided in Fig. 21.33 with photographs of the device. For each possible direction of conduction there is a combination of semiconductor layers whose state will be controlled by the signal applied to the gate terminal.

One fundamental application of the triac is presented in Fig. 21.34. In this capacity, it is controlling the ac power to the load by switching on and off during the positive and negative regions of input sinusoidal signal. The action of this circuit during the positive portion of the input signal is very similar to that encountered for the Shockley diode in Fig. 21.29. The advantage of this configuration is that during the negative portion of the input signal the same type of response will result, since both the diac and triac can fire in the reverse direction. The resulting waveform for the current through the load is provided in Fig. 21.34. By varying the resistor R the conduction angle can be controlled. There are units available today that can handle in excess of 10-kW loads.

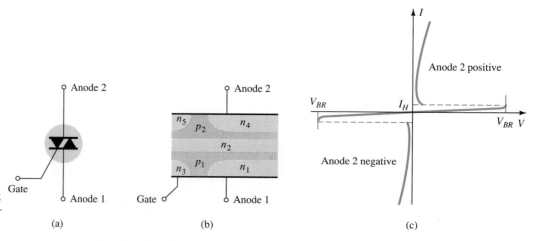

Figure 21.33 Triac: (a) symbol; (b) basic construction; (c) characteristics; (d) photographs.

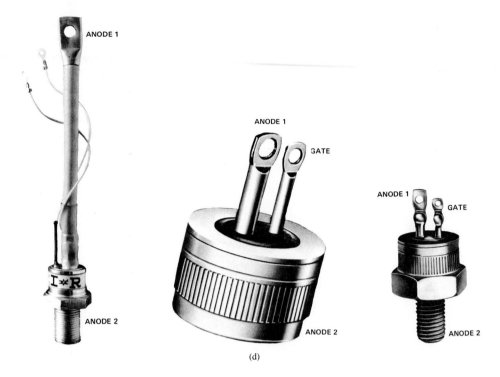

Figure 21.33 Continued.

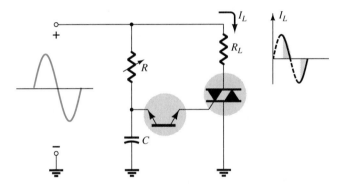

Figure 21.34 Triac application: phase (power) control.

OTHER DEVICES

21.13 UNIJUNCTION TRANSISTOR

Recent interest in the unijunction transistor (UJT) has, like that for the SCR, been increasing at an exponential rate. Although first introduced in 1948, the device did not become commercially available until 1952. The low cost per unit, combined with the excellent characteristics of the device, have warranted its use in a wide variety of applications. A few include oscillators, trigger circuits, sawtooth generators, phase control, timing circuits, bistable networks, and voltage- or current-regulated supplies. The fact that this device is, in general, a low-power-absorbing device under normal operating conditions is a tremendous aid in the continual effort to design relatively efficient systems.

The UJT is a three-terminal device having the basic construction of Fig. 21.35. A slab of lightly doped (increased resistance characteristic) n-type silicon material has two base contacts attached to both ends of one surface and an aluminum rod alloyed to the opposite surface. The p-n junction of the device is formed at the boundary of the aluminum rod and the n-type silicon slab. The single p-n junction accounts for the terminology unijunction. It was originally called a duo (double) base diode due to the presence of two base contacts. Note in Fig. 21.35 that the aluminum rod is alloyed to the silicon slab at a point closer to the base 2 contact than the base 1 contact and that the base 2 terminal is made positive with respect to the base 1 terminal by V_{BB} volts. The effect of each will become evident in the paragraphs to follow.

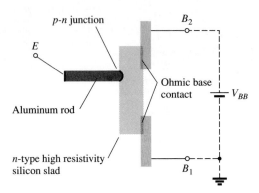

Figure 21.35 Unijunction transistor (UJT): basic construction.

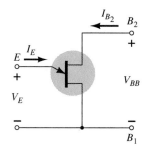

Figure 21.36 Symbol and basic biasing arrangement for the unijunction transistor.

The symbol for the unijunction transistor is provided in Fig. 21.36. Note that the emitter leg is drawn at an angle to the vertical line representing the slab of n-type material. The arrowhead is pointing in the direction of conventional current (hole) flow when the device is in the forward-biased, active, or conducting state.

The circuit equivalent of the UJT is shown in Fig. 21.37. Note the relative simplicity of this equivalent circuit: two resistors (one fixed, one variable) and a single diode. The resistance R_{B_1} is shown as a variable resistor since its magnitude will vary with the current I_E. In fact, for a representative unijunction transistor, R_{B_1} may vary from 5 kΩ down to 50 Ω for a corresponding change of I_E from 0 to 50 μA. The interbase resistance R_{BB} is the resistance of the device between terminals B_1 and B_2 when $I_E = 0$. In equation form,

$$R_{BB} = (R_{B_1} + R_{B_2})|_{I_E=0}$$

(21.2)

(R_{BB} is typically within the range of 4 to 10 kΩ.) The position of the aluminum rod of Fig. 21.35 will determine the relative values of R_{B_1} and R_{B_2} with $I_E = 0$. The magnitude of $V_{R_{B_1}}$ (with $I_E = 0$) is determined by the voltage-divider rule in the following manner:

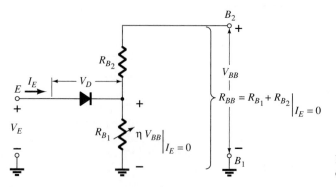

Figure 21.37 UJT equivalent circuit.

Chapter 21 *pnpn* and Other Devices

$$V_{R_{B_1}} = \frac{R_{B_1} V_{BB}}{R_{B_1} + R_{B_2}} = \eta V_{BB}\bigg|_{I_E=0} \qquad (21.3)$$

The Greek letter η (eta) is called the *intrinsic stand-off* ratio of the device and is defined by

$$\eta = \frac{R_{B_1}}{R_{B_1} + R_{B_2}}\bigg|_{I_E=0} = \frac{R_{B_1}}{R_{BB}} \qquad (21.4)$$

For applied emitter potentials (V_E) greater than $V_{R_{B_1}} = \eta V_{BB}$ by the forward voltage drop of the diode, V_D ($0.35 \rightarrow 0.70$ V) the diode will fire, assume the short-circuit representation (on an ideal basis), and I_E will begin to flow through R_{B_1}. In equation form the emitter firing potential is given by

$$V_P = \eta V_{BB} + V_D \qquad (21.5)$$

The characteristics of a representative unijunction transistor are shown for $V_{BB} = 10$ V in Fig. 21.38. Note that for emitter potentials to the left of the peak point, the magnitude of I_E is never greater than I_{EO} (measured in microamperes). The current I_{EO} corresponds very closely with the reverse leakage current I_{CO} of the conventional bipolar transistor. This region, as indicated in the figure, is called the cutoff region. Once conduction is established at $V_E = V_P$, the emitter potential V_E will drop with increase in I_E. This corresponds exactly with the decreasing resistance R_{B_1} for increasing current I_E, as discussed earlier. This device, therefore, has a *negative resistance* region which is stable enough to be used with a great deal of reliability in the areas of application listed earlier. Eventually, the valley point will be reached, and any further increase in I_E will place the device in the saturation region. In this region the characteristics approach that of the semiconductor diode in the equivalent circuit of Fig. 21.37.

The decrease in resistance in the active region is due to the holes injected into the n-type slab from the aluminum p-type rod when conduction is established. The increased hole content in the n-type material will result in an increase in the number of free electrons in the slab, producing an increase in conductivity (G) and a correspond-

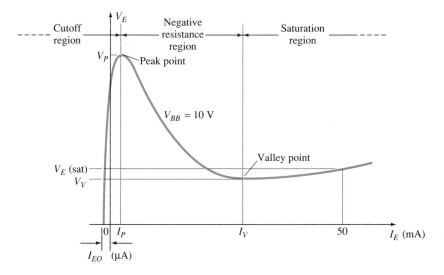

Figure 21.38 UJT static emitter-characteristic curve.

ing drop in resistance ($R \downarrow = 1/G \uparrow$). Three other important parameters for the unijunction transistor are I_P, V_V, and I_V. Each is indicated on Fig. 21.38. They are all self-explanatory.

The emitter characteristics as they normally appear are provided in Fig. 21.39. Note that I_{EO} (μA) is not in evidence since the horizontal scale is in milliamperes. The intersection of each curve with the vertical axis is the corresponding value of V_P. For fixed values of η and V_D, the magnitude of V_P will vary as V_{BB}, that is,

$$V_P \uparrow = \eta V_{BB} \uparrow + V_D$$

fixed

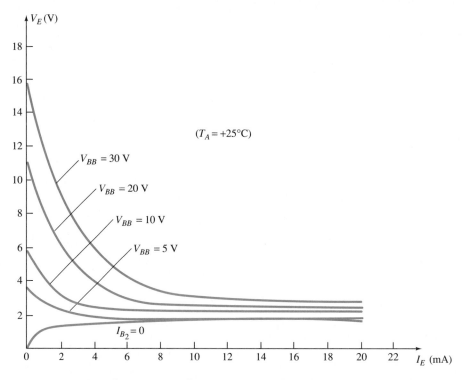

Figure 21.39 Typical static emitter-characteristic curves for a UJT.

A typical set of specifications for the UJT is provided in Fig. 21.40b. The discussion of the last few paragraphs should make each quantity readily recognizable. The terminal identification is provided in the same figure with a photograph of a representative UJT. Note that the base terminals are opposite each other while the emitter terminal is between the two. In addition, the base terminal to be tied to the higher potential is closer to the extension on the lip of the casing.

One rather common application of the UJT is in the triggering of other devices such as the SCR. The basic elements of such a triggering circuit are shown in Fig. 21.41. The resistor R_1 must be chosen to ensure that the load line determined by R_1 passes through the device characteristics in the negative resistance region, i.e., to the right of the peak point but to the left of the valley point as shown in Fig. 21.42. If the load line fails to pass to the right of the peak point, the device cannot turn on. An equation for R_1 that will ensure a turn-on condition can be established if we consider the peak point at which $I_{R_1} = I_P$ and $V_E = V_P$. (The equality $I_{R_1} = I_P$ is valid since the charging current of the capacitor, at this instant, is zero; that is, the capacitor is at this particular instant changing from a charging to a discharging state.) Then $V - I_{R_1}R_1 =$

Chapter 21 *pnpn* and Other Devices

Absolute maximum ratings (25°C):

Power dissipation	
RMS emitter current	300 mW
Peak emitter current	50 mA
Emitter reverse voltage	2 A
Interbase voltage	30 V
Operating temperature range	35 V
Storage temperature range	−65°C to +125°C
	−65°C to +150°C

Electrical characteristics (25°C):

		Min.	Typ.	Max.
Intrinsic standoff ratio		0.56	0.65	
(V_{BB} = 10 V)	η	0.56	0.65	0.75
Interbase resistance (kΩ)				
(V_{BB} = 3 V, I_E = 0)	R_{BB}	4.7	7	9.1
Emitter saturation voltage				
(V_{BB} = 10 V, I_E = 50 mA)	$V_{E(sat)}$		2	
Emitter reverse current				
(V_{BB} = 3 V, I_{B1} = 0)	I_{EO}		0.05	12
Peak point emitter current	I_P (μA)		0.04	5
(V_{BB} = 25 V)				
Valley point current				
(V_{BB} = 20 V)	I_V (mA)	4	6	

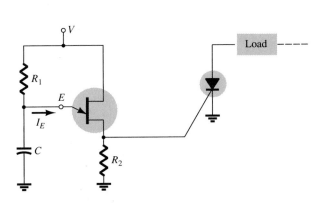

(a) (b) (c)

Figure 21.40 UJT: (a) appearance; (b) specification sheet; (c) terminal identification. (Courtesy General Electric Company.)

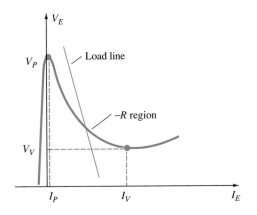

Figure 21.41 UJT triggering of an SCR.

Figure 21.42 Load line for a triggering application.

V_E and $R_1 = (V - V_E)/I_{R_1} = (V - V_P)/I_P$ at the peak point. To ensure firing,

$$R_1 < \frac{V - V_P}{I_P} \qquad (21.6)$$

At the valley point $I_E = I_V$ and $V_E = V_V$, so that

$$V - I_{R_1}R_1 = V_E$$

becomes

$$V - I_V R_1 = V_V$$

and

$$R_1 = \frac{V - V_V}{I_V}$$

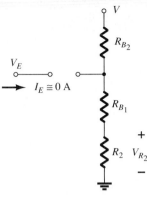

Figure 21.43 Triggering network when $I_E \cong 0$ A.

or to ensure turning off,

$$\boxed{R_1 > \frac{V - V_V}{I_V}} \tag{21.7}$$

The range of R_1 is therefore limited by

$$\boxed{\frac{V - V_V}{I_V} < R_1 < \frac{V - V_P}{I_P}} \tag{21.8}$$

The resistance R_2 must be chosen small enough to ensure that the SCR is not turned on by the voltage V_{R_2} of Fig. 21.43 when $I_E \cong 0$ A. The voltage

$$\boxed{V_{R_2} \cong \frac{R_2 V}{R_2 + R_{BB}}\bigg|_{I_E \cong 0 \text{ A}}} \tag{21.9}$$

The capacitor C will determine, as we shall see, the time interval between triggering pulses and the time span of each pulse.

At the instant the dc supply voltage V is applied, the voltage $v_E = v_C$ will charge toward V volts from V_V as shown in Fig. 21.44 with a time constant $\tau = R_1 C$.

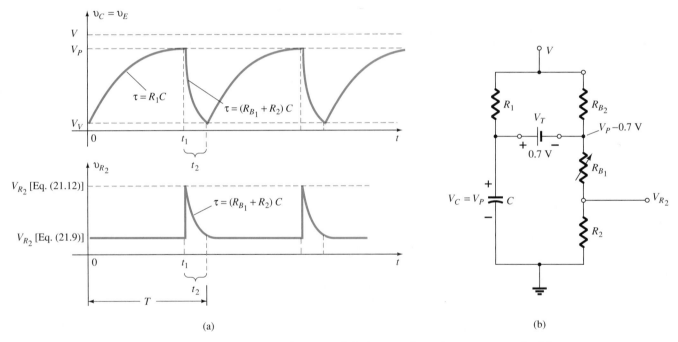

Figure 21.44 (a) Charging and discharging phases for trigger network of Fig. 21.41; (b) equivalent network when UJT turns on.

The general equation for the charging period is

$$\boxed{v_C = V_V + (V - V_V)(1 - e^{-t/R_1 C})} \tag{21.10}$$

As noted in Fig. 21.44, the voltage across R_2 is determined by Eq. (21.9) during this charging period. When $v_C = v_E = V_P$, the UJT will enter the conduction state and the capacitor will discharge through R_{B_1} and R_2 at a rate determined by the time constant $\tau = (R_{B_1} + R_2)C$.

Chapter 21   *pnpn* and Other Devices

The discharge equation for the voltage $v_C = v_E$ is the following:

$$\boxed{v_C \cong V_P e^{-t/(R_{B_1}+R_2)C}}$$ (21.11)

Equation (21.11) is complicated somewhat by the fact that R_{B_1} will decrease with increasing emitter current and the other elements of the network, such as R_1 and V, will affect the discharge rate and final level. However, the equivalent network appears as shown in Fig. 21.44 and the magnitude of R_1 and R_{B_2} are typically such that a Thévenin network for the network surrounding the capacitor C will be only slightly affected by these two resistors. Even though V is a reasonably high voltage, the voltage-divider contribution to the Thévenin voltage can be ignored on an approximate basis.

Using the reduced equivalent of Fig. 21.45 for the discharge phase will result in the following approximation for the peak value of V_{R_2}:

$$\boxed{V_{R_2} \cong \frac{R_2(V_P - 0.7)}{R_2 + R_{B_1}}}$$ (21.12)

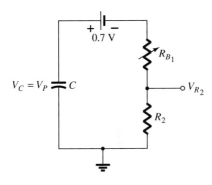

Figure 21.45 Reduced equivalent network when UJT turns on.

The period t_1 of Fig. 21.44 can be determined in the following manner:

$$v_C \text{ (charging)} = V_V + (V - V_V)(1 - e^{-t/R_1C})$$
$$= V_V + V - V_V - (V - V_V)e^{-t/R_1C}$$
$$= V - (V - V_V)e^{-t/R_1C}$$

when $v_C = V_P$, $t = t_1$ and $V_P = V - (V - V_V)e^{-t_1/R_1C}$, or

$$\frac{V_P - V}{V - V_V} = -e^{-t_1/R_1C}$$

and

$$e^{-t_1/R_1C} = \frac{V - V_P}{V - V_V}$$

Using logs, we have

$$\log_e e^{-t_1/R_1C} = \log_e \frac{V - V_P}{V - V_V}$$

and

$$\frac{-t_1}{R_1C} = \log_e \frac{V - V_P}{V - V_V}$$

with

$$\boxed{t_1 = R_1C \log_e \frac{V - V_V}{V - V_P}}$$ (21.13)

For the discharge period the time between t_1 and t_2 can be determined from Eq. (21.11) as follows:

$$v_C \text{ (discharging)} = V_P e^{-t/(R_{B_1}+R_2)C}$$

Establishing t_1 as $t = 0$ gives us

$$v_C = V_V \text{ at } t = t_2$$

and

$$V_V = V_P e^{-t_2/(R_{B_1}+R_2)C}$$

or

$$e^{-t_2/(R_{B_1}+R_2)C} = \frac{V_V}{V_P}$$

Using logs yields

$$\frac{-t_2}{(R_{B_1} + R_2)C} = \log_e \frac{V_V}{V_P}$$

and

$$t_2 = (R_{B_1} + R_2)C \log_e \frac{V_P}{V_V} \qquad (21.14)$$

The period of time to complete one cycle is defined by T in Fig. 21.44. That is,

$$T = t_1 + t_2 \qquad (21.15)$$

If the SCR were dropped from the configuration, the network would behave as a *relaxation oscillator,* generating the waveform of Fig. 21.44. The frequency of oscillation is determined by

$$f_{\text{osc}} = \frac{1}{T} \qquad (21.16)$$

In many systems $t_1 \gg t_2$ and

$$T \cong t_1 = R_1 C \log_e \frac{V - V_V}{V - V_P}$$

Since $V \gg V_V$ in many instances,

$$T \cong t_1 = R_1 C \log_e \frac{V}{V - V_P}$$

$$= R_1 C \log_e \frac{1}{1 - V_P/V}$$

but $\eta = V_P/V$ if we ignore the effects of V_D in Eq. (21.5) and

$$T \cong R_1 C \log_e \frac{1}{1 - \eta}$$

or

$$f \cong \frac{1}{R_1 C \log_e [1/(1 - \eta)]} \qquad (21.17)$$

EXAMPLE 21.1

Given the relaxation oscillator of Fig. 21.46:
(a) Determine R_{B_1} and R_{B_2} at $I_E = 0$ A.
(b) Calculate V_P, the voltage necessary to turn on the UJT.
(c) Determine whether R_1 is within the permissible range of values as determined by Eq. (21.8) to ensure firing of the UJT.
(d) Determine the frequency of oscillation if $R_{B_1} = 100$ Ω during the discharge phase.
(e) Sketch the waveform of v_C for a full cycle.
(f) Sketch the waveform of v_{R_2} for a full cycle.

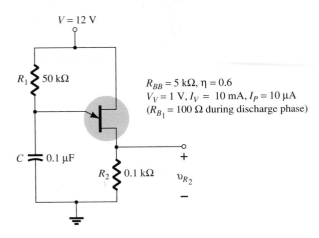

$R_{BB} = 5\ \text{k}\Omega,\ \eta = 0.6$
$V_V = 1\ \text{V},\ I_V = 10\ \text{mA},\ I_P = 10\ \mu\text{A}$
$(R_{B_1} = 100\ \Omega\ \text{during discharge phase})$

Figure 21.46 Example 21.1

Solution

(a) $\eta = \dfrac{R_{B_1}}{R_{B_1} + R_{B_2}}$

$0.6 = \dfrac{R_{B_1}}{R_{BB}}$

$R_{B_1} = 0.6 R_{BB} = 0.6(5\ \text{k}\Omega) = \mathbf{3\ k\Omega}$

$R_{B_2} = R_{BB} - R_{B_1} = 5\ \text{k}\Omega - 3\ \text{k}\Omega = \mathbf{2\ k\Omega}$

(b) At the point where $v_C = V_P$, if we continue with $I_E = 0$ A, the network of Fig. 21.47 will result where

$$V_P = 0.7\ \text{V} + \dfrac{(R_{B_1} + R_2)\ 12\ \text{V}}{\underbrace{R_{B_1} + R_{B_2}}_{R_{BB}} + R_2}$$

$$= 0.7\ \text{V} + \dfrac{(3\ \text{k}\Omega + 0.1\ \text{k}\Omega)\ 12\ \text{V}}{5\ \text{k}\Omega + 0.1\ \text{k}\Omega} = 0.7\ \text{V} + 7.294\ \text{V}$$

$$\cong \mathbf{8\ V}$$

(c) $\dfrac{V - V_V}{I_V} < R_1 < \dfrac{V - V_P}{I_P}$

$\dfrac{12\ \text{V} - 1\ \text{V}}{10\ \text{mA}} < R_1 < \dfrac{12\ \text{V} - 8\ \text{V}}{10\ \mu\text{A}}$

$1.1\ \text{k}\Omega < R_1 < 400\ \text{k}\Omega$

The resistance $R_1 = 50\ \text{k}\Omega$ falls within this range.

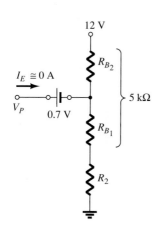

Figure 21.47 Network to determine V_P, the voltage required to turn on the UJT.

(d) $t_1 = R_1 C\ \log_e \dfrac{V - V_V}{V - V_P}$

$= (50\ \text{k}\Omega)(0.1\ \text{pF})\ \log_e \dfrac{12\ \text{V} - 1\ \text{V}}{12\ \text{V} - 8\ \text{V}}$

$= 5 \times 10^{-3}\ \log_e \dfrac{11}{4} = 5 \times 10^{-3}(1.01)$

$= 5.05\ \text{ms}$

$$t_2 = (R_{B_1} + R_2)C \log_e \frac{V_P}{V_V}$$

$$= (0.1 \text{ k}\Omega + 0.1 \text{ k}\Omega)(0.1 \text{ pF}) \log_e \frac{8}{1}$$

$$= (0.02 \times 10^{-6})(2.08)$$

$$= 41.6 \ \mu s$$

and

$$T = t_1 + t_2 = 5.05 \text{ ms} + 0.0416 \text{ ms}$$

$$= 5.092 \text{ ms}$$

with

$$f_{osc} = \frac{1}{T} = \frac{1}{5.092 \text{ ms}} \cong \mathbf{196 \ Hz}$$

Using Eq. (21.17) gives us

$$f \cong \frac{1}{R_1 C \log_e [1/(1 - \eta)]}$$

$$= \frac{1}{5 \times 10^{-3} \log_e 2.5}$$

$$= \mathbf{218 \ Hz}$$

(e) See Fig. 21.48.

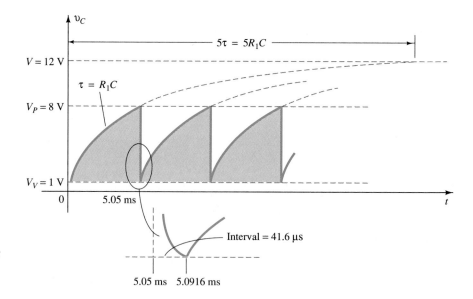

Figure 21.48 The voltage v_C for the relaxation oscillator of Fig. 21.46.

(f) During the charging phase, (Eq. 21.9)

$$V_{R_2} = \frac{R_2 V}{R_2 + R_{BB}} = \frac{0.1 \text{ k}\Omega(12 \text{ V})}{0.1 \text{ k}\Omega + 5 \text{ k}\Omega} = \mathbf{0.235 \ V}$$

When $v_C = V_P$ (Eq. 21.12)

$$V_{R_2} \cong \frac{R_2(V_P - 0.7 \text{ V})}{R_2 + R_{B_1}} = \frac{0.1 \text{ k}\Omega(8 \text{ V} - 0.7 \text{ V})}{0.1 \text{ k}\Omega + 0.1 \text{ k}\Omega}$$

$$= \mathbf{3.65 \ V}$$

The plot of v_{R_2} appears in Fig. 21.49.

Chapter 21 *pnpn* and Other Devices

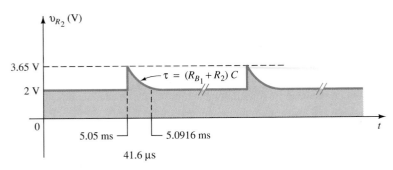

Figure 21.49 The voltage v_{R_2} for the relaxation oscillator of Fig. 21.46.

21.14 PHOTOTRANSISTORS

The fundamental behavior of photoelectric devices was introduced earlier with the description of the photodiode. This discussion will now be extended to include the phototransistor, which has a photosensitive collector–base *p-n* junction. The current induced by photoelectric effects is the base current of the transistor. If we assign the notation I_λ for the photoinduced base current, the resulting collector current, on an approximate basis, is

$$I_C \cong h_{fe}I_\lambda \qquad (21.18)$$

A representative set of characteristics for a phototransistor is provided in Fig. 21.50 with the symbolic representation of the device. Note the similarities between these curves and those of a typical bipolar transistor. As expected, an increase in light intensity corresponds with an increase in collector current. To develop a greater degree of familiarity with the light-intensity unit of measurement, milliwatts per square centimeter, a curve of base current versus flux density appears in Fig. 21.51a. Note the exponential increase in base current with increasing flux density. In the same figure a sketch of the phototransistor is provided with the terminal identification and the angular alignment.

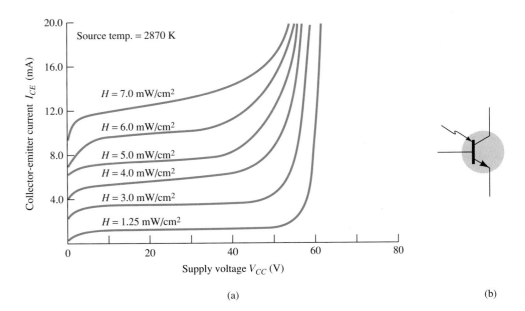

(a)

(b)

Figure 21.50 Phototransistor: (a) collector characteristics (MRD300); (b) symbol. (Courtesy Motorola, Inc.)

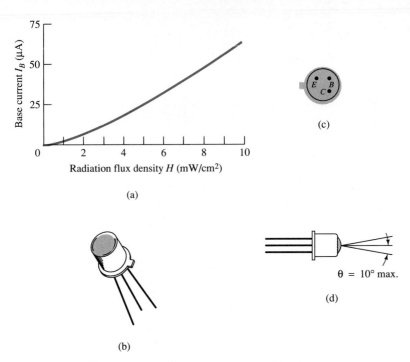

(c)

$\theta = 10°$ max.

(d)

(b)

Figure 21.51 Phototransistor: (a) base current versus flux density; (b) device; (c) terminal identification; (b) angular alignment. (Courtesy Motorola, Inc.)

Some of the areas of application for the phototransistor include punch-card readers, computer logic circuitry, lighting control (highways, etc.), level indication, relays, and counting systems.

A high-isolation AND gate is shown in Fig. 21.52 using three phototransistors and three LEDs (light-emitting diodes). The LEDs are semiconductor devices that emit light at an intensity determined by the forward current through the device. With the aid of discussions in Chapter 1, the circuit behavior should be relatively easy to understand. The terminology "high isolation" simply refers to the lack of an electrical connection between the input and output circuits.

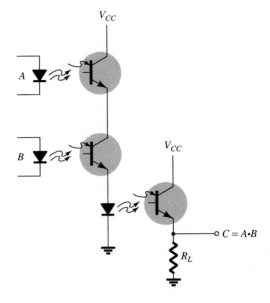

Figure 21.52 High-isolation AND gate employing phototransistors and light-emitting diodes (LEDs).

Chapter 21 *pnpn* and Other Devices

21.15 OPTO-ISOLATORS

The *opto-isolator* is a device that incorporates many of the characteristics described in the preceding section. It is simply a package that contains both an infrared LED and a photodetector such as a silicon diode, transistor Darlington pair, or SCR. The wavelength response of each device is tailored to be as identical as possible to permit the highest measure of coupling possible. In Fig. 21.53, two possible chip configurations are provided, with a photograph of each. There is a transparent insulating cap between each set of elements embedded in the structure (not visible) to permit the passage of light. They are designed with response times so small that they can be used to transmit data in the megahertz range.

ISO-LIT 1

(Top view)

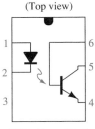

LED chip on Pin 2
PT chip on Pin 5

Pin No.	Function
1	anode
2	cathode
3	nc
4	emitter
5	collector
6	base

ISO-LIT Q1

(Top view)

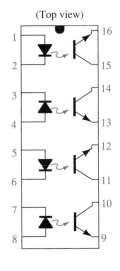

Pin No.	Function
1	anode
2	cathode
3	cathode
4	anode
5	anode
6	cathode
7	cathode
8	anode
9	emitter
10	collector
11	collector
12	emitter
13	emitter
14	collector
15	collector
16	emitter

Figure 21.53 Two Litronix opto-isolators. (Courtesy Siemens Components, Inc.)

The maximum ratings and electrical characteristics for the IL-1 model are provided in Fig. 21.54. Note that I_{CEO} is measured in nanoamperes and that the power dissipation of the LED and transistor are about the same.

The typical optoelectronic characteristic curves for each channel are provided in Figs. 21.55 through 21.59. Note the very pronounced effect of temperature on the output current at low temperatures but the fairly level response at or above room temperature (25°C). As mentioned earlier, the level of I_{CEO} is improving steadily with improved design and construction techniques (the lower the better). In Fig. 21.55 we do not reach 1 μA until the temperature rises above 75°C. The transfer characteristics of Fig. 21.56 compare the input LED current (which establishes the luminous flux) to

(a) Maximum Ratings

Gallium arsenide LED (each channel) IL-1
  Power dissipation @ 25°C   200 mW
  Derate lineraly from 25°C   2.6 mW/°C
  Continuous forward current   150 mA
Detector silicon phototransistor (each channel) IL-1
  Power dissipation @ 25°C   200 mW
  Derate linearly from 25°C   2.6 mW/°C
  Collector-emitter breakdown voltage   30 V
  Emitter-collector breakdown voltage   7 V
  Collector-base breakdown voltage   70 V
Package IL-1
  Total package dissipation at 25°C ambient (LED plus detector)   250 mW
  Derate linearly from 25°C   3.3 mW/°C
  Storage temperature   −55°C to +150°C
  Operating temperature   −55°C to +100°C

(b) Electrical Characteristics per Channel (at 25°C Ambient)

Parameter	Min.	Typ.	Max.	Units	Test Conditions
Gallium arsenide LED					
Forward voltage		1.3	1.5	V	$I_F = 60$ mA
Reverse current		0.1	10	µA	$V_R = 3.0$ V
Capacitance		100		pF	$V_R = 0$ V
Phototransistor detector					
BV_{CEO}	30			V	$I_C = 1$ mA
I_{CEO}		5.0	50	nA	$V_{CE} = 10$ V, $I_F = 0$ A
Collector-emitter capacitance		2.0		pF	$V_{CE} = 0$ V
BV_{ECO}	7			V	$I_E = 100$ µA
Coupled characteristics					
dc current transfer ratio	0.2	0.35			$I_F = 10$ mA, $V_{CE} = 10$ V
Capacitance, input to output		0.5		pF	
Breakdown voltage	2500			V	DC
Resistance, input to output		100		GΩ	
V_{sat}			0.5	V	$I_C = 1.6$ mA, $I_F = 16$ mA
Propagation delay					
$t_{D \, on}$		6.0		µs	$R_L = 2.4$ kΩ, $V_{CE} = 5$ V
$t_{D \, off}$		25		µs	$I_F = 16$ mA

Figure 21.54 Litronix IL-1 opto-isolator.

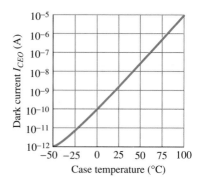

Figure 21.55 Dark current (I_{CEO}) versus temperature.

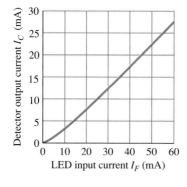

Figure 21.56 Transfer characteristics.

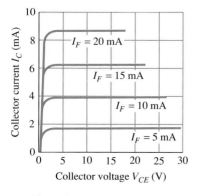

Figure 21.57 Detector output characteristics.

Chapter 21 *pnpn* and Other Devices

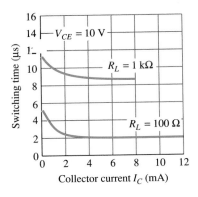

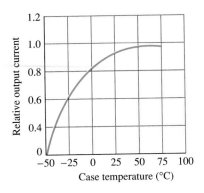

Figure 21.58 Switching time versus collector current.

Figure 21.59 Relative output versus temperature.

the resulting collector current of the output transistor (whose base current is determined by the incident flux). In fact, Fig. 21.57 demonstrates that the V_{CE} voltage affects the resulting collector current only very slightly. It is interesting to note in Fig. 21.58 that the switching time of an opto-isolator decreases with increased current, while for many devices it is exactly the reverse. Consider that it is only 2 μs for a collector current of 6 mA and a load R_L of 100 Ω. The relative output versus temperature appears in Fig. 21.59.

The schematic representation for a transistor coupler appears in Fig. 21.53. The schematic representations for a photodiode, photo-Darlington, and photo-SCR opto-isolator appear in Fig. 21.60.

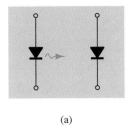

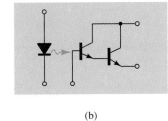

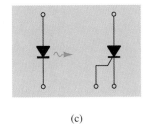

(a)   (b)   (c)

Figure 21.60 Opto-isolators: (a) photodiode; (b) photo-Darlington; (c) photo-SCR.

21.16 PROGRAMMABLE UNIJUNCTION TRANSISTOR

Although there is a similarity in name, the actual construction and mode of operation of the programmable unijunction transistor (PUT) is quite different from the unijunction transistor. The fact that the $I–V$ characteristics and applications of each are similar prompted the choice of labels.

As indicated in Fig. 21.61, the PUT is a four-layer *pnpn* device with a gate connected directly to the sandwiched *n*-type layer. The symbol for the device and the basic biasing arrangement appears in Fig. 21.62. As the symbol suggests, it is essentially an SCR with a control mechanism that permits a duplication of the characteristics of the typical SCR. The term ''programmable'' is applied because R_{BB}, η, and V_P as defined for the UJT can be controlled through the resistors, R_{B_1}, R_{B_2}, and the supply voltage V_{BB}. Note in Fig. 21.62 that through an application of the voltage-divider rule when $I_G = 0$:

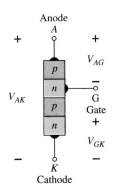

Figure 21.61 Programmable UJT (PUT).

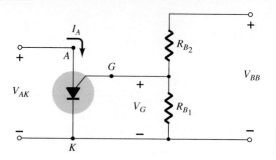

Figure 21.62 Basic biasing arrangement for the PUT.

$$V_G = \frac{R_{B_1}}{R_{B_1} + R_{B_2}} V_{BB} = \eta V_{BB} \qquad (21.19)$$

where

$$\eta = \frac{R_{B_1}}{R_{B_1} + R_{B_2}}$$

as defined for the UJT.

The characteristics of the device appear in Fig. 21.63. As noted on the diagram, the "off" state (I low, V between 0 and V_P) and the "on" state ($I \geq I_V$, $V \geq V_V$) are separated by the unstable region as occurred for the UJT. That is, the device cannot stay in the unstable state—it will simply shift to either the "off" or "on" stable states.

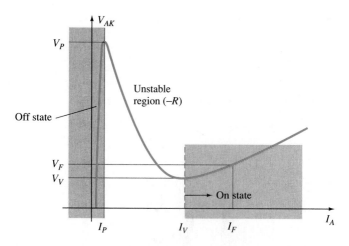

Figure 21.63 PUT characteristics.

The firing potential (V_P) or voltage necessary to "fire" the device is given by

$$V_P = \eta V_{BB} + V_D \qquad (21.20)$$

as defined for the UJT. However, V_P represents the voltage drop V_{AK} in Fig. 21.61 (the forward voltage drop across the conducting diode). For silicon V_D is typically 0.7 V. Therefore,

$$V_{AK} = V_{AG} + V_{GK}$$

$$V_P = V_D + V_G$$

and

$$V_P = \eta V_{BB} + 0.7 \text{ V} \qquad (21.21)$$
_{silicon}

We noted above, however, that $V_G = \eta V_{BB}$ with the result that

$$\boxed{V_P = V_G + 0.7 \text{ V}} \quad \text{silicon} \qquad (21.22)$$

Recall that for the UJT both R_{B_1} and R_{B_2} represent the bulk resistance and ohmic base contacts of the device—both inaccessible. In the development above, we note that R_{B_1} and R_{B_2} are external to the device permitting an adjustment of η and hence V_G above. In other words, the PUT provides a measure of control on the level of V_P required to turn on the device.

Although the characteristics of the PUT and UJT are similar, the peak and valley currents of the PUT are typically lower than those of a similarly rated UJT. In addition, the minimum operating voltage is also less for a PUT.

If we take a Thévenin equivalent of the network to the right of gate terminal in Fig. 21.62, the network of Fig. 21.64 will result. The resulting resistance R_S is important because it is often included in specification sheets since it affects the level of I_V.

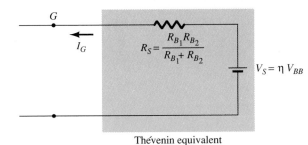

Thévenin equivalent

Figure 21.64 Thévenin equivalent for the network to the right of the gate terminal in Fig. 21.62.

The basic operation of the device can be reviewed through reference to Fig. 21.63. A device in the "off" state will not change state until the voltage V_P as defined by V_G and V_D is reached. The level of current until I_P is reached is very low, resulting in an open-circuit equivalent since $R = V$ (high)/I (low) will result in a high resistance level. When V_P is reached the device will switch through the unstable region to the "on" state, where the voltage is lower but the current higher, resulting in a terminal resistance $R = V$ (low)/I(high) which is quite small, representing short-circuit equivalent on an approximate basis. The device has therefore switched from essentially an open-circuit to a short-circuit state at a point determined by the choice of R_{B_1}, R_{B_2}, and V_{BB}. Once the device is in the "on" state, the removal of V_G will not turn the device off. The level of voltage V_{AK} must be dropped sufficiently to reduce the current below a holding level.

Determine R_{B_1} and V_{BB} for a silicon PUT if it is determined that $\eta = 0.8$, $V_P = 10.3$ V, and $R_{B_2} = 5$ kΩ.

EXAMPLE 21.2

Solution

$$\text{Eq. (21.19):} \quad \eta = \frac{R_{B_1}}{R_{B_1} + R_{B_2}} = 0.8$$

$$R_{B_1} = 0.8(R_{B_1} + R_{B_2})$$

$$0.2\,R_{B_1} = 0.8\,R_{B_2}$$

$$R_{B_1} = 4\,R_{B_2}$$

$$R_{B_1} = 4(5 \text{ k}\Omega) = \textbf{20 k}\boldsymbol{\Omega}$$

$$\text{Eq. (21.20):} \quad V_P = \eta V_{BB} + V_D$$

$$10.3 \text{ V} = (0.8)(V_{BB}) + 0.7 \text{ V}$$

$$9.6 \text{ V} = 0.8 V_{BB}$$

$$V_{BB} = \mathbf{12 \text{ V}}$$

One popular application of the PUT is in the relaxation oscillator of Fig. 21.65. The instant the supply is connected, the capacitor will begin to charge toward V_{BB} volts, since there is no anode current at this point. The charging curve appears in Fig. 21.66. The period T required to reach the firing potential V_P is given approximately by

$$T \cong RC \log_e \frac{V_{BB}}{V_{BB} - V_P} \tag{21.23}$$

or when $V_P \cong \eta V_{BB}$

$$T \cong RC \log_e \left(1 + \frac{R_{B_1}}{R_{B_2}} \right) \tag{21.24}$$

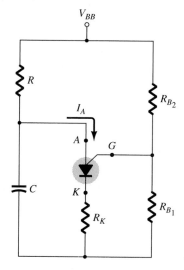

Figure 21.65 PUT relaxation oscillator.

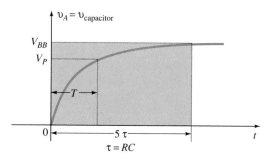

Figure 21.66 Changing wave for the capacitor C of Fig. 21.65.

The instant the voltage across the capacitor equals V_P, the device will fire and a current $I_A = I_P$ established through the PUT. If R is too large, the current I_P cannot be established and the device will not fire. At the point of transition,

$$I_P R = V_{BB} - V_P$$

and

$$R_{\max} = \frac{V_{BB} - V_P}{I_P} \tag{21.25}$$

The subscript is included to indicate that any R greater than $R_{\max}$ will result in a current less than I_P. The level of R must also be such to ensure it is less than I_V if oscillations are to occur. In other words, we want the device to enter the unstable region and then return to the "off" state. From reasoning similar to that above:

$$R_{\min} = \frac{V_{BB} - V_V}{I_V} \tag{21.26}$$

The discussion above requires that R be limited to the following for an oscillatory system:

$$R_{\min} < R < R_{\max}$$

Chapter 21 *pnpn* and Other Devices

The waveforms of v_A, v_G, and v_K appear in Fig. 21.67. Note that T determines the maximum voltage v_A can charge to. Once the device fires, the capacitor will rapidly discharge through the PUT and R_K, producing the drop shown. Of course, v_K will peak at the same time due to the brief but heavy current. The voltage v_G will rapidly drop down from V_G to a level just greater than 0 V. When the capacitor voltage drops to a low level, the PUT will once again turn off and the charging cycle repeated. The effect on V_G and V_K is shown in Fig. 21.67.

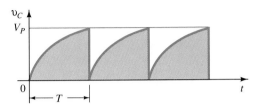

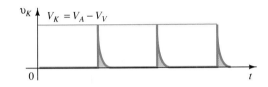

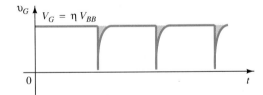

Figure 21.67 Waveforms for PUT oscillator of Fig. 21.65.

If $V_{BB} = 12$ V, $R = 20$ kΩ, $C = 1$ μF, $R_K = 100$ Ω, $R_{B_1} = 10$ kΩ, $R_{B_2} = 5$ kΩ, $I_P = 100$ μA, $V_V = 1$ V, and $I_V = 5.5$ mA, determine:

EXAMPLE 21.3

(a) V_P.
(b) R_{max} and R_{min}.
(c) T and frequency of oscillation.
(d) The waveforms of v_A, v_G, and v_K.

Solution

(a) Eq. 21.20: $V_P = \eta V_{BB} + V_D$

$$= \frac{R_{B_1}}{R_{B_1} + R_{B_2}} V_{BB} + 0.7 \text{ V}$$

$$= \frac{10 \text{ k}\Omega}{10 \text{ k}\Omega + 5 \text{ k}\Omega}(12 \text{ V}) + 0.7 \text{ V}$$

$$= (0.67)(12 \text{ V}) + 0.7 \text{ V} = \mathbf{8.7 \text{ V}}$$

(b) From Eq. (21.25): $R_{max} = \dfrac{V_{BB} - V_P}{I_P}$

$$= \frac{12 \text{ V} - 8.7 \text{ V}}{100 \text{ }\mu\text{A}} = \mathbf{33 \text{ k}\Omega}$$

From Eq. (21.26): $R_{\min} = \dfrac{V_{BB} - V_V}{I_V}$

$$= \dfrac{12 \text{ V} - 1 \text{ V}}{5.5 \text{ mA}} = \mathbf{2 \text{ k}\Omega}$$

$$R: 2 \text{ k}\Omega < 20 \text{ k}\Omega < 33 \text{ k}\Omega$$

(c) Eq. (21.27): $T = RC \log_e \dfrac{V_{BB}}{V_{BB} - V_P}$

$$= (20 \text{ k}\Omega)(1 \text{ pF}) \log_e \dfrac{12 \text{ V}}{12 \text{ V} - 8.7 \text{ V}}$$

$$= 20 \times 10^{-3} \log_e (3.64)$$

$$= 20 \times 10^{-3}(1.29)$$

$$= \mathbf{25.8 \text{ ms}}$$

$$f = \dfrac{1}{T} = \dfrac{1}{25.8 \text{ ms}} = \mathbf{38.8 \text{ Hz}}$$

(d) As indicated in Fig. 21.68.

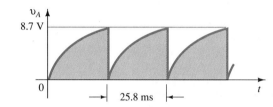

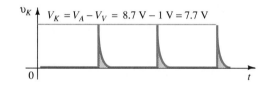

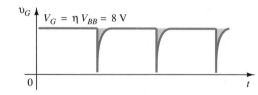

Figure 21.68 Waveforms for the oscillator of Example 21.3.

PROBLEMS

§ 21.3 Basic Silicon-Controlled Rectifier Operation

1. Describe in your own words the basic behavior of the SCR using the two-transistor equivalent circuit.

2. Describe two techniques for turning an SCR off.

3. Consult a manufacturer's manual or specification sheet and obtain a turn-off network. If possible, describe the turn-off action of the design.

§ 21.4 SCR Characteristics and Ratings

* **4.** (a) At high levels of gate current the characteristics of an SCR approach those of what two-terminal device?

Chapter 21 *pnpn* and Other Devices

(b) At a fixed anode-to-cathode voltage less than $V_{(BR)F^*}$, what is the effect on the firing of the SCR as the gate current is reduced from its maximum value to the zero level?

(c) At a fixed gate current greater than $I_G = 0$, what is the effect on the firing of the SCR as the gate voltage is reduced from $V_{(BR)F^*}$?

(d) For increasing levels of I_G, what is the effect on the holding current?

5. (a) Using Fig. 21.8, will a gate current of 50 mA fire the device at room temperature (25C°)?

(b) Repeat part (a) for a gate current of 10 mA.

(c) Will a gate voltage of 2.6 V trigger the device at room temperature?

(d) Is $V_G = 6$ V, $I_G = 800$ mA a good choice for firing conditions? Would $V_G = 4$ V, $I_G = 1.6$ A be preferred? Explain.

§ 21.6 SCR Applications

6. In Fig. 21.11b, why is there very little loss in potential across the SCR during conduction?

7. Fully explain why reduced values of R_1 in Fig. 21.12 will result in an increased angle of conduction.

* 8. Refer to the charging network of Fig. 21.13.

(a) Determine the dc level of the full-wave rectified signal if a $1:1$ transformer were employed.

(b) If the battery in its uncharged state is sitting at 11 V, what is the anode-to-cathode voltage drop across SCR_1?

(c) What is the maximum possible value of V_R ($V_{GK} \cong 0.7$ V)?

(d) At the maximum value of part (c), what is the gate potential of SCR_2?

(e) Once SCR_2 has entered the short-circuit state, what is the level of V_2?

§ 21.7 Silicon-Controlled Switch

9. Fully describe in your own words the behavior of the networks of Fig. 21.17.

§ 21.8 Gate Turn-Off Switch

10. (a) In Fig. 21.23, if $V_Z = 50$ V, determine the maximum possible value the capacitor C_1 can charge to ($V_{GK} \cong 0.7$ V).

(b) Determine the approximate discharge time (5τ) for $R_3 = 20$ kΩ.

(c) Determine the internal resistance of the GTO if the rise time is one-half the decay period determined in part (b).

§ 21.9 Light-Activated SCR

11. (a) Using Fig. 21.25b, determine the minimum irradiance required to fire the device at room temperature (25°C).

(b) What percent reduction in irradiance is allowable if the junction temperature is increased from 0°C (32°F) to 100°C (212°F)?

§ 21.10 Shockley Diode

12. For the network of Fig. 21.29, if $V_{(BR)} = 6$ V, $V = 40$ V, $R = 10$ kΩ, $C = 0.2$ μF, and V_{GK} (firing potential) $= 3$ V, determine the time period between energizing the network and the turning on of the SCR.

§ 21.11 Diac

13. Using whatever reference you require, find an application of a diac and explain the network behavior.

14. If V_{BR_2} is 6.4 V, determine the range for V_{BR_1} using Eq. (21.1).

§ 21.12 Triac

15. Repeat Problem 13 for the triac.

§ 21.13 Unijunction Transistor

16. For the network of Fig. 21.41, in which $V = 40$ V, $\eta = 0.6$, $V_V = 1$ V, $I_V = 8$ mA, and $I_P = 10$ μA, determine the range of R_1 for the triggering network.

17. For a unijunction transistor with $V_{BB} = 20$ V, $\eta = 0.65$, $R_{B_1} = 2$ kΩ($I_E = 0$), and $V_D = 0.7$ V, determine:
    (a) R_{B_2}.
    (b) R_{BB}.
    (c) $V_{R_{B_1}}$.
    (d) V_P.

* 18. Given the relaxation oscillator of Fig. 21.69.
    (a) Find R_{B_1} and R_{B_2} at $I_E = 0$ A.
    (b) Determine V_P, the voltage necessary to turn on the UJT.
    (c) Determine whether R_1 is within the permissible range of values defined by Eq. (21.8).
    (d) Determine the frequency of oscillation if $R_{B_1} = 200$ Ω during the discharge phase.
    (e) Sketch the waveform of v_C for two full cycles.
    (f) Sketch the waveform of v_{R_2} for two full cycles.
    (g) Determine the frequency using Eq. (21.17) and compare to the value determined in part (d). Account for any major differences.

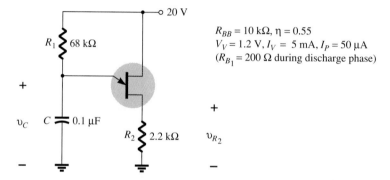

$R_{BB} = 10$ kΩ, $\eta = 0.55$
$V_V = 1.2$ V, $I_V = 5$ mA, $I_P = 50$ μA
($R_{B_1} = 200$ Ω during discharge phase)

Figure 21.69 Problem 18

§ 21.14 Phototransistors

19. For a phototransistor having the characteristics of Fig. 21.51, determine the photoinduced base current for a radiant flux density of 5 mW/cm². If $h_{fe} = 40$, find I_C.

* 20. Design a high-isolation OR-gate employing phototransistors and LEDs.

§ 21.15 Opto-Isolators

21. (a) Determine an average derating factor from the curve of Fig. 21.59 for the region defined by temperatures between $-25°C$ and $+50°C$.
    (b) Is it fair to say that for temperature greater than room temperature (up to 100°C), the output current is somewhat unaffected by temperature?

22. (a) Determine, from Fig. 21.55, the average change in I_{CEO} per degree change in temperature for the range 25 to 50°C.
    (b) Can the results of part (a) be used to determine the level of I_{CEO} at 35°C? Test your theory.

23. Determine, from Fig. 21.56, the ratio of LED output current to detector input current for an output current of 20 mA. Would you consider the device to be relatively efficient in its purpose?

* 24. (a) Sketch the maximum-power curve of $P_D = 200$ mW on the graph of Fig. 21.57. List any noteworthy conclusions.
    (b) Determine β_{dc} (defined by I_C/I_F) for the system at $V_{CE} = 15$ V, $I_F = 10$ mA.
    (c) Compare the results of part (b) with those obtained from Fig. 21.56 at $I_F = 10$ mA. Do they compare? Should they? Why?

Chapter 21 *pnpn* and Other Devices

* **25.** (a) Referring to Fig. 21.58, determine the collector current above which the switching time does not change appreciably for $R_L = 1$ kΩ and $R_L = 100$ Ω.

　　(b) At $I_C = 6$ mA, how does the ratio of switching times for $R_L = 1$ kΩ and $R_L = 100$ Ω compare to the ratio of resistance levels?

§ 21.16 Programmable Unijunction Transistor

26. Determine η and V_G for a PUT with $V_{BB} = 20$ V and $R_{B_1} = 3$ R_{B_2}.

27. Using the data provided in Example 21.3 determine the impedance of the PUT at the firing and valley points. Are the approximate open- and short-circuit states verified?

28. Can Eq. (21.24) be derived exactly as shown from Eq. (21.23)? If not, what element is missing in Eq. (21.24)?

* **29.** (a) Will the network of Example 21.3 oscillate if V_{BB} is changed to 10 V? What minimum value of V_{BB} is required (V_V a constant)?

　　(b) Referring to the same example, what value of R would place the network in the stable "on" state and remove the oscillatory response of the system?

　　(c) What value of R would make the network a 2-ms time-delay network? That is, provide a pulse v_k 2 ms after the supply is turned on and then stay in the "on" state.

---

*Please Note: Asterisks indicate more difficult problems.

22 Oscilloscope and Other Measuring Instruments

22.1 INTRODUCTION

One of the basic functions of electronic circuits is the generation and manipulation of electronic waveshapes. These electronic signals may represent audio information, computer data, television signals, timing signals (as used in radar), and so on. The common meters used in electronic measurement are the multimeter—analog or digital, to enable measuring dc or ac voltages, currents, or impedances. Most meters provide ac measurements that are correct for nondistorted sinusoidal signals only. The oscilloscope, on the other hand, displays the exact waveform, and the viewer can decide what to make of the various readings observed.

The cathode ray oscilloscope (CRO) provides a visual presentation of any waveform applied to the input terminals. A cathode ray tube (CRT), much like a television tube, provides the visual display showing the form of the signal applied as a waveform on the front screen. An electron beam is deflected as it sweeps across the tube face, leaving a display of the signal applied to input terminals.

While multimeters provide numeric information about an applied signal, the oscilloscope allows the actual form of the waveform to be displayed. A wide range of oscilloscopes are available, some suited to measure signals below a specified frequency, others to provide measuring signals of the shortest time span. A CRO may be built to operate from a few hertz up to hundreds of megahertz; CROs may also be used to measure time spans from fractions of a nanosecond (10^{-9}) to many seconds.

22.2 CATHODE RAY TUBE—THEORY AND CONSTRUCTION

The cathode ray tube (CRT) is the "heart" of the CRO, providing visual display of an input signal's waveform. A CRT contains four basic parts:

1. An electron gun to produce a stream of electrons
2. Focusing and accelerating elements to produce a well-defined beam of electrons
3. Horizontal and vertical deflecting plates to control the path of the electron beam
4. An evacuated glass envelope with a phosphorescent screen, which glows visibly when struck by the electron beam.

Figure 22.1 shows the basic construction of a CRT. We will first consider the device's basic operation. A cathode (K) containing an oxide coating is heated indi-

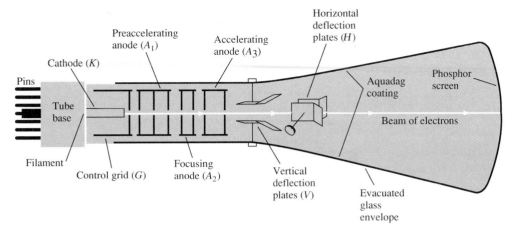

Figure 22.1 Cathode ray tube: basic construction.

rectly by a filament, resulting in the release of electrons from the cathode surface. A control grid (G) provides for control of the number of electrons passing farther into the tube. A voltage on the control grid determines how many of the electrons freed by heating are allowed to continue moving toward the face of the tube. After the electrons pass the control grid, they are focused into a tight beam and accelerated to a higher velocity by the focusing and accelerating anodes. The parts discussed so far comprise the electron gun of the CRT.

The high-velocity, well-defined electron beam then passes through two sets of deflection plates. The first set of plates is oriented to deflect the electron beam vertically, up or down. The direction of the vertical deflection is determined by the voltage polarity applied to the deflecting plates. The amount of deflection is set by the magnitude of the applied voltage. The beam is also deflected horizontally (left or right) by a voltage applied to the horizontal deflecting plates. The deflected beam is then further accelerated by very high voltages applied to the tube, with the beam finally striking a phosphorescent material on the inside face of the tube. This phosphor glows when struck by the energetic electrons—the visible glow seen at the front of the tube by the person using the scope.

The CRT is a self-contained unit with leads brought out through a base to pins. Various types of CRTs are manufactured in a variety of sizes, with different phosphor materials and deflection electrode placement. We can now consider how the CRT is used in an oscilloscope.

22.3   CATHODE RAY OSCILLOSCOPE OPERATION

For operation as an oscilloscope the electron beam is deflected horizontally by a sweep voltage, and vertically by the voltage to be measured. While the electron beam is moved across the face of the CRT by the horizontal sweep signal, the input signal deflects the beam vertically, resulting in a display of the input signal waveform. One sweep of the beam across the face of the tube, followed by a "blank" period during which the beam is turned off while being returned to the starting point across the tube face, constitutes one sweep of the beam.

A steady display is obtained when the beam repeatedly sweeps across the tube with exactly the same image each sweep. This requires a synchronization, starting the sweep at the same point in a repetitive waveform cycle. If the signal is properly synchronized, the display will be stationary. In the absence of sync the picture will appear to drift or move horizontally across the screen.

Basic Parts of a CRO

The basic parts of a CRO are shown in Fig. 22.2. We will first consider the CRO's operation for this simplified block diagram. To obtain a noticeable beam deflection from a centimeter to a few centimeters, the usual voltage applied to the deflection plates must be on the order of tens to hundreds of volts. Since the signals measured using a CRO are typically only a few volts, or even a few millivolts, amplifier circuits are needed to increase the input signal to the voltage levels required to operate the tube. There are amplifier sections for both the vertical and the horizontal deflection of the beam. To adjust the level of a signal, each input goes through an attenuator circuit which can adjust the amplitude of the display.

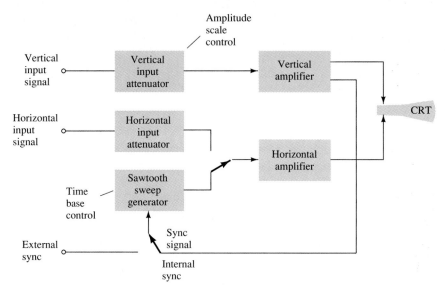

Figure 22.2 Cathode ray oscilloscope: general block diagram.

22.4 VOLTAGE SWEEP OPERATION

When the vertical input is 0 V, the electron beam may be positioned at the vertical center of the screen. If 0 V is also applied to the horizontal input, the beam is then at the center of the CRT face and remains a stationary dot. The vertical and horizontal positioning controls allow moving the dot anywhere on the tube face. Any dc voltage applied to an input will result in shifting the dot. Figure 22.3 shows a CRT face with a centered dot, and with a dot moved by a positive horizontal voltage (to the right) and a negative vertical input voltage (down from center).

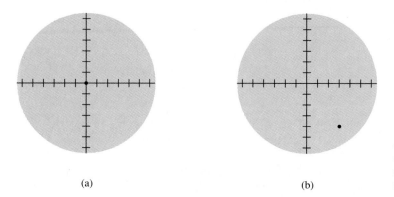

(a)   (b)

Figure 22.3 Dot on CRT screen due to stationary electron beam: (a) centered dot due to stationary electron beam; (b) off-center stationary dot.

Horizontal Sweep Signal

To view a signal on the CRT face it is necessary to deflect the beam across the CRT with a horizontal sweep signal so that any variation of the vertical signal can be observed. Figure 22.4 shows the resulting straight-line display for a positive voltage applied to the vertical input using a linear (sawtooth) sweep signal applied to the horizontal channel. With the electron beam held at a constant vertical distance, the horizontal voltage, going from negative to zero to positive voltage, causes the beam to move from the left side of the tube, to the center, to the right side. The resulting display is a straight line above the vertical center with the dc voltage properly displayed as a straight line.

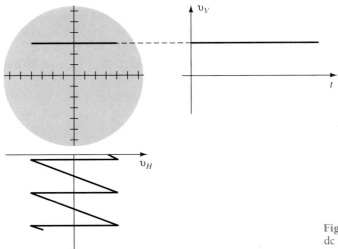

Figure 22.4 Scope display for dc vertical signal and linear horizontal sweep signal.

The sweep voltage is shown to be a continuous waveform, not just a single sweep. This is necessary if a long-term display is to be seen. A single sweep across the tube face would quickly fade out. By repeating the sweep, the display is generated over and over, and if enough sweeps are generated per second, the display appears present continuously. If the sweep rate is slowed down (as set by the time-scale controls of the scope), the actual travel of the beam across the tube face can be observed.

Applying only a sinusoidal signal to the vertical inputs (no horizontal sweep) results in a vertical straight line as shown in Fig. 22.5. If the sweep speed (frequency

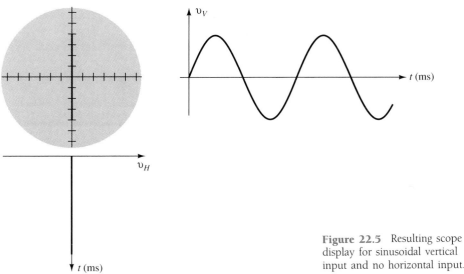

Figure 22.5 Resulting scope display for sinusoidal vertical input and no horizontal input.

22.4 Voltage Sweep Operation

of the sinusoidal signal) is reduced, it is possible to see the electron beam moving up and down along a straight-line path.

Use of Linear Sawtooth Sweep to Display Vertical Input

To view a sinusoidal signal it is necessary to use a sweep signal on the horizontal channel, so that the signal applied to the vertical channel can be seen on the tube face. Figure 22.6 shows the resulting CRO display resulting from a horizontal linear sweep and a sinusoidal input to the vertical channel. For 1 cycle of the input signal to appear as shown in Fig. 22.6a, it is necessary that the signal and linear sweep frequencies be synchronized. If there is any difference the display will appear to move (not be synchronized), unless the sweep frequency is some multiple of the sinusoidal frequency. Lowering the sweep frequency allows more cycles of the sinusoidal signal to be displayed, whereas increasing the sweep frequency results in less of the sinusoidal vertical input to be displayed, thereby appearing as a magnification of a part of the input signal.

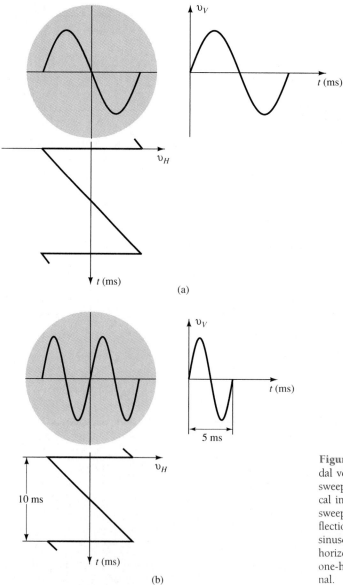

Figure 22.6 Display of sinusoidal vertical input and horizontal sweep input: (a) display of vertical input signal using linear sweep signal for horizontal deflection; (b) scope display for a sinusoidal vertical input and a horizontal sweep speed equal to one-half that of the vertical signal.

Chapter 22   Oscilloscope and Other Measuring Instruments

Determine how many cycles of a 2-kHz sinusoidal signal are viewed if the sweep frequency is:

EXAMPLE 22.1

(a) 2 kHz.
(b) 4 kHz.
(c) 1 kHz.

Solution

(a) When the two signals have the same frequency, a full cycle will be seen.
(b) When the sweep frequency is increased to 4 kHz, a half-cycle will be seen.
(c) When the sweep frequency is reduced to 1 kHz, two cycles will be seen.

Figure 22.7 shows a pulse-type waveform applied as vertical input with a horizontal sweep, resulting in a scope display of the pulse signal. The numbering at each waveform permits following the display for variation of input and sweep voltage during one cycle.

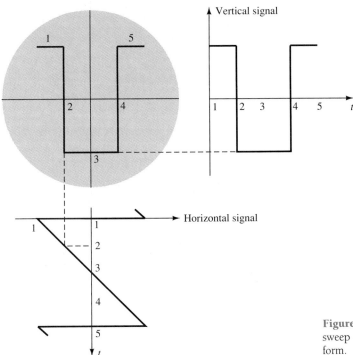

Figure 22.7 Use of the linear sweep for a pulse-type waveform.

22.5 SYNCHRONIZATION AND TRIGGERING

A CRO display can be adjusted by setting the sweep speed (frequency) to display either one cycle, a number of cycles, or part of a cycle. This is a very valuable feature of the CRO. Figure 22.8 shows the display resulting for a few cycles of the sweep signal. Each time the horizontal sawtooth sweep voltage goes through a linear sweep cycle (from maximum negative to zero to maximum positive), the electron beam is caused to move horizontally across the tube face, from left to center to right. The sawtooth voltage then drops quickly back to the negative starting voltage, with the beam back to the left side. During the time the sweep voltage goes quickly negative

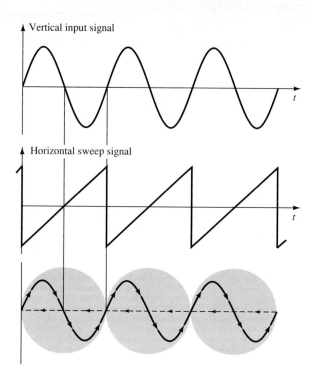

Figure 22.8 Steady scope display—input and sweep signals synchronized.

(retrace), the beam is blanked (the grid voltage prevents the electrons from hitting the tube face).

To see a steady display each time the beam is swept across the face of the tube, it is necessary to start the sweep at the same point in the input signal cycle. In Fig. 22.9 the sweep frequency is too low and the CRO display will have an apparent "drift" to the left. Figure 22.10 shows the result of setting the sweep frequency too high, with an apparent drift to the right.

It should be obvious that adjusting the sweep frequency to exactly the same as the signal frequency to obtain a steady sweep is impractical. A more practical procedure

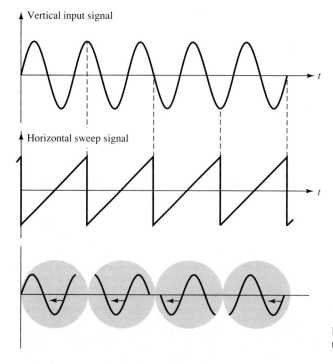

Figure 22.9 Sweep frequency too *low*—apparent drift to left.

Chapter 22 Oscilloscope and Other Measuring Instruments

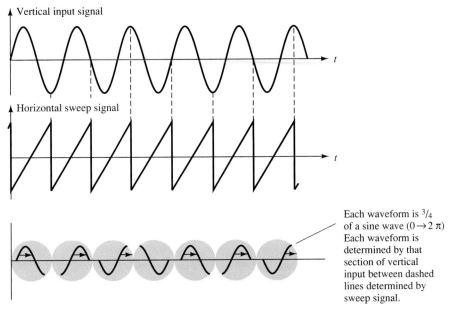

Each waveform is ³/₄ of a sine wave $(0 \rightarrow 2\pi)$ Each waveform is determined by that section of vertical input between dashed lines determined by sweep signal.

Figure 22.10 Sweep frequency too *high*—apparent drift to right.

is to wait until the signal reaches the same point in a cycle to start the trace. This triggering has a number of features, as described next.

Triggering

The usual method of synchronizing uses a portion of the input signal to trigger a sweep generator so that the sweep signal is locked or synchronized to the input signal. By using a portion of the same signal to be viewed to provide the synchronizing signal assures synchronization. Figure 22.11 shows a block diagram of how a trigger signal is derived in a single-channel display. The trigger signal source is obtained from the line frequency (60 Hz) for viewing signals related to the line voltage, from an external signal (one other than that to be viewed), or more likely, from a signal derived

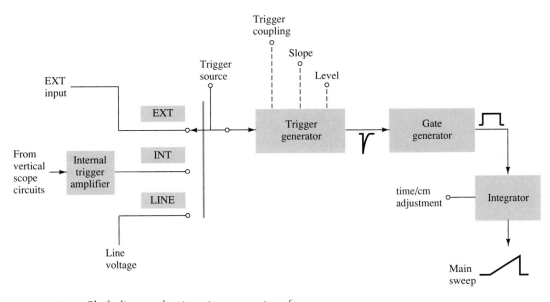

Figure 22.11 Block diagram showing trigger operation of scope.

from that applied as vertical input. The selector switch on the scope being set to INTERNAL will provide a part of the input signal to the trigger generator circuit. The output of the trigger generator is a trigger signal that is used to start the main sweep of the scope, which lasts a time set by the time/cm adjustment. Figure 22.12 shows triggering being started at various points in a signal cycle.

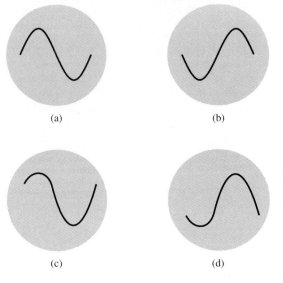

(a)   (b)

(c)   (d)

Figure 22.12 Triggering at various points of signal level (*Note*: sign starts at same point in cycle each sweep and is therefore synchronized): (a) positive-going zero level; (b) negative-going zero level; (c) positive-voltage trigger level; (d) negative-voltage trigger level.

The trigger sweep operation can also be seen by looking at some of the resulting waveforms. From a given input signal a trigger waveform is obtained to provide for a sweep signal. As seen in Fig. 22.13, the sweep is started at a time in the input signal cycle and lasts a period set by the sweep length controls. Then the scope waits until the input reaches an identical point in its cycle before starting another sweep operation. The length of the sweep determines how many cycles will be viewed, while the triggering assures that synchronization takes place.

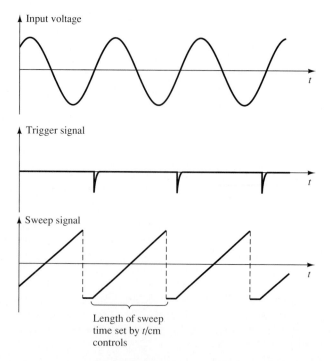

Length of sweep time set by *t*/cm controls

Figure 22.13 Triggered sweep.

22.6 MULTITRACE OPERATION

Most modern oscilloscopes provide viewing two or more traces on the scope face at the same time. This allows comparing amplitude, special waveform features, and other important waveform characteristics. A multiple trace can be obtained using more than one electron gun, with the separate beams creating separate displays. More often, however, a single electron beam is used to create the multiple images.

Two methods of developing two traces are CHOPPED and ALTERNATE. With two input signals applied an electronic switch first connects one input, then the other, to the deflection circuitry. In the ALTERNATE mode of operation the beam is swept across the tube face displaying however many cycles of one input signal are to be displayed. Then the input switches (alternates) to the second input and displays the same number of cycles of the second signal. Figure 22.14a shows the operation with alternate display. In the CHOPPED mode of operation (Fig. 22.14b), the beam repeatedly switches between the two input signals during one sweep of the beam. As long as the signal is of relatively low frequency, the action of switching is not visible and two separate displays are seen.

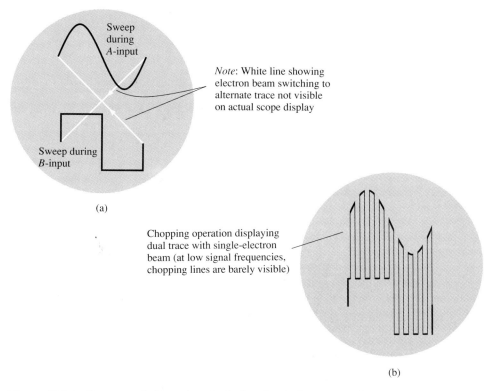

(a)

(b)

Figure 22.14 Alternate and chopped mode displays for dual-trace operation: (a) alternate mode for dual-trace using single electron beam; (b) chopped mode for dual-trace using single electron beam.

22.7 MEASUREMENT USING CALIBRATED CRO SCALES

The oscilloscope tube face has a calibrated scale to use in making amplitude or time measurements. Figure 22.15 shows a typical calibrated scale. The boxes are divided into centimeters (cm), 4 cm on each side of center. Each centimeter (box) is further divided into 0.2-cm intervals.

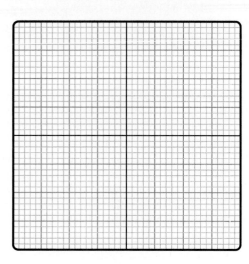

Figure 22.15 Calibrated scope face.

Amplitude Measurements

The vertical scale is calibrated in either volts per centimeter (V/cm), or millivolts per centimeter (mV/cm). Using the scale setting of the scope and the signal measured off the face of the scope, one can measure typically peak-to-peak or peak voltages for an ac signal.

EXAMPLE 22.2

Calculate the peak-to-peak amplitude of the sinusoidal signal in Fig. 22.16 if the scope scale is set to 5 mV/cm.

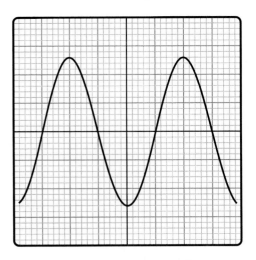

Figure 22.16 Waveform for Example 22.2.

Solution

The peak-to-peak amplitude is

$$2 \times 2.6 \text{ cm} \times 5 \text{ mV/cm} = \textbf{26 mV}$$

Note that a scope provides easy measurement of peak-to-peak values, whereas a multimeter typically provides measurement of rms (for a sinusoidal waveform).

Calculate the amplitude of the pulse signal in Fig. 22.17 (scope setting 100 mV/cm).

EXAMPLE 22.3

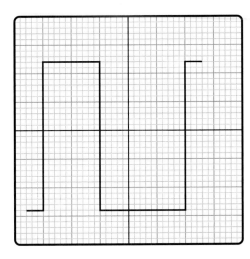

Figure 22.17 Waveform for Example 22.3.

Solution

The peak-to-peak amplitude is

$$(2.8 \text{ cm} + 2.4 \text{ cm}) \times 100 \text{ mV/cm} = \mathbf{520 \text{ mV}} = \mathbf{0.52 \text{ V}}$$

Time Measurements

PERIOD

The horizontal scale of the scope can be used to measure time, in either seconds (s), milliseconds (ms), microseconds (μs), or nanoseconds (ns). The interval of a pulse from start to end is the period of the pulse. When the signal is repetitive, the period is one cycle of the waveform.

Calculate the period of the waveform shown in Fig. 22.18 (scope setting at 20μs/cm).

EXAMPLE 22.4

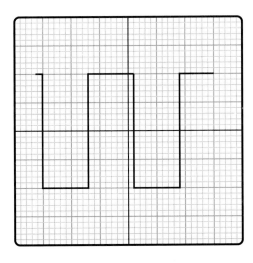

Figure 22.18 Waveform for Example 22.4.

Solution

For the waveform of Fig. 22.18,

$$\text{period} = T = 3.2 \text{ cm} \times 20 \ \mu\text{s/cm} = \mathbf{64 \ \mu s}$$

FREQUENCY

The measurement of a repetitive waveform's period can be used to calculate the signal's frequency. Since frequency is the reciprocal of the period,

$$f = \frac{1}{T} \tag{22.1}$$

EXAMPLE 22.5

Determine the frequency of the waveform shown in Fig. 22.18 (scope setting at 5 μs/cm).

Solution

From the waveform

$$\text{period} = T = 3.2 \text{ cm} \times 5 \ \mu\text{s/cm} = 16 \ \mu\text{s}$$

$$f = \frac{1}{T} = \frac{1}{16 \ \mu\text{s}}$$
$$= \mathbf{62.5 \ kHz}$$

PULSE WIDTH

The time interval that a waveform is high (or low) is the pulse width of the signal. When the waveform edges go up and down instantly the width is measured from start (leading edge) to end (trailing edge) (see Fig. 22.19a). For a waveform with edges that rise or fall over some time the pulse width is measured between the 50% points as shown in Fig. 22.19b.

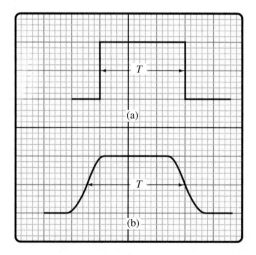

Figure 22.19 Pulse-width measurement.

Determine the pulse width of the waveform in Fig. 22.20.

EXAMPLE 22.6

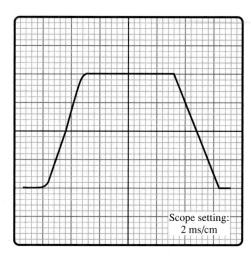

Scope setting: 2 ms/cm

Figure 22.20 Waveform for Example 22.6.

Solution

For a reading of 4.6 cm at the midpoint of the waveform the pulse width is

$$T_{PW} = 4.6 \text{ cm} \times 2 \text{ } \mu\text{s/cm} = \textbf{9.2 } \boldsymbol{\mu}\textbf{s}$$

PULSE DELAY

The time interval between pulses is called the pulse delay. For waveforms, as shown in Fig. 22.21, the pulse delay is measured between the midpoint (50% point) at the start of each pulse.

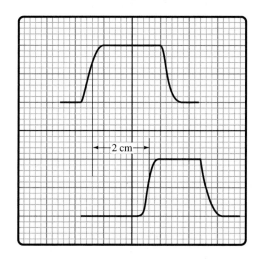

←—2 cm—→

Scope setting: 50 μs/cm

Figure 22.21 Waveform for Example 22.7.

Determine the pulse delay for the waveforms of Fig. 22.21.

EXAMPLE 22.7

Solution

From the waveforms in Fig. 22.21,

$$\text{pulse delay} = T_{PD} = 2 \text{ cm} \times 50 \text{ } \mu\text{s/cm} = \textbf{100 } \boldsymbol{\mu}\textbf{s}$$

22.8 SPECIAL CRO FEATURES

The CRO has become more sophisticated and specialized in use. The range of amplitude measurements, the scales of time measurements, the number of traces displayed, the methods of providing sweep triggering, and the types of measurements are different depending on the area of specialized scope usage.

Delayed Sweep

A useful CRO feature uses two time bases to provide selection of a small part of the signal for viewing. One time base selects the overall signal viewed on the scope, while a second permits selecting a small part of the viewed signal to be displayed in an expanded mode. The main time base is referred to as the A time base, while the second time base, referred to as B, displays the signal after a selected delay time.

Figure 22.22 provides a block diagram showing the operation of the two time bases. With front-panel controls set to operate from the A sweep, a main sweep signal is set to view a number of cycles of the input signal. The controls then allow setting the B sweep using a variable setting dial, with the B sweep usually an intensified interval which can be moved over the face of the displayed sweep. When the desired portion of the displayed sweep is set, the controls are moved to display the delayed part of the signal, which is seen at the second time base setting as a magnified display. Figure 22.23 shows a pulse-type signal first viewed using the A sweep and then the selected portion on a magnified sweep setting.

Figure 22.22 Operation of delayed sweep—block diagram.

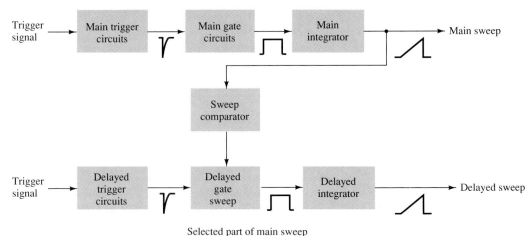

Figure 22.23 Main and delayed sweeps.

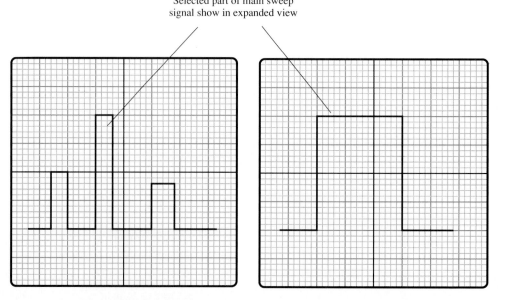

Selected part of main sweep signal show in expanded view

22.9 SIGNAL GENERATORS

A signal generator provides an ac signal of adjustable amplitude and varying frequency to use when operating an amplifier or other linear circuit. The frequency can typically be adjusted from a few hertz to a few megahertz. The signal amplitude can be adjusted from millivolts to a few volts of amplitude. While the signal is typically a sinusoidal waveform, pulse waveforms or even triangular waveforms are often available.

Waveform Generator IC(8038)

A precision waveform generator is provided by the 8038 IC unit shown in Fig. 22.24. The single 14-pin IC is capable of producing highly accurate sinusoidal, square, or triangular waveforms to use in operating or testing other equipment. Consideration of the IC's operation will help understand how any commercially available signal generator operates. This particular IC can provide output frequency which may be adjusted from less than 1 Hz up to about 300 kHz. The range of commercial units can be considerably higher. As indicated in Fig. 22.24, the IC provides three types of output waveform: all at the same frequency, the frequency being selected by the user.

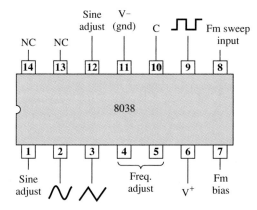

Figure 22.24 8038 Waveform generator IC.

Figure 22.25 shows the connection of the IC when used to provide an adjustable frequency output. The frequency of the unit would then be

$$f = \frac{0.15}{RC} \qquad\qquad (22.2)$$

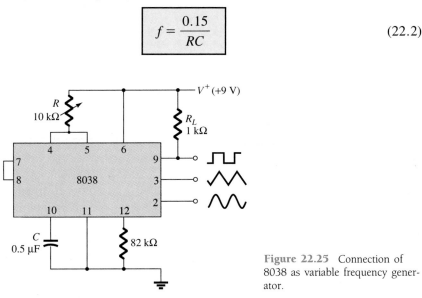

Figure 22.25 Connection of 8038 as variable frequency generator.

EXAMPLE 22.8

Determine the lowest and highest frequencies obtained when varying the 10-kΩ potentiometer from its minimum to its maximum setting.

Solution

Using Eq. (22.2), for a potentiometer set at 0, $R = 10\ \Omega$:

$$f = \frac{0.15}{(10\ \Omega)(0.5\ \mu F)} = \textbf{30 kHz}$$

For a potentiometer set to its maximum,

$$f = \frac{0.15}{(10\ k\Omega)(0.5\ \mu F)} = \textbf{30 Hz}$$

ADJUSTABLE OUTPUT AMPLITUDE

The connection of Fig. 22.26 shows how to provide adjustment of the sinusoidal waveform amplitude with the sinusoidal output provided through a buffered driver. The 310 op-amp is a unity-gain buffer providing the sinusoidal output from a low-impedance output. [The 310 has a voltage gain near unity (1), with an output impedance of about 1 Ω.] The output frequency is adjustable over a range from about 30 Hz to 30 kHz, with an amplitude adjustable up to about 9 V peak.

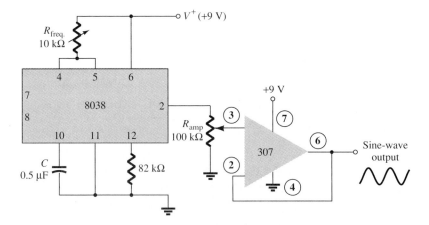

Figure 22.26 Sinusoidal waveform generator with adjustable frequency and amplitude.

5-V (TTL) PULSE GENERATOR

A circuit providing a 5-V pulse waveform for use with TTL digital circuits is shown in Fig. 22.27. The 8038 IC provides a rectangular or pulse waveform at a fixed output between 0 and +5 V. The frequency of the output can be varied from about 30 Hz to 30 kHz when adjusting the value of the 10-kΩ potentiometer. A commercial signal generator would probably include switched capacitors to provide frequency over a range of values. As long as the supply uses an IC regulator to provide the +5-V supply voltage, the output will be a well-defined value, as is typically used in TTL circuits. The 310 unity follower provides the output from a low-impedance source making it possible to connect the output to a number of loads, without affecting the amplitude or frequency of the signal waveform.

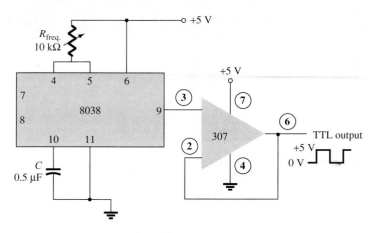

Figure 22.27 TTL signal waveform generator.

Hybrid Parameters— Conversion Equations (Exact and Approximate)

A.1 EXACT

Common-Emitter Configuration

$$h_{ie} = \frac{h_{ib}}{(1 + h_{fb})(1 - h_{rb}) + h_{ob}h_{ib}} = h_{ic}$$

$$h_{re} = \frac{h_{ib}h_{ob} - h_{rb}(1 + h_{fb})}{(1 + h_{fb})(1 - h_{rb}) + h_{ob}h_{ib}} = 1 - h_{rc}$$

$$h_{fe} = \frac{-h_{fb}(1 - h_{rb}) - h_{ob}h_{ib}}{(1 + h_{fb})(1 - h_{rb}) + h_{ob}h_{ib}} = -(1 + h_{fc})$$

$$h_{oe} = \frac{h_{ob}}{(1 + h_{fb})(1 - h_{rb}) + h_{ob}h_{ib}} = h_{oc}$$

Common-Base Configuration

$$h_{ib} = \frac{h_{ie}}{(1 + h_{fe})(1 - h_{re}) + h_{ie}h_{oe}} = \frac{h_{ic}}{h_{ic}h_{oc} - h_{fc}h_{rc}}$$

$$h_{rb} = \frac{h_{ie}h_{oe} - h_{re}(1 + h_{fe})}{(1 + h_{fe})(1 - h_{re}) + h_{ie}h_{oe}} = \frac{h_{fc}(1 - h_{rc}) + h_{ic}h_{oc}}{h_{ic}h_{oc} - h_{fc}h_{rc}}$$

$$h_{fb} = \frac{-h_{fe}(1 - h_{re}) - h_{ie}h_{oe}}{(1 + h_{fe})(1 - h_{re}) + h_{ie}h_{oe}} = \frac{h_{rc}(1 + h_{fc}) - h_{ic}h_{oc}}{h_{ic}h_{oc} - h_{fc}h_{rc}}$$

$$h_{ob} = \frac{h_{oe}}{(1 + h_{fe})(1 - h_{re}) + h_{ie}h_{oe}} = \frac{h_{oc}}{h_{ic}h_{oc} - h_{fc}h_{rc}}$$

Common-Collector Configuration

$$h_{ic} = \frac{h_{ib}}{(1 + h_{fb})(1 - h_{rb}) + h_{ob}h_{ib}} = h_{ie}$$

$$h_{rc} = \frac{1 + h_{fb}}{(1 + h_{fb})(1 - h_{rb}) + h_{ob}h_{ib}} = 1 - h_{re}$$

$$h_{fc} = \frac{h_{rb} - 1}{(1 + h_{fb})(1 - h_{rb}) + h_{ob}h_{ib}} = -(1 + h_{fe})$$

$$h_{oc} = \frac{h_{ob}}{(1 + h_{fb})(1 - h_{rb}) + h_{ob}h_{ib}} = h_{oe}$$

A.2 APPROXIMATE

Common-Emitter Configuration

$$h_{ie} \cong \frac{h_{ib}}{1 + h_{fb}} \cong \beta r_e$$

$$h_{re} \cong \frac{h_{ib}h_{ob}}{1 + h_{fb}} - h_{rb}$$

$$h_{fe} \cong \frac{-h_{fb}}{1 + h_{fb}} \cong \beta$$

$$h_{oe} \cong \frac{h_{ob}}{1 + h_{fb}}$$

Common-Base Configuration

$$h_{ib} \cong \frac{h_{ie}}{1 + h_{fe}} \cong \frac{-h_{ic}}{h_{fc}} \cong r_e$$

$$h_{rb} \cong \frac{h_{ie}h_{oe}}{1 + h_{fe}} - h_{re} \cong h_{rc} - 1 - \frac{h_{ic}h_{oc}}{h_{fc}}$$

$$h_{fb} \cong \frac{-h_{fe}}{1 + h_{fe}} \cong \frac{-(1 + h_{fc})}{h_{fc}} \cong -\alpha$$

$$h_{ob} \cong \frac{h_{oe}}{1 + h_{fe}} \cong \frac{-h_{oc}}{h_{fc}}$$

Common-Collector Configuration

$$h_{ic} \cong \frac{h_{ib}}{1 + h_{fb}} \cong \beta r_e$$

$$h_{rc} \cong 1$$

$$h_{fc} \cong \frac{-1}{1 + h_{fb}} \cong -\beta$$

$$h_{oc} \cong \frac{h_{ob}}{1 + h_{fb}}$$

APPENDIX

B

Ripple Factor and Voltage Calculations

B.1 RIPPLE FACTOR OF RECTIFIER

The ripple factor of a voltage is defined by

$$r \equiv \frac{\text{rms value of ac component of signal}}{\text{average value of signal}}$$

which can be expressed as

$$r = \frac{V_r(\text{rms})}{V_{\text{dc}}}$$

Since the ac voltage component of a signal containing a dc level is

$$\nu_{\text{ac}} = \nu - V_{\text{dc}}$$

the rms value of the ac component is

$$V_r(\text{rms}) = \left[\frac{1}{2\pi}\int_0^{2\pi} \nu_{\text{ac}}^2\, d\theta\right]^{1/2}$$

$$= \left[\frac{1}{2\pi}\int_0^{2\pi} (\nu - V_{\text{dc}})^2\, d\theta\right]^{1/2}$$

$$= \left[\frac{1}{2\pi}\int_0^{2\pi} (\nu^2 - 2\nu V_{\text{dc}} + V_{\text{dc}}^2)\, d\theta\right]^{1/2}$$

$$= [V^2(\text{rms}) - 2V_{\text{dc}}^2 + V_{\text{dc}}^2]^{1/2}$$

$$= [V^2(\text{rms}) - V_{\text{dc}}^2]^{1/2}$$

where $V(\text{rms})$ is the rms value of the total voltage. For the half-wave rectified signal,

$$V_r(\text{rms}) = [V^2(\text{rms}) - V_{\text{dc}}^2]^{1/2}$$

$$= \left[\left(\frac{V_m}{2}\right)^2 - \left(\frac{V_m}{\pi}\right)^2\right]^{1/2}$$

$$= V_m\left[\left(\frac{1}{2}\right)^2 - \left(\frac{1}{\pi}\right)^2\right]^{1/2}$$

$$\boxed{V_r(\text{rms}) = 0.385 V_m \qquad \text{(half-wave)}} \tag{B.1}$$

For the full-wave rectified signal,

$$V_r(\text{rms}) = [V^2(\text{rms}) - V_{\text{dc}}^2]^{1/2}$$

$$= \left[\left(\frac{V_m}{\sqrt{2}}\right)^2 - \left(\frac{2V_m}{\pi}\right)^2\right]^{1/2}$$

$$= V_m\left(\frac{1}{2} - \frac{4}{\pi^2}\right)^{1/2}$$

$$\boxed{V_r(\text{rms}) = 0.308V_m \qquad \text{(full-wave)}} \tag{B.2}$$

B.2   RIPPLE VOLTAGE OF CAPACITOR FILTER

Assuming a triangular ripple waveform approximation as shown in Fig. B.1, we can write (see Fig. B.2)

$$V_{\text{dc}} = V_m - \frac{V_r(\text{p-p})}{2} \tag{B.3}$$

During capacitor-discharge the voltage change across C is

$$V_r(\text{p-p}) = \frac{I_{\text{dc}}T_2}{C} \tag{B.4}$$

From the triangular waveform in Fig. B.1

$$V_r(\text{rms}) = \frac{V_r(\text{p-p})}{2\sqrt{3}} \tag{B.5}$$

(obtained by calculations, not shown).

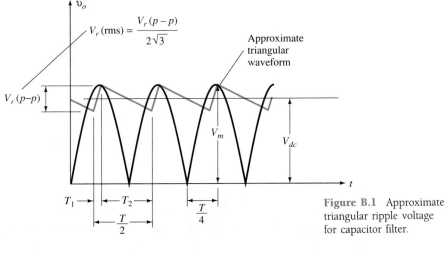

Figure B.1   Approximate triangular ripple voltage for capacitor filter.

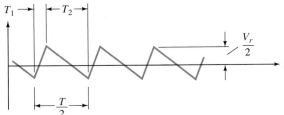

Figure B.2   Ripple voltage.

Using the waveform details of Fig. B.1 results in

$$\frac{V_r(\text{p-p})}{T_1} = \frac{V_m}{T/4}$$

$$T_1 = \frac{V_r(\text{p-p})(T/4)}{V_m}$$

Also,

$$T_2 = \frac{T}{2} - T_1 = \frac{T}{2} - \frac{V_r(\text{p-p})(T/4)}{V_m} = \frac{2TV_m - V_r(\text{p-p})T}{4V_m}$$

$$T_2 = \frac{2V_m - V_r(\text{p-p})}{V_m}\frac{T}{4} \tag{B.6}$$

Since Eq. (B.3) can be written as

$$V_{\text{dc}} = \frac{2V_m - V_r(\text{p-p})}{2}$$

we can combine the last equation with Eq. (B.6):

$$T_2 = \frac{V_{\text{dc}}}{V_m}\frac{T}{2}$$

which, inserted into Eq. (B.4), gives

$$V_r(\text{p-p}) = \frac{I_{\text{dc}}}{C}\left(\frac{V_{\text{dc}}}{V_m}\frac{T}{2}\right)$$

$$T = \frac{1}{f}$$

$$V_r(\text{p-p}) = \frac{I_{\text{dc}}}{2fC}\frac{V_{\text{dc}}}{V_m} \tag{B.7}$$

Combining Eqs. (B.5) and (B.7), we solve for $V_r(\text{rms})$:

$$\boxed{V_r(\text{rms}) = \frac{V_r(\text{p-p})}{2\sqrt{3}} = \frac{I_{\text{dc}}}{4\sqrt{3}\,fC}\frac{V_{\text{dc}}}{V_m}} \tag{B.8}$$

B.3 RELATION OF V_{dc} AND V_m TO RIPPLE, r

The dc voltage developed across a filter capacitor from a transformer providing a peak voltage, V_m, can be related to the ripple as follows:

$$r = \frac{V_r(\text{rms})}{V_{\text{dc}}} = \frac{V_r(\text{p-p})}{2\sqrt{3}\,V_{\text{dc}}}$$

$$V_{\text{dc}} = \frac{V_r(\text{p-p})}{2\sqrt{3}r} = \frac{V_r(\text{p-p})/2}{\sqrt{3}r} = \frac{V_r(\text{p})}{\sqrt{3}r} = \frac{V_m - V_{\text{dc}}}{\sqrt{3}r}$$

$$V_m - V_{\text{dc}} = \sqrt{3}rV_{\text{dc}}$$

$$V_m = (1 + \sqrt{3}r)V_{\text{dc}}$$

$$\boxed{\frac{V_m}{V_{dc}} = 1 + \sqrt{3}\,r} \qquad\qquad (B.9)$$

The relation of Eq. (B.9) applies to both half- and full-wave rectifier-capacitor filter circuits and is plotted in Fig. B.3. As example, at a ripple of 5% the dc voltage is $V_{dc} = 0.92V_m$, or within 10% of the peak voltage, where as at 20% ripple the dc voltage drops to only $0.74V_m$ which is more than 25% less than the peak value. Note that V_{dc} is within 10% of V_m for ripple less than 6.5%. This amount of ripple represents the borderline of the light-load condition.

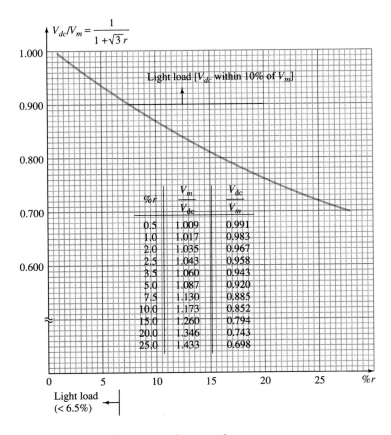

Figure B.3 Plot of V_{dc}/V_m as a function of % r.

B.4 RELATION OF V_r (RMS) AND V_m TO RIPPLE, r

We can also obtain a relation between V_r(rms), V_m, and the amount of ripple for both half-wave and full-wave rectifier-capacitor filter circuits as follows:

$$\frac{V_r(\text{p-p})}{2} = V_m - V_{dc}$$

$$\frac{V_r(\text{p-p})/2}{V_m} = \frac{V_m - V_{dc}}{V_m} = 1 - \frac{V_{dc}}{V_m}$$

$$\frac{\sqrt{3}V_r(\text{rms})}{V_m} = 1 - \frac{V_{dc}}{V_m}$$

Using Eq. (B.9), we get

$$\frac{\sqrt{3}V_r(\text{rms})}{V_m} = 1 - \frac{1}{1 + \sqrt{3}\,r}$$

$$\frac{V_r(\text{rms})}{V_m} = \frac{1}{\sqrt{3}}\left(1 - \frac{1}{1 + \sqrt{3}r}\right) = \frac{1}{\sqrt{3}}\left(\frac{1 + \sqrt{3}r - 1}{1 + \sqrt{3}r}\right)$$

$$\boxed{\frac{V_r(\text{rms})}{V_m} = \frac{r}{1 + \sqrt{3}r}} \qquad (B.10)$$

Equation (B.10) is plotted in Fig. B.4.

Since V_{dc} is within 10% of V_m for ripple $\leq 6.5\%$,

$$\frac{V_r(\text{rms})}{V_m} \cong \frac{V_r(\text{rms})}{V_{dc}} = r \qquad \text{(light load)}$$

and we can use $V_r(\text{rms})/V_m = r$ for ripple $\leq 6.5\%$.

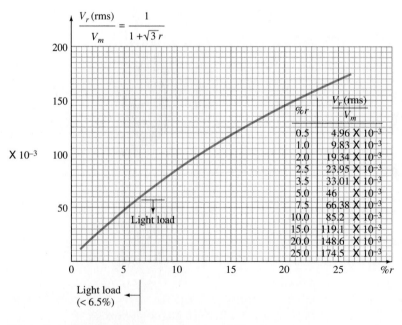

Figure B.4 Plot of $V_r(\text{rms})/V_m$ as a function of % r.

Appendix B **Ripple Factor and Voltage Calculations**

B.5 RELATION BETWEEN CONDUCTION ANGLE, % RIPPLE, AND I_{peak}/I_{dc} FOR RECTIFIER-CAPACITOR FILTER CIRCUITS

In Fig. B.1 we can determine the angle at which the diode starts to conduct, θ, as follows: Since

$$v = V_m \sin \theta = V_m - V_r(\text{p-p}) \qquad \text{at} \qquad \theta = \theta_1$$

$$\theta_1 = \sin^{-1}\left[1 - \frac{V_r(\text{p-p})}{V_m}\right]$$

Using Eq. (B.10) and $V_r(\text{rms}) = V_r(\text{p-p})/2\sqrt{3}$ gives

$$\frac{V_r(\text{p-p})}{V_m} = \frac{2\sqrt{3}V_r(\text{rms})}{V_m}$$

so that

$$1 - \frac{V_r(\text{p-p})}{V_m} = 1 - \frac{2\sqrt{3}V_r(\text{rms})}{V_m} = 1 - 2\sqrt{3}\left(\frac{r}{1 + \sqrt{3}r}\right)$$

$$= \frac{1 - \sqrt{3}r}{1 + \sqrt{3}r}$$

and

$$\boxed{\theta_1 = \sin^{-1}\frac{1 - \sqrt{3}r}{1 + \sqrt{3}r}} \qquad (B.11)$$

where θ_1 is the angle at which conduction starts.

When the current becomes zero after charging the parallel impedances R_L and C, we can determine that

$$\theta_2 = \pi - \tan^{-1} \omega R_L C$$

An expression for $\omega R_L C$ can be obtained as follows:

$$r = \frac{V_r(\text{rms})}{V_{dc}} = \frac{(I_{dc}/4\sqrt{3}\,fC)(V_{dc}/V_m)}{V_{dc}} = \frac{V_{dc}/R_L}{4\sqrt{3}\,fC}\frac{1}{V_m}$$

$$= \frac{V_{dc}/V_m}{4\sqrt{3}\,fCR_L} = \frac{2\pi\left(\dfrac{1}{1 + \sqrt{3}r}\right)}{4\sqrt{3}\,\omega CR_L}$$

so that

$$\omega R_L C = \frac{2\pi}{4\sqrt{3}\,(1 + \sqrt{3}r)r} = \frac{0.907}{r(1 + \sqrt{3}r)}$$

Thus conduction stops at an angle

$$\boxed{\theta_2 = \pi - \tan^{-1}\frac{0.907}{(1 + \sqrt{3}r)r}} \qquad (B.12)$$

(19.12)

From Eq. (16.10b) we can write

$$\frac{I_{peak}}{I_{dc}} = \frac{I_p}{I_{dc}} = \frac{T}{T_1} = \frac{180°}{\theta} \qquad \text{(full-wave)}$$

$$= \frac{360°}{\theta} \qquad \text{(half-wave)} \qquad (B.13)$$

A plot of I_p/I_{dc} as a function of ripple is provided in Fig. B.5 for both half- and full-wave operation.

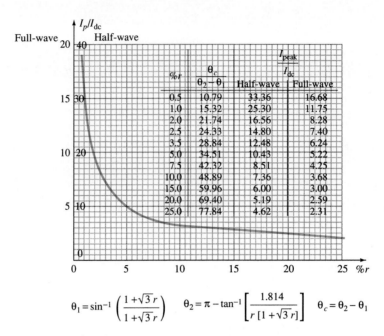

%r	θ_c $\theta_2 - \theta_1$	$\dfrac{I_{peak}}{I_{dc}}$	
		Half-wave	Full-wave
0.5	10.79	33.36	16.68
1.0	15.32	25.30	11.75
2.0	21.74	16.56	8.28
2.5	24.33	14.80	7.40
3.5	28.84	12.48	6.24
5.0	34.51	10.43	5.22
7.5	42.32	8.51	4.25
10.0	48.89	7.36	3.68
15.0	59.96	6.00	3.00
20.0	69.40	5.19	2.59
25.0	77.84	4.62	2.31

$$\theta_1 = \sin^{-1}\left(\frac{1+\sqrt{3}\,r}{1+\sqrt{3}\,r}\right) \qquad \theta_2 = \pi - \tan^{-1}\left[\frac{1.814}{r\,[1+\sqrt{3}\,r]}\right] \qquad \theta_c = \theta_2 - \theta_1$$

Figure B.5   Plot of I_p/I_{dc} versus % r half- and full-wave operation.

Charts and Tables

TABLE C.1 Greek Alphabet and Common Designations

Name	Capital	Lowercase	Used to Designate:
alpha	A	α	Angles, area, coefficients
beta	B	β	Angles, flux density, coefficients
gamma	Γ	γ	Conductivity, specific gravity
delta	Δ	δ	Variation, density
epsilon	E	ϵ	Base of natural logarithms
zeta	Z	ζ	Impedance, coefficients, coordinates
eta	H	η	Hysteresis coefficient, efficiency
theta	Θ	θ	Temperature, phase angle
iota	I	ι	
kappa	K	κ	Dielectric constant, susceptibility
lambda	Λ	λ	Wavelength
mu	M	μ	Micro, amplification factor, permeability
nu	N	ν	Reluctivity
xi	Ξ	ξ	
omicron	O	o	
pi	Π	π	Ratio of circumference to diameter = 3.1416
rho	P	ρ	Resistivity
sigma	Σ	σ	Sign of summation
tau	T	τ	Time constant, time phase displacement
upsilon	Υ	υ	
phi	Φ	ϕ	Magnetic flux, angles
chi	X	χ	
psi	Ψ	ψ	Dielectric flux, phase difference
omega	Ω	ω	Capital: ohms; lowercase: angular velocity

TABLE C.2 Standard Values of Commercially Available Resistors

Ohms (Ω)					Kilohms (kΩ)		Megohms (MΩ)	
0.10	**1.0**	**10**	**100**	**1000**	**10**	**100**	**1.0**	**10.0**
0.11	1.1	11	110	1100	11	110	1.1	11.0
0.12	**1.2**	**12**	**120**	**1200**	**12**	**120**	**1.2**	**12.0**
0.13	1.3	13	130	1300	13	130	1.3	13.0
0.15	**1.5**	**15**	**150**	**1500**	**15**	**150**	**1.5**	**15.0**
0.16	1.6	16	160	1600	16	160	1.6	16.0
0.18	**1.8**	**18**	**180**	**1800**	**18**	**180**	**1.8**	**18.0**
0.20	2.0	20	200	2000	20	200	2.0	20.0
0.22	**2.2**	**22**	**220**	**2200**	**22**	**220**	**2.2**	**22.0**
0.24	2.4	24	240	2400	24	240	2.4	
0.27	**2.7**	**27**	**270**	**2700**	**27**	**270**	**2.7**	
0.30	3.0	30	300	3000	30	300	3.0	
0.33	**3.3**	**33**	**330**	**3300**	**33**	**330**	**3.3**	
0.36	3.6	36	360	3600	36	360	3.6	
0.39	**3.9**	**39**	**390**	**3900**	**39**	**390**	**3.9**	
0.43	4.3	43	430	4300	43	430	4.3	
0.47	**4.7**	**47**	**470**	**4700**	**47**	**470**	**4.7**	
0.51	5.1	51	510	5100	51	510	5.1	
0.56	**5.6**	**56**	**560**	**5600**	**56**	**560**	**5.6**	
0.62	6.2	62	620	6200	62	620	6.2	
0.68	**6.8**	**68**	**680**	**6800**	**68**	**680**	**6.8**	
0.75	7.5	75	750	7500	75	750	7.5	
0.82	**8.2**	**82**	**820**	**8200**	**82**	**820**	**8.2**	
0.91	9.1	91	910	9100	91	910	9.1	

TABLE C.3 Typical Capacitor Component Values

pF					μF				
10	100	1000	10000		0.10	1.0	10	100	1000
12	120	1200							
15	150	1500	15000		0.15	1.5	18	180	1800
22	220	2200	22000		0.22	2.2	22	220	2200
27	270	2700							
33	330	3300	33000		0.33	3.3	33	330	3300
39	390	3900							
47	470	4700	47000		0.47	4.7	47	470	4700
56	560	5600							
68	680	6800	68000		0.68	6.8			
82	820	8200							

PSpice

SPICE is a popular electrical circuit analysis program which allows the user to enter a description of any given circuit in a simple descriptive manner, then ask for dc, ac, or graphical plots to be provided for various points (nodes) in the circuit. Knowing how to use such program can make circuit design, analysis, and even simulation much easier.

The PSpice software package used to analyze problems in the computer sections throughout this book is provided by MicroSim Corporation. An evaluation version is provided for so that educators and students can begin to discover some of the many ways the PSpice can be used to analyze various circuits in many different ways.

While the programs in this book were run on IBM-compatible machines, PSpice can be run on VAX, SUN, and MAC systems as well as others. This appendix will provide some help in using PSpice on an IBM system. For more detailed information there are many suitable educational texts,[*] company manuals, and commercial texts available.

Basic System Requirements

PSpice can be run on a suitably configured IBM PC or AT system. It can be run under DOS on a 640K system, preferably using an 80286 or faster microprocessor. Most printers can provide both text and graphical output provided by the PSpice program. The operation is most easily accomplished when running the program on a fixed disk because of the large storage needed by the program and auxiliary libraries. The latest version of PSpice includes an integrated input editor and SPICE analysis program, along with the ability to browse through the output file (containing the analysis results) and to run PROBE to view output graphical displays.

Overall Operation Using PSpice

To carry out a PSpice analysis one must proceed as follows:

1. Identify the complete circuit, marking all node points and components (component name and value).
2. Use a text editor—the one provided with PSpice or any other providing ASCII output—to enter a circuit (.CIR) file describing the given circuit.

---

[*]*SPICE: A Guide to Circuit Simulation and Analysis Using PSpice*, Prentice-Hall, Englewood Cliffs, N.J., 1988; *Computer-Aided Circuit Analysis Using SPICE*, W. Banzhaf, Prentice-Hall, Englewood Cliffs, N.J., 1989.

3. Run the PSpice analysis to produce an output (.OUT) file listing the results obtained, or a probe (PROBE.DAT) file that can be viewed using the PROBE program supplied with PSpice.

File Names

Files used with PSpice use standard three-letter extensions.

Input file describing circuit	.CIR
Output file holding results	.OUT
Probe data file to run	PROBE.DAT (only one file present at a time)

Getting a Printout

To get a hard copy of the .OUT file, use the DOS PRINT command. For example, to get a copy of HW5.OUT, type

PRINT HW5.OUT

If probe was used, type PROBE and follow directions to obtain a graphical display and, if desired, a hard copy.

E Solutions to Selected Odd-Numbered Problems

CHAPTER 1

3. Conduction in only one direction
5. (a) 150 kΩ (b) 12.5 kΩ
(c) 800 kΩ (d) 3 μΩ
$R_{Si}:R_{Cu} = 50 \times 10^9:1$
9. 18 J
21. 56.35 mA
23. (b) 1 (c) For $V = 0$ V, $e^0 = 1$ and $I_D = 0$ mA
27. 330 Ω
29. -10 V:100 MΩ, -30 V:300 MΩ
31. $R_{DC} = 76$ Ω
$r_d = 3$ Ω
$R_{DC} \gg r_d$
33. $I_D = 1$ mA, $r_d = 52$ Ω vs.
55 Ω(#32)
$I_D = 15$ mA, $r_d = 1.73$ Ω vs.
2 Ω(#32)
35. 33.3 Ω vs. 24.4 Ω(#34)
37. Using the best approximation to the curve beyond $V_D = 0.7$ V
$r_{av} = 4$ Ω
39. Decrease rapidly with increase in reverse-bias voltage
41. Log scale, $T = 25°C:I_R = 0.5$ nA
$T = 100°C:I_R = 60$ nA
Yes, at 95°C I_R increased to 64 nA.
43. $T = 25°C:P_{max} = 500$ mW, $I_{F_{max}} = 714.29$ mA
$T = 100°C:P_{max} = 260$ mW, $I_{F_{max}} = 371.43$ mA
45. (a) $V_R = -25$ V:$C_T \simeq 0.75$ pF
$V_R = -10$ V:$C_T \simeq 1.25$ pF
$\Delta C_T/\Delta V_R = 0.033$ pF/V
(b) $V_R = -10$ V:$C_T \simeq 1.25$ pF
$V_R = -1$ V:$C_T \simeq 3$ pF
$\Delta C_T/\Delta V_R = 0.194$ pF/V
(c) 0.194 pF/v:0.033 pF/V $\simeq$ 5.88:1
Increased sensitivity near $V_D = 0$ V

49. $I_F = 1$ mA, $I_R = 0.5$ mA
$t_s = 3$ ns, $t_t = 6$ ns
51. $T_1 = 129.17°$
53. 20 V:$T_C \simeq 0.06\%/°C$
5 V:$T_C \simeq -0.025\%/°C$
55. 0.2 mA: $\simeq 400$ Ω
1 mA $\simeq 95$ Ω, 10 mA $\simeq 13$ Ω
Nonlinear relationship between I_Z and dynamic impedance
57. $V_F = 2.3$ V
59. (a) $I_{peak(max)} = 37$ mA
(b) $I_{peak(max)} = 56$ mA

CHAPTER 2

1. (a) $I_{D_Q} \simeq 21.5$ mA, $V_{D_Q} \simeq 0.92$ V, $V_R = 7.08$ V
(b) $I_{D_Q} \simeq 22.2$ mA, $V_{D_Q} = 0.7$ V, $V_R = 7.3$ V
(c) $I_{D_Q} = 24.24$ mA, $V_{D_Q} = 0$ V, $V_R = 8$ V
3. $R = 0.62$ kΩ
5. (a) $I = 0$ mA (b) $I = 0.965$ A
(c) $I = 1$ A
7. (a) $V_o = 9.5$ V (b) $V_o = 7$ V
9. (a) $V_{o_1} = 11.3$ V, $V_{o_2} = 0.3$ V
(b) $V_{o_1} = -9$ V, $V_{o_2} = -6.6$ V
11. (a) $V_o = 9.7$ V, $I = 9.7$ mA
(b) $V_o = 14.6$ V, $I = 0.553$ mA
13. $V_o = 6.2$ V, $I_D = 1.55$ mA
15. $V_o = 9.3$ V
17. $V_o = 10$ V
19. $V_o = -0.7$ V
21. $V_o = 4.7$ V
23. $v_i:V_m = 6.98$ V, v_d:pos.max = 0.7 V, neg.peak = -6.98 V
i_d:pos.pulse of 2.85 mA
25. Pos.pulse, peak = 155.56 V,
$V_{dc} = 49.47$ V

27. (a) $I_{D_{max}} = 20$ mA (b) $I_{max} = 36.71$ mA (c) $I_D = 18.36$ mA (d) Yes
(e) $I_D = 36.71$ mA > $I_{D_{max}} = 20$ mA
29. Full rectified waveform, peak = 100 V; PIV = 100 V
31. Full rectified waveform, peak = 56.67 V; $V_{dc} = 36.04$ V
33. (a) Pos.pulse of 3.28 V
(b) Pos.pulse of 14.3 V
35. (a) Clipped at 4.7 V (b) Pos.clip at 0.7 V, neg.peak = -12 V
37. (a) 0 V to 40 V swing (b) -5 V to 35 V swing
39. (a) 28 ms (b) 56:1 (c) -1.3 V to -21.3 V swing
41. Network of Fig. 2.161 with battery reversed
43. (a) $R_S = 20$ Ω, $V_Z = 12$ V
(b) $P_{Z_{max}} = 2.4$ W
45. $R_S = 0.5$ kΩ, $I_{ZM} = 40$ mA
47. $V_o = 339.36$ V

CHAPTER 3

3. Forward and reverse biased
9. $I_C = 7.921$ mA, $I_B = 79.21$ μA
11. $V_{CB} = 1$ V:$V_{BE} = 800$ mV
$V_{CB} = 10$ V:$V_{BE} = 770$ mV
$V_{CB} = 20$ V:$V_{BE} = 750$ mV
Only slight
13. (a) $I_C \simeq 4.5$ mA (b) $I_C \simeq 4.5$ mA
(c) Negligible (d) $I_C \simeq I_E$
15. (a) $I_C = 3.992$ mA (b) $\alpha = 0.993$
(c) $I_E = 2$ mA
17. $A_v = 50$
21. (a) $\beta_{dc} = 117.65$ (b) $\alpha_{dc} = 0.992$
(c) $I_{CEO} = 0.3$ mA (d) $I_{CBO} = 2.4$ μA
23. (a) $\beta_{dc} = 83.75$ (b) $\beta_{dc} = 170$
(c) $\beta_{dc} = 113.33$

25. $\beta_{dc} = 116$, $\alpha_{dc} = 0.991$,
$I_E = 2.93$ mA
31. $I_C = I_{C_{max}}$, $V_{CB} = 5$ V
$V_{CB} = V_{CB_{max}}$, $I_C = 2$ mA
$I_C = 4$ mA, $V_{CB} = 7.5$ V
$V_{CB} = 10$ V, $I_C = 3$ mA
33. $I_C = I_{C_{max}}$, $V_{CE} = 3.125$ V
$V_{CE} = V_{CE_{max}}$, $I_C = 20.83$ mA
$I_C = 100$ mA, $V_{CE} = 6.25$ V
$V_{CE} = 20$ V, $I_C = 31.25$ mA
35. h_{FE}:$I_C = 0.1$ mA, $h_{FE} \cong 43$
$I_C = 10$ mA, $h_{FE} \cong 98$
h_{fe}:$I_C = 0.1$ mA, $h_{fe} \cong 72$
$I_C = 10$ mA, $h_{fe} \cong 160$
37. $I_C = 1$ mA, $h_{fe} \approx 120$
$I_C = 10$ mA, $h_{fe} \approx 160$
39. (a) $\beta_{ac} = 190$ (b) $\beta_{dc} = 201.7$
(c) $\beta_{ac} = 200$ (d) $\beta_{dc} = 230.77$ (e) Yes

CHAPTER 4

1. (a) $I_{B_Q} = 32.55$ μA (b) $I_{C_Q} =$
2.93 mA (c) $V_{CE_Q} = 8.09$ V (d) $V_C =$
8.09 V (e) $V_B = 0.7$ V (f) $V_E = 0$ V
3. (a) $I_C = 3.98$ mA (b) $V_{CC} =$
15.96 V (c) $\beta = 199$ (d) $R_B = 763$ kΩ
5. (b) $R_B = 812$ kΩ (c) $I_{C_Q} =$
3.4 mA, $V_{CE_Q} = 10.75$ V (d) $\beta_{dc} =$
136 (e) $\alpha = 0.992$ (f) $I_{C_{sat}} = 7$ mA
(h) $P_D = 36.55$ mW (i) $P_s =$
71.92 mW (j) $P_R = 35.37$ mW
7. (a) $R_C = 2.2$ kΩ (b) $R_E = 1.2$ kΩ
(c) $R_B = 356$ kΩ (d) $V_{CE} = 5.2$ V
(e) $V_B = 3.1$ V
9. $I_{C_{sat}} = 5.13$ mA
11. (a) $I_C = 2.93$ mA, $V_{CE} = 8.09$ V
(b) $I_C = 4.39$ mA, $V_{CE} = 4.15$ V
(c) %$\Delta I_C = 49.83$%, %$\Delta V_{CE} =$
48.70%
(d) $I_C = 2.92$ mA, $V_{CE} = 8.61$ V
(e) $I_C = 3.93$ mA, $V_{CE} = 4.67$ V
(f) %$\Delta I_C = 34.59$%, %$\Delta V_{CE} =$
46.76%
13. (a) $I_C = 1.28$ mA (b) $V_E = 1.54$ V
(c) $V_B = 2.24$ V (d) $R_1 = 39.4$ kΩ
15. $I_{C_{sat}} = 3.49$ mA
17. (a) $I_C = 2.28$ mA (b) $V_{CE} = 8.2$ V
(c) $I_B = 19.02$ μA (d) $V_E \approx 2.28$ V
(e) $V_B = 2.98$ V Approximate ap-
proach valid
19. (a) $R_C = 2.4$ kΩ, $R_E = 0.8$ kΩ
(b) $V_E = 4$ V (c) $V_B = 4.7$ V (d) $R_2 =$
5.84 kΩ (e) $\beta_{dc} = 129.8$
(f) 103.84 kΩ $\geq$ 58.4 kΩ (checks)
21. I. (a) $I_C = 2.43$ mA,
$V_{CE} = 7.55$ V

(b) $I_C = 2.33$ mA, $V_{CE} = 7.98$ V;
(c) Approx. approach: %$\Delta I_C = 0$%,
%$\Delta V_{CE} = 0$%
 Exact approach: %$\Delta I_C = 2.19$%,
%$\Delta V_{CE} = 2.68$%
(d) %$\Delta I_C = 2.19$% vs. 49.83% for
Prob. 11, %$\Delta V_{CE} = 2.68$% vs.
49.70% for Prob. 11
(e) Voltage-divider configuration least
sensitive
II. %ΔI_C and %ΔV_{CE} are quite small
23. (a) $I_C = 2.01$ mA (b) $V_C =$
17.54 V (c) $V_E = 3.02$ V (d) $V_{CE} =$
14.52 V
25. V_C from 5.98 V to 8.31 V
27. (a) $I_B = 13.04$ μA (b) $I_C =$
2.56 mA (c) $\beta = 196.32$
(d) $V_{CE} = 8$ V
29. (a) $I_B = 13.95$ μA (b) $I_C =$
1.81 mA (c) $V_E = -4.42$ V (d) $V_{CE} =$
5.95 V
31. (a) $I_E = 3.32$ mA (b) $V_C = 4.02$ V
(c) $V_{CE} = 4.72$ V
33. $R_B = 430$ kΩ, $R_C = 1.6$ kΩ,
$R_E = 390$ Ω
35. $R_E = 1.1$ kΩ, $R_C = 1.6$ kΩ,
$R_1 = 51$ kΩ, $R_2 = 15$ kΩ
37. $R_B = 43$ kΩ, $R_C = 0.62$ kΩ
39. (a) Open circuit, damaged
transistor
(b) Open at collector terminal, shorted
base–emitter junction
(c) Open circuit, open transistor
41. (a) $R_B \downarrow$, $I_B \downarrow$, $I_C \downarrow$, $V_C \uparrow$
(b) $\beta \downarrow$, $I_C \downarrow$ (c) Unchanged
(d) $V_{CC} \downarrow$, $I_B \downarrow$, $I_C \downarrow$
(e) $\beta \downarrow$, $I_C \downarrow$, $V_{R_C} \downarrow$, $V_{R_E} \downarrow$, $V_{CE} \uparrow$
43. (a) R_B open, $I_B = 0$ μA, $I_C =$
$I_{CEO} \cong 0$ mA, $V_C \cong V_{CC} = 18$ V
(b) $\beta \uparrow$, $I_C \uparrow$, $V_{R_C} \uparrow$, $V_{R_E} \uparrow$, $V_{CE} \downarrow$
(c) $R_C \downarrow$, $I_B \uparrow$, $I_C \uparrow$, $V_E \uparrow$
(d) Drop to relatively low voltage $\approx$
0.06 V
(e) Open base terminal
45. $V_C = -13.53$ V, $I_B = 17.5$ μA
47. (a) $S(I_{CO}) = 91$ (b) $S(V_{BE}) =$
-1.92×10^{-4} S (c) $S(\beta) = 32.56 \times$
10^{-6} A (d) $\Delta I_C = 1.66$ mA
49. (a) $S(I_{CO}) = 11.08$ (b) $S(V_{BE}) =$
-1.27×10^{-3} S (c) $S(\beta) = 2.41 \times$
10^{-6} A (d) $\Delta I_C = 0.411$ mA
51. $S(I_{CO})$ – Voltage divider less than
the other three
$S(V_{BE})$ – Voltage divider more sensitive than the other three (which have
similar levels)
$S(\beta)$ – Voltage divider least sensitive
with fixed bias very sensitive

In general, voltage divider least sensitive and fixed bias the most sensitive.

CHAPTER 5

3. (a) $V_{DS} \approx 1.4$ V (b) $r_d = 233.33$ Ω
(c) $V_{DS} \approx 1.6$ V (d) $r_d = 533.33$ Ω
(e) $V_{DS} \approx 1.4$ V (f) $r_d = 933.33$ Ω
(g) $r_d = 414.81$ Ω (h) $r_d = 933.2$ Ω
(i) In general, yes
11. (a) $I_D = 9$ mA (b) $I_D = 1.653$ mA
(c) $I_D = 0$ mA (d) $I_D = 0$ mA
15. $I_{DSS} = 12$ mA
17. $V_{DS} = 25$ V, $I_D = 4.8$ mA
$I_D = 10$ mA, $V_{DS} = 12$ V
$I_D = 7$ mA, $V_{DS} = 17.14$ V
19. Yes
21. $I_D = 4$ mA (exact match)
29. $I_{DSS} = 11.11$ mA
31. $V_{DS} = 25$ V
35. $V_T = 2$ V, $k = 5.31 \times 10^{-4}$
$I_D = 5.31 \times 10^{-4}$ $(V_{GS} - 2$ V$)^2$
37. $V_{GS} = 27.36$ V

CHAPTER 6

1. (c) $I_{D_Q} \approx 4.7$ mA, $V_{DS_Q} = 6.36$ V
(d) $I_{D_Q} = 4.69$ mA, $V_{DS_Q} = 6.37$ V
3. (a) $I_D = 3.125$ mA (b) $V_{DS} = 9$ V
(c) $V_{GG} = 1.5$ V
5. $V_D = 18$ V
7. $I_{D_Q} \approx 2.6$ mA, $V_{GS} = -1.95$ V
9. (a) $I_{D_Q} = 3.33$ mA (b) $V_{GS_Q} \approx$
-1.7 V (c) $I_{DSS} = 10.06$ mA (d) $V_D =$
11.34 V (e) $V_{DS} = 9.64$ V
11. $V_S = 1.4$ V
13. (a) $I_{D_Q} \approx 5.8$ mA, $V_{GS_Q} \approx$
-0.85 V, $I_{D_Q} \uparrow$, $V_{GS_Q} \downarrow$ (b) 216 Ω
15. (a) $I_{D_Q} = 2.7$ mA, $V_{GS_Q} = -2$ V
(b) $V_{DS} = 8.12$ V, $V_S = 2$ V
17. (a) $I_{D_Q} = 2.9$ mA, $V_{GS_Q} = -1.2$ V
(b) $V_{DS} = 9.27$ V, $V_D = 10.52$ V
19. (a) $I_{D_Q} \approx 8.25$ mA (b) $V_{GS_Q} =$
$V_{DS_Q} = 7.9$ V (c) $V_D = 12.1$ V, $V_S =$
4.21 V (d) $V_{DS} = 7.89$ V
21. (a) $V_G = 3.3$ V (b) $V_{GS_Q} =$
-1.25 V, $I_{D_Q} = 3.75$ mA (c) $I_E =$
3.75 mA (d) $I_B = 23.44$ μA (e) $V_D =$
11.56 V (f) $V_C = 15.88$ V
23. $R_S = 0.43$ kΩ, $R_D = 1.3$ kΩ
25. $R_D = 0.75$ kΩ, $R_G = 10$ MΩ
27. D-S short-circuited; actual I_{DSS} or
V_P or combination thereof larger in
magnitude than specified.
29. (a) $I_{D_Q} = 3$ mA, $V_{GS_Q} = 1.55$ V
(b) $V_{DS} = -9.87$ V (c) $V_D = -11.4$ V

31. $I_{D_Q} = 4.68$ mA vs. 4.69 mA of #1, $V_{DS_Q} = 6.38$ V vs. 6.37 V of #1
33. $I_{D_Q} = 3.3$ mA (same), $V_{GS_Q} = -1.47$ V vs. -1.5 V of #12

CHAPTER 7

1. (a) 0 (b) Clipping (c) 80.4%
3. 1 kHz: $X_C = 15.92$ Ω
100 kHz: $X_C = 0.1592$ Ω
Yes, better at 100 kHz
7. (a) $Z_o = 50$ kΩ (b) $I_L = 5.747$ mA
9. (a) $I_i = 8$ μA (b) $Z_i = 500$ Ω
(c) $V_o = -720$ mV (d) $I_o = 1.41$ mA
(e) $A_i = 176.25$ (f) $A_i = 176.47$
11. (a) $r_e = 15$ Ω (b) $Z_i = 15$ Ω
(c) $I_c = 3.168$ mA (d) $V_o = 6.97$ V
(e) $A_v = 145.21$ (f) $I_b = 32$ μA
13. (a) $r_e = 8.571$ Ω (b) $I_b = 25$ μA
(c) $I_c = 3.5$ mA (d) $A_i = 132.84$
(e) $A_v = -298.89$
19. (a) $V_o = -160\,V_i$ (b) $I_b = 9.68 \times 10^{-4}\,V_i$ (c) $I_b = 1 \times 10^{-3}\,V_i$ (d) 3.2%
(e) Valid first approximation
21. (a) $V_o = -180\,V_i$ (b) $I_b = 2.32 \times 10^{-4}\,V_i$ (c) $I_b = 2.5 \times 10^{-4}\,V_i$
(d) 7.2% (e) Yes, less than 10%
23. (a) $h_{fe} = 100$ (b) $h_{ie} = 120$
25. (a) $h_{ie} = 1.5$ kΩ (b) $h_{ie} = 6.5$ kΩ
27. $h_{fe} = 100$, $h_{ie} = 2$ kΩ
29. $r_e = 15$ Ω, $\beta = 100$, $r_o = 30.3$ kΩ
31. (a) 75% (b) 70%
33. (a) $h_{oe} = 200$ μS (b) 5 kΩ, not a good approximation
35. (a) h_{fe} (b) h_{oe}
(c) Maximum: $h_{oe} \simeq 30$ (normalized)
Minimum: $h_{oe} \simeq 0.1$ (normalized)
At low levels of I_C
(d) Midregion

CHAPTER 8

1. (a) $Z_i = 497.47$ Ω, $Z_o = 2.2$ kΩ
(b) $A_v = -264.74$, $A_i = 60$
(c) $Z_i = 497.47$ Ω, $Z_o = 2.09$ kΩ
(d) $A_v = -250.94$, $A_i = 56.74$
3. (a) $I_B = 23.85$ μA, $I_C = 2.38$ mA, $r_e = 10.79$ Ω
(b) $Z_i = 1.08$ kΩ, $Z_o = 4.3$ kΩ
(c) $A_v = -398.52$, $A_i = 100$
(d) $A_v = -359.83$, $A_i = 90.38$
5. $V_{CC} = 30.68$ V
7. (a) $r_e = 5.34$ Ω
(b) $Z_i = 118.37$ kΩ, $Z_o = 2.2$ kΩ
(c) $A_v = -1.81$, $A_i = 97.39$
(d) Same results

9. (a) $r_e = 5.34$ Ω
(b) $Z_i = 746.17$ Ω, $Z_o = 2.2$ kΩ
(c) $A_v = -411.99$, $A_i = 139.73$
(d) $Z_i = 746.17$ Ω, $Z_o = 2.11$ kΩ
$A_v = -394.57$, $A_i = 133.83$
11. (a) $r_e = 8.72$ Ω, $\beta r_e = 959.2$ Ω
(b) $Z_i = 142.25$ kΩ, $Z_o = 8.69$ Ω
(c) $A_v \simeq 0.997$, $A_i = -52.53$
(d) $Z_i = 137.88$ kΩ, $Z_o = 8.69$ Ω
$A_v \simeq 0.997$, $A_i = -54.33$
13. (a) $I_B = 4.61$ μA, $I_C = 0.922$ mA
(b) $r_e = 28.05$ Ω
(c) $Z_i = 7.03$ kΩ, $Z_o = 27.66$ Ω
(d) $A_v = 0.986$, $A_i = -3.47$
15. $A_v = 163.2$, $A_i = 0.9868 \simeq 1$
17. $R_C = 1.6$ kΩ, $R_F = 33.59$ kΩ, $V_{CC} = 5.28$ V
19. (a) $Z_i = 0.62$ kΩ, $Z_o = 1.66$ kΩ
(b) $A_v = -209.82$, $A_i = 72.27$
(c) $Z_i = 0.62$ kΩ, $Z_o = 1.6$ kΩ
$A_v = -201.77$, $A_i = 69.5$
21. (a) $r_e = 8.31$ Ω
(b) $h_{fe} = 60$, $h_{ie} = 498.6$ Ω
(c) $Z_i = 497.47$ Ω, $Z_o = 2.2$ kΩ
(d) $A_v = -264.74$, $A_i = 60$
(e) $Z_i = 497.47$ Ω, $Z_o = 2.09$ kΩ
(f) $A_v = -250.90$, $A_i = 56.73$
23. (a) $Z_i = 9.38$ Ω, $Z_o = 2.7$ kΩ
(b) $A_v = 283.43$, $A_i \simeq -1$
(c) $\alpha = 0.992$, $\beta = 124$, $r_e = 9.45$ Ω, $r_o = 1$ MΩ
25. (a) $Z_i = 816.21$ Ω, $Z_i' = 814.8$ Ω
(b) $A_v = -357.68$
(c) $A_i = 132.70$, $A_i' = 132.43$
(d) $Z_o = 72.94$ kΩ, $Z_o' = 2.14$ kΩ
27. (a) No! (b) R_2 disconnected at base

CHAPTER 9

1. $g_{m0} = 6$ mS
3. $I_{DSS} = 8.75$ mA
5. $I_{DSS} = 12.5$ mA
7. $g_{mo} = 2.4$ mS
9. $Z_o = 40$ kΩ, $A_v = -180$
11. (a) $g_{mo} = 4$ mS, (b) $g_m = 2.8$ mS, (c) $g_m = 2$ mS
13. $g_m = 0.75$ mS, $r_d = 100$ kΩ
15. $A_v = -5.2$
17. (a) $Z_i = 10$ MΩ, (b) $Z_o = 2.9$ kΩ
19. (a) $A_v = -10.7$, (b) $A_v = -2.5$
21. $A_v = -5.9$
23. $A_v = -9.7$
25. $A_v = -33$
27. $g_m = 6.1$ mS
29. $Z_i = 9$ MΩ, $Z_o = 198.8$ Ω

31. $A_v = 7.4$, $Z_i = 344$ Ω, $Z_o = 3.3$ kΩ
33. $Z_i = 277.8$ Ω, $Z_o = 2.2$ kΩ, $A_v = 5.7$
35. $A_v = -4.8$
37. $Z_i = 1.97$ MΩ, $Z_o = 7.7$ kΩ, $A_v = -10.1$

CHAPTER 10

1. (a) $A_{v_{NL}} = -326.22$, $Z_i = 1.01$ kΩ, $Z_o = 3.3$ kΩ
(c) $A_v = -191.65$
(d) $A_i = 41.18$
(e) The same
3. $R_L = 4.7$ kΩ: $A_v = -191.65$
$R_L = 2.2$ kΩ: $A_v = -130.49$
$R_L = 0.5$ kΩ: $A_v = -42.42$
As $R_L \downarrow$, $A_v \downarrow$
5. (a) $A_{v_{NL}} = -557.36$, $Z_i = 616.52$ Ω, $Z_o = 4.3$ kΩ
(c) $A_v = -214.98$, $A_{v_s} = -81.91$
(d) $A_i = 49.04$
(e) $A_{v_s} = -120.12$, As $R_L \uparrow$, $A_{v_s} \uparrow$
(f) $A_{v_s} = -118.67$, As $R_s \downarrow$, $A_{v_s} \uparrow$
(g) Unaffected
7. (b) $A_v = -160$ vs. -162.4(#6)
9. (a) $A_{v_{NL}} = -3.61$, $Z_i = 81.17$ kΩ, $Z_o = 3$ kΩ
(c) $A_v = -2.2$, $A_{v_s} = -2.18$
(d) None
(e) A_v – none, $A_{v_s} = -2.17$, As $R_s \uparrow$, $A_{v_s} \downarrow$ (only slightly for moderate changes in R_s since Z_i typically so large)
11. (a) $Z_i = 10.74$ Ω, $Z_o = 4.7$ kΩ, $A_{v_{NL}} = 435.59$
(c) $A_v = 236.83$, $A_{v_s} = 22.97$
(d) The same
(e) $A_v = 138.88$, $A_{v_s} = 2.92$, A_{v_s} very sensitive to increase in R_s due to small Z_i, $R_L \downarrow$, $A_v \downarrow$, $A_{v_s} \downarrow$
13. (a) $A_{v_{NL}} = 0.737$, $Z_i = 2$ MΩ, $Z_o = 0.867$ kΩ
(c) $A_{v_s} \simeq A_v = 0.529$
(d) $A_{v_s} \simeq A_v = 0.622$, $R_L \uparrow$, $A_{v_s} \simeq A_v \uparrow$
(e) Little effect since $R_i \gg R_{sig}$
(f) No effect on Z_i or Z_o
15. (a) $A_{v_1} = -97.67$, $A_{v_2} = -189$
(b) $A_{v_T} = 18.46 \times 10^3$, $A_{v_{s(T)}} = 11.54 \times 10^3$
(c) $A_{i_1} = 97.67$, $A_{i_2} = 70$
(d) $A_{i_T} = 6.84 \times 10^3$
(e) No effect
(f) No effect
(g) In phase

CHAPTER 11

1. (a) 3, 1.699, -0.151 (b) 6.908, 3.912, -0.347 (c) Results differ by magnitude of 2.3

3. (a) Same: 13.98 (b) Same: -13.01 (c) Same: 0.699

5. $G_{dBm} = 43.98$ dBm

7. $G_{dB} = 67.96$ dB

9. (a) $G_{dB} = 69.83$ dB (b) $G_v = 82.83$ dB (c) $R_i = 2$ kΩ (d) $V_o = 1385.64$ V

11. (a) $|A_v| = 1/\sqrt{1 + (1950.43 \text{ Hz}/f)^2}$
(b) 100 Hz: $|A_v| = 0.051$
    1 kHz: $|A_v| = 0.456$
    2 kHz: $|A_v| = 0.716$
    5 kHz: $|A_v| = 0.932$
    10 kHz: $|A_v| = 0.982$
(c) $f_1 = 1950.43$ Hz

13. (a) 10 kHz (b) 1 kHz (c) 5 kHz (d) 100 kHz

15. (a) $r_e = 28.48$ Ω
(b) $A_{v_{mid}} = -72.91$
(c) $Z_i = 2.455$ kΩ
(d) $A_{v_s} = -54.68$
(e) $f_{L_S} = 103.4$ Hz, $f_{L_C} = 38.05$ Hz, $f_{L_E} = 235.79$ Hz
(f) $f_1 \simeq f_{L_E}$

17. (a) $r_e = 30.23$ Ω
(b) $A_{v_{mid}} \simeq 0.983$
(c) $Z_i = 21.13$ kΩ
(d) $A_{v_s \, mid} \simeq 0.955$
(e) $f_{L_S} = 71.92$ Hz, $f_{L_C} = 193.16$ Hz
(f) $f_1 \simeq f_{L_C}$; $f_1 \simeq 210$ Hz (PSpice)

19. (a) $V_{GS_Q} = -2.45$ V, $I_{D_Q} = 2.1$ mA
(b) $g_{mo} = 2$ mS, $g_m = 1.18$ mS
(c) $A_{v_{mid}} = -2$
(d) $Z_i = 1$ MΩ
(e) $A_{v_s} \simeq A_v = -2$
(f) $f_{L_G} = 1.59$ Hz, $f_{L_C} = 4.91$ Hz, $f_{L_S} = 32.04$ Hz
(g) $f_1 \simeq 32$ Hz

21. (a) $V_{GS_Q} = -2.55$ V, $I_{D_Q} = 3.3$ mA
(b) $g_{mo} = 3.33$ mS, $g_m = 1.91$ mS
(c) $A_{v_{mid}} = -4.39$
(d) $Z_i = 51.94$ kΩ
(e) $A_{v_s \, mid} = -4.27$
(f) $f_{L_G} = 2.98$ Hz, $f_{L_C} = 2.46$ Hz, $f_{L_S} = 41$ Hz
(g) $f_1 \simeq f_{L_S} = 41$ Hz
Z_i considerably less but still sufficiently greater than R_{sig} to result in minimum effect on A_{v_s}; reduced Z_i, however, can raise level of f_{L_G}

23. (a) $f_{H_i} \simeq 293$ kHz, $f_{H_o} = 3.22$ MHz
(b) $f_\beta = 8.03$ MHz, $f_T = 883.3$ MHz

25. (a) $f_{H_i} \simeq 584$ MHz, $f_{H_o} = 2.93$ MHz
(b) $f_\beta = 5.01$ MHz, $f_T = 400.8$ MHz

27. (a) $g_{mo} = 3.33$ mS, $g_m = 1.91$ mS
(b) $A_{v_{mid}} = -4.39$, $A_{v_s \, mid} = -4.27$
(c) $f_{H_i} = 1.84$ MHz, $f_{H_o} = 3.68$ MHz

29. $f_2' = 1.09$ MHz

31. (a) $v = 12.73 \times 10^{-3}$ [sin $2\pi(100 \times 10^3)t + \frac{1}{3}$ sin $2\pi(300 \times 10^3)t + \frac{1}{5}$ sin $2\pi(500 \times 10^3)t + \frac{1}{7}$ sin $2\pi(700 \times 10^3)t + \frac{1}{9}$ sin $2\pi(900 \times 10^3)t$] (b) BW = 500 kHz (c) $f_{L_o} \simeq 3.53$ kHz

CHAPTER 12

1. $V_G = 0$ V, $V_S = 1.4$ V, $V_D = 9.86$ V

3. $V_G = 0$ V, $V_S = 1.4$ V, $V_D = 10.3$ V

5. $Z_i = 10$ MΩ, $Z_o = 2.1$ kΩ

7. $A_{V1} = -75.8$, $A_{V2} = -311.9$, $A_V = 23{,}642$

9. $V_B = 2.55$ V, $V_E = 1.85$ V, $V_C = 2.7$ V, $I_C = 0.84$ mA

11. $Z_i = 10$ MΩ, $Z_o = 2.7$ kΩ

13. $A_V = -214$, $V_o = -2.14$ V

15. $V_{E2} = 8.06$ V, $I_{E2} = 15.8$ mA

17. $V_{B1} = 4.88$ V, $V_{C2} = 5.58$ V, $I_C = 104.2$ mA

19. (a) Q_1 on, Q_2 on, Q_3 off, Q_4 off
(b) Q_1 off, Q_2 off, Q_3 on, Q_4 on
(c) Q_1 on, Q_2 off, Q_3 on, Q_4 off

21. $I_D = 6$ mA

23. $I = 3.67$ mA

25. $I = 2$ mA

27. $I_C = 1$ mA, $V_C = 0$ V

29. $V_o = 1.89$ V

CHAPTER 14

1. CMRR = 75.56 dB

3. $V_o = -18.75$ V

5. $V_i = -40$ mV

7. $V_o = -9.3$ V

9. V_o ranges from 5.5 V to 10.5 V

11. $V_o = -3.39$ V

13. $V_o = 0.5$ V

15. $V_2 = -2$ V, $V_3 = 4.2$ V

17. $V_o = 6.4$ V

19. $I_{IB}^+ = 22$ nA, $I_{IB}^- = 18$ nA

21. $A_{CL} = 80$

23. V_o(offset) = 105 mV

CHAPTER 15

1. $V_o = -175$ mV, rms

3. $V_o = 412$ mV

7. $V_o = -2.5$ V

11. $I_L = 6$ mA

13. $I_o = 0.5$ mA

15. $f_{OH} = 1.45$ kHz

17. $f_{OL} = 318.3$ Hz, $f_{OH} = 397.9$ Hz

CHAPTER 16

1. $P_i = 10.4$ W, $P_o = 640$ mW

3. $P_o = 2.1$ W

5. R(eff) = 2.5 kΩ

7. $a = 44.7$

9. %$\eta = 37$%

13. (a) Maximum $P_i = 49.7$ W
(b) Maximum $P_o = 39.06$ W
(c) Maximum %$\eta = 78.5$%

17. (a) $P_i = 27$ W (b) $P_o = 8$ W
(c) %$\eta = 29.6$% (d) $P_{2Q} = 19$ W

19. %$D_2 = 14.3$%, %$D_3 = 4.8$%, %$D_4 = 2.4$%

21. %$D_2 = 6.8$%

23. $P_D = 25$ W

25. $P_D = 3$ W

CHAPTER 17

9. $V_o = 13$ V

13. Period = 204.8 μs

17. $f_o = 60$ kHz

19. $C = 133$ pF

21. $C_1 = 300$ pF

CHAPTER 18

1. $A_f = -9.95$

3. $A_f = -14.3$, $R_{if} = 31.5$ kΩ, $R_{of} = 2.4$ kΩ

5. Without feedback: $A_v = -303.2$, $Z_i = 1.18$ kΩ, $Z_o = 4.7$ kΩ
With feedback: $A_{vf} = -3.82$, $Z_{if} = 45.8$ kΩ

7. $f_o = 4.2$ kHz

9. $f_o = 1.05$ MHz

11. $f_o = 159.2$ kHz

CHAPTER 19

1. Ripple factor = 0.028

3. Ripple voltage = 24.2 V

5. $V_r = 1.2$ V

7. $V_r = 0.6$ V rms, $V_{dc} = 17$ V
9. $V_r = 0.12$ V rms
11. $V_m = 13.7$ V
13. $\%r = 7.2\%$
15. $\%r = 8.3\%$, $\%r' = 3.1\%$
17. $V_r = 0.325$ V rms
19. $V_o = 7.6$ V, $I_Z = 3.66$ mA
21. $V_o = 24.6$ V
25. $I_{dc} = 225$ mA
27. $V_o = 9.9$ V

CHAPTER 20

3. 30:1 or better is typical, short period of time, casing design
5. 124% increase, $V_R \simeq 25$ V
7. (a) $C_T = 41.85$ pF (b) $k \simeq 71 \times 10^{-12}$

9. (a) $C = 5.17$ pF (b) Graph, $C \simeq 5$ pF
11. $T_1 = 50°C$
13. $Q = 26.93$, Q drops significantly with increase in frequency
19. $I_T = 5$ mA, $V_T = 60$ mV $I_T = 2.8$ mA, $V_T = 900$ mV
21. $f_p \simeq 2228$ Hz
23. (a) 3750 Å $\rightarrow$ 7500 Å (b) $\simeq 8400$ Å (c) BW = 4200 Å
25. (a) Silicon (b) Orange
27. (a) $\simeq 0.9$ Ω/fc (b) $\simeq 380$ Ω/fc (c) $\simeq 78$ kΩ/fc Low-illumination region
29. $V_i = 21$ V
31. As fc increases, t_r and t_d decrease exponentially
33. (a) $\Phi \simeq 5$ mW (b) 2.27 lm
35. $\Phi = 3.44$ mW
41. Lower levels

45. $R \simeq 20$ kΩ
47. $R(\text{thermistor}) = 90$ Ω

CHAPTER 21

5. (a) Yes (b) No (c) No (d) Yes, no
11. (a) $\simeq 0.7$ mW/cm^2 (b) 82.35%
17. (a) $R_{B_2} = 1.08$ kΩ (b) $R_{BB} = 3.08$ kΩ (c) $V_{R_{B_1}} = 13$ V (d) $V_P = 13.7$ V
19. $I_B = 25$ μA, $I_C = 1$ mA
21. (a) For decreasing temperatures, $-0.53\%/\text{C}°$ (b) Yes
23. $I_C/I_F = 0.44$ Relatively efficient
25. (a) $I_C \geq 3$ mA (b) $\Delta R : \Delta t \simeq 2.3 : 1$
27. $Z_P = 87$ kΩ, $Z_V = 181.8$ Ω, To a degree
29. (a) Yes, 8.18 V (b) $R < 2$ kΩ (c) $R = 1.82$ kΩ

Index

Y

Yield rate, 571, 577

Z

Zener constant current source, 545
Zener diodes, 35–38, 87–94, 830, 843
  applications, 87–94

characteristics, 35–36
equivalent circuit, 36
specification sheet, 36–37
symbol, 38
temperature coefficient, 37–38
Zener region, 14–15
Zone-refining apparatus, 571–72

FREE SOFTWARE

PSpice, produced by MicroSim Corporation, is an integrated package for the analysis of electronic and electrical circuitry. To receive an evaluation copy, please complete and detach the postage paid card, and return it to MicroSim Corporation.

PSpice
 MicroSim Corporation

☐ Yes, please send me more information about PSpice and a free, updated Evaluation Version of PSpice on:

       ☐ 5 1/4" PC DOS diskettes
       ☐ 3 1/2" Macintosh diskettes

☐ Please add my name to your mailing list.

Mr / Ms / Prof _____

Address _____

Address _____

City, State _____

Zip, Country _____

Phone _____

PSpice is a registered trademark of MicroSim Corporation.

[FOLD HERE]

RB

NO POSTAGE
NECESSARY
IF MAILED
IN THE
UNITED STATES

BUSINESS REPLY MAIL

First Class Mail   Permit No. 9677   Irvine, CA

Postage will be paid by addressee

MicroSim Corporation

20 Fairbanks
Irvine, CA 92718-9905 U.S.A.

11 BJT and FET Frequency Response $\log_e a = 2.3 \log_{10} a$, $\log_{10} 1 = 0$, $\log_{10} a/b = \log_{10} a - \log_{10} b$, $\log_{10} 1/b = -\log_{10} b$, $\log_{10} ab = \log_{10} a + \log_{10} b$, $G_{\mathrm{dB}} = 10 \log_{10} P_2/P_1$, $G_{\mathrm{dBm}} = 10 \log_{10} P_2/1 \text{ mW}|_{600\Omega}$, $G_{\mathrm{dB}} = 20 \log_{10} V_2/V_1$, $G_v = G_{v_1} + G_{v_2} + G_{v_3} + \cdots + G_{v_n}$, $P_{O_{\mathrm{HPF}}} = 0.5 P_{o_{\mathrm{mid}}}$, $\mathrm{BW} = f_1 - f_2$; low frequency (BJT): $f_{L_S} = 1/2\pi(R_s + R_i)C_S$, $f_{L_C} = 1/2\pi(R_o + R_L)C_C$, $f_{L_E} = 1/2\pi R_e C_E$, $R_e = R_E \| (R_s'/\beta + r_e)$, $R_s' = R_s \| R_1 \| R_2$, FET: $f_{L_G} = 1/2\pi(R_{\mathrm{sig}} + R_i)C_G$, $f_{L_C} = 1/2\pi R_o C_C$, $f_{L_S} = 1/2\pi R_{\mathrm{eq}} C_S$, $R_{\mathrm{eq}} = R_S \| 1/g_m (r_d \simeq \infty \ \Omega)$; Miller effect: $C_{M_i} = (1 - A_v)C_f$, $C_{M_o} = (1 - 1/A_v)C_f$; high frequency (BJT): $f_{H_i} = 1/2\pi R_{\mathrm{Th}_1} C_i$, $R_{\mathrm{Th}_1} = R_s \| R_1 \| R_2 \| R_i$, $C_i = C_{W_i} + C_{be} + C_{M_i}$, $f_{H_o} = 1/2\pi R_{\mathrm{Th}_2} C_o$, $R_{\mathrm{Th}_2} = R_C \| R_L \| r_o$, $C_o = C_{W_o} + C_{ce} + C_{M_o}$, $f_\beta \simeq 1/2\pi \beta_{\mathrm{mid}} r_e (C_{be} + C_{bc})$, $f_T = \beta_{\mathrm{mid}} f_\beta$; FET: $f_{H_i} = 1/2\pi R_{\mathrm{Th}_1} C_i$, $R_{\mathrm{Th}_1} = R_{\mathrm{sig}} \| R_G$, $C_i = C_{W_i} + C_{gs} + C_{M_i}$, $f_{H_o} = 1/2\pi R_{\mathrm{Th}_2} C_o$, $R_{\mathrm{Th}_2} = R_D \| R_L \| r_d$, $C_o = C_{W_o} + C_{ds} + C_{M_o}$; multistage: $f_1' = f_1/\sqrt{2^{1/n} - 1}$, $f_2' = (\sqrt{2^{1/n} - 1})f_2$; square-wave testing: $f_{H_i} = 0.35/t_r$, % tilt $= [(V - V')/V] \times 100\%$, $f_{L_o} = (P/\pi)f_s$, $P = (V - V')/V$

12 Compound Configurations   Differential voltage gain: $A_v = \beta R_C/2r_i$; common-mode voltage gain: $\beta R_C/[r_i + 2(\beta + 1)R_E]$

13 Discrete and IC Manufacturing Techniques $R = R_s l/w$, $C_T = \epsilon A/W$

14 Opamps $\mathrm{CMRR} = A_d/A_c$; $\mathrm{CMRR(log)} = 20 \log_{10}(A_d/A_c)$; constant-gain multiplier: $V_o/V_1 = -R_f/R_1$; noninverting amplifier: $V_o/V_1 = 1 + R_f/R_1$; unity follower: $V_o = V_1$; summing amplifier: $V_o = -[(R_f/R_1)V_1 + (R_f/R_2)V_2 + (R_f/R_3)V_3]$; integrator: $v_o(t) = -(1/R_1 C_1)\int v_1 dt$

15 Opamp Applications   Constant-gain multiplier: $A = -R_f/R_1$; noninverting: $A = 1 + R_f/R_1$; voltage summing: $V_o = -[(R_f/R_1)V_1 + (R_f/R_2)V_2 + (R_f/R_3)V_3]$; high-pass active filter: $f_{oL} = 1/2\pi R_1 C_1$; low-pass active filter: $f_{oH} = 1/2\pi R_1 C_1$

16 Power Amplifiers
Power in: $P_i = V_{CC} I_{CQ}$
power out: $P_o = V_{CE} I_C = I_C^2 R_C = V_{CE}^2/R_C$ rms
$\qquad = V_{CE} I_C/2 = (I_C^2/2)R_C = V_{CE}^2/(2R_C)$ peak
$\qquad = V_{CE} I_C/8 = (I_C^2/8)R_C = V_{CE}^2/(8R_C)$ peak-to-peak
efficiency: $\%\eta = (P_o/P_i) \times 100\%$